Studies
in the History of Mathematics and
Physical Sciences

10

Sources in the History of Mathematics and Physical Sciences

Vol. 1: G.J. Toomer (Ed.), **Diocles on Burning Mirrors: The Arabic Translation of the Lost Greek Original,** Edited, with English Translation and Commentary by G.J. Toomer

Vol. 2: A. Hermann, K.V. Meyenn, V.F. Weisskopf (Eds.), **Wolfgang Pauli: Scientific Correspondence I: 1919 – 1929**

Vol. 3: J. Sesiano, **Books IV to VII of Diophantus'** *Arithmetica*: **In the Arabic Translation of Qustā ibn Lūqā**

Vol. 4: P.J. Federico, **Descartes on Polyhedra: A Study of the** *De Solidorum Elementis*

Vol. 5: O. Neugebauer, **Astronomical Cuneiform Texts**

Vol. 6: K. von Meyenn, A. Hermann, V.F. Weisskopf (Eds.), **Wolfgang Pauli: Scientific Correspondence II: 1930 – 1939**

Vol. 7: J.P. Hogendijk, **Ibn Al-Haythan's** *Completion of the Conics*

Studies in the History of Mathematics and Physical Sciences

Vol. 1: O. Neugebauer, **A History of Ancient Mathematical Astronomy**

Vol. 2: H. Goldstine, **A History of Numerical Analysis from the 16th through the 19th Century**

Vol. 3: C.C. Heyde/E. Seneta, **I.J. Bienaymé: Statistical Theory Anticipated**

Vol. 4: C. Truesdell,**The Tragicomical History of Thermodynamics, 1822 – 1854**

Vol. 5: H.H. Goldstine, **A History of the Calculus of Variations from the 17th through the 19th Century**

Vol. 6: J. Cannon/S. Dostrovsky, **The Evolution of Dynamics: Vibration Theory from 1687 to 1742**

Vol. 7: J. Lützen, **The Prehistory of the Theory of Distributions**

Vol. 8: G.H. Moore, **Zermelo's Axiom of Choice**

Vol. 9: B. Chandler and W. Magnus, **The History of Combinatorial Group Theory**

Vol. 10: N.M. Swerdlow/O. Neugebauer, **Mathematical Astronomy in Copernicus's De Revolutionibus**

N. M. Swerdlow
O. Neugebauer

Mathematical Astronomy in Copernicus's De Revolutionibus

Part 2

Springer-Verlag
New York Berlin Heidelberg Tokyo

N. M. Swerdlow
Department of Astronomy
and Astrophysics
University of Chicago
5640 S. Ellis Avenue
Chicago, IL 60637
U.S.A.

O. Neugebauer
Department of the History
of Mathematics
Brown University
Box 1900
Providence, RI 02912
U.S.A.

AMS Classifications: 01A20, 01A40, 05-13

Library of Congress Cataloging in Publication Data
Swerdlow, N. M. (Noel M.), 1941–
Mathematical astronomy in Copernicus's *De revolutionibus*.
(Studies in the history of mathematics and physical sciences; 10)
Bibliography: p.
Includes index.
1. Astronomy—Mathematics—History. 2. Copernicus, Nicolaus, 1473–1543. De revolutionibus orbium coelestium.
I. Neugebauer, O. (Otto), 1899– . II. Title.
III. Series.
QB47.S86 1984 521.1 83-20081

Typeset by Composition House Ltd., Salisbury, England.
Printed and bound by R. R. Donnelley & Sons, Harrisonburg, Virginia.
Printed in the United States of America.

9 8 7 6 5 4 3 2 1

ISBN 0-387-90939-7 Springer-Verlag New York Berlin Heidelberg Tokyo
ISBN 3-540-90939-7 Springer-Verlag Berlin Heidelberg New York Tokyo

Table of Contents

Part 2

PART 2

Notation and Symbols

The following list, which is not complete, gives the usual meaning of the notation we have used, not without some inconsistencies, throughout this study. Under the heading Spherical Astronomy, we have also included the notation for precession, parallax, and apparent radii.

1. Spherical Astronomy

α right ascension
β latitude
γ angle between ecliptic and meridian, meridian angle
δ declination
δ_P equation of precession
ε obliquity of ecliptic
ζ zenith distance, $\bar{\zeta} = 90° - \zeta$ altitude
ϑ anomaly of obliquity and precession
λ longitude
ν angle between ecliptic and horizon, horizon angle
ξ angle between ecliptic and circle of altitude
π precession, $\bar{\pi}$ mean precession
ρ oblique ascension
φ geographical or terrestrial latitude, $\bar{\varphi} = 90° - \varphi$ co-latitude
a ortive amplitude
h *arcus visionis*; equal hours, h' seasonal hours
l geographical or terrestrial longitude
n correction or difference of ascension
p parallax, p_λ in longitude, p_β in latitude, p' difference of lunar and solar parallax
r' apparent radius (of sun, moon, shadow)
s semidiurnal arc, half-length of daylight
t time, in hours or degrees
D descendant, setting point of ecliptic
G horoscopus, ascendant, rising point of ecliptic

M midheaven, culminating point of ecliptic
N nonagesimal, $G - 90°$
P horizontal parallax
S sun

Zodiacal Signs

♈	Aries	♋	Cancer	♎	Libra	♑	Capricorn
♉	Taurus	♌	Leo	♏	Scorpio	♒	Aquarius
♊	Gemini	♍	Virgo	♐	Sagittarius	♓	Pisces

2. Solar, Lunar, and Planetary Theory

$\bar{}$ mean motion, longitude, time, etc.
$_0$ epoch value
α anomaly
β geocentric latitude, β_1, β_2, β_3 components of latitude, β' apparent latitude
δ difference of longitude
Δ elongation of inferior planet from mean sun
ΔE equation of time
ε correction angle
η elongation of moon from mean sun, η' from true sun
ι inclination of lunar orbit to ecliptic
κ eccentric anomaly
λ geocentric tropical longitude, λ^* sidereal longitude, λ' apparent longitude, $\bar{\lambda}'$ adjusted mean longitude
λ_A tropical (λ_A^* sidereal) longitude of apogee
μ mean distance
ν lunar nodal distance
ρ geocentric distance
ω argument of latitude from northern limit, ω' from ascending node
b heliocentric latitude
c correction, equation; column number
d day
e eccentricity
h hour
i inclination of orbital plane to ecliptic
r radius, r' radius of small epicycle, ^{r} rotations of 360°
s distance in terrestrial radii (of sun, moon, vertex of shadow)
t time
tr terrestrial radii

v	angular velocity
y	year, usually Egyptian year of 365^d
A	apogee
C	center of epicycle or orbit
E	equant
M	midpoint of eccentric
N	midpoint of eccentric in model for second anomaly; northern limit of latitude
O	observer, earth
P	planet, moon
R	radius of eccentric
S	sun, $\bar{S}$ mean sun, center of earth's orbit
☊	ascending node, ☋ descending node

Planetary Symbols

☉	sun	♄	Saturn	♀	Venus
☾	moon	♃	Jupiter	☿	Mercury
		♂	Mars		

Parameters

The following list of parameters is intended to show the results of Copernicus's derivations in N. It does not necessarily contain the parameters underlying the correction tables, which are partially copied from the *Alfonsine Tables* and the *Almagest* even if Copernicus's parameters differ slightly, and it gives only one value for each element, where Copernicus sometimes uses alternates. These cases, and other parameters derived in M, are discussed in the preceding study. Positions are given for three epochs:

Alexander	−323 Nov 12	noon	$\Delta t = 5,23^{y}\ 2,10\tfrac{1}{2}^{d}$
Christ	1 Jan 1	midnight	25, 0 8
1500	1500 Jan 0	midnight	

All are for the meridian of Frauenburg–Cracow used by Copernicus, and the epoch of 1500 has been added to facilitate computation of positions in the sixteenth century. In the line Ecc–rad for the planets are listed all eccentricities, epicyclic radii, and orbital radii in each model where $R = 10{,}000$, and the line Equations gives the corresponding maximum equations. For the sun, we list the limits of its variable eccentricity and equation, but for Mars and Venus only the values applicable to the time of Copernicus. The uncertainties of the motion of the apsidal lines of Mars and Mercury are discussed in the preceding sections on each planet.

1. Precession and Obliquity

Precession	$\bar{\pi}$ 0;0,50,12,5$^{\circ/y}$	0;0,0,8,15$^{\circ/d}$	
Anomaly	ϑ 0;6,17,24,9	0;0,1,2,2	
Epochs	$\bar{\pi}_0$	ϑ_0	
Alexander	1;2°	5,32;52°	
Christ	5;32	6;45	
1500	26;27,3	2,24;0,11	
Equations	δ_P 1;10°	$\Delta\varepsilon$ 0;24°	
Obliquity	ε_{max} 23;52°	ε_{min} 23;28°	$\bar{\varepsilon}$ 23;40°

2. Solar Theory

Sidereal longitude	$\bar{\lambda}^*$	5,59;44,49,7,4°/y	0;59,8,11,22°/d
Tropical longitude	$\bar{\lambda}$	5,59;45,39,19,9	0;59,8,19,37
Anomaly	$\bar{\alpha}$	5,59;44,24,46,50	0;59,8,7,22
Apogee	λ^*_A	0;0,24,20,14	—

Epochs	$\bar{\lambda}^*_0$	$\bar{\lambda}_0$	$\bar{\alpha}_0$	$\bar{\lambda}^*_{A0}$
Alexander	3,45;37°	3,46;38°	2,46;31°	59;6°
Christ	4,32;31	4,38;2	3,31;14	1, 1;17
1500	4,20;51,31	4,47;18,4	3, 9;26,34	1,11;25

Eccentricities	e_1 369	e_2 48	e_{max} 417	e_{min} 321
Equations	$c_{3(\vartheta)}$ 7;28°	$c_{(\alpha,\vartheta)}$ 2;23° to 1;51°		

Distance	max 1179tr	mean 1142tr	min 1105tr
Parallax	0;2,55°	0;3°	0;3,7°
Appar. diam.	0;31,40°	—	0;33,54°

3. Lunar Theory

Elongation	$\bar{\eta}$	2, 9;37,22,36,25°/y	12;11,26,41,31°/d
Anomaly	$\bar{\alpha}$	1,28;43,9,7,15	13;3,53,56,30
Arg. of Lat.	$\bar{\omega}$	2,28;42,45,17,21	13;13,45,39,22

Epochs	$\bar{\eta}_0$	$\bar{\alpha}_0$	$\bar{\omega}_0$
Alexander	5,10;44°	1,25;41°	2,16;57°
Christ	3,29;58	3,27;7	2, 9;45
1500	5,41;54,43	3,10;26,12	1,44;27,18

Epic. rad.	r_1 1097	r_2 237	Equa. $c_{3(2\bar{\eta})}$ 12;28°

	Syzygy $\bar{\eta} = (0°, 180°)$	Quadrature $\bar{\eta} = \pm 90°$
Epic. rad.	r_{min} 860	r_{max} 1334
Equations	$c_{(\alpha, 2\bar{\eta})}$ 4;56°	$c_{(\alpha, 2\bar{\eta})}$, 7;40°
Distance	65;30tr to 55;8tr	68;20tr to 52;17tr
Parallax	0;52,24° to 1;2,21°	0;50,18° to 1;5,45°
Appar. diam.	0;30° to 0;35,38°	0;28,45° to 0;37,34°
Diam. of shadow	1;20,36° to 1;35,44°	—

Inclination	ι	5°

4. Planetary Theory

Saturn

Anomaly	$\bar{\alpha}$	5,47;32,3,9,40$^{\circ/y}$	0;57,7,44,5$^{\circ/d}$
Longitude	$\bar{\lambda}^*$	12;12,45,57,24	0;2,0,27,17
Apogee	λ^*_A	0;0,36	—

Epochs	$\bar{\alpha}_0$	$\bar{\lambda}^*_0$	λ^*_{A0}
Alexander	2,28;1°	1,17;36°	3,41;48°
Christ	3,25;49	1, 6;42	3,45;2
1500	3,47;15,38	33;35,53	4, 0;2

Ecc–rad	$2e$ 1139	e_1 854	r' 285	r 1090
Equations	$c_{3(\bar{\kappa})}$ 6;31°	$c_{(\alpha,\bar{\kappa})}$ 5;55° to 6;38°		
Latitude	ω_A −50°	i_o 2;44°	i_c 2;16°	Δi 0;28°

Jupiter

Anomaly	$\bar{\alpha}$	5,29;25,8,15,6$^{\circ/y}$	0;54,9,3,49$^{\circ/d}$
Longitude	$\bar{\lambda}^*$	30;19,40,51,58	0;4,59,7,33
Apogee	λ^*_A	0;0,12	—

Epochs	$\bar{\alpha}_0$	$\bar{\lambda}^*_0$	λ^*_{A0}
Alexander	2,18;10°	1,27;27°	2,32;50°
Christ	1,38;16	2,54;15	2,33;55
1500	5,13;55,29	5, 6;56,2	2,38;55

Ecc–rad	$2e$ 917	e_1 687	r' 229	r 1916
Equations	$c_{3(\bar{\kappa})}$ 5;15°	$c_{(\alpha,\bar{\kappa})}$ 10;33° to 11;35°		
Latitude	ω_A +20°	i_o 1;42°	i_c 1;18°	Δi 0;24°

Mars

Anomaly	$\bar{\alpha}$	$2,48;28,30,36,4^{\circ/y}$	$0;27,41,40,22^{\circ/d}$
Longitude	$\bar{\lambda}^*$	3,11;16,18,31,0	0;31,26,31,0
Apogee	λ_A^*	nonuniform	—

Epochs	$\bar{\alpha}_0$	$\bar{\lambda}_0^*$
Alexander	2, 0;39°	1,44;58°
Christ	3,58;22	34;9
1500	3,54;48,34	26;2,57

Ecc–rad	$2e$ 1960	e_1 1460	r' 500	r 6580
Equations	$c_{3(\bar{\kappa})}$ 11;8°	$c_{(\alpha,\bar{\kappa})}$ 36;54° to 46;43°		
Latitude	ω_A 0°	i_o 1;51°	i_c 0;9°	Δi 1;42°

Venus

Anomaly	$\bar{\alpha}$	$3,45;1,45,3,40^{\circ/y}$	$0;36,59,28,35^{\circ/d}$
Longitude	$\bar{\lambda}^*$	$=\bar{\odot}$	$=\bar{\odot}$
Apogee	λ_A^*	fixed	—

Epochs	$\bar{\alpha}_0$	$\bar{\lambda}_0^*$	λ_{A0}^*
Alexander	1,21;52°	$=\bar{\odot}$	48;20°
Christ	2, 6;45	$=\bar{\odot}$	48;20°
1500	5,55;27,25	$=\bar{\odot}$	48;20°

Ecc–rad	$2e$ 350	e_1 246	e_2 104	r 7193
Equations	$c_{3(\bar{\kappa})}$ 2;0°	$c_{(\alpha,\bar{\kappa})}$ 45;10° to 46;51°		
Latitude	i_1 2;30°	i_2 3;30°	i_3 +0;10°	

Mercury

Anomaly	$\bar{\alpha}$ $3^{r} + 53;57,23,6,30^{\circ/y}$	$3;6,24,13,40^{\circ/d}$
Longitude	$\bar{\lambda}^*$ $=\bar{\odot}$	$=\bar{\odot}$
Apogee	λ^*_{A} (1) 0;1,14,11 (2) 0;0,57,20	—

Epochs	$\bar{\alpha}_0$	$\bar{\lambda}^*_0$	(1) λ^*_{A0}	(2)
Alexander	3,33;3°	$=\bar{\odot}$	2,53;50°	3, 2;24°
Christ	46;24	$=\bar{\odot}$	3, 0;30	3, 7;33
1500	5, 5;52,55	$=\bar{\odot}$	3,31;28	3,31;26

Ecc–rad	e 524	e_1 736	e_2 212	$\bar{r}$ 3763	r' 190
Equations	$c_{3(\bar{\kappa})}$ 3;1°	$c_{(\alpha,\bar{\kappa})}$ 19;3° to 23;52°			
Latitude	i_1 6;15°	i_2 7;0°	i_3 −0;45°		

Tables

Tables to Chapter 3

Table 1. Observations of Fixed Stars.

No.	Observer	Star	λ°	Date in Text		Correct Julian Date	Julian Day
				Original	E.A.		
1.	Timocharis	β Sco	212	Call 1,36	30	−294 Dec 21	1614,029
2.	Timocharis	α Vir	172;20	Call 1,36	30	−293 Mar 9	1614,107
3.	Timocharis	α Vir	172;30	Call 1,48	[42]	−282 Nov 9	1618,370
4.	Hipparchus	α Leo	119;50	Call 3,50	196	−128/127	
5.	Menelaus	α Vir	176;15	Trajan 1 A.D. 99	422	+98 Jan 11	1756,863
6.	Menelaus	β Sco	215;55	Trajan 1 A.D. 99	422	98 Jan 14	1756,866
7.	Ptolemy	α Leo	122;30	Antoninus 2	462	139 Feb 23	1771,881
8.	Ptolemy	α Vir	176;30	Antoninus 2	462		
9.	Ptolemy	β Sco	216;20	Antoninus 2	462		
10.	Battānī	α Leo	134;5		1202	879/80	
11.	Battānī	β Sco	227;50		1202	879/80	
12.	Copernicus	α Vir	197;14	A.D. 1515	[1839]	1515	
13.	Copernicus	α Vir	197;21	A.D. 1525	1849	1525 Apr 16?	2278,170

Table 2. Observations of the Sun.

No.	Observer	Type	Date and Time in Text	Julian Date and Time		Julian Day
1.	Hipparchus	A.	E.A. 177 Epag 3/4, midnight	−146 Sep 26/27,	0^h	1668,000/1
2.	Hipparchus	V.	E.A. 178 Mech 27, sunrise	−145 Mar 24,	6	1668,179.25
3.	Ptolemy	A.	E.A. 463 Ath 9, 1^h after sunrise	139 Sep 26,	7	1772,096.292
4.	Ptolemy	V.	E.A. 463 Pach 7, 1^h after noon	140 Mar 22,	13	1772,274.542
5.	Battānī	A.	E.A. 1206 Pach 7/8, $7\frac{2}{5}^h$ after sunset	882 Sep 19,	1;24	2043,470.058
6.	Copernicus	V.	(from III,16) 186;5,30^d before 7	1515 Mar 12,	4;18	2274,482.179
7.	Copernicus	A.	1515 XVIII Kal. Oct E.A. 1840 Phao 6, $\frac{1}{2}^h$ after sunrise	1515 Sep 14,	6;30	2274,668.271
8.	Copernicus	♏15°	(from III,16) 45;16^d after 7	1515 Oct 29,	12;54	2274,713.538
9.	Copernicus	V.	1516 V Id. Mar [E.A. 1840 Phar 5], $4\frac{1}{3}^h$ after midnight	1516 Mar 11	$4;20^h$	2274,847.181

Table 3

Observations	(1) Δt^{ty}	(2) Δt^{y-d-h}	(3) $ty \cdot \frac{1}{4}^{d}$	(4) Deficit (3) − $(2)^{d-h}$	(5) Deficit (4)/(1)	(6) ty $365\frac{1}{4}^{d}$ − (5)
1.–3.	285^{ty}	285^{y} 70^{d} $7\frac{1}{5}^{h}$	71^{d} 6^{h}	$19/20^{d}$	$1^{d}/300^{y}$	$365;14,48^{d}$
2.–4.	285	285 70 $7\frac{1}{5}$	71 6	$19/20^{d}$	1 /300	365;14,48
3.–5.	743	743 178 $17\frac{3}{5}$	$185\frac{3}{4}$	7^{d} $\frac{2}{5}^{h}$	1 /106	365;14,26 $= 365^{d}$ $5;46,24^{h}$
5.–7.	633	633 153 $6\frac{3}{4}$	158 6	5^{d} − $1\frac{1}{4}^{h}$	1 /128	[365;14,32]
3.–7.	1376	1376 332 $\frac{1}{2}$	344	$\sim 12^{d}$	1 /115	[365;14,29]
4.–9.	1376	1376 332 $16\frac{1}{4}$	344	[$\sim 11^{d}$ 8^{h}]	[1 /121]	[365;14,30]

Table 4

No.	Observer	Julian Date and Time	λ Obs.	λ Comp.
1.	Hipparchus	−146 Sep 26, 23^{h}	180°	180°
2.	Hipparchus	−145 Mar 24, 5	0	0;1
3.	Ptolemy	139 Sep 26, 6	180	180
4.	Ptolemy	140 Mar 22, 12	0	0;3
5.	Battānī	882 Sep 18, 23;44	180	179;50
6.	Copernicus	1515 Mar 12, 4;18	0	0;14
7.	Copernicus	1515 Sep 14, 6;30	180	179;59
8.	Copernicus	1515 Oct 29, 12;54	225	224;59
9.	Copernicus	1516 Mar 11, 4;20	0	359;59

Tables to Chapter 4

Table 1. Summary of Mean Motions.

	Source		$\bar{\eta}$	$\bar{\alpha}$	$\bar{\omega}$
1.	Hip.	°/y	129;37,21,28,29°	88;43, 8,40,20°	148;42,46,20, 3°
2.	Hip.	°/d	12;11,26,41,20,18	13; 3,53,56,29	13;13,45,39,40
3.	Ptol.	corr.	0; 0	−0; 0, 1,11,39	+0; 0, 0,52,41
4.	Ptol.	°/y 1 + 3	129;37,21,28,29	88;43, 7,28,41	148;42,47,12,44
5.	N	corr.	+0; 0, 1, 7,56	+0; 0, 0,26,55	−0; 0, 1, 2,42
6.	N	°/y 1 + 5	129;37,22,36,25	88;43, 9, 7,15	148;42,45,17,21
7.	M	corr.	+0; 0, 1, 2,49	+0; 0, 0,24,49	−0; 0, 1, 1,44
8.	M	°/y	129;37,22,32,40	88;43, 9, 5, 9	148;42,44,31, 6
9.	N, M	°/d	12;11,26,41,31	13; 3,53,56,30	13;13,45,39,22

Table 2. Observations of the Moon.

No.	Date and Time	Julian Day	Description
1.	Hadrian 17 Pay 20/21 133 May 6 22;15^h	1769,762.927	E. total
2.	Hadrian 19 Choi 2/3 134 Oct 20 22;0^h	1770,294.917	E. 10′ north
3.	Hadrian 20 Phar 19/20 136 Mar 6 3;0^h	1770,797.125	E. 6′ north
4.	1511 Oct 6/7 0;35^h	2273,230.024	E. total
5.	1522 Sep 5/6 1;20^h	2277,217.056	E. total
6.	1523 Aug 25/26 4;25^h	2277,517.184	E. total
7.	Ptol. Phil. 7 Phar 27/28 −173 May 1 1;20^h	1657,990.056	E. 7′ north
8.	1509 Jun 2 23;45^h	2272,373.990	E. ~8′ south
9.	1500 Nov 6 2;20^h	2269,243.097	E. 10′ north
10.	E.A. 197 Pay 17 −126 Jul 7 15;10^h	1675,224.632	near octant
11.	1522 Sep 27 17;40^h	2277,238.736	near horizon
12.	1524 Aug 7 18;0^h	2277,918.75	near horizon
13.	1497 Mar 9 23;36^h	2267,905.983	occult. α Tauri
14.	[Hadrian 20 Ath 13] 135 Oct 1 16;50^h	1770,640.701	Ptolemy's parallax

Table 3

No.	Year	C. $\bar{\lambda}'_\odot$	C. $\lambda_\odot$	$\bar{\lambda}_\odot$	$\bar{\lambda}'_\odot$	$\lambda_\odot$	$\alpha_\odot$	$\alpha_\odot - \bar{\lambda}_\odot$
1.	133	42;21°	43;15°	43;8°	42;22°	43;18°	40;42°	−2;26°
2.	134	206;43	205;10	207;29	206;43	205;9	203;15	−4;14
3.	136	341;44	344;5	342;29	341;43	344;4	345;23	+2;54
4.	1511	204;13	202;25	203;39	204;13	202;27	200;44	−2;55
5.	1522	173;59	172;12	173;27	173;59	172;12	172;50	−0;37
6.	1523	163;2	161;21	162;30	163;2	161;22	162;48	+0;18
7.	−173	—	6th° ♉	34;47	35;4	36;13	33;51	−0;56
8.	1509	—	81;0	79;54	80;29	80;59	80;12	+0;18
9.	1500	—	233;16	233;57	234;34	233;18	230;52	−3;5
10.	−126	102;3	100;40	101;49	102;5	100;42	101;38	−0;11
11.	1522	[195;20]	193;29	194;49	195;21	193;31	192;24	−2;25
12.	1524	[145;33]	144;14	145;4	145;36	144;15	146;33	+1;29
13.	1497	—	358;30	356;2	356;39	358;30	358;37	+2;35
14.	135	[187;31]	185;28	188;19	187;32	185;28	185;0	−3;19

Table 4

No.	Year	$\Delta\bar{t}$	$\bar{\eta}$	$+c_☾$	η	$-c_\odot$	η'	Error
1.	133	166,364.4220^d	177;27°	+3;25°	180;52°	−0;54°	179;58°	−0;2°
2.	135	166,896.4108	182;47	−4;21	178;26	+1;34	180;0	0;0
3.	136	167,398.6389	185;19	−2;44	182;35	−2;21	180;14	+0;14
4.	1511	669,831.5221	177;53	+0;16	178;9	+1;46	179;55	−0;5
5.	1522	673,818.5597	182;51	−4;40	178;11	+1;47	179;58	−0;2
6.	1523	674,172.6907	179;59	−1;39	178;20	+1;40	180;0	0;0
7.	−173	54,591.5588	182;40	−1;24	181;16	−1;9	180;7	+0;7
8.	1509	668,975.4838	182;8	−1;45	180;23	−0;30	179;53	−0;7
9.	1500	665,844.5945	174;16	+4;27	178;43	+1;16	179;59	−0;1

Table 5. Components of Parallax.

Taurus t	p'_λ	p'_β
−6;59^h	+0;17°	−0;48°
−6;0	+0;18	−0;47
−5;0	+0;18	−0;45
−4;0	+0;15	−0;43
−3;0	+0;11	−0;40
−2;0	+0; 6	−0;37
Nonagesimal	[0;0]	[−0;35]
−1;0	−0; 4	−0;33
Noon	−0;12	−0;30
+1;0	−0;19	−0;27
+2;0	−0;27	−0;24
+3;0	−0;35	−0;20
+4;0	−0;40	−0;21
+5;0	−0;43	−0;22
+6;0	−0;45	−0;23
+6;59	−0;45	−0;24

Tables to Chapter 5

Table 1. Mean Motions in V,1.

	Saturn	Jupiter	Mars	Venus	Mercury
(a) Version in N					
A^r	57^r	65^r	37^r	5^r	145^r
$N^y + c^d$	$59^y + \sim 1;7,18^d$	$71^y - 5;45,23^d$	$79^y + 2;23,45^d$	$8^y - 2;26,44^d$	$46^y + 0;25^d$
$R^r + c'^{\circ}$	$2^r + \sim 1;5,50^{\circ}$	$6^r - 5;42,32^{\circ}$	$42^y + 2;21,48^{\circ}$	$13^r - 2;23,29^{\circ}$	$191^r + 0;21,53^{\circ}$
P^d_{α}	$378;5,32,42^d$	$398;53,2,58^d$	$779;56,13,55^d$	$583;55,17,50^d$	$115;52,38,53^d$
$\bar{\alpha}^{\circ/y}$	$347;32,3,9,40^{\circ}$	$329;25,8,15,6^{\circ}$	$168;28,30,36,4^{\circ}$	$225;1,45,3,40^{\circ}$	$3^r + 53;57,23,6,30^{\circ}$
$\bar{\alpha}^{\circ/d}$	$0;57,7,44,5^{\circ}$	$0;54,9,3,49^{\circ}$	$0;27,41,40,22^{\circ}$	$0;36,59,28,35^{\circ}$	$3;6,24,13,40^{\circ}$
$\bar{\lambda}^{*\circ/y}$	$12;12,45,57,24^{\circ}$	$30;19,40,51,58^{\circ}$	$191;16,18,30,36^{\circ}$	$= \bar{\odot}$	$= \bar{\odot}$
(b) Version in M					
A^r	57^r	65^r	37^r	5^r	145^r
$N^y + c^d$	$59^y + \sim 1;6,48^d$	$71^y - 5;45,27^d$	$79^y + 2;27,3^d$	$8^y - 2;26,46^d$	$46^y + 0;34,23^d$
$R^r + c'^{\circ}$	$2^r + \sim 1;6,6^{\circ}$	$6^r - 5;41,2\frac{1}{2}^{\circ}$	$42^r + 2;24,56^{\circ}$	$13^r - 2;24,40^{\circ}$	$191^r + \sim 0;34,23^{\circ}$
P^d_{α}	$378;5,32,11^d$	$398;53,2,56^d$	$779;56,19,7^d$	$583;55,17,24^d$	$115;52,42,12^d$
$\bar{\alpha}^{\circ/y}$	$347;32,2,54,12^{\circ}$	$329;25,8,15,6^{\circ}$	$168;28,29,13,12^{\circ}$	$225;1,48,54,30^{\circ}$	$3^r + 53;56,46,54,40^{\circ}$
$\bar{\alpha}^{\circ/d}$	$0;57,7,44,[2]^{\circ}$	$0;54,9,3,49^{\circ}$	$0;27,41,40,8^{\circ}$	$0;36,59,28,35^{\circ}$	$3;6,24,7,43^{\circ}$
$\bar{\lambda}^{*\circ/y}$	$12;12,46,12,52^{\circ}$	$30;19,40,51,58^{\circ}$	$191;16,19,53,52^{\circ}$	$= \bar{\odot}$	$= \bar{\odot}$

Table 2. Observations of Saturn.

No.	Date and Time	Julian Day	$\bar{\lambda}^*_{\odot}$	λ^*_p	Cop. λ^*_p	π	λ_p	Ptol. λ_p
1.	Hadrian 11 Pach 7 127 Mar 26 17^h	1767,529.708	354;41°	174;41°	~174;40°	6;33°	181;14°	181;13°
2.	Hadrian 17 Epi 18 133 June 3 15^h	1769,790.625	63;3	243;3	243;3	6;37	249;40	249;40
3.	Hadrian 20 Mes 24 136 Jul 8 11^h	1770,921.45	97;37	277;37	277;37	6;39	284;16	284;14
4.	1514 May 5 $22;48^h$	2274,171.95	25;24	205;24	205;24	27;13	232;37	—
5.	1520 Jul 13 12^h	2276,432.5	93;25	273;25	273;25	27;17	300;42	—
6.	1527 Oct 10 $6;24^h$	2279,077.267	180;7	0;7	0;7	27;21	27;28	—
7.	1514 Feb 24 5^h	2274,101.208	315;41	—	209;0	27;13	236;13	—

Table 3. Observations of Jupiter.

No.	Date and Time	Julian Day	$\bar{\lambda}^*_\odot$	λ^*_p	Cop. λ^*_p	π	λ_p	Ptol. λ_p
1.	Hadrian 17 Epi 1							
	133 May 17 22^h	1769,773.917	46;35°	226;35°	226;33°	6;37°	233;12°	233;11°
2.	Hadrian 21 Phao 12							
	136 Aug 31 21^h	1770,975.875	151;15	331;15	331;16	6;39	337;54	337;54
3.	Antoninus 1 Ath 20							
	137 Oct 8 4^h	1771,378.167	187;45	7;45	7;45	6;39	14;24	14;23
4.	1520 Apr 30 11^h	2276,358.458	20;27	200;27	200;28	27;16	227;43	—
5.	1526 Nov 28 3^h	2278,761.125	228;32	48;32	48;34	27;21	75;53	—
6.	1529 Feb 1 19^h	2279,557.792	293;44	113;44	113;44	27;22	141;6	—
7.	1520 Feb 18 6^h	2276,286.25	309;17	—	205;9	27;16	232;25	—

Table 4. Observations of Mars.

No.	Date and Time	Julian Day	$\bar{\lambda}^*_\odot$	λ^*_p	Cop. λ^*_p	π	λ_p	Ptol. λ_p
1.	Hadrian 15 Tyb 26							
	130 Dec 15 0^h	1768,889.0	254;24°	74;24°	74;20°	6;35°	80;59°	81;0°
2.	Hadrian 19 Phar 6							
	135 Feb 21 20^h	1770,418.833	322;13	142;13	142;10	6;38	148;51	148;50
3.	Antoninus 2 Epi 12							
	139 May 27 21^h	1771,974.875	55;52°	235;52	235;54	6;41	242;33	242;34
4.	1512 Jun 5 1^h	2273,472.042	55;34	235;34	235;33	27;12	262;46	—
5.	1518 Dec 12 20^h	2275,853.833	243;5	63;5	63;2	27;16	90;21	—
6.	1523 Feb 22 5^h	2277,386.208	313;24	133;24	133;20	27;18	160;42	—
7.	1512 Jan 1 6^h	2273,316.25	262;1	—	191;20	27;12	218;32	—

Table 5. Observations of Venus.

No.	Observer	Egyptian Date and Time		E.A.	Time	Julian Date	Julian Day
1.	Theon	Hadrian 16 Phar 21	E	455	18^h	132 Mar 8	1769,338.75
2.	Ptol.	Antoninus 4 Tho 12	M	464	4;30	140 Jul 30	1772,404.188
3.	Theon	Hadrian 4 Ath 20	M	443	5;30	119 Oct 12	1764,807.229
4.	Ptol.	Hadrian 21 Mech 9	E	460	18	136 Dec 25	1771,091.75
5.	Theon	Hadrian 13 Epi 3	M	452	4	129 May 20	1768,315.167
6.	Ptol.	Hadrian 21 Tyb 2	E	460	16	136 Nov 18	1771,054.667
7.	Ptol.	Hadrian 18 Phar 2	M	457	5;30	134 Feb 18	1770,050.229
8.	Ptol.	Antoninus 3 Phar 4	E	463	16;30	140 Feb 18	1772,241.688
9.	Tim.	Ptol. Phil. 13 Mes 18	M	52	5;30	−271 Oct 12	1622,360.229
10.	Ptol.	Antoninus 2 Tyb 30	M	462	3;45	138 Dec 16	1771,812.156
11.	Cop.	— [Phar 9]	E	[1853]	19;30	1529 Mar 12	2279,596.813

Table 6. Observations of Venus (*Continued*)

No.	Cop. $\bar{\lambda}^*_{\odot}$	$\bar{\lambda}^*_{\odot}$	π	$\bar{\lambda}'_{\odot}$	Ptol. $\bar{\lambda}_{\odot}$	Obs. λ_p	Cop. λ^*_p	Δ
1.	337;41°	337;41°	6;36°	344;17°	344;15°	31;30°	[24;56°]	+47;15°
2.	119;0	119;0	6;42	125;42	125;45	78;30	[71;45]	−47;15
3.	191;13	191;23	6;28	197;51	197;52	150;20	[143;41 or ;51]	−47;32
4.	265;[27]	265;27	6;40	272;7	272;4	319;36	[312;59]	+47;32
5.	48;50	48;50	6;34	55;24	55;24	10;36	4;[2]	−44;48
6.	228;54	228;54	6;39	235;33	235;30	282;50	276;10	+47;16 or ;20
7.	318;50	318;55	6;38	325;33	325;30	281;55	275;15	−43;35
8.	318;50	318;50	6;42	325;32	325;30	13;50	7;10	+48;20
9.	194;23	194;27	2;34	197;1	197;3	154;10	151;30	−42;53
10.	255;30	255;30	6;40	262;10	262;9	216;30	209;46	−45;44 or ;39
11.	332;11	332;12	27;22	359;34	—	36;36	9;12	+37;1

Table 7. Observations of Mercury.

No.	Observer	Egyptian Date and Time		Julian Date and Time		Julian Day	Δt since Epoch
1.	Ptol.	Antoninus 1 Epi 20	E	138 Jun 4	17^h	1771,617.708	137^y $188;42,30^d$
2.	Ptol.	Antoninus 4 Pham 19	M	141 Feb 2	4;48	1772,591.2	140 ~ 67;12
3.	Ptol.	Hadrian 19 Ath 15	M	134 Oct 3	5	1770,277.208	[133 308;12,30]
4.	Ptol.	Hadrian 19 Pach 19	E	135 Apr 5	16;48	1770,461.7	[134 127;42]
5.	Theon	Hadrian 14 Mes 18	E	130 Jul 4	18	1768,725.75	129 216;45
6.	Ptol.	Antoninus 2 Mes 21	M	139 Jul 5	4;48	1772,013.2	138 219;12
7.	?	Ptol. Phil. 21 Tho 19	M	−264 Nov 15	6	1624,951.25	59 17;45 (E.A.)
8.	Walther			1491 Sep 9	5	2265,897.208	1491 258;12,30
9.	Walther			1504 Jan 9	6;30	2270,402.271	[1504 18;16,15]
10.	Walther			1504 Mar 18	19;30	2270,471.813	[1504 87;48,45]

Table 8. Observations of Mercury (*Continued*)

No.	Cop. $\bar{\lambda}^*_{\odot}$	$\bar{\lambda}^*_{\odot}$	π	$\bar{\lambda}'_{\odot}$	Ptol. $\bar{\lambda}_{\odot}$	Obs. λ_p	Cop. λ^*_p	Δ
1.	63;50°	63;51°	6;40°	70;31°	70;30°	97°	90;20°	+26;30°
2.	303;19	303;20	6;42	310;2	310;0	283;30	~276;49	−26;30
3.	182;38	182;38	6;38	189;16	189;15	170;12	163;35	−19;3
4.	4;28	4;28	6;38	11;6	11;5	34;20	27;43	+23;15
5.	93;30	93;30	6;35	100;5	100;5	126;20	119;45	+26;15
6.	93;39	93;39	6;41	100;20	100;20	80;5	73;24	−20;15
7.	228;8	228;11	2;38	230;49	230;50	213;20	210;40	−17;28 or ;30
					Cop. $\bar{\lambda}'_{\odot}$		Cop. λ_p	
8.	149;48	149;45	26;59	176;44	176;47	163;23	163;30	~ −13;15 or ;24
9.	[270;2]	269;59	27;6+	297;5	297;7	273;15	273;20	−23;47 or ;52
10.	[338;34]	338;31	27;6+	5;38	5;39	26;30	26;55	+21;17 or 20;51

Table 9. Additional Observations from the *Almagest*.

Planet	No	Julian Date and Time		Julian Day	$\bar{\lambda}^*_\odot$	π	$\bar{\lambda}'_\odot$	Ptol. $\bar{\lambda}_\odot$	Ptol. λ_p	Δ
Saturn	8.	138 Dec 22	19^h	1771,818.792	262;2°	6;40°	268;42°	268;41°	309;4°	
Saturn	9.	−228 Mar 1	18	1637,841.75	333;10	3;0	336;10	336;10	159;30	
Jupiter	8.	139 Jul 11	4	1772,019.167	99;32	6;41	106;13	106;11	75;45	
Jupiter	9.	−240 Sep 4	5	1633,645.208	157;1	2;53	159;54	159;56	97;33	
Mars	8.	139 May 30	20	1771,977.833	58;47	6;41	65;28	65;27	241;36	
Mars	9.	−271 Jan 18	6	1622,093.25	291;19	2;33	293;52	293;54	212;13	
Mercury	11.	132 Feb 2	18	1769,303.75	303;11	6;36	309;47	309;45	331;0	+21;15°
Mercury	12.	134 Jun 4	5	1770,156.208	63;23	6;38	70;1	70;0	48;45	−21;15
Mercury	13.	139 May 17	18	1771,964.75	45;54	6;41	52;35	52;34	77;30	+24;56

Table 10. Superior Planets.

No.	λ^*_p	λ_p	Obs. λ_p	Error
		Saturn		
1.	174;40°	181;13°	181;13°	0;0°
2.	243;3	249;40	249;40	0;0
3.	277;36	284;15	284;14	+0;1
4.	205;21	232;34	232;37	−0;3
5.	273;21	300;38	300;42	−0;4
6.	0;4	27;25	27;28	−0;3
7.	208;56	236;9	236;13	−0;4
8.	302;25	309;5	309;4	+0;1
9.	156;31	159;31	159;30	+0;1
		Jupiter		
1.	226;37	233;14	233;11	+0;3
2.	331;16	337;55	337;54	+0;1
3.	7;45	14;24	14;23	+0;1
4.	200;25	227;41	227;43	−0;2
5.	48;28	75;49	75;53	−0;4
6.	113;41	141;3	141;6	−0;3
7.	205;8	232;24	232;25	−0;1
8.	69;3	75;44	75;45	−0;1
9.	95;9	98;2	97;33	+0;29
		Mars		
1.	74;19	80;54	81;0	−0;6
2.	142;8	148;46	148;50	−0;4
3.	235;53	242;34	242;33	+0;1
4.	235;39	262;51	262;46	+0;5
5.	63;13	90;29	90;21	+0;8
6.	133;22	160;40	160;42	−0;2
7.	191;27	218;39	218;32	+0;7
8.	234;57	241;38	241;36	+0;2
9.	209;33	212;6	212;13	−0;7

Table 11. Inferior Planets.

No.	Δ	Obs. Δ	Error	λ_p	Obs. λ_p	Error
			Venus			
1.	+46;52°	+47;15°	−0;23°	31;9°	31;30°	−0;21°
2.	−46;15	−47;15	+1;0	79;27	78;30	+0;57
3.	−47;23	−47;32	+0;9	150;28	150;20	+0;8
4.	+47;27	+47;32	−0;5	319;34	319;36	−0;2
5.	−43;58	−44;48	+0;50	11;26	10;36	+0;50
6.	+46;8	+47;20	−1;12	281;41	282;50	−1;9
7.	−43;33	−43;35	+0;2	282;0	281;55	+0;5
8.	+48;18	+48;20	−0;2	13;50	13;50	0;0
9.	−42;53	−42;53	0;0	154;8	154;10	−0;2
10.	−45;29	−45;39	+0;10	216;41	216;30	+0;11
11.	+37;4	+37;1	+0;3	36;38	36;36	+0;2
			Mercury			
1.	+26;6	+26;30	−0;24	96;37	97;0	−0;23
2.	−26;28	−26;30	+0;2	283;34	283;30	+0;4
3.	−18;47	−19;3	+0;16	170;29	170;12	+0;17
4.	+23;19	+23;15	+0;4	34;25	34;20	+0;5
5.	+26;11	+26;15	−0;4	126;16	126;20	−0;4
6.	−20;11	−20;15	+0;4	80;9	80;5	+0;4
7.	−17;25	−17;30	+0;5	213;24	213;20	+0;4
8.	−13;36	−13;24	−0;12	163;11	163;23	−0;12
9.	−23;45	−23;52	+0;7	273;23	273;15	+0;7
10.	+21;18	+20;51	+0;27	26;57	26;30	+0;27
11.	+21;3	+21;15	−0;12	330;50	331;0	−0;10
12.	−21;3	−21;15	+0;12	48;58	48;45	+0;13
13.	+25;20	+24;56	+0;24	77;54	77;30	+0;24

Table 12. Distances of the Planets.

Planet	Battānī				Copernicus	
	m	M	μ	$\mu/\mu_{\odot}$	$\mu/\mu_{\odot}$	μ
Mercury	64;10[tr]	166[tr]	115[tr]	0;6,14	0;22,35	430[tr]
Venus	166	1070	618	0;33,28	0;43,10	822
Sun or Earth	1070	1146	1108	1;0	1;0	1142
Mars	1146	8022	4584	4;8,14	1;31,11	1736
Jupiter	8022	12924	10473	9;27,8	5;13,9	5960
Saturn	12924	18094	15509	13;59,50	9;10,27	10477

Figures

Figures to Chapter 1

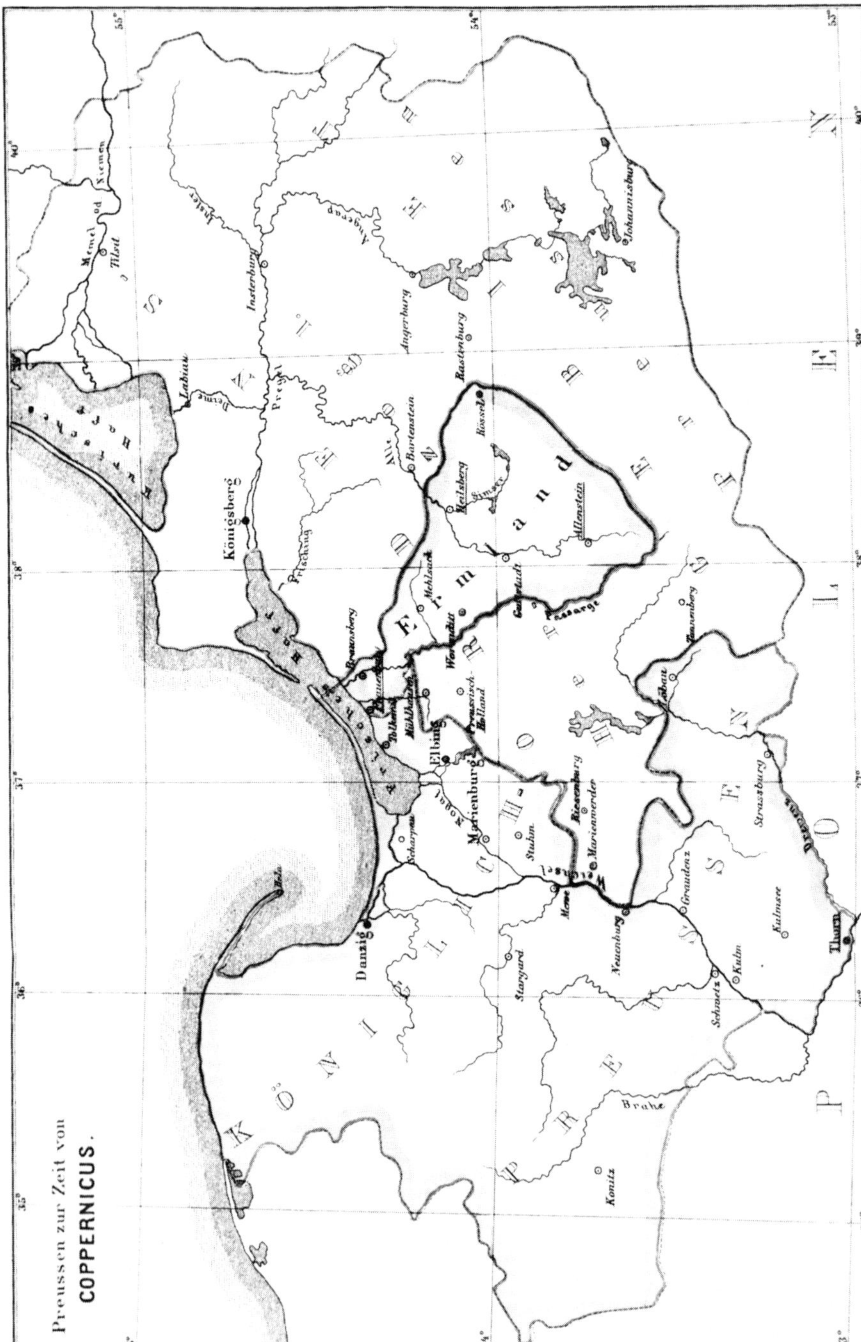
Preussen zur Zeit von
COPPERNICUS.
Königsberg
Danzig
Elbing
Marienburg
Thorn
Ermland

Fig. 1

Fig. 2. Venice, Bibl. Naz. Mar. Sign.: 138c. 4., 26.

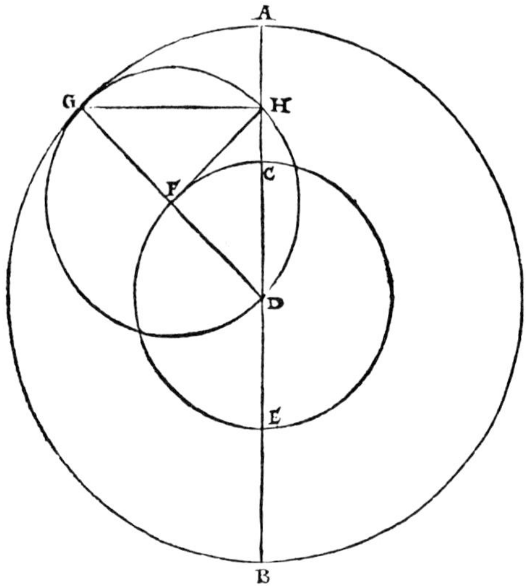

Fig. 3. Copernicus, De revolutionibus, Nuremberg ed., f. 67v.

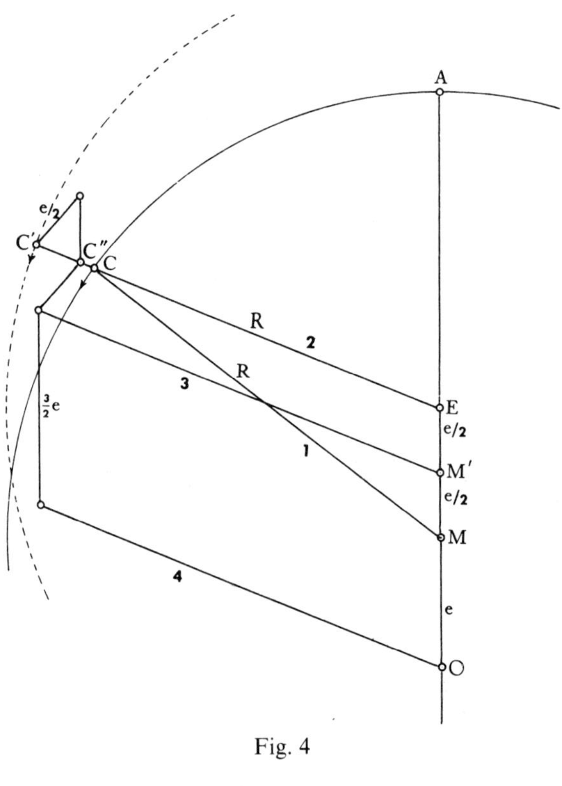

Fig. 4

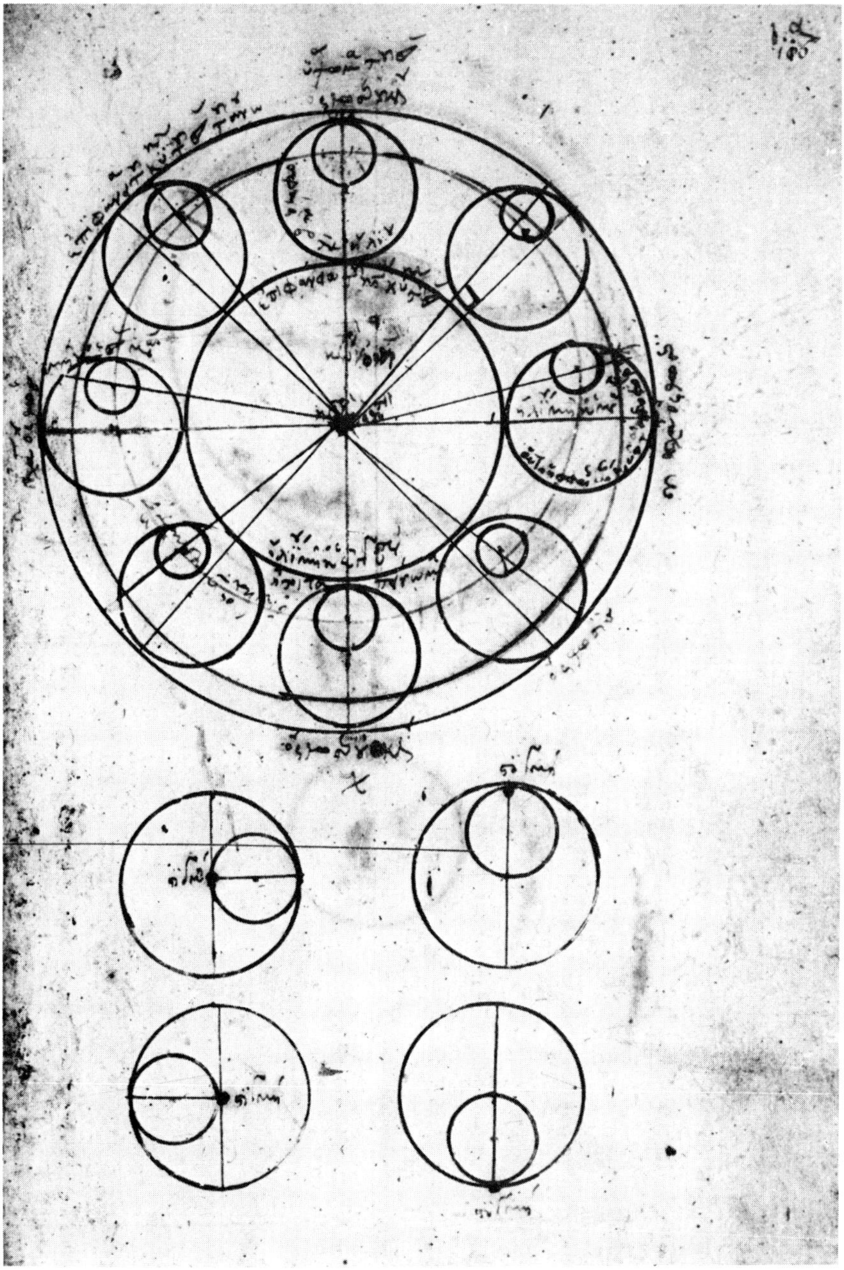

Fig. 5. Vat. Gr. 211, f. 116r.

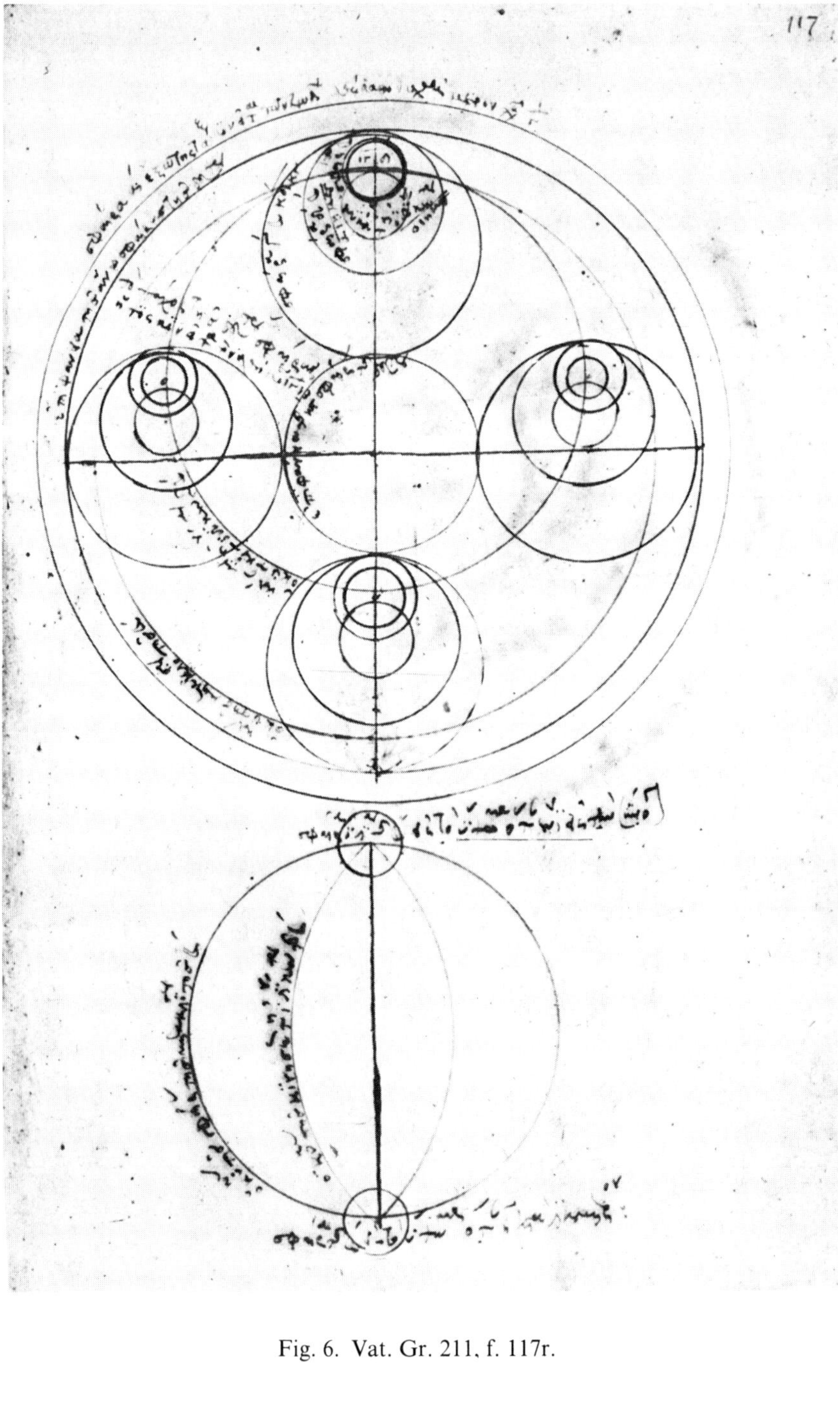

Fig. 6. Vat. Gr. 211, f. 117r.

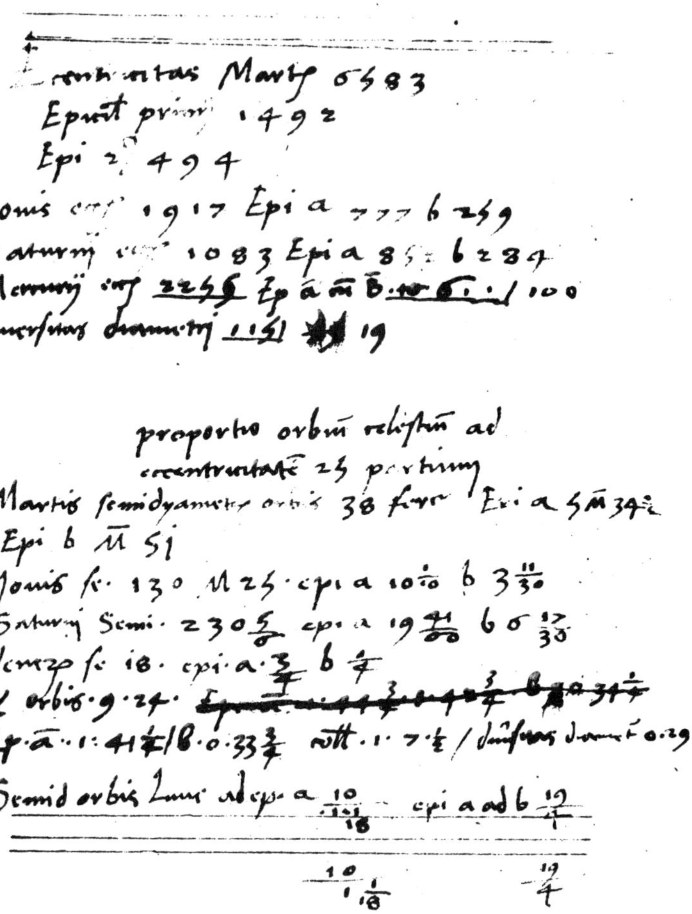

Fig. 7. Uppsala Univ. Bibl. 34.VII.65, f. 284v.

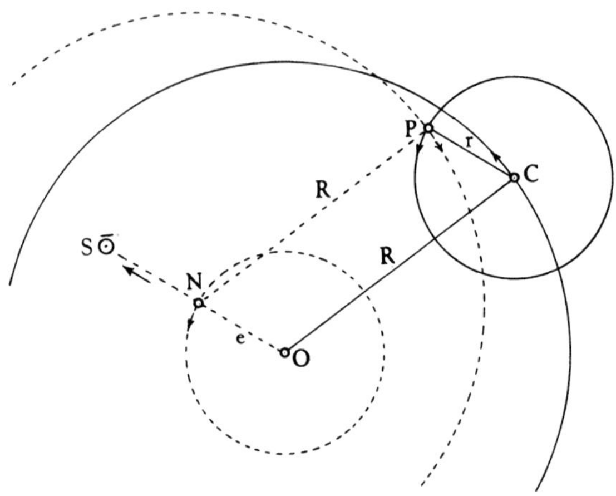

(a)

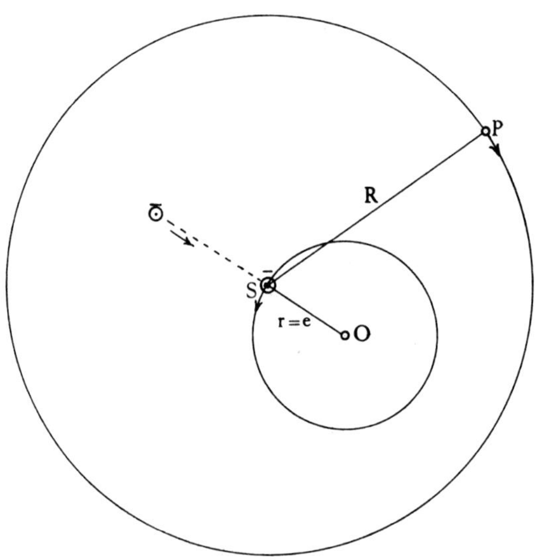

(b)

Fig. 8

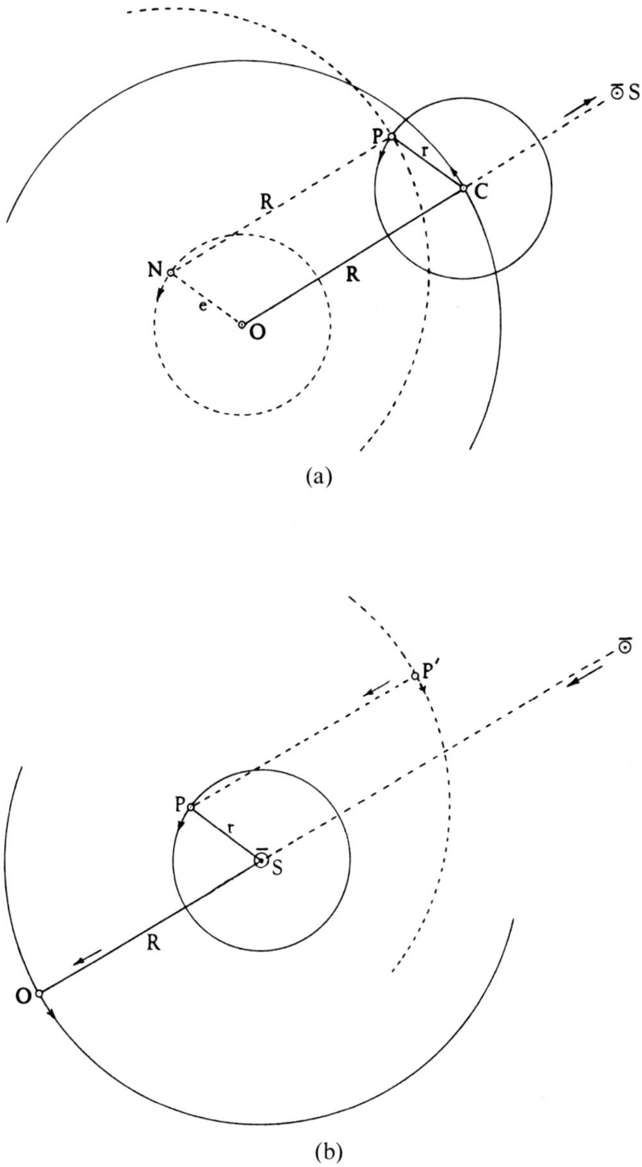

Fig. 9

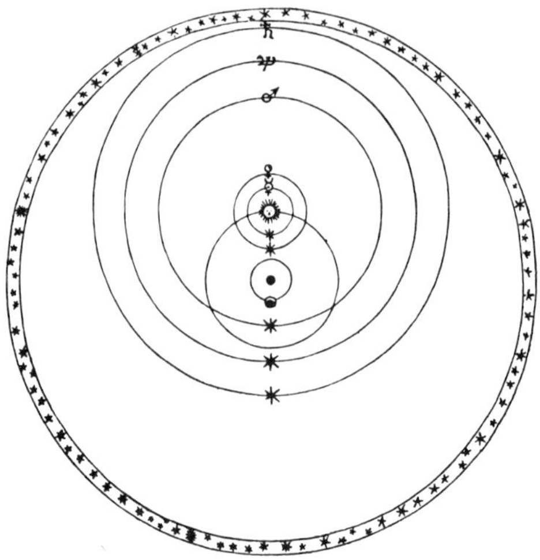

Fig. 10. Brahe, De mundi aetherei recentioribus phaenomenis, p. 189.

Fig. 11(a). Copernicus, De revolutionibus, autograph ms., f. 9v.

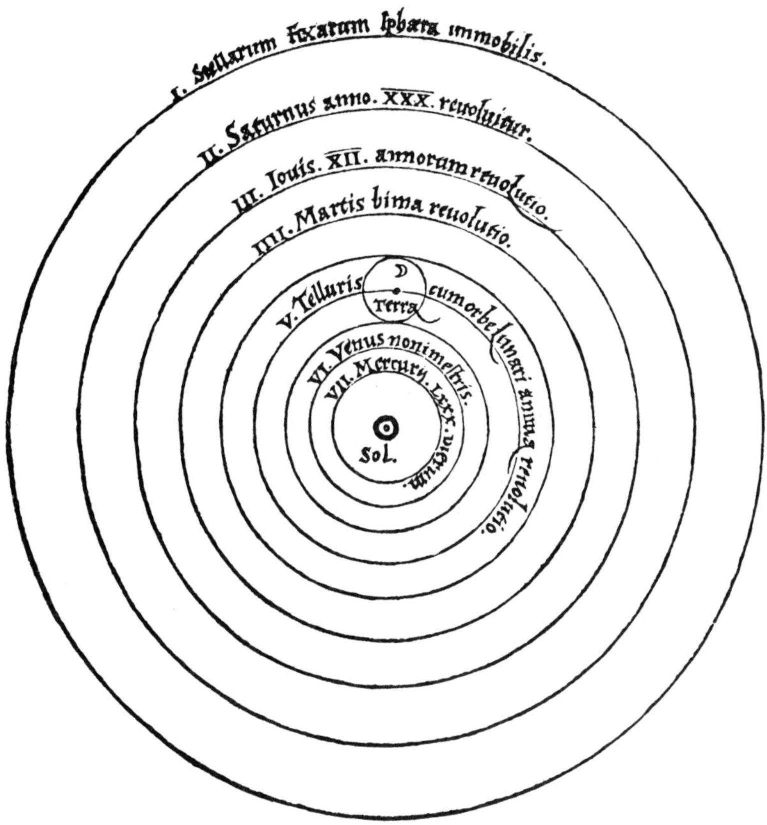

Fig. 11(b). Copernicus, De revolutionibus, Nuremberg ed., f. 9v.

Figures to Chapter 2

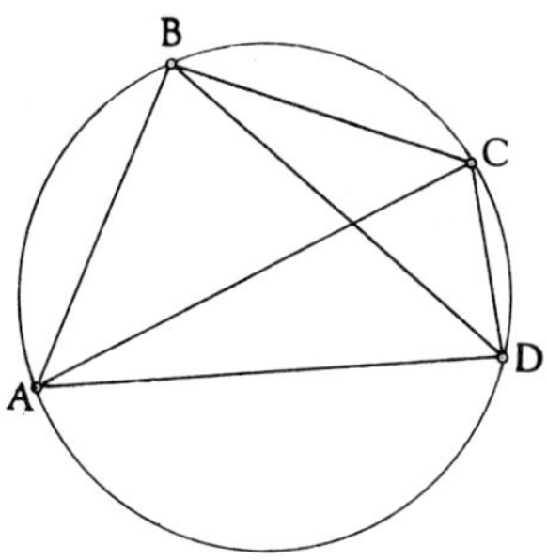

Fig. 1

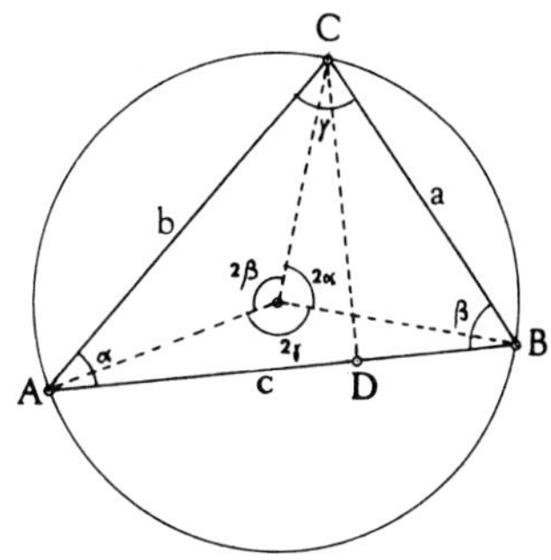

Fig. 2

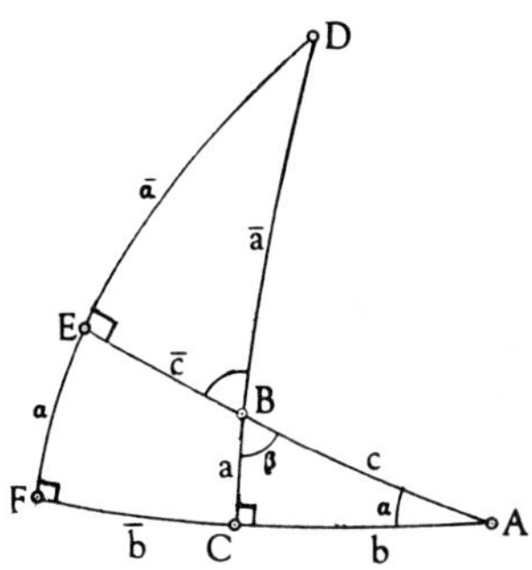

Fig. 3

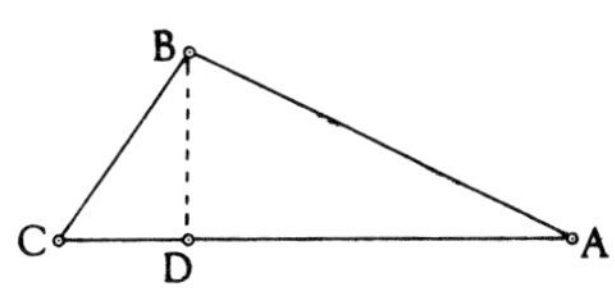

(a)

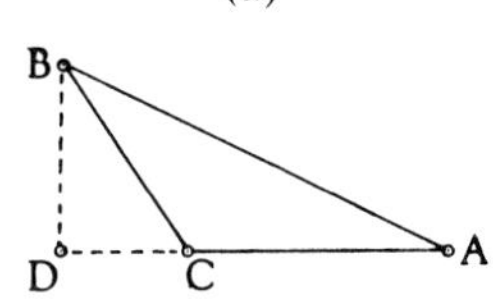

(b)

Fig. 4

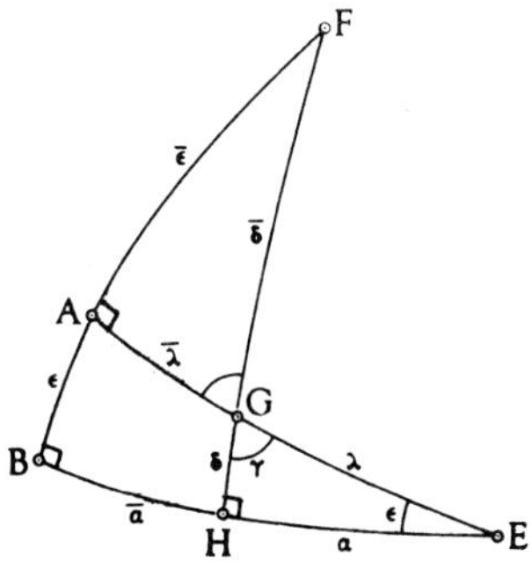

Fig. 5

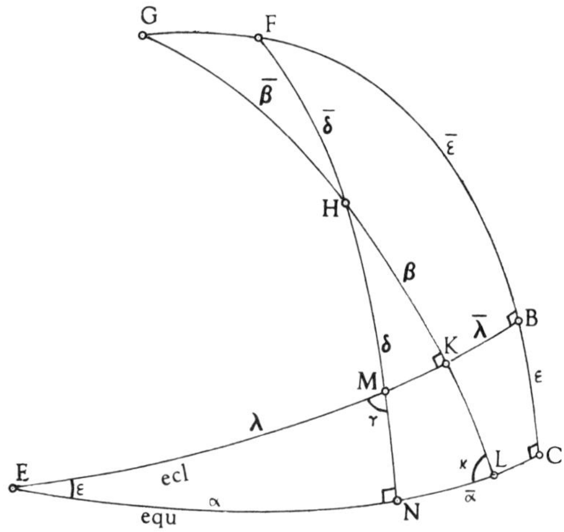

Fig. 6

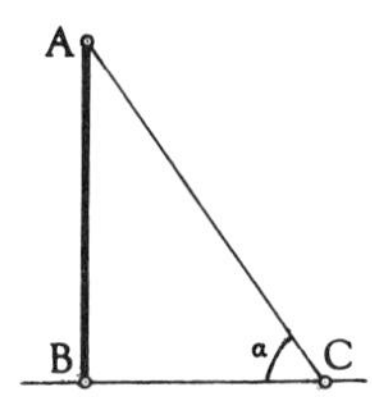

Fig. 7

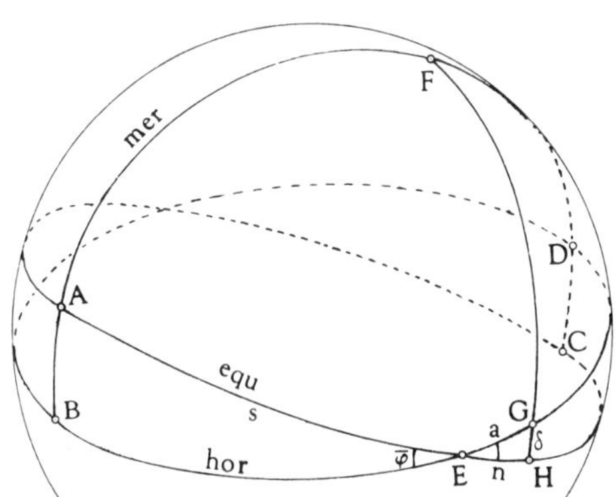

Fig. 8

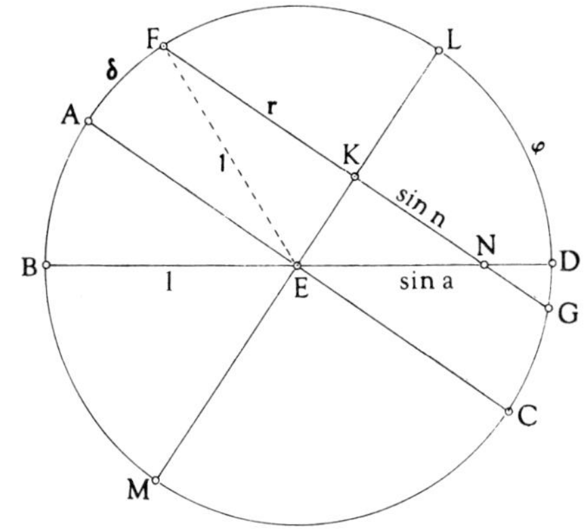

Fig. 9

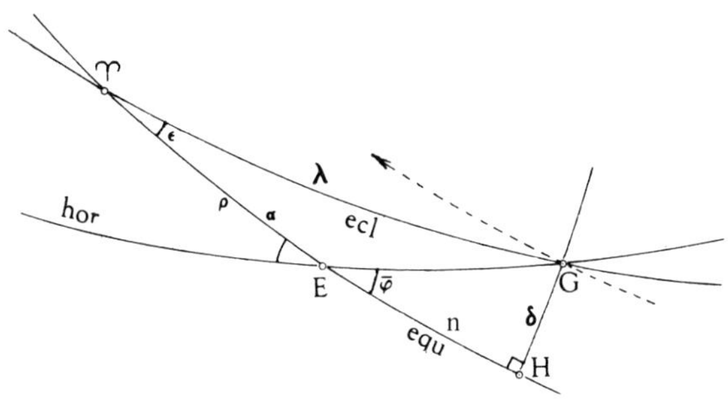

Fig. 10

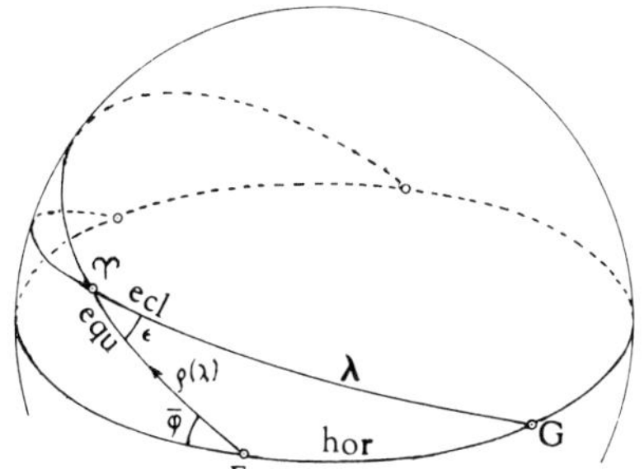

Fig. 11

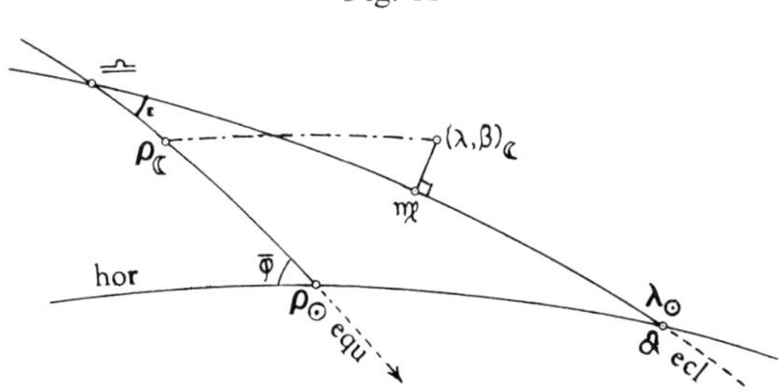

Fig. 12

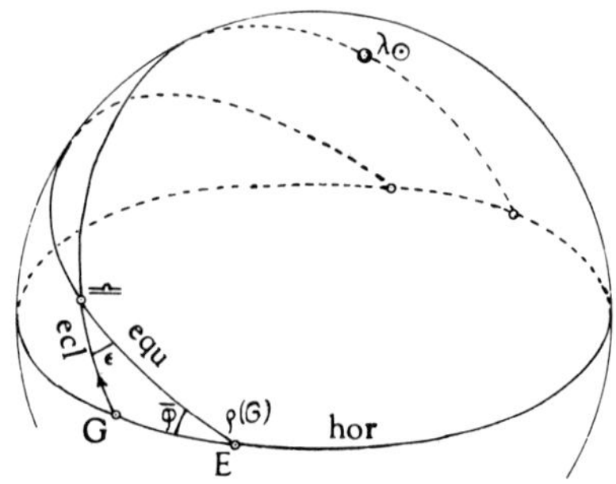

Fig. 13

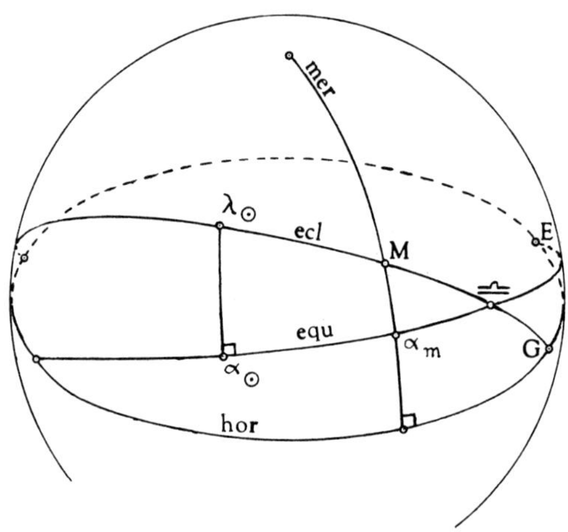

Fig. 14

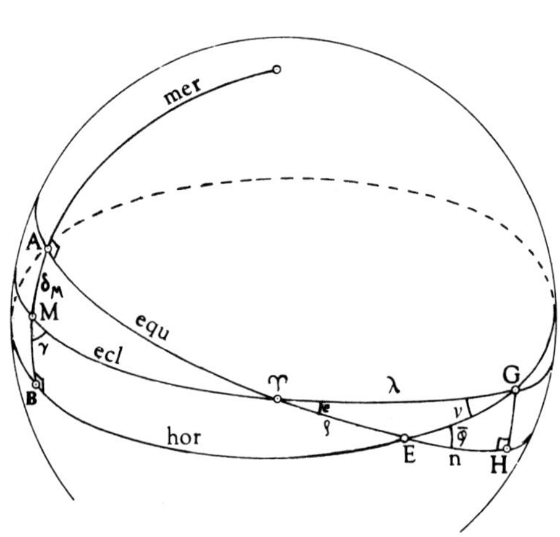

Fig. 15

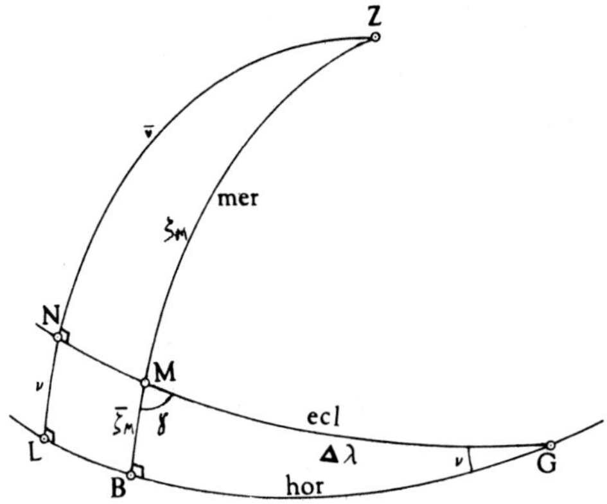

Fig. 16

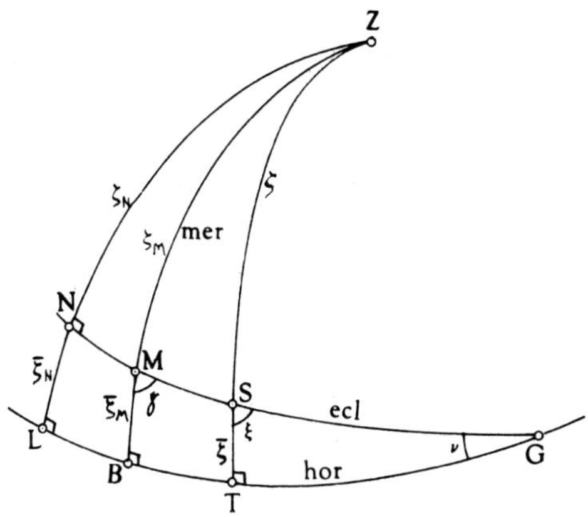

Fig. 17

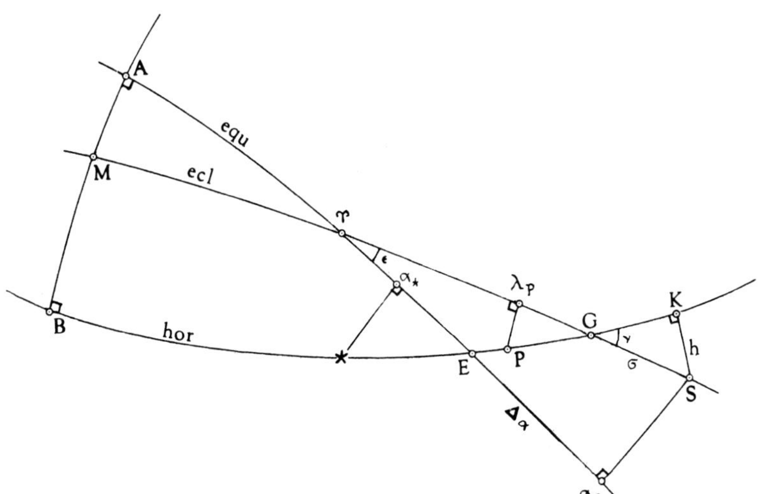

Fig. 18

Figures to Chapter 3

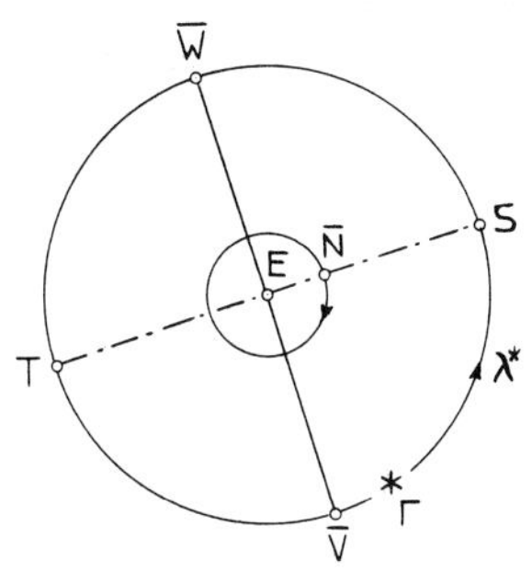

Fig. 1

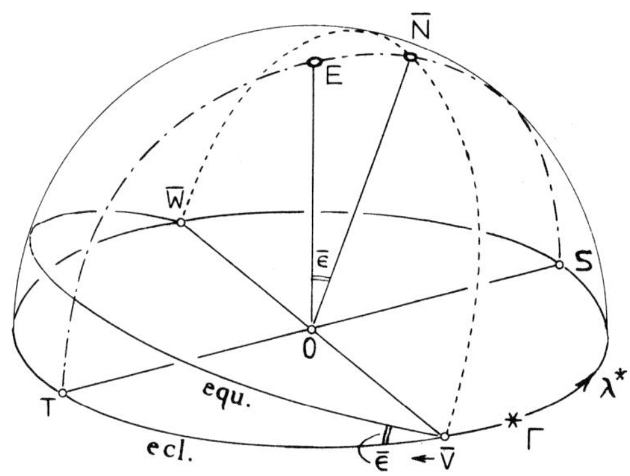

Fig. 2

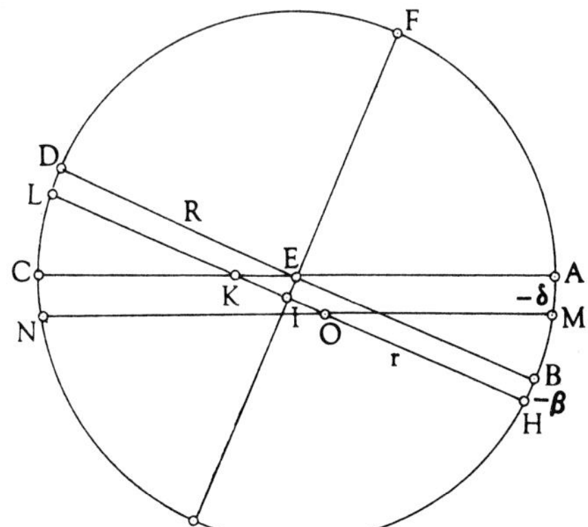

Fig. 3

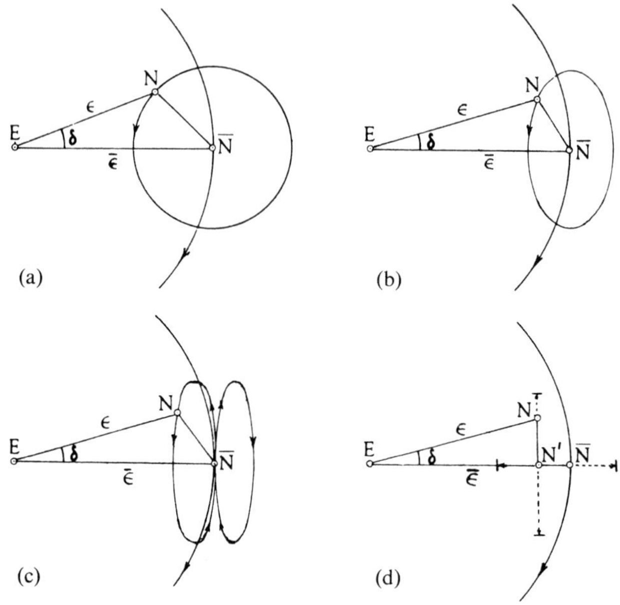

Fig. 4

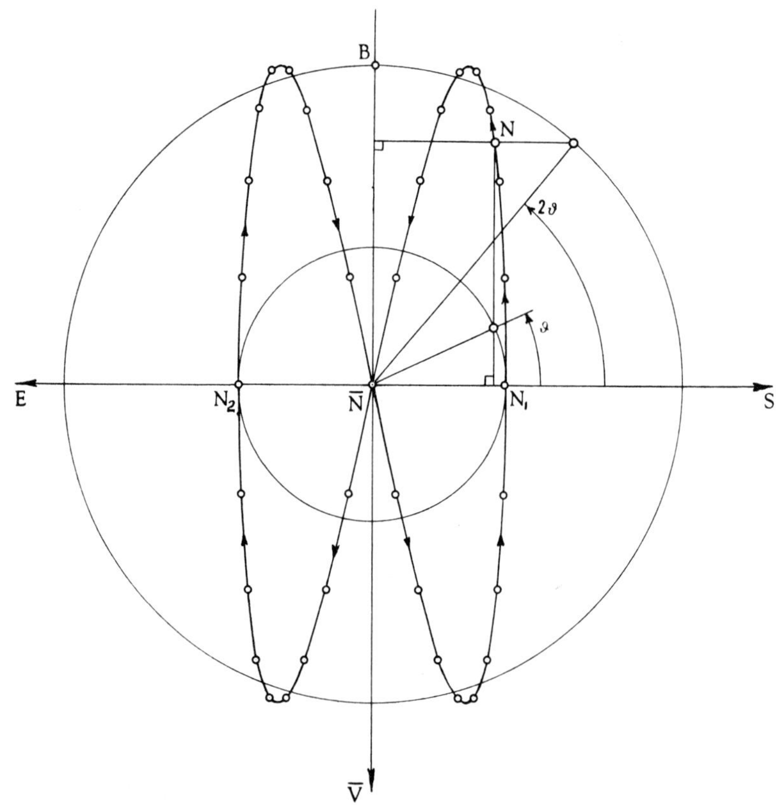

Fig. 5

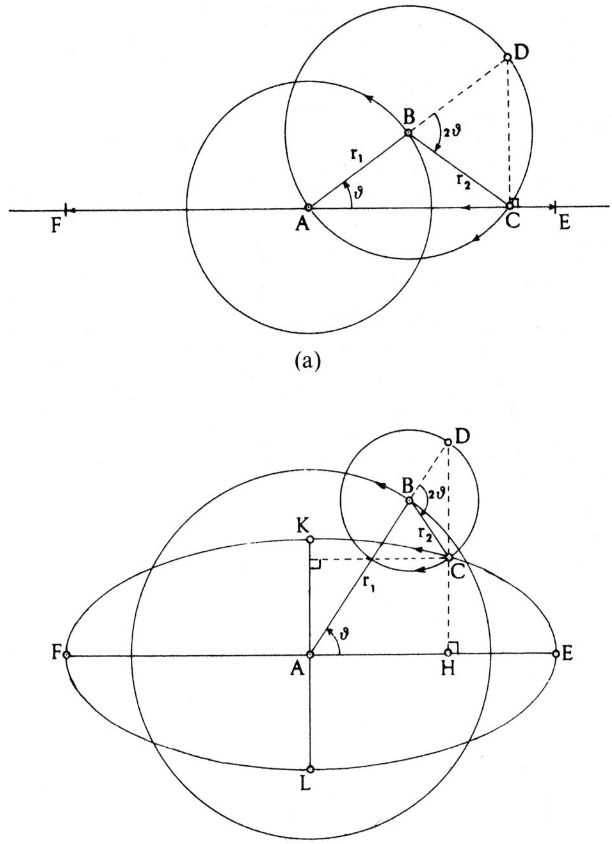

(a)

(b)

Fig. 6

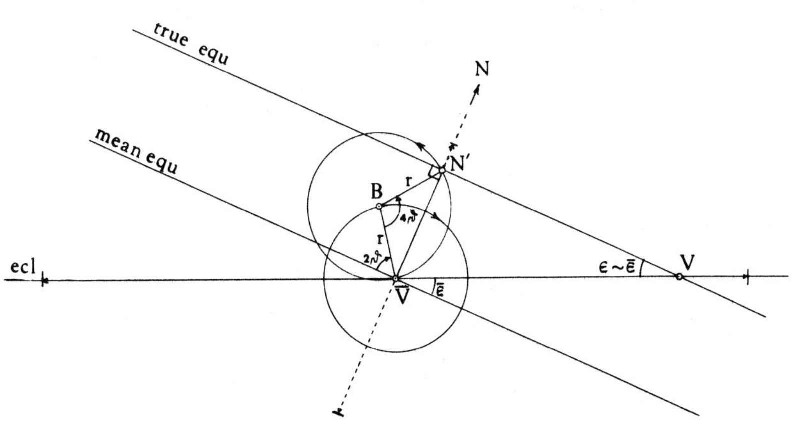

Fig. 7

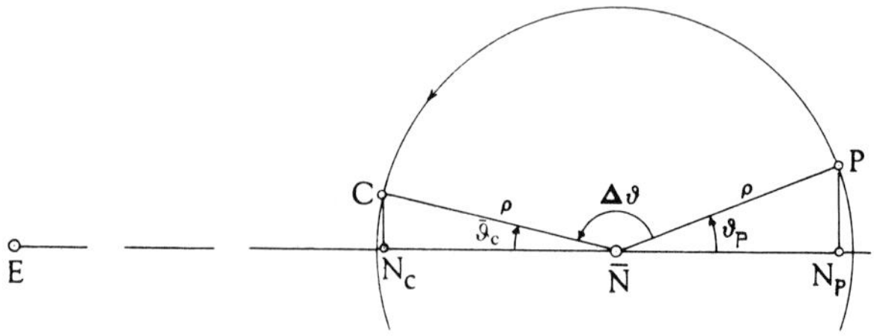

Fig. 8

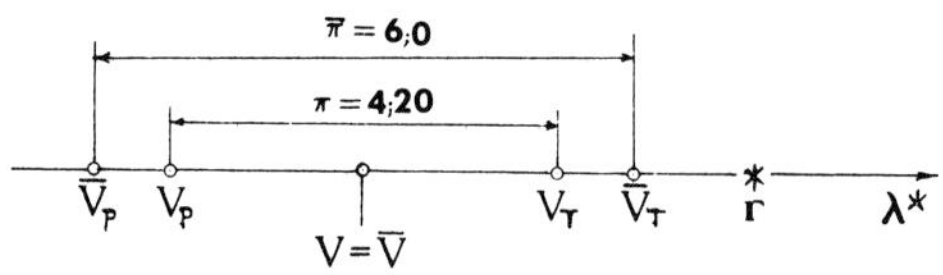

Fig. 9

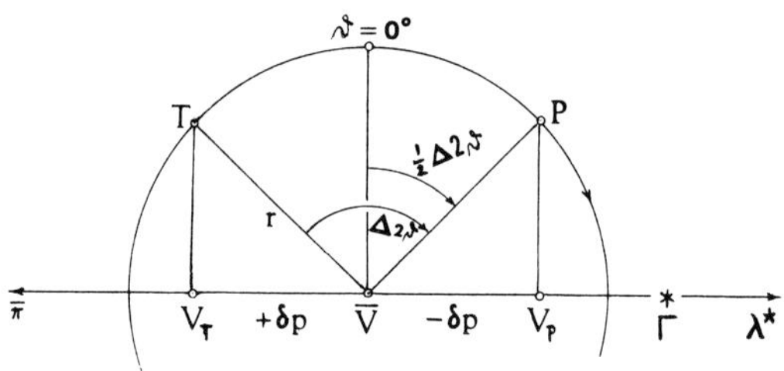

Fig. 10

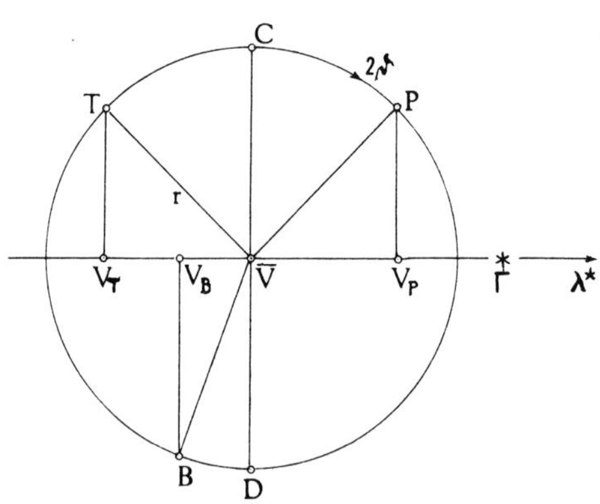

Fig. 11

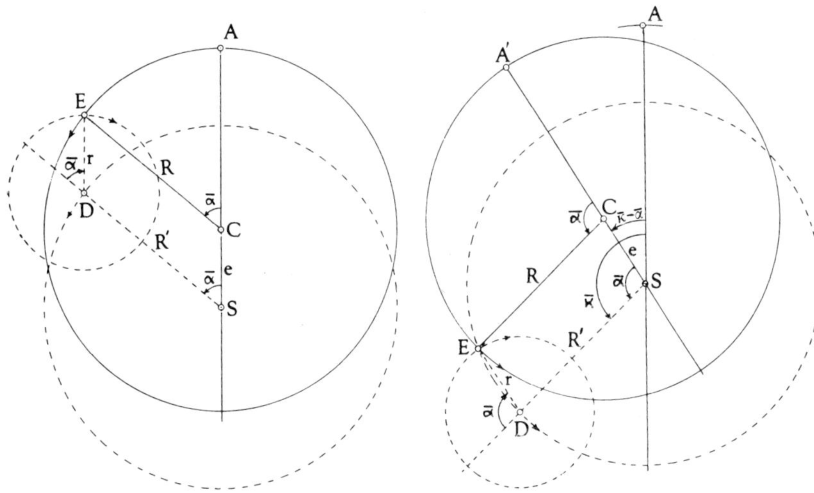

(a) (b)

Fig. 12

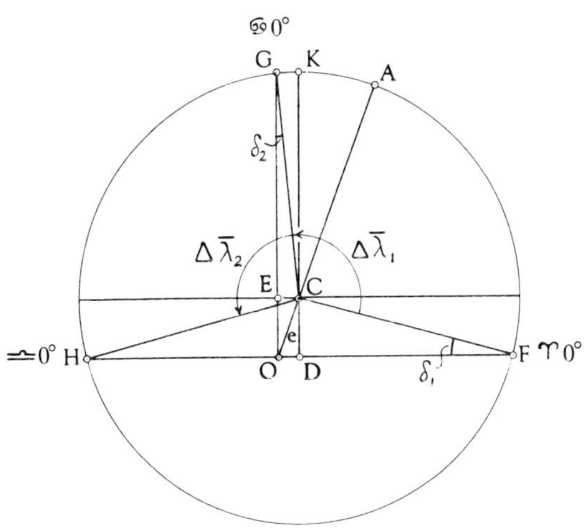

Fig. 13

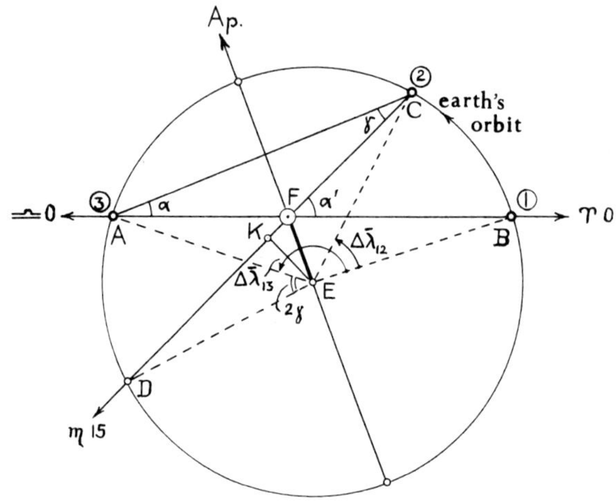
Ap.
earth's orbit
F
C
A
B
K
E
D
α
α'
γ
2γ
$\Delta\bar{\lambda}_{12}$
$\Delta\bar{\lambda}_{13}$
♎ 0
♈ 0
♏ 15

Fig. 14

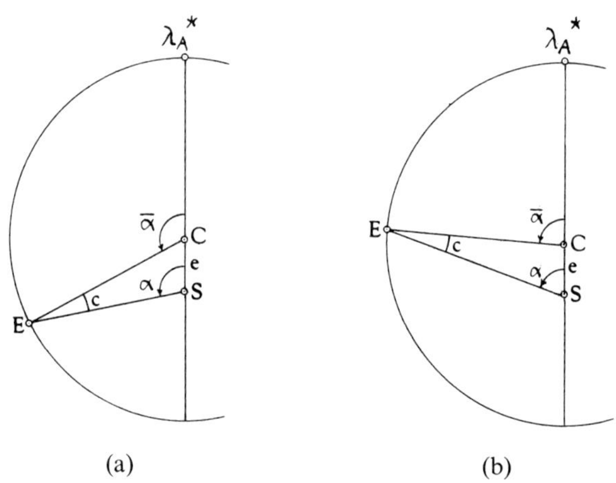
λ_A^*
C
S
E
e
c
$\bar{\alpha}$
α
(a)
λ_A^*
C
S
E
e
c
$\bar{\alpha}$
α
(b)

Fig. 15

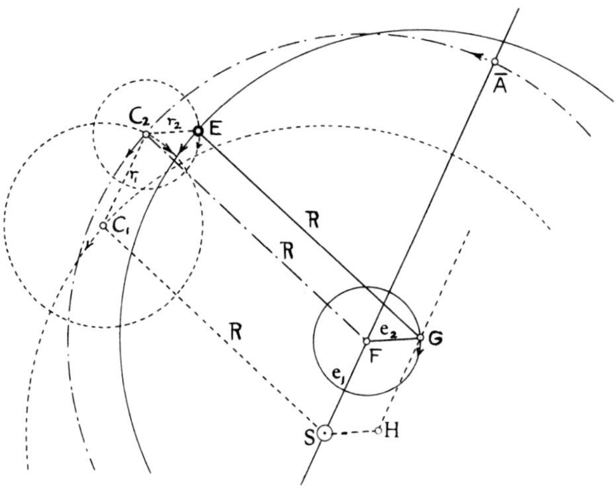

Fig. 16

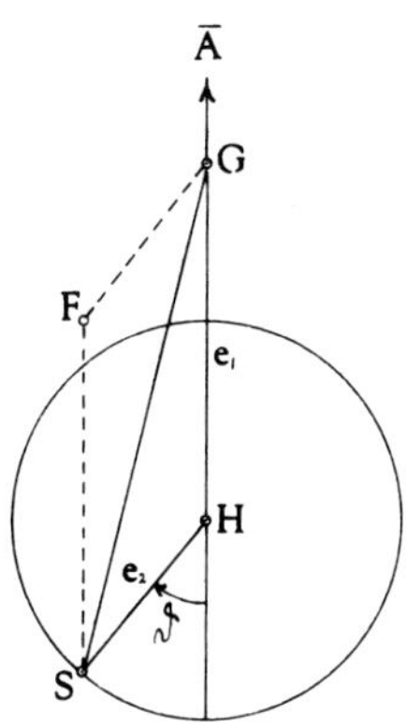

Fig. 17

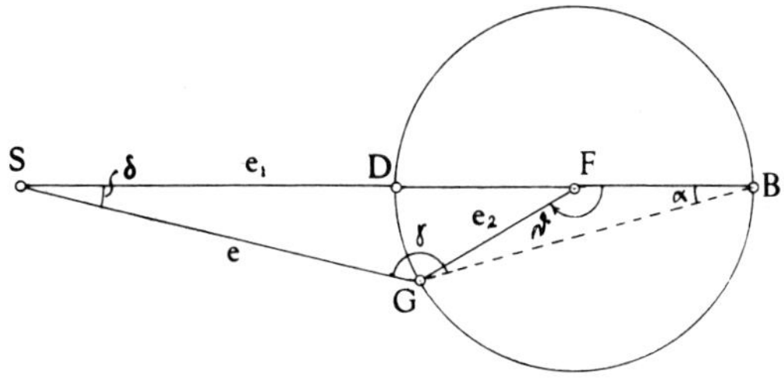

Fig. 18

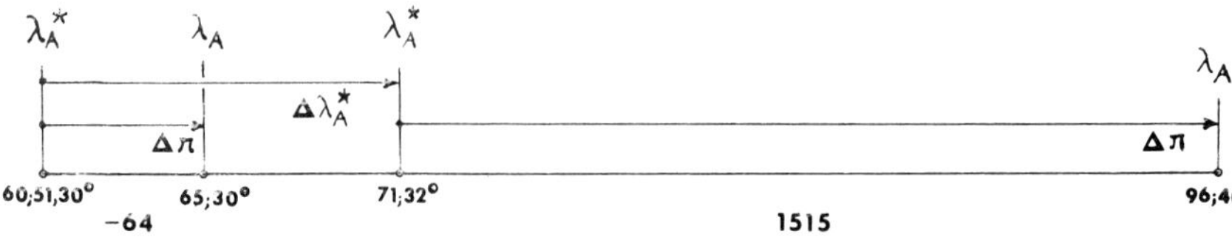
λ_A^*
λ_A
λ_A^*
λ_A
$\Delta\lambda_A^*$
$\Delta\pi$
$\Delta\pi$
60;51,30°
65;30°
71;32°
96;40°
-64
1515

Fig. 19

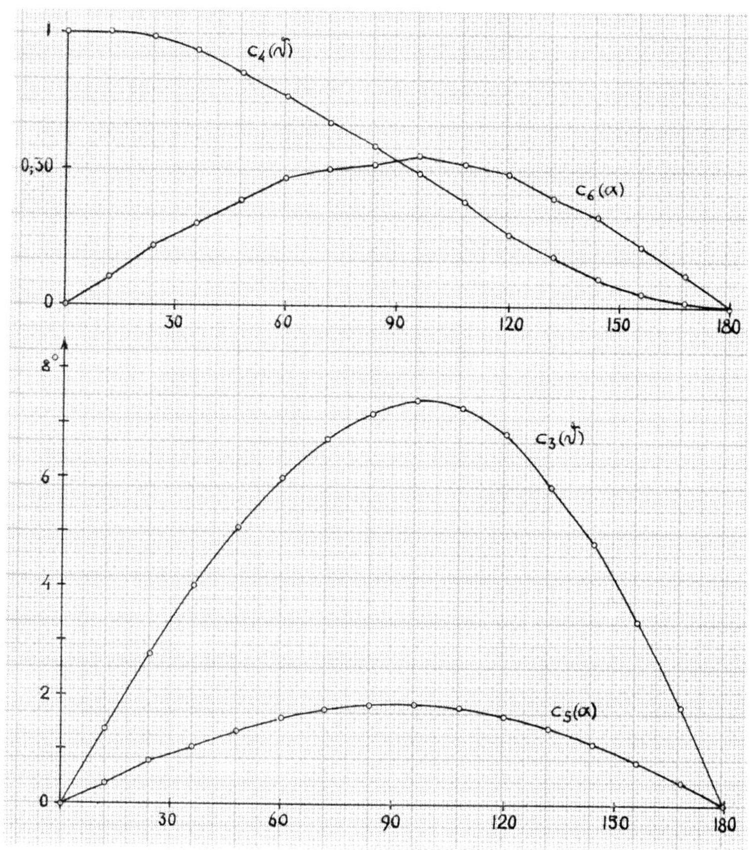
1
0;30
0
30
60
90
120
150
180
$c_4(\vartheta)$
$c_6(\alpha)$
8°
6
4
2
0
$c_3(\vartheta)$
$c_5(\alpha)$

Fig. 20

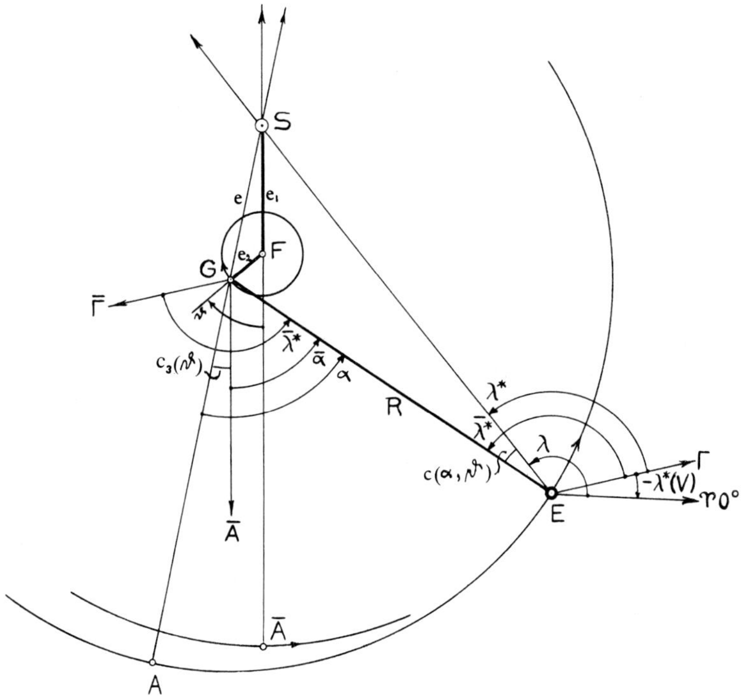
S
e
e_1
F
G
e_2
$\bar{\Gamma}$
$\bar{\lambda}^*$
$\bar{\alpha}$
α
$c_3(\vartheta)$
R
λ^*
$\bar{\lambda}^*$
λ
$c(\alpha, \vartheta)$
Γ
$-\lambda^*(V)$
♈0°
E
$\bar{A}$
$\bar{A}$
A

Fig. 21

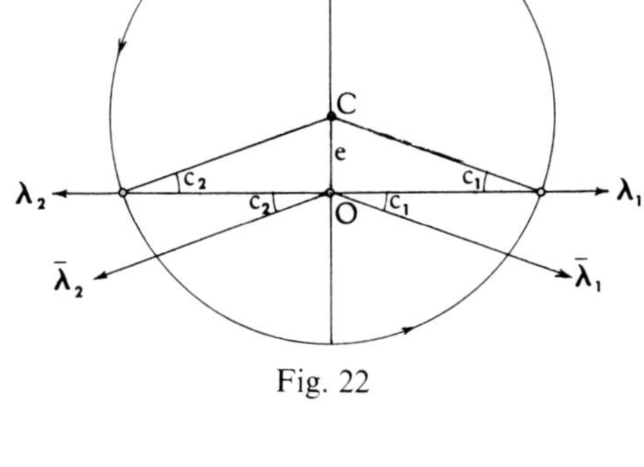
A
C
e
λ_2
c_2
c_2
O
c_1
c_1
λ_1
$\bar{\lambda}_2$
$\bar{\lambda}_1$

Fig. 22

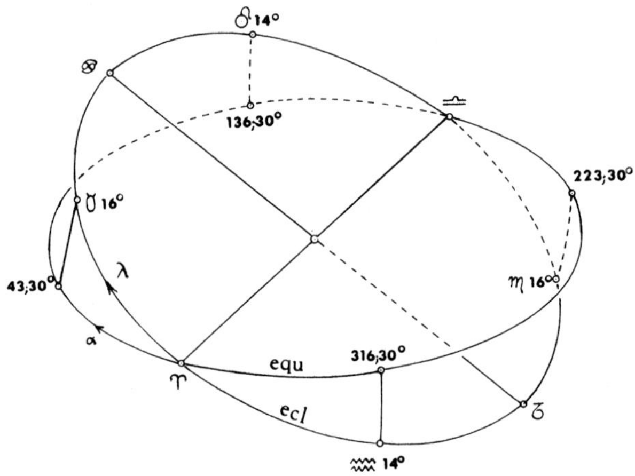
14°
136;30°
223;30°
16°
43;30°
λ
α
m 16°
equ
316;30°
ecl
14°

Fig. 23

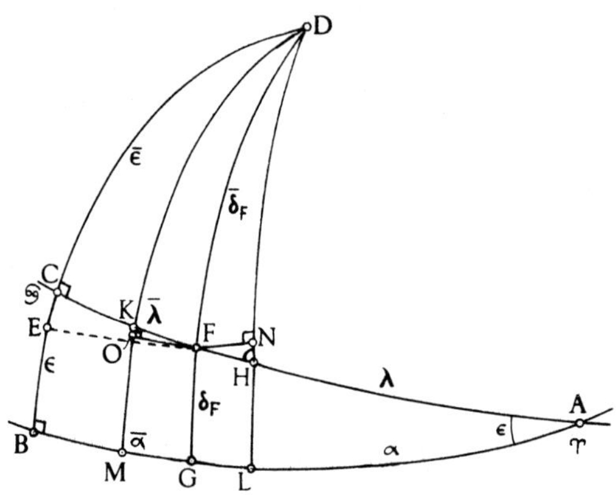
D
C
K
E
F
N
O
H
λ
A
B
M
G
L
α
ε

Fig. 24

Figures to Chapter 4

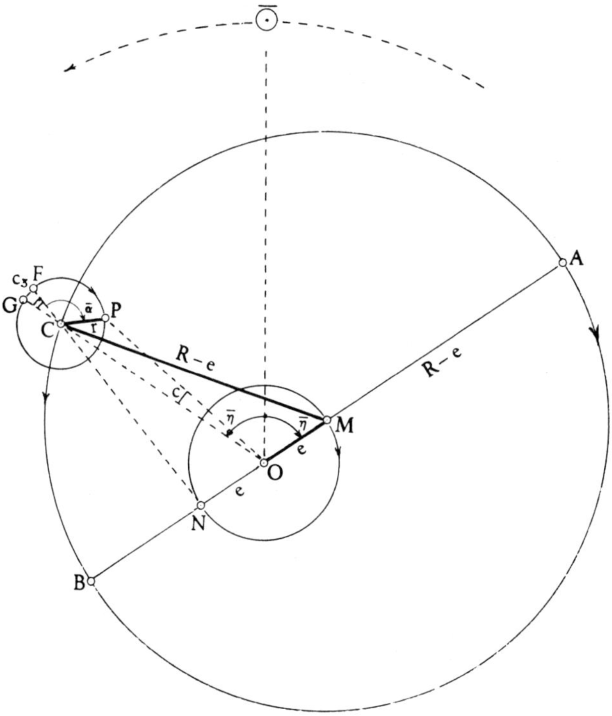

Fig. 1

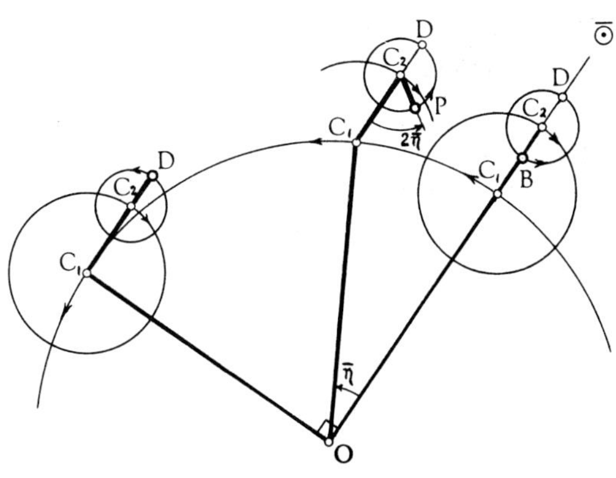

Fig. 2

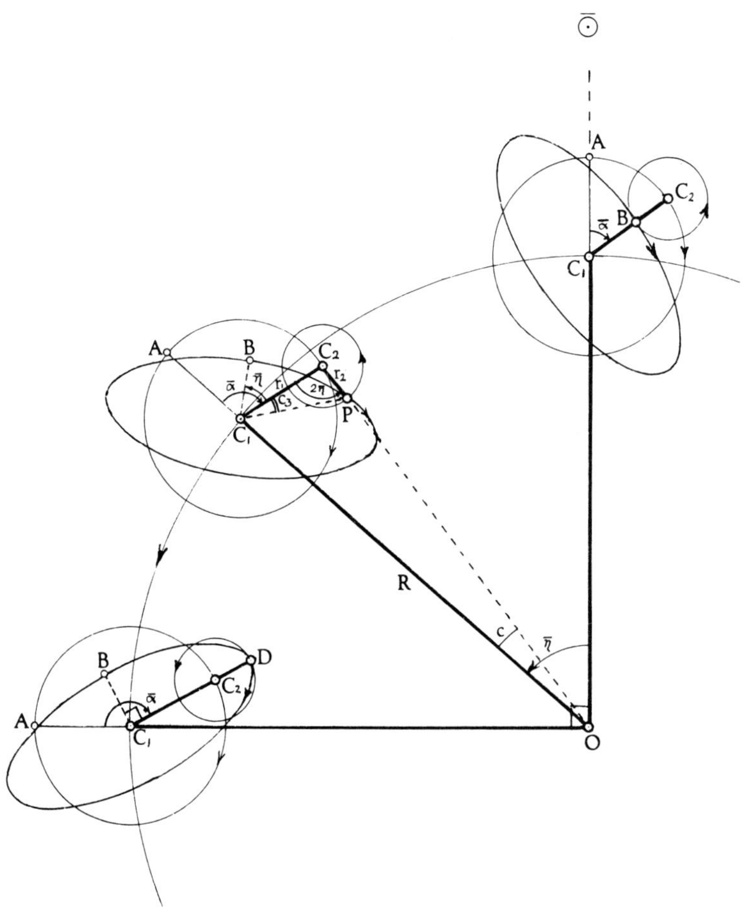

Fig. 3

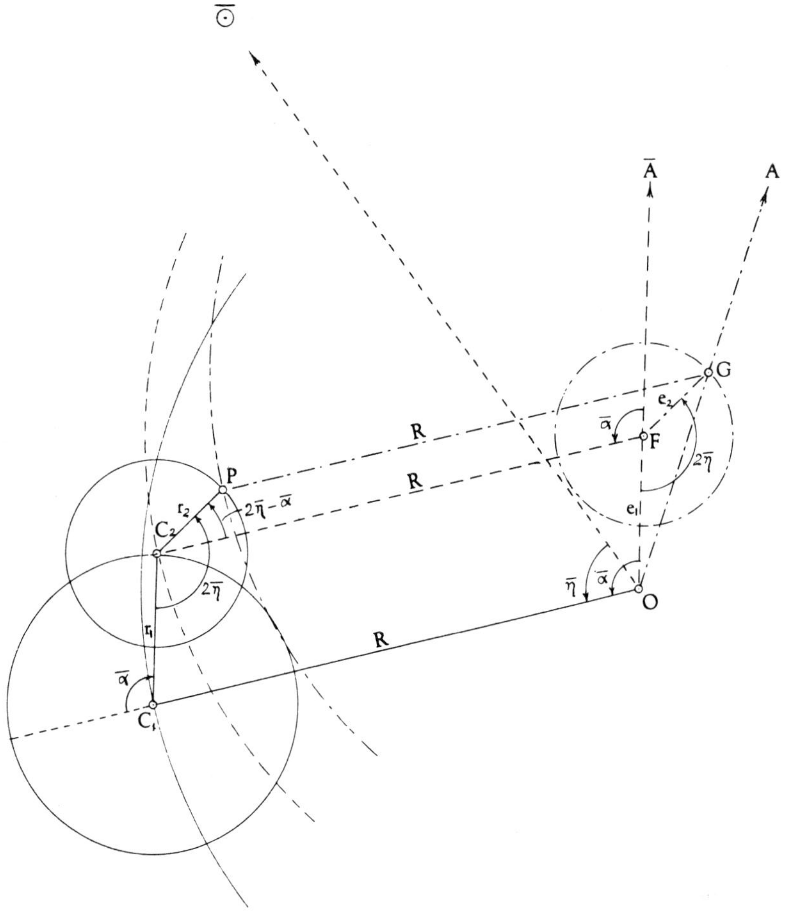

Fig. 4

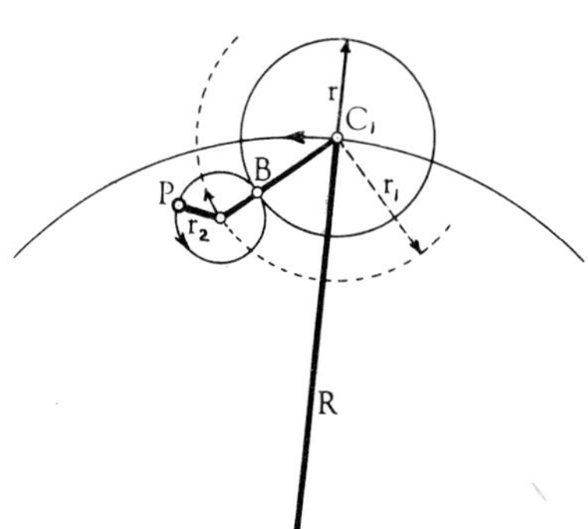

Fig. 5

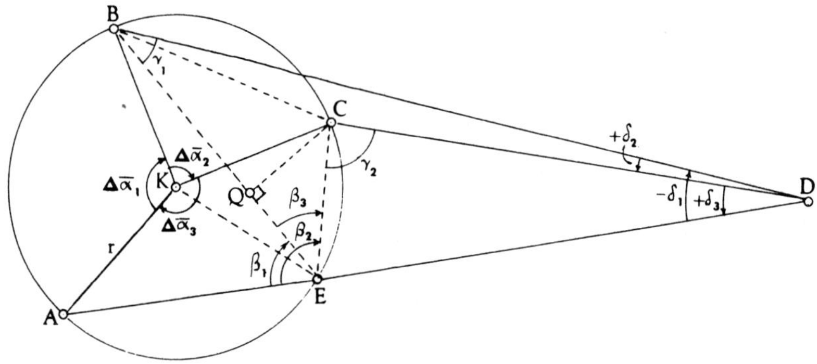
B
C
D
E
A
K
Q
r
γ_1
γ_2
β_1
β_2
β_3
$\Delta\bar{\alpha}_1$
$\Delta\bar{\alpha}_2$
$\Delta\bar{\alpha}_3$
$+\delta_2$
$-\delta_1$
$+\delta_3$

(a)

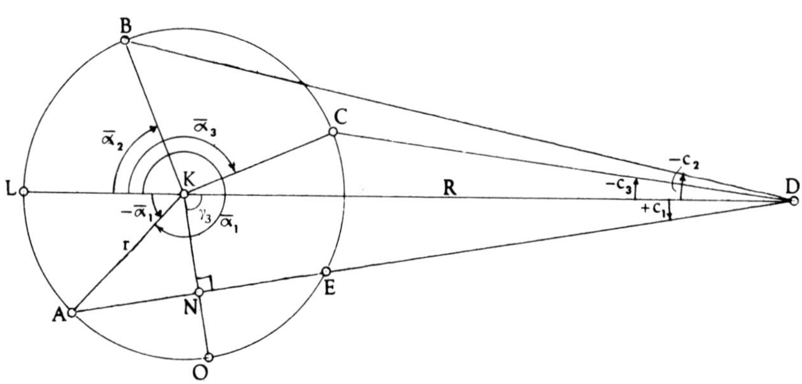
B
C
D
E
A
K
L
N
O
R
r
$\bar{\alpha}_2$
$\bar{\alpha}_3$
$-\bar{\alpha}_1$
$\bar{\alpha}_1$
γ_3
$-c_2$
$-c_3$
$+c_1$

(b)

Fig. 6

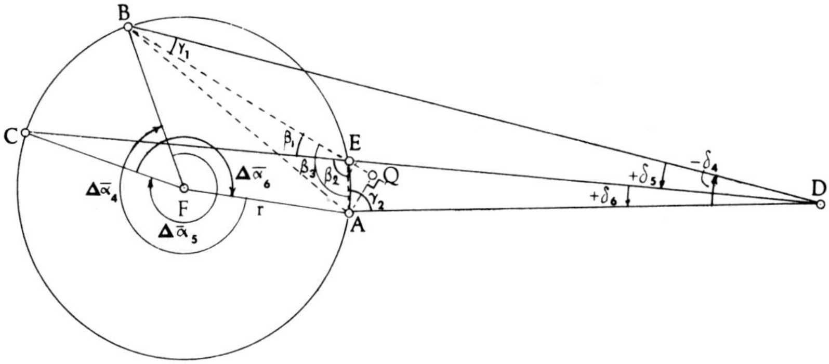
B
C
E
Q
D
F
A
r
γ_1
β_1
β_2
β_3
γ_2
$\Delta\bar{\alpha}_4$
$\Delta\bar{\alpha}_5$
$\Delta\bar{\alpha}_6$
$+\delta_5$
$+\delta_6$
$-\delta_4$

(a)

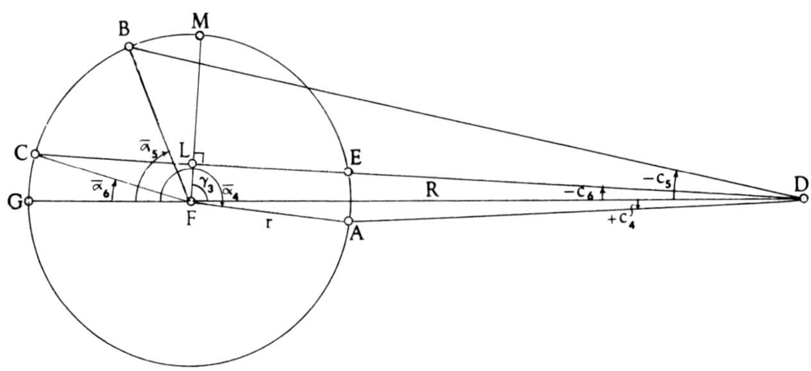
B
M
C
L
E
G
F
A
D
R
r
$\bar{\alpha}_5$
$\bar{\alpha}_6$
$\bar{\alpha}_4$
γ_3
$-c_6$
$-c_5$
$+c_4$

(b)

Fig. 7

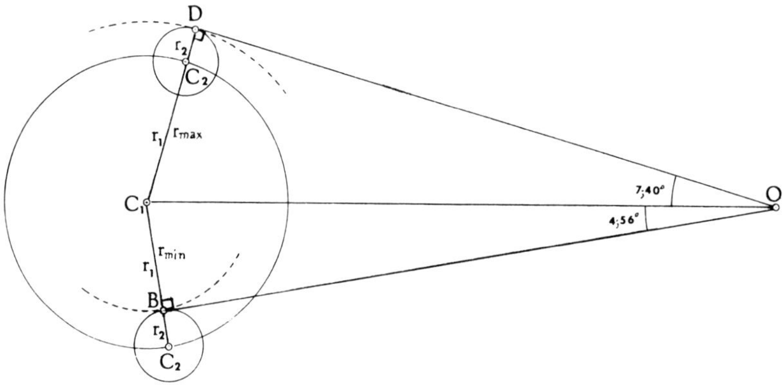
D
r_2
C_2
r_1
r_{max}
C_1
7;40°
4;56°
O
r_{min}
r_1
B
r_2
C_2

(a)

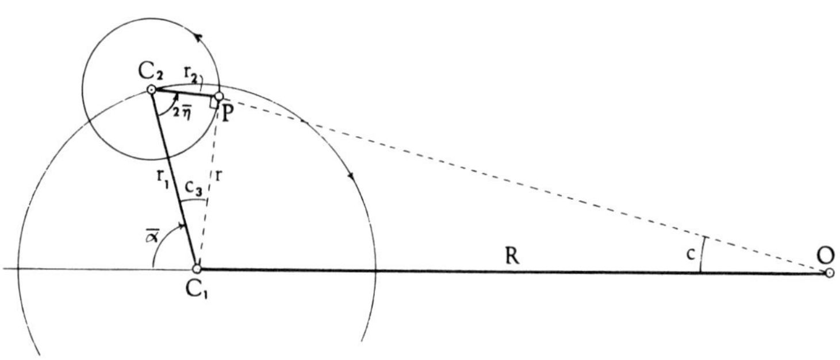
C_2
r_2
$2\bar{\eta}$
P
r_1
c_3
r
$\bar{\alpha}$
R
c
O
C_1

(b)

Fig. 8

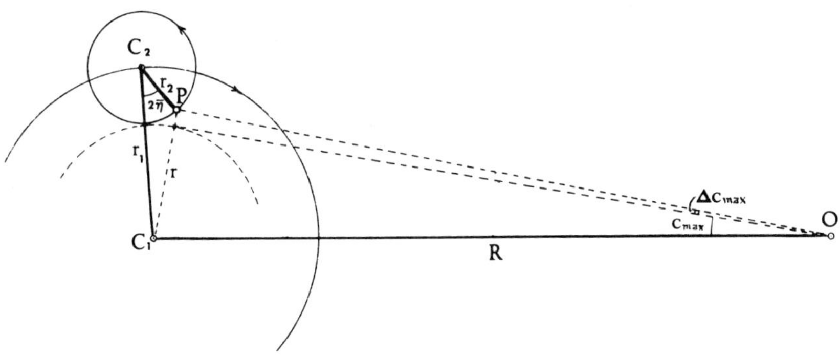

(a)

(b)

Fig. 9

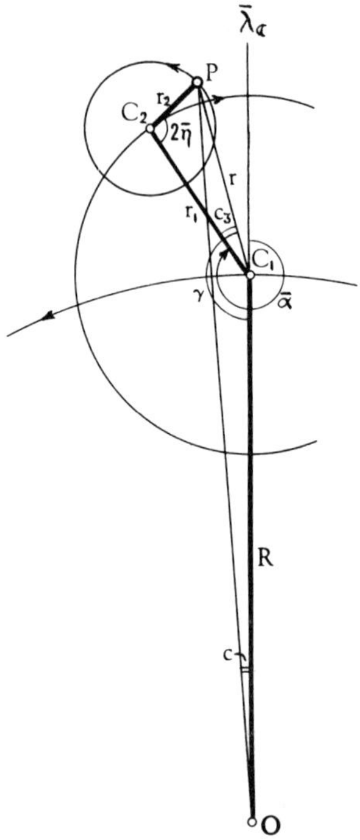

Fig. 10

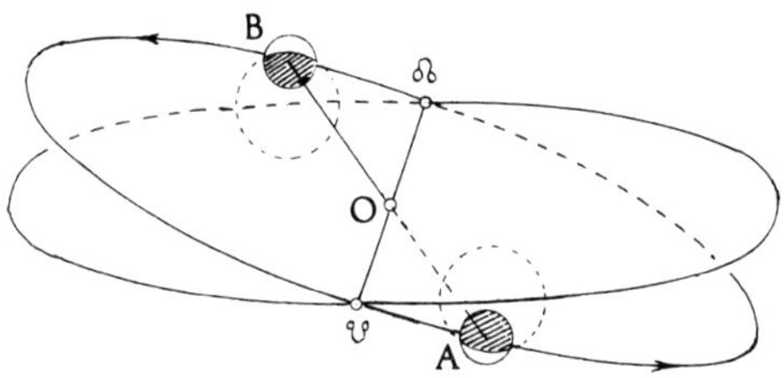

Fig. 11

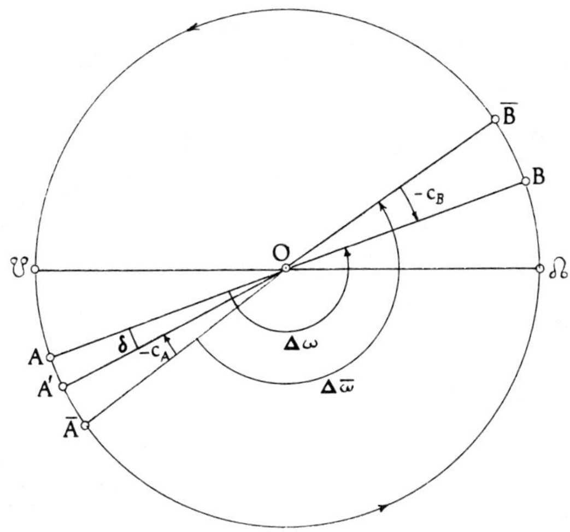

Fig. 12

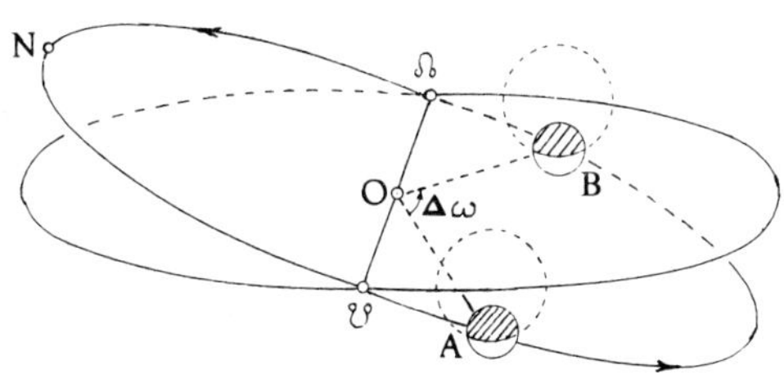

Fig. 13

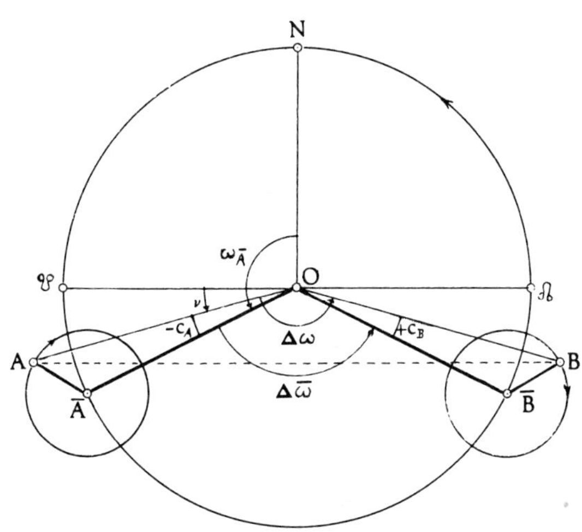

Fig. 14

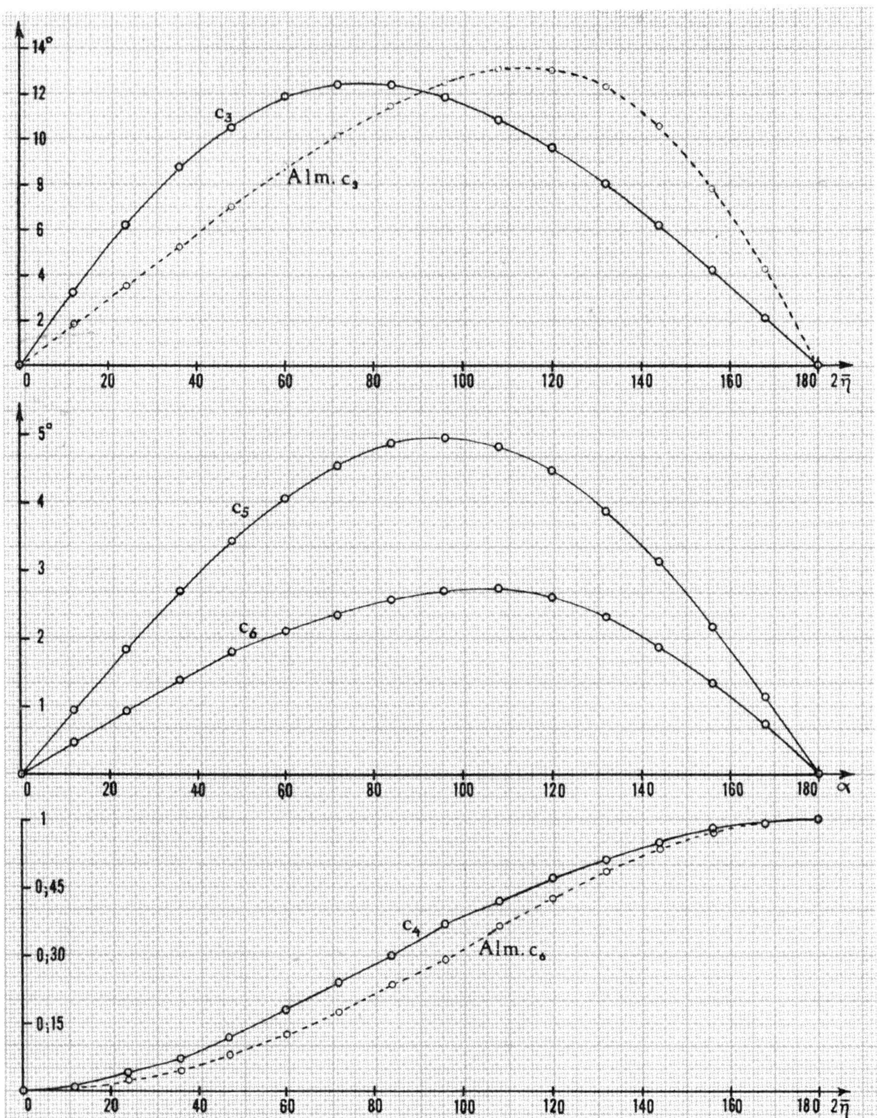
14°
12
10
8
6
4
2
0
20
40
60
80
100
120
140
160
180
2η̄
c3
Alm. c3
5°
4
3
2
1
c5
c6
α
1
0;45
0;30
0;15
c4
Alm. c6

Fig. 15

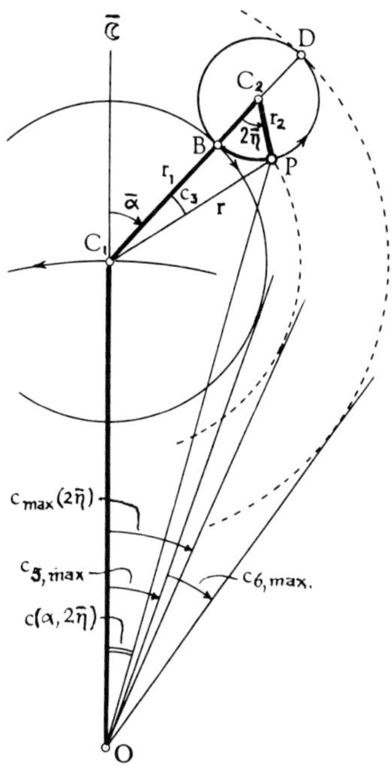
$\bar{\mathfrak{C}}$
D
C_2
r_2
$2\bar{\eta}$
B
P
r_1
$\bar{\alpha}$
c_3
r
C_1
$c_{max}(2\bar{\eta})$
$c_{5,max}$
$c_{6,max}$
$c(\alpha, 2\bar{\eta})$
O

Fig. 16

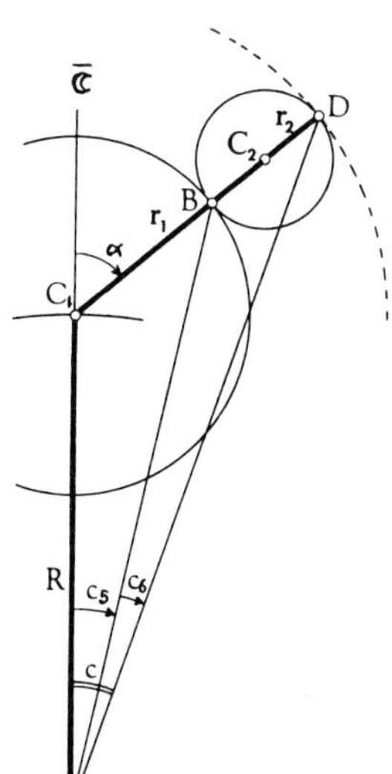
$\bar{\mathfrak{C}}$
D
r_2
C_2
B
r_1
α
C_1
R
c_5
c_6
c
O

Fig. 17

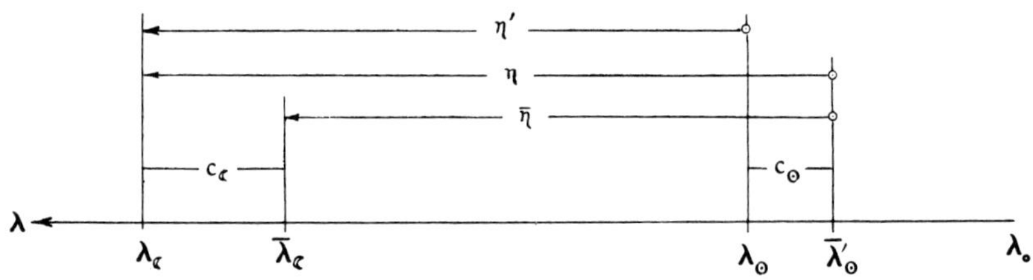

Fig. 18

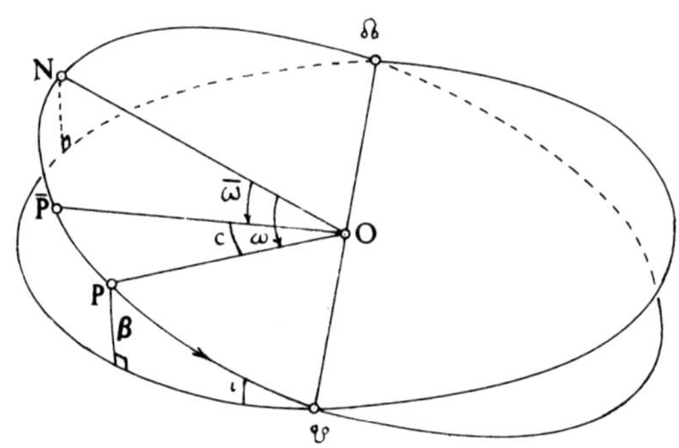

Fig. 19

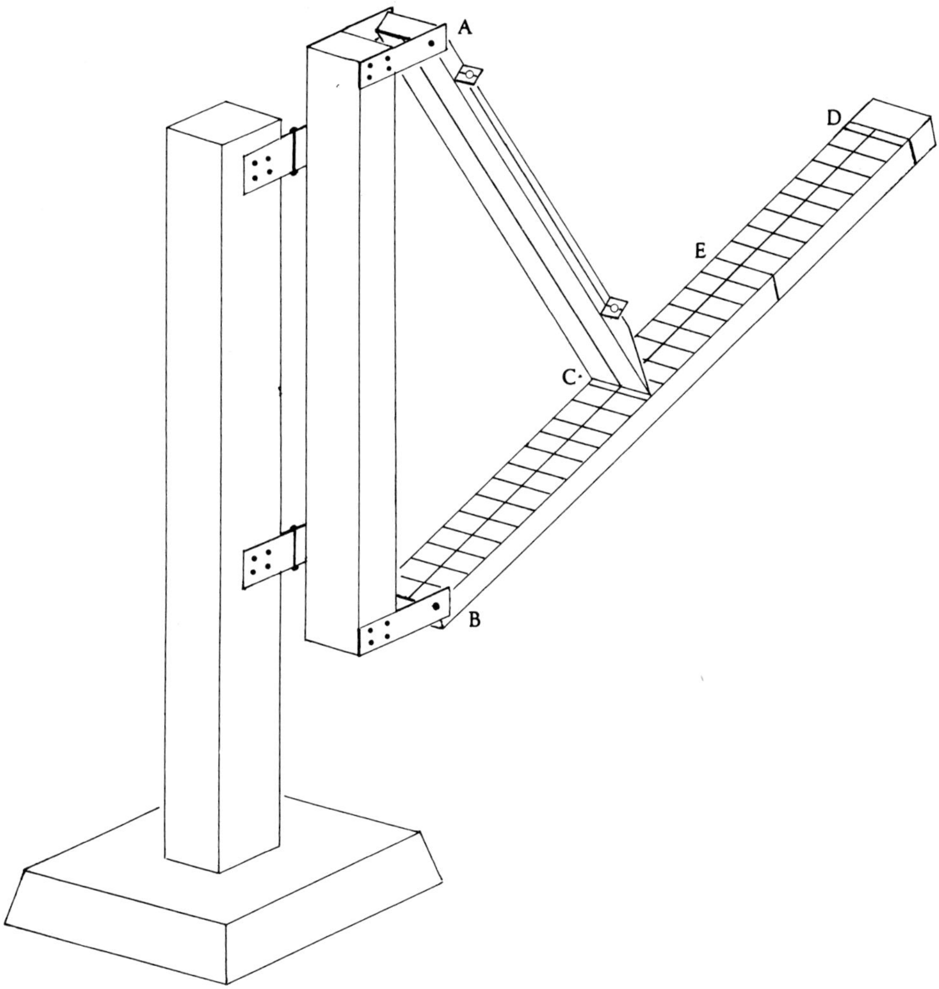

(a)

Fig. 20

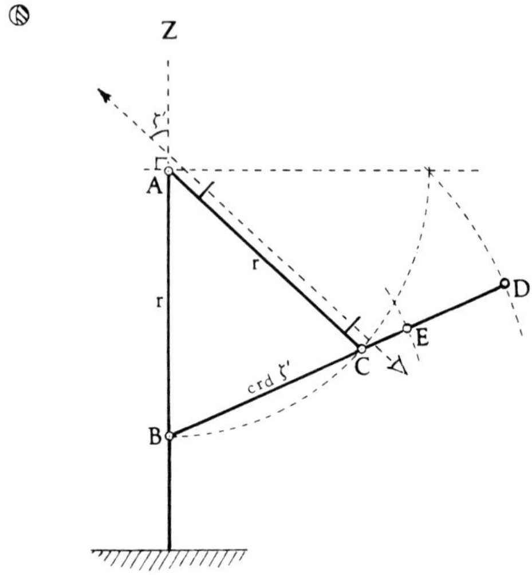

(b)

Fig. 20

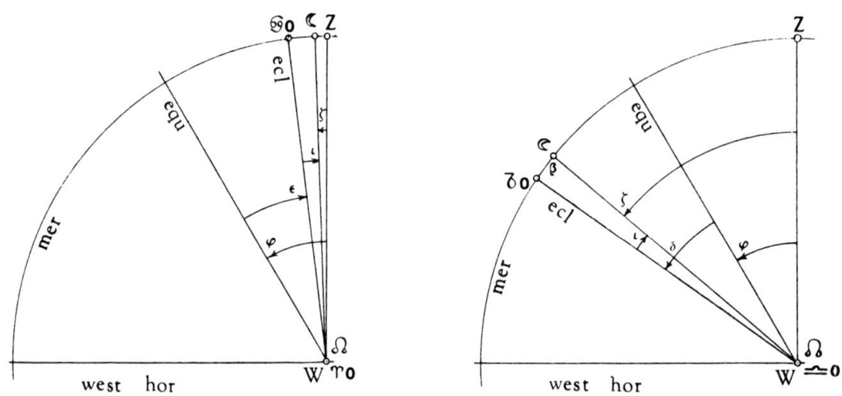

(a) (b)

Fig. 21

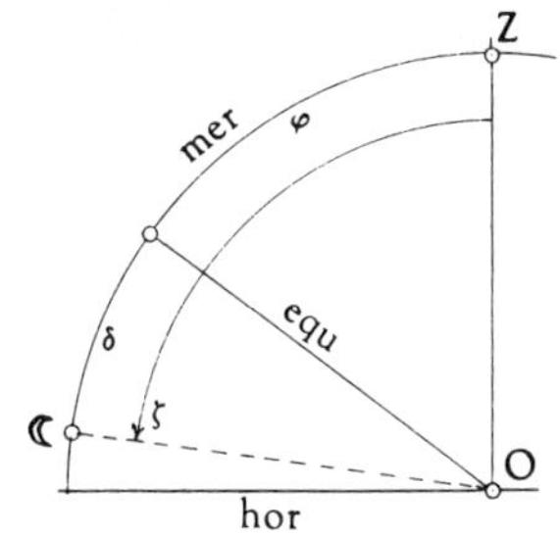

Fig. 22

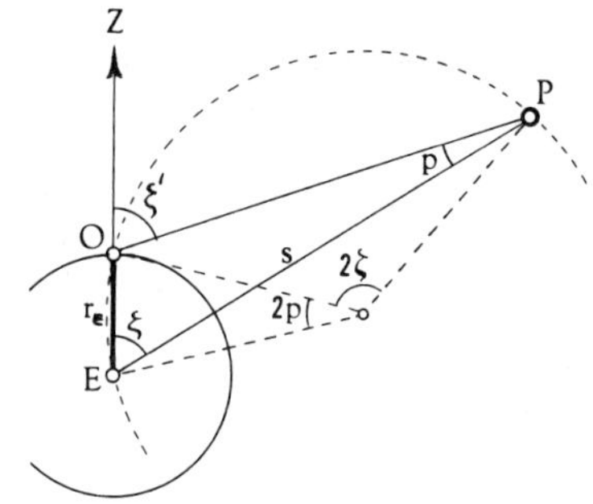

Fig. 23

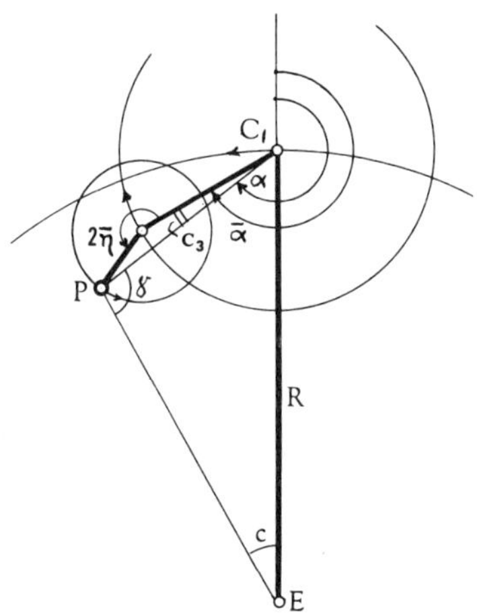

Fig. 24

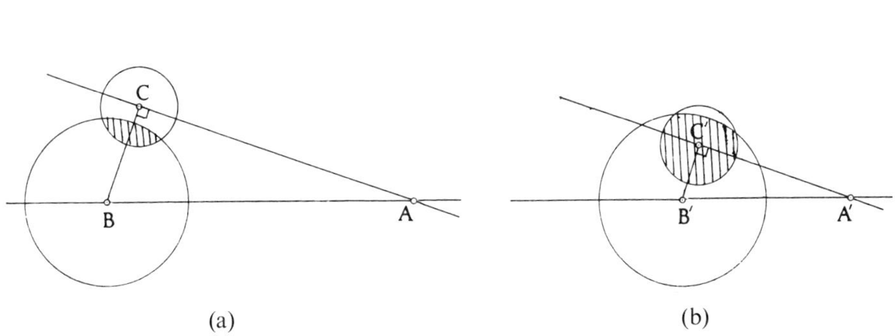

Fig. 25

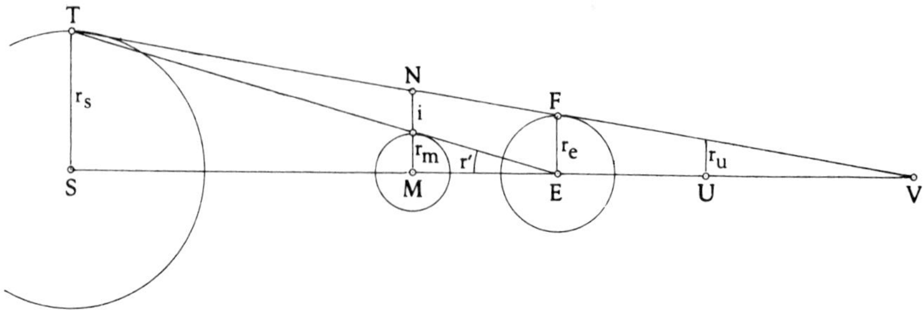

Fig. 26

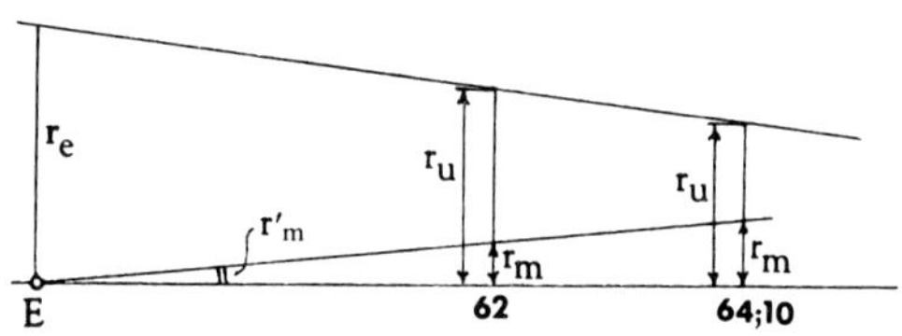

Fig. 27

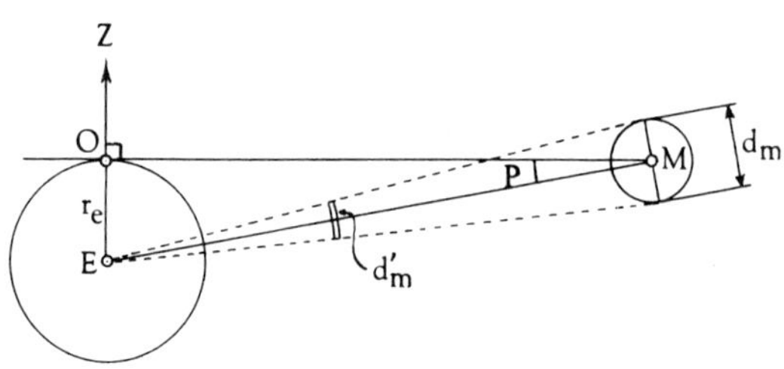

Fig. 28

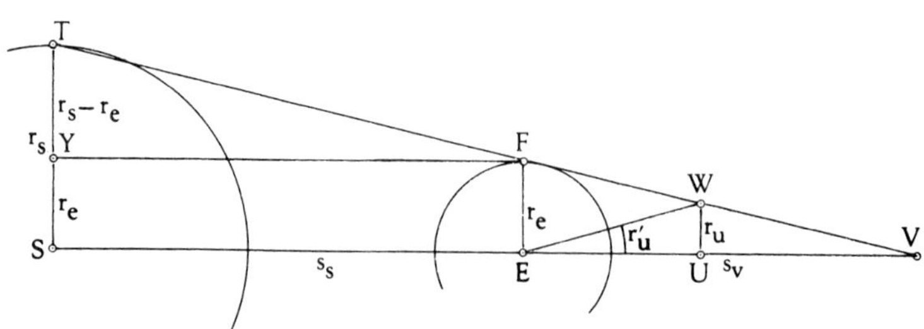

Fig. 29

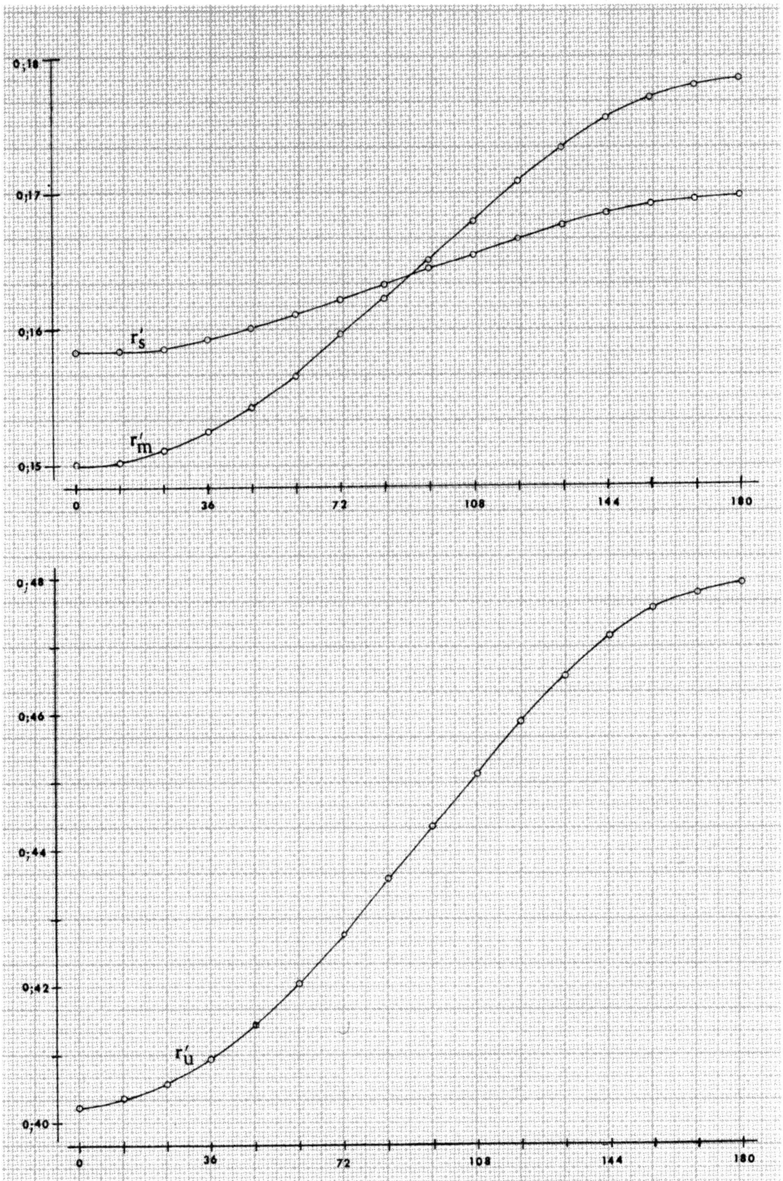
0;18
0;17
0;16
0;15
r's
r'm
0
36
72
108
144
180
0;48
0;46
0;44
0;42
0;40
r'u
0
36
72
108
144
180

Fig. 30

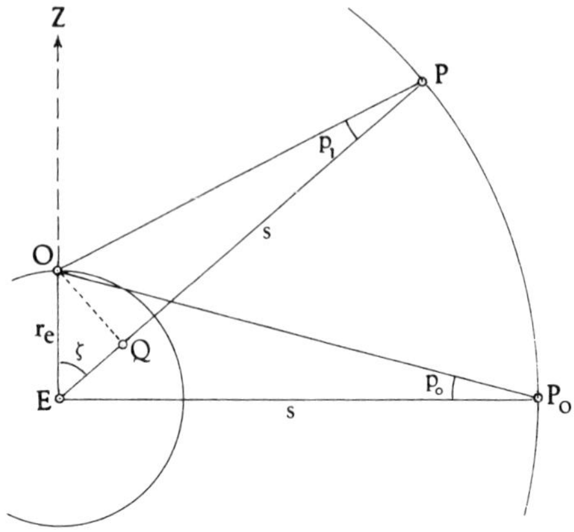

Fig. 31

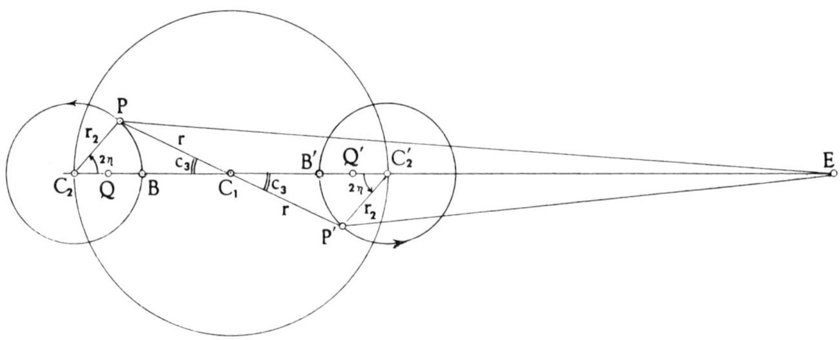

Fig. 32

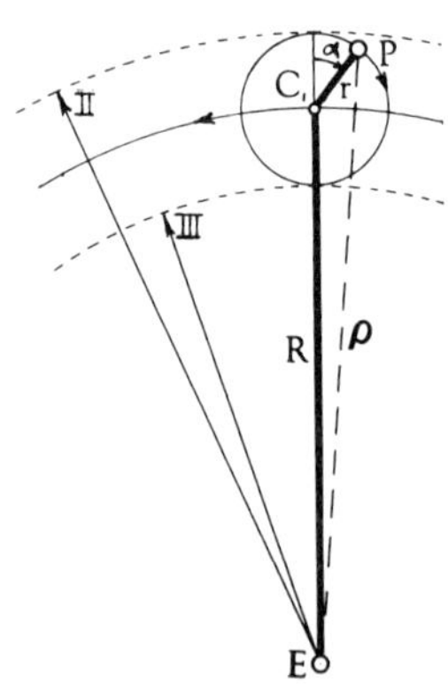

Fig. 33

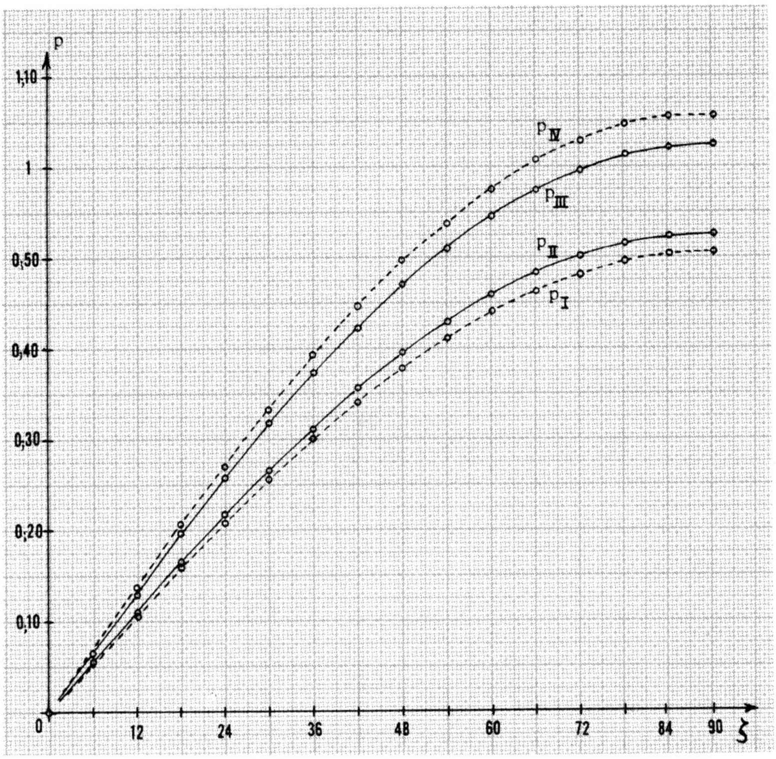
p
1,10
1
0,50
0,40
0,30
0,20
0,10
0
12
24
36
48
60
72
84
90
ζ
p_{IV}
p_{III}
p_{II}
p_{I}

Fig. 34

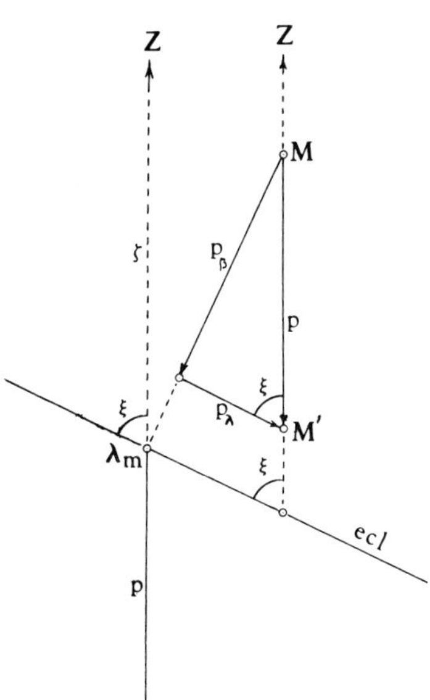
Z
Z
M
ζ
p_β
p
ξ
ξ
p_λ
M′
λ_m
ξ
ecl
p

Fig. 35

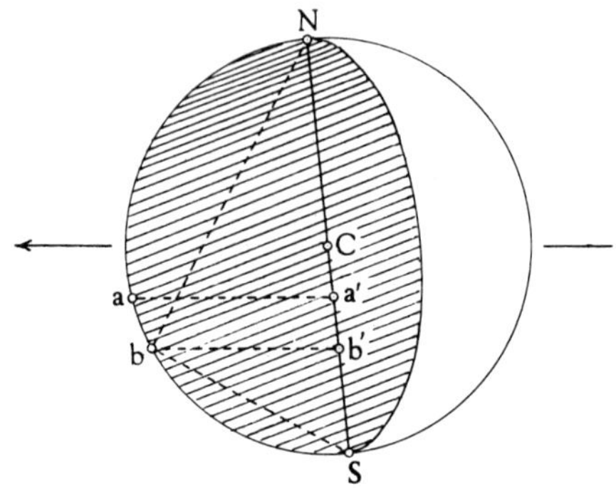

Fig. 36

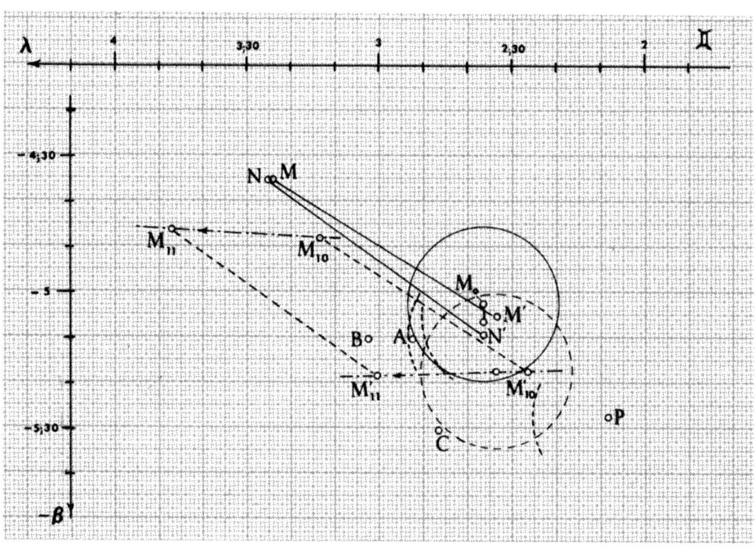

Fig. 37

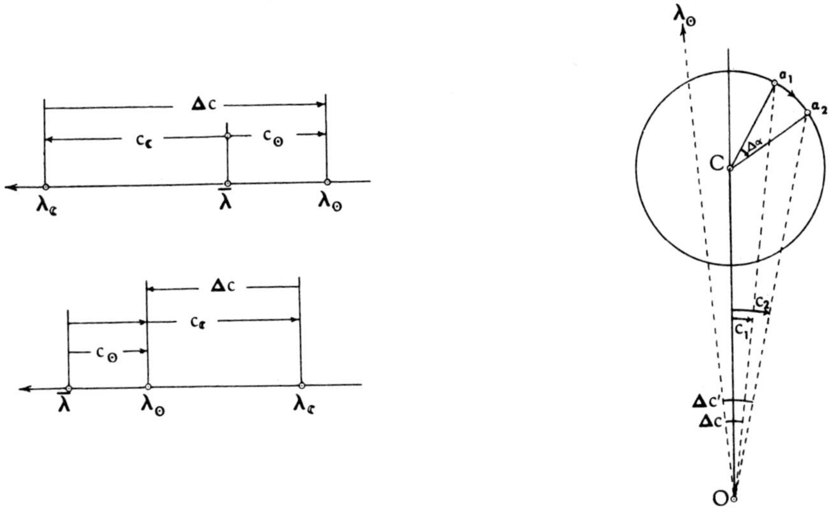

Fig. 38

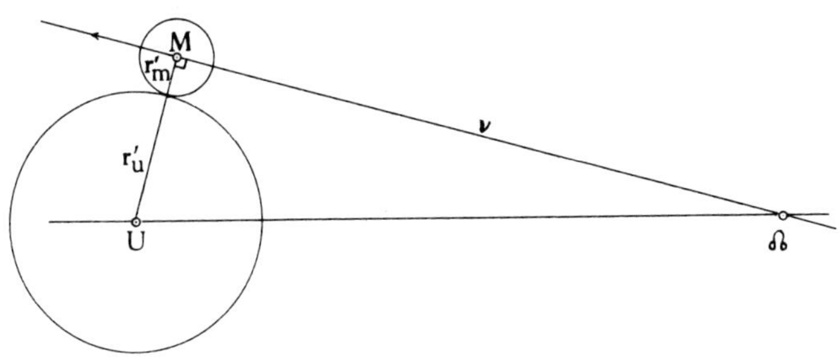

Fig. 39

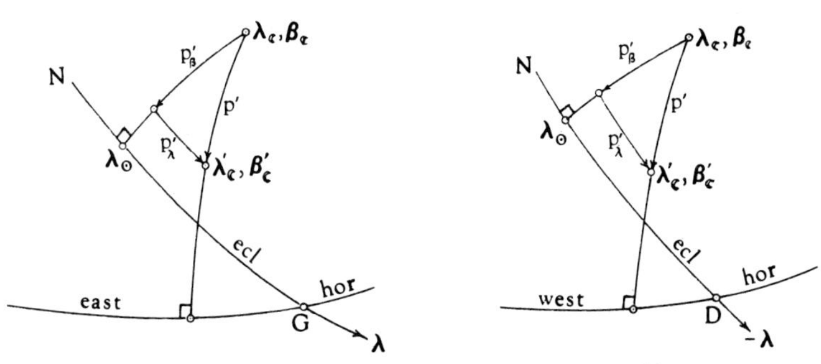

Fig. 40

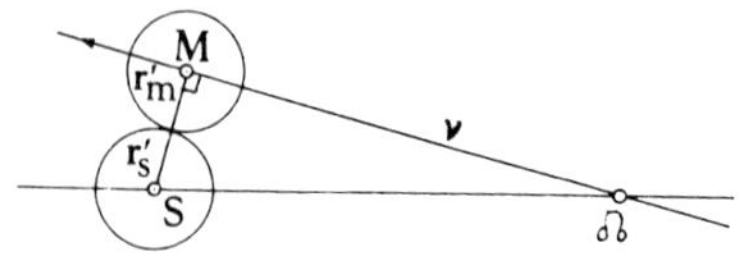

Fig. 41

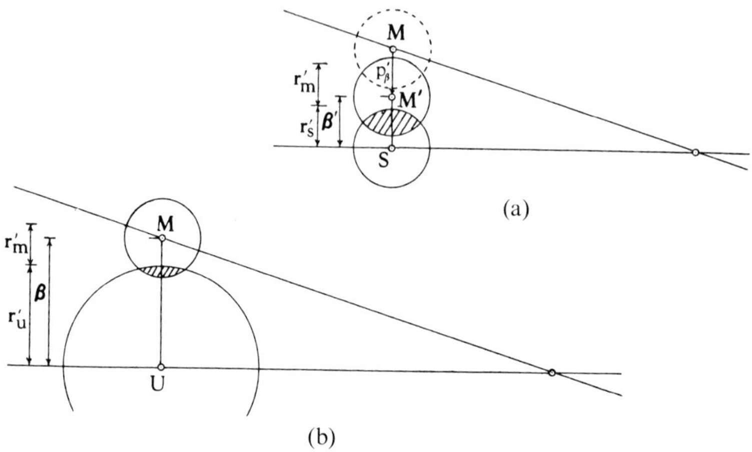

Fig. 42

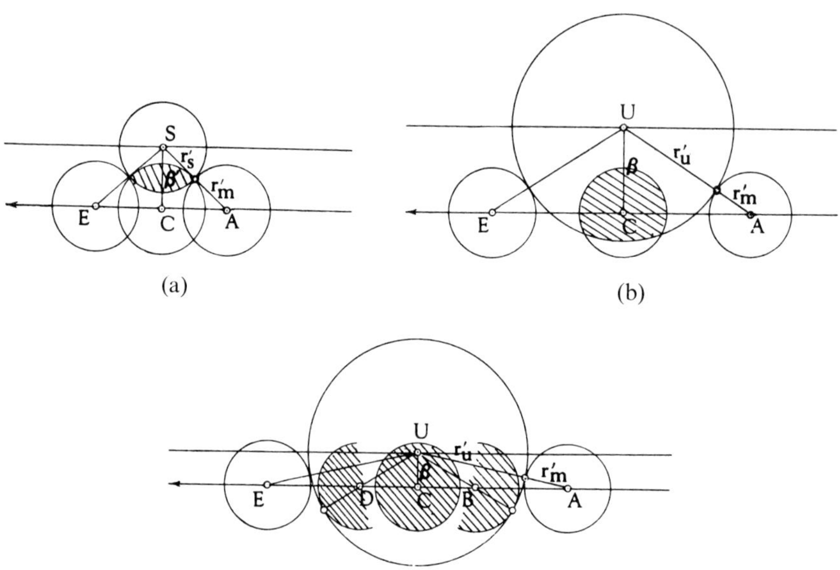

Fig. 43

Figures to Chapter 5

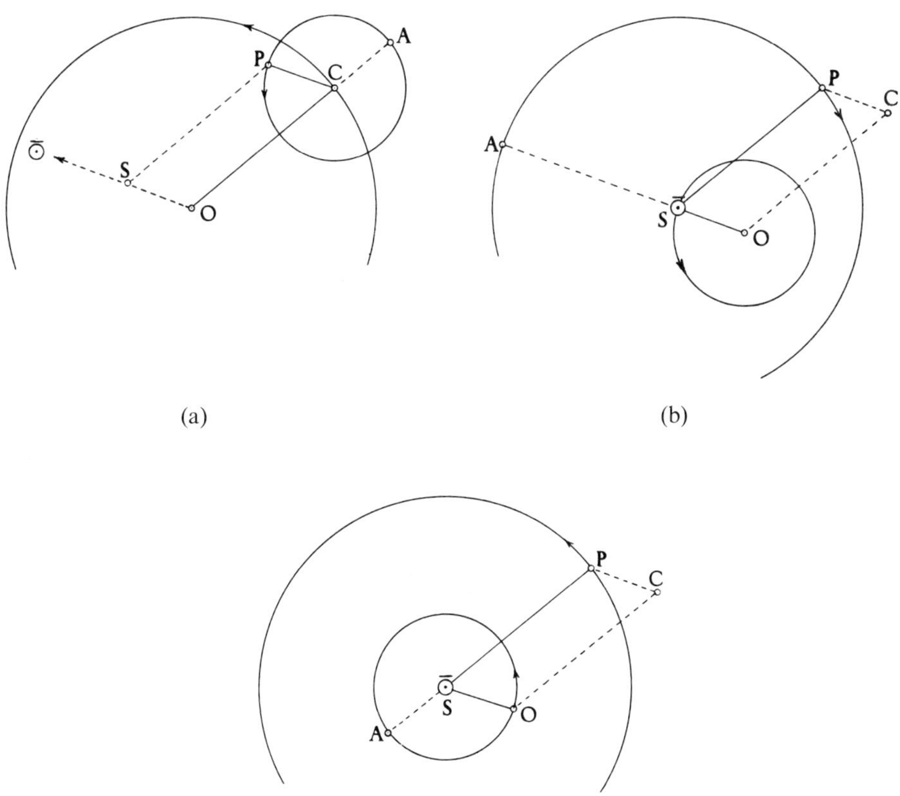

(a) (b)

(c)

Fig. 1

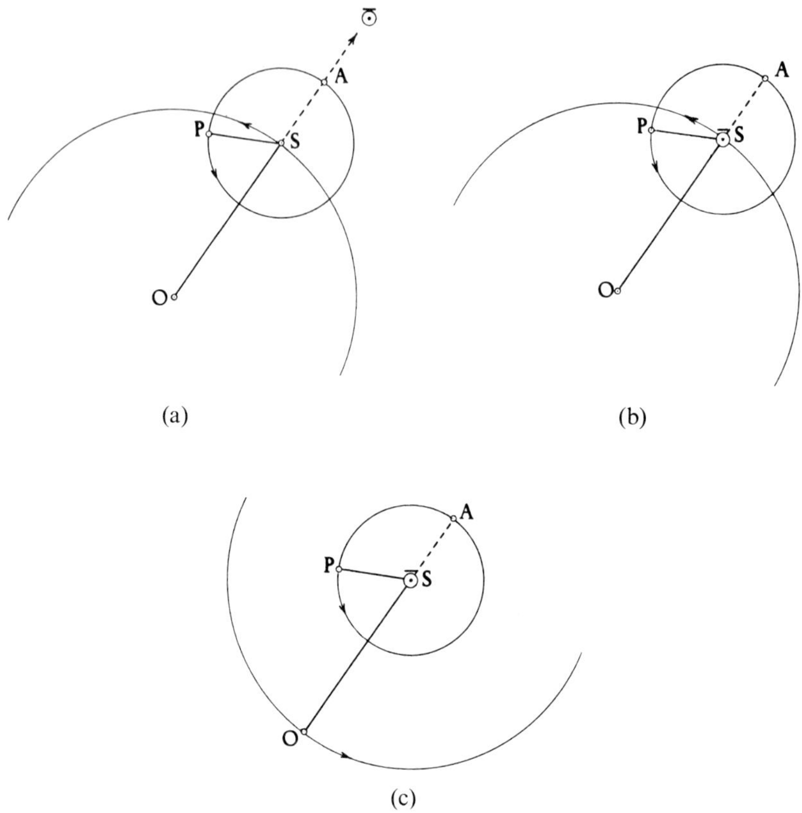

Fig. 2

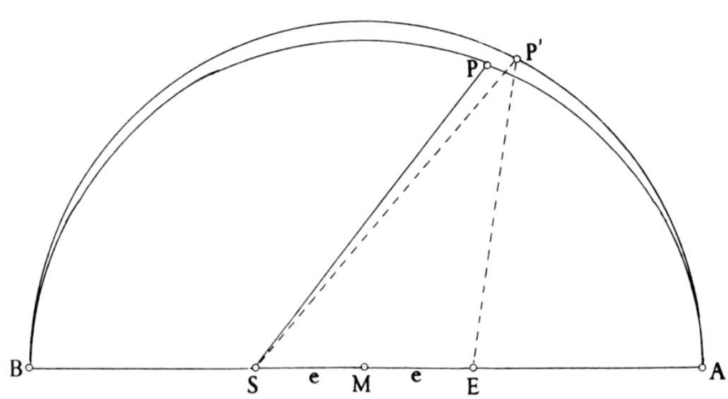

Fig. 3

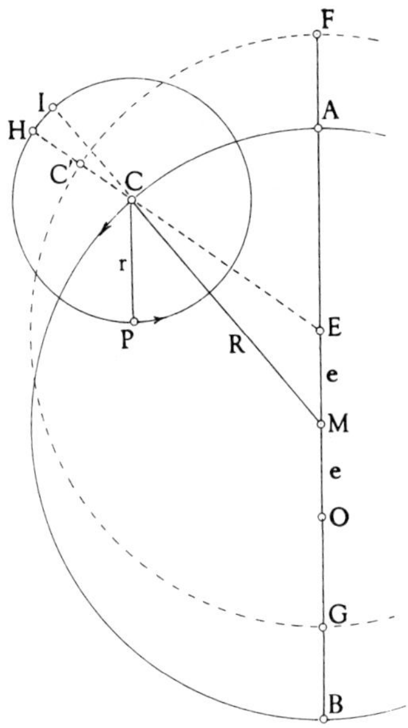

Fig. 4

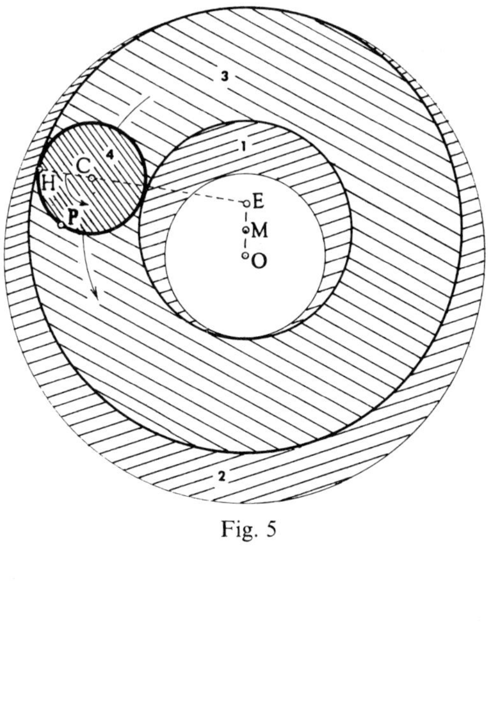

Fig. 5

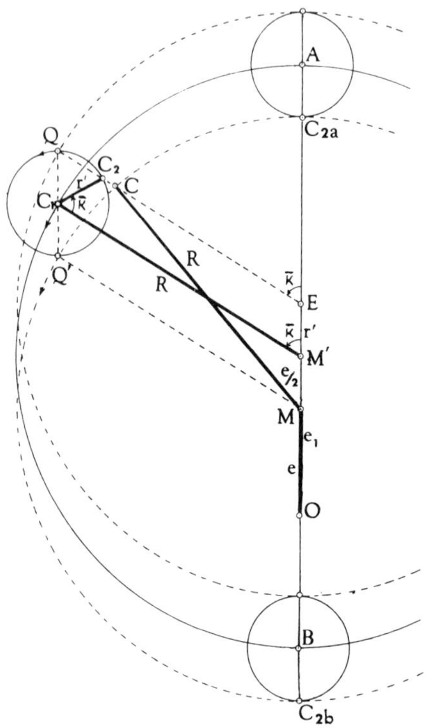
A
C_{2a}
Q
C_2
C
r'
C_1
$\bar{\kappa}$
R
R
Q'
$\bar{\kappa}$
E
$\bar{\kappa}$
r'
M'
e/2
M
e_1
e
O
B
C_{2b}

Fig. 6

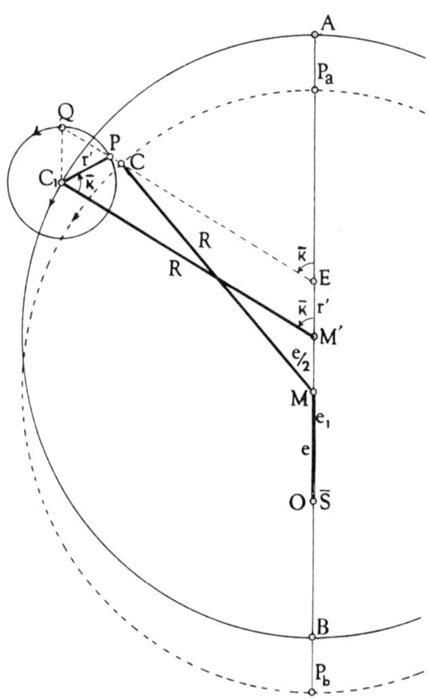
A
P_a
Q
P
C
r'
C_1
$\bar{\kappa}$
R
R
$\bar{\kappa}$
E
$\bar{\kappa}$
r'
M'
e/2
M
e_1
e
O
$\bar{S}$
B
P_b

Fig. 7

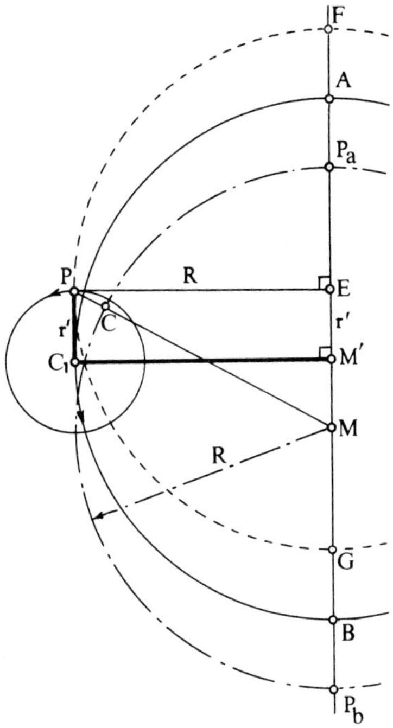

Fig. 8

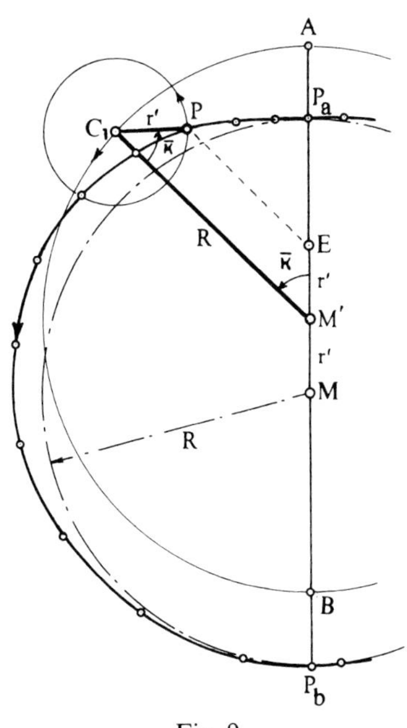

Fig. 9

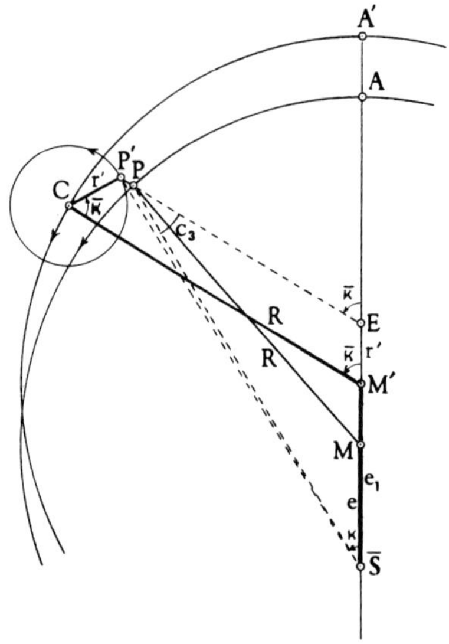

Fig. 10

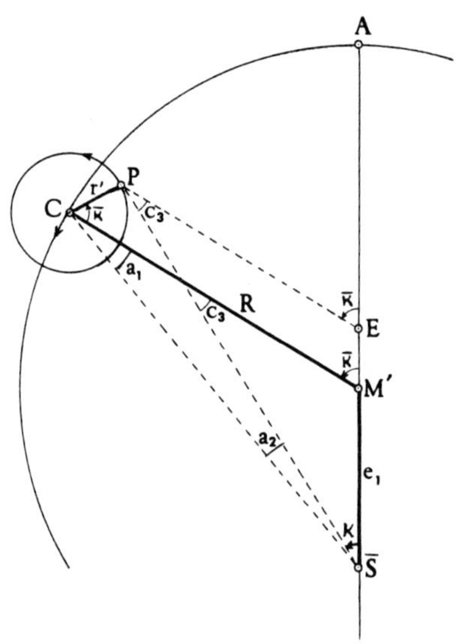

Fig. 11

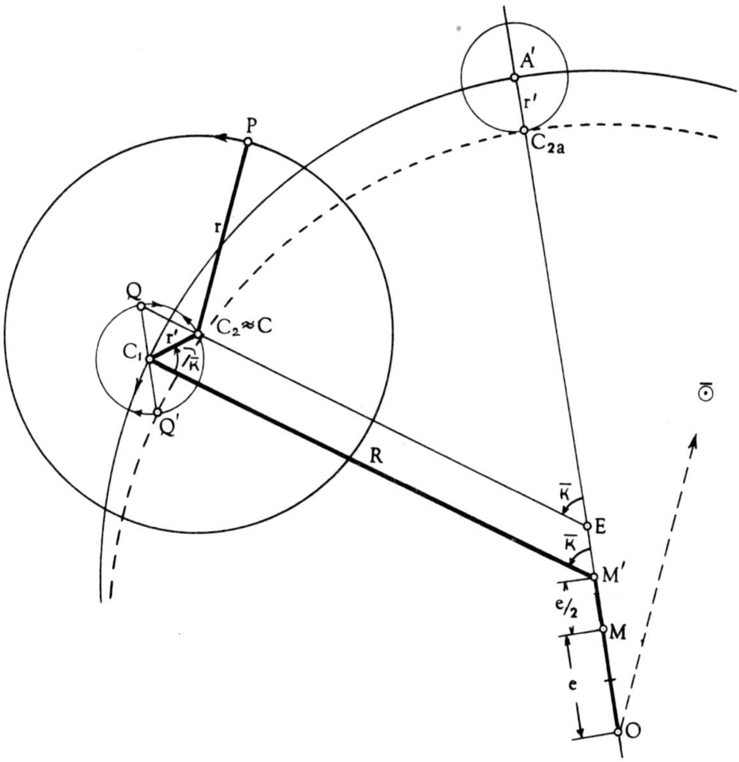
A'
r'
C2a
P
r
Q
C1
r'
C2≈C
κ̄
Q'
R
κ̄
E
κ̄
M'
e/2
M
e
O

Fig. 12

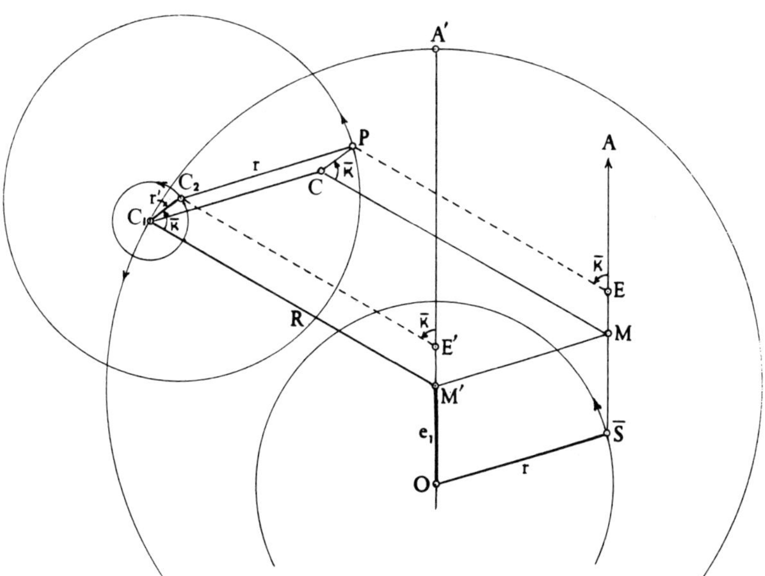
A'
P
A
r
C2
r'
C
κ̄
C1
κ̄
κ̄
E
R
κ̄
E'
M
M'
e1
S̄
O
r

Fig. 13

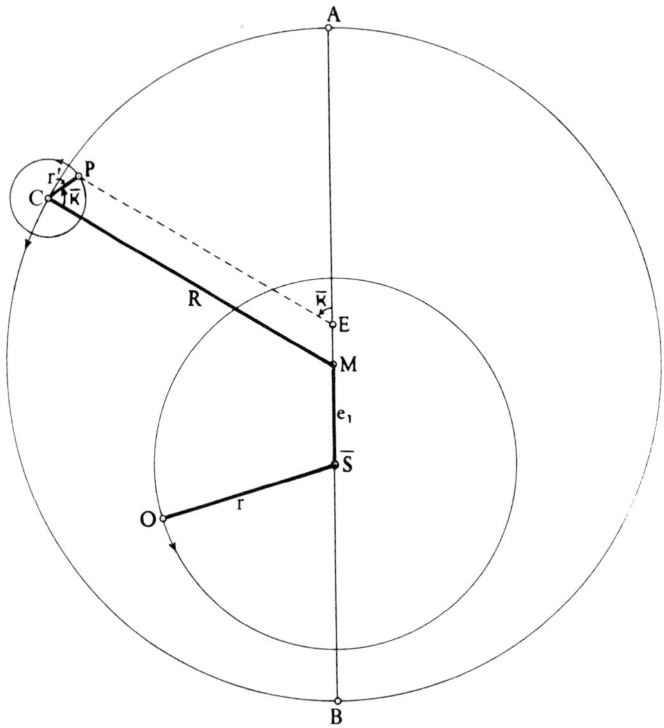
A
r'
P
C
$\bar{\kappa}$
R
$\bar{\kappa}$
E
M
e_1
$\bar{S}$
O
r
B

Fig. 14

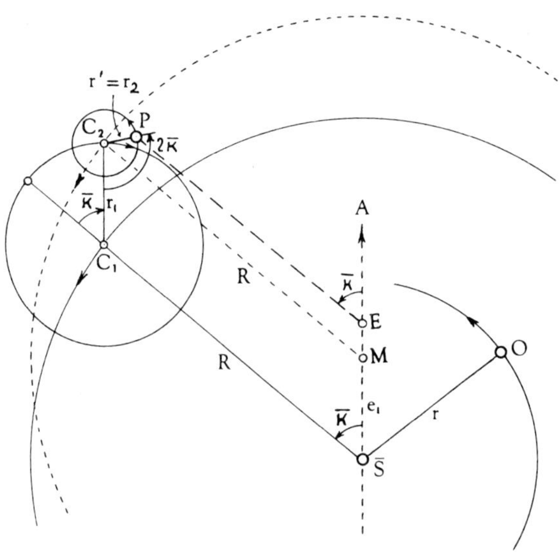
r' = r_2
C_2
P
$2\bar{\kappa}$
$\bar{\kappa}$
r_1
C_1
R
A
$\bar{\kappa}$
E
M
O
R
$\bar{\kappa}$
e_1
r
$\bar{S}$

Fig. 15

(a)

(b)

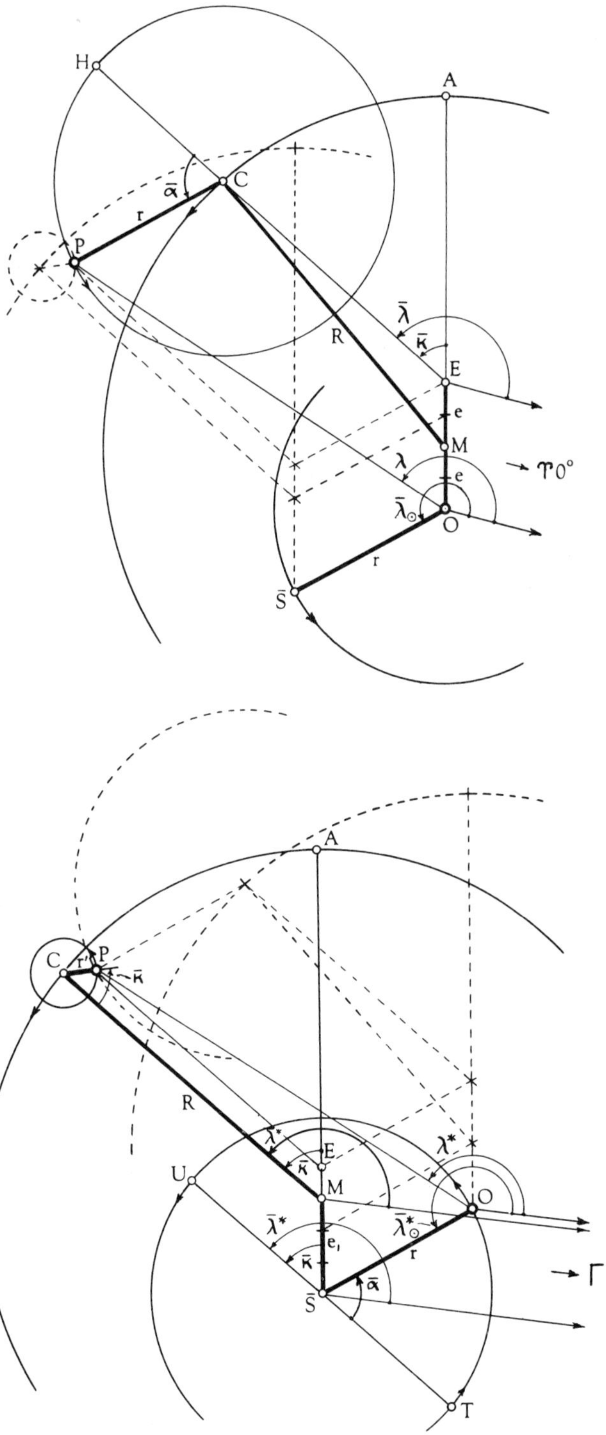
H
A
C
$\bar{\alpha}$
r
P
R
$\bar{\lambda}$
$\bar{\kappa}$
E
e
M
λ
e
→ ♈0°
$\bar{\lambda}_{\odot}$
O
r
$\bar{S}$
A
C
r′
P
$\bar{\kappa}$
R
λ^*
E
$\bar{\kappa}$
M
U
$\bar{\lambda}^*$
e_1
$\bar{\kappa}$
$\bar{S}$
$\bar{\alpha}$
$\bar{\lambda}^*_{\odot}$
r
O
λ^*
→ Γ
T

Fig. 16

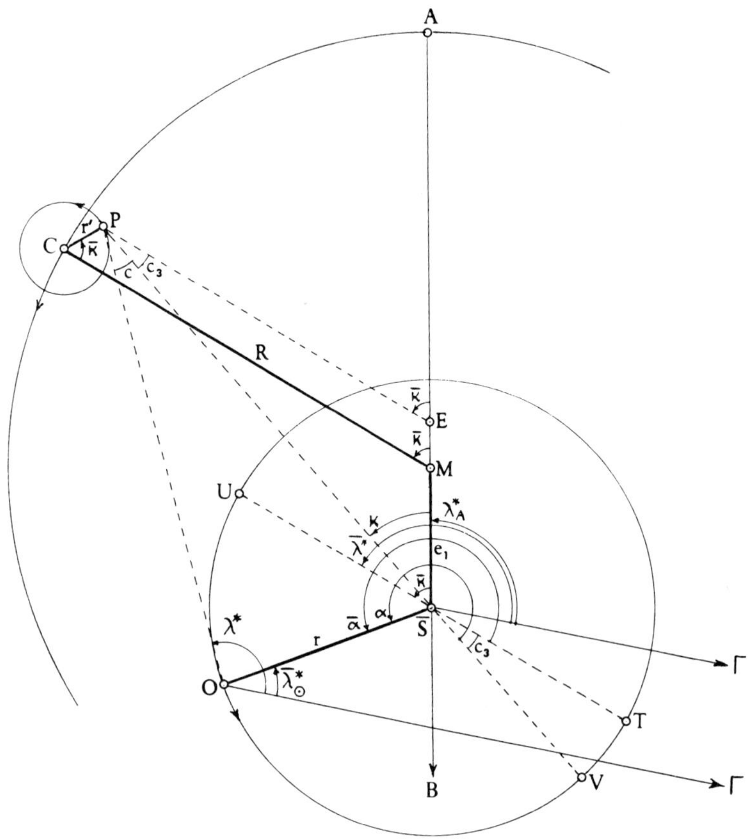

Fig. 17

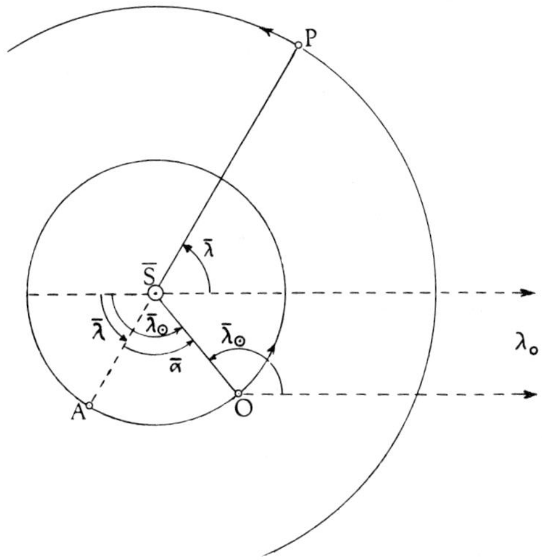
P
$\bar{\lambda}$
$\bar{S}$
$\bar{\lambda}$
$\bar{\lambda}_{\odot}$
$\bar{\lambda}_{\odot}$
$\bar{\alpha}$
λ_o
A
O

(a)

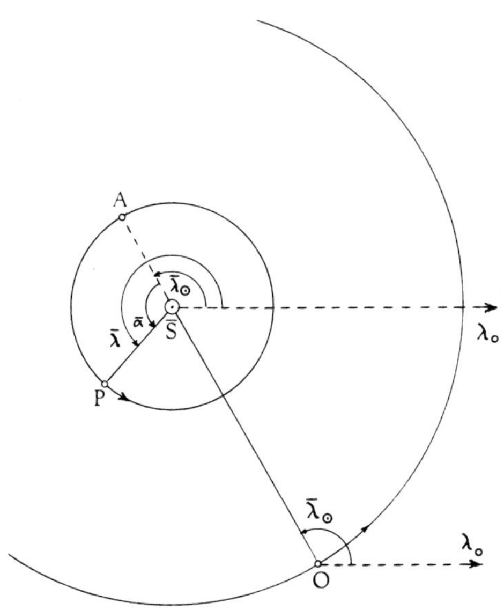
A
$\bar{\lambda}_{\odot}$
$\bar{\alpha}$
$\bar{\lambda}$
$\bar{S}$
λ_o
P
$\bar{\lambda}_{\odot}$
λ_o
O

(b)

Fig. 18

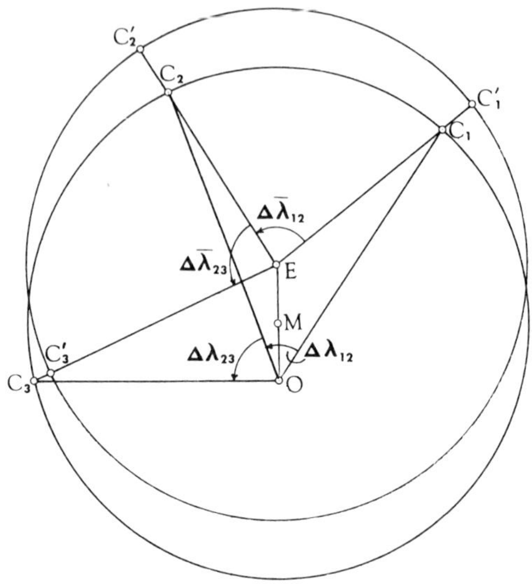
C'2
C2
C'1
C1
Δλ̄12
Δλ̄23
E
M
Δλ23
Δλ12
C'3
C3
O

Fig. 19

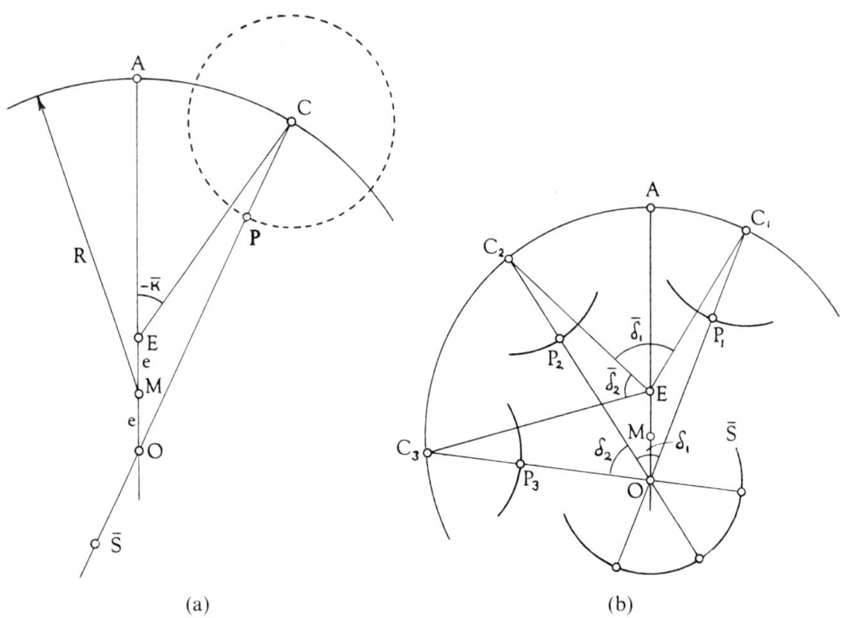
A
C
R
P
-κ̄
E
e
M
e
O
S̄
A
C1
C2
P1
P2
δ̄1
δ̄2
E
M
S̄
δ2
δ1
C3
P3
O
(a)
(b)

Fig. 20

(a)

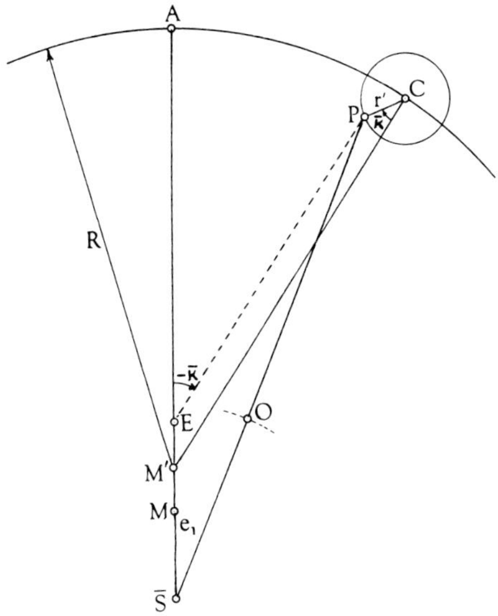

(b)

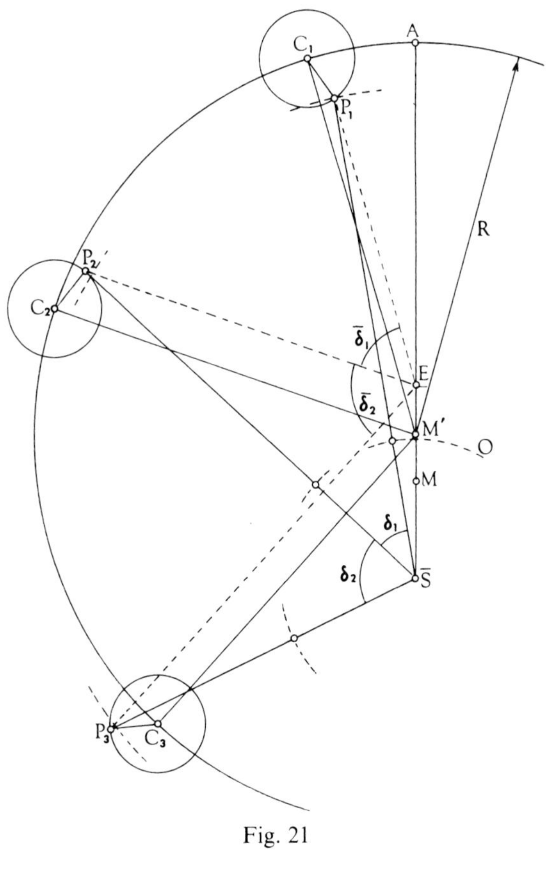

Fig. 21

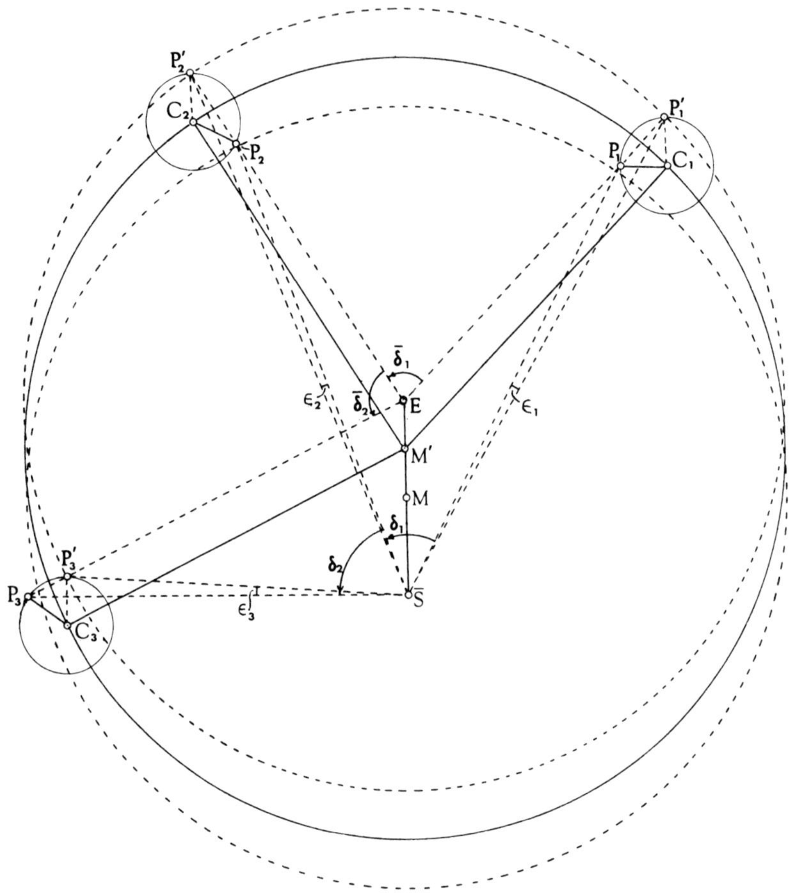

Fig. 22

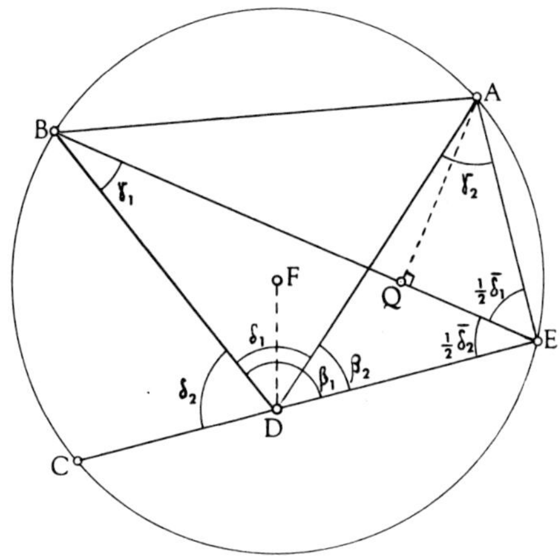
A
B
C
D
E
F
Q
γ_1
γ_2
$\frac{1}{2}\bar{\delta}_1$
$\frac{1}{2}\bar{\delta}_2$
δ_1
δ_2
β_1
β_2

(a)

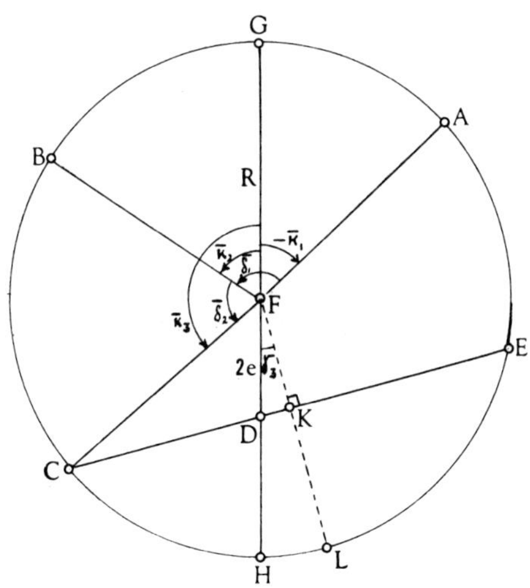
G
A
B
R
$-\bar{\kappa}_1$
F
E
2e
K
D
C
H
L

(b)

Fig. 23

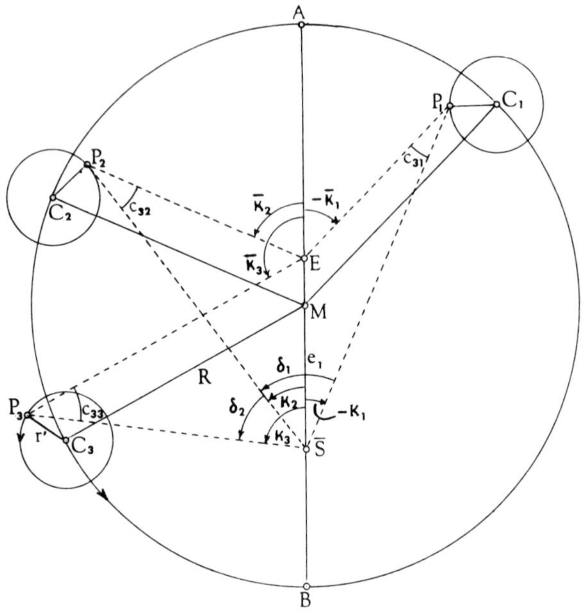
A
P_1
C_1
P_2
C_2
c_{32}
c_{31}
$\bar{\kappa}_2$
$-\bar{\kappa}_1$
$\bar{\kappa}_3$
E
M
R
δ_1
e_1
δ_2
κ_2
$-\kappa_1$
κ_3
P_3
c_{33}
r′
C_3
$\bar{S}$
B

Fig. 24

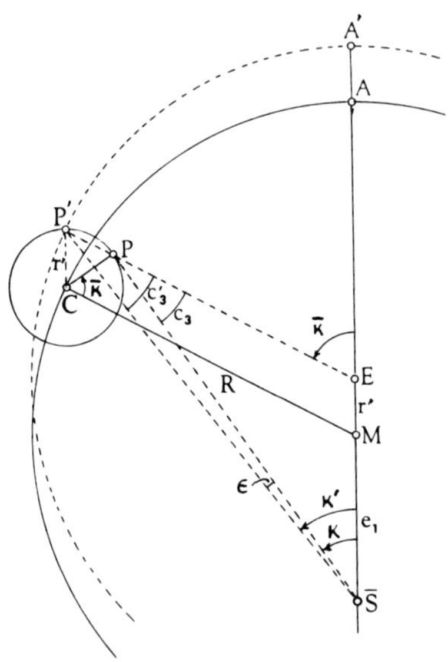
A′
A
P′
P
r′
C
$\bar{\kappa}$
c'_3
c_3
$\bar{\kappa}$
E
r′
R
M
ε
κ′
e_1
κ
$\bar{S}$

Fig. 25

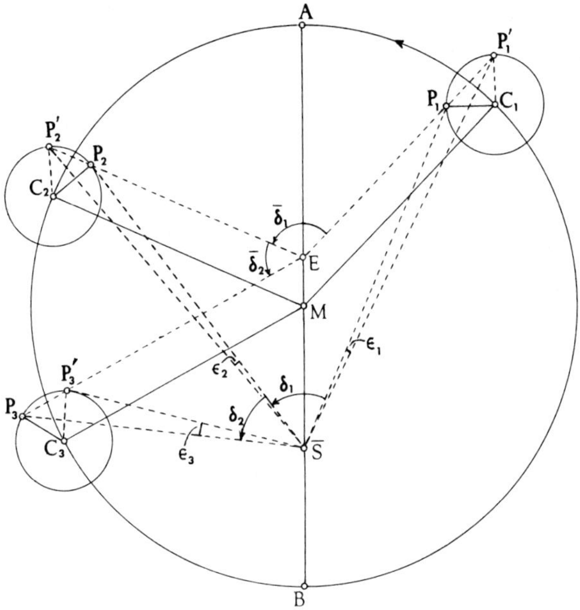
A
P′1
P1
C1
P′2
P2
C2
E
M
P′3
P3
C3
S̄
B
ε1
ε2
ε3
δ1
δ2
δ̄1
δ̄2

Fig. 26

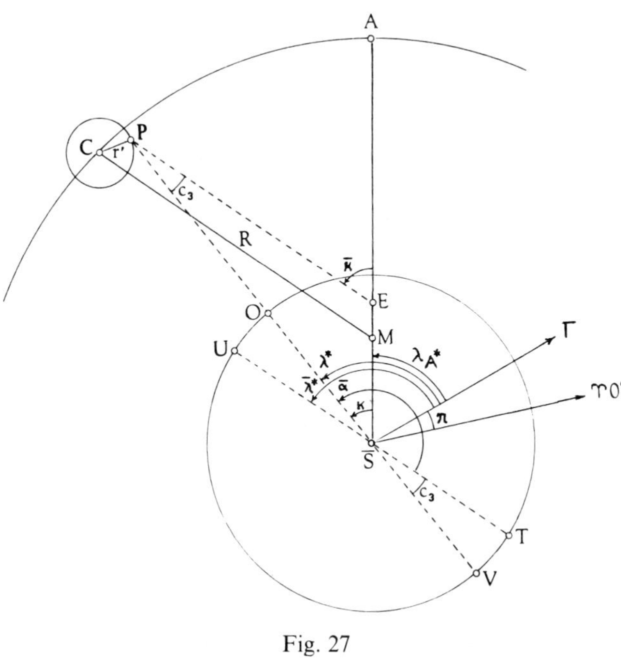
A
C
P
r′
c3
R
κ̄
E
O
M
U
λ*
λ̄*
ᾱ
κ
λA*
Γ
♈0°
π
S̄
c3
T
V

Fig. 27

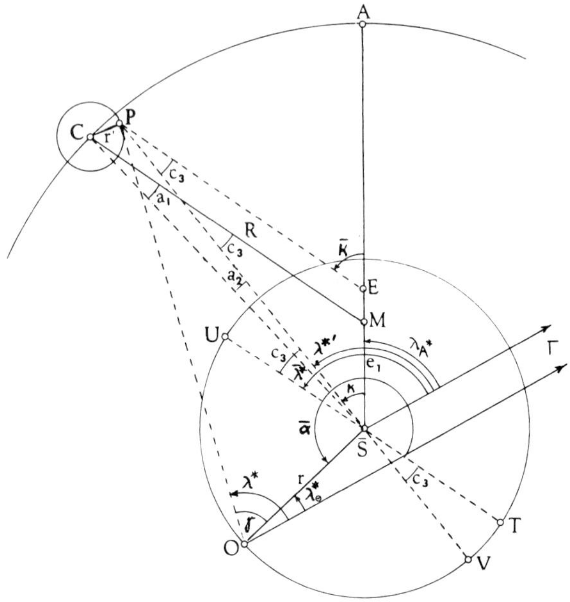

Fig. 28

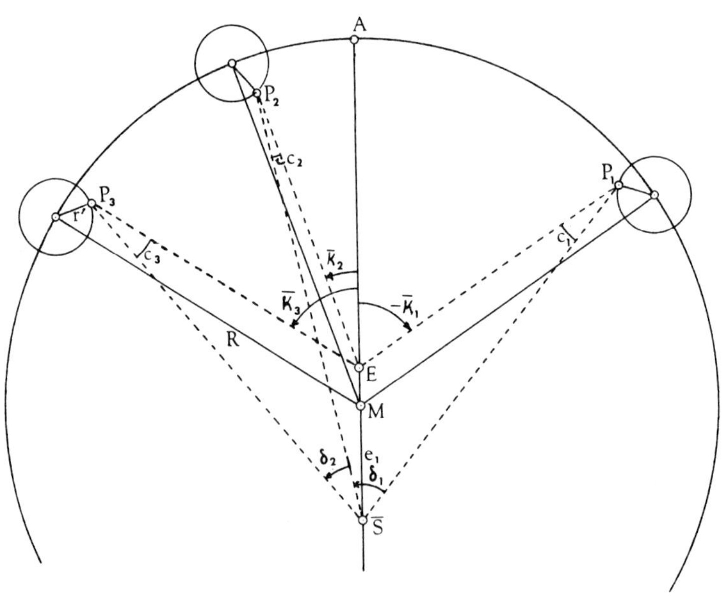

Fig. 29

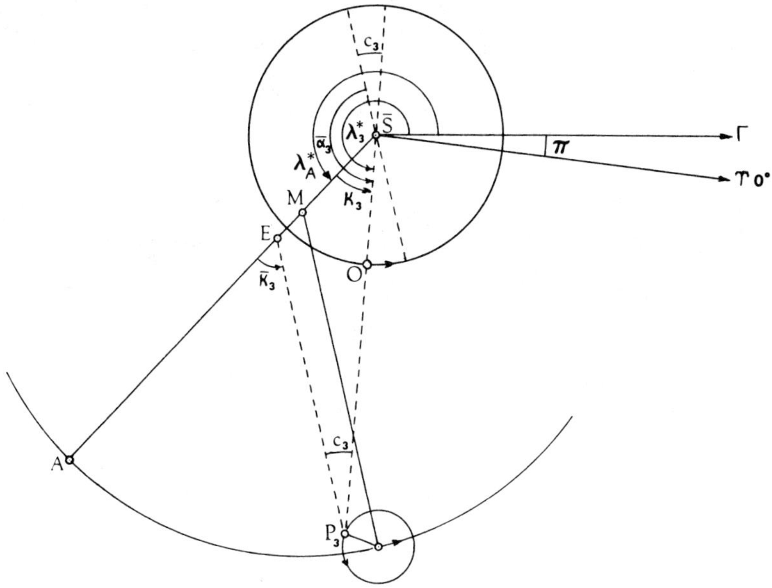

Fig. 30

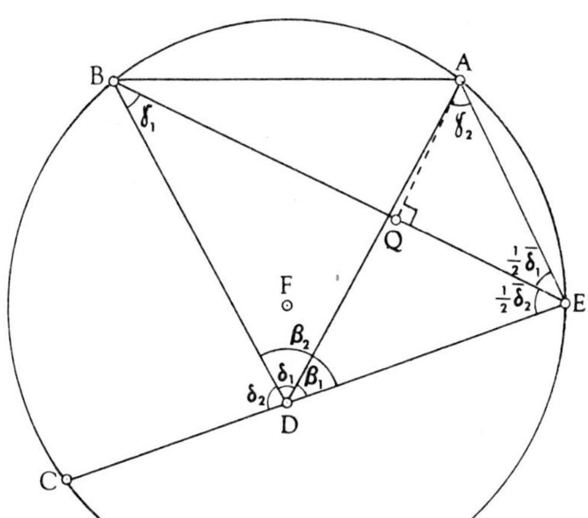

(a)

Fig. 31

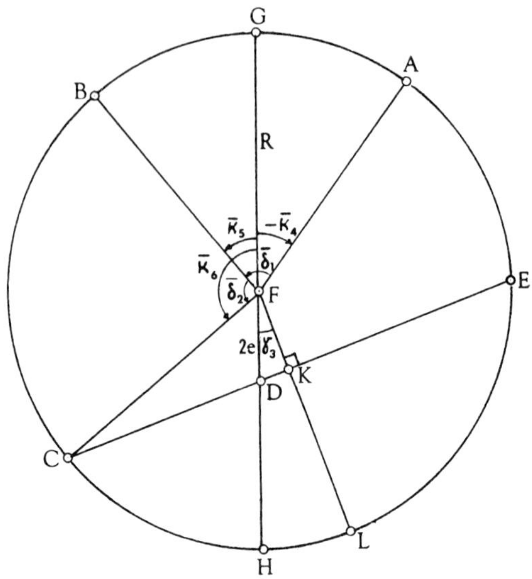

(b)

Fig. 31

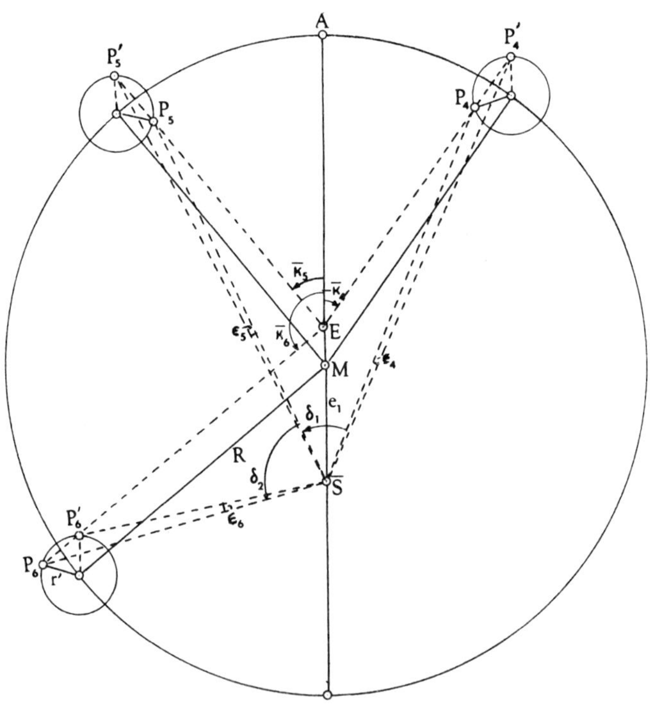

Fig. 32

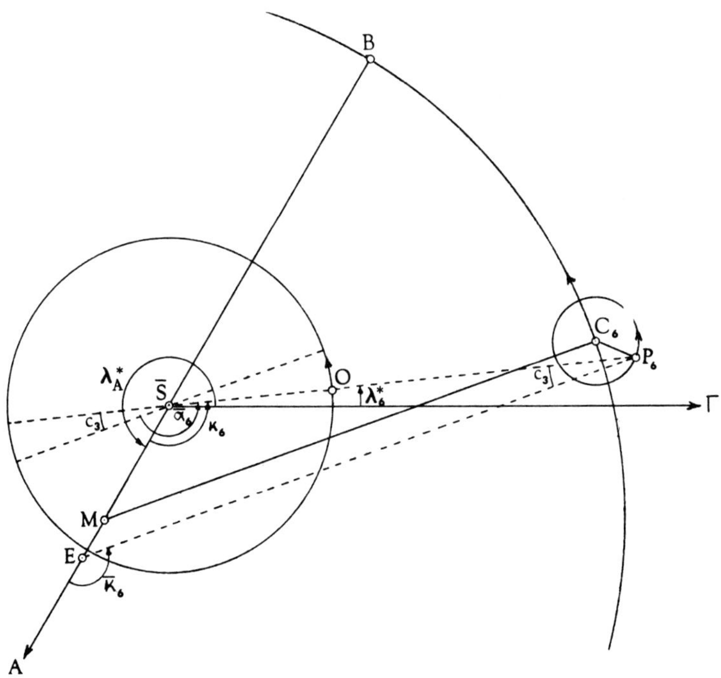

Fig. 33

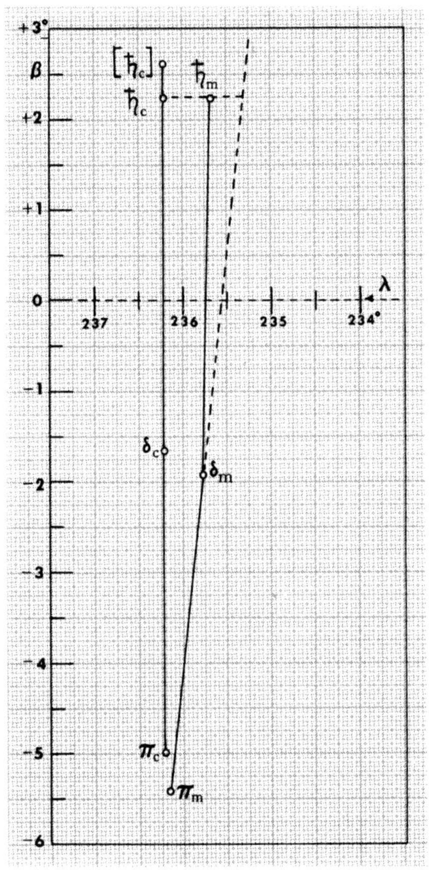

Fig. 34

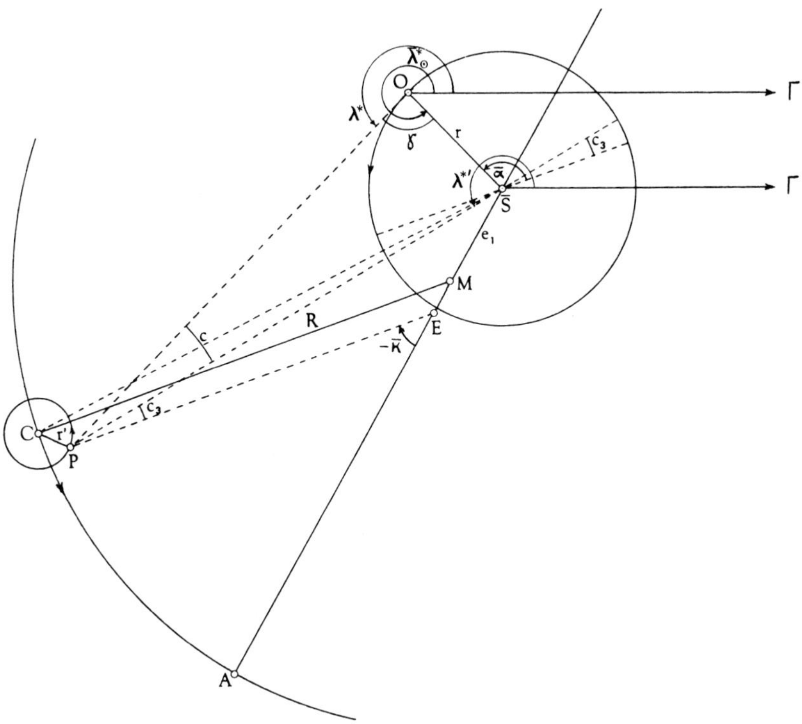

Fig. 35

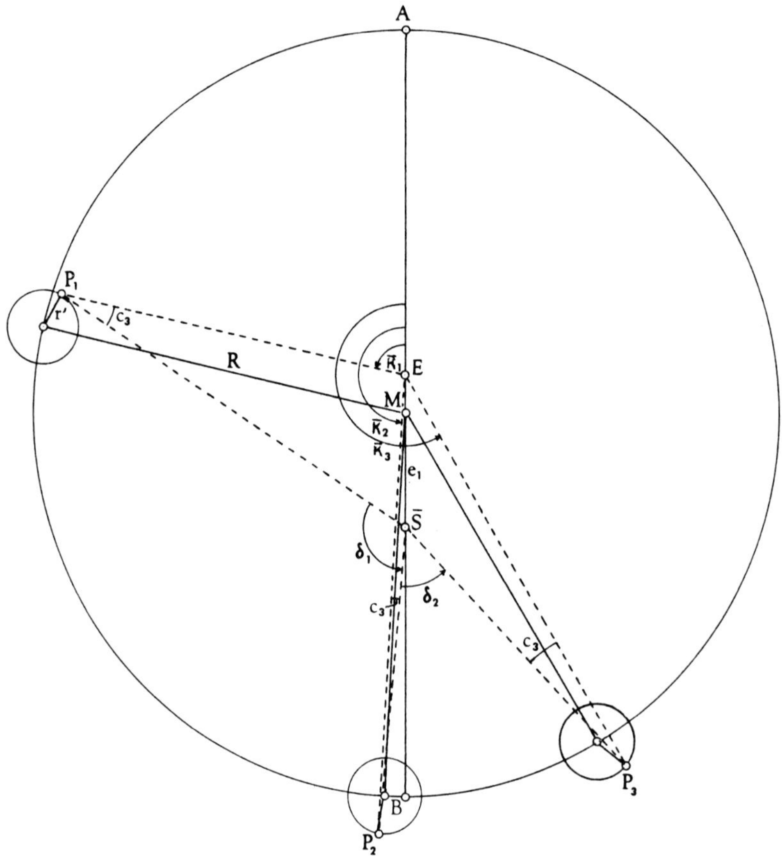

Fig. 36

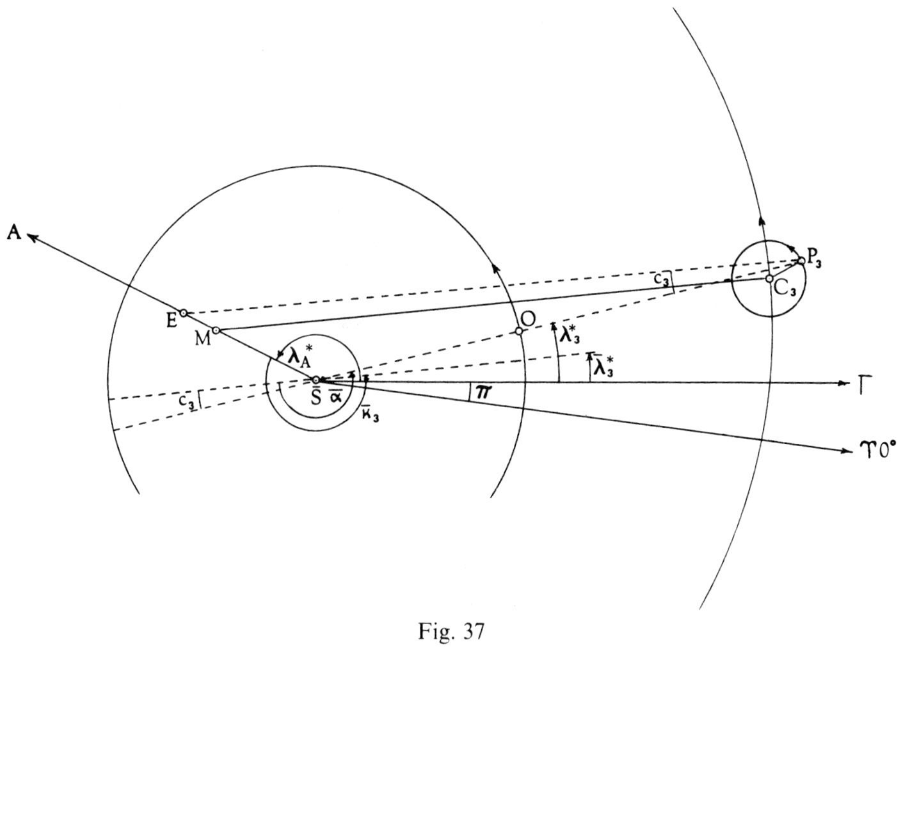
A
E
M
O
λ_A^*
λ_3^*
$\bar{\lambda}_3^*$
S
$\bar{\alpha}$
$\bar{\kappa}_3$
π
c_3
c_3
P₃
C₃
Γ
♈0°

Fig. 37

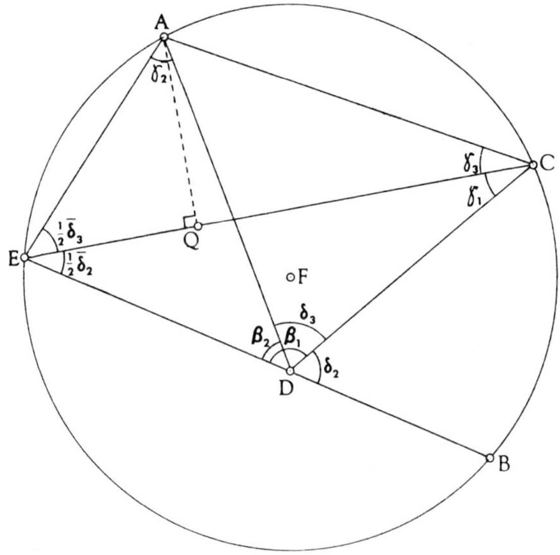

(a)

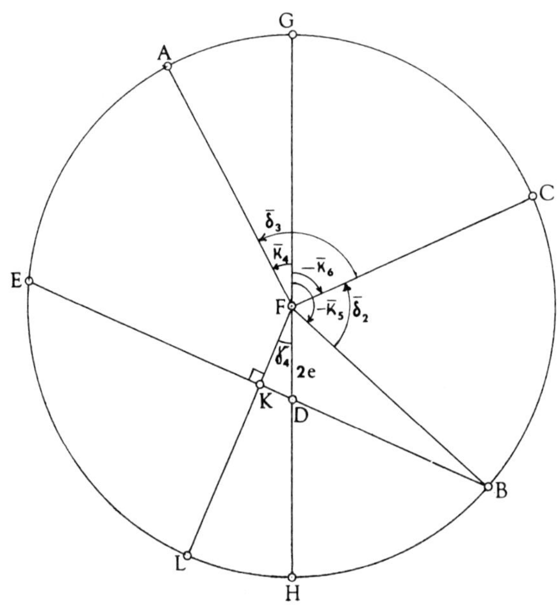

(b)

Fig. 38

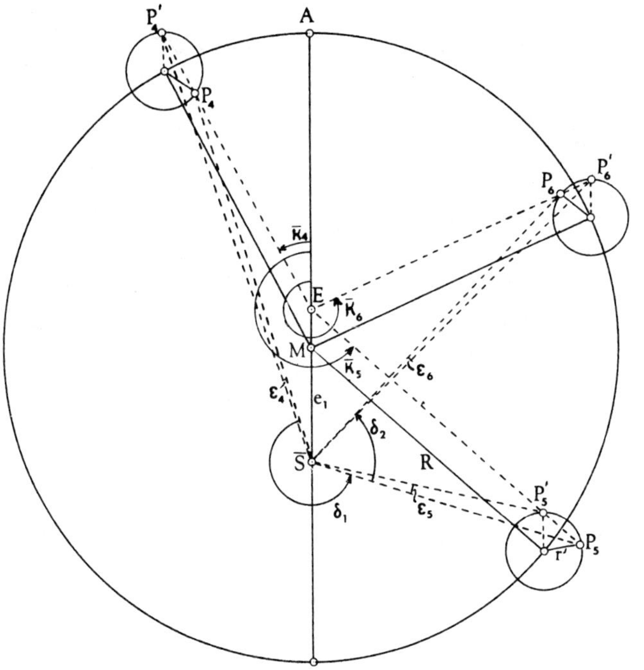

Fig. 39

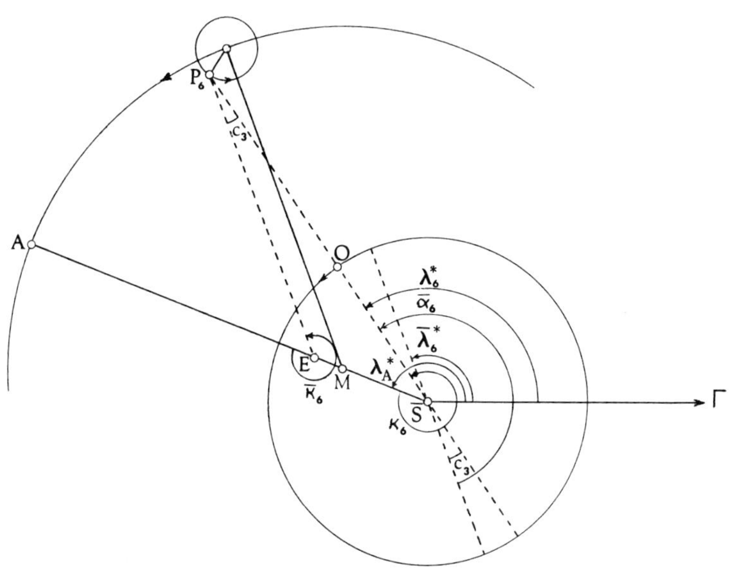

Fig. 40

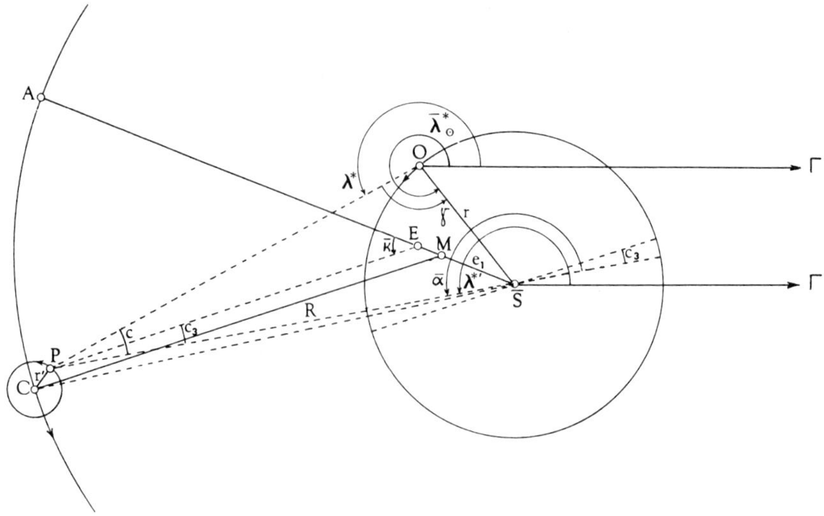
A
O
Γ
Γ
r
E
M
e_1
R
P
C
r'
$\bar{S}$
c
c_3
c_3
$\bar{\kappa}$
$\bar{\alpha}$
λ^*
$\bar{\lambda}^*_\odot$
$\lambda^{*\prime}$
γ

Fig. 41

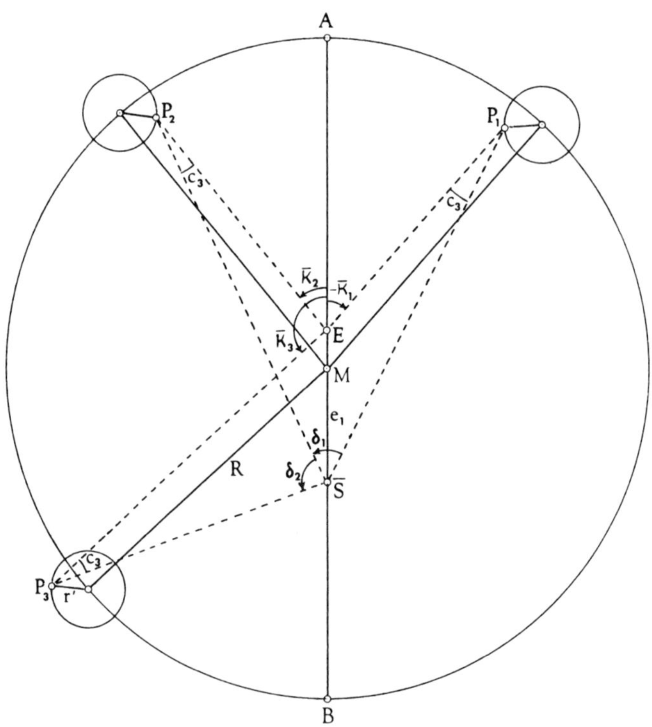
A
B
P_1
P_2
P_3
E
M
$\bar{S}$
e_1
R
r'
c_3
c_3
c_3
$\bar{\kappa}_2$
$-\bar{\kappa}_1$
$\bar{\kappa}_3$
δ_1
δ_2

Fig. 42

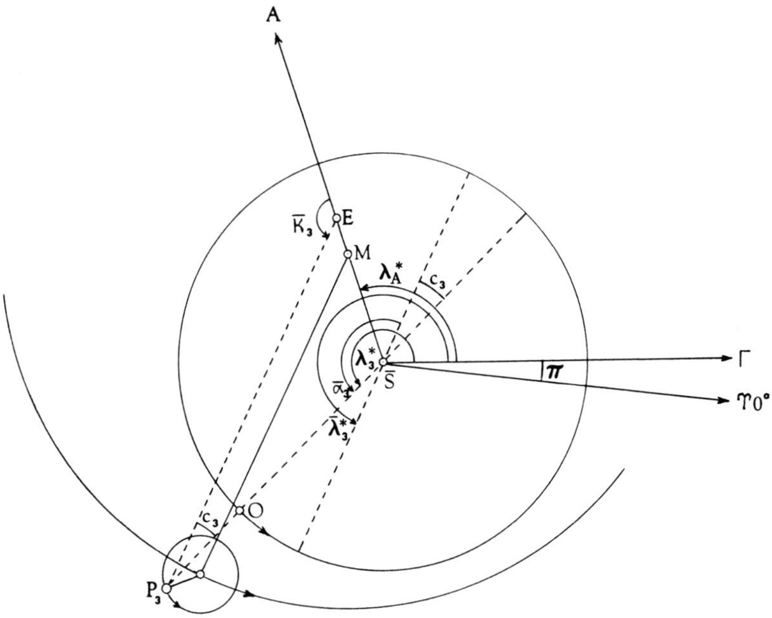

Fig. 43

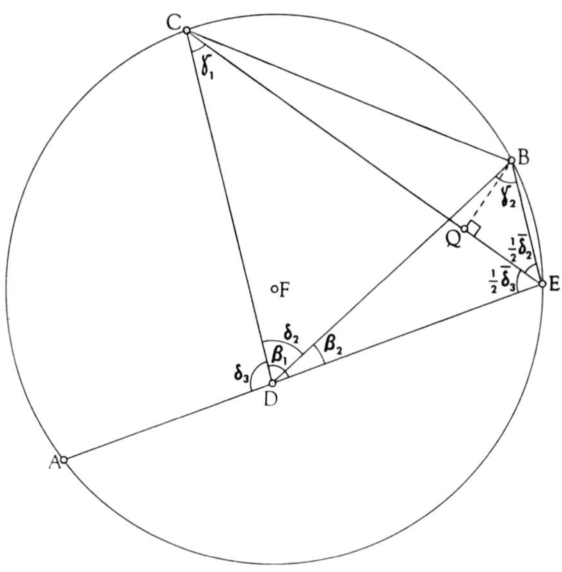

(a)

Fig. 44

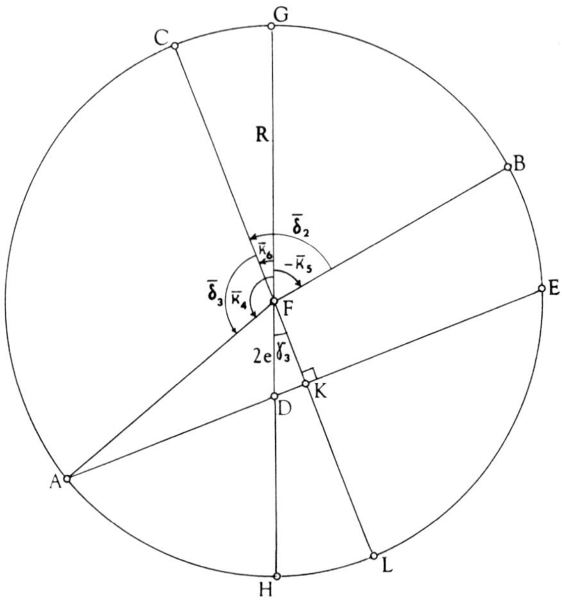

(b)

Fig. 44

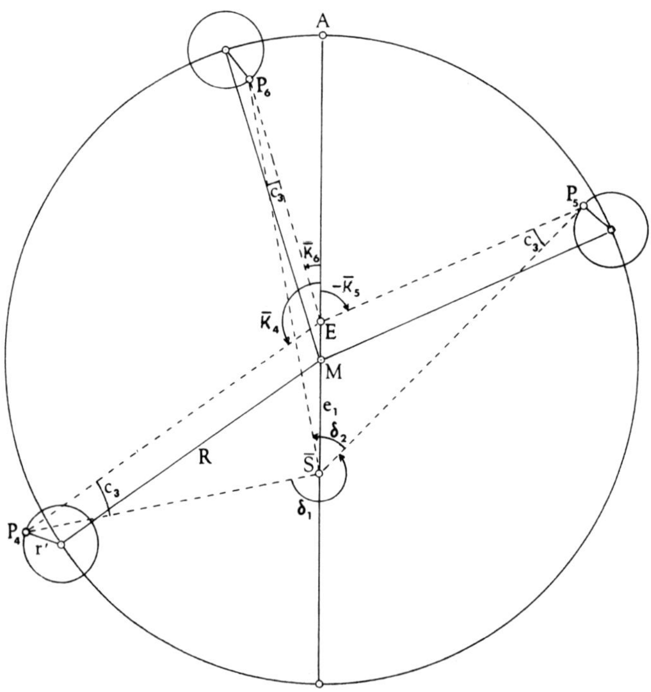

Fig. 45

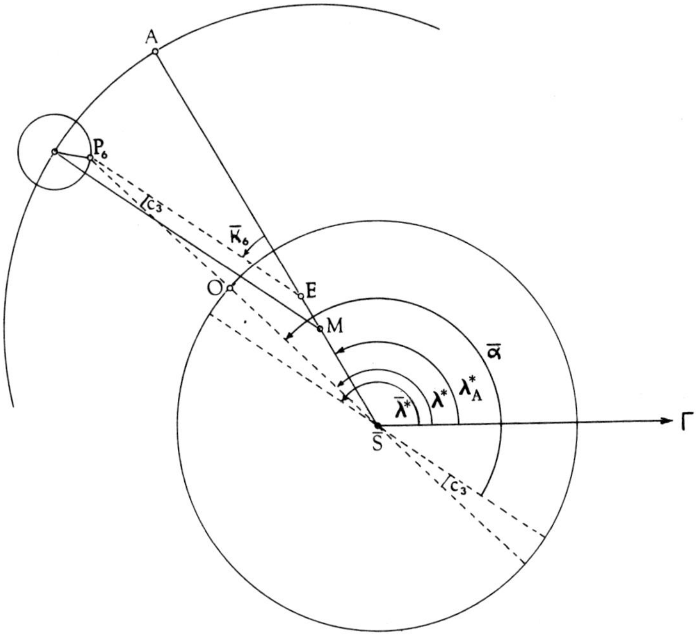

Fig. 46

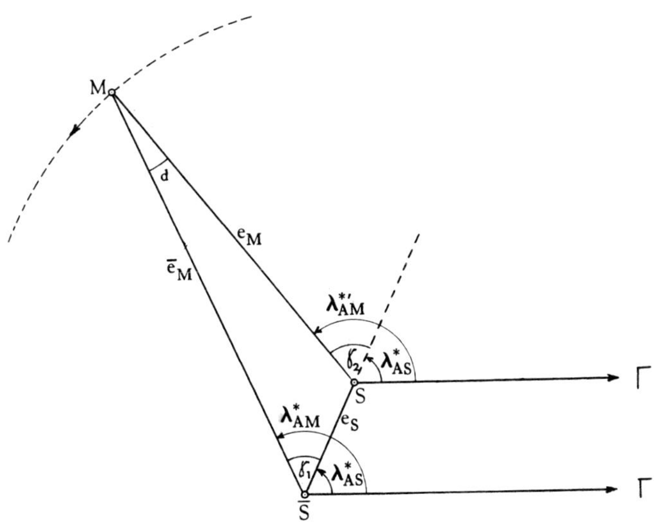

Fig. 47

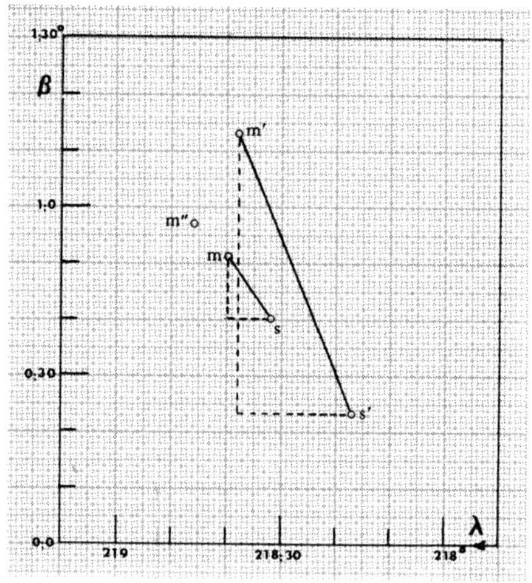
1;30°
β
1;0
0;30
0;0
219
218;30
218°
λ
m'
m"
m
s
s'

Fig. 48

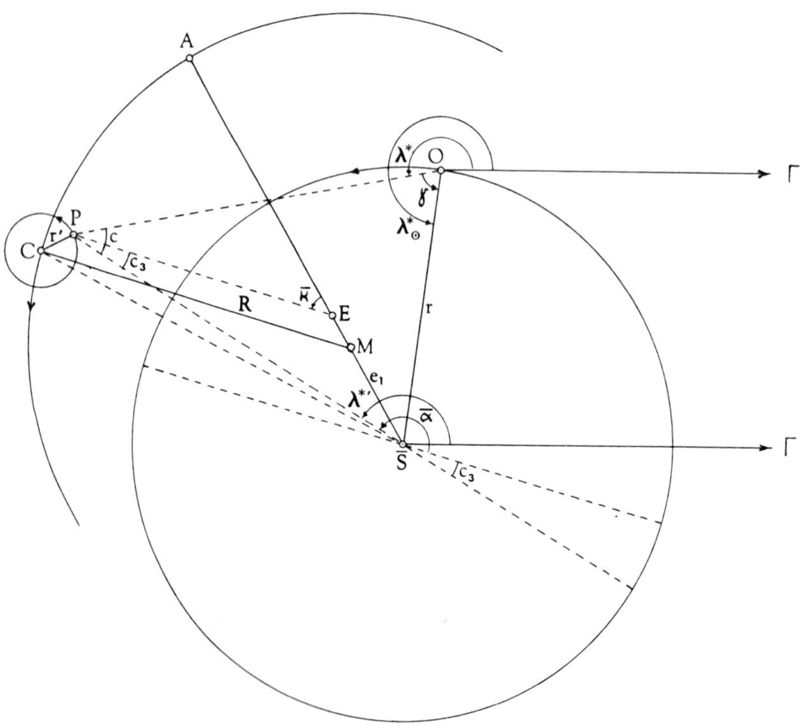
A
O
Γ
Γ
P
C
r'
c
R
E
M
e1
r
S
c3
c3
λ*
λ*⊙
λ*'

Fig. 49

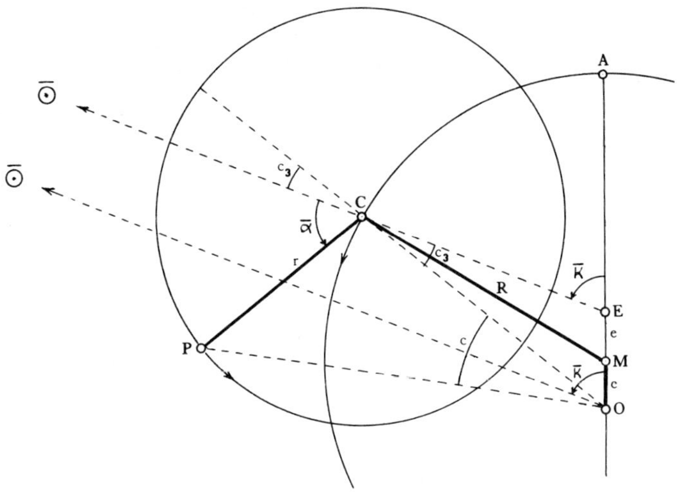
A
C
$\bar{\alpha}$
r
R
c_3
c_3
$\bar{\kappa}$
$\bar{\kappa}$
E
e
c
P
M
c
O

Fig. 50

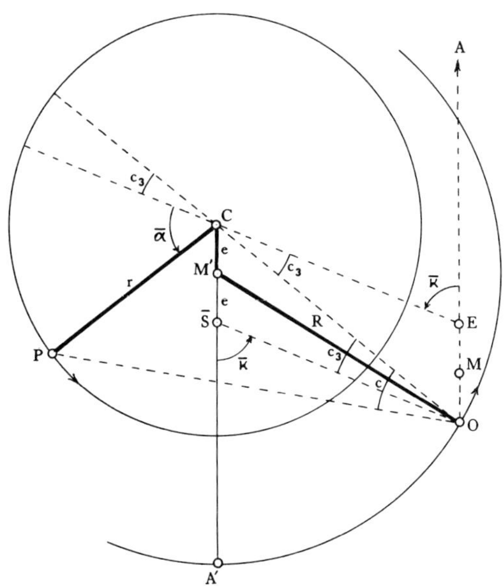
A
c_3
C
$\bar{\alpha}$
e
M′
c_3
$\bar{\kappa}$
r
e
E
$\bar{S}$
R
P
$\bar{\kappa}$
c_3
M
c
O
A′

Fig. 51

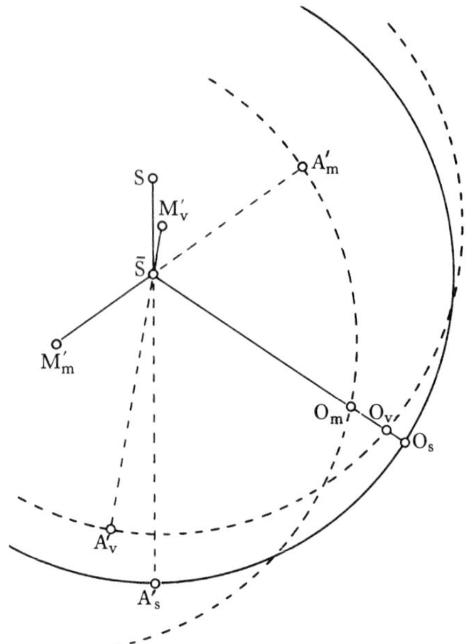

Fig. 52

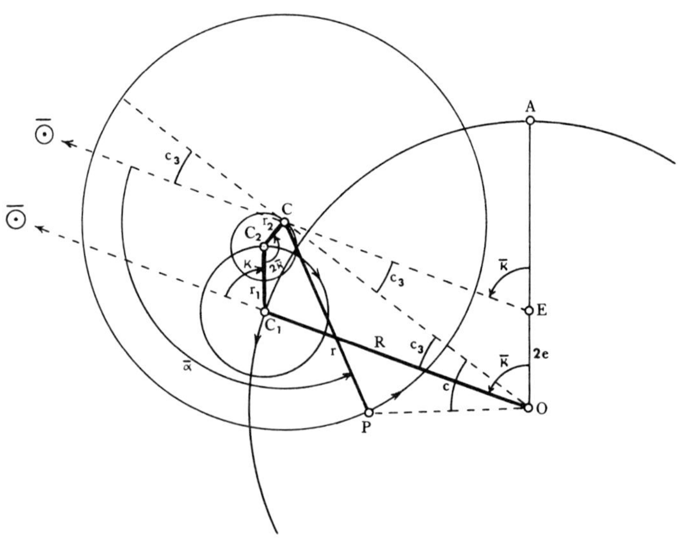

Fig. 53

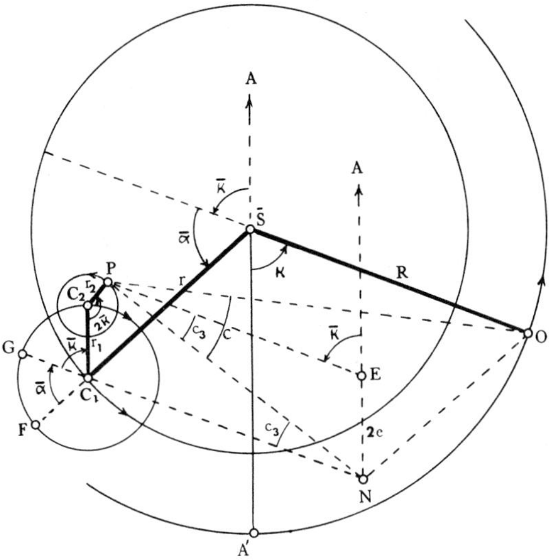
A
A
S̄
κ̄
ᾱ
κ̄
κ
R
P
r
C₂
2κ̄
c₃
c
O
G
κ̄
r₁
E
ᾱ
C₁
F
c₃
2e
N
A′

Fig. 54

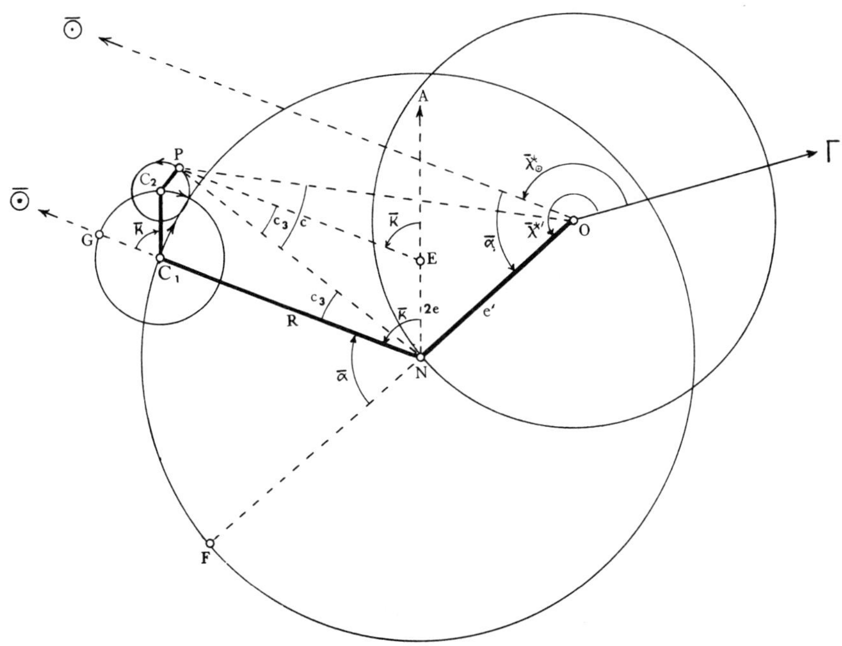
⊙̄
A
P
Γ
C₂
⊙̄
c₃
c
κ̄
ᾱ
O
G
κ̄
E
C₁
c₃
κ̄
2e
e′
R
N
ᾱ
F

Fig. 55

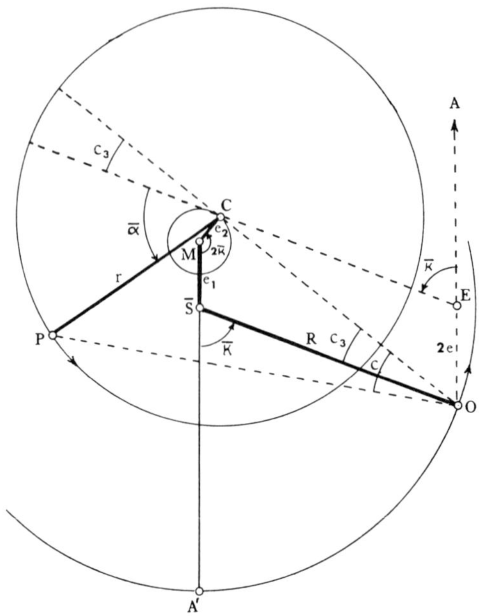
A
C
c_3
$\bar{\alpha}$
e_2
M
$2\bar{\kappa}$
e_1
$\bar{\kappa}$
E
r
$\bar{S}$
P
R
c_3
$\bar{\kappa}$
2e
c
O
A'

(a)

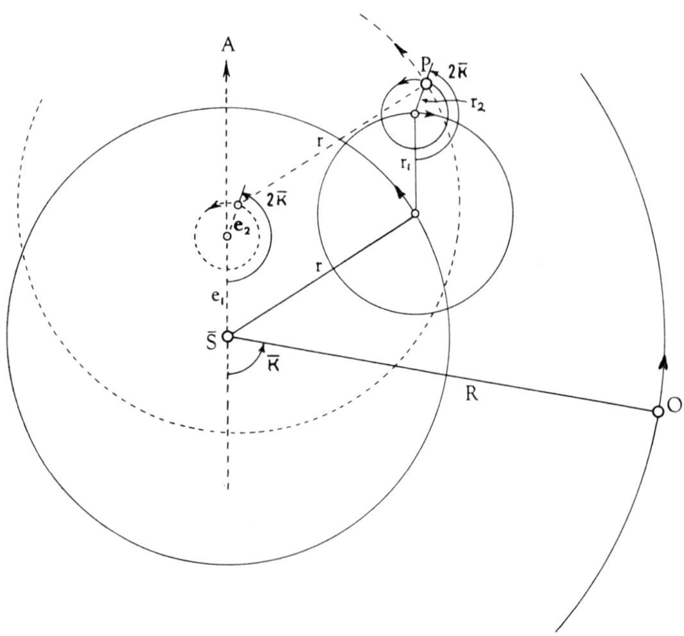
A
P
$2\bar{\kappa}$
r_2
r
r_1
$2\bar{\kappa}$
e_2
r
e_1
$\bar{S}$
$\bar{\kappa}$
R
O

(b)

Fig. 56

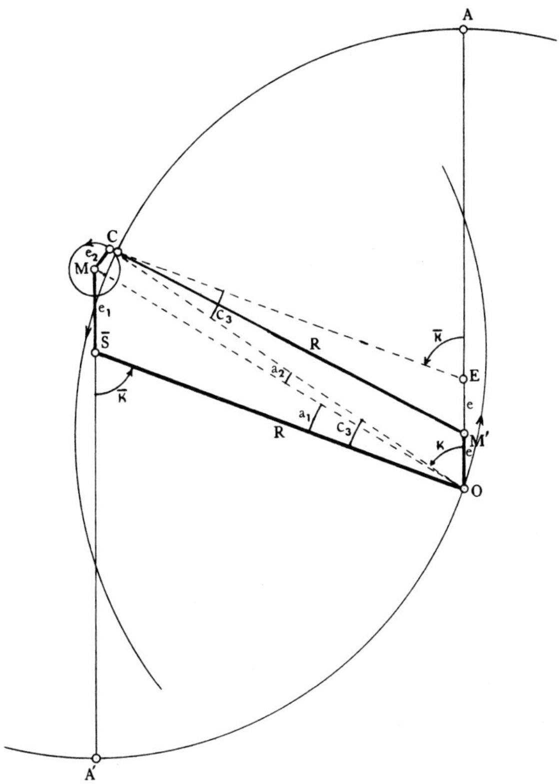

Fig. 57

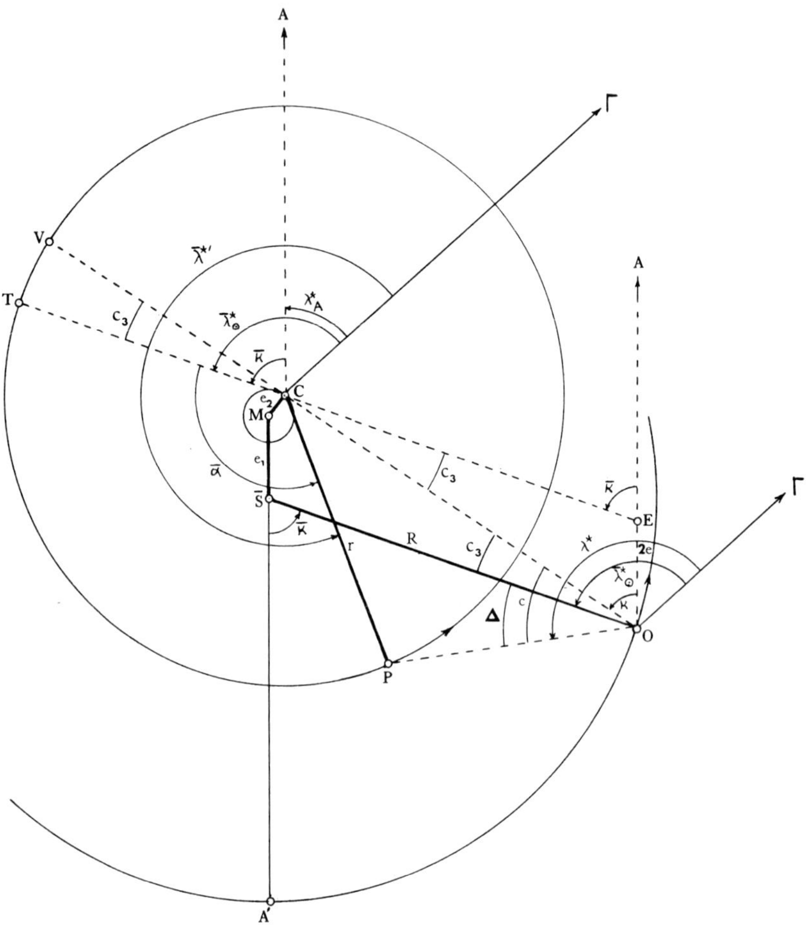
A
Γ
V
T
$\bar{\lambda}^{*\prime}$
c_3
$\bar{\lambda}^*_\odot$
λ^*_A
$\bar{\kappa}$
C
e_2
M
e_1
$\bar{\alpha}$
$\bar{S}$
$\bar{\kappa}$
r
R
c_3
c_3
c
Δ
P
A
$\bar{\kappa}$
E
λ^*
2e
$\lambda^*_\odot$
$\bar{\kappa}$
O
Γ
A'

Fig. 58

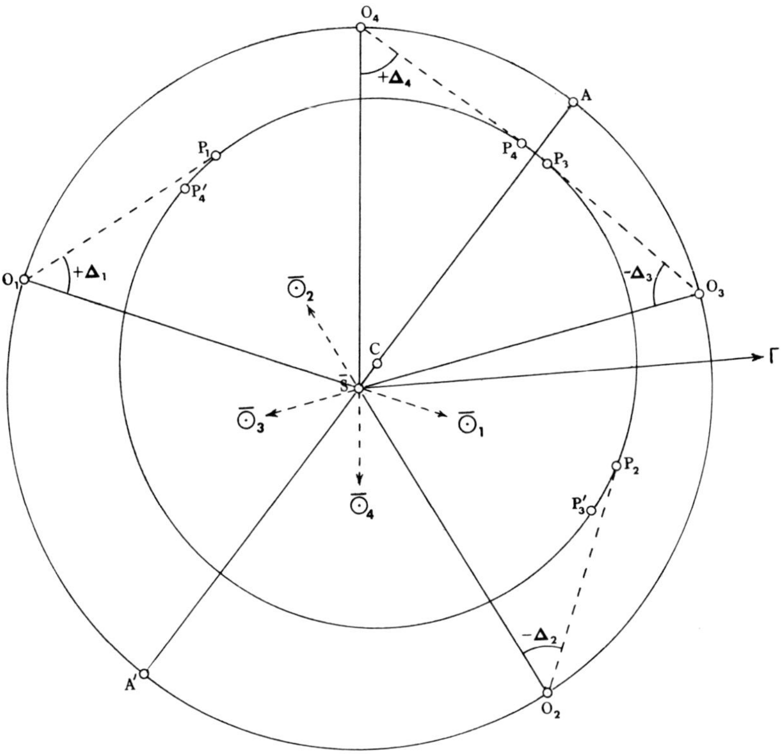
O_4
+Δ_4
A
P_1
P_4
P_3
P'_4
O_1
+Δ_1
-Δ_3
O_3
Θ_2
C
Γ
S̄
Θ_3
Θ_1
P_2
Θ_4
P'_3
-Δ_2
A'
O_2

Fig. 59

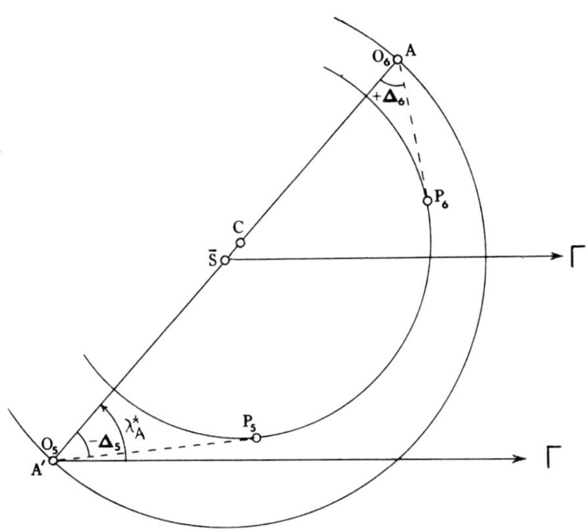
O_6
A
+Δ_6
P_6
C
S̄
Γ
λ*_A
P_5
O_5
-Δ_5
A'
Γ

Fig. 60

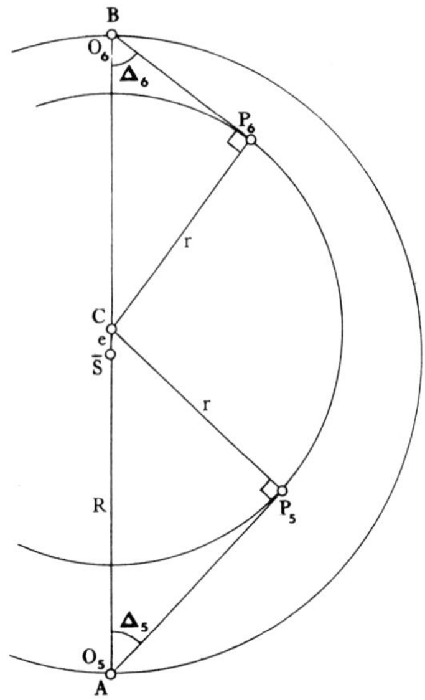

Fig. 61

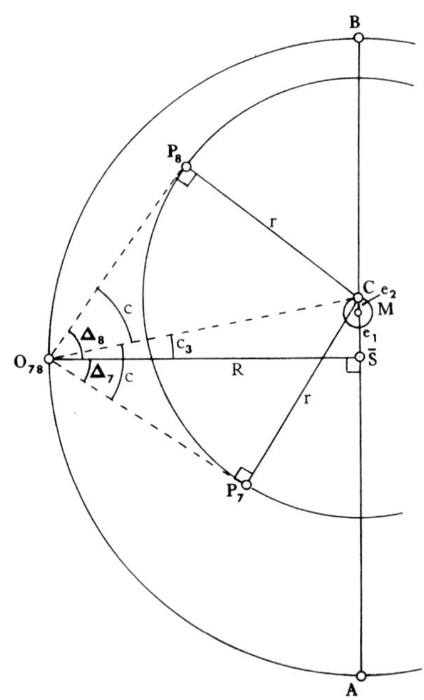

Fig. 62

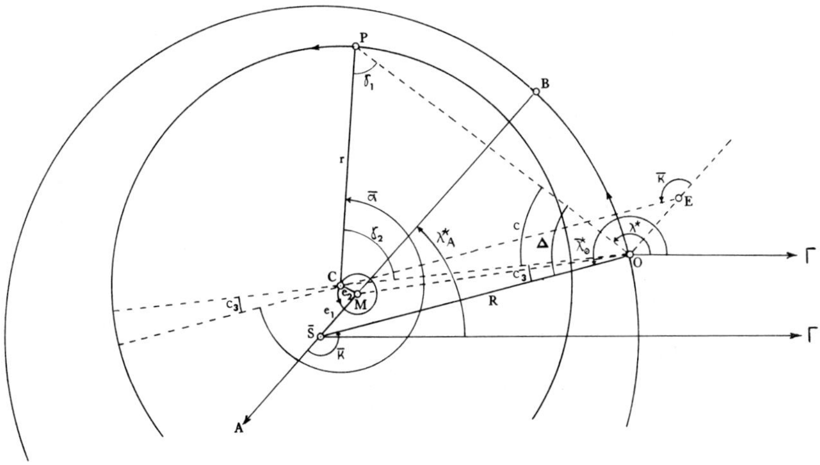
P
B
r
δ1
σ̄
γ2
λ*A
c
Δ
λ*⊙
λ*
κ̄
E
O
C
c3
e2
M
e1
c3
R
S̄
κ̄
A
Γ
Γ

Fig. 63

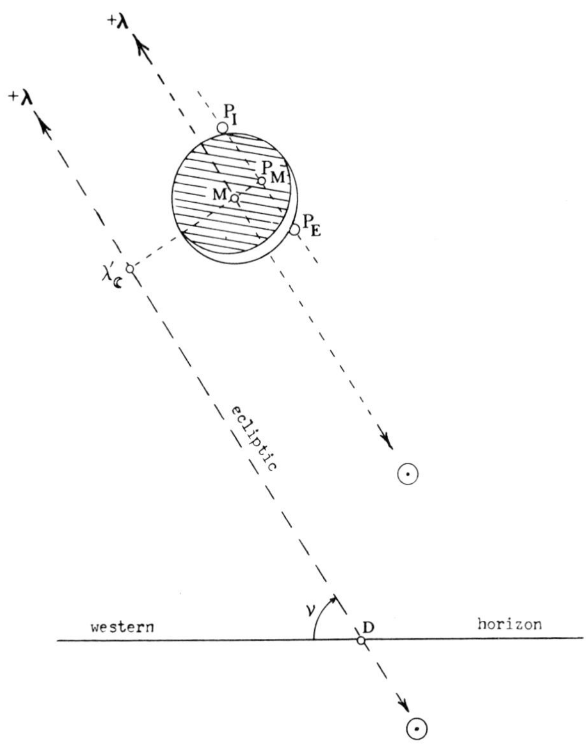
+λ
+λ
PI
PM
M
PE
λ'☾
ecliptic
γ
western
D
horizon

Fig. 64

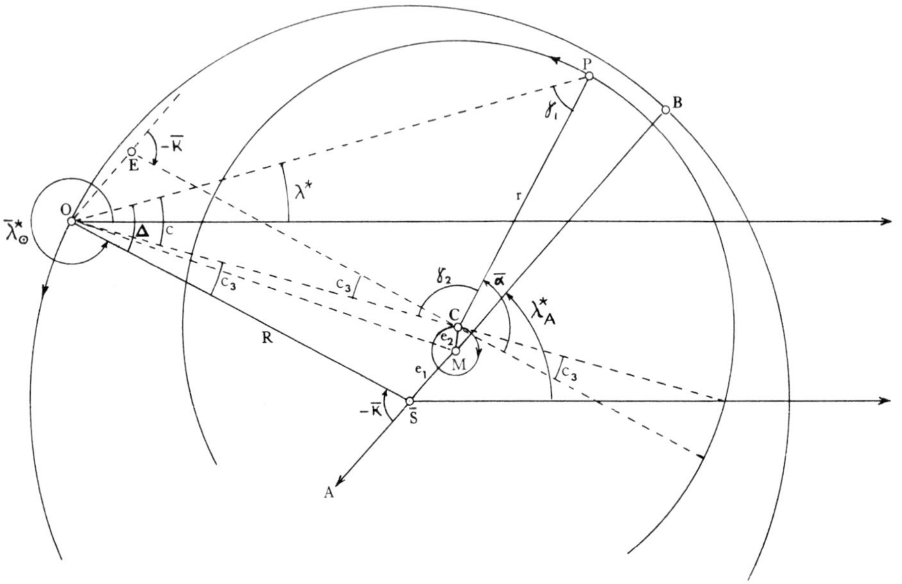
P
B
γ_1
E
$-\bar{\kappa}$
λ^*
O
r
Γ
$\bar{\lambda}^*_\odot$
Δ
c
c_3
c_3
γ_2
$\bar{\alpha}$
λ^*_A
C
e_2
M
R
e_1
c_3
$-\bar{\kappa}$
$\bar{S}$
Γ
A

Fig. 65

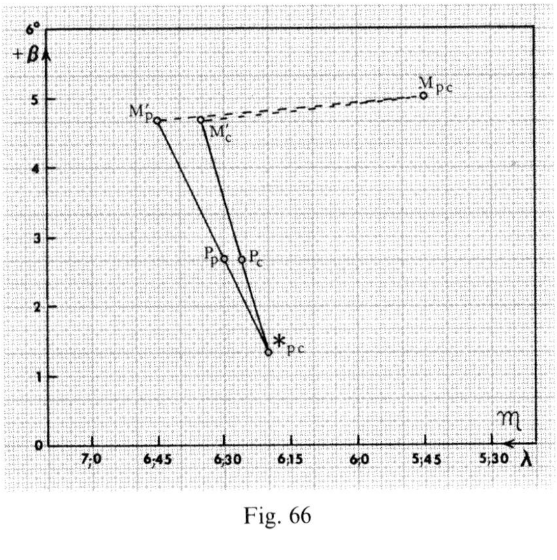
6°
+β
5
4
3
2
1
0
M'_p
M'_c
M_{pc}
P_p
P_c
$*_{pc}$
♏
7;0
6;45
6;30
6;15
6;0
5;45
5;30
λ

Fig. 66

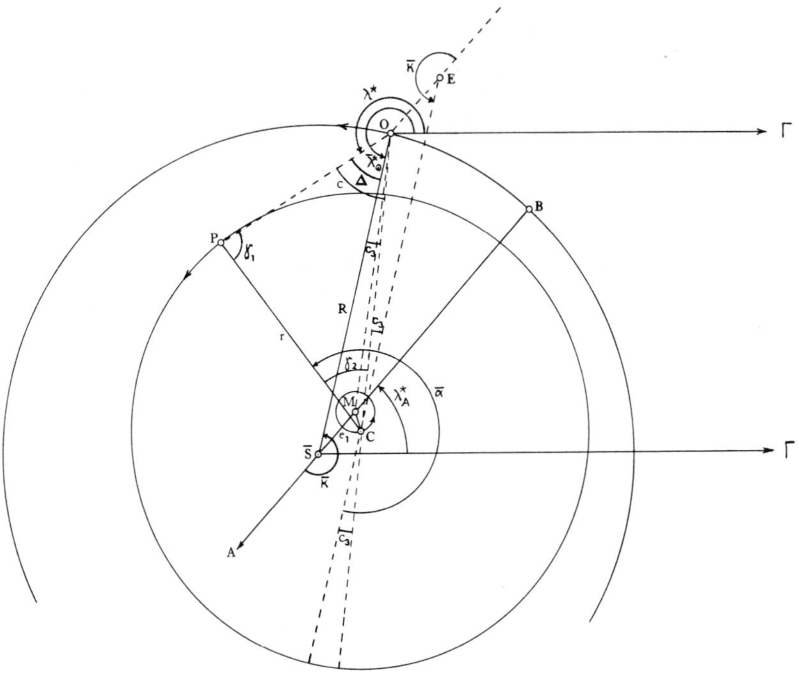

Fig. 67

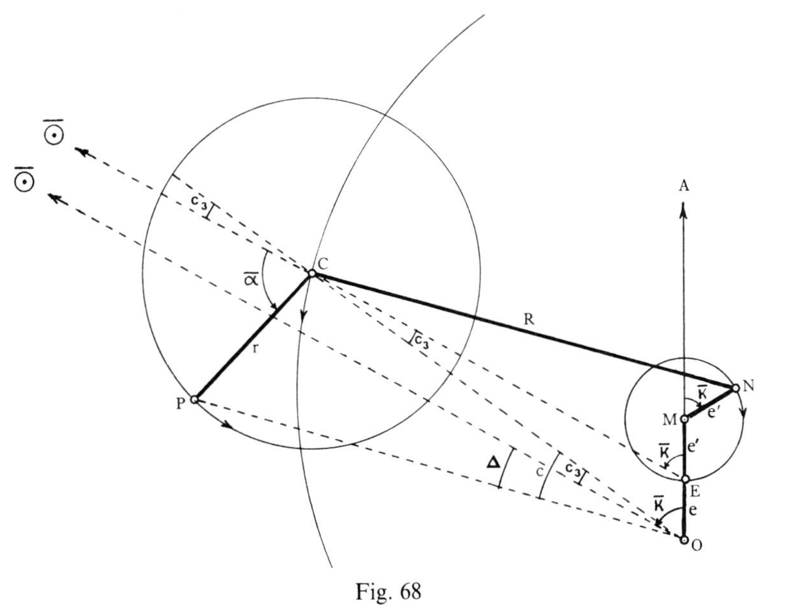

Fig. 68

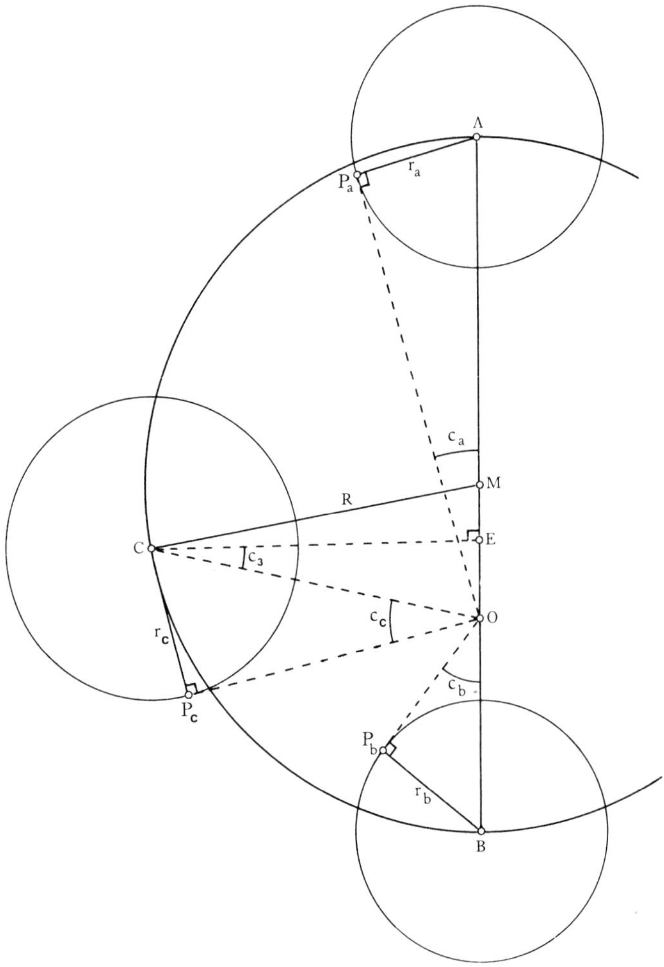

Fig. 69

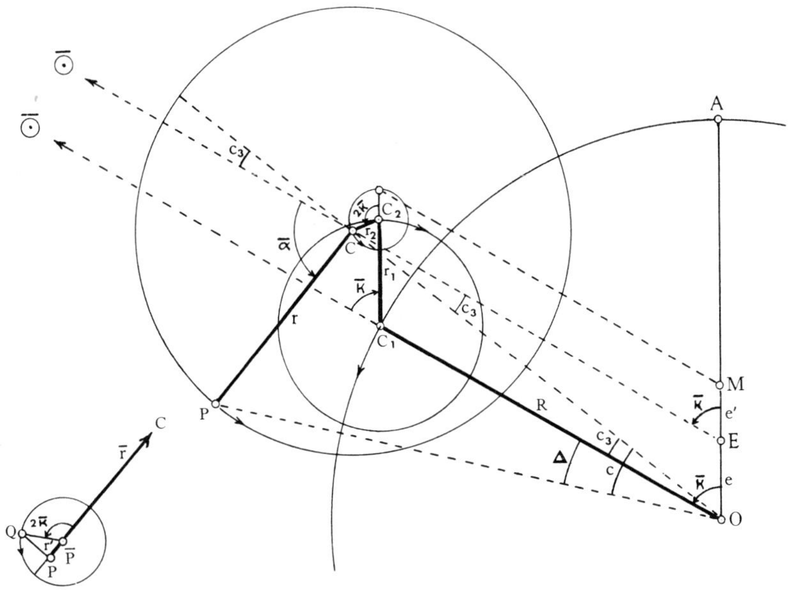

Fig. 70

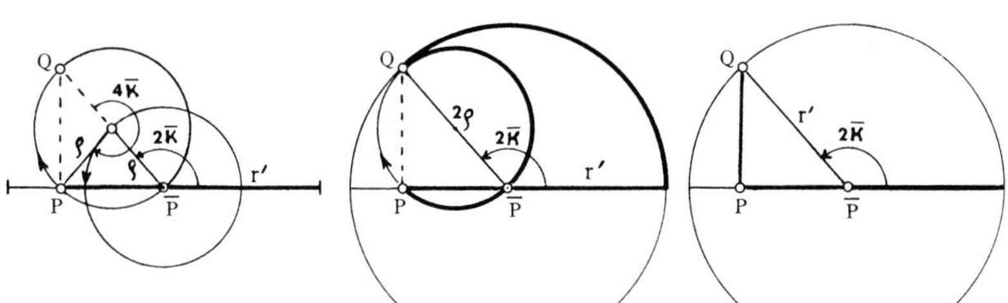

(a) (b) (c)

Fig. 71

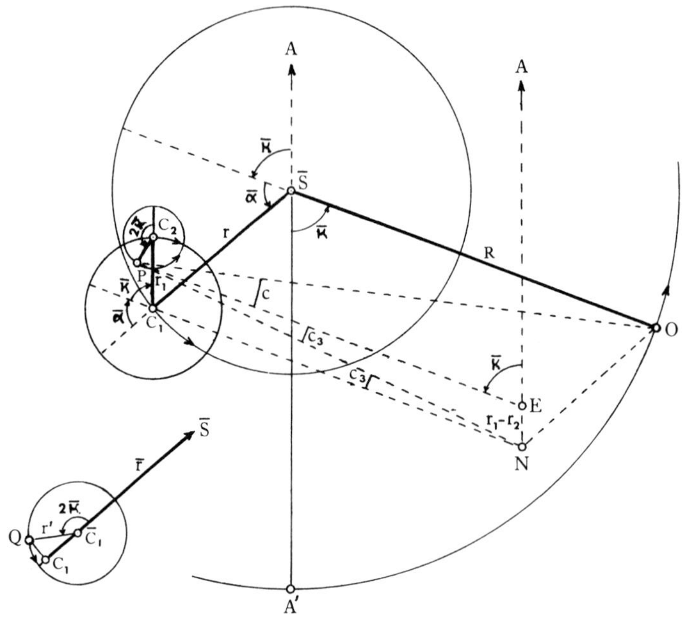

Fig. 72

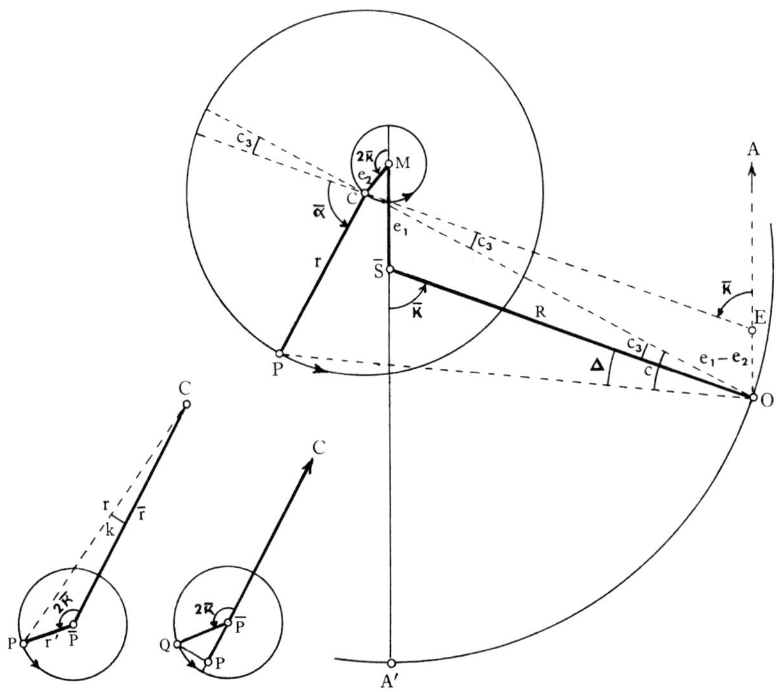

(a)

Fig. 73

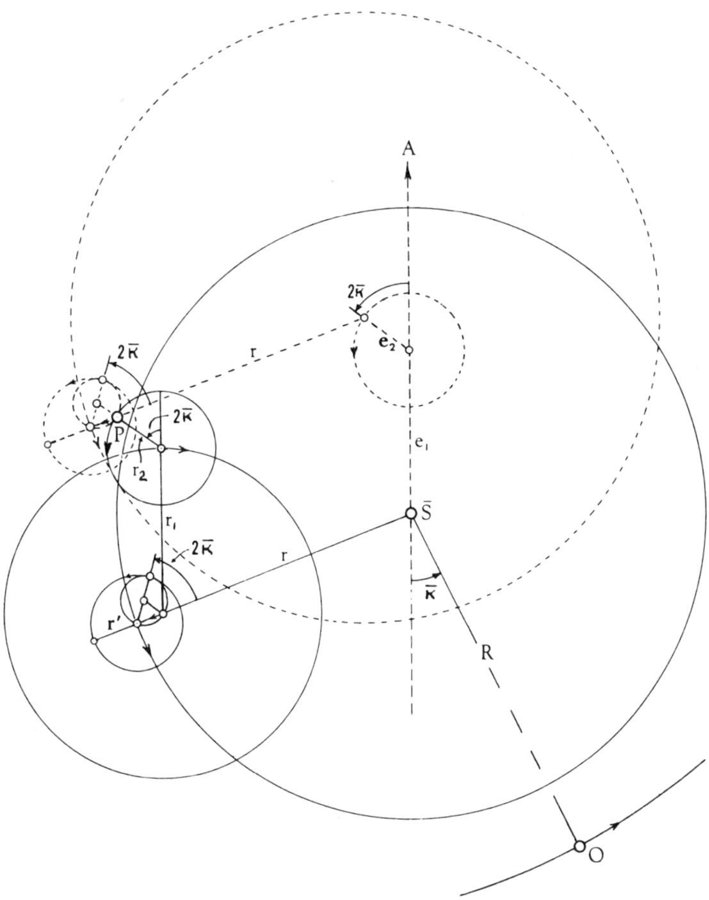

(b)

Fig. 73

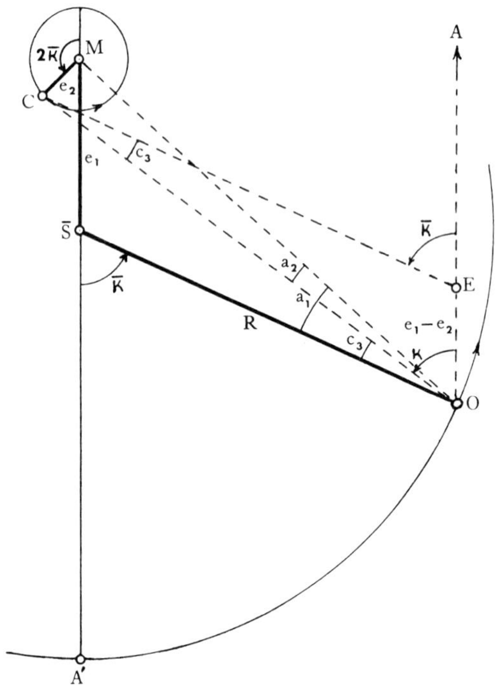

Fig. 74

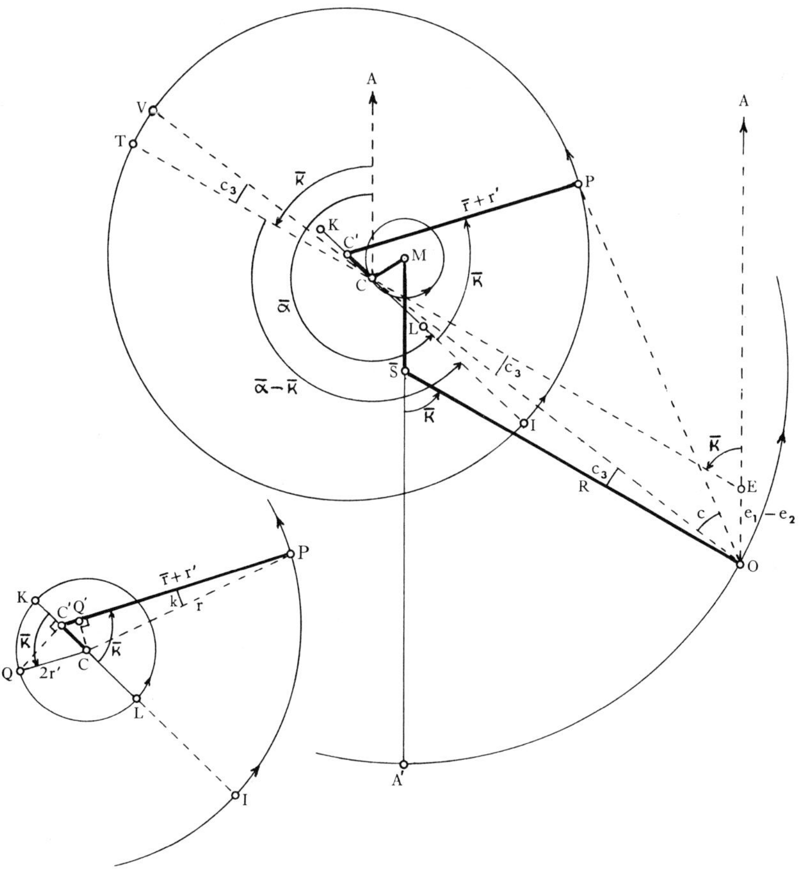

Fig. 75

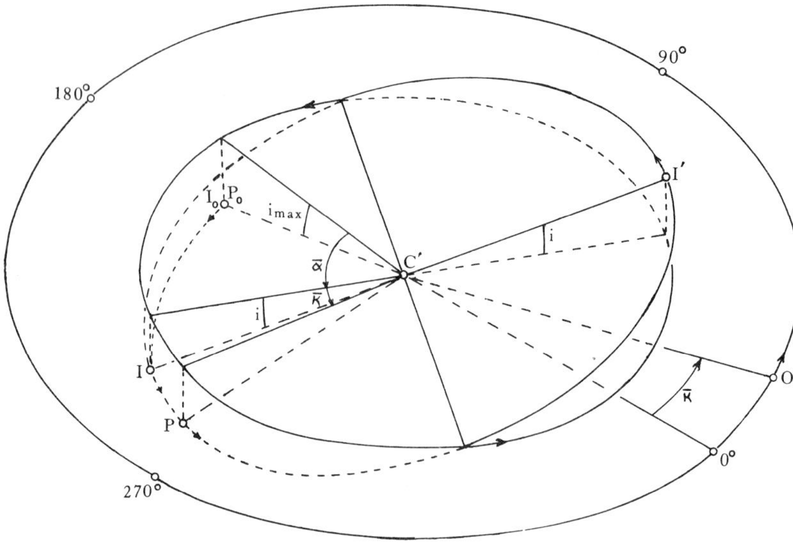
90°
180°
I'
I_0 P_0
i_{max}
i
C'
$\bar{\alpha}$
$\bar{\kappa}$
i
I
O
$\bar{\kappa}$
P
0°
270°

Fig. 76

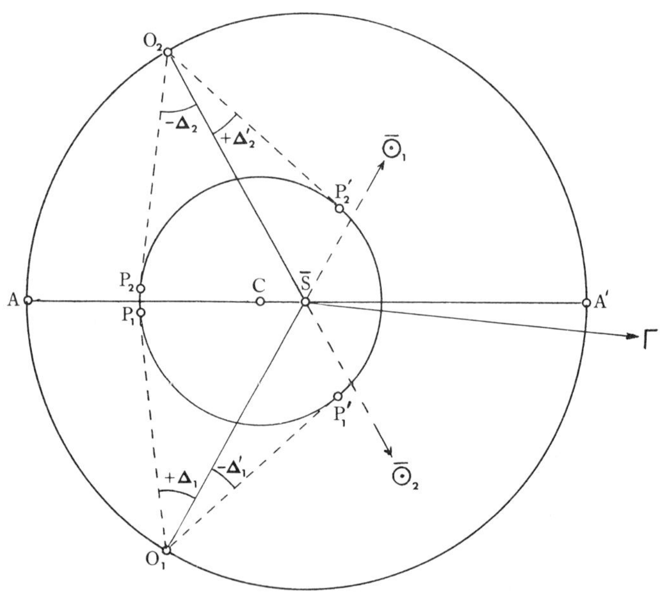
O_2
$-\Delta_2$
$+\Delta_2'$
$\bar{\odot}_1$
P_2'
P_2
A
C
$\bar{S}$
A'
P_1
Γ
P_1'
$\bar{\odot}_2$
$+\Delta_1$
$-\Delta_1'$
O_1

Fig. 77

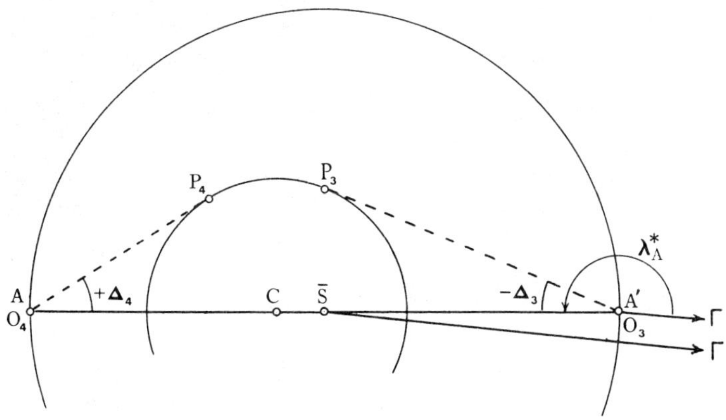

Fig. 78

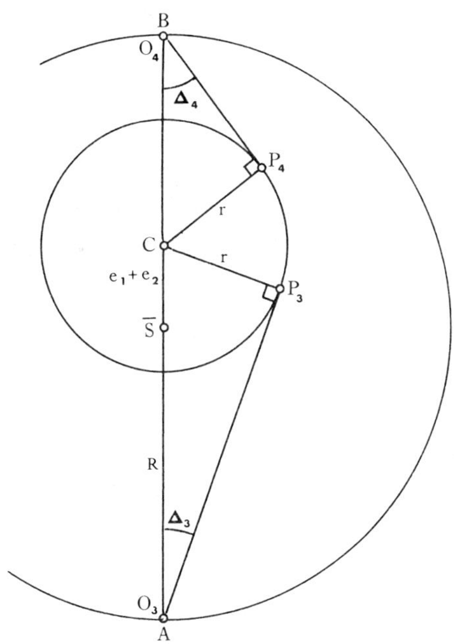

Fig. 79

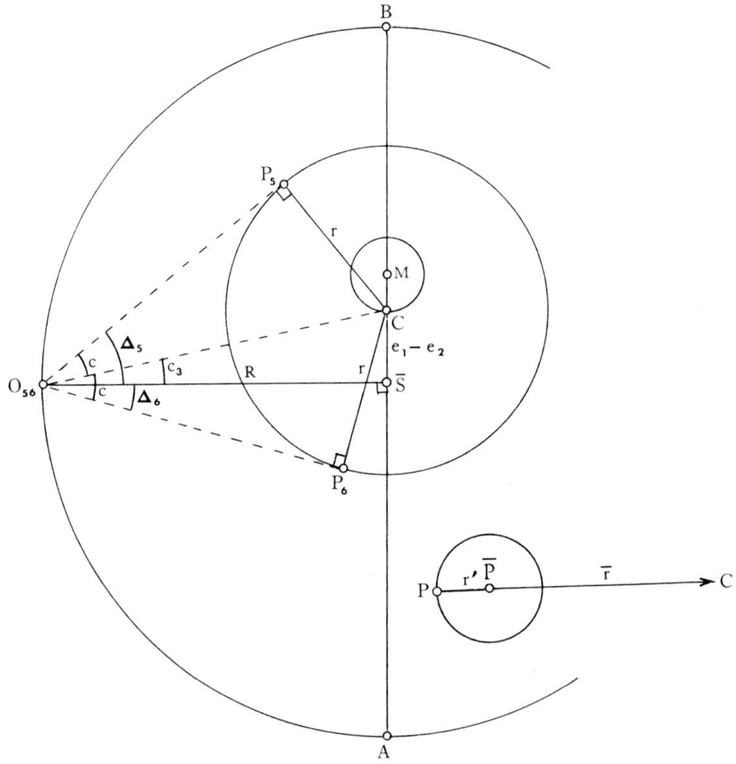

Fig. 80

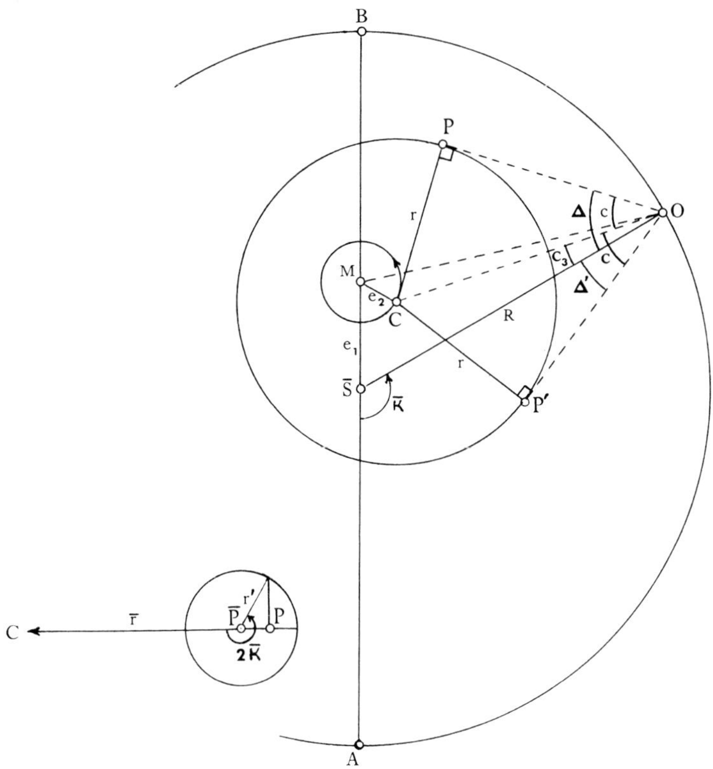

Fig. 81

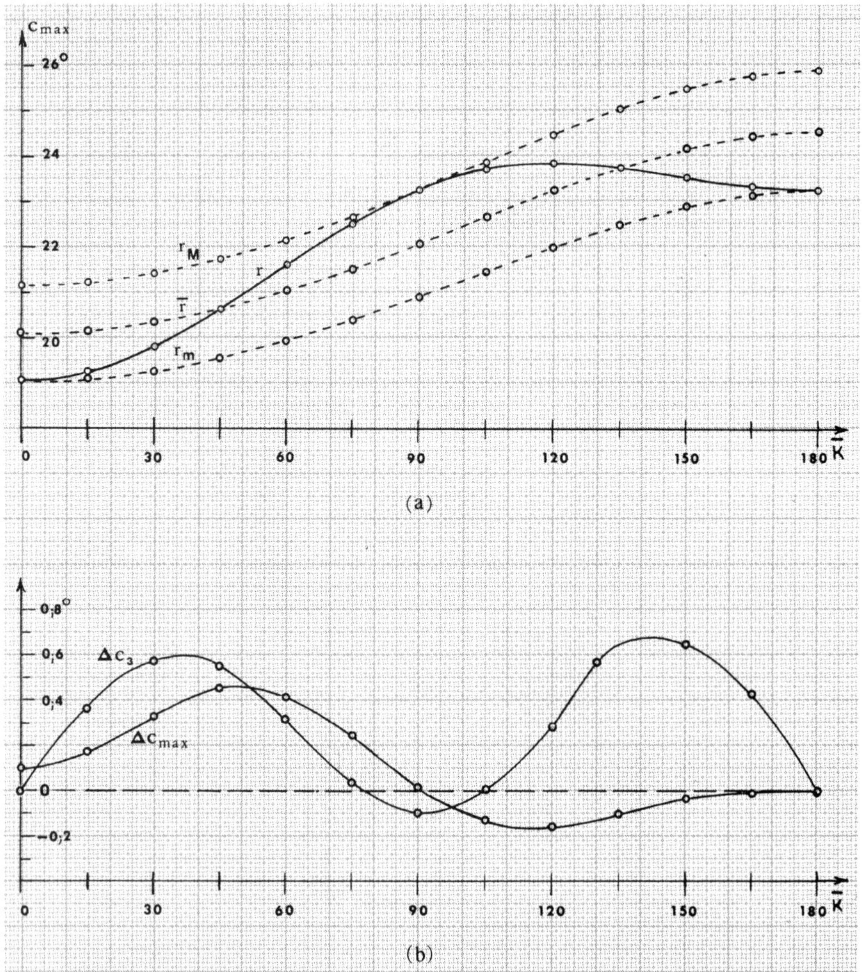

c max
26°
24
22
20
r M
r
r̄
r m
0
30
60
90
120
150
180
K̄
(a)
0;8°
0;6
0;4
0
−0;2
Δc₃
Δc max
0
30
60
90
120
150
180
K̄
(b)

Fig. 82

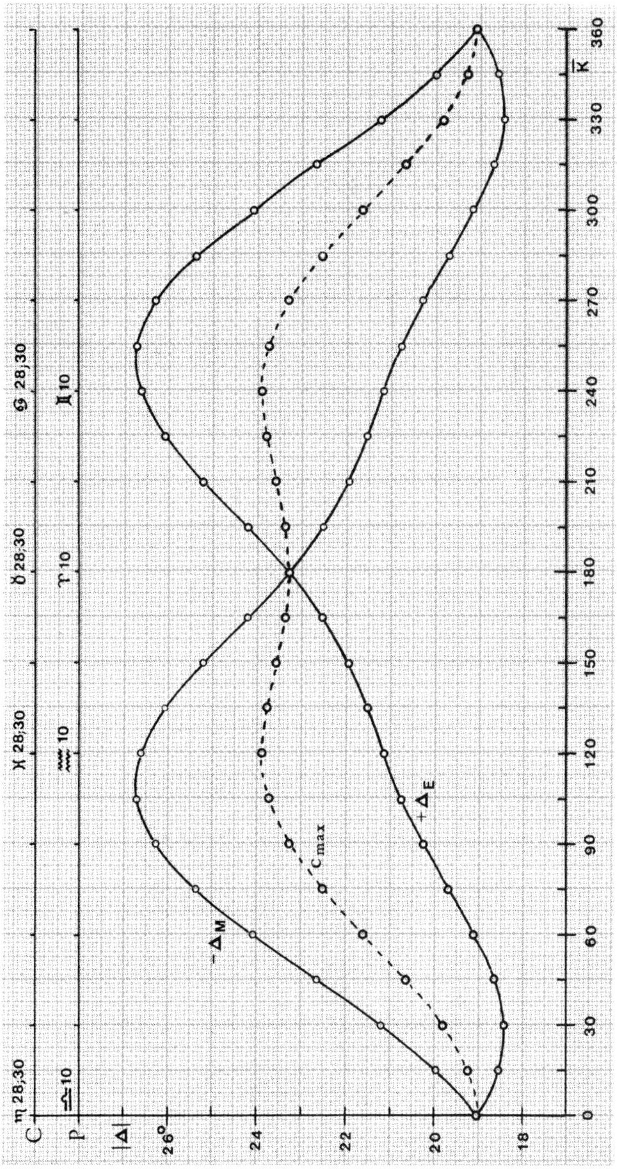
C
P
|Δ|
♍ 28;30
♎ 10
♓ 28;30
♒ 10
♉ 28;30
♈ 10
♋ 28;30
♊ 10
$-\Delta_M$
c_{max}
$+\Delta_E$
26°
24
22
20
18
0
30
60
90
120
150
180
210
240
270
300
330
360
$\bar{\kappa}$

Fig. 83

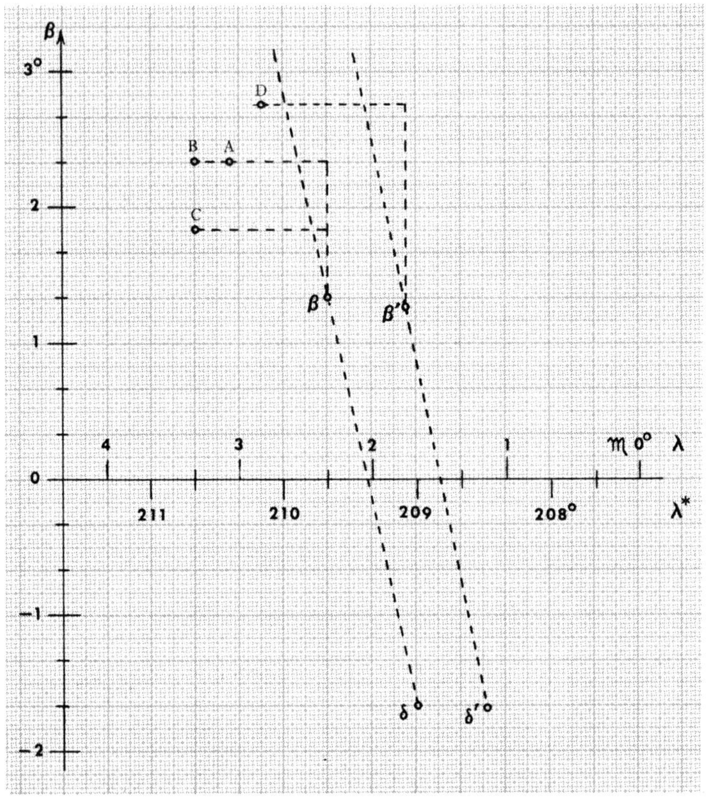
β
3°
2
1
0
−1
−2
D
B
A
C
β
β′
δ
δ′
4
3
2
1
♏ 0°
λ
211
210
209
208°
λ*

Fig. 84

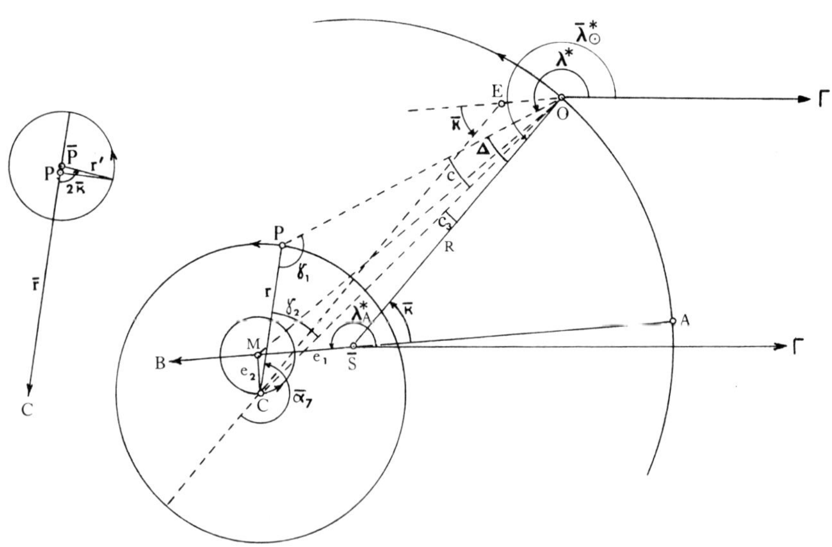
λ̄*☉
λ*
E
O
Γ
κ̄
Δ
c
ς
R
P̄
P
r′
2κ̄
r̄
C
P
γ1
r
γ2
λ*A
κ̄
A
M
B
e1
S̄
e2
C
ᾱ7

Fig. 85

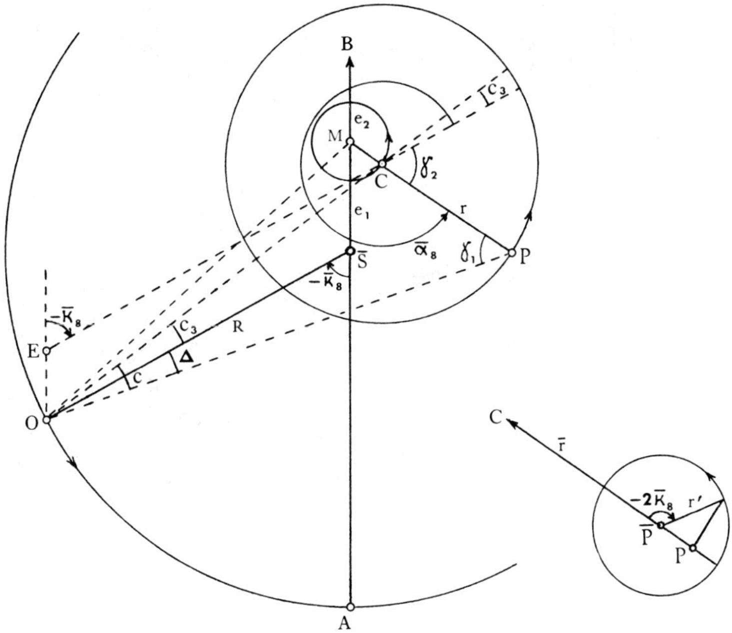
B
e_2
M
C
γ_2
r
c_3
e_1
$\bar{\alpha}_8$
γ_1
P
$\bar{S}$
$-\bar{\kappa}_8$
$-\bar{\kappa}_8$
E
c_3
R
Δ
c
O
A
C
$\bar{r}$
$-2\bar{\kappa}_8$
r′
$\bar{P}$
P

Fig. 86

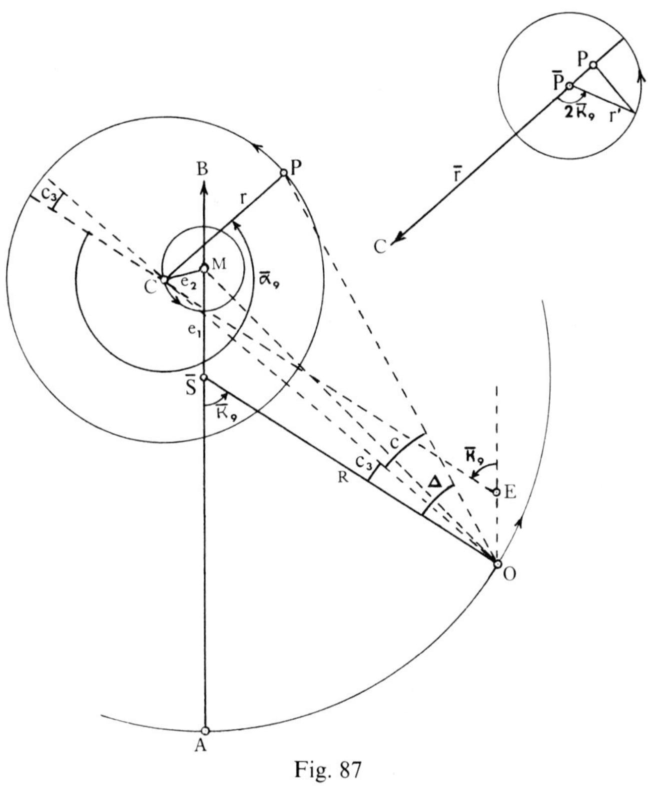
P
$\bar{P}$
$2\bar{\kappa}_9$
r′
$\bar{r}$
C
B
P
r
c_3
M
C
e_2
$\bar{\alpha}_9$
e_1
$\bar{S}$
$\bar{\kappa}_9$
c
$\bar{\kappa}_9$
c_3
R
Δ
E
O
A

Fig. 87

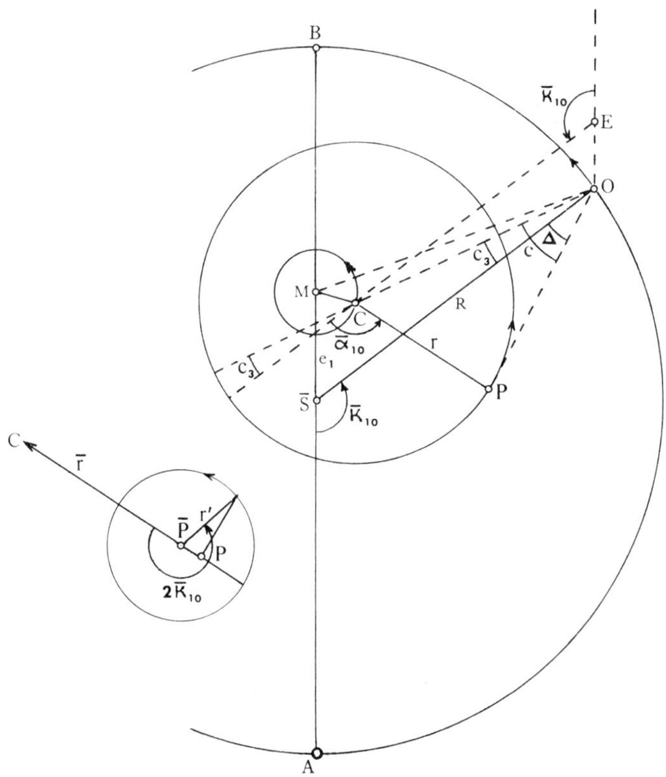

Fig. 88

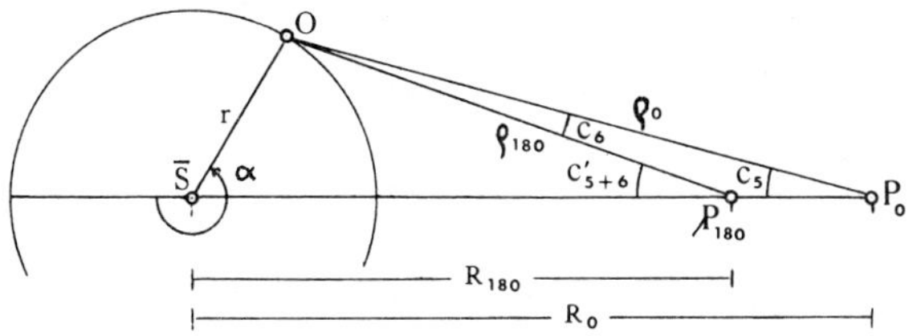
O
r
ρ180
c6
ρ0
S̄
α
c′5+6
c5
P0
P180
R180
R0

(a)

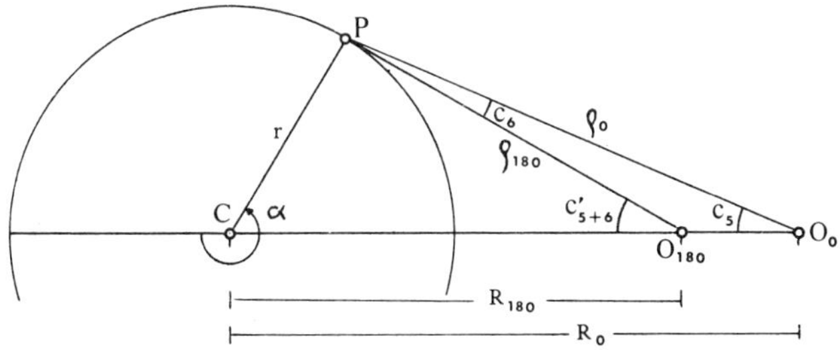
P
r
c6
ρ0
ρ180
C
α
c′5+6
c5
O0
O180
R180
R0

(b)

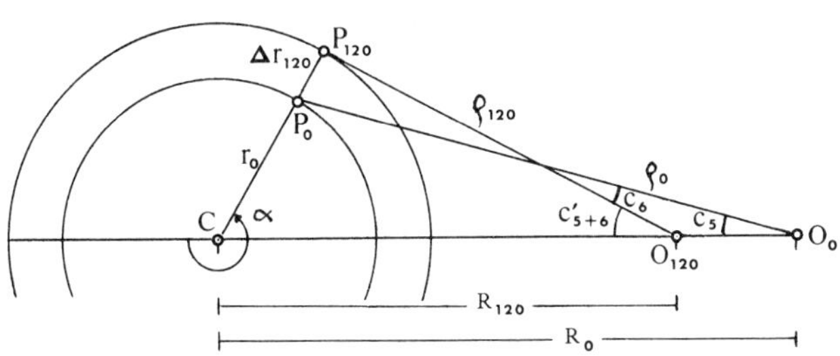
P120
Δr120
ρ120
P0
r0
ρ0
C
α
c′5+6
c6
c5
O0
O120
R120
R0

(c)

Fig. 89

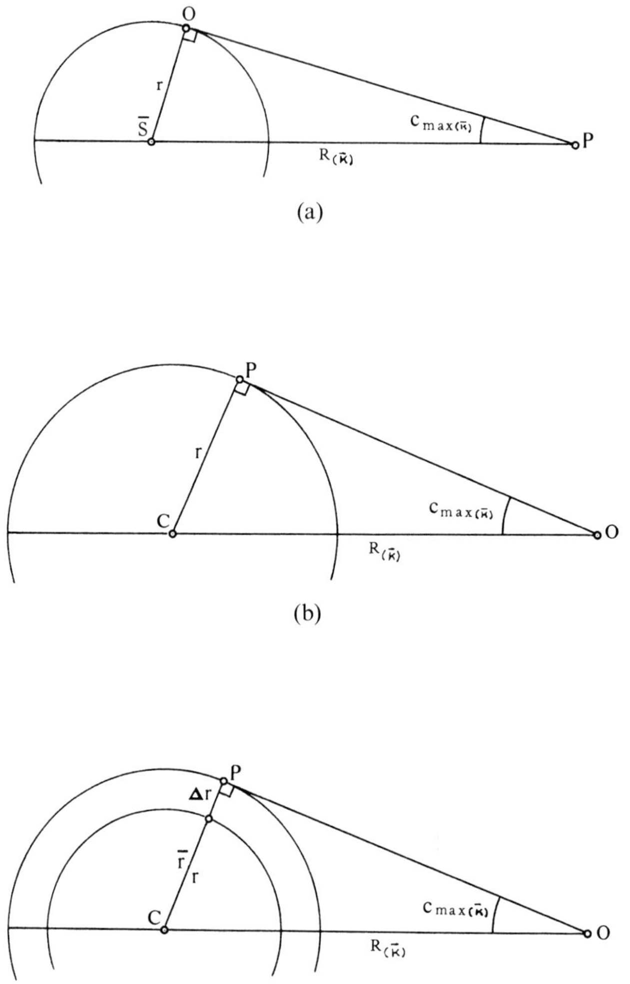

Fig. 90

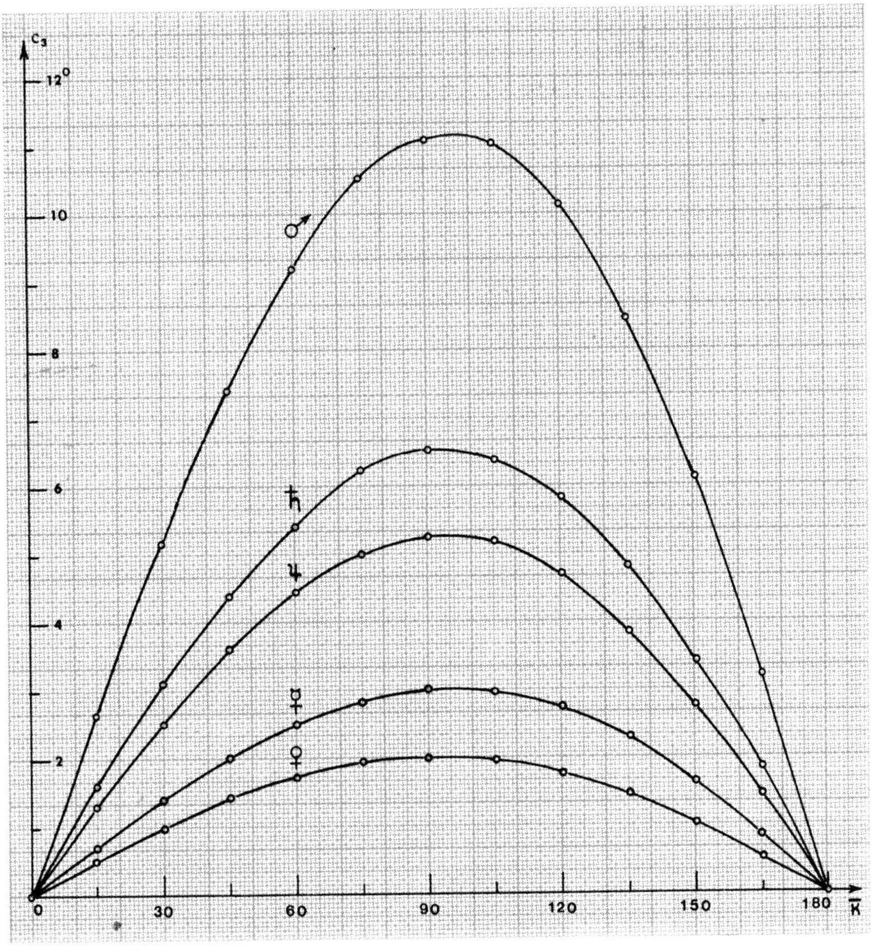

Fig. 91

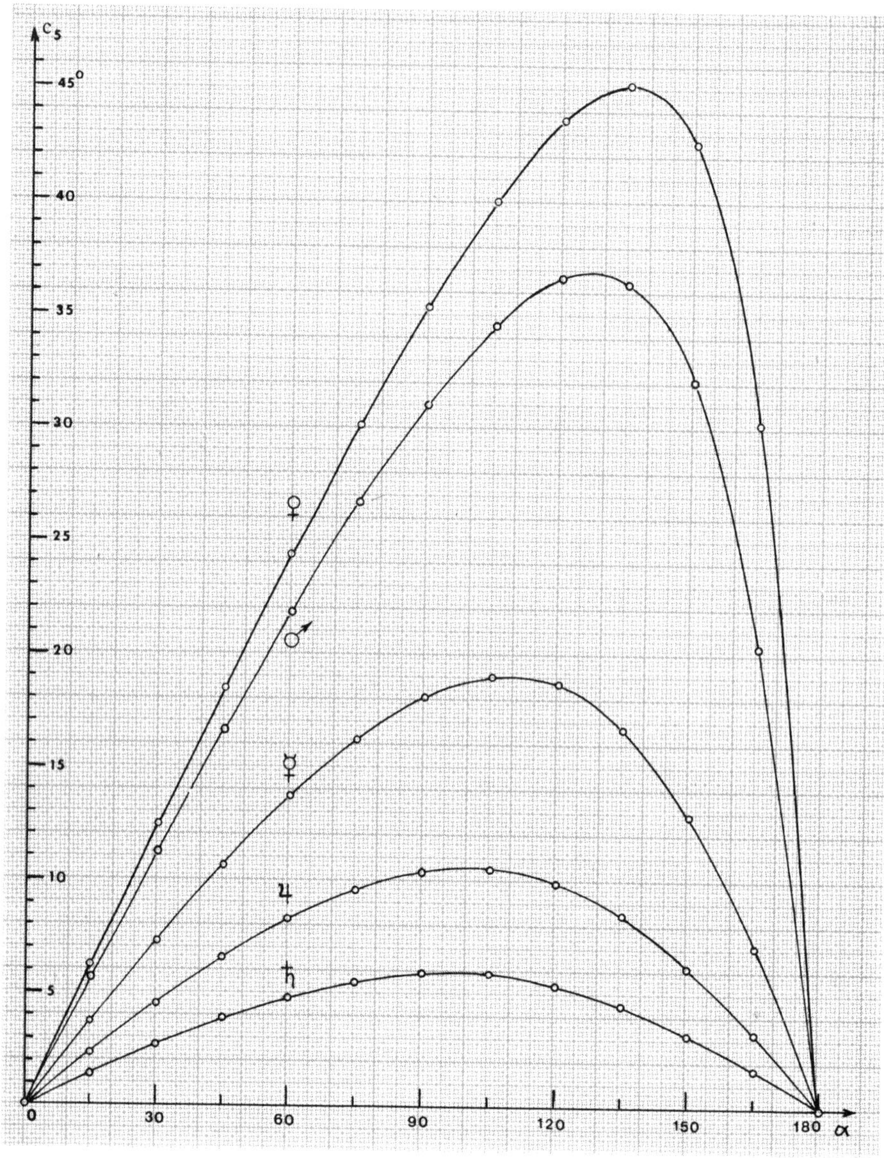

Fig. 92

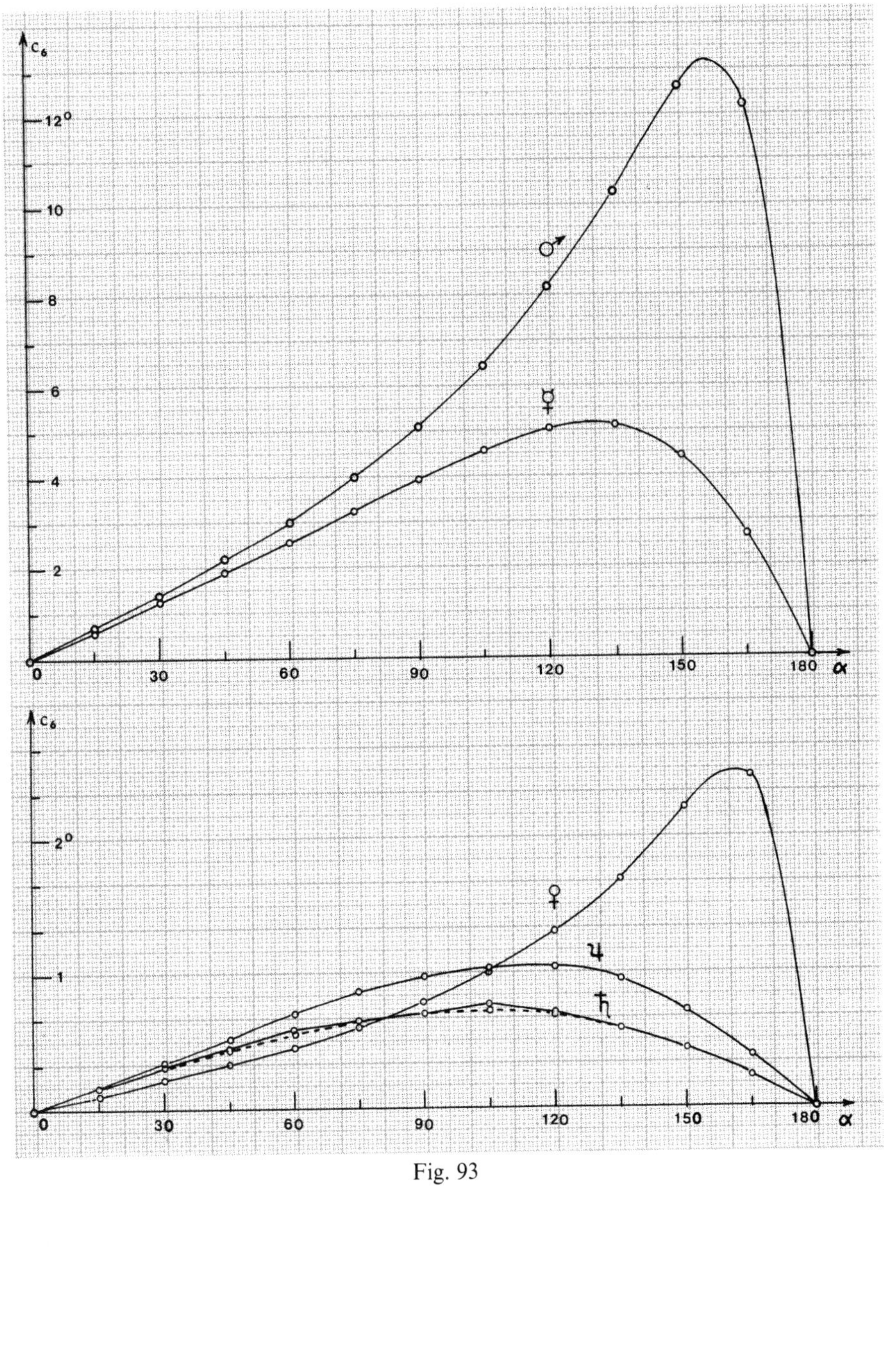

Fig. 93

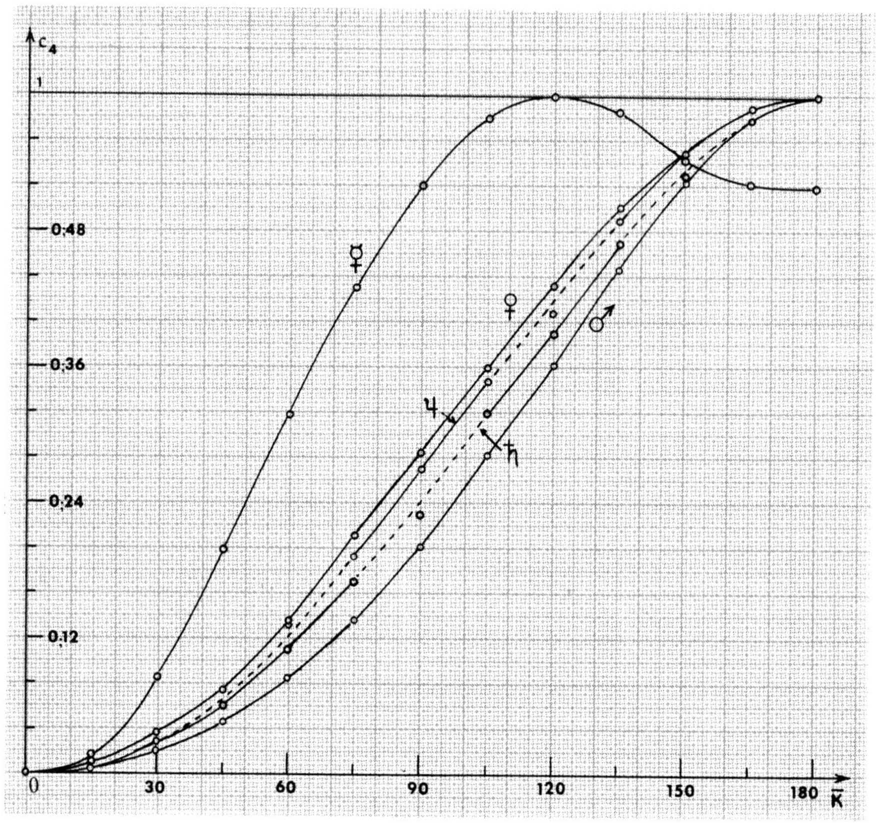

Fig. 94

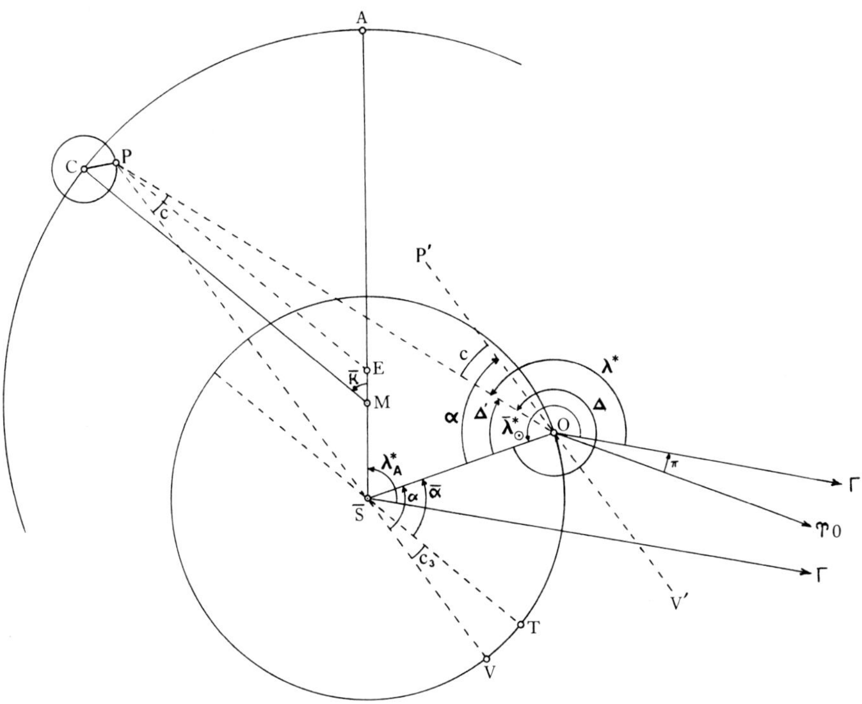

Fig. 95

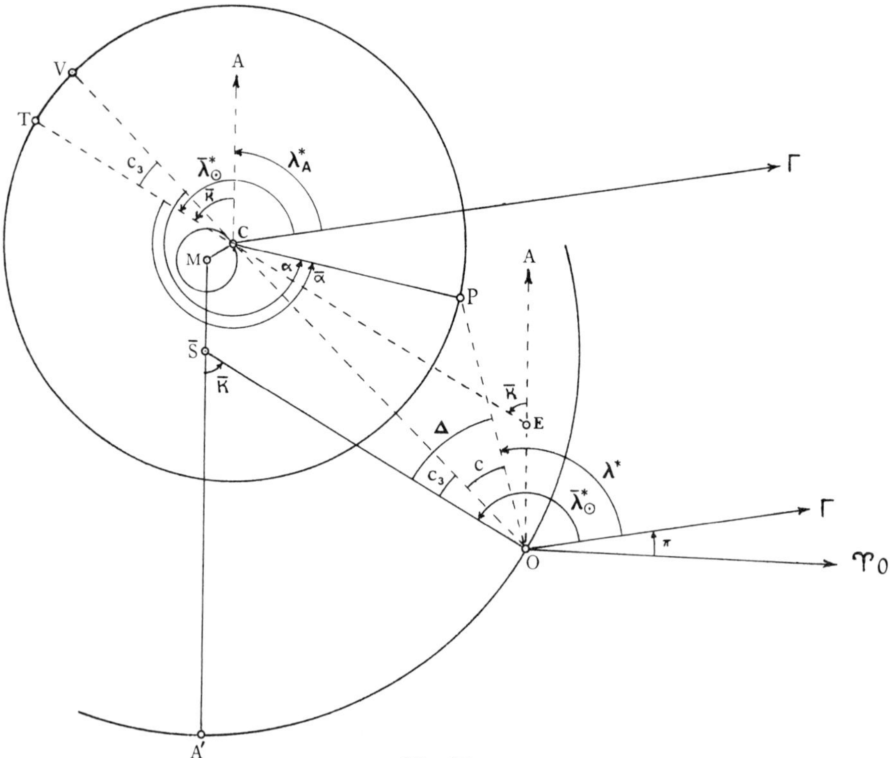
A
V
T
c_3
$\bar{\lambda}^*_\odot$
λ^*_A
Γ
$\bar{\kappa}$
C
M
α
$\bar{\alpha}$
P
A
$\bar{S}$
$\bar{\kappa}$
$\bar{\kappa}$
E
Δ
c
λ^*
c_3
$\bar{\lambda}^*_\odot$
Γ
π
O
Υ_0
A'

Fig. 96

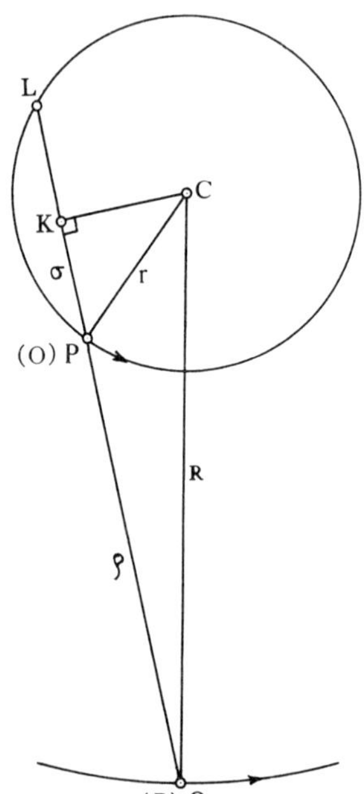
L
C
K
σ
r
(O) P
R
ρ
(P) O

Fig. 97

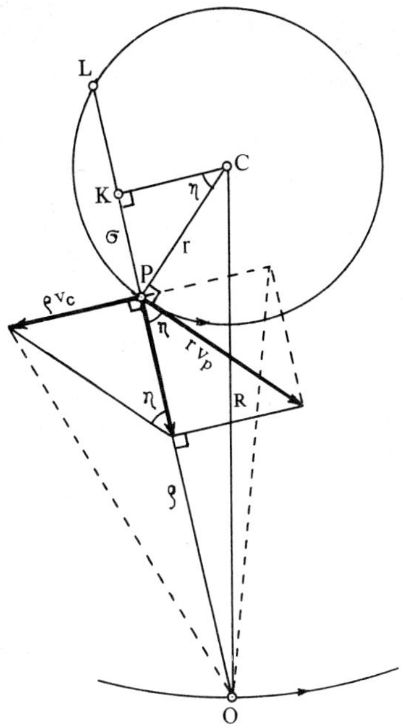

Fig. 98

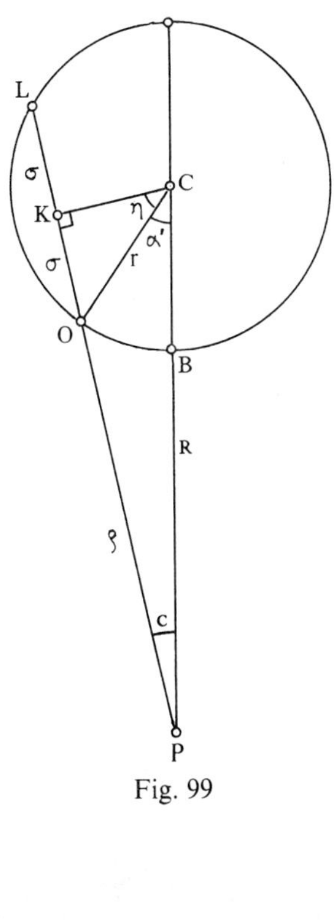

Fig. 99

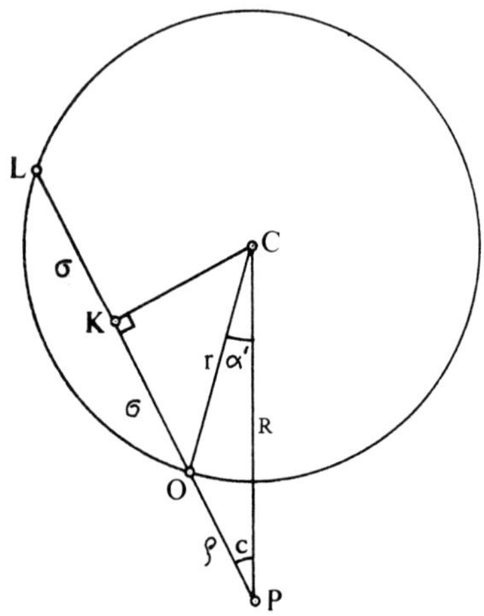

Fig. 100

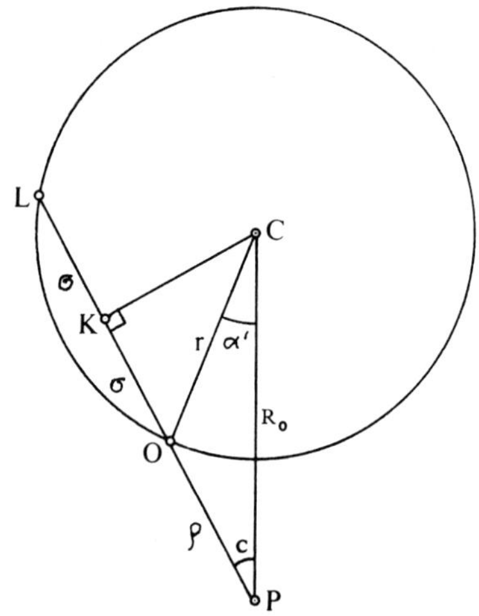

Fig. 101

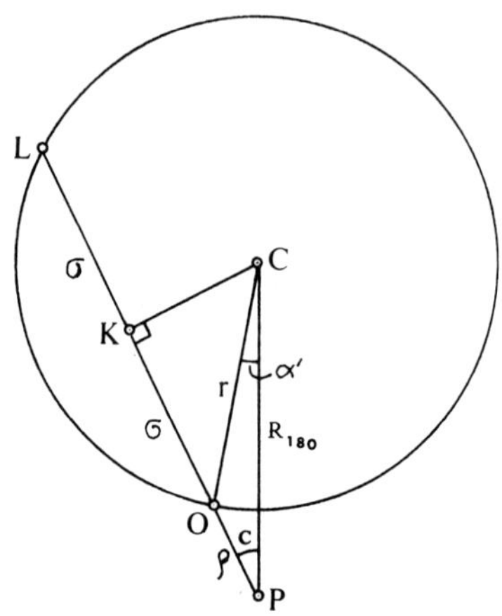

Fig. 102

Figures to Chapter 6

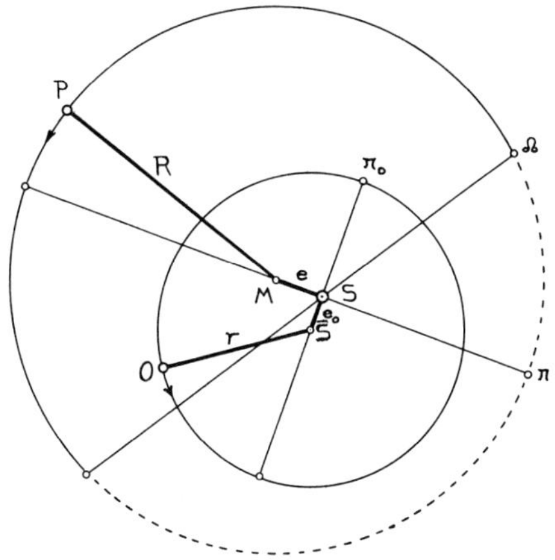

(a)

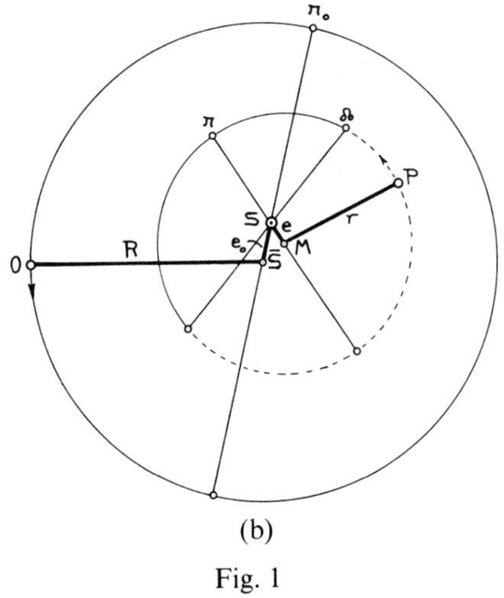

(b)

Fig. 1

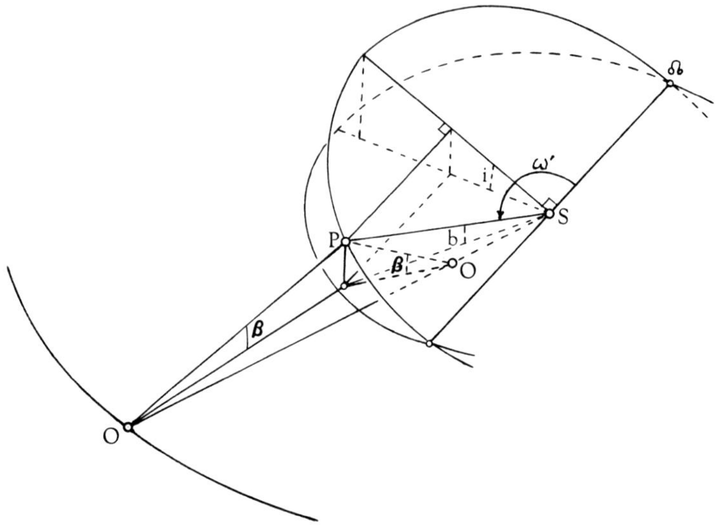

Fig. 2

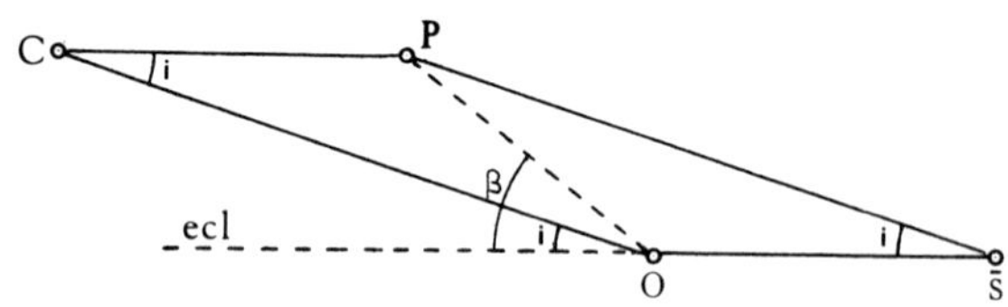

Fig. 3

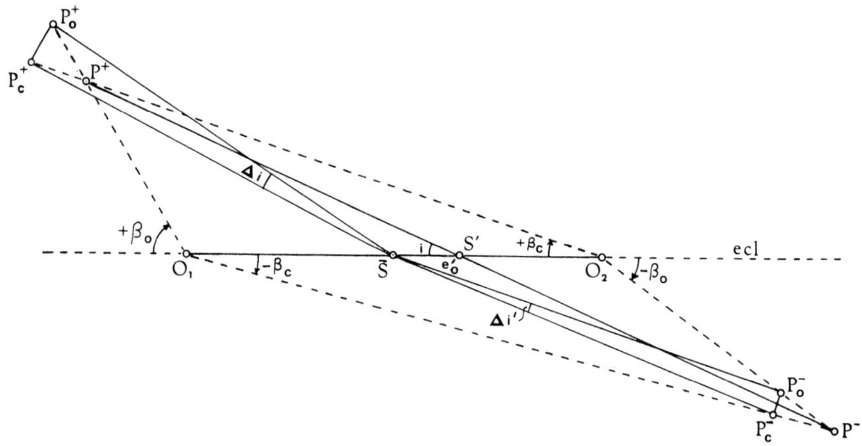

Fig. 4

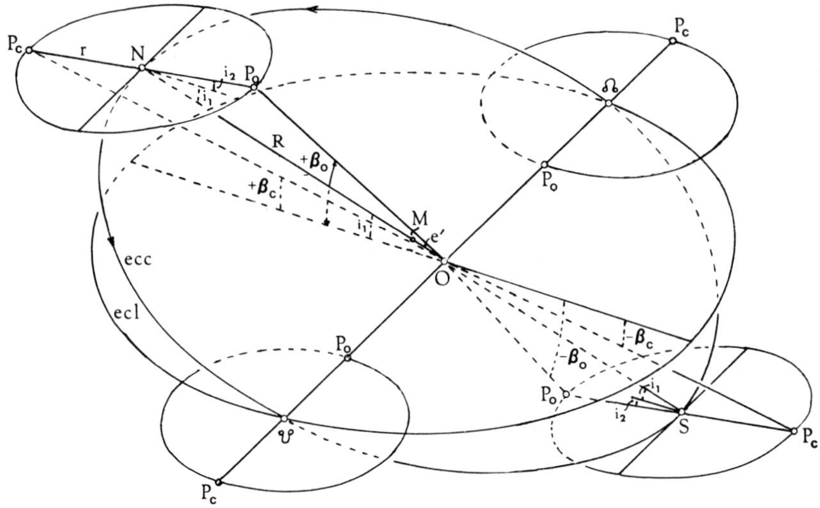
P_c
r
N
i_2
P_o
i_1
R
+β_o
+β_c
M
e'
ecc
ecl
O
P_c
P_o
-β_c
-β_o
P_o
i_1
i_2
S
P_c
P_o
P_c

Fig. 5

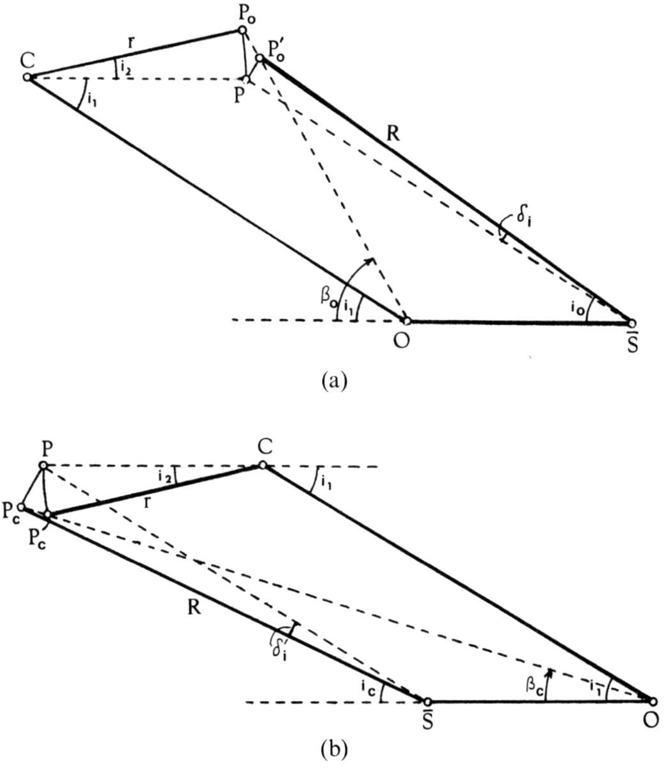
P_o
r
P'_o
C
i_2
i_1
P
R
δ_i
β_o
i_1
i_o
O
S̄

(a)

P
C
i_2
i_1
P_c
r
P'_c
R
δ'_i
i_c
β_c
i_1
S̄
O

(b)

Fig. 6

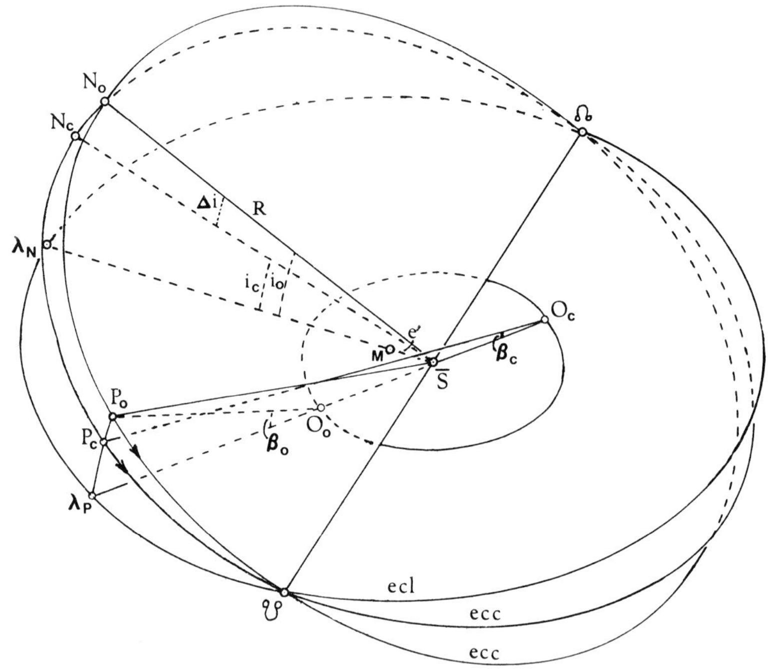

N₀
N_c
λ_N
Δi
R
i_c
i_o
e'
M
O_c
β_c
S̄
P_o
P_c
O_o
β_o
λ_P
ecl
ecc
ecc

Fig. 7

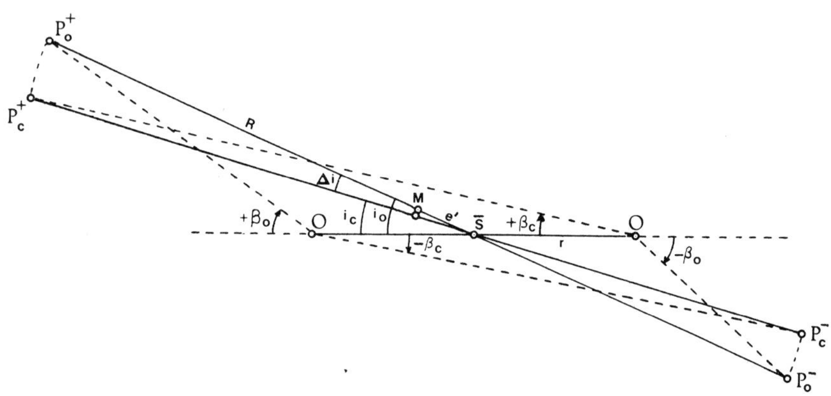

P_o^+
P_c^+
R
Δi
M
+β_o
O
i_c
i_o
e'
S̄
+β_c
O
−β_c
r
−β_o
P_c^-
P_o^-

Fig. 8

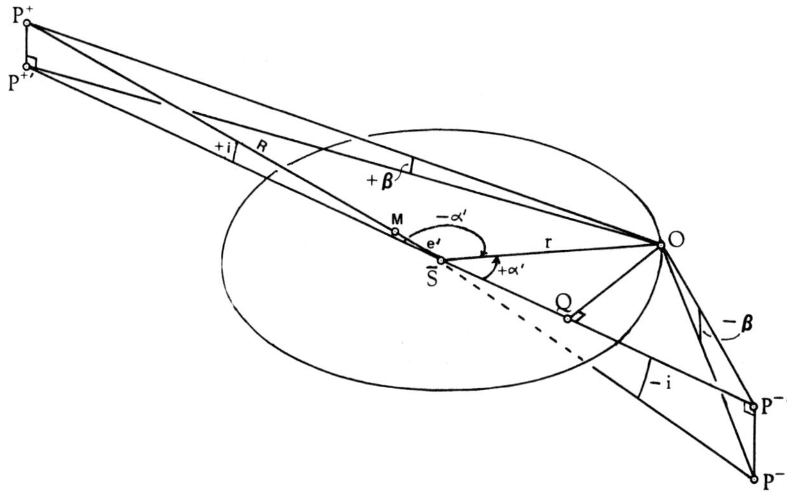

Fig. 9

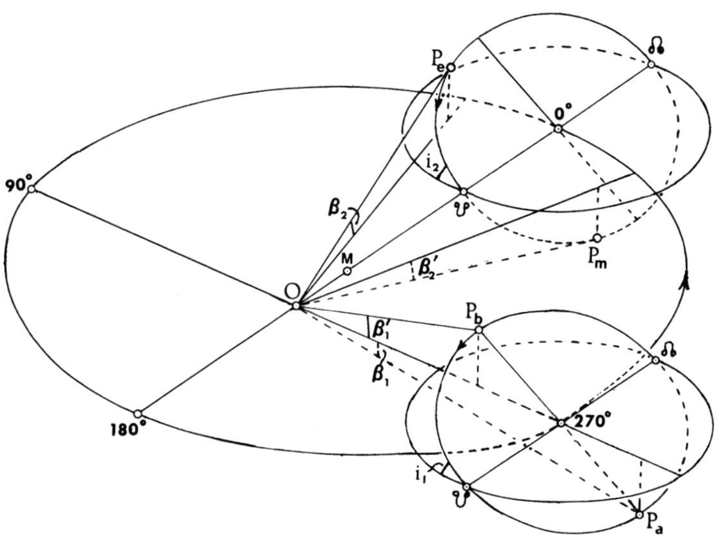

Fig. 10

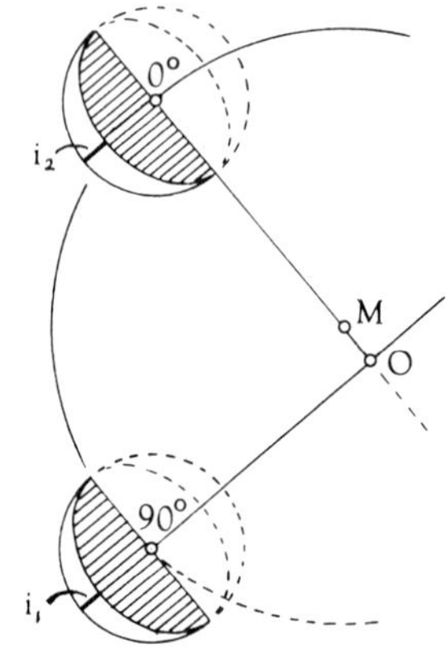

Fig. 11

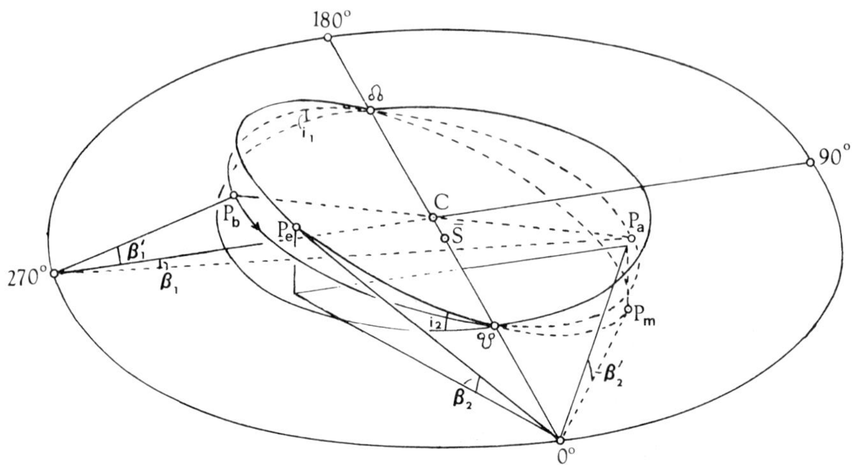

Fig. 12

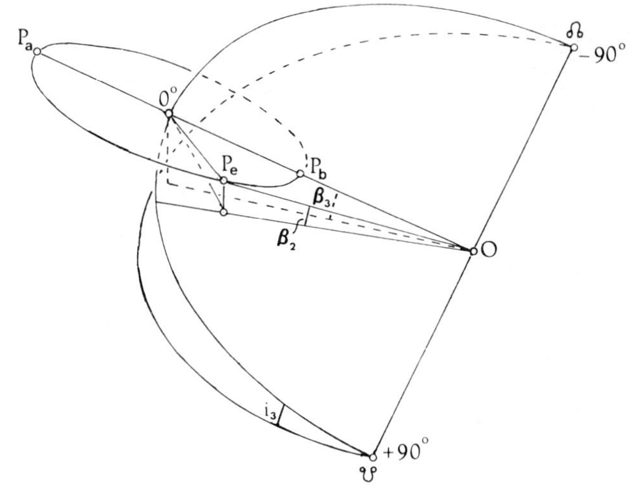

Fig. 13

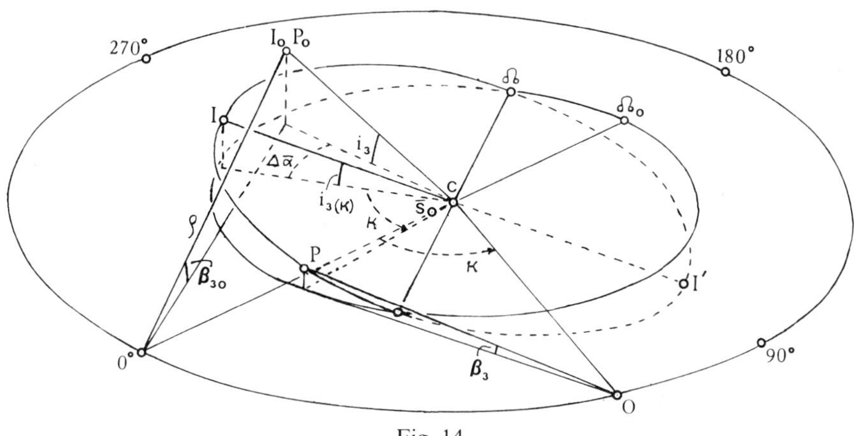

Fig. 14

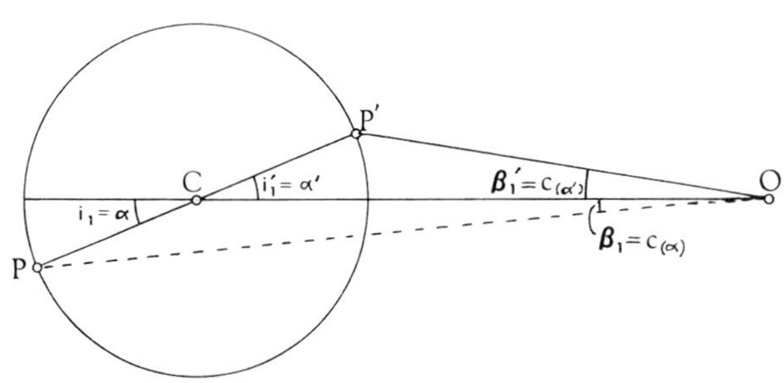

Fig. 15

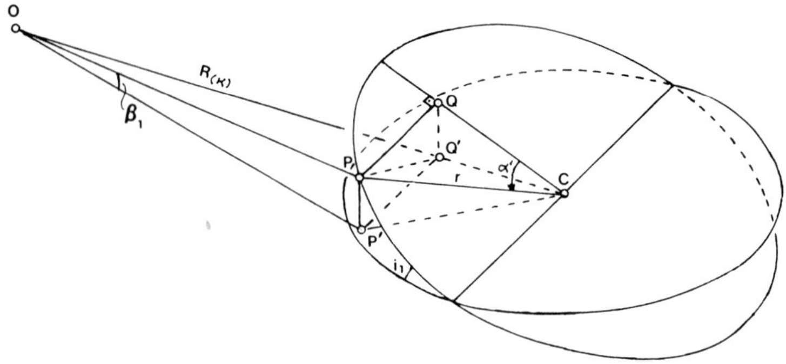

Fig. 16

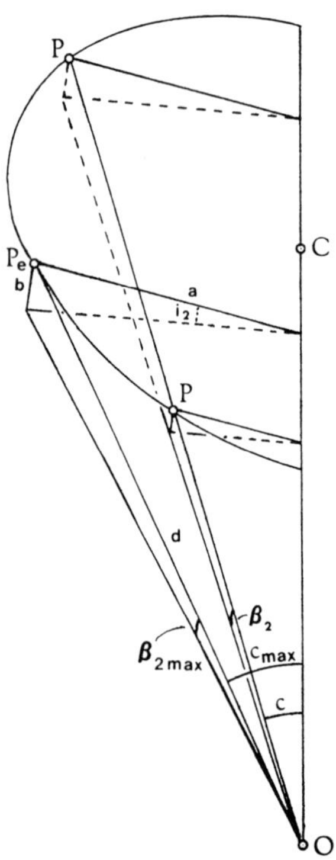

Fig. 17

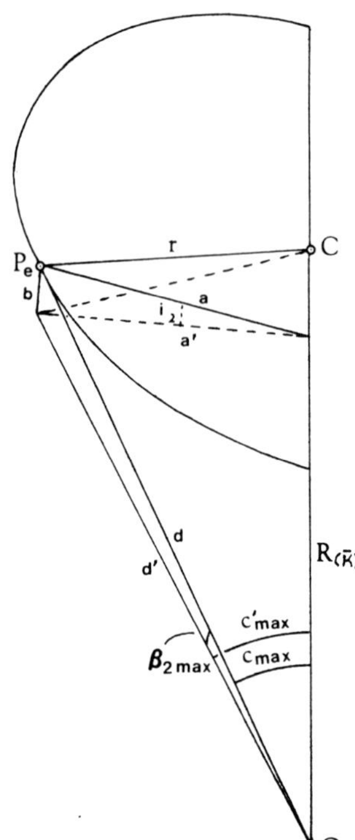

Fig. 18

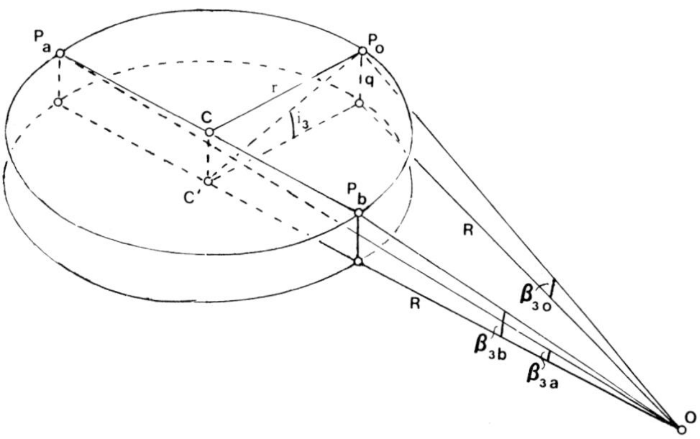

Fig. 19

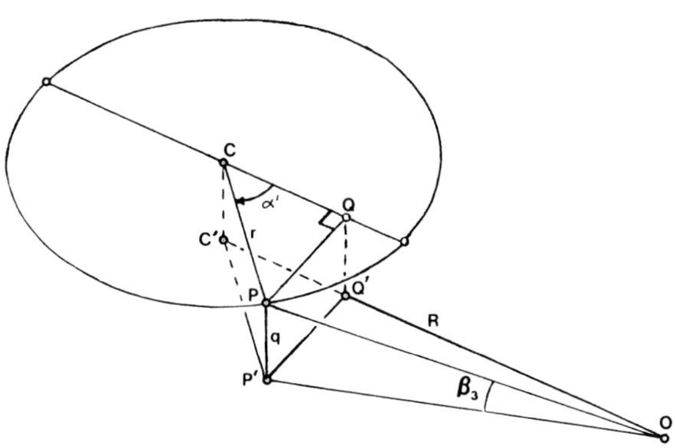

Fig. 20

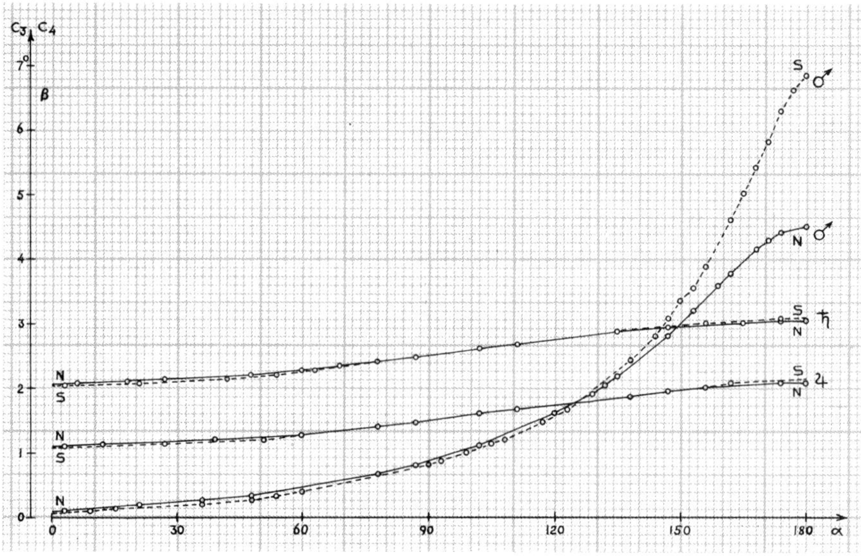
C_3, C_4
7°
β
6
5
4
3
2
1
0
0
30
60
90
120
150
180
α
N
S
N
S
N
S
N
S
N
S
N
♂
♂
♄
♃

Fig. 21

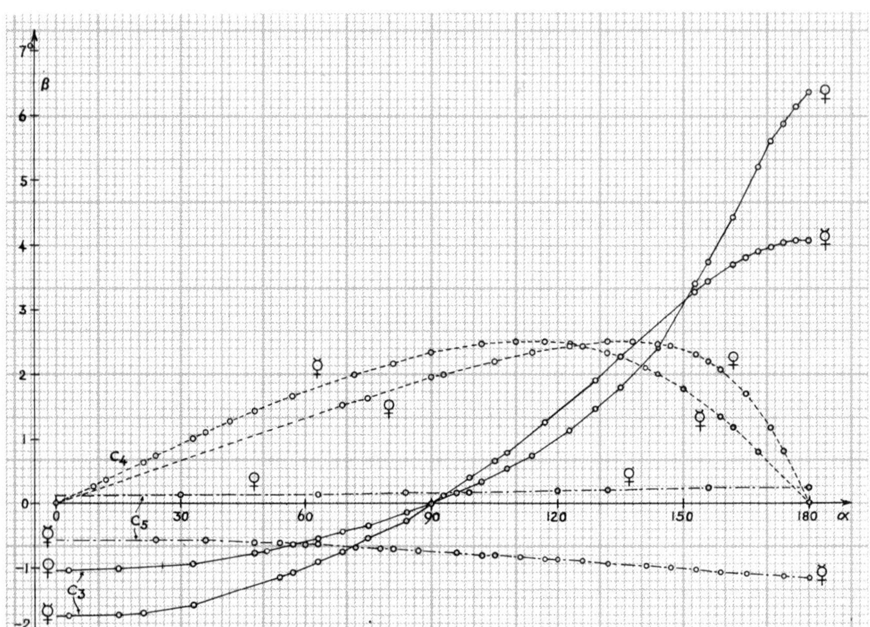
7°
β
6
5
4
3
2
1
0
-1
-2
0
30
60
90
120
150
180
α
C_4
C_5
C_3
♀
☿
♀
☿
♀
☿
♀
☿
♀
☿
♀
☿

Fig. 22

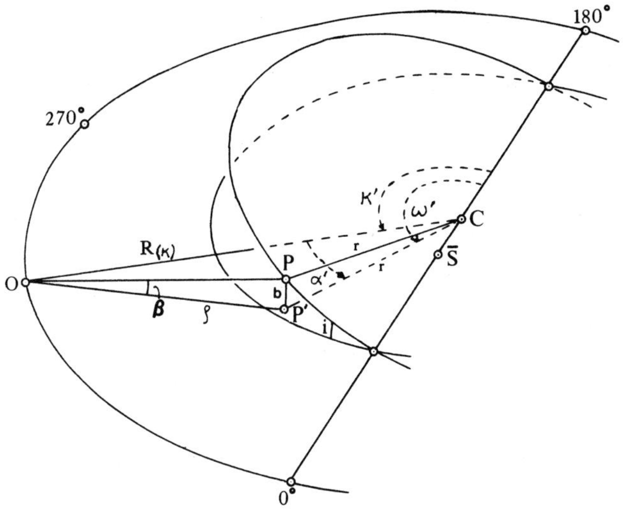
180°
270°
κ'
ω'
C
R(κ)
r
P
S̄
O
α'
r
b
β
ϱ
P'
i
0°

Fig. 23

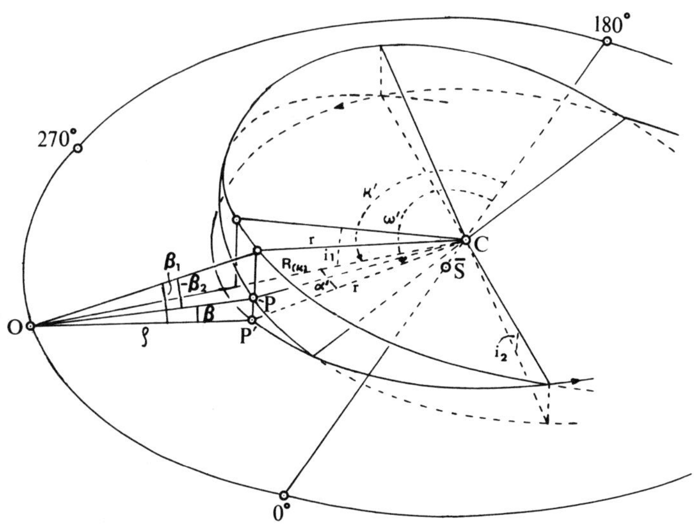
180°
270°
κ'
ω'
r
C
β₁
-β₂
i₁
R(κ)
S̄
α'
r
P
β
O
ϱ
P'
i₂
0°

Fig. 24

Bibliographical Abbreviations

The following list is in no sense a comprehensive, or even a selective, bibliography of writings on Copernicus. It is restricted to those works referred to or consulted in the course of this study, and a few additional introductory works that are of interest, for example, Dreyer (1906), Pannekoek (1947–48), Armitage (1957), and Kuhn (1957). There are some large bibliographies concerning Copernicus. Surely the most extensive is H. Baranowski, *Bibliografia kopernikowska* (Warsaw, 1958), which covers the period 1509 to 1955 and, at nearly 4000 entries in over 400 pages, is not a little daunting. Baranowski has published continuations for 1956–71 (Warsaw, 1973) and 1972–75 (*Studia Copernicana* 17, 1977), the latter containing the deluge of publications provoked by the 500th anniversary commemorations of 1973. Somewhat more manageable are the bibliographies for 1939–58 and 1959–70 in Rosen (1971), which contain short descriptions showing that few of the nearly 1100 publications need be read.

Aaboe, A. 1955. On the Babylonian Origin of Some Hipparchian Parameters. Centaurus 4 (1955–56), 122–25.

Abbud, F. 1962. The Planetary Theory of Ibn al-Shāṭir: Reduction of the Geometric Models to Numerical Tables. Isis 53, 492–99.

AHES. Archive for History of Exact Sciences.

AIHS. Archives internationales d'histoire des sciences.

Alfonso X. 1492. Tabule astronomice Alfonsi Regis. Venice. Copernicus's copy: Uppsala Univ. Bibl. 34.VII.65.

1518/21. Tabule Astronomice Divi Alfonsi Regis Romanorum et Castelle. Venice.

Almagest. See Ptolemy.

Almagesti minoris libri VI. Nürnberg Stadtbibl. MS Cent. VI,12, ff. 1–66. Copy by Regiomontanus.

Amico, G. B. 1536. De motibus corporum coelestium iuxta principia peripatetica sine eccentricis et epicyclis. Venice.

Apianus, P. 1534. Instrumentum primi mobilis. Nuremberg. Contains also Jābir (1534).

Armitage, A. 1957. Copernicus, the Founder of Modern Astronomy. New York.

Al-Battānī. 1537. De motu stellarum. Trans. Plato of Tivoli. Nuremberg. Contains also al-Farghānī (1537). See also Nallino.

Benjamin, F. S. and G. J. Toomer. 1971. Campanus of Novara and Medieval Planetary Theory. Madison.

Bialas, V. 1973. Die Planetenbeobachtungen des Copernicus; Zur Genauigkeit der Beobachtungen und ihrer Funktion in seinem Weltsystem. Philosophia Naturalis 14, 328–52.

Birkenmajer, A. 1972. Le huit du cavalier d'Eudoxe et la couronne tordue de Copernic. SC 4, 659–73.

Birkenmajer, L. A. 1900. Mikolaj Kopernik. Cracow.

Biskup M. 1973. Regesta Copernicana (Calendar of Copernicus' Papers). SC 8.

1977. Biography and Social Background of Copernicus. SC 17, 137–52.

Biskup, M. and J. Dobrzycki. 1972. Copernicus, Scholar and Citizen. Warsaw.

Brahe, T. 1913–29. Tychonis Brahe Dani Opera omnia, ed. I. L. E. Dreyer. 15 vols., Copenhagen.

1946. Tycho Brahe's Description of his Instruments and Scientific Work as given in Astronomiae Instauratae Mechanicae. Trans. and ed. H. Raeder, E. Strömgren, B. Strömgren. Copenhagen.

Braunmühl, A. von. 1900. Vorlesungen über Geschichte der Trigonometrie I. Leipzig.

Burmeister, K. H. 1967–68. Georg Joachim Rheticus, 1514–1574, Eine Bio-Bibliographie. 3 vols. Wiesbaden.

1969. Georg Joachim Rheticus as a geographer and his contribution to the first map of Prussia. Imago Mundi 23, 73–76.

Caraci, G. 1927. Heinrich Zell, G. Gastaldi und einige der ältesten Karten von Deutschland. Petermanns Mitteilungen 73, 200–05, Tafel XII.

Cardano, H. 1563. Opera. Lyons.

Carmody, F. J. 1960. The Astronomical Works of Thabit b. Qurra. Berkeley and Los Angeles.

Carra de Vaux. 1893. Les spheres célestes selon Nasir-Eddin Attûsî. Appendice VI in P. Tannery. Recherches sur l'histoire de l'astronomie ancienne. Paris.

Censorinus. 1867. De die natali liber, ed. F. Hultsch, Leipzig.

Copernicus, N. Commentariolus. See Prowe (1883–84), Curtze (1878), (1882), Lindhagen (1881), Müller (1899), Rossmann (1948), Zinner (1951), Rosen (1971), Swerdlow (1973), Hugonnard-Roche, Rosen, Verdet (1975).

1542. De lateribus et angulis triangulorum. Wittenberg.

1543. De revolutionibus orbium coelestium libri VI. Nuremberg. Repr. Paris, 1927; Turin-Amsterdam, 1943; Leipzig, New York-London, 1965; Brussels, 1966.

1566. Nicolai Copernici Torinensis de revolutionibus orbium caelestium, Libri VI. Basel. Repr. Prague, 1971.

1617. Astronomia instaurata, Libris sex comprehensa, qui de Revolutionibus orbium coelestium inscribuntur. Ed. N. Muler, Amsterdam.

1873. Nicolai Copernici Thorunensis De revolutionibus orbium caelestium libri VI . . . Accedit Georgii Ioachim Rhetici de libris revolutionum Narratio prima. Thorn.

1879. Über die Kreisbewegungen der Weltkörper. Übers. u. mit Anmerk. von C. L. Menzzer. Thorn. Repr. Leipzig, 1939.

1944. Nikolaus Kopernikus Gesamtausgabe I, Opus de revolutionibus caelestibus manu propria, Faksimile-Wiedergabe. Munich-Berlin.

1949. Nikolaus Kopernikus Gesamtausgabe II, De revolutionibus orbium caelestium, Textkritische Ausgabe Ed. F. Zeller, K. Zeller. Munich.

1952. On the Revolutions of the Heavenly Spheres. Trans. C. G. Wallis. Great Books of the Western World 16. Chicago, etc.

1959. Über die Kreisbewegungen der Weltkörper. Ed. G. Kraus, comm. A. Birkenmajer. Berlin. I,1–11 in Menzzer's translation.
1973. Nicolai Copernici opera omnia I, De revolutionibus, codicis propria auctoris manu scripti imago phototypa, prolegomenis instruxit: G. Zathey. Warsaw-Cracow.
1975. Nicolai Copernici opera omnia II, De revolutionibus libri sex. Ed. R. Gansiniec, J. Domanski, J. Dobrzycki, comm. A. Birkenmajer, J. Dobrzycki. Warsaw–Cracow.
1976. On the Revolutions of the Heavenly Spheres. Trans. A. M. Duncan with introduction and notes. Newton Abbot-London-Vancouver, New York.
1978. Nicolas Copernicus. On the Revolutions, ed. J. Dobrzycki, trans. and comm. E. Rosen. Warsaw–Cracow.
See also Koyré (1970).
Curtze, M. 1874. Reliquiae Copernicanae I & II. ZMP 19, 76–82, 432–58.
1875. Reliquiae Copernicanae III. ZMP 20, 221–48.
1878. Inedita Copernicana. MCVT 1.
1882. Ergänzungen zu den "Inedita Copernicana." MCVT 4, 1–9.
Czartoryski, P. 1978. The Library of Copernicus. SC 16, 355–96.
Czwalina, A. 1959. Ptolemaeus, die Bahnen der Planeten Venus und Merkur. Centaurus 6, 1–35.

Delambre, J. B. J. 1817. Histoire de l'astronomie ancienne. 2v. Paris.
1819. Histoire de l'astronomie du Moyen Age. Paris.
1821. Histoire de l'astronomie moderne. 2v. Paris.
Devreesse, R. 1965. Le Fonds Grec de la Bibliotèque Vaticane des Origines a Paul V. Studi e Testi 244, Città del Vaticano.
Dobrzycki, J. 1963. Katalog gwiazd w De revolutionibus. Studia i Materialy z Dziejow Nauki Polskiej, C 7, 109–53.
1965. Theoria Precesji w Astronomii Sredniowiecznej. Studia i Materialy z Dziejow Nauki Polskiej, C 11, 3–47.
See also Biskup and Dobrzycki (1972), Copernicus (1975), (1978).
Doppelmayr, J. G. 1730. Historische Nachricht von den Nürnbergischen Mathematicis und Künstlern. Nuremberg.
Dreyer, J. L. E. 1890. Tycho Brahe. A Picture of Scientific Life and Work in the Sixteenth Century. Edinburgh. Repr. New York, 1963.
1906. History of Planetary Systems from Thales to Kepler. Cambridge. Repr. as A History of Astronomy from Thales to Kepler. New York, 1953.
See also Brahe (1913–29).
DSB. Dictionary of Scientific Biography.

Eusebius. 1483. Chronicon. Trans. Jerome. Venice.
1529. Chronicon. Trans. Jerome. Basel.

Al-Farghānī. 1537. Rudimenta astronomica. Trans. John of Seville. Nuremberg. Contains also al-Battānī (1537).
Folkerts, M. 1976. Johann Werner. DSB 14, 272–77.

Gingerich, O. 1971. The Theory of Mercury from Antiquity to Kepler. Actes de XII Congrès International d'histoire des sciences III A, 57–64.

1973. The Role of Erasmus Reinhold and the Prutenic Tables in the Dissemination of Copernican Theory. SC 6, 43–62, 123–25.

1975. Remarks on Copernicus's Observations. In Westman (1975), 99–107.

1978. Early Copernican Ephemerides. SC 16, 403–17.

1981. An Early Tradition of an Extended Errata List for Copernicus's "De Revolutionibus." JHA 12, 47–52.

Goldstein, B. R. 1964. On the Theory of Trepidation according to Thabit b. Qurra and al-Zarqāllu and its implications for homocentric planetary theory. Centaurus 10, 232–47.

1967. The Arabic Version of Ptolemy's Planetary Hypotheses. Trans. Amer. Philos. Soc. N.S. 57,4.

1977. Ibn Muᶜādh's Treatise on Twilight and the Height of the Atmosphere. AHES 17, 97–118.

Goldstein, B. R. and F. W. Sawyer. 1977. Remarks on Ptolemy's Equant Model in Islamic Astronomy. Appendix: On Ptolemy's Determination of the Apsidal Line for Venus. ΠΡΙΣΜΑΤΑ. Festschrift für Willy Hartner. Ed. Y. Maeyama and G. Saltzer. Wiesbaden. 165–81.

Goldstine, H. H. 1973. New and Full Moons 1001 B.C. to A.D. 1651. Memoirs Amer. Philos. Soc. 94. Philadelphia.

Grafton, A. 1973. Michael Maestlin's Account of Copernican Planetary Theory. Proc. Amer. Philos. Soc. 117, 523–50.

HAMA. See Neugebauer (1975).

Hamann, G. ed. 1980. Regiomontanus-Studien. Sitz. d. öst. Akad. d. Wiss., phil-hist. Kl. 364. Vienna.

Harrassowitz, O. publ. 1971. Deutsche Buchdrucker des fünfzehnten Jahrunderts. Wiesbaden.

Hartner, W. 1955 The Mercury Horoscope of Marcantonio Michiel of Venice: A Study in the History of Renaissance Astrology and Astronomy. Vistas in Astronomy 1, 84–138. Also in Hartner (1968), 440–95.

1964. Medieval Views on Cosmic Dimensions and Ptolemy's Kitāb al-Manshūrāt. Mélanges Alexandre Koyré. Paris. I, 254–82. Also in Hartner (1968), 319–48.

1968. Oriens-Occidens. Ausgewählte Schriften. Hildesheim.

1969. Nasir al-Dīn al-Ṭūsī's Lunar Theory. Physis 11, 287–304.

1974a. Ptolemy, Azarquiel, Ibn al-Shāṭir, and Copernicus on Mercury. A Study of Parameters. AIHS 24, 5–25.

1974b. Ptolemy's and Copernicus' Mercury Models. An Accuracy Test. AIHS 24, 367–69.

1980. Ptolemäische Astronomie im Islam und zur Zeit des Regiomontanus. In Hamann (1980), 109–24.

Heiberg, J. L. See Ptolemy, 1898, 1903, 1907.

Henderson, J. A. 1973. On the Distances between Sun, Moon, and Earth according to Ptolemy, Copernicus and Reinhold. Yale University Dissertation.

1975. Erasmus Reinhold's Determination of the Distance of the Sun from the Earth. In Westman (1975), 108–29.

Herz, N. 1887–94. Geschichte der Bahnbestimmung von Planeten und Kometen. 2v. Leipzig.

1897. Allgemeine Einleitung in die Astronomie. In W. Valentiner, Handwörterbuch der Astronomie. 3v. in 4. Breslau. 1, 1–163.

Hill, G. W. 1900. Ptolemy's Problem. Astronomical Journal 21, 33–35. Also in Collected Mathematical Works 4 (1907), 207–11.
Hipler, F. 1876. Die Chorographie des Joachim Rheticus. ZMP. Hist.-lit. Abteil. 21, 125–50.
Horn, W. 1951. Sebastian Münster's Map of Prussia and its Variants. Imago Mundi 7, 66–73.
Hugonnard-Roche, H., E. Rosen, J.-P. Verdet. 1975. Introductions à l'astronomie Copernic. Paris.
Hugonnard-Roche, H. and J.-P. Verdet. 1982. Rheticus, Narratio prima. Ed. crit., trad. franc. et comm. SC 20.

Jābir Ibn Aflaḥ. 1534. Gebri Filii Affla Hispalensis, de astronomia libri IX. Nuremberg. Contained in Apianus (1534).
Jarzebowski, L. 1971. Biblioteka Mikolaja Kopernika. Torun.
JHA. Journal for the History of Astronomy.
JHAS. Journal for the History of Arabic Science.

Kamienski, M. 1963. Obserwacje Kopernika w Swietle Astronomii Wspolczesnej. Studia i Materialy z Dziejow Nauki Polskiej. Seria C, Z. 7, 85–108.
Kennedy, E. S. 1966. Late Medieval Planetary Theory. Isis 57, 365–78.
1983. Studies in the Islamic Exact Sciences. Beirut.
Kennedy, E. S. and I. Ghanem. 1976. The Life and Work of Ibn al-Shāṭir. An Arab Astronomer of the Fourteenth Century. Aleppo.
Kennedy, E. S. and V. Roberts. 1959. The Planetary Theory of Ibn al-Shāṭir. Isis 50, 227–35.
Kepler, J. 1858–71. Opera omnia. Ed. Ch. Frisch. 8v. in 9. Frankfurt–Erlangen.
1937–. Gesammelte Werke. Munich.
Kesten, H. 1945. The World of Copernicus. Trans. from German by E. B. Ashton and M. Guterman. New York.
Al-Khwārizmī. See Suter (1914), Neugebauer (1962).
King, D. A. 1975. Ibn al-Shāṭir. DSB 12, 357–64.
Koestler, A. 1959. The Sleepwalkers, a History of Man's Changing Vision of the Universe. New York.
Koyré, A. 1961. La révolution astronomique. Paris.
1970. Des Révolutions des Orbes Célestes. Paris. Trans. of De rev. I,1–11 with intro. and notes.
Kremer, R. L. 1980. Bernard Walther's Astronomical Observations. JHA 11, 174–91.
1981. The Use of Bernard Walther's Astronomical Observations: Theory and Observation in Early Modern Astronomy. JHA 12, 124–32.
1983. Walther's Solar Observations: A Reply to R. R. Newton. Quarterly Journal of the Royal Astronomical Society 24, 36–47.
Kuhn, T. S. 1957. The Copernican Revolution, Planetary Astronomy in the Development of Western Thought. Cambridge, Mass.

Langermann, T. 1979. Ibn al-Haytham's Hayʾat al-ʿālam. Harvard University Dissertation.
Lindhagen, A. 1881. Nicolai Copernici de hypothesibus motuum coelestium a se constitutis commentariolus. Bihang till K. Svenska Vet. Akad. Handlinger, Band 6 N:o 12.
List, M. 1978. Marginalien zum Handexemplar Keplers von Copernicus: De revolutionibus orbium coelestium (Nürnberg, 1543). SC 16, 443–60.

Manitius, K. See Ptolemy (1912–13).

MCVT. Mitteilungen des Coppernicus-Vereins für Wissenschaft und Kunst zu Thorn.

Menzzer, C. L. See Copernicus (1879).

Millás Vallicrosa, J. M. 1942. Las traducciones orientales en los manuscritos de la Biblioteca Catedral de Toledo. Madrid.

1943–50. Estudios sobre Azarquiel. Madrid-Granada.

Moesgaard, K. P. 1968. The 1717 Egyptian Years and the Copernican Theory of Precession. Centaurus 13, 120–38.

1974. Success and Failure in Copernicus' Planetary Theory. AIHS 24, 73–111, 243–318.

1974a. Thābit Ibn Qurra between Ptolemy and Copernicus: An Analysis of Thābit's Solar Theory. AHES 12, 199–216.

Müller, A. 1899. Nicolai Copernici de hypothesibus motuum coelestium a se constitutis commentariolus. Zeitschrift fur die Geschichte und Altertumskunde Ermlands 12, 359–82.

Müller, Johannes of Königsberg. See Regiomontanus.

Nallino, C. A. 1899–1907. Al-Battānī sive al-Battenii Opus Astronomicum. Pubblicazioni del Reale Osservatorio di Brera in Milano 40, 3v. Milan. See also al-Battānī (1537).

Neugebauer, O. 1962. The Astronomical Tables of al-Khwārizmī. Hist. Filos. Skr. Dan. Vid. Selsk. 4, no. 2. Copenhagen. See also Suter (1914).

1962a. Thâbit ben Qurra "On the Solar Year" and "On the Motiȯn of the Eighth Sphere." Proc. Amer. Philos. Soc. 106, 264–99.

1968. On the Planetary Theory of Copernicus. Vistas in Astronomy 10, 89–103.

1968a. Three Copernican Tables. Centaurus 12, 97–106.

1975. A History of Ancient Mathematical Astronomy (HAMA). 3 pts. Springer-Verlag, Berlin–Heidelberg–New York.

Neugebauer, O. and D. Pingree. 1971. The Pañcasiddhāntikā of Varāhamihira. Hist. Filos. Skr. Dan. Vid. Selsk. 6,1. 2 pts. Copenhagen.

Neugebauer, P. V. Tafeln zur astronomischen Chronologie. Leipzig:

1912. I Sterntafeln von 4000 v. Christ bis zur Gegenwart.

1914. II. Tafeln für Sonne, Planeten und Mond.

1922. III. Hilfstafeln zur Berechnung von Himmelserscheinungen.

1929. Astronomische Chronologie. 2v. Berlin–Leipzig.

1938. Tafeln zur Berechnung der jährlichen Auf- und Untergänge der Planeten zwischen Äquator und 40° nördlicher und südlicher Breite. Astronomische Nachrichten 264, No. 6331, 313–22.

Nunes, P. 1943. Obras II. De crepusculis. Lisbon.

Oppolzer, Th. R. von. 1887. Canon der Finsternisse. Denkschr. d. kais. Akad. d. Wiss., math.-naturwiss. Kl. 52. Vienna. Repr. New York, 1962.

Pannekoek, A. 1945. A remarkable place in Copernicus' "De Revolutionibus" Bulletin of the Astronomical Institutes of the Netherlands 10 (No. 366), 68–69.

1947–48. Planetary Theories. Popular Astronomy 55–56.

Pedersen, O. 1974. A Survey of the Almagest. Odense.

Peurbach, G. ca. 1472. Theoricae novae planetarum. Nuremberg. Printed by Regiomontanus. Repr. in Regiomontanus (1972).

1514. Tabulae eclipsium. Vienna.
See also Regiomontanus.
Pingree, D. 1970. On the Classification of Indian Planetary Tables. JHA 1, 95–108.
1978. History of Mathematical Astronomy in India. DSB 15, 533–633.
Price, D. J. de S. 1959. Contra-Copernicus . . . Critical Problems in the History of Science, ed. M. Clagett, Madison, 197–218.
Prowe, L. 1883–84. Nicolaus Coppernicus. 2v. in 3. Berlin. Repr. Osnabrück, 1967.
Ptolemy. 1482. Cosmographia. Ulm.
1515. Almagestum Cl. Ptolemaei. Venice. Trans. Gerard of Cremona.
1551. Omnia, que extant, opera, Geographia excepta. Basel. Contains trans of Almagest by George of Trebizond.
Opera quae extant omnia. Ed. J. L. Heiberg. Leipzig:
1898–1903. I. Syntaxis mathematica. 2 pts.
1907. Opera astronomica minora.
1912–13. Ptolemäus, Handbuch der Astronomie. Trans. K. Manitius. 2v. Leipzig. Repr. Leipzig, 1963.
1984. Ptolemy's Almagest. Trans. and annot. by G. J. Toomer. London–New York.

Ragep, F. J. 1982. Cosmography in the "Tadhkira" of Naṣīr al-Dīn al-Ṭūsī. Harvard University Dissertation. Ed., trans., and comm. on Book I and Book II, chap. 1–11.
Regiomontanus, J. ca. 1474. Disputationes contra cremonensia deliramenta. Nuremberg. Printed by Regiomontanus. Repr. in Regiomontanus (1972).
1490. Tabulae directionum et profectionum. Augsburg. Copernicus's copy Uppsala Univ. Bibl. 34.VII.65.
1533. De triangulis omnimodis libri quinque. Nuremberg. Repr. in Regiomontanus (1972).
1541. Compositio tabularum sinuum rectorum. Nuremberg.
1972. Opera collectanea. Ed. F. Schmeidler. Osnabrück.
See also Schoener (1544).
Regiomontanus, J. and G. Peurbach. ca. 1463. Epitome in Almagestum Ptolemaei. Venice Bibl. Naz. Mar. MS Lat. 328, ff. 1–117.
1496. Epytoma Joannis de monte regio In almagestum ptolemaei. Venice. Repr. in Regiomontanus (1972).
Reinhold, E. Commentarius in opus revolutionibus Copernici. Berlin, Deutsche Staatsbibl. MS fol. lat. 391.
1551. Prutenicae Tabulae Coelestium Motuum. Tübingen.
Rheticus, G. J. 1540. Ad clarissimum virum D. Ioannem Schonerum, de libris revolutionum . . . D. Doctoris Nicolai Copernici Torunnaei . . . narratio prima. Danzig. Repr. Osnabrück, 1967. See also Copernicus (1873); Prowe (1883–84) 2; Kepler (1937–) 1; Zeller (1943); Rosen (1971); Hugonnard-Roche, Rosen, Verdet (1975); Hugonnard-Roche and Verdet (1982).
Riddell, R. C. 1978. The Latitudes of Venus and Mercury in the Almagest. AHES 19, 95–111.
1980. Parameter Disposition in pre-Newtonian Planetary Theories. AHES 23, 87–157.
Roberts, V. 1957. The Solar and Lunar Theory of Ibn ash-Shāṭir: A Pre-Copernican Copernican Model. Isis 48, 428–32.
1966. The Planetary Theory of Ibn al-Shāṭir: Latitudes of the Planets. Isis 57, 208–19.

Rome, A. 1927. L'Astrolabe et le Météoroscope d'apres le commentaire de Pappus sur le 5e livre de l'Almagest. Annales de la Société scientifique de Bruxelles, sér A, Sci. math. 47, 1927,2. Mémoires, 77–102.
Rosen, E. 1943. The Authentic Title of Copernicus's Major Work. Journal of the History of Ideas 4, 457–74.
1971. Three Copernican Treatises. 3rd. ed. New York.
1977. When Did Copernicus Write the "Revolutions"? Sudhoffs Archiv 61, 144–55.
1982. The Earliest Translation of Copernicus' "Revolutions" into German. Sudhoffs Archiv 66, 301–05.
See also Copernicus (1978), Hugonnard-Roche, Rosen, Verdet (1975).
Rossmann, F. 1948. Nicolaus Kopernicus, Erster Entwurf seines Weltsystems. Munich. Repr. Darmstadt, 1966.

Sabra, A. I. 1967. The Authorship of the *Liber de crepusculis*. Isis 58, 77–85.
1972. Ibn al-Haytham. DSB 6, 189–210.
1978. An Eleventh-Century Refutation of Ptolemy's Planetary Theory. SC 16, 117–31.
1979. Ibn al-Haytham's Treatise: Solution of Difficulties Concerning the Movement of Iltifāf. JHAS 3, 388–422.
Sabra, A. I. and N. Shehaby. 1971. Ibn al-Haytham, Al-Shukūk ʿalā Baṭlamyūs. Cairo.
Saliba, G. 1979a. The Original Source of Qutb al-Dīn al-Shīrāzī's Planetary Model JHAS 3, 4–18.
1979b. The First Non-Ptolemaic Astronomy at the Maraghah School. Isis 70, 571–76.
1980. Ibn Sīnā and Abū ʿUbayd al-Jūzjānī: The Problem of the Ptolemaic Equant. JHAS 4, 376–404.
Samuel, A. E. 1972. Greek and Roman Chronology. Handbuch der Altertumswissenschaft I, 7. Munich.
SC. Studia Copernicana. Wroclaw–Warsaw–Cracow–Gdansk.
Schoener, J. 1533. Opusculum geographicum ex diversorum libris ac cartis summa cura et diligentia collectum. ex urbe Norica (Nuremberg).
1536. Tabulae astronomicae, quas uulgo . . . resolutas vocant. Nuremberg.
1544. Scripta clarissimi mathematici M. Ionnis Regiomontani. Nuremberg. Repr. in Regiomontanus (1972).
Schram, R. 1908. Kalendariographische und chronologische Tafeln. Leipzig.
Shipman, J. C. 1967. Johannes Petreius, Nuremberg Publisher of Scientific Works, 1524–1550. Homage to a Bookman. Essays on Manuscripts, Books and Printing written for Hans P. Kraus on his 60th Birthday, Oct. 12 1967. Berlin.
Suchodolski, B. ed. 1973. Poland, the Land of Copernicus. Wroclaw.
Suter, H. 1914. Die astronomischen Tafeln des Muhammed Ibn Mūsā al-Khwārizmī. D. Kgl. Vidensk. Selsk. Skrifter, 7 Raekke, Hist. og Filos. Afd. III, 1. Copenhagen. See also Neugebauer (1962).
Swerdlow, N. M. 1972. Aristotelian Planetary Theory in the Renaissance: Giovanni Battista Amico's Homocentric Spheres. JHA 3, 36–48.
1973. The Derivation and First Draft of Copernicus's Planetary Theory. A Translation of the Commentariolus with Commentary. Proc. Amer. Philos. Soc. 117, 423–512.
1973a. Al-Battānī's Determination of the Solar Distance. Centaurus 17, 97–105.
1974. The Holograph of De Revolutionibus and the Chronology of its Composition. JHA 5, 186–98.

1975. On Copernicus' Theory of Precession. In Westman (1975), 49–98.
1975a. The Planetary Theory of François Viète. 1. The Fundamental Planetary Models. JHA 6, 185–208.
1975b. Copernicus's Four Models of Mercury. SC 13, 141–55.
1976. Pseudodoxia Copernicana . . . AIHS 26, 108–58.
1980. Long-period Motions of the Earth in De revolutionibus. Centaurus 24, 212–45.
1981. On Establishing the Text of "De Revolutionibus." JHA 12, 35–46.

Toomer, G. J. 1968. A Survey of the Toledan Tables. Osiris 15, 5–174.
1969. The Solar Theory of az-Zarqāl. A History of Errors. Centaurus 14, 306–36.
1980. Hipparchus' Empirical Basis for His Lunar Mean Motions. Centaurus 24, 97–109.
See also Ptolemy (1984).
Tropfke, J. 1923. Geschichte der Elementar-Mathematik V (Trigonometrie). Berlin–Leipzig.
Tuckerman, B. 1962. Planetary, Lunar, and Solar Positions 601 B. C. to A.D. 1 at Five-day and Ten-day Intervals. Memoirs Amer. Philos. Soc. 56. Philadelphia.
1964. Planetary, Lunar, and Solar Positions A.D. 2 to A.D. 1649 at Five-day and Ten-day Intervals. Memoirs Amer. Philos. Soc. 59. Philadelphia.

Valla, G. 1501. De expetendis et fugiendis rebus . . . 2v. Venice.
Viète, F. 1646. Opera mathematica. Ed. F. van Schooten. Leyden. Repr. Hildesheim, 1970.
Vogt. H. 1920. Des Kalender des Claudius Ptolemäus. S. B. d. Heidelberger Akad. d. Wiss., phil.-hist. Kl. 1920 Abh. 15.

Werner, J. 1522. Libellus Ioannis Verneri Nurembergensis super vigintiduobus elementis conicis . . . eiusdem Ioannis, de motu octauae Sphaerae, Tractatus duo. Nuremberg.
Westman, R. ed. 1975. The Copernican Achievement. Berkeley.
Whiteside, D. T. 1974. Keplerian Planetary Eggs, Laid and Unlaid, 1600–1605. JHA 5, 1–21.
Wilson, C. 1972. The Inner Planets and the Keplerian Revolution. Centaurus 17, 205–48.

Zeller, K. 1943. Des Georg Joachim Rheticus erster Bericht über die 6 Bücher des Kopernikus von den Kreisbewegungen der Himmelsbahnen. Munich–Berlin.
Zeller, M. C. 1946. The Development of Trigonometry from Regiomontanus to Pitiscus. Univ. of Michigan Diss. 1944. Ann Arbor.
Zinner, E. 1937. Arbeiten zur Geschichte der Astronomie XIII. Vierteljahrschrift der astronomischen Gesellschaft 72, 56–58.
1943. Entstehung und Ausbreitung der Coppernicanischen Lehre. S.B. d. Phys.-med. Sozietät zu Erlangen 74.
1951. Astronomie: Geschichte ihrer Probleme. Munich.
1968. Leben und Wirken des Joh. Müller von Königsberg, genannt Regiomontanus. Zweite Auflage. Osnabrück.
ZMP. Zeitschrift für Mathematik und Physik.

Subject Index

The index is arranged analytically, that is, subjects are gathered under general headings, e.g. planetary theory of longitude, spherical astronomy, under which may be found specific entries, e.g. equation of center, oblique ascension, and cross-references are provided for larger subjects. Frequently occurring names, e.g. Ptolemy, Regiomontanus, Copernicus, and titles of books, e.g. the *Almagest*, *Epitome of the Almagest*, *De revolutionibus*, are indexed only selectively, while incidental references to names and books, and secondary works referred to in footnotes, are not indexed.

Sources in the History of Mathematics and Physical Sciences

Vol. 1: G.J. Toomer (Ed.), **Diocles on Burning Mirrors: The Arabic Translation of the Lost Greek Original,** Edited, with English Translation and Commentary by G.J. Toomer

Vol. 2: A. Hermann, K.V. Meyenn, V.F. Weisskopf (Eds.), **Wolfgang Pauli: Scientific Correspondence I: 1919 – 1929**

Vol. 3: J. Sesiano, **Books IV to VII of Diophantus'** *Arithmetica*: **In the Arabic Translation of Qustã ibn Lũqã**

Vol. 4: P.J. Federico, **Descartes on Polyhedra: A Study of the** *De Solidorum Elementis*

Vol. 5: O. Neugebauer, **Astronomical Cuneiform Texts**

Vol. 6: K. von Meyenn, A. Hermann, V.F. Weisskopf (Eds.), **Wolfgang Pauli: Scientific Correspondence II: 1930 – 1939**

Vol. 7: J.P. Hogendijk, **Ibn Al-Haythan's** *Completion of the Conics*

Studies in the History of Mathematics and Physical Sciences

Vol. 1: O. Neugebauer, **A History of Ancient Mathematical Astronomy**

Vol. 2: H. Goldstine, **A History of Numerical Analysis from the 16th through the 19th Century**

Vol. 3: C.C. Heyde/E. Seneta, **I.J. Bienaymé: Statistical Theory Anticipated**

Vol. 4: C. Truesdell,**The Tragicomical History of Thermodynamics, 1822 – 1854**

Vol. 5: H.H. Goldstine, **A History of the Calculus of Variations from the 17th through the 19th Century**

Vol. 6: J. Cannon/S. Dostrovsky, **The Evolution of Dynamics: Vibration Theory from 1687 to 1742**

Vol. 7: J. Lützen, **The Prehistory of the Theory of Distributions**

Vol. 8: G.H. Moore, **Zermelo's Axiom of Choice**

Vol. 9: B. Chandler and W. Magnus, **The History of Combinatorial Group Theory**

Vol. 10: N.M. Swerdlow/O. Neugebauer, **Mathematical Astronomy in Copernicus's De Revolutionibus**

Studies in the History of Mathematics and Physical Sciences

10

N. M. Swerdlow
O. Neugebauer

Mathematical Astronomy in Copernicus's De Revolutionibus

In Two Parts

With 222 Figures

Springer-Verlag
New York Berlin Heidelberg Tokyo

N. M. Swerdlow
Department of Astronomy
and Astrophysics
University of Chicago
5640 S. Ellis Avenue
Chicago, IL 60637
U.S.A.

O. Neugebauer
Department of the History
of Mathematics
Brown University
Box 1900
Providence, RI 02912
U.S.A.

AMS Classifications: 01A20, 01A40, 05-13

Library of Congress Cataloging in Publication Data
Swerdlow, N. M. (Noel M.), 1941–
Mathematical astronomy in Copernicus's *De revolutionibus.*
(Studies in the history of mathematics and physical sciences; 10)
Bibliography: p.
Includes index.
1. Astronomy—Mathematics—History. 2. Copernicus, Nicolaus, 1473–1543. De revolutionibus orbium coelestium.
I. Neugebauer, O. (Otto), 1899– . II. Title.
III. Series.
QB47.S86 1984 521.1 83-20081

Typeset by Composition House Ltd., Salisbury, England.
Printed and bound by R. R. Donnelley & Sons, Harrisonburg, Virginia.
Printed in the United States of America.

9 8 7 6 5 4 3 2 1

ISBN 0-387-90939-7 Springer-Verlag New York Berlin Heidelberg Tokyo
ISBN 3-540-90939-7 Springer-Verlag Berlin Heidelberg New York Tokyo

To
S. Chandrasekhar
with admiration and gratitude

Preface

When I first laid out the framework for *A History of Ancient Mathematical Astronomy*, I intended to carry the discussion down to the last applications of Greek astronomical methodology, i.e. Copernicus, Brahe, and Kepler. But as the work proceeded, it became evident that this plan was much too ambitious, and so I decided to terminate my *History* with late antiquity, well before Islam.

Nevertheless, I did not discard the running commentary that I had prepared when studying *De revolutionibus* in its relation to the methodology of the *Almagest*. Only recently, E. S. Kennedy and his collaborators had opened access to the "Marāgha School" (mainly Ibn ash-Shāṭir), revealing close parallels to Copernicus's procedures. Accordingly, it seemed useful to make available a modern analysis of *De revolutionibus*, and thus in 1975 I prepared for publication "Notes on Copernicus."

In the meantime, however, Noel Swerdlow, also starting from Greek astronomy, not only extended his work into a deep analysis of *De revolutionibus*, but also systematically investigated its sources and predecessors (Peurbach, Regiomontanus, etc.). I was aware of these studies through his publications as well as from numerous conversations on the subject at The Institute for Advanced Study and at Brown University. It became clear to me that my own investigations lay at too superficial a level, and I therefore withdrew my manuscript and suggested to Swerdlow that he undertake a thoroughgoing revision and amplification of my "Notes." His acceptance of my proposal initiated the present publication.

The addition of my name to the title page is nothing more than a reflection of this prehistory. In the present publication Swerdlow has not only substantially extended the investigation of Copernicus's work in breadth and depth, but has also completely rewritten the whole manuscript. I am happy to see that my own study has thus been replaced by a work of real competence that represents a significant step in our attempts to evaluate the position of ancient science in the evolutionary process of mathematical astronomy.

O. N.

The *Almagest* has been the subject of two recent expositions, by Pedersen and Neugebauer, the appearance of which has made it reasonable to devote similar analyses to other works of mathematical astronomy in the Ptolemaic tradition. The most famous of these is *De revolutionibus*, the last important work to employ the full range of Ptolemaic methods, in its use of heliocentric models showing their nearly universal applicability, and in its failure to go beyond Ptolemy in descriptive and numerical accuracy showing their limitations. Some years ago, I began a rather ambitious project of preparing a translation of *De revolutionibus* with a technical commentary and translation of Peurbach's and Regiomontanus's *Epitome of the Almagest*, Copernicus's principal guide to Ptolemy's astronomy. The appearance of two new translations of *De revolutionibus* in 1976 and 1978, whatever their quality, made a third superfluous, and the *Epitome* deserves a proper edition, translation, and commentary in its own right, for it is not merely ancillary, but, along with *De revolutionibus*, the most important astronomical work of the Renaissance.

However, the commentary still seemed worth pursuing. While a number of papers on special topics have been published in recent years, the last technically proficient exposition of *De revolutionibus* as a whole was a chapter of 40 pages written by Norbert Herz in 1894. Hence, when Neugebauer showed me his original manuscript, it seemed to present the opportunity to undertake an analysis of *De revolutionibus* on about the same scale as the analyses of the *Almagest*. The result has actually ended up somewhat longer than anticipated because of the necessity of explaining the background to Copernicus's work, much of which he reviews himself, from the *Almagest*, the *Epitome*, and a number of other sources, and the decision to set out Copernicus's numerical work in the derivation of parameters—the subject that occupies the greatest part of *De revolutionibus*—reconstructing, as far as possible, the derivations he omits. As the work continued to increase in size, it appeared that it could well end up being more on the subject of Copernicus's mathematical astronomy than anyone cares to know. But it seemed useful to include whatever could reasonably be accomplished within the limits of a general exposition in order to provide a foundation for others who may wish to correct or extend our work without having to start over from the beginning. Even so, the study is far from exhaustive. Certain subjects, such as trigonometry, spherical astronomy, the catalogue of stars, basic properties of eccentric and epicyclic models, to which Copernicus's own contribution is minimal, we have treated briefly, and the first chapters of Book I, which have frequently been commented on at length by scholars whose qualifications far exceed our own, we have omitted entirely. In any case, the study is quite long enough as it is.

It is with gratitude that I acknowledge the support and encouragement of the Department of Astronomy and Astrophysics and the Division of the Physical Sciences of The University of Chicago. They have not only granted me the freedom and best possible conditions for research, but have held firmly to the principle that historical studies are a proper part of scientific research, and for this I am especially grateful.

There are several people who have helped with this work in a number of ways. Mrs. Maria D'Onofrio of the Sciences Library, Mrs. Madeline Gross-Millin of the Rockefeller Library, and Mrs. Mary Russo of Special Collections of Brown University have always been of the greatest help. My debt to the Department of History of Mathematics of Brown University is not to be measured. Professor Gerald Toomer has read the manuscript of this study, caught any number of errors, and made many suggestions, always correct. Professor David Pingree has always been ready to listen, and answer questions on the most obscure points. The late Professor Abraham Sachs and Mrs. Janet Sachs have watched over me for many years, offering their guidance on many occasions when it was very much needed. And I am especially grateful to Molly Schen, my best and closest friend through four summers in Providence. Of course, my greatest debt is to my co-author, who has taught me whatever I know.

Providence, Summer 1983 N. S.

Table of Contents

Part 1

PART 1

1

General Introduction

Der Narr will die ganze Kunst Astronomiae umkehren.

M. LUTHER (1539)

1. Life of Nicolaus Copernicus

The biography of Copernicus is imperfectly known and provides little information pertinent to his work in astronomy.[1] He was born in the city of Torun on the Vistula in West or Royal Prussia, according to his horoscope on 19 February 1473,[2] the youngest of four children of Niclas Koppernigk, a prosperous merchant who had come to Torun from Cracow in the 1450's, and Barbara Watzenrode (or Watczelrode), a daughter of one of the city's leading merchant families. Torun was at the time an important inland city of the Hanseatic League, and Copernicus, as a son of the merchant patriciate, grew up in surroundings of relative affluence. His father died at some time between 1483 and 1485, and he came under the protection of his mother's brother Lucas Watzenrode (1447–1512), according to tradition a stern and overbearing man—even worse, *ein herber, finsterer Mann*, Prowe calls him, *niemand will ihn lachen gesehen haben*—who had a profound influence on his nephew's future, and perhaps on his character. Watzenrode had studied at the universities of Cracow and Köln, and in 1473 received a doctorate in canon law from the University of Bologna. He advanced by stages in both the Church and State, from canonries of Kulm and Warmia to Bishop of Warmia in 1489. This was a position of considerable power that he used to the end of preserving the independence of his own rule by opposing

---

[1] What must have been the most valuable source for the life of Copernicus was a biography written by Rheticus that is unfortunately lost. The principal sources for this selective biography, concentrating on his scientific work and making no claim to originality, are: Prowe (1883–84), cited here as P., the standard, comprehensive biography, a bit out of date where new sources have been discovered, but a splendid work of scholarship; Biskup (1973), document numbers from which are cited here as B., a calendar of all known documents pertaining to Copernicus's life, paraphrased with full references to sources and literature, a very useful work second only in importance to Prowe; Biskup and Dobrzycki (1972), particularly good on Copernicus's administrative activities in Warmia; Burmeister (1967–68), supersedes all previous work on Rheticus and on the period he was with Copernicus. In addition to the sources mentioned in the notes, we have consulted the standard study of Zinner (1943), and the popular biographies of Kesten (1945), Koestler (1959), and Rosen (1971).

There is a great variety in the language and orthography of the names of persons and places in the literature on Copernicus, sometimes for reasons of nationalism, in which we have no interest. We shall generally use the German or Latin forms, which are in most cases closest to the forms used in the documents in B., and for places to the 1542 map of Prussia shown here in Fig. 2. Numerous contemporary spellings can be found in the index of B.; there are no less than 16 for Frauenburg and 37 for Copernicus, including Cappernitz, Cupperinckenn, and Koppernikel.

[2] B. 15 and pl. 22. See also below, pp. 454–457.

the Teutonic Order in East Prussia, which virtually surrounded Warmia, and promoting when advantageous the authority of the Polish Crown in Royal Prussia, the land of prosperous merchant cities to the west. He was by all accounts a prince of the Church of whom Machiavelli would have written approvingly, and fortunately for the young Copernicus, he fully understood the advantages of nepotism in assuring the future of his realm.

There is speculation, but no certain knowledge, concerning Copernicus's early education in Torun. In 1491 he, along with his older brother Andreas, enrolled in the University of Cracow, doubtless under the guidance and with the support of their Uncle Lucas, who evidently intended for both youths an ecclesiastical and political career like his own. Copernicus is thought to have remained at Cracow until 1495, but there is no record of his taking a degree and his course of study is unknown. Nevertheless, it is reasonably presumed that he took some of the courses in astronomy and mathematics offered at the time of his attendance, among them lectures on spherical astronomy, planetary theory, eclipse tables, the *Tabulae resolutae*, geography, and astrology.[3] Two volumes that he kept for the rest of his life appear to have been acquired by him in Cracow, as shown by the style of their bindings.[4] One contains the 1492 Venice edition of the *Alfonsine Tables* and the 1490 Augsburg edition of Regiomontanus's *Tabulae directionum*, the two books forming what was at the time the most up-to-date and more-or-less complete collection for astronomical calculation, the *Tabulae directionum* for spherical astronomy and the *Alfonsine Tables* for planetary theory and eclipses. On a number of pages bound in at the end of the volume, he copied parts of Peurbach's *Tabulae eclipsium* and a large set of planetary latitude tables related to those in the *Tabulae resolutae*—both useful as supplements to the *Alfonsine Tables*—and various smaller tables and notes made over a period of years.[5] The other volume from Cracow contains the 1482 Venice edition of Euclid's *Elements* with Campanus's commentary, for two centuries the most commonly used version of Euclid, and the 1485 Venice edition of *In iudiciis astrorum* by ʿAlī ibn Abī ʾr-Rijāl, one of the most comprehensive and influential Arabic astrological treatises to be translated into Latin.

It can be seen from the four books of this period that Copernicus was already acquiring a library of fundamental works in spherical and planetary astronomy, geometry, and astrology, and these are four books that just happen to survive. Assuming that he did in fact attend lectures on these subjects, he appears to have left Cracow at the age of about 22 with at least the competence of university training in mathematics, astronomy, and astrology. In August of 1495 Uncle Lucas obtained for him one of the sixteen canonries of the Chapter of Warmia, which, however, he did not

---

[3] P. I,1, 141–42; Jarzebowski (1971), 8.
[4] Czartoryski (1978), 365–66.
[5] Much of this material is printed, not altogether accurately, in P. II, 206–44.

formally take possession of until 1497, and in the account book of the Chapter of Warmia, where he appears in 1495 and 1496 among canons delinquent in required payments, he is designated *Nicolaus de Thorn, nepos episcopi.*[6] It is possible that for about a year he travelled to various German universities, perhaps attending lectures, but not formally registering. However, in the autumn of 1496 he enrolled at the University of Bologna as a student of canon law and a member of the German Nation, where he was joined two years later by Andreas, both brothers again following in the footsteps of their Uncle Lucas.[7] At Bologna Copernicus must also have studied civil law, since he is referred to in a document from 1499 as a "student in both laws," and it was probably in Bologna that he undertook the study of Greek, a subject in which he became proficient, although not distinguished. Also in 1499 Nicolaus and Andreas ran short of money, as students are wont to do, borrowing 100 ducats from a bank to be repaid in four months time. But after one month, Bernard Scultetus, who had guaranteed the loan, sent an urgent request to Uncle Lucas that he repay the money more quickly to avoid a greater financial loss to his *nepotes* and their guarantor.

During this period Copernicus did not neglect astronomy. According to Rheticus, he lived with Dominico Maria di Novara (1454–1504), and was "not so much the student as the assistant, and witness to the observations, of the learned man," whatever this may mean. Dominico Maria occupied the chair of astronomy at Bologna, was a prominant astrologer, and held the interesting theory that an increase of 1;10° in the latitudes of a number of places in Italy, compared with their locations in Ptolemy's *Geography*, indicated a geographical shift of the earth's axis of rotation, an opinion that Copernicus does not appear to have followed.[8] Copernicus's earliest dated observation in *De revolutionibus*, an occultation of Aldebaran on 9 Mar 1497, was made in Bologna, perhaps in the presence of Dominico Maria, and on 9 Jan and 4 Mar of 1500 he observed conjunctions of the moon and Saturn. Rheticus also reports that in about 1500, when Copernicus was about twenty-seven years old, he lectured on mathematics in Rome before a large crowd of students, important men, and experts in this subject. Exactly what this means about Copernicus's reputation at the time is not clear, but in the latter part of 1500 he was definitely in Rome, where he observed a lunar eclipse on 6 November.

Copernicus had left Bologna without taking a degree—unlike Uncle Lucas—and in July of 1501 appeared before the Chapter in Frauenburg where he received permission to continue his studies for two years, but now in medicine in order to serve as medical advisor to the Bishop and members of the Chapter.[9] He returned to Italy and commenced his medical studies at

---

[6] B. 23–24.
[7] For the period in Bologna and Rome, B. 27–37.
[8] According to Magini; cf. P. I,1, 240–42 n.
[9] For the period in Padua and Ferrara, B. 38–44.

the University of Padua, at the time the most eminent institution in Europe for the study of medicine and all scientific subjects. Unfortunately, nothing is known about Copernicus's studies at Padua, and once again he left without a degree. Perhaps it was at this time that he learned of the planetary theory of the Marāgha astronomers that later provided the foundation of much of his own planetary theory. There is in any case evidence that Marāgha theory was known in Italy, and specifically in Padua, in the late fifteenth and early sixteenth centuries, and if Copernicus learned of it while in Padua, it would be the most important contribution of his years in Italy to his later work.[10]

At some time during the first half of 1503, Copernicus travelled from Padua to Ferrara where, on 31 May, he finally received a degree, a doctorate in canon law from the University of Ferrara. It has been suggested that the reason he took the degree from Ferrara, rather than Bologna where he had studied for four years, was that the requisite expenses were less in Ferrara. In the mean time Uncle Lucas had been looking out for his nephew's interest, and by the beginning of 1503 had added to his Warmia Canonry a scholastery of the Church of the Holy Cross in Breslau, a benefice that Copernicus held until 1538.[11] Thus, when he returned to Warmia in the latter part of 1503, his living was assured, as were his responsibilities to the Chapter and Uncle Lucas, and aside from brief trips into Prussian and Polish territory, Copernicus never again left what he called in the dedication of *De revolutionibus* "this remote corner of the earth."

Initially he went into the service of the Bishop, residing at the episcopal palace, really more of a fortress, in Heilsberg.[12] There he acted as Uncle Lucas's personal physician and, more significantly, assisted the Bishop in his administrative duties, travelling with him to audiences with the King of Poland and to assemblies of the Royal Prussian Estates and its Council, of which the Bishop of Warmia was the permanent chairman. There seems little doubt that the nephew was being groomed as his uncle's successor—Watzenrode had reached the office at the age of forty-two, and Copernicus was now in his middle thirties—and not unrelated to such advancement was a grant of permission from Pope Julius II, doubtless obtained through Uncle Lucas's intercession, allowing Copernicus to acquire two further benefices. This, however, Copernicus did not do, and in about 1510 he left the Bishop's service in Heilsberg and moved to Frauenburg, the general headquarters of the Chapter of Warmia.

Some interesting conclusions have been drawn from Copernicus's departure from Heilsberg in 1510,[13] namely, that he had decided to decline a higher ecclesiastical and political career—probably because he had already

---

[10] See below, pp. 47–48.

[11] B. 42, 387, 395.

[12] For the period in Heilsberg, B. 45–63.

[13] Rosen (1971), 334–35, an important insight into Copernicus's life and work.

devised his new planetary theory and resolved to devote more attention to astronomy than such a career would permit—and that consequently he had a falling-out with Uncle Lucas, who would hardly wish to see the years of education, training, and influence he had expended on behalf of his nephew go to waste. This interpretation has much to recommend it, for it is compatible with the other evidence for the period of Copernicus's discovery of his new theory—between 1508 and 1514, more likely before 1512—and even provides further evidence for fixing the date at about 1510. In any case, it is clear that it would require something of great importance to break Copernicus's ties with his stern uncle—unless he was finally just sick of being ordered about—and it is hard to think of anything more important in his life than his discovery of the heliocentric theory.

In 1509, while still in Heilsberg, Copernicus had printed in Cracow his first, and for over thirty years his only, publication, a Latin version of the *Moral, Pastoral, and Amatory Epistles* of Theophylactus Simocatta (7th century), the Greek text of which had been printed by Aldus in 1499.[14] The dullness, if not insipidity, of Copernicus's choice of these undistinguished Byzantine pseudepigrapha has occasioned much wonder among later historians, but since he writes admiringly of them in his dedication to the Bishop or Warmia, we can only conclude that they were congenial to his taste, or at least to the taste of Uncle Lucas, to whom in the dedication Copernicus addresses the ominous line from Ovid to Germanicus Caesar, "[My] talent stands or falls at your look." Whether this should be interpreted as the conventional sentiment of a dedication or as something more personal is hard to know. But whatever the relations between the uncle and nephew had become, within two years after Copernicus left Heilsberg, Bishop Watzenrode died suddenly in Torun at the end of March of 1512 while returning to Warmia from the coronation of Queen Barbara Zapolya in Cracow, and Copernicus was now for all time released from the protection, and perhaps the oppression, of his somber patron. The death of Watzenrode was doubtless a relief to the Teutonic Order in East Prussia, whose depredations in Warmia increased in the following years under his more amiable successor Fabian von Lossainen, a canon who had been a fellow student with Copernicus in Bologna. And even the King of Poland, Sigismund I, who had had his own problems with the imperious Watzenrode, made sure that in the future he would maintain greater control over Warmia by initially refusing to recognize Lossainen until the Chapter acknowledged the King's right to confirm the election of the Bishop.

During the decade following his return from Italy, Copernicus continued work on astronomy, and appears to have achieved some distinction. In a prefatory poem to the translation of Theophylactus written probably in 1508 by Laurentius Corvinus (Rabe), a humanist and geographer in Breslau whom Copernicus may have known earlier while a student in Cracow, he is

[14] Text in P. II, 45–137; facsimile of 1509 edition, Warsaw, 1953.

commended for treating the motions of the sun, moon, and planets, and for seeking "the hidden causes of things with wonderful principles."[15] What this means is, to say the least, obscure, but it does speak well for Copernicus's reputation and indicates that he already may have been doing something out of the ordinary, perhaps working with Marāgha rather than Ptolemaic models, although not yet having arrived at the heliocentric theory.

There is even evidence that Copernicus was known, or still remembered, in Italy. During the Fifth Lateran Council, Pope Leo X introduced the recurring question of reforming the ecclesiastical calendar, issuing in 1514 a request for the advice of astronomers throughout Europe. The request seems to have reached Copernicus through Paul of Middelburg, the Bishop of Fossombrone and a competent astronomer who, in his *Secundum Compendium correctionis Calendarii*, published in Rome in 1516, mentions a letter on the subject from Copernicus, although its contents are not described.[16] Nevertheless, the incident must have been of significance to Copernicus, for in the dedication of *De revolutionibus* to Pope Paul III he says that it was on the advice of Paul of Middelburg that he has since that time devoted attention to more accurate determination of the length of the year and month and to the motion of the sun and moon.

However, Copernicus's investigations into planetary theory had already taken an unexpected turn, and he had written a short account of his discoveries. In the catalogue of the library of Matthew of Miechow, a Cracow physician and professor of medicine, dated 1 May 1514, there is an entry that reads; *Item sexternus Theorice asserentis Terram moveri, Solem vero quiescere*, "Next a quire of six leaves of a *Theorica* maintaining that the earth moves while the sun is at rest."[17] Since it is all but unthinkable that another work holding this remarkable theory could have reached Cracow by 1514, the entry can only refer to Copernicus's *Commentariolus*, as it is called, the first description of his new heliocentric planetary theory, which, during the author's life circulated anonymously and not widely. Copernicus makes no reference to it in *De revolutionibus*—perhaps by then it was something he wished to disown—no manuscript of it survives among his own books and papers, and it probably would have perished entirely had not Tycho received a copy in 1575 and later had further copies prepared and distributed. At present, three manuscripts of it are known—in Vienna, Stockholm, and Aberdeen—and all are descended from Tycho's copy, which has itself disappeared.[18]

---

[15] B. 57; P. II, 48. The phrase about Copernicus's treating the "alternating movements" of the moon's "brother" (*celerem lunae cursum alternosque meatus Fratris*) has been reasonably explained (cf. Zinner (1943), 185; Rosen (1971), 339) as referring to the seasonal variation of the declination of the sun, the moon's brother, and not to a motion of the earth.

[16] B. 103; P. I,2, 65–72.

[17] B. 91.

[18] On editions and translations, see below, p. 92.

The *Commentariolus*, containing errors that show signs of haste, may have been written in a momentary burst of enthusiasm, perhaps immediately upon devising the heliocentric theory. In essence, the planetary theory of longitude is a fairly straightforward modification of Marāgha models, specifically those of Ibn ash-Shāṭir, to a heliocentric arrangement referred to the center of the earth's orbit, with numerical parameters largely adapted from the *Alfonsine Tables*, while the latitude theory is an entirely original, although not altogether satisfactory, heliocentric conversion of Ptolemy's models. Although the fundamental principles of Copernicus's later work are already present, the *Commentariolus* is in many ways but a rough outline, requiring considerable amplification to be either convincing or useful. Copernicus evidently had something further in mind, for he remarks that "mathematical demonstrations," which were omitted for the sake of brevity, are intended for a larger volume. What he intended at this time is difficult to know—the addition of mathematical demonstrations to the *Commentariolus* hardly describes *De revolutionibus* of thirty years later—and Copernicus must soon have realized that a far more extensive treatise was required to present his new theory convincingly, and that this would require many additional years of work. While the *Commentariolus* has a number of shortcomings, two in particular stand out that at the time were acceptable to Copernicus, but that upon further reflection he realized had to be corrected, the correction requiring far more than supporting the theory of the *Commentariolus* with mathematical demonstrations of whatever kind were originally intended.

First, the heliocentric theory and the motion of the earth were presented as a series of postulates, although there is no doubt that Copernicus considered them true. This was not really objectionable, and was in fact entirely reasonable, because Copernicus knew that at the time he had no way of proving that the earth in fact moves. But he must also have realized that eventually he had to provide proof, or at least considerably more evidence, for so startling an innovation, and it is well known that the difficulty—indeed, impossibility—of his doing so later made him reluctant to publish his theory at all. The other problem was that the parameters of the models had merely been adapted from existing sources, principally the *Alfonsine Tables*. This too was not objectionable for demonstrative purposes in the *Commentariolus*, where Copernicus's goal was to show that his new models "agree with computations and observations," which can only mean with the sort of checks that could be carried out with whatever tables were at hand. But he must already have known, or soon came to know, that the *Alfonsine Tables* and its descendents were not sufficiently accurate to agree with observations, even quite crude observations of the times of conjunctions and occultations. Equally important, they could not come anywhere close to reproducing both contemporary observations and Ptolemy's observations in the *Almagest*, the discrepancies indicating the possibility of long-period alterations of parameters, something already well established for the rate of precession and the length of the tropical years. Therefore, it would be necessary to derive new parameters, or confirm old parameters, directly from observation, just as Ptolemy had done in the *Almagest*, and this immense labor would require many years of observation, delaying the composition of Copernicus's larger book until after all the observations were made and all the parameters derived, itself the work of many years. Thus, what may originally have been intended as an expansion of the *Commentariolus*—perhaps "mathematical demonstrations" of the compatibility of the new heliocentric models and derivative parameters with existing theory and tables, something Copernicus could expect to do in a comparatively short time—became a far more ambitious work, a new *Almagest* in fact, and Copernicus was, almost of necessity, to devote the rest of his life to an arduous task, the difficulty of which nearly defeated him.

The commonplaces about Copernicus's extreme caution and secretiveness, which it must be confessed his dedication to *De revolutionibus* has done much to encourage, must therefore be abandoned, or at least relegated to a late stage of his life when he was despairing of successfully completing the great labor he had set himself. Considering the magnitude of his undertaking and the time that it necessarily required, it can be seen that he carried out his work about as rapidly as could be expected. It is probably not coincidental that the earliest of Copernicus's planetary observations in *De revolutionibus* are from 1512, about the time of or, more likely, shortly after the writing of the *Commentariolus*. And aside from three lunar eclipses and the 1497 occultation of Aldebaran, all of Copernicus's observations in *De revolutionibus* fall between 1512 and 1529, which is, if anything, a relatively short time for making the necessary observations. During this period his principal astronomical work was patiently making as many observations as time and circumstances allowed—certainly far more than the number reported in *De revolutionibus*—and he was not yet in a position to begin seriously the writing of his book. Further, there is considerable evidence for Copernicus's administrative activities in the Chapter during these years, and they were sufficiently numerous and diverse to suggest that the time he could devote to astronomy, particularly to difficult theoretical work and extensive calculation, was limited.

At this point it is informative to consider Copernicus's occupation as a canon of the Chapter of Warmia.[19] Warmia (Ermland) is a small territory of about 4000 square kilometers, roughly triangular in shape, with a base of about 80 km running north-east, and the apex, about 100 km to the north, forming a coast of less than 20 km along the Frisches Haff, a long fresh-water lagoon along the southern coast of the Baltic. The land is almost entirely surrounded by East Prussia except for a short border with Royal Prussia of about 20 km in its north-west corner near Frauenburg (Frombork), the primary residence of the Chapter. Warmia and both Prussias are illustrated

---

[19] A good account is in Biskup and Dobrzycki (1972), 60–82.

in a modern map in Fig. 1 that shows the principal towns pertinent to Copernicus's activities, while Fig. 2 shows part of a very detailed contemporary map of Prussia published in Nuremberg in 1542 by Heinrich Zell.[20] Copernicus's residence, Frauenberg, is about 70 km from Danzig (Gdansk) to the west and Königsberg to the east, the largest cities of Royal Prussia and East Prussian respectively, while the Bishop's residence, Heilsberg (Lidzbark Warminski) is about 70 km to the south east, and Allenstein (Olsztyn), the secondary residence of the Chapter, about 80 km to the south where the most extensive of the Chapter's three land holdings was located.

Following the Treaty of Torun in 1466, Warmia and both Prussias were technically dependencies in varying degrees of the Polish Crown, but each maintained its local independence for most purposes: Warmia as an ecclesiastical state under the rule of the Bishop, Royal Prussia through the extensive trade and wealth of its merchant cities, and East Prussia under the rule of the Teutonic Order, owing its primary allegiance to Pope and Emperor, and, not incidentally, armed to the teeth. Warmia was a rather poor and sparsely populated agricultural region with no sizable towns—Braunsberg (Braniewo), 10 km to the east of Frauenburg, was the largest—comparing unfavorably with the prosperous trade and agriculture of Royal Prussia, and having more in common both economically and politically, with conditions in the surrounding territory of East Prussia which, under the Teutonic Order, was also an ecclesiastical state of sorts. The population of Warmia consisted mostly of some thousands of peasants of Old Prussian and Polish descent living in small villages, tending their oats, barley, and pigs, and bound by various degrees of servitude to the land, about one-third of which belonged to the Chapter and most of the rest to the Bishop. The administration of justice, the collection of rents, and the exaction of services were in the hands of the sixteen canons of the Chapter and the Bishop's other agents, the whole arrangement reminiscent of Voltaire's description of Paraguay under the Jesuits, "Los Padres y ont tout, et les peuples rien; c'est le chef-d'oeuvre de la raison et de la justice."

Most of the canons were, like Copernicus, sons of wealthy merchant families of Royal Prussia and university educated, principally at Cracow and Leipzig. Each was entitled to a residence within the walls of Frauenburg, additional estates, called *curiae*, in the surrounding country, and the income from some number of land holdings, the rights to all of which could be bought, sold, leased, or traded. With his income, each canon was required to maintain at least two servants and three horses, and when called upon by the Bishop, to bear arms—although not at Chapter meetings—perhaps a relic of the founding of the Diocese of Warmia by the Teutonic Order in the thirteenth century. Essentially, the canons of the Chapter held much the

[20] The map in Fig. 1 is from P. I,2, and the map in Fig. 2 is probably based upon a map or description of Prussia originally made by Copernicus and Alexander Scultetus, on which see below, p. 28.

same position in Warmia as the nobility in the more feudal secular states of the period, and while the canonries could hardly be hereditary, nepotism did what it could to keep them in the same families of the Royal Prussian patriciate.

Administrative duties, some more responsible and demanding than others, were assigned and re-assigned among the canons in November of each year, and Copernicus occupied a number of positions over the years between 1510 and about 1530, after which his obligations seem to have been reduced. From 1510 to 1512, and again in 1520, 1524–25, and 1529, he held the office of Chancellor, responsible for the Chapter's official correspondence and for keeping some of the accounts. In 1513 he was the head of the provisioning fund, supervising the Chapter's bakery, brewery, and mills. From Nov 1516 to Nov 1519, and again for part of 1521, he served as the Administrator of Benefices, residing at Allenstein and travelling through the Chapter's estates and over 100 villages, in charge of the collection of revenues and enforcement of services, the administration of justice, and the settling of the succession of peasants to abandoned or repossessed lands, the records of these dispositions being entered into a register.[21] In addition, he participated in Chapter meetings, attended assemblies of the Warmian Estates and the Royal Prussian Estates, and served on diplomatic missions to representatives of the Polish Crown and East Prussia. Copernicus's highest administrative office appears to have been reached in 1523 when, following the death of Bishop Lossainen at the end of January, and until the confirmation of his successor Maurice Ferber in October, he served as the Administrator General of the Diocese of Warmia, residing in Heilsberg and executing the administrative duties of the Bishop.

Copernicus even had the opportunity to bear arms, or at least to serve in a war, for the Bishop.[22] During Lossainen's tenure, relations between Warmia and the Teutonic Order in East Prussia, never very good, were deteriorating, and it came to the point that robbers from East Prussia were raiding in Warmia under the protection of the Order. The increased danger followed upon the accession in 1511 of Albrecht von Hohenzollern (1490–1568), a leader of strength, intelligence, and great ambition, who became Grand Master of the Order at the remarkable age of twenty. East Prussia had technically been a fief of the Polish Crown since the Treaty of Torun concluded the Thirteen Years' War in 1466, and Albrecht himself was the grandson on his mother's side of Casimir IV, and was thus related to the reigning Sigismund I. Nevertheless, he continued to assert the independence of East Prussia, as had previous Grand Masters with the support of the Emperor and Pope, and had designs on Warmia, leading to a series of disputes with the King and the Bishop. Old Watzenrode might have been a match for Albrecht; Lossainen was not.

---

[21] Published in facsimile by Biskup, Olsztyn, 1970; cf. B. 110–195 *passim*, 225–30.

[22] B. 200–23 *passim*; Biskup and Dobrzycki (1972), 68–78.

After a series of incidents and provocations, war broke out between East Prussia and Poland, and on 31 December 1519 Albrecht invaded Warmia, seizing the largest town, Braunsberg, which he refused to evacuate when envoys, including Copernicus, were sent by the Bishop. Then on 23 January Frauenburg was attacked, the town itself and the curiae of the canons outside the walls of the cathedral being destroyed by fire. The cathedral was saved by defending Polish troops, but the canons fled, some to Danzig and Elbing in Royal Prussia, but others, including Copernicus, to Allenstein, which remained relatively safe for the moment. (In February, April, and July Copernicus made observations of Jupiter and Saturn.) By autumn, however, much of Warmia had been occupied and devastated by East Prussian troops, the Bishop was under siege in Heilsberg, and Allenstein was threatened. Copernicus appealed for aid to Sigismund, but his request was intercepted by the Order. Although troops were eventually sent, the danger continued, and in early 1521 Copernicus requested and received arms, ammunition, and food from Elbing. Finally, in April a four-year truce was concluded between the Order and Poland, but the Order continued to occupy a considerable part of Warmia, including Braunsberg. By May Copernicus was once again confirming the succession of peasants to land, and by July he was back in Frauenburg where he held an office called Commissioner of Warmia.

The final resolution of the conflict, which does not directly involve Copernicus, is nevertheless of interest.[23] The issues were referred to the Emperor Charles V, but this produced no results and Albrecht went to Germany to seek aid for continuing the war. In 1522 he came to Nuremberg where he met with one Andreas Osiander (1498–1552)—his original name was Hosemann and his schoolmates called him Hosen Enderle[24]—who had just adopted the Lutheran Reform and all but convinced the Grand Master of the Teutonic Order to do the same. Albrecht then travelled to Wittenberg to see Luther himself, who advised him to disband the archaic Order, get married, and found an hereditary duchy in his own name, all of which Albrecht thought a good idea. After various negotiations with Sigismund, and solemn assurances to the Pope that he would not join the Reformers, it was done, and in 1525 the Treaty of Cracow established the Duchy of Prussia under a Hohenzollern dynasty as a fief of the Polish Crown. Duke Albrecht immediately did everything in his power to further the religion of Luther, and married a Danish princess the very next year. He also continued to maintain a close friendship with Osiander, whom he seems to have regarded as his favorite Reformer, and both were to play a role in the later publication of Copernicus's work.

Relations between Warmia and Ducal Prussia improved after the settlement of 1525—even with the differences in religion, Bishop Ferber

---

[23] P. I,2, 188–92; *Ency. Brit.* 11th ed. (1910) 1, 497.

[24] Doppelmayr (1730), 58 n. *ll.*

had kindly sent Albrecht additional troops to put down the peasant uprisings—and it was now possible to do something about a problem affecting Royal Prussia and Warmia that had concerned Copernicus for several years, namely, the improvement of coinage.[25] Already at the beginning of the century, the silver coinage in Royal Prussia had been debased to the point where better coins were being melted for their bullion value, leaving in circulation only the most worthless coins, particularly issues of very low silver content from East Prussia. The problem evidently became acute after Albrecht's accession in 1511 when an even more debased coinage minted by the Teutonic Order was introduced into circulation, leading to a protest by the Royal Prussian Estates in 1516. All this would of course be particularly irritating to the Bishop and Chapter of Warmia since the peasants could pay their rents in the debased coins. On the other hand, in Royal Prussia the mints of Torun, Danzig, and Elbing benefited from issuing coins with a metal content far below their face value, so that the councils of these cities were understandably reluctant to give up such dependable profits, and nothing could be done about East Prussia so long as hostilities continued.

As early as 1517 Copernicus drew up a set of recommendations for reform, and he continued to revise his tract in a German version of 1519, read to the Assembly of the Estates of Royal Prussia after the war in 1522, and a final Latin version of 1528, apparently prepared for a meeting of the Prussian Estates with envoys of Duke Albrecht for the purpose of standardizing the coinage of Royal and Ducal Prussia.[26] In the final version, called *Monete cudende ratio* (Method of Minting Coinage), he points out that in addition to the three generally recognized disasters that bring about the ruin of kingdoms, namely, civil strife, fatal disease, and famine (*discordia, mortalitas, terre sterilitas*), there is a fourth, debased coinage or worthless money (*monete vilitas*), that is known to few, and only to the wisest, because it does not overthrow the state suddenly, but gradually and imperceptibly. But the effects are nevertheless severe. Coinage of low silver content causes old coinage of higher content to be melted down and pass out of circulation, which increases prices and destroys foreign trade so that commerce and manufacture suffer, the people grow lazy, and the land sinks into decay and ruin.

The tract is both interesting and admirable (although not so original as frequently claimed) in that it shows research into the history of Prussian coinage as far back as the beginning of the fifteenth century, specifically chronicling its decline in value, a careful analysis of the purpose and principles of a precious metal medium of exchange, and a series of practical recommendations for regulating the relation of Prussian and Polish coinage. Copernicus recommends that the metal content of Prussian coinage be standardized with the stronger coinage of Poland at twenty marks to a pound

[25] Biskup and Dobrzycki (1972), 83–97; P. I,2, 193–201.

[26] Texts in P. II, 21–44.

of pure silver, less cost of manufacture, so that three Royal Prussian shillings and one Ducal Prussian groschen shall equal one Polish groschen, which shall define the Prussian mark. He recommends further that the minting of the new coins be restricted to one location in each Prussia, that the profit to the mints be limited to the cost of manufacture, and that the old coinage be demonetized, withdrawn from circulation, and converted to the new according to its precious metal value at the mint. Copernicus's plan seems to have had some effect on the reforms that were eventually adopted—the coinage of both Prussias was at least standardized with that of Poland—but the other suggestions were not followed, most notably the restriction of the mints' profits to cost of manufacture, there were problems in withdrawing the old coins, and no less than four mints were in operation after 1530. Copernicus turned his attention to another, more local, economic subject in 1531 when he prepared a schedule regulating the price of bread in Warmia.[27] In this case, not unlike his suggested regulation of minting, he directed that the price of bread be set equal to the price of its constituent grain, the bakers' profits being limited to the value of the by-products of the milling.

One of Copernicus's constant occupations was that of physician, for which he had earlier received training at Padua.[28] He served as physician to the Bishop and members of the Chapter, consulted when called upon for dignitaries in both Prussias, and according to a later, idealized biographical tradition, made his services available to the people of Warmia and was highly regarded. He owned a number of medical books, mostly older works printed in the fifteenth century, in some of which he made notes or wrote down recipes for medications. But none of this is very informative, and since he wrote nothing on medicine we are really entirely ignorant of his thoughts on the subject or the methods of his practice. While he resided in Heilsberg, until 1510, he was the personal physician to Bishop Watzenrode, but he was not present at Uncle Lucas's death in Torun in 1512, and neither apparently was any other physician. Brother Andreas had contracted some kind of loathsome disease before 1508, described in 1512 as *contagioso leprae morbo infectus*, and twice sought treatment in Rome where, by 1519, he is believed to have died.[29] Bishop Lossainen died in 1523 of syphilis—according to the Warmian Church Chronicler, God had punished the poor Bishop for being too tolerant of the Lutheran heresy by inflicting him *a morbo Gallico*—and there was obviously little Copernicus could do about that.[30] His successor Maurice Ferber was not going to make the same mistake, issuing an edict in 1526 expelling all Lutherans from Warmia and in general fighting

[27] P. I,2, 213–16; B. 322.

[28] Numerous documents cited in B; P. esp. I,2, 293–320. On Copernicus's medical books, see also Czartoryski (1978).

[29] P. I,2, 26–32, quotation at 30 n.

[30] P. I,2, 156 n.

a rearguard action against the Reformers, who were making great progress in Royal Prussia and were already dominant in Ducal Prussia. Despite these precautions he suffered from every kind of ailment—stomach and bladder trouble, fainting spells, and eventually paralytic strokes—time and again calling upon Copernicus to rush to Heilsberg, and when things got too serious enlisting for consultation Lorenz Wille, physician to Duke Albrecht, and Jan Solpha, professor of medicine at Cracow and personal physician to King Sigismund. But it was all to no avail, for after years of miserable health, he finally died of another stroke in 1537, and Copernicus, who had again hastened to Heilsberg, could do no more than arrange a funeral, inventory his possessions, and transport the Bishop's corpse and money back to Frauenburg.

As we have mentioned, during the years of his heaviest administrative responsibilities, Copernicus's principal astronomical work was carrying out the observations required for deriving the parameters of his solar, lunar, and planetary theory, a task that extended over about seventeen years from 1512 to 1529. His only writing on an astronomical subject since the *Commentariolus* had been a critical evaluation of Johann Werner's *De motu octavae sphaerae* (Nuremberg, 1522) that he wrote at the request of Bernard Wapowski, a distinguished cartographer and secretary to King Sigismund in Cracow. Werner (1468–1522) was an eminent Nuremberg mathematician, whom Wapowski would have known from his important earlier publication in 1514 of an extensively annotated translation of Book I of Ptolemy's *Geography* with much additional material on map projections.[31] However, for the new work, which concerned the variation of the precession of the equinoxes and the obliquity of the ecliptic, he required the opinion of an astronomer, and so sent a copy of the book to Copernicus. On 3 June 1524 Copernicus sent Wapowski a letter strongly criticizing Werner's treatment of these subjects, and so well written that it appears to have been intended for public circulation.[32] In particular, he took Werner to task for the serious error of misdating Ptolemy's observation of Regulus by eleven years, from 139 to 150, and for not trusting the accuracy of ancient observations, a subject upon which Copernicus was later to change his mind, as is known from conversations reported by Rheticus.[33] He did not set forth his own views on the precession, remarking that they are intended for another place, which shows that he was already concerned with the subject. But it is of interest that he later incorporated material from Werner's book into his own treatment of the precession and obliquity in *De revolutionibus*, although without any reference to Werner.

By the early 1530's knowledge of Copernicus's new theory was circulating in Europe, even reaching the high and learned circles of the Vatican. In 1533

---

[31] On Werner, see Doppelmayr (1730), 31–35; Folkerts (1976).

[32] On editions and translations, see below, p. 92.

[33] Below, p. 20.

the papal secretary Johann Albrecht Widmanstadt was presented with a valuable Greek manuscript of Alexander of Aphrodisias and Proclus by Pope Clement VII for explaining to him, in the presence of two cardinals, a bishop, and a physician in the Vatican gardens, the "Copernican theory of the motion of the earth."[34] Following the death of Clement VII in 1535, Widmanstadt became secretary to Cardinal Nicolaus Schoenberg (1472–1537), and must likewise have told him of Copernicus's work. In a letter dated from Rome on 1 November 1536, Schoenberg wrote to Copernicus, requesting that he communicate his discovery to the learned, and that he send him his *lucubrationes* on the sphere of the world together with tables and anything else of pertinence, all of which Schoenberg instructed Dietrich von Reden, a Warmian canon resident in Rome, to have copied and sent to him, presumably at the Cardinal's expense. The letter contains a clear and correct outline of the location of the sun, earth, and moon in the heliocentric theory, a statement that Copernicus had already written an account of his theory and had computed tables, and the Cardinal's enthusiasm to learn more is obviously genuine. Copernicus's immediate response to this generous invitation is not known, but he retained the letter and had it printed in *De revolutionibus* as a commendation of his work by an influential, and now deceased, dignitary of the Church.[35]

Copernicus's work was also known in Cracow at this time, as shown by an even more interesting letter.[36] On 15 October 1535 Bernard Wapowski sent to Sigismund von Herberstein in Vienna "something new and long awaited by learned men, an almanach with the truest and most correct motions of the planets, which differ greatly from the common almanach, calculated from new tables" drawn up by Copernicus. He explains that Copernicus is a great mathematician who has held for many years that in order to correct the motions of the planets it is necessary to grant some motion to the earth, although the earth's motion is insensible. Wapowski wishes copies of the almanach to be sent to almanach makers in Germany, so that they may acknowledge their errors and the errors of their tables, or that it be printed in Vienna so that the astronomers of Europe will acknowledge their errors and diligently seek more correct motions of the planets, because neither meteorological prognostications nor annual horoscopes can be done correctly without true motions and aspects of the planets. He adds that Copernicus and many others desire this for the common benefit of all. Wapowski's letter is very important because it gives more specific information about the progress of Copernicus's work than Schoenberg's letter of a year later, and is definitely based upon first-hand knowledge. Evidently by 1535 Copernicus had proceeded far enough in his work to complete a set of solar, lunar, and planetary tables— although not necessarily

---

[34] B. 339; P. I,2, 274 n.; Birkenmajer (1900), 538.

[35] B. 359; P. I,2, 274–78.

[36] B. 345 and pl. 16–17; text in Zinner (1937), 56–58.

identical to those in *De revolutionibus*—and as a specimen of the accuracy of his work had computed an almanach, presumably in standard form giving daily positions and aspects for a full year (one would guess 1536). Wapowski says that the aspects were copied incompletely and with some errors, but can easily be corrected since the planetary positions were copied correctly. Unfortunately Herberstein did not act on Wapowski's request, Wapowski died on 21 November, scarcely a month after sending the letter and the almanach, and Copernicus's almanach has disappeared.

From the existence of a set of tables in 1535, it can be concluded that most of the preparatory work for *De revolutionibus*, specifically the derivation of parameters, had been completed at this time, although much still remained to be done, including the writing of at least the latter part of the book and substantial revisions of the earlier parts—as shown by the watermark dating of the paper in Copernicus's manuscript—some possibly altering parameters underlying the tables. By 1539, when Rheticus came to visit Copernicus, the work was more-or-less complete, for Rheticus read a version of the text close, but not identical, to Copernicus's surviving holograph (which, however, continued to receive important, and still later, revisions).

But at some time before Rheticus's arrival, Copernicus's courage failed him, and he came close to giving up the publication of his book entirely. It is possible that his declining to comply with Cardinal Schoenberg's request in 1536 is already a sign of this, but it is also possible that Copernicus wrote the Cardinal—he must have sent some kind of response—that his work was not yet completed, and this would have been true. Since Schoenberg died the following year, his patronage, for that is what his invitation amounted to, could no longer be enlisted. Likewise, Copernicus's eagerness to publish an almanach in 1535 has been taken to show that he was willing to release an almanach, and perhaps even tables, but not his theoretical work. This is entirely possible, and receives some later confirmation from Rheticus, but it is also possible that Copernicus believed that an accurate new almanach would create a more favorable reception for the later publication of his controversial theory, and this would have been a reasonable assumption. Here one may guess that it was Wapowski's death that frustrated Copernicus's intentions, whatever they may have been.

Nevertheless, during the late 1530's, perhaps even while he continued to work on his book, Copernicus became very reluctant to publish. For this there may have been a number of reasons, among them that it seems to have been a dismal period in his life. The events are well known, and are reported at length in most biographies of Copernicus since they introduce a bit of scandal into an otherwise obscure personal life.[37] In 1538 Johannes Flachsbinder (1485–1548), called Dantiscus and the son of a brewer from

[37] On Dantiscus, P. I,2, 321 ff., on Scultetus, 347 ff., on Anna Schillings, 363 ff. and B. 403–27 *passim*, 507–08. Cf. the accounts in Kesten (1945), Koestler (1959), Rosen (1971).

Danzig, became Bishop of Warmia. A protege of Sigismund, for whom he had served as ambassador to the courts of the Emperors Maximilian in Austria and Charles V in Spain, and a noted poet and correspondent with humanists and learned men, his tenure as Bishop saw a further surrendering of the independence of Warmia to the Polish Crown and an intensification of action against the Reformers, with some of whom, it should be noted, Dantiscus was on personally friendly terms. There was not much that Dantiscus or Sigismund could do about the Reformation in Ducal Prussia or even the larger cities of Royal Prussia—the King had earlier executed some radical Reformers in Danzig, but then acquiesced in an uneasy, unofficial tolerance—but clerical discipline could be tightened in Warmia where, as in the rest of Prussia, it was traditionally lax. It was not unusual for the canons of the Chapter to keep, in addition to their horses and servants, concubines, and Copernicus had living with him a younger woman named Anna Schillings, variously described as his housekeeper or cook. Several years earlier, in 1531, Bishop Ferber had reprimanded Copernicus for having a divorced woman spend a single night in his house.[38] But the relation with Anna Schillings was more serious and a source of greater scandal. And now Copernicus's predicament was shared by two other canons, Leonard Niederhoff and Alexander Scultetus, the latter a learned and close friend of Copernicus, who thought highly of him—they had earlier worked together on a map or topographical description of Prussia—but a long-time enemy of Dantiscus, whose original nomination to a Warmian canonry Scultetus had opposed. All three canons were ordered to get rid of their women, which they did, but throughout much of 1539 the situation was quite ugly, with spies and informers reporting to Dantiscus rumors of secret assignations, the women terrorized by the threat of some kind of trial and expulsion from Warmia, and Scultetus, the principal object of the Bishop's wrath, named as a Lutheran heretic, an accusation that, beginning in 1540, cost him years of trials and imprisonments in Rome.

How these humiliating accusations affected Copernicus's determination to complete his work is difficult to know; they could hardly have been helpful. But it is possible, even likely, that his reluctance to publish had come in an earlier period, so that there was really no connection here between his personal and intellectual life. For there were other, internal obstacles to his work for which there is better evidence. We have noted that after writing the *Commentariolus*, Copernicus set himself two tasks that greatly altered and delayed his intended "larger volume." One was the derivation of new parameters, upon which he then expended at least seventeen years of observation and several more years of computation, until by 1535 he had composed what he considered to be satisfactory new tables and an almanach. But Copernicus did not long rest content with his work, for the holograph of *De revolutionibus* shows substantial alterations continuing

[38] B. 315.

through later years of parameters that already may have superseded those of 1535. Further, in remarks reported by Rheticus in the preface to his *Ephemerides novae* of 1551, and thus admittedly years after the conversations took place, Copernicus told Rheticus that he would rejoice if only he could reach an accuracy of one-sixth of a degree (in either his theory or his observations), that the ancient positions of fixed stars, from which positions of planets must be measured, are not better, and worse still, that many observations of the ancients were not genuine, but were adapted to fit their particular theories.[39] Some years earlier Copernicus had criticized Werner for believing that the ancients could even err in their observations, but now he had reached the more disturbing conclusion that they had deliberately modified them.

All these doubts must have undermined Copernicus's confidence that he had achieved a planetary theory accurate for his own time and, equally important, accurate in representing the observations of antiquity and the long-period alteration of parameters indicated by those observations. These reservations, however, are never expressed in *De revolutionibus*—our knowledge depends entirely upon Rheticus's later testimony—and although they are by no means trivial, they do not appear to have been the reason that nearly led Copernicus to suppress his treatise. Copernicus could at least have confidence that his tables, being based upon modern observations, and thus containing improved modern epochs of the mean motions, were necessarily more accurate, or less inaccurate, than any other tables in use at his time. In any case, the testing of the accuracy of his work was something that would have to be left to the future, and in revising the book for publication his principal object was to make his work numerically as internally consistent as possible, whatever doubts he may have had about its foundation.

But his other goal, to provide proof, or at least convincing evidence, for the heliocentric theory and the motion of the earth, proved more intractable—as we can well understand—and in the dedication to Paul III he expresses in no uncertain terms his fear that few readers will be convinced by his arguments. It was this very natural fear that he would be unable to convince what he unfortunately called the "drones among bees" that nearly led him to give up his work. "Therefore when I pondered these difficulties with myself, the scorn, which I had reason to fear on account of the novelty and preposterousness of my opinion, nearly persuaded me that I should entirely abandon the work already begun." This of course was Copernicus's great fear, and it only makes sense if he realized, as he did, that he had not been able to prove the motion of the earth, but only argue with greater or lesser persuasiveness for its plausibility, a distinction that is crucial to understanding his difficulty. Copernicus was no fool. He knew what he could and could not do, and little service has been done to his reputation

---

[39] B. 515; text in P. II, 387–96, esp. 390–92.

by the common biographical tradition that he had thoroughly proved his case and merely feared that the rest of the world would be too stupid to understand. He was in the situation—not infrequent in the sciences, in scholarship, in law—of being certain that he was right, but lacking conclusive proof. And to make matters worse, he believed he was right about something so unusual that others would find it, not merely uncertain or doubtful, but impossible and even absurd. This was the difficulty that accounted for his reluctance to publish, and for the controversial solution that accompanied the published book.

There are few claims of authors more suspect than the prefatory commonplace that "I should never have consented to release my work for publication were it not for the urging of friends and other learned men of quality." In Copernicus's case, the claim is probably sincere. Of course since he was now old, and since his livelihood never depended upon his work in astronomy, he had nothing to lose by publishing. But for the same reasons he had nothing to gain except the ridicule of his contemporaries—Martin Luther did call him a *Narr*—and, if he turned out to be correct, a posthumous fame surpassing that of all his contemporaries except the same Martin Luther. Among the friends that urged him to publish, Copernicus names only Cardinal Schoenberg, "a man distinguished in every kind of knowledge," and next to him, "the man most friendly to me, Tiedemann Giese, Bishop of Kulm, so devoted is he to sacred and all secular learning."

Tiedemann Giese (1480–1550), Copernicus's fellow canon and closest friend, came from a wealthy Danzig family and had begun his studies at the University of Leipzig at the remarkable age of twelve. He studied further in Basel and Italy, and after he became a canon in 1502 or 1504, often worked together with Copernicus on various duties for the Chapter. Giese appears to have taken a genuine and knowledgeable interest in Copernicus's astronomical work—according to Rheticus he owned some astronomical instruments of the best quality—and more important, he seems to have believed that Copernicus was correct. And the astronomer was equally sympathetic to his friend's contributions to theology. In 1523, when Albrecht was busily encouraging Reform, Georg von Polentz, the Bishop of Samland in East Prussia, had printed in Königsberg a collection of one hundred and ten propositions, "Little Flowers" (*flosculi*, i.e. choice passages) he called them, very favorable to Luther. The following year Bishop Ferber published an edict anathemizing everyone infected with the "Lutheran pestilence," while only a few days earlier Bishop Polentz had published an edict directing his clergy to pay close attention to Luther's writings. (The sly Luther later reprinted both edicts with his own annotations.) Giese made his contribution to the dispute in the form of a reply to the *Flosculi* called the *Anthelogikon* (title in Greek), published in Cracow in 1525, that took an extremely conciliatory tone, rejecting conflict and pleading only for a reconciliation of the Lutheran and Roman faiths. In a prefatory letter dated 8 April 1524, Giese says that he was persuaded to have his work printed by Nicolaus Copernicus, "in other matters a man of sound judgment," from which it has been concluded that Copernicus shared his friend's moderate sentiments.[40]

When Copernicus published his book, it was only fair of Giese to return the favor of nearly twenty years earlier by lending his name as a recommendation. But Giese did far more than that. In 1538, when Dantiscus became Bishop of Warmia, Giese had succeded him as Bishop of Kulm, moving to Löbau in Royal Prussia where Copernicus and Rheticus visited him in the summer of 1539. In the *Borussiae encomium*, appended to the *Narratio prima*, Rheticus gives a wonderful account of discussions between the two old friends, the insistent Bishop and the reluctant astronomer, which Rheticus says he learned of from friends, an expression that could include Giese and even Copernicus.[41] Giese urged Copernicus to publish. Very well, replied Copernicus, he would imitate the Alfonsine astronomers rather than Ptolemy by publishing tables without demonstrations—an echo, perhaps of the 1535 almanach—from which learned men would be able to reconstruct the underlying principles, while common mathematicians would have a correct calculation of the motions. (This is a curious claim since Copernicus's tables in no way show that his theory is heliocentric rather than geocentric. Quite the contrary, they really suggest the latter.) Thus there would be no disturbance among philosophers, and, as was observed by the Pythagoreans, the secrets would be reserved for the initiates of mathematical philosophy. But, Giese argued, tables alone would be incomplete. One should follow the example of Ptolemy by setting out full demonstrations rather than compel assent on mere authority, which has no place in mathematics. And if the tables were indeed successful, hardly anyone would bother to investigate their principles, especially since they are diametrically opposed to the hypotheses of the ancients (and, we may add, would be undetectable in any case). Further, true philosophers would wish to examine Aristotle's arguments for the immobility of the earth, and if they are deficient, will see that a true method of astronomy must be adopted, even if it leads to the questioning of many arguments by which Aristotle refuted the opinions of Timaeus and the Pythagoreans concerning the motion of the earth. But if the philosophers are too stubborn to be convinced, one should remember that even Ptolemy, the king of this science, was condemned (wrongly) by Averroës, and as for the uneducated, the Bishop concluded,

---

[40] P. I,2, 167–87, gives a full account of this interesting episode based upon the extensive research of F. Hipler. In 1536 Giese completed a larger work, *De regno Christi*, that he sent to Erasmus and Melanchthon for their comments. After Giese's death, Stanislaus Hosius, the founder of the Jesuit College in Braunsberg and a leader of the Counter Reformation in Poland, condemned it as containing *horrendas haereses*, and it has since been lost.

[41] Text in P. II, 372–75, and in Kepler (1937–) 1, 129–31; English translation in Rosen (1971), 192–95. The dating of these conversations is uncertain. They appear to have taken place before Rheticus arrived, but it is not impossible that he was a witness to them (who else could have been?). That Giese is referred to as Bishop is probably not significant for dating.

their clamor should be ignored since virtuous men do not undertake labors for their sake. By such arguments, according to Rheticus, Giese prevailed upon Copernicus to allow learned men and posterity to pass judgment upon his work, for which, Rheticus says, virtuous men and those devoted to mathematics will be, like himself, very grateful to the Bishop of Kulm.

Rheticus says nothing of his own role in convincing Copernicus to publish, and Copernicus does not acknowledge his assistance in his dedication to the Pope, perhaps because Rheticus was a Lutheran or, as Giese later explained the omission, "due to apathy and a kind of carelessness, as he was less attentive to all that was not philosophical."[42] It is of course possible that Giese had already convinced Copernicus to publish before Rheticus arrived, but there is no doubt that Copernicus's only disciple played a crucial role in the publication of *De revolutionibus*.

Georg Joachim Rheticus (1514–1574) was born in Feldkirch—conventionally, although unfairly, referred to as a *Dorf*—in Vorarlberg, at present the western-most province of Austria.[43] His father, Georg Iserin, was a physician, astrologer, alchemist, and perhaps something of a charlatan, who was executed in 1528 on a charge of witchcraft. The son came to share his father's interests and his mercurial, even Faustian, temperament, but fortunately not his fate, although he had trouble enough throughout his restless life. He studied in Zürich, met the remarkable Theophrastus Bombast von Hohenheim, called Paracelsus, by whom the impressionable youth was profoundly affected, and then attended the University of Wittenberg, where he was encouraged to study mathematics by Philip Melanchthon himself. In 1536 he presented for his candidacy disputation the question "whether the laws condemn astrological prognostications," the answer to which was no, and in the same year he was appointed to the chair for lower mathematics, arithmetic and geometry. The chair for higher mathematics, that is, astronomy, was held by Erasmus Reinhold (1511–1553), but teaching responsibilities were not fixed, and in the following years Rheticus also lectured on astronomy and astrology. He was also known for riotous living within a circle of dissolute friends.

In the autumn of 1538 Rheticus set out on an educational journey to visit distinguished mathematicians and scholars. It was to last for three years. He went first to Nuremberg to see Johann Schoener (1477–1547), a noted astronomer, astrologer, and geographer, who at the time possessed many, if not most, of the manuscripts of Regiomontanus, Bernhard Walther (ca. 1430–1504), and Johann Werner (1468–1522). While studying with Schoener for several months, he must have met the distinguished printer of technical and scholarly books, Johann Petreius (1497–1550), and perhaps also Duke Albrecht's old friend Osiander, the minister of St. Lorenz Kirche, who

---

[42] B. 503; Burmeister (1967–68) 3, 55; P. II, 420.

[43] Burmeister (1967–68) is the most important source on Rheticus and his relation with Copernicus. For *einem Dorff*, see Doppelmayr (1730), 59 n. *yy*.

was associated with Petreius and had some kind of interest in mathematics.[44] He then travelled to Ingolstadt to see Peter Apian (1495–1552), best known for his work in geography and his elaborate *Astronomicum Caesareum* (1540), prepared for Charles V, which Kepler called an extravagant waste of time and ingenuity.[45] His next stop was Tübingen, where he met the famous Greek scholar Joachim Camerarius (1500–1574), who later procured for him the chair of mathematics at the University of Leipzig. In April of 1539 he went briefly to Feldkirch to see his old teacher Achilles Pirmin Gasser (1503–1577), a physician with interests in astronomy and geography, and in May he finally arrived in Frauenburg, where he was to spend the better part of the next two years and three months with Copernicus, his *Dominus Praeceptor*, as he called him in the *Narratio prima*.

Perhaps to ingratiate himself with the old canon, and perhaps for business reasons, Rheticus brought as a gift a number of books published in the last few years, mostly by Petreius. Those surviving at present are:[46] Regiomontanus, *De triangulis omnimodis* (1533), Apianus, *Instrumentum primi mobilis* (1534) together with Jābir ibn Aflaḥ, *De astronomia* (1534), Vitelo, *Perspectiva* (1535), all printed by Petreius, and two important first printings in Greek, Euclid's *Elements* with Proclus's commentary on Book I (Basel, 1533) and Ptolemy's *Almagest* with Theon's commentary (Basel, 1538). It is particularly notable that all the Latin texts were printed by Petreius, and were perhaps presented as a demonstration to Copernicus of the quality of his work, indicating that he and Rheticus may already have discussed the desirability of Petreius's printing whatever Copernicus may have written. There were four other publications of Petreius that could also have been of interest to Copernicus, namely, Regiomontanus's *Problemata XXIX saphaeae instrumenti astronomici* (1534), Camerarius's first Greek edition and partial translation of Ptolemy's *Tetrabiblos* (1535), Schoener's edition of the *Tabulae resolutae* (1536), and perhaps most interesting of all, Schoener's edition of al-Farghānī and al-Battānī with Regiomontanus's *Oration* on the mathematical sciences (1537). If, however, Rheticus also brought these, they have since disappeared.

Whether Rheticus went to Frauenburg on Schoener's behest, as has often been surmised since the *Narratio prima* was addressed to him, or whether it was his original intention to seek out Copernicus, is not known. By the time of his visit, it appears to have been common knowledge among astronomers and scholars in Germany that the canon of Frauenburg had invented some kind of planetary theory based on the belief that the earth moves, but much more than this was not known. It was to be Rheticus's

---

[44] Doppelmayr (1730) is still of great value on the Nuremberg mathematicians and artisans, including all those mentioned here. For Regiomontanus, the most comprehensive study is Zinner (1968), and there is an important collection of papers in Hamann (1980). A list of Petreius's printings, and it is quite impressive, is given in Shipman (1967).

[45] Kepler (1937–) 3, 142.

[46] Descriptions in P. I,2, 406–12; Czartoryski (1978), 367–68.

self-appointed or assigned role to provide more information. Curiously, Copernicus must have been happy to oblige—Giese may already have convinced him not to withhold his work—for he handed over to the twenty-five years old professor from Wittenberg his manuscript of *De revolutionibus*, and must have given him permission to write and publish an account of his work. This not incidentally shows a high degree of trust in Rheticus's ability by someone who feared misunderstanding and ridicule of his theories even when written in his own carefully chosen words. Rheticus worked fast. He claims to have studied the treatise for scarcely ten weeks, leaving it aside when he and Copernicus stayed for several weeks with Giese in Löbau. Rheticus had fallen ill and Copernicus could attempt to escape the continuing recriminations about Anna Schillings, which, as it turned out, followed him even to Löbau.

Back in Frauenburg, by 23 September Rheticus had completed the *Narratio prima* in the form of a letter to Schoener, which was printed in Danzig by March of 1540 and reprinted in Basel in 1541 with a prefatory letter by Gasser.[47] If the purpose of the *Narratio* was to inform the learned world and create a greater interest in Copernicus's work, it succeeded splendidly. It is brilliantly written—eloquent, learned, quite detailed and clear—above all enthusiastic, and even after four hundred years the most informative popularization of Copernicus's more difficult book. While most of the theory of *De revolutionibus* is at least summarized, the treatment of all subjects is not equally complete, and Rheticus planned a *Narratio altera*, which, however, never appeared. In any case, by the time the *Narratio prima* was written, Copernicus had consented to release his own work—Rheticus says so in no uncertain terms—and to judge from surviving evidence, Rheticus had already performed his task of arousing interest in his teacher's promised publication.

Arrangements for publication appear to have been soon underway, if they had not already begun. In August of 1540 Petreius in Nuremberg addressed the preface of an astrological work, *De judiciis nativitatum* by Antonio de Montulmo (late 14th century), to Rheticus, commending the excellence of the *Narratio* and expressing the wish that Rheticus would communicate to him Copernicus's observations.[48] Evidently, Rheticus's visit to Frauenburg was recognized as an important event, and it is difficult to escape the conclusion that Petreius was offering to publish Copernicus's work, if not advertising by this notice that he was already committed to do so. In the mean time, Copernicus had set to work revising the book for publication. The revision was extensive, taking the form both of alterations in the surviving holograph and then of many further changes incorporated in the fair copy from which the book was printed. The work continued, although how steadily is not known, at least into the summer of 1541, for in June of

---

[47] B. 428, 433–35, 467–68. For editions and translations, see below, p. 92.

[48] B. 442; Burmeister (1967–68) 3, 19–21. The name also appears as Monte Ulmi.

that year Rheticus wrote in a letter that his teacher was in good health and writing much.[49] Throughout most of the period from May of 1539 to September of 1541, Rheticus was staying with Copernicus, and it is possible, although far from certain, that he assisted with the revision.

The revision, although extensive, was largely technical and numerical, hardly touching the first eleven chapters of Book I where Copernicus attempted to argue for the heliocentric theory and the motion of the earth. Here he did little more than eliminate a pseudepigraphical letter from one Lysis to one Hipparchus (not the astronomer) on the Pythagorean doctrine of secrecy, which Copernicus was now violating anyway, and perhaps he eliminated his original preface, none of these changes affecting his arguments. But the problem of believing that he was correct about the motion of the earth, and knowing that he was not able to prove it, was still a source of concern, and on 1 July 1540 he wrote to, of all people, Osiander, apparently expressing his worry or even asking what could be done about the anticipated opposition of philosophers and theologians.[50] Almost a year later, on 20 April 1541, Osiander wrote to Copernicus that he considered astronomer's *hypotheses*, i.e. models, not to be articles of faith, but only bases of computation that, even if false, exactly reproduce the phenomena. He suggested that some words to this effect in the preface would disarm the criticism of peripatetics and theologians. On the same day, he wrote much the same to Rheticus, and added the interesting observation that potential critics, induced in this way to inquiring, will be more fair-minded and, after seeking in vain for something better, "will eventually give their support to the opinion of the author."

This was Osiander's plan for the presentation of Copernicus's work and for the eventual acceptance of its new theories, which he seemed to consider at least as true as any in existence, even if equally unprovable. At the time this was a reasonable, perhaps exceedingly favorable, judgment, and doubtless a sensible way of getting a fair hearing for Copernicus's theories. But Copernicus and Rheticus apparently did not agree, believing instead that, even in the absence of proof, it was better to present the new theories without qualification as a true description of the planetary system. And this was also understandable since it had always been Copernicus's object to demonstrate the correct physical structure of the planetary system, and not to devise geometrical models suitable for computation, for Ptolemy's models,

---

[49] B. 459; Burmeister (1967–68) 3, 27.

[50] B. 431, 453–54; P. I,2, 521–23. The correspondence is discussed, and Osiander's letters quoted, by Kepler in *Apologia Tychonis contra Ursum*, edited by Frisch in Kepler (1858–71) 1, 245–46, and there is an extensive literature on this subject. In our opinion, Kepler possessed only the letters from Osiander, from which he learned the date of Copernicus's letter and little or nothing more, for if he had any pertinent letters by Copernicus (or Rheticus), he surely would have quoted from them and discussed their contents, while in fact he does neither. See List (1978), 452–56.

provided with accurate parameters, already fulfilled this requirement adequately. Thus, there was no point in even presenting the new theory unless it was also considered to be true.

In fact, the two positions were both in a way correct and at the same time irreconcilable. Copernicus could no more prove the motion of the earth than he could prove the semiannual variation of the radius of Mercury's orbit or the synodic oscillations of the orbital planes of the superior planets, but even without such proof he wished his theories to be judged right or wrong, and there is certainly nothing unusual about scientific theories being accepted or rejected on the basis of incomplete, even contradictory, evidence. Osiander's opinion was equally sensible, if somewhat more cautious, namely, if you cannot prove something, you must call it hypothetical, and leave others to judge its truth by seeing whether they can come up with anything better, which is also not an uncommon path to acceptance in the history of science. Stated in this way, the opinion is valid whether Copernicus's models be considered physically or only geometrically, for even the geometry requires proof. And as it turned out, both positions did play a role in the eventual acceptance of the Copernican theory, for by the second half of the seventeenth century, even before Newton, it was almost universally acknowledged as true by the philosophers and scientists of Europe on the basis of still incomplete evidence and the discrediting or implausibility of all other theories.

In late August of 1541, Rheticus enlisted the influential assistance of Duke Albrecht, to whom Giese had the previous year sent a copy of the *Narratio prima.*[51] Copernicus had spent much of April in Königsberg giving medical attention to the Duke's advisor Georg von Kunheim, and now Rheticus dedicated to Albrecht a German Chorography, which he sent along with a map of Prussia drawn up with the help of friends, and a small instrument for determining the length of daylight. At the same time he asked that Albrecht write to the Elector of Saxony and the University of Wittenberg requesting that he be granted a further leave to deliver Copernicus's book to the printer (*das opus D. praeceptoris mei in den truk zu geben*). Albrecht was grateful to Copernicus and Rheticus, even though he found the instrument of poor workmanship and useless, and immediately dispatched the requested letters. In October Rheticus returned to Wittenberg to resume his duties. He may already have taken with him the completed revision of *De revolutionibus*, or Copernicus may have sent it to him in the course of the year. What he certainly had was a copy of the chapters on plane and spherical trigonometry (I,13–14), which he had printed in Wittenberg before June of 1542 as *De lateribus et angulis triangulorum*, Copernicus's first publication

---

[51] On these events: B. 437–38, 447–57 *passim*, 461–64, 471–76; Burmeister (1967–68) 1, 51–66 and 3, 28–42. Rheticus's German Chorography (*Chorographia tewsch.*) is published in Hipler (1876).

since the letters of Theophylactus more than thirty years earlier.[52] Rheticus added a seven-place sine table at one-minute intervals that appears to be a slight modification of Regiomontanus's table, which he could have had copied in Nuremberg when visiting Schoener, who himself had Regiomontanus's table printed by Petreius in 1541.

Indirectly, Copernicus probably had another publication in 1542, namely, the map of Prussia sent by Rheticus to Duke Albrecht, and this fairly recent discovery is worth a short digression.[53] In 1529 Bishop Ferber asked Alexander Scultetus to enlist Copernicus's help in preparing a map or description of Prussia. Exactly what the two canons did at the time is not known, but they must have carried out their assignment, for it is more than likely, and perhaps certain, that the map prepared by Rheticus, "with the help of some good gentlemen and friends, as far as was possible to me as a foreigner" (*mit hulffe, etlicher gutter herren vnd frunde, so weit mir als fromden moglich gewesen ist*), was based on the map or description of Scultetus and Copernicus. Rheticus himself was hardly in a position to assemble the information required for such a map during only two years in Prussia, and he had to rely on earlier descriptions. But there is more. In 1542 Heinrich Zell published in Nuremberg a large (48 × 70 cm) and very detailed map of Prussia and surrounding territories with over 140 towns depicted, a part of which, containing Warmia, is shown here as Fig. 2. Zell had been a student of Rheticus in Wittenberg during 1538, apparently accompanied him to Prussia in 1539, and supervised the printing of the *Narratio prima* in Danzig in 1540. It is most unlikely that the map he published in Nuremberg in 1542 is independent of the map sent to Duke Albrecht, indeed, the conclusion is inescapable that it is the same map, and was taken to Nuremberg by Rheticus or Zell in 1542. Consequently, the map of Prussia published in Nuremberg was based, at least in part, on the work done earlier by Scultetus and Copernicus. It is the first detailed map of the region ever made, and is an excellent example of mid-sixteenth century *chorography*, that is, the mapping of particular regions or countries, as *geography* is the mapping of the whole earth. Note in particular the large cities of Königsberg, Danzig, Elbing, and Torun, the towns of Frauenburg,

[52] There is a dedicatory preface by Rheticus to Georg Hartmann (1489–1564), a mathematician he met at Nuremberg who, he says, was acquainted with Copernicus's brother Andreas in Rome, and this is followed by an eighteen-line epigram, often said to be the one written by Dantiscus for *De revolutionibus* (B. 465). This is hardly possible since the epigram is definitely for a brief treatment of the elements of a *doctrina—prius est doctrina tenenda, Quam breuiter* (text: *brevirer*) *tradunt haec elementa tibi*—namely, trigonometry, preliminary to both astronomy and astrology, while *De revolutionibus* is neither brief nor elementary. Let us just admit that the authorship of this epigram is unknown and Dantiscus's epigram is lost.

[53] On the map, see Burmeister (1969) and references in n. 51. The unique copy of the Nuremberg printing is Venice, Bib. Naz. Mar., Sign.: 138c. 4., 26, and reproductions may be found in *Petermanns Mitteilungen* 73 (1927), Tafel 12, of which our Fig. 2 is a partial enlargement, and in *Imago Mundi* 7 (1951), 66. The map is sufficiently interesting and splendid to deserve a full-size reproduction.

Heilsberg, and Allenstein, and not far from Löbau, the scene of the great Battle of Tannenberg of 1410 in which the Teutonic Order was severely defeated by the combined armies of Poland and Lithuania.

In May of 1542 Rheticus delivered the fair copy of *De revolutionibus* to Petreius in Nuremberg—who actually prepared the fair copy is unknown—and by June the printing was underway with Rheticus correcting proof. In the same month Copernicus completed the dedication to Pope Paul III, which he must then have sent separately to Nuremberg.[54] Rheticus in the meantime had gone briefly to Feldkirch, but was back by early July and remained until October when he took up his new professorship at Leipzig that had been arranged for him by Camerarius and Melanchthon. On his departure, the supervision of the printing of Copernicus's book was apparently turned over to Osiander, who added the famous, unsigned prefatory note "To the reader concerning the hypotheses of this work," in which he praised Copernicus, but expressed his opinion on the uncertainty of astronomical hypotheses in language considerably stronger than in his letters to Copernicus and Rheticus. In fact, he pretty much said that astronomy is filled with absurdities, that it is essentially impossible for astronomical hypotheses to reach true causes—unless they are divinely revealed—and that anyone who takes them as true will depart from astronomy a greater fool than he entered. Whatever, if anything, had previously been agreed upon, this was definitely going too far. After the book appeared, which occurred by 21 March 1543, Rheticus and Giese became enraged at Petreius and Osiander for what they rightfully considered a breach of trust, and Giese attempted to bring an action before the Nuremberg Town Council to compel Petreius to issue a corrected edition, as Giese informed Rheticus in a letter of 26 July. Giese wished Rheticus to add a short preface, cleansing the copies already distributed from the "crime of deceit," and also add his elegantly written biography of Copernicus and his short work defending the motion of the earth against contradiction with Scripture. Unfortunately, the Town Council rejected Giese's action, and Rheticus's additions never appeared.[55]

Copernicus's reaction to Osiander's note, if he ever saw it, is uncertain, although it is not likely that he would have been pleased. According to a letter written in 1609 by Johann Praetorius, who had known Rheticus in Cracow forty years earlier, several first pages of the book were sent to Copernicus, but a little after that he died before he could see the entire work. The "first pages" would have been the first pages printed, not the unnumbered pages of the front matter, including the offending preface, which were generally printed last. Praetorius then writes, "Rheticus, however, earnestly affirmed that the preface of Osiander was clearly displeasing to Copernicus, indeed, he was not a little exasperated," meaning that he was

[54] B. 486, 481.

[55] B. 494, 503, 506; Burmeister (1967–68) 3, 54–55.

very angry, which we can well understand.[56] This is a good story—and more likely than the oft-told tale that the sight of Osiander's preface gave Copernicus's the apoplexy that brought on his death—but it was told and written so long after the event that one wonders whether Rheticus might have spoken in the conditional in attributing his own displeasure to Copernicus, who doubtless would have shared it.

It appears more likely that Copernicus never saw, or comprehended, what had been added to his book. By early December of 1542 he became seriously ill, suffering a cerebral hemorrhage and paralysis of his right side, in which condition he lingered on for several months. According to Giese, only on the day of his death, 24 May 1543, did he see his completed book, although for many days before that he had been without memory and "liveliness of mind" (*vigore mentis*), that is, he was comatose.[57]

During the last months of Copernicus's illness, he was looked after by Georg Donner, a canon to whom Giese had written on 8 December 1542, asking that he take care of him because he does not have many friends who are concerned about his health. This Donner did, and after Copernicus's death, he sent to Duke Albrecht a copy of *De revolutionibus*, later writing about the book in a letter to Albrecht of 3 August 1543:[58]

> Vnd mochte wol dasselbe D. Nicolai getichte der Schwanen Gesenge vergleichet werden, welche Im sterbenn myt den szuessen thoenen beslissen vnd aufgebenn Ir lebenn. Vnd ist wol wert der vnmehren ader vngemeynen erudition halben, das es auffgehobenn vnd behalten werde.

Chronology of Copernicus's Life

1473 Feb. 19	Born in Torun to Niclas Koppernigk and Barbara Watzenrode.
ca. 1483–85	Death of father; comes under protection of Uncle Lucas Watzenrode, who becomes Bishop of Warmia in 1489.
1491–95	Attends University of Cracow.
1496–1500	Studies canon, and probably civil, law at University of Bologna; lives with and assists Dominico Maria di Novara.

[56] Partial text in Zinner (1943), 454, and discussion by Koestler and Zinner in Koestler (1959), 169–71. Their conclusion, that Copernicus saw Osiander's note, rests on the doubtful assumption that Copernicus saw all the proof sheets of his book and was still sentient as the last ones arrived. Praetorius's letter is, to say the least, confusing. There is an extensive literature on Osiander's note—more extensive, it seems, than the serious literature on Copernicus's planetary theory— and it is not our intention to add to it. Praetorius also reports that Copernicus intended the title of his book to be *De revolutionibus orbium mundi*, which Osiander changed to *orbium coelestium*, while Doppelmayr (1730), 60 n. *a*, says that Copernicus intended simply *De revolutionibus*, to which Osiander added *orbium coelestium*. This subject also has been much discussed; cf. Rosen (1943), Copernicus (1975), 355, and Copernicus (1978), 330 as representative.

[57] B. 490, 492, 498, 503. The date 21 May has also been suggested.

[58] B. 505; P. I,2, 558.

1500 by Nov.	In Rome; delivers a lecture on mathematics.
1501–03	Studies medicine at University of Padua.
1503 May 31	Receives doctorate in canon law from University of Ferrara.
1504–10	In Heilsberg with Bishop Watzenrode.
1509	Translation of Theophylactus Simocatta published in Cracow.
ca. 1510	Invents new heliocentric planetary theory and writes *Commentariolus*; leaves Heilsberg and moves to Frauenburg.
1512	Earliest observations of planets used in *De revolutionibus*.
Mar. 30	Death of Bishop Watzenrode in Torun.
ca. 1514	Advice on calendar solicited by Paul of Middelburg.
1516 Nov.–1519 Nov.	Administrator of Benefices; resides at Allenstein.
1520 Jan.–1521 Apr.	War with East Prussia; defense of Allenstein.
1521 ca. Jul.	Returns to Frauenburg; Commissioner of Warmia.
1523 ca. Jan.–Oct.	Administrator General of Warmia; resides at Heilsberg.
1524 before Apr. 8	Encourages Tiedemann Giese to publish *Anthelogikon*.
Jun. 3	Sends to Bernard Wapowski criticism of Johann Werner's *De motu octavae sphaerae*.
1528	Final version of treatise on coinage, *Monete cudende ratio*.
1529	Latest observations, of Jupiter and Venus, in *De revolutionibus*.
1533	Johann Widmanstadt explains Copernican theory to Pope Clement VII.
1535 Oct. 15	Bernard Wapowski sends to Vienna Copernicus's almanach computed from new tables.
1536 Nov. 1	Cardinal Schoenberg invites Copernicus to communicate his new discovery, and send his studies, tables, etc.
1538 Autumn	Rheticus travels to Nuremberg to visit Johann Schoener; also meets Johann Petreius and others.
1539	Problems with Bishop Dantiscus concerning Anna Schillings.
May	Rheticus arrives in Frauenburg.
Summer	In Löbau with Rheticus and Giese.
Sep. 23	Rheticus completes *Narratio prima* in Frauenburg.

1540 by Mar.	*Narratio prima* published in Danzig.
Jul. 1	Writes to Osiander, apparently about opposition of philosophers and theologians.
Aug. 1	Petreius publicly addresses to Rheticus an invitation to communicate Copernicus's observations.
1541 Apr. 20	Osiander writes to Copernicus and Rheticus suggesting that astronomical hypotheses need not be considered true.
Jun. 2	Writing, presumably revising *De revolutionibus*.
Aug. 28	Rheticus sends map of Prussia to Duke Albrecht.
by Oct.	Rheticus returns to Wittenberg, possibly with revised text of *De revolutionibus*, certainly with trigonometry.
1542 May	Rheticus brings fair copy of *De revolutionibus* to Petreius in Nuremberg; printing begins with Rheticus correcting proof.
Jun.	Writes dedication to Pope Paul III.
before Jun. 20	*De lateribus* published in Wittenberg.
?	Heinrich Zell publishes map of Prussia in Nuremberg.
by Oct.	Rheticus departs for Leipzig; supervision of printing presumably turned over to Osiander, who adds anonymous preface on hypotheses.
before Dec. 8	Falls ill, suffering cerebral hemorrhage and paralysis.
1543 before Mar. 21	*De revolutionibus* published in Nuremberg.
May 24	Dies in Frauenburg.
July–Aug.	Giese brings action to compel Petreius to issue corrected front matter to *De revolutionibus*; action fails.

2. The Astronomy of Copernicus

Copernicus divitiarum suarum ipse ignarus Ptolemaeum sibi exprimendum omnino sumpsit, non rerum naturam, ad quam tamen omnium proxime accesserat.

J. Kepler, *Astronomia nova xiv* (1609)

Copernicus made one fundamental innovation in planetary theory, the consequences of which only became evident in the work of Kepler and Newton. In the remainder of his astronomy, he was one of the last representatives of a tradition extending from Hipparchus, or better Ptolemy, to his most direct predecessor, Regiomontanus, whose *Epitome of the Almagest* was his guide to the astronomy of Ptolemy, and may have provided the crucial step to the heliocentric theory. The tradition of Ptolemaic astronomy received, in the course of nearly fourteen centuries, many additions and modifications, of non-Ptolemaic Greek, Indian, Arabic, and last of all, European origin. Copernicus was heir to some fraction of these, but fundamentally his astronomy, in common with the most sophisticated astronomy of the intervening period, rests upon the work of Ptolemy. And even the principal ways in which he differs from Ptolemy—except for the heliocentric theory—are part of an Arabic tradition concerned more with internal problems in Ptolemy's work than with new descriptions of the motions of the planets, something that did not occur until the observational and theoretical innovations of Tycho and Kepler. The background to Copernicus's astronomy is of course the entire accumulation of observations, procedures, models, and parameters since the time of Ptolemy, in so far as they were transmitted to Copernicus. But out of this large and diverse body of material, what is the most important to consider here are the general principles of Ptolemy's mathematical and physical astronomy, the interesting modifications in the latter made by the astronomers of Marāgha in the thirteenth and fourteenth centuries, and the rebirth of a true understanding of Ptolemaic astronomy in Europe through the work of Regiomontanus.

Ptolemy's Astronomy in the *Almagest* and the *Planetary Hypotheses*

It is now a commonplace to observe that there are two aspects to Ptolemy's astronomy: the mathematical, represented in its theoretical form by the *Almagest* and its practical form by the *Handy Tables*, and the physical, which is the subject of the second book of the *Planetary Hypotheses*.[1] It should be added that there are two further divisions of Ptolemy's work in astronomy, namely, what one might call, for lack of a better term, the metaphysical, in the third book of the *Harmonics*, and the astrological in the *Tetrabiblos*, although these two, as important as they may be, are not here of concern. Somewhat more difficult to classify is a fifth aspect, likewise not directly of interest here, concerning the substances and material properties of the heavens and the earth, and the causes of motion, not unlike the treatment of these subjects by Aristotle. Ptolemy considers them only briefly, in the first book of the *Almagest* and the second book of the *Planetary Hypotheses*, and regards them rather as the fundamental principles of a science, call it natural philosophy or Aristotelian physics, that he accepts as established, without making any further contribution of his own.

[1] There are expositions of the *Almagest* in Pedersen (1974) and *HAMA*, which also contains treatments of the *Handy Tables* and the *Planetary Hypotheses*.

The *Almagest* is a comprehensive treatise on every aspect of mathematical astronomy, and its very thoroughness has left it isolated from whatever treatments of the subject preceded it, for the works of Ptolemy's predecessors fell rapidly into disuse, and have largely disappeared. Yet the more one studies the *Almagest* and compares it with the surviving fragments of pre-Ptolemaic astronomy, whether in Greek or, with some caution, in Indian sources, the more evident it becomes that even in its own day the *Almagest* was like no other work, and much of what is considered characteristic of Greek astronomy is in fact uniquely Ptolemy. Among the original contributions of Ptolemy are some that are fundamental, e.g. the development of spherical astronomy on the basis of the rigorous application of spherical trigonometry, the construction of tables of mean motions and of corrections computed rigorously by trigonometry from the geometrical models, perhaps the very invention of the equation of time, a complete strategy for the calculation of eclipses including the effects of lunar parallax, the consistent use of longitude and latitude for the cataloguing of stars, and doubtless others that, in our ignorance, we assume to be earlier Greek practice. Naturally all of these basic contributions remained a part of astronomy at the time of Copernicus, and in some instances continue in use in our day. But the part of Ptolemy's work that is of principal concern in this selective review is the progressive structure of the *Almagest*, in which each subject is built upon what has gone before, and Ptolemy's use of observations both for determining the structure of models and for deriving numerical parameters.

The most sophisticated aspect of Ptolemy's work is his use of observations, and this takes two forms, one explicit in the *Almagest*, the other not. The explicit use is for the derivation of parameters from observations in carefully selected locations that isolate the effect of one or more elements from complicating factors that cannot otherwise be eliminated. For the sun the procedure, which had previously been used by Hipparchus, is quite simple, although the observations are not, taking the time between the sun's entry into the equinoxes and summer solstice 90° apart. The observations are basic, requiring previous knowledge only of the observer's latitude in order to determine the equinox—when the zenith distance of the sun at meridian transit is equal to the latitude—but the time of the solstice is difficult to specify since the zenith distance of the sun changes hardly at all from day to day. The derivation of the parameters is a special case of the problem: given three points on the circumference of a circle, to find a point, within or without the circle, from which lines drawn to the three points will contain given angles. A geometrical solution for this problem was probably invented by Apollonius of Perga (ca. 200 B.C.), and about the middle of the next century a trigonometric solution was applied to the sun and moon by Hipparchus, who may have been the first to do so. From the observations of equinoxes and solstices, the given angles are right angles, and the intervals of time between the observations, along with an estimate of the length of the year, give the arcs of the circle, the uniform or mean motion, between the observations (see Fig. 3.13).[2] The solution for the required elements—the eccentricity, the direction of the apsidal line, and the mean distance of the sun from the apsidal line at any one observation—is then straightforward.

Only after the solar theory is established can the moon be considered. Because the moon is only about 60 terrestrial radii from the earth, it displays a parallax that can reach about 1° in the horizon, so that its apparent position, depending upon its altitude, is in general not the same as its true geocentric position. One cannot, therefore, observe the moon at points 90° apart and apply the same demonstration used for the sun, for the geocentric positions will differ in each case from the apparent positions by unknown amounts, since the lunar parallax is not yet known. In order to find geocentric positions unaffected by parallax, Hipparchus, and following him, Ptolemy, estimated the times of the midpoints of lunar eclipses when the geocentric position of the moon in longitude is almost exactly opposite the sun. Two points are here worthy of notice. First, by this method the single observational quantity wanted, geocentric longitude, has been isolated from the unwanted and, at present, unknown effects of parallax. Second, the method requires a computed true solar longitude, and thus presupposes that the solar theory is already developed. This sort of progression, from one completed theory to the next, is characteristic of Ptolemy's method, and determines the systematic order of presentation in the *Almagest*. The addition of 180° to the computed true longitudes of the sun then gives the true geocentric longitudes of the moon. But eclipses may occur in any part of the zodiac, and therefore it is necessary to solve the unrestricted problem of three points located anywhere on a circle. Further, the eclipses should be as close together in time as possible because it is necessary to estimate the mean motions of the moon in longitude and anomaly with the least possible cumulative error. Hipparchus and Ptolemy use Babylonian period relations to find the mean motions, and then solve the unrestricted problem for either an eccentric or epicyclic lunar model (Fig. 4.6).

The parameters of the lunar model were established using oppositions of the moon, but Ptolemy discovered that computed positions outside of opposition and conjunction showed various kinds of errors when compared with observations. These observations either included some kind of correction for parallax or, more likely, were restricted to the *nonagesimal*, the point of the ecliptic 90° from the horizon, where the ecliptic is parallel to the horizon, so that the moon's parallax in longitude is zero or at least very small. These are special observations, made only for a particular purpose, and they bring us to the other kind of observations mentioned earlier, those that are implicit in the *Almagest*. The implicit observations formed part of Ptolemy's preliminary work in determining the apparent motions to be accounted for and the structure of the appropriate model, and they are

[2] I.e., Chapter 3, Fig. 13. We shall use this notation to refer to figures in later chapters.

not directly reported, although the characteristic features of the motions are described and the model justified on their basis. Ptolemy had already done this in *Almagest* III,4 for the sun and in IV,5 for the first lunar model where he was presumably, but not certainly, repeating Hipparchus's analysis. But now in V,2 he explains that he has discovered the behavior of the second lunar inequality, and distinguished it from the first, by a very careful analysis of his own. He then uses the results of that analysis to determine the structure of his refined lunar model, and gives a briefer justification in V,5 for a further modification of the model. Standing behind Ptolemy's cryptic remarks is the analysis of what must have been a substantial number of selected observations from which the new inequalities could be isolated, so that the specific characteristics of the apparent motions could then determine the required model. Once this was done—and it must have been far from easy—only four carefully selected observations are necessary to derive the additional parameters of the refined model.

Why does Ptolemy omit reporting his analysis in detail? Here we can only speculate, but the reason that seems most likely is that the *Almagest* was intended as an exposition of final results reached by the most direct demonstrations, and not of preliminary analyses, which must have been less elegant, filled with numerical approximations of no permanent value, and highly redundant—much the sort of thing Kepler does report in the *Astronomia nova*. It is striking that except for mean motions, Ptolemy never gives provisional estimates of parameters, although he must certainly have used them in earlier stages of his work, but only final results, even for the first lunar theory. Ptolemy must also have believed that through his analyses he had described the phenomena accurately and had reached models that correctly reproduced them, so that this part of his work was completed once and for all, and no later astronomer would ever have to repeat it. But here he was not necessarily correct, for it was only by the kinds of analyses that Ptolemy had carried out in his preliminary work, and not by the elegant derivations of parameters reported at length, that it would be possible to improve upon Ptolemy's own descriptions of apparent motions and models in order to effect any substantial improvement in astronomy. This is an important point for understanding the relation of Copernicus to Ptolemy, for underlying all of Copernicus's work is the belief that Ptolemy's descriptions of phenomena are correct and the models that represent them at least theoretically accurate. And this belief was also held by medieval Arabic astronomers. Only later did Tycho and Kepler understand that it was really necessary to start over from the beginning without such assumptions.

Ptolemy's refined lunar model (Fig. 4.1) accounts for the second inequality by placing the lunar epicycle on a circle of rather large eccentricity that rotates such that the epicycle is at the apogee at syzygy, conjunction and opposition, and at the perigee at quadrature. A curious consequence of the model is a variation of the lunar distance in the ratio of nearly two to one between syzygy and quadrature, which should produce a corresponding variation in the apparent diameter and parallax of the moon, which is not seen to occur. Nevertheless, Ptolemy takes the model seriously, at least as regards parallax, for his parallax tables show a maximum parallax about $\frac{3}{4}^{\circ}$ too large at quadrature, and the incorrect effect on the apparent size of the moon was later frequently remarked on. A further refinement of the model is to give the mean apogee of the epicycle a *prosneusis*, an inclination, toward a point (N in Fig. 4.1) symmetrical to the center of the eccentric. And even though the center of the epicycle is located on the eccentric circle, its motion is uniform with respect to the center of the earth. Both of these properties of the model were later found objectionable.

Having established the complete lunar theory, useful at any elongation from the sun, Ptolemy next determines the apparent diameter, distance, and parallax of the moon and sun, which give all that is necessary to set out the theory of eclipses in Book VI. Next, by observing with an armillary the longitudinal separation from the moon of some number of zodiacal stars, and computing the apparent position of the moon from the lunar and parallax theories, Ptolemy establishes the longitudes of fundamental stars. Using earlier observations of Timocharis, Agrippa, and Menelaus, he confirms Hipparchus's estimate of the rate of precession of 1° per century, which is too low. Then, also using the armillary, by measuring the longitudinal distances of other stars from the fundamental stars, and by reading latitudes directly off the armillary, he compiles the catalogue of about a thousand stars in Books VII–VIII. Note again that each step depends upon what has previously been established. Only at this point can Ptolemy turn his attention to the planets because observations of the planets are made by measuring, with the armillary, their distances from zodiacal stars of known longitude, and from the apparent position of the moon.

The same technique of deriving both the model and parameters from observation is continued in the planetary theory in longitude of Books IX–XI. In IX,5 Ptolemy sets out the principal phenomena of the planets, justifying the model to which they lead in IX,6, but an important part of this analysis does not come until X,6 when he explains the motivation for the bisection of the ecentricity of the superior planets, the most remarkable, and ingenious feature of his planetary theory. He found, as he states it, that the eccentricity derived from the maximum equation of zodiacal anomaly is about twice the eccentricity derived from the length of the retrograde arc at greatest and least distance. What this means is that he carried out two different examinations of the eccentricity—necessarily of Mars, the only superior planet with an eccentricity large enough to show an obvious variation in the length of retrograde arc—one using oppositions, much as was done for the moon, the other using the variation of the length of the retrograde arc, which is strongly affected by any variation in distance. The reason for using oppositions is that the planet is at the perigee of the epicycle at opposition to the mean sun (Fig. 5.20(a)), so that the observed position of the planet also gives the location of the center of the epicycle. From three locations of the center of the epicycle, the eccentricity may be found, while the variation of the length of the retrograde arc gives a completely different method of finding the eccentricity. Ptolemy found that the eccentricity derived from the oppositions was about twice the eccentricity derived from the retrograde arcs. This is another, and very important, example of deducing the structure of the model directly from observations.

The model arrived at in the course of this analysis is shown in Fig. 5.4, in which O is the earth, C the center of the epicycle, P the planet, and the two eccentricities are respectively $OE = 2e$ derived from the oppositions, and $OM = e$ derived from the variation of the retrograde arcs. The motion of the center of the epicycle is such that its angular motion with respect to E is uniform, while its distance from M is constant, that is, the centers determining direction and distance have been separated so that C now moves nonuniformly in the circle about M. This property of the model was later found objectionable, but in fact there is a good reason for it. The point E, called in the middle ages the "equant point," corresponds to the empty focus of the ellipse in Kepler's model for planetary motion (Fig. 5.3), with respect to which the motion of the planet is very nearly uniform, while the orbital motion of the planet with respect to the center of the ellipse is truly, and not merely optically, nonuniform. Ptolemy's model, in both direction and distance, was the best earlier approximation to Kepler's first two laws of planetary motion.

But the model immediately creates a serious difficulty. The numerical values of the eccentricities derived in the preliminary analysis were quite crude since they were based, in the case of OE on the false assumption that C moves in a circle about E, as though it were located at C' in the figure, and in the case of OM on conventional estimates of the length and time of retrograde arcs, the latter being very difficult to determine with any accuracy. Thus, the initial analysis was used only for the information that one eccentricity was twice the size of the other, that is, to establish the model, and accurate values were yet to be found. This is again done using oppositions, but no longer directly by the method used previously because C' describes arcs uniformly on a circle about E—later called the "equant circle"—while the center of the epicycle C is located on a circle about M. Ptolemy's solution to this problem was an iterative procedure of some difficulty. One first assumes, falsely, that C is located at C', that the epicycle moves on the circle about the equant, and solves by the previous method for the double eccentricity $OE = 2e$. Next, this eccentricity is bisected into $OM = e$, from which one finds the location of C on the circle about M, and computes a corrected location of C' on the circle about E. This in turn is used to find a new value of $2e$, and the procedure is continued until the resulting elements can reproduce correctly the apparent motion of C between the oppositions. The convergence is rapid—two, three, or four solutions suffice to an accuracy of $0;1°$—although the computation is not, and the final results are the eccentricity, the direction of the apsidal line, and the distance of the center of the epicycle from the apsidal line at one of the oppositions.

For the inferior planets this method is inapplicable, since it depends upon the observation of oppositions, and Ptolemy had to use a less satisfactory procedure in several steps based upon inherently less certain observations of greatest elongations of the planet with the mean sun in selected locations, which, however, has the interesting property of allowing each of the eccentricities to be found separately. In the course of analyzing the greatest elongations of Mercury, he made the remarkable, and in this case incorrect, discovery that the sum of Mercury's greatest morning and evening elongations was greater when the center of the epicycle was 120° from the apogee than at the expected distance of 180°. Once again he derived the appropriate model from the observations—this time unfortunately—inventing a special, and rather complicated, model for Mercury, the peculiar properties of which were dutifully reproduced by later astronomers, including Copernicus, until the early seventeenth century. The basic principle of the model (Fig. 5.68) is that the center of the eccentric moves in a small circle such that the epicycle is closest to the earth at $\pm 120°$ from the direction of the apogee, thus accounting for the larger elongations.

Having derived the parameters of the models, Ptolemy then makes up tables of corrections between mean and true positions, and also sets out tables of mean motions, although the precise origin of the mean motions underlying the tables is not completely clear. Book XII is devoted to the method of computing the lengths and times of retrograde arcs, a rather archaic subject, and to the greatest elongations of the inferior planets. Ptolemy's procedure for finding the location of the stationary points that define the retrograde arc on the epicycle is based upon Apollonius's Theorem, but subject to the considerable complication that the motion of the center of the epicycle is no longer uniform and its distance no longer constant.

The last subjects treated, in Book XIII, are the latitudes and visibility phases of the planets, the latter again, like the retrograde arcs, an archaic topic. The most remarkable feature of the latitude theory is the variable inclination of the plane of the epicycle, which, for the superior planets, lies in the plane of the ecliptic when the center of the epicycle crosses the node, but is inclined to the plane of the ecliptic in all other locations, reaching the maximum inclination at the northern and southern limits of the inclined eccentric (Fig. 6.5). In the case of the inferior planets (Fig. 6.10), the epicycle has two separate, variable inclinations, one, called the *inclination*, in the line of sight, the other, called the *slant*, perpendicular to the line of sight, and in addition the plane of the eccentric has a variable inclination of its own. Here again, and unfortunately so, the model was virtually compelled by the observations, for the most part conventional estimates of extremal latitudes in integer degrees and of no great accuracy. Those for the superior planets were too large at opposition and too small near conjunction, while the latitudes of the inferior planets were underestimated near both inferior and superior conjunction, where the planet is very difficult to observe or entirely invisible. Ptolemy must later have discovered these errors through his own observations, for the latitude theory is modified and successively improved in the *Handy Tables* and the *Planetary Hypotheses*. In the latter, the epicycles of the superior planets remain parallel to the plane of the ecliptic, while those of the inferior planets hold a single, fixed inclination, the result being an essentially correct latitude theory by modern standards. Unfortunately, Ptolemy's modifications appear to have had little influence on later astronomers, who either used the latitude theory of the *Almagest* directly or attempted, as Copernicus did, to reproduce it in their own models.

The *Planetary Hypotheses* is later than the *Almagest*, to which it refers, and even later than Ptolemy's version of the *Handy Tables*, to judge by changes in parameters and particularly the evolution of the latitude theory. The part of Book I that survives in Greek is not, as is usually said, a summary of the models in the *Almagest*, but a quite different sort of work, concerned with the making of replicas of the models—the word "hypothesis" means "model"—out of wood or metal, with graduated circles for the analogue computation of the corrections from mean to true positions and for the direct representation of latitudes. They are early versions of what were later called "equatoria," and were accompanied by tables of mean motions, now lost, similar to those in the *Handy Tables*. In fact, the replicas are an extension of the graphical representation of corrections already described in Ptolemy's Introduction to the *Handy Tables*. The remainder of Book I, surviving only in Arabic and Hebrew, contains the theory of distances and sizes of the planets, which is based upon the assumption that spheres of successive planets are contiguous such that the greatest distance of one planet is equal to the least distance of the next higher planet—except for a space between Venus and the sun—and so on up to the sphere of the fixed stars at a distance of 20,000 terrestrial radii. After the distances have been determined, a set of apparent diameters of the planets—all too large—as fractions of the apparent diameter of the sun, allows the computation of their true diameters and volumes, taking the diameter and volume of the earth as the unit. The largest planets and stars turn out to be about $4\frac{1}{2}$ times the diameter, or 90 times the volume, of the earth.

In Book II, which likewise survives only in Arabic and Hebrew, Ptolemy describes physical models for the motions of the heavens, composed either of spheres or of sections (*manshūrāt*) of spheres cut off about the equators. A cross section of the model for a superior planet is shown in Fig. 5.5, in which the eccentric sphere, marked 3, carries within itself the epicyclic sphere, marked 4, and two surrounding spheres, 1 and 2, make the innermost and outermost surfaces of the spheres concentric to the earth so that they nest one surrounding another. The spheres are supposed to rotate just as the motions of the models in the *Almagest*, so sphere 3 must rotate uniformly with respect to E, although its axis passes through M, and sphere 4 is supposed to rotate so that the planet moves uniformly with respect to line ECH, where H is the mean apogee of the epicycle. Ptolemy does not raise the issue, but it is hard to see how the eccentric sphere can rotate uniformly about E if its axis passes through M, for the consequence is that the rotation of the sphere about its axis is nonuniform. This, in its most fundamental physical sense, is the violation of uniform circular motion that has been so celebrated, and that later proved important to the Marāgha astronomers and to Copernicus as the motivation to seek alternatives to Ptolemy's models.

From this brief and highly selective review of Ptolemy's astronomy, there are three points that are of importance for later astronomy in general and for Copernicus in particular:

1. The *Almagest* contains an explicit use of observations, with complete demonstrations, for the purpose of deriving numerical parameters. These demonstrations were to some extent understood and repeated by later astronomers.
2. The *Almagest* contains an implicit use of observations, alluded to only briefly, for the purpose of describing apparent motions and deriving the appropriate models. The very existence of these observations and procedures was on the whole ignored by later astronomers, who took as given the properties of Ptolemy's models and—if necessary, with corrected parameters—their agreement with apparent motions.
3. The *Planetary Hypotheses* contains physical representations of the models that are supposed to exist in the heavens and produce the apparent motions of the planets, but these lead to difficulties, the most notable being the violation of uniform circular motion by spheres that are required to rotate uniformly with respect to points not located on their axes, and thus rotate nonuniformly about their axes.

Arabic Astronomy and the Marāgha School

In the course of the ninth century, many of Ptolemy's writings were translated into Arabic, including the *Almagest* and the *Planetary Hypotheses*, and substantial parts of the *Handy Tables*, which now survive in Arabic only so far as they were incorporated into medieval tables. Ptolemy's writings form one source, and that the most advanced, of medieval Arabic astronomy, while other, earlier sources were translations of Indian works, in some cases transmitted through Persian intermediaries, which are lost except for citations by Arabic writers. There are a few relics of Indian astronomy that even found their way into *De revolutionibus*, in the lunar theory of Book IV, but in all other regards it is the development of Ptolemaic astronomy among Arabic writers that is here of concern. The history of Arabic astronomy is a subject of great wealth and diversity—there are far more astronomical texts in Arabic than the total number in Greek and medieval Latin—and most of the important developments in medieval astronomy were the work of Arabic writers unknown in Europe. We shall, however, restrict our review to the handful of sources that were, directly or indirectly, pertinent to the work of Copernicus. These fall into two groups, first those writers whose works were translated into Latin, mostly in the twelfth century, and became part of late medieval European astronomy, and second the astronomy of the Marāgha school that appears to have reached Europe, Italy in particular, in the fifteenth century through Byzantine Greek intermediaries.

The first group is pertinent principally to Copernicus's precession and solar theories. During the ninth century, a number of astronomers redetermined the fundamental parameters of Ptolemy's solar theory, i.e. the obliquity of the ecliptic, the length of the tropical year, the rate of precession, and the eccentricity and tropical longitude of the apsidal line of the solar model. Almost without exception they found that, compared with Ptolemy's values, the obliquity had decreased, the tropical year had decreased, the precession had increased, the eccentricity had decreased, and the longitude of the apsidal line had increased. Now, Ptolemy had considered all these parameters to have been constant, at least during the three hundred years between Hipparchus and himself, and thus in the ninth century astronomers were confronted with a difficult question. Had these parameters really changed in the intervening seven hundred or so years, or were Ptolemy's determinations simply erroneous? While this is something of an over-simplification of a considerable diversity of opinion, the conclusion that had the greatest effect on medieval European astronomy was the incorrect one: Ptolemy had not erred, for the parameters were subject to long-period variations that were necessarily periodic.

The names most closely associated with these theories are Thābit ibn Qurra (836–901), who lived in Baghdad, and az-Zarqāl (d. 1100), who lived in Toledo and Cordoba, although the most well-known works containing models and tables for the long-period variations, such as *De motu octavae sphaerae* and the *Toledan Tables*, are of uncertain attribution. *De motu octavae sphaerae* contains a model for a nonuniform, periodic "trepidation" of the sphere of the fixed stars that simultaneously changes both the rate of precession, and thus the length of the tropical year, and the obliquity. The tables from this work are contained in the *Toledan Tables*, which also contains, although in a corrupt form, a table for a variable eccentricity of the sun. Theories of variable precession underwent modifications in the *Alfonsine Tables*, in which the precession has both a uniform and periodic component, and models for such theories were known to Copernicus through the more recent works of Peurbach, Regiomontanus, and Johann Werner. The usual assumption to account for the increase of the tropical longitude of the sun's apsidal line was that it is, like the planetary apsidal lines, sidereally fixed, as in the *Toledan Tables* and *Alfonsine Tables*. Az-Zarqāl, on the other hand, gave it an independent, uniform sidereal motion, and also developed a model for the variation of the solar eccentricity, of which there is a cryptic, and misleading, account in the *Epitome*. Al-Battānī (d. 929), who lived in Raqqa, and Thābit, in the presumably authentic *De anno solis*, both find the solar eccentricity smaller than Ptolemy's and an increased tropical longitude of the apsidal line, but assume only that the apsidal line is sidereally fixed, following rates of uniform precession considerably more rapid than Ptolemy's.[3]

It should be noted that while the models for a nonuniform precession or variation of the solar eccentricity have nothing corresponding to them in Ptolemy's work, the methods by which they were discovered were not original. Rather, they were found by applying Ptolemy's procedures for the derivation of parameters, or slight modifications of them, to new observations of the sun or a few zodiacal stars, and finding results not in agreement with the *Almagest*. The theories were then invented to account for the discrepancies. To judge by Ptolemy's treatment in *Almagest* III,1 of Hipparchus's suspicion of short-term inequalities in the length of the tropical year, he would have been more skeptical, and more likely to question the accuracy of the observations, even if they were his own. Further, the models were more-or-less decided upon in advance, and then the parameters chosen to fit some number of observations, usually from the time of either Hipparchus or Ptolemy and the period of the inventor, with not much concern for what happens before, in between, or after. What is missing is Ptolemy's practice of deriving the model from the observations, rather than imposing the model upon them. The tables for precession in both the *Toledan Tables* and the *Alfonsine Tables*, which are otherwise quite different, both indicate that at some time in the future, before A.D. 2000 in the former and about A.D. 3000 in the latter, the precession will be negative, and therefore the tropical year longer than the sidereal year. Needless to say, no observational evidence could be offered for this surprising occurrence. (Perhaps it was believed that the world would not survive that long, which appears at the moment to be true.)

The other body of Arabic astronomy that is pertinent to the work of Copernicus, and in this case it is fundamental, is that of astronomers associated with the observatory of Marāgha in north-western Iran. Here the motivation was entirely different from the previous group, for it was concerned only secondarily with parameters, but primarily with the physical problems of Ptolemy's models, particularly in the *Planetary Hypotheses*. We have remarked that when Ptolemy's planetary model is physically represented as complete spheres, a problem arises in that the sphere carrying

[3] Ragep (1982) 1, 219–29, has presented evidence for believing that *De motu octavae sphaerae* is not by Thābit, but originated in Spain, where its tables were incorporated into the *Toledan Tables*. The short text has been published several times, e.g. Millás Vallicrosa (1943–50), Carmody (1960), is translated in Neugebauer (1962a), and there is much literature on it, some of awesome complexity. On the *Toledan Tables*, see Toomer (1968), and on az-Zarqāl's solar theory, Toomer (1969). Thabit's *De anno Solis* is published in Carmody (1960), translated in Neugebauer (1962a), and there is an analysis in Moesgaard (1974a). All aspects of Battānī's work are, of course, treated in Nallino (1899–1907).

the epicycle must rotate uniformly about a point, the equant point, not on its axis, so that the motion of the sphere with respect to its axis is nonuniform. The serious physical, or mechanical, problem is that there is no way of compelling a sphere, a "simple body" in the Aristotelian sense, to do this all by itself. This objection, and a good many others, were raised in the early eleventh century by Ibn al-Haytham (965–ca. 1040), who worked principally in Cairo, in a treatise called *Al-Shukūk ʿalā Baṭlamyūs* (Doubts concerning Ptolemy), a rather ill-tempered diatribe against what he perceived to be errors, contradictions, and impossibilities in the *Almagest*, the *Planetary Hypotheses*, and, rather briefly, the *Optics*.[4] Ibn al-Haytham's objections, which may not all be original, extend from quibbles to complex, and often obscure, examinations of physical difficulties in Ptolemy's models. In some cases they show his own misunderstanding of Ptolemy's intentions, as when he is disturbed by differences between the latitude theories of the *Almagest* and the *Planetary Hypotheses*—it does not occur to him that Ptolemy might have legitimate reasons for revising his work—but in others his objections are right on the mark if Ptolemy's models are to be taken seriously as physical bodies in the heavens.

And there is no doubt that Ibn al-Haytham and other astronomers wished to do so, that is, they were not content with a mathematical representation of the apparent motions of the planets using models that were to be taken only as geometry, but wished to understand the structure of the physical mechanisms, composed of rotating spherical bodies, that carried the visible bodies of the planets through their apparent motions. At some earlier time Ibn al-Haytham had written a work called *Hayʾat al-ʿālam* (Configuration of the World), in which he set out detailed descriptions of the arrangement and motions of the spheres in essentially Ptolemaic form, and without raising any objections.[5] Essentially, this was no different from what Ptolemy attempted in Book II of the *Hypotheses*. However, he must later have decided that by strict physical standards Ptolemy had failed to provide models compatible with the only possible motion of spherical bodies, uniform axial rotation. Instead he had devised models that, when examined in detail, contained nonuniform rotations of spheres, back-and-forth motions of lines directing the apogee and perigee of the epicycle toward the equant point or prosneusis point, and, if the latitude theory of the *Almagest* is considered physically, unaccountable oscillations of orbital

[4] Text edited in Sabra and Shehaby (1971); a translation by D. Voss is in preparation. On Ibn al-Haytham, and the *Shukūk* in particular, see Sabra (1972), (1978), and the class of literature of which the *Shukūk* is representative is discussed in Ragep (1982) 1, 102–07.

[5] Text, translation, and commentary in Langermann (1979). It appears that Ibn al-Haytham wrote this work before he had seen the *Planetary Hypotheses*. In a brief appendix, he sets out the basic physical principles of the motion of bodies in the heavens, but without mentioning that the models just described violate these principles. The appendix appears in only one Arabic manuscript of the text, and is not in the Latin translations.

planes. All of this, and more, is pointed out in the *Shukūk*, where Ibn all-Haytham insists that one must do better than Ptolemy by devising a physically correct planetary theory.

Note that Ibn al-Haytham's objections are physical, and are not concerned with the geometrical representation of planetary motion by Ptolemy's models—in so far as he can make such a distinction—that is, considered mathematically, Ptolemy's models are unobjectionable. Curiously, he says nothing about the great variation of distance in the lunar theory, the one case in which the model, as shown by the absence of a corresponding variation in the apparent size of the moon, is obviously incorrect. It was, of course, realized by some astronomers that Ptolemy's parameters were not perfect, that cumulative errors in mean motions should be corrected and, at least for the sun, that all parameters should be redetermined. But there was no questioning the theoretical soundness of Ptolemy's models—although eventually the problem of lunar distance was noticed—of his description of planetary phenomena, and of his procedures for the derivation of parameters. Thus, Ptolemy's models, at least in theory, were taken as given, and the problem was to find the correct mechanisms that they were representing. To take the example of the planetary model, the object was to produce by uniformly rotating spheres a motion of the center of the epicycle that was uniform with respect to the equant and at a constant distance from the center of the eccentric of Ptolemy's model, but to do physically what Ptolemy had done only schematically.

How influential Ibn al-Haytham's treatise was is not certain. It was probably not the first, and was certainly not the last, to raise such objections.[6] But even if it had little effect in itself, it was part of a tradition that, more than two hundred years later, was to produce the most innovative planetary theory of the middle ages. What ultimately happened is now well known, although much yet remains to be learned about this interesting subject. A number of astronomers associated with the observatory of Marāgha, founded in 1259 by Hulagu, the Ilkhānid conqueror of Persia, took up these problems, several of them providing physically admissible models essentially equivalent to Ptolemy's. They are now loosely referred to as the "Marāgha School," and while the initial interest in them was due to the resemblance of their planetary theory to that of Copernicus, research into their astronomy has now taken on a life of its own, which is certainly well deserved. The most noted was the Sūfī philosopher and mathematician Naṣīr ad-Dīn aṭ-Ṭūsī (1201–74). The other writers whose works have thus far been investigated are Mūʾayyad ad-Dīn al-ʿUrḍī (d. 1266), who supervised the

[6] An early, and unsuccessful, attempt to correct the problem of equant motion was made by Abū ʿUbayd al-Juzjānī (mid eleventh century) who, according to Shīrāzī, "disgraced himself." The short text has been edited and translated in Saliba (1980). Ibn al-Haytham himself wrote a work on physical models for latitude theory; see Sabra (1979).

construction of the observatory, Ṭūsī's most prominent student Quṭb ad-Dīn ash-Shīrāzī (1236–1311), and a later astronomer of the fourteenth century in Damascus, Ibn ash-Shāṭir (1304–75).[7]

Before considering Marāgha planetary theory, we shall examine a well-known invention of Ṭūsī's called by Kennedy (1966) the "Ṭūsī couple," that has various applications and exists in two forms, one for spheres with parallel axes and the other for spheres with inclined axes. It is shown in a plane representation in Fig. 3, which is, appropriately, Copernicus's diagram from *De revolutionibus* III,4, and is nearly identical to Ṭūsī's own illustration in the *Tadhkira fī ʿilm al-hayʾa*.[8] If the two equal circles with center D and F rotate such that the circle about F rotates opposite to and with twice the speed of the circle about D, the point H on its circumference will oscillate along line AB. If the axes through D and F are parallel, and the circles AB and GD equatorial great circles of spheres, the device will produce a rectilinear oscillation of H that Ṭūsī uses in his planetary and lunar models for longitude. If the axes through D and F are inclined and intersect at the center of two concentric spheres, the circles CE and GD will be small circles near their poles, and H will oscillate approximately along the arc of a great circle—provided the oscillation is small—and this Ṭūsī applies to the oscillations of the epicyclic planes in the models for latitude.

The method of the Marāgha planetary models was to break up the equant motion in Ptolemy's models into two or more components of uniform circular motion, physically the uniform rotation of spheres, that together control the direction and distance of the center of the epicycle, so that it comes to lie in nearly the same position it would have in Ptolemy's model, and always moves uniformly with respect to the equant. In a sense, Marāgha planetary theory is based upon generalizing the principle of the Ṭūsī couple to unequal, and even unconnected, circles, and in modern terms shows a keen understanding of vector addition. Four models are shown superimposed in vector form in Fig. 4, which is adapted from Kennedy (1966), in which the rotating vectors are to be understood as radii of the spheres that produce the motions. The earth is at O, the center of the eccentric in Ptolemy's model is M, the equant point E, and a point C' moves uniformly about E in a circle of radius $EC' = R$. The epicycle of the second anomaly is omitted.

---

[7] The literature on the subject of Marāgha astronomy has grown quite rapidly, and complete texts are beginning to appear. The principal publications to date are: Carra de Vaux (1893), Roberts (1957), (1966), Kennedy and Roberts (1959), Abbud (1962), Kennedy (1966)—the last five collected in Kennedy and Ghanem (1976) and in Kennedy (1983)—Hartner (1969), (1974a), (1980), Saliba (1979a), (1979b), Ragep (1982). The last is an edition of the astronomical parts of Ṭūsī's *Tadhkira*, with translation and extensive commentary, and is thus far the most penetrating analysis of the physical motivation of Marāgha planetary theory. G. Saliba has completed an edition of ʿUrḍī's *Kitab al-hayʾah*, which will be followed by a translation and commentary.

[8] Cf. Ṭūsī's illustration from Istanbul, Ms. Laleli 2116 in Hartner (1980).

The four models are as follows:

1. Ptolemy's model in which the center of the epicycle C lies at the intersection of EC' and a circle of radius R about M, where $OM = e$ and $OE = 2e$.
2. Ṭūsī's model in which C'' oscillates about C' on radius EC' over a distance $\pm e$ by the uniform rotation of two circles of radii $e/2$.
3. ʿUrḍī's and Shīrāzī's model in which C'' is placed on EC' by the uniform rotation of a radius R from M', where $OM' = \frac{3}{2}e$, and an epicycle of radius $e/2$.
4. Ibn ash-Shāṭir's model in which C'' is placed on EC' by the uniform rotation of a radius R from O, a larger epicycle of radius $\frac{3}{2}e$, equal to the eccentricity OM', and a smaller epicycle of radius $e/2$.

The common property of models 2–4 is that the same point C'' moves uniformly with respect to the equant E and lies on EC' just beyond C, while C and C'' coincide in the apsidal line.

What does all this have to do with Copernicus? Rather a lot. In *De revolutionibus* he uses the form of Ṭūsī's device with inclined axes for the inequality of the precession and the variation of the obliquity of the ecliptic, and in both the *Commentariolus* and *De revolutionibus* he uses it for the oscillation of the orbital planes in the latitude theory. In the *Commentariolus* he uses the form with parallel axes for the variation of the radius of Mercury's orbit, and by implication does the same in *De revolutionibus* although without giving a description of the mechanism. The planetary models for longitude in the *Commentariolus* are all based upon the models of Ibn ash-Shāṭir—although the arrangement for the inferior planets is incorrect—while those for the superior planets in *De revolutionibus* use the same arrangement as ʿUrdī's and Shīrāzī's model, and for the inferior planets the smaller epicycle is converted into an equivalent rotating eccentricity that constitutes a correct adaptation of Ibn ash-Shāṭir's model. In both the *Commentariolus* and *De revolutionibus* the lunar model is identical to Ibn ash-Shāṭir's and finally in both works Copernicus makes it clear that he was addressing the same physical problems of Ptolemy's models as his predecessors. It is obvious that with regard to these problems, his solutions were the same.

The question therefore is not whether, but when, where, and in what form he learned of Marāgha theory. There is evidence for the transmission of some of this material, although how much is uncertain, to Italy in the fifteenth century by way of Byzantine sources. MS Vat. Gr. 211, ff. 106v–117r, contains a short treatise with figures that is a Greek translation made about 1300 by Gregory Chioniades from an Arabic original that gives, among other things, descriptions of spherical models. The models for the sun and planets are Ptolemy's, but the lunar model is Ṭūsī's, and there are illustrations of Ṭūsī's device for rectilinear oscillation (f. 116r, lower part) reproduced here as Fig. 5, and of Ṭūsī's lunar model from the *Tadhkira* (drawn a bit inaccurately on f. 117r, upper figure) reproduced here as Fig. 6.[9] The Vatican manuscript shows only the direction of the transmission, not the source used by Copernicus, and the Marāgha planetary models are not present in Chioniades's treatise.

It is, however, certain that Ṭūsī's device was known in Italy and applied to planetary theory, as shown by a curious treatise by one Giovanni Battista Amico called *On the Motions of the Heavenly Bodies according to Peripatetic Principles without Eccentrics or Epicycles* (Venice, 1536). Amico, who was murdered at the age of 26 in 1538 "by an unknown assassin, it is believed, out of envy of his learning and virtue," was a strict Aristotelian at the University of Padua who believed that only homocentric spheres were permissible in the heavens. His models, which really do not work at all, are based entirely on Ṭūsī's device applied to concentric spheres with inclined axes for every conceivable purpose in both longitude and latitude theory.[10]

Since Copernicus spent the years 1501–03 in Padua, one might guess that it was there that he became concerned about physical objections to Ptolemy's models, and learned of Ṭūsī's device and whatever other Marāgha theory had been transmitted to Italy. Evidently it was the sort of thing that was of interest to Paduan Aristotelians, who were themselves concerned about physical problems of uniform rotation of spheres, and even homocentric spheres. It is not known what form Copernicus's information took, but it is less likely that he would have studied Greek manuscripts than that he had access to some presently unknown account in Latin, or simply had it explained to him by someone who had studied the Greek sources. A direct transmission of the Arabic is of course extremely unlikely.

European Astronomy and Regiomontanus

Although fragments of a translation of Ptolemy's *Handy Tables* dating from the sixth century, the so-called *Praeceptum canonis Ptolemaei*, are extant, there is really no astronomy of significance in Europe until the period of translation in the twelfth and thirteenth centuries made available texts such as the *Almagest*, Jābir ibn Aflaḥ's commentary on (and criticism of) the *Almagest*, and the *Toledan Tables*, translated by Gerard of Cremona, and al-Battānī's *Zīj aṣ-Ṣābi*, translated, apparently without its tables, by Plato of Tivoli. The same period saw the translation of works attributed to Thābit ibn Qurra and az-Zarqāl, a number of elementary astronomical and astrological texts, such as those of al-Farghānī and al-Qabīṣī, and some very

---

[9] The same treatise with figures is also in Vat. Gr. 1058, and without figures in Laur. Gr. 28/17. Chioniades's treatise is being edited by D. Pingree for the series *Corpus des Astronomes Byzantins*. Vat. Gr. 211 is definitely, and 1058 possibly, in the inventory of Vatican Latin and Greek manuscripts (Vat. Lat. 3954, ff. 1–75) made by B. Platyna in 1475 under Sixtus IV; cf. Devreesse (1965), 59–60. Figure 5 should be compared with Ṭūsī's figure from the *Tadhkira* in Hartner (1980).

[10] See Swerdlow (1972).

substantial astrological works such as the *In iudiciis astrorum* that Copernicus acquired while in Cracow. Still, the vast body of important Arabic astronomical writings was hardly touched during this period, and has only become known in the west through the historical investigations of the past two centuries.

The earliest works of any sophistication written in Europe, except Spain, are probably the *Theorica planetarum* by Campanus of Novara and a summary of the first six books of the *Almagest*, of unknown authorship, called the *Almagesti minoris libri VI*, which also contains additional information from the *Toledan Tables*, Jābir, Thābit, and Battānī. Both date from the late thirteenth century, as does a translation of Ibn al-Haytham's *Hayʾat al-ʿālam*, which appears to be the first Latin text containing descriptions of physical models composed of spheres.[11]

Perhaps the most widely used astronomical work introduced into Europe in this period was the *Alfonsine Tables*, originally composed from a mass of heterogeneous material—some as old as the *Handy Tables*, mostly by way of the *Toledan Tables*, and some more recent—at the court of Alfonso X in Toledo in the late thirteenth century. By the early fourteenth century the *Alfonsine Tables* reached northern Europe, where it was frequently arranged into different forms. One, which, contrary to prevailing opinion, we believe to be the original version made in Toledo, gives the mean motions sexagesimally to five or six places for single days so that they may be used with a variety of calendars and epochs, no less than ten of which are given for the meridian of Toledo, most of them never used in Europe for any purpose whatever. This form was first printed in Venice in 1483 in an early fourteenth century version with canons by John of Saxony. Copernicus owned a somewhat expanded version printed in Venice in 1492, differing in a number of particulars, with detailed instructions by Johann Santritter. More commonly, however, the mean motions were arranged into collected years, years, months, days, and hours, adjusted to various local meridians in Europe. The earliest version of this kind was also made in Toledo. Another, called the *Tabulae resolutae*, of unknown attribution and dating from the late fourteenth century, was very common in northern Europe, particularly in Germany and Poland. It was known to Copernicus while he was at Cracow, for lectures were given on it, and he copied a set of latitude tables of the sort found in the *Tabulae resolutae* into his volume of the *Alfonsine Tables*. Johann Schoener edited a version for the meridian of Nuremberg that was printed by Petreius in 1536, who printed yet another version in 1542.

Thus far we have been enumerating translations from Arabic and a few standard works of medieval European astronomy. The beginning of something new comes in the middle of the fifteenth century. A translation of the

[11] Campanus's *Theorica planetarum* has been published with a translation and commentary in Benjamin and Toomer (1971). The *Almagesti minoris libri VI* is not published, and our remarks are based upon Nürnberg MS. Cent VI,12, ff. 1–66, copied by Regiomontanus. The Latin version of *Hayʾat al-ʿālam* has been edited in Millás Vallicrosa (1942), 285–312.

Almagest from the Greek was made in 1451 by George of Trebizond, although it had little circulation until it was printed in 1528, after which it displaced the older translation from the Arabic by Gerard of Cremona, printed only once in 1515. Copernicus seems to have relied on both versions, although his copy of the 1528 edition has not survived, and he finally received the 1538 Basel edition of the Greek text from Rheticus in 1539. In Ferrara about the middle of the century, Giovanni Bianchini composed extensive tables for both spherical astronomy and planetary theory, the latter based on the *Alfonsine Tables*, but completely rearranged in format, and wrote a work called *Flores Almagesti*, which is partly an introduction and partly a commentary to the *Almagest*.

The most influential new work, however, began at the University of Vienna with Georg Peurbach (1423–61) and more so, his student and associate Johannes Müller of Königsberg, called Regiomontanus (1436–76).[12] Peurbach's first important, and by far his most popular, work was the *Theoricae novae planetarum*, originally written in 1454, but supplemented later with a description of the motion of the eighth sphere attributed to Thābit. The *Theoricae novae* contains detailed descriptions of physical models composed of spheres, rather like those described by Ibn al-Haytham, followed by geometrical descriptions using circles for the purpose of elucidating the structure and use of tables. It was very popular as a school text, and was printed dozens of times in the late fifteenth and early sixteenth centuries following its first printing by Regiomontanus in Nuremberg in about 1472. Lectures on planetary theory were given in Cracow while Copernicus was there, at which time the planetary theory must have been that of the *Theoricae novae*. The physical objections to Ptolemy's models expressed by Copernicus in the *Commentariolus* and in *De revolutionibus* apply above all to the spherical representations of the models in the *Theoricae novae*, which was doubtless Copernicus's source for spherical models in general. In about 1459 Peurbach completed an enormous set of tables for computing eclipses, the *Tabulae eclipsium*, first printed in Vienna in 1514, although Copernicus had earlier copied parts of them into his volume of the *Alfonsine Tables*, preumably while he was in Cracow. The eclipse tables are based upon the solar and lunar theory of the *Alfonsine Tables*, but Peurbach recomputed everything required for eclipses in greatly expanded form, particularly, extensive tables for finding the time of true conjunction or opposition, and parallax tables—for northern Europe only—for finding the time of apparent conjunction and the apparent longitude and latitude of the moon in solar eclipses.

In 1460 Cardinal Johannes Bessarion (ca. 1403–72) came to Vienna as a legate of Pope Pius II, among other reasons, to seek aid for a crusade against

[12] Königsberg is a small town in Lower Franconia, about 30 km north-west of Bamberg. On Peurbach and Regiomontanus, the most comprehensive study is Zinner (1968), and there are a number of interesting papers in Hamann (1980).

the Turks to recapture Constantinople. There he met Peurbach and Regiomontanus, and enlisted their assistance in his more modest project to prepare a new translation of the *Almagest*. Bessarion did not like George of Trebizond's translation, and he did not like George of Trebizond either. They came from the same town on the coast of the Black Sea, and worse, Bessarion was a Platonist who did not approve of Trebizond's interpretations of Plato, which were critical and Aristotelian. Bessarion also wished to produce an abridgement of the *Almagest* for instructional use, and Peurbach, who Regiomontanus says knew the *Almagest* almost by heart, began writing the abridgement, but died after completing only the first six books. On his deathbed, he made Regiomontanus promise to complete the rest, which was done, and by 1463 Regiomontanus, now in Italy, dedicated the completed *Epitome of the Almagest* to Bessarion. The *Epitome* was printed for the first time in Venice in 1496 in a rather poor edition, and was later reprinted in Basel in 1543 and Nuremberg in 1550. Although it was not served well by its printers, the printing and wide distribution of the *Epitome* was of the greatest importance for all serious astronomy in the sixteenth century, and particularly for Copernicus. While the *Almagest* had been available since the late twelfth century, there is little to show that Ptolemy's work was in any real sense understood until Regiomontanus, for it is to him that most of the credit belongs, prepared this lucid exposition of the entirety of Ptolemy's mathematical astronomy. Although Copernicus's own copy of the book has not been discovered, it is clear that he used the *Epitome* constantly, sometimes in preference to the *Almagest*, and during the early period of his work he does not appear to have known the *Almagest*. Thus, the general features of the *Epitome* are worth a description.

Except for the introductory section of Book I, the first six books of the *Epitome* follow rather closely the *Almagesti minoris libri VI*, leading to the suspicion that Peurbach stayed very close to this work and any differences are probably due to Regiomontanus. The text of the *Almagest* underlying all of the *Epitome*, except the first propositions of Book I, is the Gerard of Cremona translation, while the beginning propositions are based upon the Greek text, and must have been the work of Regiomontanus, perhaps with help from Bessarion, when he was in Italy.[13] Regiomontanus also did what he could to emend errors in the text of Gerard of Cremona, but apparently without reference to the Greek text and not always successfully. As in the *Almagestum minus*, the *Epitome* breaks up the chapters of the *Almagest* into propositions in which a model is described, a theorem proved, a parameter derived, and such, and procedures are often explained in general form where Ptolemy gives only his numerical work. This is

[13] The earliest manuscript of the *Epitome*, Venice, lat. 329, omits the propositions prior to the trigonometry, and preserves a version of the text preceding that found in other manuscripts and the 1496 printing, while Venice, lat. 328 contains a later revision presented by Regiomontanus to Bessarion. We have used both lat. 328 and the 1496 Venice edition.

particularly notable in the spherical astronomy of Book II and in the derivation of parameters for the solar, lunar, and planetary theories, and for didactic purposes the *Epitome* is often clearer than the *Almagest*. Occasionally Regiomontanus finds something in the *Almagest* of particular interest or worthy of a more detailed treatment, an example of some significance here being the description of the eccentric model for the second anomaly of the planets, mentioned briefly by Ptolemy in *Almagest* XII,1, and expanded considerably in *Epitome* XII,1–2. There is also a fair amount of material from later sources, most of all from Jābir ibn Aflaḥ, but also from Thābit, Battānī, az-Zarqāl, the *Toledan Tables*, some of which, in Books I–VI, was already in the *Almagestum minus*.

This is not the place to review Regiomontanus's many contributions to astronomy, some of a very practical sort, and we shall confine our comments to only a few points relating to Copernicus. In 1467, during his period in Hungary, Regiomontanus prepared a large set of tables, called the *Tabulae directionum*, for spherical astronomy and particulary for those parts of astrology, such as the determination of the twelve houses, that depend upon spherical astronomy. Copernicus acquired, while in Cracow, the 1490 Augsburg edition of the *Tabulae directionum*, which is bound in the same volume with his 1492 Venice edition of the *Alfonsine Tables*, and he used the *Tabulae directionum* in Book II of *De revolutionibus*. Regiomontanus also computed a seven-place sine table at intervals of one minute that appears to be the basis of the sine table Rheticus had printed with Copernicus's *De lateribus et angulis triangulorum*, which is the separate printing of *De revolutionibus* I, 13–14 that appeared in Wittenberg in 1542. By far the largest and most important work on trigonometry was Regiomontanus's *De triangulis omnimodis libri quinque*, first printed by Petreius in 1533 and brought to Copernicus by Rheticus. Books I–II are on plane triangles, Books III–V on spherical triangles, and it is difficult to imagine a more comprehensive treatment of trigonometry as it was once understood purely geometrically. Copernicus is thought to have used Regiomontanus's book in revising his own chapters on trigonometry, and while this is not impossible, errors and a general inferiority of Copernicus's exposition show that he could not have studied *De triangulis* all that carefully.

In 1471 Regiomontanus moved to Nuremberg, and there set about an ambitious plan for the improvement of astronomy and mathematics: to publish good editions and translations of the principal astronomical and mathematical works, both of antiquity and the modern era; to manufacture astronomical instruments of improved design; to undertake a comprehensive series of observations of the sun, moon, and planets; to derive improved elements on the basis of the observations; to compute from the improved elements new tables and ephemerides.

The plan for printing editions and translations, often characterized as an example of humanism in the sciences, was the most comprehensive venture in scientific publishing yet undertaken, and even to this day not all of the more than fifty titles listed in Regiomontanus's prospectus from about

1474 have been printed. Among the ancient works are a new translation of the *Almagest*, Theon's commentary, a new translation of the *Geography*, the *Tetrabiblos*, *Optics*, *Harmonics* with Porphyry's commentary, treatises on spherical astronomy by Menelaus and Theodosius, the works of Archimedes with Eutocius's commentaries, Apollonius's *Conics*, and Aristotle's *Mechanical Problems*. (It is not altogether clear whether all of these were to be editions or translations.) Among the modern works are Peurbach's *Theoricae novae*, the *Epitome*, and a number of Regiomontanus's own works including *De triangulis omnimodis*, the *Tabulae directionum*, *Tabulae primi mobilis* (double entry tables for direct solutions of right spherical triangles), a criticism of George of Trebizond's translation of and commentary on the *Almagest*, commentaries on Ptolemy's *Geography* and on the works of Archimedes lacking commentaries by Eutocius, and ephemerides for no less than thirty-two years. The whole enterprise was terminated by Regiomontanus's death in 1476, by which time only a few works had appeared, among them Peurbach's *Theoricae novae*, Manilius's *Astronomica*, Regiomontanus's *Disputatio contra Cremonensia deliramenta* (a criticism in dialogue form of the old *Theorica planetarum* attributed to Gerard of Cremona), and *Ephemerides* for the years 1475–1506, giving daily positions and aspects of the sun, moon, and planets. After Regiomontanus's death, his manuscripts were in the possession of his associate Bernhard Walther (ca. 1430–1504), and following Walther's death they were used by Johann Werner (1468–1522), Willibald Pirkheimer (1470–1530), and Johann Schoener (1477–1547), and a number of works from the manuscripts were published in Nuremberg, principally through the efforts of Schoener. Some of the manuscripts have since been dispersed or lost, but many are still in the Nuremberg Stadtbibliothek at the present day.

The program of observations had barely begun at the time of Regiomontanus's departure in July of 1475 for Rome, where he died the following year. They were, however, continued with great dedication by Bernhard Walther for fully thirty years, from 1475 to 1504, the earliest surviving series of nearly continuous observations. At first Walther observed only with a parallactic instrument, with which he took hundreds of zenith distances of the sun at meridian transit, and a *rectangulus* or *radius astronomicus*, an instrument for the measurement of angular separations of up to a few degrees, which he used for hundreds of observations of distances of planets from fixed stars. Only in 1488 did he begin using an armillary for the direct measurement of the longitude and latitude of planets. The earliest known use of Walther's observations is by Copernicus, who used three observations of Mercury from 1491 and 1504 for confirming the location of its apsidal line and correcting its mean motion. Walther's observations were first published in Nuremberg in 1544 by Schoener, along with earlier observations of Regiomontanus and Peurbach.

Since Regiomontanus did not carry out his plan for observations, new elements could not be derived, nor could the new tables and ephemerides be computed. The ephemerides published by Regiomontanus were presumably based upon the *Alfonsine Tables* in one form or another, or on Bianchini's adaptation, and they were continued in the same way by Johann Stoeffler and Jacob Pflaum for the years 1507–31, and in a second series for 1532–51. These ephemerides were reprinted many times—no astrologer could be without them—and Copernicus received after 1530 a 1492 Augsburg edition for the years 1491–1506 and a 1499 Ulm edition for 1499–1531.[14]

Regiomontanus had a clear and practical understanding of what was necessary for the improvement of astronomy, beginning with the basic subjects of trigonometry and spherical astronomy, and working on through observations to the derivation of improved elements and the computation of new tables and ephemerides. But most important of all, by the writing of the *Epitome*, the mathematical astronomy of Ptolemy was, in a sense, reborn. Through the eventual publication of the *Epitome* and his works in trigonometry and spherical astronomy, it became possible to do what Regiomontanus intended, but did not even begin: to take up planetary theory that, at least in Europe, was still much as it had been left by Ptolemy some fourteen centuries earlier. It is at this point that we turn to the work of Copernicus.

Early Period to the Writing of the *Commentariolus*

We begin with a summary of the period in Cracow and Italy:

Although the subjects studied by Copernicus at the University of Cracow during the period 1491–95 are not known, it is reasonable to presume that he attended some of the lectures given on spherical astronomy, planetary theory (probably Peurbach's *Theoricae novae*), the *Tabulae resolutae*, eclipse tables (again probably Peurbach's), geography, and astrology. What is certain, however, is that while in Cracow he acquired copies of the *Alfonsine Tables*, the *Tabulae directionum*, Euclid's *Elements*, and ar-Rijāl's *In iudiciis astrorum*. There is no reason to believe that he had not become proficient in the subjects of these books, i.e. spherical astronomy, planetary and eclipse calculation, geometry, and astrology. On pages bound into the volume containing the *Alfonsine Tables* and the *Tabulae directionum*, he copied out parts of Peurbach's *Tabulae eclipsium* and a large set of planetary latitude tables related to those in the *Tabulae resolutae*, for those in the *Alfonsine Tables* are very brief. He also made a number of notes on these

[14] There are facsimiles of Regiomontanus's prospectus in Zinner (1968), Abb. 45, Harrassowitz (1971), 30, and Regiomontanus (1972), 533, the last a collection that also contains facsimiles of the *Epitome* (Venice, 1496), *De triangulis* (Nuremberg, 1533), Schoener, *Scripta* (Nuremberg, 1544), Peurbach, *Theoricae novae* (Nuremberg, ca. 1472), *Ephemerides* for 1475 (Nuremberg, 1474), and other works. The observations of Walther, Regiomontanus, and Peurbach are printed in Schoener (1544), which also gives descriptions of the parallactic instrument, 23–26, and the *rectangulus* or *radius astronomicus*, 34v–35v. On Walther's observations, see Kremer (1980), (1981), (1983). Copernicus's copies of the *Ephemerides* for 1491–1506 and 1499–1531 came from the estate of Hildebrand Ferber; cf. Czartoryski (1978), 371.

pages, one page of which, certainly written several years later, is directly related to the *Commentariolus*.

Of Copernicus's concern with astronomy during his years in Italy, 1496–1503, we know only that he assisted Dominico Maria di Novara in Bologna, gave some kind of lecture on mathematics in Rome, and made some observations recorded in his notes or later used in *De revolutionibus*. Nevertheless, there is evidence that some account of Marāgha planetary theory was transmitted to Italy in the fifteenth century through Byzantine Greek sources, that it reached both Rome and Padua, and that it was of interest to Aristotelians concerned with physical problems of uniform circular motion of spheres, and even with homocentric spheres. It is thus probable that whatever Copernicus learned of Marāgha theory, and perhaps also his concern with physically permissible planetary models, were the result of his period in Italy, more specifically in Padua during 1501–03, which thus made a substantial contribution to his later work.

The evidence for his early work in planetary theory comes entirely from the *Commentariolus*, written certainly before 1514 and probably about 1510, and from the page of notes just mentioned in the volume of the *Alfonsine Tables*.[15] Copernicus explains in the *Commentariolus* that his concern in planetary theory was the representation of apparent motions by uniform circular motions, a principle violated by Ptolemy's models, which, however, are satisfactory for purposes of computation. This criticism is directed principally against the physical representation of Ptolemy's models as complete spheres in Peurbach's *Theoricae novae planetarum*, the form in which Copernicus and his contemporaries envisioned Ptolemy's models. Hence, his initial interest was in the same physical problems that guided the Marāgha astronomers, and not with problems of observations and the numerical accuracy of the astronomy of his time, whatever he might then have known about the failings of the *Alfonsine Tables* and its derivatives. In one sense, the solution to his problem was already at hand in the Marāgha theory, but Copernicus did a good deal more, and made the models in the *Commentariolus* heliocentric, as originally stated in a series of seven postulates that assume the heliocentric arrangement of the planetary system. But there is another document that contains an earlier statement of the heliocentric theory, or something very close to it.

The page of notes in the volume of the *Alfonsine Tables*, shown here in Fig. 7, is prior to the *Commentariolus* and provides the principal evidence for how Copernicus reached the heliocentric theory. It contains two sets of numbers. The lower, with the heading *proportio orbium celestium ad eccentricitatem 25 partium* (proportion of the heavenly spheres to an eccentricity of 25 parts), is the source of the parameters of the models in the *Commentariolus*, that is, the radii of the spheres and the two epicycles of each planet

---

[15] The following summary of Copernicus's derivation of the heliocentric theory and the contents of the *Commentariolus* is based on Swerdlow (1973).

and of the moon. The upper set of numbers is the source of the lower. It gives the radii of the epicycles and a quantity called the *eccentricitas* for each planet—although Venus is omitted—that are derived from the *Alfonsine Tables*, for the epicycles from the maximum equations of center, and for the *eccentricitas* from the maximum equations of the anomaly. The number called the *eccentricitas* is what would usually be called the radius of the epicycle that determines the equation of the anomaly in Ptolemy's models. But that Copernicus calls it an *eccentricity* indicates that he was at the time working with an alternate, eccentric model of the second anomaly that Ptolemy refers to briefly in *Almagest* XII,1, but that Regiomontanus describes in detail and supplements in *Epitome* XII,1–2. The model, in the two forms described by Regiomontanus, for the superior and inferior planets, leads respectively to the Tychonic and Copernican forms of the heliocentric theory, and Copernicus's use of the eccentric model in the upper set of numbers indicates the analysis that he followed, doubtless beginning from Regiomontanus's description, to the heliocentric theory.

Copernicus probably undertook an investigation of the second anomaly, and of the eccentric model, because even with the Marāgha solution to the first anomaly, the uniform motion of the planet on the epicycle must still be measured from the mean apogee lying on a line directed to the equant (see Fig. 5.53 for Venus). Thus, technically there is still a violation of uniform circular motion, or in physical terms, of the uniform rotation of the epicyclic sphere.

The epicyclic and eccentric forms of the model for the superior planets are shown in Fig. 8(a). The fundamental property of the epicyclic model is that the radius CP of the epicycle to the planet is always parallel to the direction OS from the earth to the mean sun. By completing the parallelogram $OCPN$, ON will be parallel to CP and lie in the direction from the earth to the mean sun, and NP will be parallel to OC. Describing circles with radii ON and NP, P will lie on the circumference of an eccentric with center N, and the epicyclic radius r has become the eccentricity e, referred to in the upper numbers in Copernicus's notes. The transformation to the heliocentric model in its Tychonic form is shown in Fig. 8(b), in which it is only necessary to assume that the mean sun S actually lies on the circle about O. Consequently, as in Copernicus's lower set of numbers, the eccentricity $e = 25$ parts becomes the radius r of the mean sun's orbit about a fixed earth, and R is the radius of the planet's orbit, which is now measured in terms of e. This is advantageous because it gives the relative distances of the superior planets in terms of the distance between the earth and the mean sun.

In *Almagest* XII,1 Ptolemy says that the eccentric model is only possible for the superior planets, but in *Epitome* XII,2 Regiomontanus describes a corresponding model for the inferior planets, which is shown here in Fig. 9(a). The mean sun S now lies in the extended direction OC from the earth to the center of the epicycle. By completing the same parallelogram $OCPN$ as in the previous model, P lies on the circumference of an eccentric with center N, and the radius of the epicycle r again becomes the eccentricity e, referred to for Mercury in the upper set of numbers. In order to transform this model to its heliocentric form, this time Copernican as shown in Fig. 9(b), the earth and the planet are moved in parallel through the distance R, and the mean sun in the same direction to the center of the orbits of the planet and the earth. In accordance with Copernicus's terminology in the lower numbers, the radius of the earth's orbit R is equal to the eccentricity $e = 25$ parts, as in the model for the superior planets, while the eccentricity e of the model in Fig. 9(a) becomes the radius r of the planet's orbit. Once again, the radius of the planet's orbit is measured in units of the distance between the earth and the mean sun.

These transformations, however, have given two different models, the Tychonic for the superior planets and the Copernican for the inferior, and Copernicus's notes offer no evidence either for how he made his choice, or even whether he had yet done so. Thus, in the lower numbers $e = 25$ parts could still refer to the orbit of the mean sun about the earth, which would be r in Fig. 8(b) for the superior planets and R in Fig. 9(b) for the inferior. In order to find a motivation for his decision, we must return to Copernicus's physical concerns about the proper motions of spheres. The spheres must not only rotate uniformly about their axes, but they must also be free to do so without interference from each other, as in the nested sphere arrangement in Ptolemy's *Planetary Hypotheses* and all later descriptions of the spheres of the planetary system. Now, there is a peculiarity of the Tychonic system quite evident in any illustration, namely, that the orbit of Mars intersects the orbit of the mean sun. This is shown in Fig. 10, which is Tycho's illustration of his new system from *De mundi aetherei recentioribus phaenomenis* (1588), his book on the comet of 1577.[16] The intersection follows necessarily because the radius of Mars's orbit is about $\frac{3}{2}$ the radius of the sun's orbit, and thus Mars must cross over to the other side of the earth and reach opposition, which Mercury and Venus do not, but not clear the sun's orbit around the earth, as Jupiter and Saturn do. Consequently, the orbits of Mars and the mean sun must intersect.

This was not objectionable to Tycho *after* his analysis of the motion of the comet of 1577 convinced him that there were no solid spheres in the heavens, as he had previously believed. But until that time, Tycho did not believe such an arrangement possible precisely because of what he called the "absurdity" of the intersection of the spheres of Mars and the mean

[16] Brahe (1913–29) 4, 158. Note that Mars is shown at opposition so that it is inside the sun's orbit around the earth. In a system of solid spheres, Venus and Mercury would not intersect the sphere of the sun since the epicyclic sphere of Venus contains the epicycle of Mercury and the mean and true sun, the latter on an epicycle. However, the intersection of the spheres of Mars and the mean sun—actually the entire sphere carrying the epicycles of the sun, Mercury, and Venus—cannot be avoided. The intersection of spheres in the Tychonic theory, and Kepler's opinion that Copernicus's belief in the "reality of the spheres" precluded his considering it, is further discussed in Swerdlow (1976), 134–37.

sun, which are not merely the circles shown in Fig. 10, but solid bodies filling all the space within the circles, and even encompassing the epicyclic spheres of Venus and Mercury about the sun. What was objectionable to Tycho as long as he believed in solid spheres would likewise have been objectionable to Copernicus, for whom the intersection of rigid, material spheres, of solid spheres, would also be an absurdity. Thus, by his very concern for a correct representation of the physical structure of the heavens, Copernicus was forced to reject the Tychonic arrangement in favor of the motion of the earth around the sun, with all the terrestrial, physical problems that arise from the diurnal rotation of the earth, which is impossible to avoid if the earth is moving about the sun.

There is a related point here that may at first seem surprising. The Tychonic theory is not only incompatible with solid spheres, but even more so with the nested spheres of the Ptolemaic system, since part of the sphere of Mars would overlap, and cut through, the sphere carrying the epicycles of Venus, Mercury, and the sun. Ptolemy had assumed that the spheres were contiguous for two reasons, first that it made sense physically that there should be no empty spaces, "or any meaningless and useless thing," and second because it provided a reasonable assumption for finding the distances of the planets. One of the principal reasons Copernicus adopted the heliocentric theory is that it gives the distances of the planets from the mean sun unambiguously, and it gives them without making any assumptions about nested spheres. Even so, it appears that Copernicus still believed, but entirely for physical reasons, in the contiguity of the spheres of successive planets. The evidence for this is not in the *Commentariolus*, but in *De revolutionibus* (although it is hardly possible that Copernicus only thought of nested spheres when writing the later work), and it comes from the well-known diagram of the entire planetary system in I,10. This is shown in Fig. 11(a) from Copernicus's manuscript and in Fig. 11(b) from the Nuremberg edition.[17] In each case there are a series of concentric circles about the sun with captions for each planet and the stars, but by counting, it is evident that there are too many circles for the captions to apply to the circles. Therefore, the captions apply to the spaces between the circles, and these spaces are a series of contiguous spheres—not drawn to scale—of the successive planets, just as in the Ptolemaic system of the *Planetary Hypotheses* and later medieval cosmology. Hence for the physical reasons, that the solid bodies of the spheres must be able to turn freely without intersecting, and that the spheres are contiguous to each other with no empty spaces, Copernicus could not consider the Tychonic system, even though he must have been aware of it as one transformation of the eccentric model of the second anomaly.

---

[17] The captions for the stars and the superior planets in Fig. 11(b) are one space too high, leaving an empty space above the earth. Presumably this was an error in making the woodcut. Cf. Swerdlow (1976), 127–29.

The analysis set out here shows, so far as the evidence permits, *how* Copernicus arrived at the heliocentric theory and the motion of the earth. It does not, however, explain *why* he chose to adopt it as a true representation of the planetary system. This, of course is an old question on which much has been written, mostly without concern for how Copernicus reached his theory in the first place. Some rather far-fetched answers have been given, with a lot of hand-waving in the direction of Neoplatonism, Hermes Trismegistus, and (in Warmia under the eyes of Uncle Lucas?) sun-worshipping. Although one could perhaps say that anyone in 1510 who was capable of believing that the earth moved was capable of believing anything—and there is no telling what strange things Copernicus might have believed—it seems to us that there is no foundation for these claims. Among other reasons, they are based on the highly anachronistic belief that the heliocentric theory and the motion of the earth were entirely obvious and there for the taking if only one had the correct metaphysical or mystical faith. But this is simply untrue. Copernicus arrived at the heliocentric theory by a careful analysis of planetary models—and as far as is known, he was the only person of his age to do so—and if he chose to adopt it, he did so on the basis of an equally careful analysis.

His reasons are not unknown, for they are mentioned in *De revolutionibus*, in part in the *Commentariolus*, and he must have comprehended most of them in the course of investigating the consequences of his new planetary theory. As they have been pointed out so many times before, we need only review them briefly. The heliocentric theory—philosophical quibbles aside—gives the order and distances of the planets unambiguously and under the reasonable assumption that the equation of the anomaly shows the ratio of the radii of the planet's and earth's orbits. In so doing, it makes the planetary system into a single whole in which no parts can be arbitrarily rearranged. By contrast, in the geocentric theory the radii of the eccentrics and epicycles are known only relatively, one planet at a time, and only by additional assumptions, such as the contiguity of successive spheres, can the order and distances of the planets be determined. Certain features of planetary models that are unexplained in geocentric theory are understood as the necessary consequences of the transformation from a heliocentric to a geocentric arrangement: why the radii of the epicycles of the superior planets remain parallel to the direction from the earth to the mean sun; why the centers of the epicycles of the inferior planets lie in the direction of the mean sun; why planets closer to the earth's orbit, Mars and Venus, have longer synodic periods, longer periods of invisibility (at least for Mars), larger equations of the anomaly (i.e. relatively larger epicycles), and larger retrograde arcs (although this is actually more complicated) than planets farther from the earth's orbit, Jupiter and Saturn. All of these reasons are given by Copernicus, mostly in *De revolutionibus* I,10 and some of them in the explanation of the second anomaly of the superior planets and Venus in the *Commentariolus*.

Of course one question still remains, a fairly obvious one, which is why these reasons convinced Copernicus when, until the following century, they convinced hardly anyone else. One can immediately suggest that it is easier to convince yourself than to convince others. But this can be amplified slightly to an answer that is not so trivial as it appears, namely, that the new planetary theory that Copernicus worked out himself did just what he wanted. Copernicus's goal was, like the Marāgha astronomers, to give a complete and physically correct description of the planetary system. With the models of the Marāgha astronomers he was able to go so far towards this goal, and with the invention of the heliocentric arrangement he was able to go a good deal farther, and neither Ptolemy or Ṭūsī would have objected to his holding that—physical objections to the motion of the earth aside—what worked and fit together so well could not be so far wrong. And that is what, correctly as it turned out, Copernicus himself believed.

The *Commentariolus* is very concise, and gives no information about how Copernicus developed his theories. After stating his objections to the violation of uniform circular motion in Ptolemy's models, he says that it can be preserved, and every apparent irregularity will follow, if seven postulates are granted. These have nothing to do with the problem stated, but instead set forth the basic assumptions of the heliocentric theory, and assumptions they indeed are because Copernicus had no way of proving them, although he certainly believed them to be true. The best he can do is say that there will be considerable evidence for the motion of the earth in the explanation of the circles, that is, that the models are their own best evidence. This doubtless is what convinced him, but it entirely avoids the issue of proving the motion of the earth against the physical objections that can be raised by Aristotelians, and the astronomical arguments given by Ptolemy, that the earth is at rest at the center of the world. His answer to the latter, in the fourth postulate, is that the sphere of the fixed stars is so remote that none of the effects of removing the earth from the center will be detectable. This, of course, is only an assumption.

After enumerating the order and periods of the planetary spheres, he takes up the three motions of the earth: the annual revolution of the Great Sphere (*orbis magnus*), the daily rotation, and the precession, which he calls the "motion of the inclination" (*motus declinationis*). The model for the annual motion of the earth is a simple eccentric with a maximum equation of $2\frac{1}{6}°$, taken from the *Alfonsine Tables*, and a sidereally fixed apsidal line 10° west of either Castor or Pollux, probably the former, although the solar apogee in the *Alfonsine Tables* is 11;47° west of Castor. The eccentricity derived from the maximum equation is rounded from $\frac{1}{26;28}$ to $\frac{1}{25}$, and assigning the great sphere a radius of 25 parts, Copernicus then measures all the planetary spheres and epicycles in these units. This unified measure of all parts of the system is not repeated consistently in *De revolutionibus*, where eccentricities and epicycles are given only in terms of the unit radius of the eccentric.

The daily rotation follows the annual motion, doubtless because Copernicus reached it as a consequence of the heliocentric motion of the earth, and did not begin his investigations into planetary theory with a prior assumption that the earth rotates. For physical reasons, it is a rather unpleasant consequence, and Copernicus does not attempt to deal with it until *De revolutionibus*. The last of the earth's motions, the precession, accounts for the slow westward motion of the equinoxes with respect to the fixed stars, and thus for the difference between the sidereal and tropical years. In the next section, Copernicus treats the variation in the length of the tropical years given by Ptolemy, al-Battānī, and the *Alfonsine Tables*, concluding that in every case the corresponding value of the precession will give a constant sidereal year of about 365 days $6\frac{1}{6}$ hours. He points out further that the sidereally fixed apsidal lines of the planets show that all motions should be measured sidereally rather than tropically. Here again, his reasoning has a physical basis. The sidereal rotations of the great sphere and of the planetary spheres are their true motions, while the motions with respect to the equinox are accidental, and depend upon the momentary position of the earth's equator, which must in any case move nonuniformly. However, his model for the precession, a westward motion of the earth's axis completing a rotation in a tropical year while the great sphere rotates eastward in a sidereal year, will give only a uniform precession. Copernicus says that he has not yet worked out the rule of this motion, meaning a complete theory of the precession, and the simple model in the *Commentariolus* is a long way from the complex mechanism for the nonuniform precession and variation of the obliquity in *De revolutionibus*.

Ptolemy's lunar model (Fig. 4.1) is criticized both for its violation of uniform circular motion and for its unnaturally large variation of distance, which would cause variations in the apparent size and parallax of the moon that are not observed. Copernicus's own double-epicycle model (Fig. 4.2), which greatly reduces the variation of distance and preserves uniform motion, is that of Ibn ash-Shāṭir, although the parameters are derived, with an error in notation, from the maximum equations of 4;56° and 7;36°, at syzygy and quadrature, in the *Alfonsine Tables*. The regression of the nodes in the lunar latitude theory is accounted for in the usual way by giving the axis of an outermost sphere a westward motion completed in just under nineteen years.

The model for the motion of the superior planets in longitude (Fig. 5.15) is again that of Ibn ash-Shāṭir using two epicycles, but adapted to a heliocentric arrangement referred to the mean sun. The radii of the spheres and epicycles are derived from the maximum equations of the anomaly and of center in the *Alfonsine Tables*, and the apsidal lines are sidereally fixed, as in Ptolemy and the *Alfonsine Tables*. Copernicus's locations, given with respect to zodiacal stars, are within about a degree of the Alfonsine positions for Jupiter and Mars, although over 31° away for Saturn, suggesting that he made an error of one zodiacal sign in locating the apogee with respect to a star in the Alfonsine catalogue. The second anomaly, due to the motion of the earth, is described briefly with regard to its effects on direct and retrograde motion, stations, and the relative size of the equation of the anomaly. In a sense, this explanation, and the corresponding section for Venus, constitute Copernicus's principal argument or evidence for the heliocentric theory in the *Commentariolus*.

Copernicus next takes up the latitudes of the superior planets, and here he is entirely on his own in adapting Ptolemy's models to a heliocentric arrangement. Since the epicycles of the superior planets correspond to the orbit of the earth, which is confined to the plane of the ecliptic, he must convert the oscillations of the epicyclic planes in Ptolemy's models (Fig. 6.5) to oscillations of the orbital planes of the planets (Fig. 6.7) that are completed in each planet's true synodic period with respect to the earth. He produces the oscillations using Ṭūsī's device (Fig. 3) applied to spheres with inclined axes, but since the oscillations must correspond to true, not mean, synodic periods, the uniform rotation of the spheres is necessarily violated. The directions of the nodal lines, given with respect to zodiacal stars, are all within a degree or so of their Alfonsine locations—even Saturn—and the minimum and maximum inclinations of the orbital planes are about what can be derived from the latitude tables Copernicus copied into his volume of the *Alfonsine Tables*.

The longitude model for Venus (Fig. 5.54) is again an adaptation of Ibn ash-Shāṭir's model (Fig. 5.53), but in this case the adaptation is done incorrectly, so that it is necessary to measure the equation of center from an external point (*N* in Fig. 5.54), when it should be measured from the earth (Fig. 5.56(a)). The difficulty was probably due to Copernicus's originally using the eccentric model for the second anomaly, in which the external point is in fact the center of the planet's eccentric (Fig. 5.55). The problem is corrected, for both Venus and Mercury, in *De revolutionibus*. As in the *Alfonsine Tables*, the apsidal line coincides with the sun's and the equations of center of Venus and the sun are the same. Hence, the radii of the epicycles follow from the maximum equation of center, and the radius of the sphere from the maximum equation of the anomaly, in the *Alfonsine Tables*. The section on longitude concludes with a description of the effects of the motion of the earth on Venus's retrogradations and greatest elongations.

Perhaps the most difficult problem that confronted Copernicus in working out his planetary theory was the conversion of Ptolemy's latitude models for the inferior planets, something that was still giving him trouble in the latest stages of his work on *De revolutionibus*. At the time he wrote the *Commentariolus*, he believed that the first two components of latitude, the *inclination* and *slant* of the plane of the epicycle in Ptolemy's model (Fig. 6.10), could be accounted for entirely by the motion of the earth about the planet's interior orbit, the inclination being seen when the fixed inclination of the orbital plane is viewed in the line of sight, the slant when viewed across the line of sight. This was not sufficient, and in *De revolutionibus* he gave the orbit a variable inclination (Fig. 6.12). The third component, the variable inclination of the eccentric in Ptolemy's model (Fig. 6.13), is represented by a variable inclination and rotation of the planet's orbital plane (Fig. 6.14) that is not at all equivalent because it introduces the additional effect of the distance of the planet from the earth. Copernicus takes account of this in *De revolutionibus*, but appears unaware of it in the *Commentariolus*.

Ptolemy's model for Mercury (Fig. 5.68) differed from that for the other planets in that it was designed to produce the greatest sum of opposite elongations at $\pm 120°$ from the apogee rather than at 180°. Ibn ash-Shāṭir accounted for this by applying Ṭūsī's device to make the radius of Mercury's epicycle variable (Fig. 5.70), and Copernicus adapts the same mechanism to Mercury's orbit (Fig. 5.72), although his model has the same problem as for Venus of measuring the equation of center from an external point rather than from the earth. The location of the apsidal line, the radius of the sphere, and its variation all follow correctly from the *Alfonsine Tables*, but Copernicus had trouble finding the radii of the epicycles, and his results are simply erroneous. The latitude theory of Mercury is given only a brief description since it is the same as that of Venus, except that the directions in different parts of the orbit are reversed. In conclusion, Copernicus says that for the entire "choric dance" (*chorea*) of the planets 34 circles are sufficient, a number about which much foolishness has been written.

The *Commentariolus* has a number of problems, some quite serious and showing that Copernicus did not yet understand all the necessary details of adapting geocentric models to his new theory. Perhaps these are due to haste in writing the treatise very soon after he had devised the heliocentric theory. But in general it accomplishes what Copernicus set out to do, that is, to give a complete and physically admissible description of the planetary system using models that theoretically, specific parameters aside, produce essentially the same apparent motions as Ptolemy's. At least, Copernicus thought so at the time. This was the goal of the Marāgha astronomers, and it was likewise Copernicus's goal. Of the problems, some, such as the inadequacy of the precession theory to account for a nonuniform precession, were solved in *De revolutionibus*, others, particularly the violation of uniform circular motion by the spheres producing the oscillation of the orbital planes in the latitude theory, were inherent in the models, and remained uncorrected, and unacknowledged, to the end. Copernicus had said that he was omitting mathematical demonstrations as they were intended for a larger book which, as we have noted, appears to mean some kind of more detailed treatment of the material in the *Commentariolus* that could be completed in a reasonable time. However, he did not long rest content with such a plan. First of all, there was the fact that the heliocentric theory was stated as a series of postulates, assumptions, while clearly so fundamental an innovation in planetary theory required proof, or at least considerably more evidence than had thus far been presented. It is reasonable to believe that Copernicus wished to produce such proof or evidence, although ultimately he could not do so. Second, the parameters in the *Commentariolus* were adapted from sources at hand, principally the *Alfonsine Tables*, and this was acceptable enough for purposes of demonstration. But Copernicus must soon have decided that there was no point in setting out a physically correct system unless it was also numerically accurate—Kepler would have agreed—and thus he undertook the difficult task of deriving new elements of the orbits, requiring many years of observations that began by 1512 and continued until at least 1529.

His task, however, was limited in an important way because his intention was to derive parameters for the models already invented, or as it turned out, slight modifications of them, but not to determine on the basis of observations what models were actually appropriate to the motions of the planets, as Ptolemy had done and as Kepler was later to do. This, of course, is the limitation that could be called, to borrow Kepler's phrase, representing Ptolemy rather than nature, and while it saved Copernicus a good deal of trouble, it also precluded his going beyond Ptolemy to a more profound understanding of planetary theory and a greater accuracy in representing the apparent motions. But even the derivation of new parameters was an extremely difficult, perhaps unprecedented, undertaking that, including the period of observations, was to occupy Copernicus's attention for the better part of the next twenty-five or thirty years.

The Years of Observation

It is commonly said that Copernicus's observations are neither numerous nor accurate, and compared with Tycho, or even Walther, this is true. But the observations for which there are records—about sixty, of which twenty-seven are cited in *De revolutionibus*[18]—must represent only a fraction of those made in the period of about 1512–29 during which he was assembling the observations necessary for deriving new elements. Although only a handful of observations were actually used for the derivations, Copernicus himself indicates that these were selected from many more. He says in III,6 that his value of the obliquity was based on "frequent observation for thirty years," in III,16 that he applied his attention to equinoxes and other locations of the sun "for ten and more years,"[19] in V,5 that he has taken

[18] Birkenmajer (1900), 317–18, lists 63, which Zinner (1943), 408–18, esp. 415–16, reduces to 61, and the number could be reduced a bit more. The 27 in *De revolutionibus* are listed by Menzzer in Copernicus (1879), Anm. 66, and are treated in Chapters 3–5 of this study.

[19] In a curious passage in the *Narratio prima* (Kepler (1937–) 1, 94:5–10; trans. Rosen (1971), 125), Rheticus says that Copernicus observed "eclipses" and the motion of the sun for about forty years for determining the solar apogee, and compared them with all the "eclipses" in Ptolemy to find the motion of the apogee. While it is remotely possible, with a great deal of effort and a fully developed lunar and parallax theory, to find the solar apogee using eclipses, this certainly was not Copernicus's method. In III,20 Copernicus says that he used solar and lunar eclipses to confirm, in some unspecified way, his own location of the solar apogee, and that it is probably what Rheticus has in mind.

three oppositions of Saturn "from new observations" that he was compelled to make, and in V,11 and V,16 that the oppositions of Jupiter were observed "with the greatest care" and of Mars "not without care," both of which indicate a reason for selection from a larger number than the three actually used. In the case of oppositions in particular, it is reasonable that he would have observed every one during this period, weather and his responsibilities to the Chapter permitting, and an opposition is not a single observation, but an interpolation between two or more. He would likewise have observed every lunar eclipse visible in Warmia, and we have his word that he made solar observations for at least ten years. Hence, the number of observations taken for possible use is nowhere near twenty-seven or sixty, but more like a few hundred, and in addition to these are observations of interesting phenomena such as occultations or unusually close conjunctions or alignments of planets and stars that he could not but notice.

Whether Copernicus kept a journal for the recording of observations during this period is not known. Those observations not quoted in *De revolutionibus*, and a few that are, are noted by Copernicus in various books, the larger number in Johann Stoeffler, *Calendarium Romanum magnum* (Oppenheim, 1518) and Stoeffler and Pflaum, *Almanach nova in annos 1499–1531* (Ulm, 1499), a continuation of Regiomontanus's *Ephemerides* for 1475–1506, and a few more in his volumes of the *Alfonsine Tables* (Venice, 1492) and Regiomontanus's *Calendarium* and *Ephemerides* (Augsburg, 1492).[20] Most of these "occasional" observations are of conjunctions and alignments of planets with stars and the moon, and of eclipses of the sun and moon, although some of the notations to the eclipse diagrams in Stoeffler's *Calendarium* concern observations made by someone in Cracow, and the corrections to the times could in some cases be made by computation rather than observation.

There are only seven recorded observations prior to 1512. Four appear in *De revolutionibus*, of which three are lunar eclipses—of 1500 observed in Rome, 1509 in Heilsberg or possibly Cracow, and 1511 in Frauenburg—and one the 1497 occultation of Aldebaran observed in Bologna. Three that do not appear are two conjunctions of the moon and Saturn observed in Bologna in 1500, and a 1504 conjunction of Jupiter and Saturn noted by

---

[20] Facsimiles in Biskup (1973), plates 1–2, 12–15; transcriptions in Prowe (1883–84) I,2, 268 n.; Birkenmajer (1900), 460, 515–21, 551–55. There have been differences of opinion concerning the handwriting of some of the notes, but we see no reason to doubt that they are Copernicus's own records. Probably the most interesting is one in Regiomontanus's *Ephemerides* for 9 Mar. 1497 reporting a conjunction of the moon and Aldebaran at $10\frac{1}{3}^{h}$ after noon, $0;40^{h}$ earlier than the time of the occultation reported in IV,27, and in fact the correct time of the occultation in Bologna. See below, pp. 266–269.

In the *Narratio prima* (Kepler (1937–) 1, 114:24–26; trans. Rosen (1971), 163), Rheticus reports that Copernicus always had in sight "the observations of all ages along with his own, collected in order as in catalogues." This would appear to be a catalogue of fully reduced observations, i.e. date, time, longitude, etc., rather than a journal of unreduced observations, e.g. meridian zenith distances and such. In addition to the observations of all ages, one wonders how many of his own it contained, just those in *De revolutionibus* or a good many more?

Martin Biem of Ilkusch to have occurred on 12 May *secundum Copernicum*. Aside from the lunar eclipses, which are noticeable as a matter of course and predicted in almanachs, these are not the sort of observations that Copernicus needed for deriving parameters in *De revolutionibus*. The observations of the moon with Saturn and Aldebaran could be of use for testing the Alfonsine lunar theory and Ptolemy's lunar parallax—although the test would be neither easy nor certain—and Copernicus does use the occultation of Aldebaran as a test of his own lunar parallax in IV,27, but one cannot be certain about what Copernicus did with them at so early a date. There are twenty recorded observations from after 1529—if indeed the notations in the eclipse diagrams of Stoeffler's *Calendarium* all concern observational rather than computational corrections—and of this number, two are of lunar eclipses, five of solar eclipses, and thirteen, entered in Stoeffler's *Almanach nova* and Regiomontanus's *Ephemerides*, are of alignments and conjunctions of planets and stars. The latter, dating from 1537–38, after Copernicus had essentially finished deriving parameters, might be useful for testing a completed theory, although one really does not know what Copernicus intended to do with them.

In order to consider in general terms the kinds of observations Copernicus made and how he may have made them, we shall divide them into four categories:[21]

1. Eclipses.
2. Zenith distances or altitudes in the meridian.
3. Oppositions.
4. Alignments, conjunctions, and occultations.

1. *Eclipses*. Copernicus uses five lunar eclipses that he observed for finding the parameters of the lunar model in *De revolutionibus*, and three more, from 1525, 1530, and 1534, are entered into Stoeffler's *Calendarium*. In observing an eclipse, one wishes to note the time of first and last contact, of the beginning and end of totality if the eclipse is total, the magnitude of the eclipse in digits (twelfths of the lunar diameter) if it is partial, and the direction from which the shadow crosses the moon. Copernicus never gives all of this information, but always includes the magnitude and the time of mid-eclipse, presumably estimated from the time of first and last contact. What is crucial in observing an eclipse is the measuring of time. Copernicus gives no information about how he measured time—and we shall not advance any speculation—but even if one has an accurate measure of time,

[21] A subject of little concern to us here, or in general, is the accuracy of Copernicus's observations, although in order to analyze the alignments and occultations, we shall compute the circumstances according to modern theory. Comparisons of Copernicus's observations with modern computations can be found in Kamienski (1963), Bialas (1973), Moesgaard (1974), Gingerich (1975), and there are some notably large errors, e.g. of about $1\frac{1}{4}°$ and $2\frac{1}{4}°$ in the oppositions of Mars from 1518 and 1523.

the moments of first and last contact are notoriously difficult to estimate, for one is likely to be late in the former and early in the latter.

2. *Zenith distances or altitudes in the meridian.* These are among the most useful of observations because, with the aid of spherical astronomy and solar theory, and in most cases some measurement of time, they may be converted to right ascension and declination, which may in turn be converted by transformation of coordinates to longitude and latitude. In II,2 Copernicus describes a quadrant engraved on a piece of metal or stone aligned in the meridian, upon which a gnomon throws a shadow, from which the zenith distance of the sun can be read as it crosses the meridian. By observing at the solstices, one may find the arc between the tropics and the obliquity of the ecliptic, something Copernicus says he did for thirty years. The mean of the zenith distances at both solstices is, in principle, the observer's terrestrial latitude, which Copernicus reports in III,2 as 54;19,30° for Frauenburg. If the obliquity is taken as 23;28,30°, as Copernicus also reports in III,2, then he observed zenith distances of 30;51° at the summer solstice and 77;48° at the winter solstice (where refraction would decrease the true zenith distance by about 0;4°). By interpolating between observations on successive days around the equinox, one may find the time of the equinox, when the sun's zenith distance is equal to the observer's latitude, and with spherical trigonometry any solar zenith distance may be converted to longitude in order to find, e.g. when the sun reaches Scorpio 15°, which Copernicus uses in III,16.

In IV,15 Copernicus describes the construction of the "parallactic instrument," similar to the instrument described by Ptolemy in *Almagest* V,12, that gives the chord of the zenith distance, which is converted to arc by inverse use of chord or sine tables. In IV,16 Copernicus uses it to measure rather large zenith distances of the moon taken in 1522 and 1524 that were used for the purpose of finding the lunar parallax and distance in terrestrial radii. In III,2 he gives measurements of the altitude of Spica in 1515 and 1525 without specifying how the observations were made, although the measurement of zenith distances with the parallactic instrument is a possibility, and the altitudes were used to find the longitude of Spica for use in the precession theory.

3. *Oppositions.* In order to find the eccentricity and direction of the apsidal line of a superior planet, the longitudes of three oppositions to the mean sun are required, and these should be distributed as widely as possible around the planet's orbit, or at least through half the orbit. In the case of Saturn, with a zodiacal period of nearly thirty years, this obviously takes time. Copernicus uses observations from 1514, 1520, and 1527, covering a bit less than half the orbit, but optimally distributed six and seven years apart. For Jupiter and Mars time is not as serious a constraint, so Copernicus can select what he considers to be the most accurately observed and well-distributed oppositions over the period 1512–29 or so, which may be what

he means when he says that they were taken with *diligentia* (although two of the oppositions of Mars are very inaccurate). In both cases, he uses oppositions distributed through about half the orbit, with one about midway between the other two. This is a very good distribution since the equation of center will change considerably between each of them. The oppositions of Jupiter are from 1520, 1526, 1529, and of Mars from 1512, 1518, 1523.

An opposition is "observed" by finding the true longitude of the planet on successive nights, weather permitting, near the anticipated time of opposition, about in the middle of the planet's retrograde arc, and then interpolating for the time that the planet is exactly 180° from the computed longitude of the mean sun. This cannot be done until a mean solar motion has been established, so Copernicus may not have been able to compute the moment of opposition until years after the observations were made. At least two, but probably several, observations are required, in the case of Saturn and Jupiter spread over about a month, and the planet will be moving retrograde through all of them. The apparent velocity of the planet changes considerably in the retrograde arc, but the most rapid and detectable motion occurs near the opposition. There are various ways of finding the longitude. One is simply by measuring the zenith distance at the time of meridian transit using the parallactic instrument, and finding the right ascension and declination, which can then be transformed to longitude and latitude by spherical trigonometry. A difficulty with this method is that it requires a precise measure of time. An error of $0;4^h$ in the timing of the transit will produce an error of 1° in the right ascension and an error of the same order in the longitude, which is rather serious since the planets are moving only a few minutes per day. Thus, the errors of about $1\frac{1}{4}°$ and $2\frac{1}{4}°$ in the 1518 and 1523 oppositions of Mars could arise from consistent errors of only about $0;5^h$ and $0;9^h$ in timing meridian transits, although this is not necessarily the cause of Copernicus's errors.

Another way of finding the longitude is by using an armillary, or spherical astrolabe, described by Ptolemy in *Almagest* V,1 and by Copernicus in II,14. The instrument consists essentially of a graduated ecliptic ring, with rotating sights attached to a graduated ring perpendicular to the ecliptic for measuring latitude. When the ecliptic ring is set to coincide with the true ecliptic in the sky, by aligning it with the true observed or computed position of the sun, the tropical longitude and the latitude of a star or planet fixed in the sights may then be read on the graduated rings. The observations of Mercury by Bernhard Walther that Copernicus uses in V,30 were made in this way. Alternatively, the distance of the planet from a star of known longitude can be read on the ecliptic ring, and added to the longitude of the star. Finally, by a third method the distance of a star or planet may be measured from the observed or computed apparent longitude of the moon. This is how Ptolemy made his observation of Regulus reported in *Almagest* VII,2 and by Copernicus in II,14. The accuracy of using the armillary depends upon the accuracy of aligning the ecliptic ring, and then on the accuracy of one's star catalogue or lunar theory. However, the measurement of time is not as crucial as in the method of meridian transits, particularly if one measures the distance from a fixed star.

It is not altogether clear how Copernicus measured the longitudes for the oppositions. He says in V,4 that oppositions are found with "astrolabic instruments," which appears decisive, except that he does not say that he himself used one. For most of the period from November of 1516 to June of 1521, about $4\frac{1}{2}$ years, Copernicus was in Allenstein rather than Frauenburg, and three of the oppositions, one for each planet, date from 1518 and 1520. Did Copernicus have two armillaries, or drag one back and forth between Frauenburg and Allenstein? Did he take one along when he fled Frauenburg after it was sacked by Albrecht's troops in January of 1520? Obviously we cannot answer these questions, but we suspect that some, if not all, the oppositions were taken without an armillary, for there is no positive evidence that Copernicus owned one. While he certainly describes the instrument and its use, principally for Ptolemy's observation of Regulus, this is not the same as saying that he used it himself, while he does specifically say in IV,16 that he used a parallactic instrument, which Tycho later recovered and described. Still, there is one observation, mentioned in the next section, that appears to be made with an armillary, although a simpler instrument for measuring small angles cannot be ruled out. One can only wish that Copernicus had given more information on this subject.

4. *Alignments, conjunctions, and occultations.* These are rather interesting observations since Copernicus gives something more than a bare longitude, and we have recomputed them in the course of this study, sometimes with curious results. The most striking—Copernicus calls it a "wonderful sight"—is the 1529 occultation of Venus through which, by computing the apparent position of the moon, Copernicus finds the true longitude and latitude of Venus. The 1497 occultation of Aldebaran is also remarkable, and Copernicus uses it to confirm, with some difficulty, his lunar theory and parallax. The three observations of the superior planets outside of opposition are of alignments and conjunctions with stars. Interestingly, each is made within a few months of the first of each set of oppositions. This is reasonable for preventing any cumulative error in mean motions, but it is not apparent what else to make of what could be a more significant fact. Saturn was observed in 1514 to be in a straight line with π and δ Scorpii, and on 1 Jan. 1512 Mars was found $\frac{1}{4}^{\circ}$ from α Librae, "inclined toward the solstitial rising." Both of these observations, like the occultations, appear to have been made without instruments. Copernicus simply noticed the alignment and conjunction, and estimated, rather badly as it turns out, the distance of the latter. On the other hand, the observation of Jupiter in 1520 is a very precise measurement "with an instrument" that Jupiter was 4;31° from β Scorpii. In this case, the instrument, giving a direct measure of the difference of longitude, would appear to be an armillary (and in Allenstein during the war).

There were, however, other, simpler instruments that could be used to measure the angular distance between two stars or planets, although not directly the difference in longitude, and Copernicus may have used one for some of the 1537–38 observations of alignments. Here are some examples:[22]

1. 1537 Sep. 8. Mars in a straight line following (east) the heads of the Gemini (α and β Gem.).
2. Oct. 10, Wednesday. Venus and Saturn were equally distant from the hind foot of Leo (υ Leo), Venus preceding (west), Saturn following (east).
3. Nov. 7, Wednesday. Mars followed (east) by one digit the straight line of the sixth and eighth (stars) of Leo (γ and α Leo), distant from Regulus (α) by 2° and more.
4. Nov. 15, Thursday, $8\frac{1}{4}$ hours (after noon?). The moon followed (east) Jupiter by $3\frac{1}{10}°$.

The first two are alignments requiring no instrument, but the second two suggest the use of some instrument for measuring small angular separations, although the distance "one digit" is curious. In any case, these are a very common sort of observation, although one does not know how many Copernicus made. As noted earlier, Bernhard Walther made hundreds of them between 1475 and early 1488, both noting alignments and using the *rectangulus* or *radius astronomicus*, after which he began using an armillary. By graphing such configurations, the coordinates of the planets may be found, their accuracy of course depending upon the accuracy of the stellar coordinates, but it is not known what Copernicus may have done with them. And of course one can also ask why Copernicus did not use an armillary.

De revolutionibus

Having completed his observations, or perhaps during the last years they were being made, Copernicus could turn his attention to reducing them for practical use, selecting which appeared the most promising, and beginning the labor of deriving parameters. This had necessarily to proceed in a fixed order. The determination of the times of equinoxes and other locations of the sun required first settling on values of the latitude of Frauenburg and the obliquity of the ecliptic, and these presumably Copernicus reached during the period of his observations. Since he believed the tropical year to be variable and the sidereal year constant, with the variable precession as their difference, he also had to work out the precession theory before he could take up solar theory, which was based upon observations of the sun with tropical coordinates. Only then could he find the mean and true sidereal and tropical motions of the sun that are required for his lunar and planetary theories. Finding the longitudes of the eclipses that are the foundation of

[22] Biskup (1973), plate 12; Prowe (1883–84) I,2, 268 n.

the lunar theory required computing true tropical positions of the sun, while interpolating in observations of superior planets for the time of opposition required mean sidereal positions of the sun, at least as presented in the texts of Book IV and V. Just as in the *Almagest*, the relation of one part building upon the results of another determined the order of the demonstrations in *De revolutionibus*, except that the catalogue of stars, being adapted from Ptolemy's with no additional observations, falls entirely outside this order. From the letter of Wapowski of 15 October 1535 concerning Copernicus's new tables and almanach, it appears that he had, provisionally at least, completed his derivations by that time, although many revisions could still have been made in the following years.

Except for the placement of the star catalogue at the end of Book II, the first two books of *De revolutionibus* also correspond to the ordering of subjects in the *Almagest* and the *Epitome*. The first eleven chapters of Book I are devoted to a general description of the universe and of the location and motions of the earth. It is here that Copernicus attempts to defend the motion of the earth, specifically the daily rotation, against physical objections. His argument, spread through several chapters, is that the earth is a spherical body with a single center of magnitude and of gravity; that a uniform rotation about its axis is natural to a sphere, a "simple" body in the Aristotelian sense, by virtue of its form (this principle also applies to the celestial spheres that carry the planets); that no violent motions of bodies on the surface of the earth, or in the surrounding water and air, will result from the natural rotation. He also presents, principally in I,10, the reasons for believing that the heliocentric theory is a correct description of the planetary system, and in I,11 he briefly describes the three motions of the earth, in the order of the daily rotation, the annual motion, and the precession. It would be nice to say that Copernicus proved his case, but, alas, he did not, and he knew it. For however plausible his reasons for adopting the heliocentric theory may seem—to us, who are already convinced—he feared that his contemporaries would not be persuaded. And he was right, for hardly any of them were.

Trigonometry is treated in three chapters: I,12 on the measurement of chords in a circle, followed by the table of sines, which Copernicus calls half-chords of the double arc; I,13 on the solutions of plane triangles; I,14 on the solutions of spherical triangles. Chapters 1–12 of Book II are on spherical astronomy, and the presentation, based more on the *Epitome* than the *Almagest*, is quite thorough despite its brevity. I,13 is a very brief, and not very useful, treatment of visibility phases, heliacal risings and settings, of stars and planets, a rather complicated subject treated by Ptolemy in *Almagest* VIII and a special work called the *Phaseis* for the stars, and in *Almagest* XIII, with extensive tabulation in the *Handy Tables*, for the planets. Book II ends with the catalogue of stars, which is adapted from Ptolemy's catalogue by subtracting 6;40° from the longitudes in order to measure "sidereal" longitudes from the first star listed in Aries, γ Arietis.

In the first part of Book III Copernicus takes up the nonuniform precession and the diminution of the obliquity since the time of Ptolemy. These subjects had been treated before, most recently in Johann Werner's *De motu octavae sphaerae* (1522), and Copernicus draws upon Werner for part of his own treatment, most notably his analemma for finding the longitude of Spica from its declination, and a record of recent values of the obliquity, among them the value 23;28,30° used by Copernicus and Werner, and attributed by Werner to Regiomontanus. A difficulty with earlier treatments of the precession and obliquity, among them Werner's, is that they employed motions of the axes of spheres in small circles that would simultaneously affect both the rate of precession and the obliquity in rather complex ways. But already since writing the *Commentariolus*, where he had used it in the latitude theory of the planets, Copernicus had at his disposal Ṭūsī's device (Fig. 3) for producing oscillation (approximately) along a great circle. This could be applied to the problem at hand, for the variation of the obliquity and the inequality of the precession could be treated as perpendicular oscillations of the earth's equator. When transferred to the earth's pole, these gave two oscillations of the earth's axis of rotation, one, tangent to the circle of the axis's precessional motion, producing the inequality of the precession, and a second, perpendicular to the first, varying the obliquity. Copernicus assumed that the period of the variation of the obliquity was exactly twice the period of the inequality of the precession, and that the moment of the maximum obliquity and the slowest precession coincided, as indeed appeared to be true from Ptolemy's large obliquity and slow precession.

What remained was to establish the parameters of these motions, and here Copernicus's account is neither a complete nor an accurate record of what he did. The following reconstruction is, however, supported by the evidence he gives. He began by assuming that the obliquity varied periodically over a range of 0;24°, from 23;52° to 23;28°, and that the maximum occurred before Ptolemy while the minimum was yet to occur. Combining this assumption with the obliquity at the time of Ptolemy and at his own time, he found that half the period of the obliquity, from maximum to minimum, was 1717 Egyptian years (of 365 days), which is therefore the entire period of the inequality of the precession. He also found—how is not really clear—that the period of the mean precession is 25816 Egyptian years, which is very nearly correct. The mean precession, combined with longitudes of stars from the observations of Timocharis (−293) and Ptolemy (139), and the assumption that the slowest point of the precession fell exactly midway between them, allowed Copernicus to find the maximum equation of the precession of 1;10°. This assumption, however, was shown to be incorrect by longitudes of stars observed by al-Battānī (880), and so by some kind of trial and error, he found that he could fit all the observations by moving the slowest point of the precession to a date falling in −64. Finally, in a chapter added in his manuscript, he demonstrates that the obliquity varies between 23;52° and

23;28°, the very assumption he started out with (and a wonderful exercise in circular reasoning). Nevertheless, as flawed as his presentation is, the precession theory agrees very well with the observations, and the variation of the obliquity is at least close.

The principal problems in the solar theory, occupying the second part of Book III, also concern long-period, nonuniform variation of parameters. Copernicus first reviews the variation in the length of the tropical year shown by observations of equinoxes by Hipparchus, Ptolemy, al-Battānī, and himself. Therefore, he will follow Thābit ibn Qurra in assuming that the sidereal year is of constant length since the length of the tropical year and the rate of precession vary in a fixed proportion that indicates a constant sidereal year. But not all the variation in the length of the year is accounted for by the variation of the rate of precession for, Copernicus says, the apparent motion of the sun is affected by two further long-period inequalities, a diminution of the eccentricity and a nonuniform motion of the earth's apsidal line. Indeed, reviewing Hipparchus's and Ptolemy's derivation of these parameters, and then carrying out his own, he shows that the eccentricity has diminished by about one-fourth, from 415 to 323 where the radius of the eccentric is 10,000, and the tropical longitude of the apogee has increased about 31° from ♊5;30° to ♋6;40°, exceeding the precession by about 11° in the nearly 1400 years since Ptolemy. However, in the period of nearly 1700 years since Hipparchus, the sidereal longitude of the apogee increased only about 8°, indicating a retrograde sidereal motion of the apsidal line of about 3° between Hipparchus and Ptolemy. Further confirmation of the decrease of the eccentricity and motion of the apsidal line is adduced from the parameters of al-Battānī and, so Copernicus believes, az-Zarqāl, although what appears here to be a retrograde tropical motion of the apsidal line is really a confusion of sidereal and tropical longitudes in the *Epitome*, Copernicus's source of information about Battānī and az-Zarqāl.

Relying to some extent on the *Epitome's* account of az-Zarqāl, Copernicus develops a model for this motion (Fig. 3.16), in which the center of the earth's orbit, the mean sun, moves in a small circle about a point removed from the true sun by a mean eccentricity. This will both vary the eccentricity and add an inequality to the motion of the apsidal line. The next task is to find the parameters, but confronted here by rather difficult geometry that was not treated by Ptolemy, Copernicus assumes almost everything to be found in advance, and ends up solving only trivial problems. He first assumes that the period of the variation of the eccentricity is the same as that of the obliquity, 3434 years, and further, that the maximum eccentricity coincided with the maximum obliquity in −64, again as Ptolemy's large eccentricity and obliquity seem to indicate. He next assumes, and it is this that makes the problem trivial, that the maximum eccentricity is 417, only slightly larger than Ptolemy's. Then, using his own eccentricity of 323 for 1515, he finds that the mean eccentricity is 369 and the radius of the small circle 48. Like the precession, the treatment of solar theory is filled with problems, but the final theory agrees very well with the equinox observations of Hipparchus, Ptolemy, and, one erroneous time excepted, Copernicus, although less well with those of al-Battānī.

The principal source of difficulties throughout the precession and solar theories of Book III was the variation of parameters over long periods. This would be difficult enough if the parameters were correct, but mostly they were not, resulting as they did from very sensitive derivations from less than accurate observations. Faced with this confusion of information, in which little more than general trends were evident, Copernicus could either apply rather sophisticated mathematical techniques—or lacking that, trial and error—to derive elements fitting a larger part of the earlier observations and theory, or simply ignore as much as possible in order to simplify the problems. In the case of the precession, as far as it was possible for him, Copernicus attempted the first course, by trial and error, but for the solar theory, perhaps because the problems were more difficult and the record contradictory, he chose the second. His object in both cases was to reproduce Ptolemy's observations and parameters at Ptolemy's time, and his own at his own time, and, except for an attempt to take account of Battānī in the precession, pretty much let the intervening period fall where it may. On the whole he succeeded because he derived the parameters of the models on the basis of Ptolemy's and his own observations, and considering the little he knew about the period in between, this was probably the most reasonable decision.

The lunar theory of Book IV is more straightforward, and more directly an application of the methods of the *Almagest* and the *Epitome*, because the problem of long-period variation of parameters is not present. Copernicus first describes Ptolemy's model (Fig. 4.1) and its problems of violating uniform circular motion and producing too large a variation of distance, and then his own (Fig. 4.2), which is identical to Ibn ash-Shāṭir's model already used in the *Commentariolus*. He then sets out provisional mean motions based on the Babylonian period relations used by Hipparchus and Ptolemy, following these with corrections and tabulations based upon his own demonstrations. The first of these concerns the radius of the lunar epicycle at syzygy, which is found by the same method used by Ptolemy from three lunar eclipses (Fig. 4.7). Copernicus reviews Ptolemy's demonstration in full, and then presents his own. In the first case, where the radius of the deferent is 100,000, the radius of the epicycle is 8706, and in the second 8604 which, Copernicus remarks, agrees with most astronomers since Ptolemy. What he means here is that the radius 8604 produces a maximum equation at syzygy of 4;56°, the value found in the *Alfonsine Tables* that Copernicus had used in the *Commentariolus* and, although he could not know this, goes back to Indian sources nearly a thousand years earlier. He does not consider the difference between Ptolemy's result and his own to show a real reduction in the size of the epicycle, but regards his own result, presumably because it agrees with the Alfonsine value, as correct, and Ptolemy's, by implication, as slightly inaccurate (which is not true). Hence, he need not consider any long-period variation of lunar parameters, and he later treats Ptolemy's values of the lunar distance, parallax, and apparent diameter in the same way, as more-or-less incorrect, since they were so closely related to the incorrect distances in Ptolemy's model. In his own treatment of these subjects, he relies more upon later, medieval procedures reported in the *Epitome*.

Ptolemy had discovered the second lunar inequality by a painstaking examination of his own and Hipparchus's observations, only alluded to in *Almagest* V,2 and *Epitome* V,3, and the additional parameters, two eccentricities that turn out to be equal, are derived from two observations of the moon near quadrature and two near octant. Copernicus does none of this, but simply takes over from Ptolemy a maximum equation at quadrature of 7;40° which, along with the equation of 4;56° at syzygy, immediately dictates the radii of the two epicycles and the correction of the anomaly that is approximately equivalent to Ptolemy's *prosneusis* of the mean apogee of the epicycle. The latter appears a rather clear case of forcing a model on the phenomena, although it happens that the two corrections are about equal at octant, where Ptolemy's observations were taken. In any case, Copernicus's lunar theory excellently reproduces all but one of the observations used to establish it, and the one exception is actually an error in Ptolemy's timing of an eclipse. The lunar correction table is partly adapted from the *Alfonsine Tables* and partly computed for Copernicus's model.

The next part of Book IV is concerned with the distance, parallax, and apparent diameter of the sun and moon. Here Copernicus uses a number of later methods from the *Epitome*, and the smaller variation of lunar distance in his model makes the treatment of these subjects a distinct improvement over Ptolemy's. Nevertheless, it is one of the most confusing sections of *De revolutionibus*, containing many errors and internal contradictions, due to an inconsistent revision of an originally flawed exposition. The lunar parallax and distance are taken from two observations of the moon near quadrature with enormous zenith distances of more than 80°, and the parallaxes are found to be 0;50° and 1°, where Ptolemy's theory would have given parallaxes of 1;18° and 1;35°. This proves that the close approach of the moon at quadrature in Ptolemy's model is in error, although Copernicus's observations and parallax are actually filled with problems due to incorrect lunar latitude, neglect of refraction so close to the horizon, and what is most remarkable, that the moon, which is supposed to be in the meridian, is actually several degrees away. Nevertheless, Copernicus manages to find reasonable distances of about 68^{tr} and 57^{tr} (terrestrial radii), while Ptolemy's model implied distances of only 44^{tr} and 36^{tr}. The total variation of distance in Copernicus's model is confined within 55^{tr} and $65\frac{1}{2}^{tr}$ at syzygy and 52^{tr} and 68^{tr} at quadrature, compared with a possible variation of from 64^{tr} to $33\frac{1}{2}^{tr}$ at syzygy and quadrature respectively in Ptolemy's model.

Copernicus determines the solar distance by the same method used by Ptolemy and, in principle, al-Battānī, and the results are nearly the same—Ptolemy 1210^{tr}, Battānī 1146^{tr}, Copernicus 1179^{tr}—although Copernicus's demonstration is one of the strangest examples of inconsistently throwing together an original and revised version. And these same inconsistencies continue through his exposition and tabulation of the variation of the apparent diameters of the sun, moon, and the earth's shadow. Since Copernicus's lunar theory differs so much from Ptolemy's in its distances, he computes a new table of lunar parallaxes, and his final test, of both the lunar theory and parallax, is the recomputation of the occultation of Aldebaran that he observed in Bologna in 1497. The agreement between the observation and computation comes out perfectly, but only because Copernicus first altered the time of the observation by $0;40^h$ and the longitude of the star by $0;10°$.

The last subject in Book IV is the computation of eclipses. Copernicus's treatment is brief, and the only table provided is one of the motion of the moon and sun in synodic months to aid in finding the time of mean syzygy. Here we shall merely outline the order of eclipse computation, and in the part of this study concerned with eclipse theory we shall expand the exposition considerably, with an example of each step. The first is to find the time of mean syzygy, which can be done using the table of the moon's mean elongation from the sun. For the time found, one computes the mean lunar anomaly and argument of latitude, and the mean solar anomaly. The next step is to find the time of true syzygy, for which the true velocities of the sun and moon are required. Copernicus shows how to find true velocities, and also has his own method for finding true syzygy that differs from the methods in his sources and is in fact superior.

At this point, one can test the conditions of the syzygy to see whether an eclipse is possible. The criterion for a lunar eclipse is that the true latitude of the moon must be less than the sum of the apparent radii of the moon and the earth's shadow. The criterion for a solar eclipse is that the apparent latitude of the moon must be less than the sum of the apparent radii of the sun and moon. This condition is quite complicated because the apparent latitude of the moon and the time of apparent conjunction depend upon the lunar parallax, which can change considerably in the course of an eclipse. Copernicus calls finding the time of apparent conjunction "more than enough trouble," and gives a summary of a procedure that is intended to be used with standard parallax tables, such as those in the *Alfonsine Tables*.

If an eclipse turns out to be possible, one next determines its magnitude in digits, and finally the times and durations of the phases of the eclipse. As an afterthought, Copernicus says that some people measure eclipses by twelfths of the area rather than the diameter of the eclipsed body, an archaic procedure that Ptolemy says is used principally for prognostications (of weather) from eclipses. Copernicus reviews the method of computing area digits from the *Almagest* and the *Epitome*, and then says that this much will suffice about eclipses, which "others" have treated in more detail, for he wishes to hasten on to the planets. Here he is referring to the extensive treatment of eclipses in the *Almagest*, the *Epitome*, to Peurbach's monumental *Tabulae eclipsium*, and perhaps to the *Alfonsine Tables*.

The planetary theory of longitude in Book V, particularly for the superior planets, is Copernicus's most admirable, and most demanding, accomplishment. The theoretical work on this subject, the development of heliocentric models, in all essentials equivalent to Ptolemy's, but preserving the uniform rotation of the component spheres demanded by physical principles, had been worked out already in the *Commentariolus*. It was above all the decision to derive new elements for the planets that delayed for nearly half a lifetime Copernicus's continuation of his work—nearly twenty years devoted to observation and then several more to the most tedious kind of computation—and the result was recognized by his contemporaries as the equal of Ptolemy's own accomplishment, which was surely the highest praise for an astronomer. And Copernicus's elements could not but be superior to Ptolemy's in the sixteenth century in being free of the cumulative errors in mean motions that made any direct use of the tables in the *Almagest* unthinkable, and made even the *Alfonsine Tables*, then nearly three hundred years old, highly suspect.

While Copernicus's original goal in the *Commentariolus* was to devise physically correct heliocentric models equivalent to Ptolemy's, his object now was to derive elements accurate in his own time, and take account of any variation of parameters since the time of Ptolemy. At no point, however did he question the soundness of Ptolemy's models for representing the apparent motions of the planets, and so at no time did he carry out the sort of analysis that Ptolemy had, and that Kepler did later, to determine what really constituted an appropriate model for the planets. Copernicus has frequently been criticized for following Ptolemy rather than making discoveries that later appeared to be almost direct consequences of the heliocentric theory, but this criticism is really not fair, for the analysis required to discover these consequences was very difficult and required observations of a sort that it never even occurred to Copernicus to make. In the absence of such an analysis, altogether beyond his powers or his inclination, there was no way of discovering, for example, that the apsidal lines and eccentricities of the planets should be referred to the true rather than the mean sun, although the difference actually produces a considerable effect on the apparent motion of a planet seen from the earth. And the discovery of the much smaller effects of the approximation by equant motion of correct planetary motion according to Kepler's first two laws was completely out of the question. Copernicus represented Ptolemy rather than nature, as Kepler remarked, because it never really occurred to him to do anything else.

The model for the superior planets in *De revolutionibus* contains one alteration of the earlier model in the *Commentariolus*. There Copernicus had used a double epicycle for the first inequality (Fig. 5.15) exactly like Ibn ash-Shāṭir, but now he substituted for the larger epicycle an equivalent eccentricity (e_1 in Fig. 5.15), so that his model was like that of ʿUrḍī and Shīrāzī (Fig. 5.12), although one assumes that he made the transformation on his own. His reason was that he found that the apsidal lines of the superior planets were not sidereally fixed, as he had earlier supposed when relying upon Ptolemy and the *Alfonsine Tables*, but possessed independent motions, and further, the eccentricity of Mars had decreased. These changes could perfectly well be represented by the double-epicycle model, but it was didactically clearer and, so Copernicus seems to have believed, physically correct to separate the eccentricity, that is subject to the shift of the apsidal line and the diminution, from the smaller epicycle that can be left unaffected by these changes.

Copernicus's derivations of the orbital elements of the superior planets were his most laborious undertaking, requiring well over a thousand separate calculations just for the demonstrations presented in the text, which we shall later review, and God knows how many thousands more in unsuccessful trials. He uses the same iterative procedure as Ptolemy, requiring three oppositions, for finding the double eccentricity and the direction of the apsidal line. Likewise, the same method Ptolemy used for finding the radius of the epicycle from a single observation of the planet outside of opposition, Copernicus uses for finding the ratio of the radii of the earth's and planet's orbits. The results can be summarized briefly: In all cases he found that the ratio of the radii of the orbits of the earth and planet was essentially equal to Ptolemy's radius of the epicycle, so the radii of the planets' spheres were not subject to long-period variations, the possibility of which he was probably not ready to consider in any case. The double eccentricity of Saturn and Jupiter had remained unchanged, and the double eccentricity of Mars had diminished slightly, since Ptolemy's time, from which he concluded that the center of the earth's orbit was approaching the center of Mars's orbit, but that the motion had no detectable effect upon Saturn and Jupiter. The apsidal lines of Saturn and Jupiter had shifted respectively about 14° and 4;40° with respect to the fixed stars, and Copernicus treated these as uniform motions. The apsidal line of Mars had shifted by about 11°, but he did not attempt to determine how much was due to the rotation of the apsidal line and how much to the motion of the mean sun with respect to the center of Mars's orbit. He claims, however, that the distance, although not the direction, of the center of Mars's sphere from the true sun has remained unchanged, and an analysis using Copernicus's and Ptolemy's eccentricities and apsidal directions shows that this is very nearly true. The conclusion Copernicus appears to draw from this is a cosmological one, that the true sun does not move, and should therefore be regarded as the unmoving center of the universe. He does not, however, appear to consider the possibility that it should also be treated as the center of the planetary system for the determination of eccentricities and the directions of apsidal lines.

The treatment of the inferior planets is much less satisfactory, and whatever reasons there may be for this, the one that takes precedence is that Copernicus could not obtain the observations required to derive elements of his own using Ptolemy's methods. In the case of Mercury, he remarks in V,30 that fogs off the Vistula and the planet's long periods of invisibility, even at greatest elongation from the sun, due to the low inclination of the ecliptic to the horizon at northern latitudes, made the necessary observations unobtainable. While he says nothing of difficulties in observing Venus, and does cite one observation of his own, it happens that the required observations of Venus were also unobtainable, but for a different reason having nothing to do with local conditions.

Ptolemy's procedure for deriving the parameters of the inferior planets uses observations in which the mean sun occupies specified positions with respect to the apsidal line—symmetrical to the apsidal line for determining its direction, in the apsidal line for deriving the simple eccentricity, perpendicular to the apsidal line for finding the double eccentricity—while the planet is as close as possible to opposite, morning and evening, greatest elongations. The heliocentric periods of Venus and the earth have what is called a resonance, a ratio of small numbers, such that Venus completes 13 revolutions in very nearly 8 years, which therefore contain 5 synodic periods of the planet as seen from the earth. Consequently, the greatest elongations of Venus will distribute themselves in 8 years about only 5 locations of the sun in the ecliptic approximately $360°/5 = 72°$ apart. But Ptolemy's procedures require locations 90° apart, and within a period of 30 or 40 years these simply will not occur. Ptolemy had to compromise by using the correct locations of the mean sun, which are indispensable, but Venus as much as twenty or so days from greatest elongation, which is less critical. During the early sixteenth century, the conditions for finding Venus near greatest elongation with the mean sun at perigee of Venus's eccentric were extremely unfavorable, worse than at Ptolemy's time, and this in itself eliminated the possibility of Copernicus's finding the eccentricity by Ptolemy's method. And it would be unfair, and uncharacteristic, to expect him to devise a different method.

The equation of center in Ptolemy's models for all planets is determined by the distance of the center of the epicycle from the apsidal line. When transformed to a heliocentric arrangement, this corresponds, for the superior planets, to the distance of the planet from the apsidal line, with the equation of center measured at the mean sun, but for the inferior planets, to the distance of the earth from the apsidal line, the equation of center now being measured at the earth. This, of course, looks peculiar in Copernicus's models for Venus and Mercury, and has occasioned some surprise that he did not question so remarkable, and apparently coincidental, an arrangement. However, as we have noted before, since Copernicus's object was to find physically permissible heliocentric models that would reproduce, as far as possible, the apparent motions of Ptolemy's models, he was not motivated to question this coincidence or carry out the analysis—by no means easy—required to determine whether it was in fact true. Hence this dependence of the equation of center of Venus and Mercury on the distance of the earth from their apsidal lines is a fundamental property of the models in both the *Commentariolus* and *De revolutionibus*.

However, the models for both planets in *De revolutionibus* differ from those in the *Commentariolus* in that the two epicycles are moved from the circumference to the center of the planet's sphere, to form two eccentricities (see Figs. 5.56(b) and 5.73(b)). This was a correction of an error in the earlier models (Figs. 5.54 and 5.72) in which the equation of center had to be measured from an external point rather than from the earth, a difficulty possibly due, as noted earlier, to Copernicus's using the eccentric model for the second anomaly of the inferior planets, in which the external point is the center of the planet's eccentric (Fig. 5.55). The corrected models (Figs. 5.56(a) and 5.73(a)), with the eccentricities at the center of the planet's sphere, are then exactly the same as Ibn ash-Shāṭir's models (Fig. 5.53 and 5.70), and the equation of center is measured, as it should be, at the earth.

Because he could not obtain the proper observations of Venus or Mercury, Copernicus relies upon Ptolemy's to derive the parameters of his models, the results being essentially adaptations of Ptolemy's parameters. The procedure is straightforward for Venus, but rather more intricate for Mercury, where the variation of the radius of the orbit using Ṭūsī's device takes the place of the variation of the distance of the center of the epicycle in Ptolemy's model. He does, however, make some changes for both planets, of less than clear origin, that apply to his own time, although he assumes that the parameters derived from Ptolemy's observations are correct for the time of Ptolemy.

In the case of Venus, he reduces the double eccentricity from 416 to 350, giving as his reason "as many observations teach us" without specifying what kind of observations. It had become standard in medieval Arabic sources, originally under Indian influence, to make the double eccentricity of Venus equal to the eccentricity of the sun, whatever the value of the solar eccentricity, and make their apsidal lines coincident, practices followed in the *Alfonsine Tables* and so also in the *Commentariolus*. In *De revolutionibus*, Copernicus retains Ptolemy's sidereally fixed apsidal line of Venus, which does not coincide with the moving apsidal line of the earth, although the reduction of Venus's double eccentricity to 350 is rather close to the earth's 323, and originally the reduction had been to 332 or 333, closer still. But since he says the reduction was either derived or confirmed by many observations, there is probably more to what he is doing than just following a traditional practice. The observations used by Ptolemy to find Venus's double eccentricity, the distance from the earth to the equant point, are those with the planet near greatest elongation and the mean sun 90° from the apsidal line. Conditions for such observations at Copernicus's time were not ideal, but they were no worse than they were for Ptolemy, so it is possible that he derived this parameter by applying Ptolemy's method to his own observations. His only observation of Venus in the text is the 1529 occultation that is used to correct the mean motion by comparing it with a very old observation by Timocharis from -271, although originally an observation of Ptolemy's from 138 was used for this purpose.

For Mercury Copernicus does the opposite, holding the eccentricity constant since Ptolemy—or the equivalent of Ptolemy's eccentricity in Copernicus's slightly different model—but advancing the apsidal line by the large amount of 28° of sidereal longitude. The distinctive feature of Ptolemy's model for Mercury, that the sum of opposite greatest elongations is greater at $\pm 120°$ from the apogee than at 180°, is taken over without demonstration, and Copernicus even fails to present half of Ptolemy's observational evidence that motivated the model in the first place. In order both to correct the mean motion and confirm the shift of the apsidal line, he attempts a demonstration using three observations by Bernhard Walther, one from 1491 and two from 1504. The first observation is used to derive a value of the mean anomaly under the assumption that all of Ptolemy's parameters have remained constant except for the new location of the apsidal line. The two observations from 1504 are then used as a test of the mean anomaly and the shifted apsidal line, the idea being to keep shifting the apsidal line by trial and error, and repeating the whole procedure, until the observations are correctly reproduced. However, Copernicus does not manage to do this, and ends up simply altering Walther's observations to agree, more-or-less, with the results of what appears to be only his first attempt.

Following the derivations for each planet, Copernicus places the tables of corrections. These are mostly computed from Copernicus's models and parameters, but for Jupiter and Saturn some of the corrections are copied or adapted from the corresponding corrections in the *Alfonsine Tables* or the *Almagest*, something that Copernicus had done previously in the lunar correction table. For Mars and Venus, however, Copernicus's parameters are not the same as earlier sources, and for Mercury the model is not quite equivalent to Ptolemy's—although it is remarkably close—so he had to compute all the corrections himself. The last subject taken up in Book V is retrograde arcs. Here Copernicus's treatment is very brief, being confined to a proof of Apollonius's Theorem on the location of the stationary points, and a single worked example for Mars at mean distance. Originally, in his manuscript, he had also computed the retrograde arc and time for Mars at greatest and least distance, but he made a number of errors and, presumably for this reason, cancelled the demonstrations.

Of all the parts of *De revolutionibus*, it is the latitude theory of Book VI that shows the least advance over the *Commentariolus*, for even though Copernicus made a number of changes, the unmanageably complicated theory of the inferior planets remains in a state of confusion. The problem was that Copernicus was attempting to reproduce the latitudes of Ptolemy's models and tables in the *Almagest*, but his own models would not do it.

New observations of the sort required for the latitude theory would have been difficult to obtain even if Copernicus had been inclined to try, which of course he was not. Kepler was referring to the latitude theory when he made his famous remark about Copernicus's representing Ptolemy rather than nature. It was, in more ways than he knew, his source of greatest difficulty.

The latitude theory of the superior planets is identical to that of the *Commentariolus* except that Copernicus does a more careful job of adapting parameters, the variable inclinations of the orbital planes, from the limiting latitudes for opposition and conjunction in Ptolemy's tables. However, his model cannot reproduce the extrema for Mars, so he alters them to fit the narrower range of which the model is capable. Since the extreme latitudes in the *Almagest* were none too accurate anyway, the alterations are harmless. Aside from the violation of uniform circular motion in the production of the oscillations of the orbital planes, which Copernicus does not acknowledge, both the model and the adaptation of parameters for the superior planets are straightforward.

The inferior planets are another matter. Here we shall repeat some of what was said earlier about Ptolemy's theory and Copernicus's original theory in the *Commentariolus*. Ptolemy's latitude theory consists of three components, three separate inclinations, all of them varying from a maximum value to zero. The first, the *inclination*, is an inclination of the plane of the epicycle in the line of sight; the second, the *slant*, is an inclination of the plane of the epicycle across the line of sight; the third is an inclination of the plane of the eccentric. When transformed into a heliocentric arrangement, the inclination and the slant become two different ways of viewing the planet's inclined, interior orbit, in or across the line of sight, and this much Copernicus reached in the *Commentariolus*. However, the numerical values of the inclination and slant are not the same, and therefore it is necessary to vary the inclination of the planet's orbit in accordance with the motion of the earth, rather like the equation of center of the inferior planets. But in the *Commentariolus*, Copernicus believed this to be unnecessary—he thought that the motion of the earth in itself was adequate—for he gave the orbit a single fixed inclination for both components. In *De revolutionibus* he corrects this by giving the orbital plane a variable inclination with exactly Ptolemy's parameters. However, he did not do this until very late, for Rheticus's account of the latitude theory in the *Narratio prima* agrees with the *Commentariolus* in giving the orbital plane a fixed inclination and attributing the effects of the inclination and slant solely to the motion of the earth. Even with the correction, however, a problem remains, for the model is not equivalent to Ptolemy's.

The third component, intended to substitute for Ptolemy's variable inclination of the plane of the eccentric, is a variable inclination of the orbital plane arranged such that the inclination is maximum, and the point of greatest inclination coincides with the planet, when the earth is in the planet's apsidal line, the inclination is zero when the earth is 90° from the apsidal line, and the point of maximum inclination moves with the speed of the mean anomaly of the planet. The model is not equivalent to Ptolemy's because the distance of the planet from the earth has a pronounced effect on the latitude, which is not the case in Ptolemy's model. In the *Commentariolus* Copernicus overlooked this, but in *De revolutionibus* he takes the effect of distance into account, and computes additional columns for the latitude tables, although with an error for Venus that greatly reduces the effect of distance on the apparent latitude.

The latitude tables are in the same form as Ptolemy's, and more than is warranted, Copernicus copies directly from Ptolemy's tables. The columns for Saturn and most of Jupiter are copied, with interpolation where necessary, from the tables in the *Almagest*, while Mars is computed independently since Copernicus's model does not reach the extrema of Ptolemy's. The columns for the inclination and slant of Venus and Mercury are copied, again with interpolation, from the *Almagest*, but this is not correct since the models are not equivalent. In Ptolemy's model there are two separate variable inclinations of the epicycle on two perpendicular nodal lines, one in and one across the line of sight, while in Copernicus's model the orbital plane has a single variable inclination. The differences in the apparent latitudes of the two models can amount to about 1° for Venus, but only about 0;12° for Mercury. Finally, additional columns are computed in order to take account of the effect of distance on the third component of latitude of the inferior planets, and these have nothing corresponding in the *Almagest*. Actually this is rather interesting, for what Copernicus has done is to take seriously the initially unintended consequences of a model that was supposed to behave differently, and then impose these consequences on the planets in the absence of any further justification.

Conclusions

In summarizing the astronomy of Ptolemy, we singled out three aspects of his work that are significant for understanding later astronomy, and particularly the astronomy of Copernicus.[23]

The last of the three, the physical problems of Ptolemy's models when represented as spheres, was the first to concern Copernicus, in the *Commentariolus*, where his object was to develop a physically correct planetary theory based upon the principle of uniformly rotating spheres. With the assistance of some kind of account of Marāgha theory, he was able to reach his goal—although some hidden difficulties remained—and with the assistance of Regiomontanus's description in the *Epitome* of the eccentric model of the second anomaly, he was able to go beyond his predecessors. For by

[23] Above, p. 41.

making his theory heliocentric, he achieved a complete physical system in which the order and distances of the planets were unambiguously determined, and all the peculiarities of the second anomaly in geocentric models were given a single explanation in the annual motion of the earth.

The first aspect of Ptolemy's work mentioned was the derivation of numerical parameters from observation. Here too Copernicus was greatly assisted by Regiomontanus's exposition of Ptolemy's, sometimes quite difficult, procedures, and here too he accomplished far more than his predecessors, deriving a complete set of elements for the sun, moon, and planets, by a consistent application of Ptolemy's methods and, as far as he could obtain the required observations, independent of earlier elements. This was the most difficult part of his work, occupying, as we have noted before, nearly twenty years of observation and several more years of computation, and by far the greatest part of *De revolutionibus* is, like the *Almagest*, devoted to setting out these derivations. That the results were not as numerically consistent or accurate as they could have been, does not detract from Copernicus's accomplishment as (probably) the first person since Ptolemy to carry out this travail de bénédictin.

But the most fundamental part of Ptolemy's astronomy, the derivation of models from observation, Copernicus never approached. Even when attempting to account for long-period variations of parameters—a problem not treated by Ptolemy—his method was, like his medieval predecessors, to invent a model in advance, and then force the earlier observations or parameters into it by trial and error, or just assumption. For the rest, the theoretical correspondence between Ptolemy's models and the apparent motions was accepted as established, while Copernicus devoted his attention first to correct physics of spheres, and then to correct parameters. Now, for this Ptolemy must carry much of the responsibility for not setting out the analyses by which he developed his models, which would at least have shown later astronomers, including Copernicus, that this was a necessary—indeed, as Kepler was to show, the most necessary—part of astronomy. And the subject is passed over just as briefly by Regiomontanus, who freely enlarges upon anything that strikes him as significant, so he too appears to have believed that the models were sound, and thus the only remaining functions of observation were the derivation and testing of parameters, not the determination of the fundamental structures of planetary theory. Copernicus may have advanced beyond Regiomontanus with regard to heliocentric conversion of models, but his understanding of Ptolemy was certainly not more profound than that of his Cicerone, so he can hardly be blamed for not undertaking to investigate what Ptolemy and Regiomontanus both considered to be settled.

The features of Copernicus's theory for which he is most frequently criticized are the result of this single cause. The measurement of eccentricities and apsidal directions from the mean rather than the true sun, the intersection of orbital planes in the mean sun, the dependence of the equation of center of the inferior planets on the motion of the earth, the oscillation of the orbital planes of the planets in accordance with the motion of the earth—all follow from Copernicus's belief that Ptolemy's models are correct, and that his task was to preserve their effects, which, as far as possible, he did. This is what Kepler called representing Ptolemy, not nature, and given his limitations, Copernicus had no other choice and, in any case, did not intend to do anything else. Kepler intended his remark as, not so much a criticism, as an explanation, and it remains the fairest, and the shortest, evaluation of Copernicus's astronomy.

3. Texts, Editions, and Translations

The Text of *De revolutionibus*

The principal documents for establishing the text of *De revolutionibus* are Copernicus's autograph manuscript, which we shall designate M, and the 1543 Nuremberg edition, which we call N. There are in addition some authentic sources of emendations to N, namely, a printed errata sheet, that we call E, attached to many surviving copies of N, and several copies of N containing a larger number of hand-written corrections, related to those in E, that appear to be based upon a comparison of N with a prior and authoritative textual source. This cannot be M, for hundreds of differences between N and M that are intentional revisions are not marked for correction, but must be either the fair copy or perhaps a working copy that Copernicus made in the course of his revision. We shall briefly consider each source.[1]

M: *Codex Cracoviensis Bibliothecae Iagellonicae 10,000.*[2]

This is Copernicus's holograph, and it is a document of great value—probably the earliest surviving holograph of a major scientific work—for it records Copernicus's composition and revision of his book over a period of a number of years. M was substantially complete in 1539, when Rheticus read Copernicus's book, but continued to receive revisions, perhaps as late as 1541. Many further revisions of the book were not entered in M, but were contained in the fair copy from which N was printed. Consequently, M

[1] There is a more detailed discussion of the relation and use of these sources in Swerdlow (1981), and a list of the corrections in copies of N, omitting those already in E, in Gingerich (1981).

[2] There is an excellent facsimile in Copernicus (1944), and a still better facsimile in Copernicus (1973), with an extensive discussion of the paper, handwriting, and history of the ownership of the manuscript by J. Zathey. Following Copernicus's death, the manuscript passed, presumably, to Giese, who died in 1550, but probably before that he gave it to Rheticus, who held it until his death in 1574. From Rheticus, the manuscript passed to his young associate Valentine Otho (ca. 1545–ca. 1603), who completed and published Rheticus's *Opus palatinum de triangulis* in 1596. The entire history of the manuscript can be found in Copernicus (1973), 41–49.

cannot be used as the fundamental text of Copernicus's work since it preserves a version containing many errors and inconsistencies that Copernicus later corrected in his final revision that appears only in N. Nevertheless, M can be used to correct some typographical errors in N—but only in cases where N is demonstrably wrong and M demonstrably right—and is of course invaluable for the information it gives on the composition and earlier stages of *De revolutionibus* (which would probably not please its author).

N: *Nicolai Copernici Torinensis De revolutionibus orbium coelestium libri VI . . . Norimbergae apud Ioh. Petreium, Anno M. D. XLIII.*[3]

This is the first edition, and the fundamental text of *De revolutionibus* in its final form, differing from M in hundreds of readings, both significant and trivial. The final revision of the text may have been the work of Copernicus alone, or he may have been assisted by Rheticus, who was with him for most of the period from May of 1539 until September of 1541.[4] But however the work may have been divided, there is no reason to believe that the entire revision did not have Copernicus's full approval. Thus N is the principal witness to the version of *De revolutionibus* that Copernicus wished to have printed, and must be taken as the fundamental text for any critical edition or translation. However, N is far from perfect, containing somewhere between two and three hundred typographical errors, and these must be corrected on the basis of M, the corrections in E and the corrected copies of N, or by careful emendation, always remembering that some errors may belong to the author.

E: Errata Sheet.[5]

The correction sheet, which claims to be based upon a collation of N with the "autograph" (*autographum*)—a manuscript in the hand of the author, but not, in this case, M—contains 108 corrections extending only to f. 146r of N, and of these 80 agree with M and 28 do not. It is possible that some of the 28 are actually late revisions made by Copernicus after the fair copy had gone to the printer. The corrections in E are, perhaps without exception, sensible and authoritative, and should be adopted into a corrected text of N.

Corrected Copies of N.

Owen Gingerich (1981) has found 94 corrections, in addition to those in E, entered into several copies of N. As far as we have been able to tell

[3] There have been four publications of facsimiles of N: Paris, 1927; Turin–Amsterdam, 1943, which includes E; Leipzig, New York–London, 1965, which is of a copy that includes the handwritten corrections; Brussels, 1966. We have used the 1965 reprint.

[4] The opinion that has sometimes been expressed, that the revisions in N were the work of Rheticus alone, is very unlikely, and that they were, as has also been suggested, due to unauthorized tampering by someone unnamed is simply not possible. The revisions show sufficient concern with and understanding of small numerical details, with errors and inconsistencies in M, that only Copernicus himself, or someone working in close collaboration with him, could have made them.

[5] There are facsimiles of E in the 1943 Turin–Amsterdam reprint of N and in Swerdlow (1981).

from our own work, virtually all of these should be adopted, and quite a number agree with M in any case.

Note on the Dating of M

The chronology of Copernicus's work on *De revolutionibus* is inseparable from the chronology of the writing of M, which contains virtually all of the evidence for the progress of his work. The dates of observations are not of much help since the date of a certain observation shows that the part of M using the observation was written later than the observation, *but not how much later*, since years can elapse between making an observation, applying it, and writing the section of the book containing it. Hence, it is only the latest of the observations that provide evidence for an earliest possible date of parts of the book. One observation that is of importance for dating is the 1525 observation of Spica in III,2 used in the precession theory. Since the solar, lunar, and planetary theories of Books III–V all use the complete precession theory, all of Books III–V must have been written later than 1525, but this in itself does not tell us how much later. Hence, it cannot be maintained, for example, that the solar theory was written shortly after the 1515–16 observations of equinoxes, for in no case are observations prior to 1525 of any interest for the dating of M.[6]

The principal evidence for dating parts of M is the dating of its paper by watermarks, either from external documents on the same paper or from the content of the pages. Watermark dating is, of course, notoriously uncertain since papers with similar watermarks were manufactured for many years, and it is difficult to know how long even a single batch of paper was in use. There are four papers used in M, conventionally designated C, D, E, and F (A and B are merely flyleaves), and because M is a composite document with layers of revision, the papers do not occur exactly in order, although the sequence is in general first C, then D, then D and E together, then E, and last F.[7] Papers E and F have been dated from external documents, dated letters by Copernicus, but papers C and D have not, and so depend upon their contents in M. The papers are approximately dated as follows:

C. Paper C is definitely the earliest layer of M, found only in the first four sheets of Book I (Preface to beginning of I,10), the star catalogue of

---

[6] The chronology of M presented here is shorter than that due to L. Birkenmajer and A. Birkenmajer, which uses early observations for dating and thereby extends the writing of M back to about 1515, directly following Copernicus's promise of a "larger volume" in the *Commentariolus*. For this long chronology, see J. Zathey's introduction to Copernicus (1973) and Rosen (1977), a rather exasperated vituperation against the shorter chronology presented by Swerdlow (1974), who remains unrepentant.

[7] The watermarks are illustrated, and there are good charts of the distribution of the papers in each fascicle, by Zathey in Copernicus (1973).

Book II with an early version of II,14, and the precession theory and the beginning of the solar theory of Book III. The use in Book III must be after 1525, the date of the observation of Spica in III,2, but the first pages of Book I show that C must be later still. Edward Rosen points out in the notes to his translation that in I,7 (M f. 5v), on why the ancients thought the earth was at rest in the center of the universe, "Copernicus' language unmistakably resembles the corresponding passage in George of Trebizond's translation of PS" (Rosen's abbreviation for the *Almagest*, a word he avoids), and this translation was first published in 1528.[8] Rosen has made a very important discovery, for it shows that the introductory chapters of Book I must have been written after 1528. On the other hand, the star catalogue and the precession theory appear to be based on the Gerard of Cremona translation, and are probably earlier than I,7.[9]

D. Paper D is found throughout M. In Books I and II, it could in part be a substitution for earlier sheets of C, while in Books III–VI it appears to be original, the transition from C to D occurring in the last pages of the precession and the first pages of the solar theories. D constitutes almost the entirety of Book IV, is used, apparently simultaneously with E, in Book V, and last occurs at the beginning of Book VI. The principal evidence for its dating is its interleaving with E, the sheets of which do not appear to be replacements, showing that D was still in use in the later 1530's, while its independent use in Book IV must be somewhat earlier.

E. Paper E is also found throughout M. In Books I–III (and perhaps IV) it is usually a substitution for earlier sheets in a series of extensive revisions, while in V–VI it is original. E is dated externally by three letters written on it by Copernicus to Bishop Dantiscus in 1537 and 1539. Thus, parts of the holograph using E are dated to the later 1530's, including the period after Rheticus's arrival in 1539 when much of the revision using E took place.

F. Paper F is used only for two sheets in Books I and VI that are very late revisions since F is dated by a letter from Copernicus to Duke Albrecht in June of 1541.

In summation, the internal and external dating of the paper of M indicates a period extending from some time after 1525 or 1528 for the earliest use of paper C, to the later 1530's for paper E, with D falling somewhere in between, and even 1541 for paper F. It is unlikely that Copernicus acquired George of Trebizond's *Almagest* and wrote I,7 immediately after the publication of the translation in Venice in 1528. Further, since the observations in *De revolutionibus* end in 1529, and since about one-third of M is written on paper E,

[8] Copernicus (1978), 350–51, notes to 14:34 and 14:40. There are two passages, and while some doubts are possible, we concur with Rosen's judgment. It is, of course, hardly possible that Copernicus used a manuscript of Trebizond's translation.
[9] See below, pp. 123, 133, n. 11.

dated to the later 1530's, it appears most likely that Copernicus began the writing of M only after his observations had ended, so that the composition of *De revolutionibus* began in the early 1530's or even somewhat later. As noted earlier, the book was substantially complete by the time Rheticus read it in 1539, but continued to receive revisions for about two more years.

Editions and Translations

The 1543 Nuremberg edition of *De revolutionibus* was reprinted, essentially unaltered, in Basel in 1566 along with Rheticus's *Narratio prima* (a facsimile of this edition was published in Prague in 1971). In 1617 Nicolaus Muler (or Muller) edited in Amsterdam a "corrected," annotated, and somewhat modernized edition, under the title *Astronomia instaurata, Libris sex comprehensa, qui de Revolutionibus orbium coelestium inscribuntur*, with specimen computations, a catalogue of observations used in the book, and a certain amount of supplementary material reflecting the development of astronomy as of the early seventeenth century (although Muler's perceptions of this hardly coincide with our own). It is, in fact, an intelligently executed and useful edition that corrects a number of numerical errors in N, for Muler may well have had a better technical understanding of Copernicus's astronomy than any subsequent editor of the text. The value of the edition was clearly recognized by J. Baranowski, an astronomer of the Warsaw observatory and the editor of the 1854 Warsaw edition, the next to be published, for he follows the Amsterdam edition, which he calls "the most accurate and suitable of all," a reasonable enough judgment, although not leading to a critical edition of Copernicus's text. Baranowski was the first editor to consult M, from which he printed the original preface, but apparently he otherwise made little use of it. He also printed the *Narratio prima*, the letters of Theophylactus Simocatta, a number of Copernicus's letters, including the criticism of Werner, the treatise on coinage, and other texts pertinent to Copernicus, all texts accompanied by a Polish translation.

Baranowski's edition has been little used, for it was followed within twenty years by the first of the three so-called critical editions that have utilized M in establishing the text. It is here that the trouble began that has continued up to the present time, for all the editors have assumed that M was the authoritative version of *De revolutionibus*, since it is in the author's hand, while N was considered the result of unauthorized tampering, by Rheticus or worse. Consequently, Copernicus's final revision of his text has been disregarded in favor of superseded, erroneous, and sometimes internally contradictory readings from M. However, the editing has never been done with much consistency, and the results have been more-or-less random pastiches of M and N, with few selections of readings based upon the detailed understanding and extensive recomputation required for editing a technical work like *De revolutionibus*.

The first of these critical editions, which also includes the *Narratio prima*, was the sumptuous Thorn edition of 1873, published by the Societas Copernicana Thorunensis, financed by and dedicated to Kaiser Wilhelm I, in which the establishment of the text was largely the work of M. Curtze. The next edition, by the Kopernikus-Kommission der deutschen Forschungsgemeinschaft, was published in Munich in 1949, five years after the same organization published the 1944 facsimile of M. The text was edited by F. Zeller and K. Zeller, and there is a more complete reporting of the variants of M and N, and of the corrections in M, than in the Thorn edition. The most recent edition was published in Warsaw and Cracow in 1975 by the Polish Academy of Sciences, also following an earlier publication of a facsimile of M in 1973. The text was partially edited by R. Gansiniec, and posthumously revised and completed by J. Domanski and J. Dobrzycki. The reporting of corrections in M is more complete than in previous editions, but entire columns of tables that differ in M and N, some of which had been printed in the 1949 Munich edition, are omitted. There is extensive annotation, to I,1–11 by A. Birkenmajer, to the rest of the text by J. Dobrzycki, and much of the annotation is interesting.

Since all of these editions have preferred readings from M to the revisions in N, although with no great consistency, none is a satisfactory version of Copernicus's final text, which should be the object of a critical edition. Any serious work on *De revolutionibus* should in any case be done from M and N directly, but that is no obstacle since facsimiles are readily available. Nevertheless, yet another critical edition is in progress in Germany, likewise as part of a multi-volume publication, and this one, it is reported, will take N as the basic text. That, of course, is the right thing to do, but after three unsuccessful attempts, we must observe that the availability of facsimiles would seem to make such an edition superfluous for scholarly purposes, and one might reasonably doubt whether a casual reader has much use for a critical edition with an extensive apparatus.

There have been a number of translations of *De revolutionibus* into various languages—German, English, Polish, Russian, Hindi, and more are on the way—of which only those in German and English are known to us. The earliest of all was a German version made before 1589 by Nicolaus Raymarus Ursus for Jost Bürgi, the noted instrument maker to Wilhelm IV, the Landgraf of Hesse.[10] This translation has not been published, although it survives in Graz, Univ. Bibl. MS 560, and it would be of interest to see what Ursus did with Copernicus's technical terms. The first published translation was the Polish version by Baranowski included in the 1854 Warsaw edition of the text. The next to appear was a German version by C. L. Menzzer published by Der Coppernicus-Verein für Wissenschaft und Kunst in Thorn in 1879 with subvention by Kaiser Wilhelm I. Menzzer's version is not inferior to more recent translations—it is conventional to observe, sometimes

[10] Rosen (1982), a clever piece of detective work.

at length, that he mistranslates the title—and it is amusing to note the dutiful reproduction of some of Menzzer's errors in the later translations. In general he follows the Thorn edition, although correcting some, but not nearly enough, of its incorrect readings, and he provides extensive annotation, checking many simpler computations, for the most part correctly. The mathematical notation is slightly modernized and run in with the text.

No less than three English translations have appeared. The first, by C. G. Wallis in the *Great Books of the Western World*, vol. 16 (1952), mostly follows the Thorn edition, sets the mathematics off from the text in modern notation, and contains essentially no annotation. The second, by A. M. Duncan (1976), is correctly based upon the Nuremberg edition, which in itself makes it preferable to the other translations, although little is done to correct the edition's faults, and is on the whole both literal and clear, bringing the reader quite close to the original Latin. For some strange reason, the figures are taken from the 1617 Amsterdam edition, and are so greatly reduced in size that they are difficult to use. The mathematical notation is quite literal—although the notation of sexagesimal fractions is curious—and run in with the text. The most recent, by Rosen (1978), is a fairly drastic rewriting of the text into a modern American-English of short sentences rather far removed from Copernicus's style, although it is no more inaccurate than previous translations. The underlying text is eclectic, combining readings from M and N more-or-less at random, and having some resemblance to the 1975 Warsaw–Cracow edition. The mathematical notation is modernized after a fashion and run in with the text, and omitted numbers are often filled in, although not always correctly. The annotation is extensive, that for I,1–11 being quite detailed, with references to many obscure sources, while the more technical remainder of the work is not treated as thoroughly. Many of the simplest computations are checked, often at great length and not always correctly, but the checking stops short of even applications of the law of cosines.

The principal problem with all of the translations is that they are frequently inaccurate where a correct translation depends upon a technical understanding of the astronomy of both Ptolemy and Copernicus. Since later translations have tended to follow earlier ones in passages of any difficulty, there is really little to chose between them in the matter of accuracy, although in the matter of text and literal approach to the original, Duncan has the advantage over the others. Perhaps one may best summarize the quality of the editions and translations of *De revolutionibus*, and the quality of the annotation provided with both, by observing that all the publications fall far short of the standard established by Nallino in his edition of al-Battānī (1899–1907) and by Manitius in his translation of the *Almagest* (1912–13). The critical editions should be used only with facsimiles of M and N at hand, and the translations should not be used for serious purposes without a copy of the Latin text, for all technical passages and numbers must be checked carefully.

The standard edition of the *Commentariolus* is that in Prowe (1883–84) 2, 184–202, which was based upon earlier publications by Curtze (1878), (1882), and Lindhagen (1881) of the Vienna and Stockholm manuscripts. Prowe's edition contains many erroneous readings and no variants, and should be used only along with the earlier publications and the more recent translations, all of which contain emendations. The same is true of the Latin text in Rossmann (1948). However, a new edition is being prepared in Poland, also using the Aberdeen manuscript, that should supersede the earlier texts. The *Commentariolus* has been translated many times: into German in Müller (1899), Rossmann (1948), which is adapted in Zinner (1951); into English in Rosen (1971) and Swerdlow (1973); into French in Hugonnard-Roche, Rosen, and Verdet (1975); and other translations will doubtless appear. Copernicus's letter to Wapowski on Werner's *De motu octavae sphaerae* is published in Prowe (1883–84) 2, 145–54, 169–83, collated with additional manuscripts in Birkenmajer (1900), 492–509, and translated into English in Rosen (1971), 93–106. The text of Rheticus's *Narratio prima* is available in a facsimile of the 1540 Danzig edition (Osnabrück, 1967), in Copernicus (1873), Prowe (1883–84) 2, and in Kepler (1937–) 1. It has been translated into German in Zeller (1943), English in Rosen (1971), and French in Hugonnard-Roche, Rosen, and Verdet (1975). Finally, there has just appeared a new edition of the *Narratio prima*, with a French translation and extensive commentary by Hugonnard-Roche and Verdet in *Studia Copernicana* 20 (1982).

Editions of Copernicus's Sources

It is usually not difficult to determine which editions of earlier texts Copernicus used, either because his copies survive in the University Library of Uppsala or because only one edition was available in his time. In this section, we shall list his principal sources and whatever editions he may have used, along with other editions and translations we have consulted in this study.

Ptolemy, *Almagest*. The Gerard of Cremona translation was published in Venice in 1515—we shall frequently refer to this as the 1515 *Almagest*—and a copy annotated by Copernicus is at Uppsala. He also used the translation by George of Trebizond, first published in Venice in 1528, although his copy has not been discovered, and we have used a 1551 Basel printing of this translation. In 1539 Rheticus brought Copernicus the 1538 Basel edition of the Greek text with Theon's commentary, but the use Copernicus could make of this at so late a date was limited. We have also used Heiberg's edition of the Greek text (1898–1903) and Manitius's translation (1912–13). A new translation of the *Almagest* with extensive annotation by G. J. Toomer has just appeared (1984).

Peurbach and Regiomontanus, *Epitome of the Almagest*. One presumes that Copernicus used the 1496 Venice edition, although his copy has not been found, and we have used both this edition, which is rather poor, and the better manuscript Venice, lat. 328.

Alfonsine Tables. Copernicus's copy of the 1492 Venice edition, with his annotations, is at Uppsala. We have used a film of this copy and also a later 1518/21 Venice printing.

Regiomontanus, *Tabulae directionum*. Copernicus had the 1490 Augsburg edition bound with the 1492 Venice edition of the *Alfonsine Tables*, and we have used a film of this copy and likewise of Copernicus's notes at the end of the volume.

Giorgio Valla, *De expetendis et fugiendis rebus*, Venice, 1501. This is the only printing of Valla's encyclopedia, of which Books XVI–XVIII are on astronomy, and Book XVII contains Ptolemy's star catalogue.

The preceding are the most important of Copernicus's sources, and of course the *Almagest* and the *Epitome* take precedence over all the others. Also of great importance at an early stage of Copernicus's work was some account of Marāgha theory, but whatever form this may have taken is at present unknown. The following is a list of other sources used by Copernicus that are, if marked with an asterisk, among his books at Uppsala. The list is not complete, for in the course of the following study we shall mention some others, and we have taken no notice of a number of sources used in the beginning parts of Book I of *De revolutionibus*.[11]

Werner, *De motu octavae sphaerae*, Nuremberg, 1522.
Peurbach, *Theoricae novae planetarum*, many early editions.
Censorinus, *De die natali*, several early editions.
Eusebius (trans. Jerome), *Chronicon*, several early editions.
*Ptolemy, *Cosmographia*, Ulm, 1486.
*Pliny, *Historia naturalis*, Venice, 1487.
*Euclid, *Elementa geometriae*, Venice, 1482.
**Scriptores astronomici veteres* (Aratus, Manilius, Firmicus, etc.), Venice, 1499.
*Euclid, *Elements*, with Proclus's commentary on Book I, Greek text, Basel, 1533.
*Regiomontanus, *De triangulis omnimodis*, Nuremberg, 1533.
*Jābir ibn Aflaḥ, *De astronomia*, Nuremberg, 1534.
*Regiomontanus, *Ephemerides*, 1491–1506, Augsburg, 1492.
*Stoeffler and Pflaum, *Ephemerides*, 1499–1531, Ulm, 1499.
*Stoeffler, *Calendarium Romanum magnum*, Oppenheim, 1518.

[11] All the surviving books owned or annotated by Copernicus are listed in Czartoryski (1978). Many sources presumed to be used by Copernicus in the beginning parts of Book I, some of them quite obscure, are identified in the notes to Copernicus (1975) and Copernicus (1978).

Purpose and Limitations of This Study

What follows is intended as an examination of Copernicus's mathematical astronomy from a technical, rather than a philosophical or sociological, point of view. While everything in *De revolutionibus* may not be mathematical astronomy, most of it certainly is, and thus this limited exposition may provide a foundation of elementary knowledge to assist scholars who wish to study more profound subjects for which we are not qualified. We have nothing to say about whether Copernicus was a good German or a good Pole, of whether he was a Priest, a Pythagorean, a Platonist, a Neoplatonist, a Hermetist, of the spelling of his name, the title of his book, of whether he liked or disliked common people, and how such things may have affected his astronomy. Historians and philosophers have already written more than we have been able to read about the meaning of Osiander's preface, the aesthetics of the heliocentric theory, whether the theory is heliocentric or heliostatic or geokinetic, about the influence of ancient philosophers whose writings were entirely lost, of medieval philosophers whose writings were entirely neglected, of Renaissance philosophers who thought the sun beautiful (and who does not?), and we have neither the desire nor the learning to add to this extensive literature. All of what we have to say about Copernicus's life is contained in the preceding biography, and all of what we have to say about his physical astronomy is contained in the preceding general essay. It is our, perhaps naive, belief that when Copernicus wrote that mathematics is written for mathematicians, he meant it, and that his work may best be understood within the history of mathematical astronomy.

Our study is primarily an exposition of *De revolutionibus* as presented in Copernicus's final revision of the text contained in the 1543 Nuremberg edition (N). Copernicus's holograph (M), containing an earlier version of the text is used constantly, but we discuss its contents only where it either clarifies material in N or shows earlier stages of Copernicus's work that appear to us to be significant or interesting, and many details, corrections, and variant readings in M pass without comment. Thus, while every demonstration in N is set out and recomputed, we usually report only final results of earlier versions in M. We have used as our texts the 1965 Leipzig, New York–London reprint of N, and both the 1944 Munich and 1973 Warsaw–Cracow reprints of M. While we have done all our work directly from the texts, we have attempted to give credit to any interesting points mentioned in the notes to the various modern editions and translations, although not recording the inevitable duplication of noting the same variants and corrections in M and N and checking the same calculations.

One of the more difficult and subjective issues in setting out an older text such as *De revolutionibus* is that of anachronism in notation and terminology. Our mathematical notation is modern, and for the notation of sexagesimals we have used commas to separate places, and semicolons to separate integers and fractions. In analyzing models and procedures, we have sometimes used modern formulations, e.g. tangent functions, that Copernicus does not use, but we have confined our recomputations to his own more limited procedures. Likewise, we have constantly used terms like *equant* and *orbit*, the former used only twice by Copernicus, the latter never, simply because one has constantly to refer to the equant property of models and because orbit is the most convenient term for the circular or nearly circular path of a moving point or body. In any case, particularly in the planetary theory, we have discussed Copernicus's own technical terms and their history.

A great deal of *De revolutionibus* is taken up with calculation, and while hardly the principal point of our analysis, we have checked every number in the demonstrations, although not in all the tables. There are few things more tedious, more detrimental to a straightforward exposition, than the reporting of computational errors, most of which are of no particular significance. But in analyzing Copernicus's derivations of parameters in detail, the reporting of errors is an unavoidable, if regrettable, consequence (which the reader may ignore if he wishes). There being no point in wasting space on correct computations, we give our result only if it differs from the text, and we have been flatly inconsistent about reporting rounding differences of ± 1 in the last place. Mercifully, we do not show the arithmetic of adding up mean motions or intervals of time, or subsidiary steps in computing the equation of center, and such, which the curious reader may find set out at length, along with many other simple calculations, in the notes to Copernicus (1879) and Copernicus (1978). The computations have been done both by hand and, for checking tables or final results of long derivations, with a programmable calculator—without which the recomputation of the omitted iterations for the parameters of the superior planets would have been, to say the least, impractical—and the steps in all cases have been those actually or probably used by Copernicus. Finally, we have at least double-checked every calculation, and have been as meticulous as we could in proof-reading the thousands of numbers in this study in every stage of manuscript and printing. Yet, in such an *ingentem numerorum multitudinem*, as Copernicus would say, it is not possible to be certain that all errors have been detected, for which we ask the reader's indulgence in checking any number that appears suspicious.

While we have attempted to make this exposition self-contained, it would naturally be helpful to read it with a copy of *De revolutionibus* at hand, preferably a facsimile of N.

2

Trigonometry and Spherical Astronomy

In the first eleven chapters of Book I, Copernicus sets out a general description of the planetary system and of the movement of the earth. It is here that he presents his principal arguments for the heliocentric theory and attempts to answer astronomical and physical objections to the motion of the earth. These chapters are so well known, and have been commented upon at such length by so many historians and philosophers, who have devoted particular attention to just this part of Copernicus's work, that anything we might have to say, after the hundreds if not thousands of pages that have already been written, would be bringing owls to Athens.[1] Therefore, we shall pass immediately to the more practical, if less celebrated, expositions of trigonometry and spherical astronomy that occupy the greater part of Books I and II. It is of course true that Copernicus made no original contributions to either subject, but his presentation is on the whole admirable, and the rest of his work is either unintelligible, or of severely reduced intelligibility, without them.

1. Trigonometry (I,12–14)

The exposition of trigonometry in Chapters 12–14 of Book I is both more extensive and more didactic than the corresponding sections of the *Almagest* and *Epitome.*[2] The utility of Copernicus's presentation was shown by the separate publication of I,13–14 as *De lateribus et angulis triangulorum* (Wittenberg, 1542), with a preface by Rheticus and a set of seven-place ($R = 10{,}000{,}000$) sine tables at intervals of 0;1°, which are more practical than the five-place ($R = 100{,}000$) tables at intervals of 0;10° included in I,12. The textual differences between the two publications are not significant. Copernicus's trigonometry rests exclusively on the chord and the sine, which he never calls by this name, but denotes as "half-chord of the double arc." The computation of the table of half-chords in I,12 closely follows the procedure described in the *Almagest* and *Epitome*. It is divided into six theorems and one problem:

1. From the inscription of the sides of regular polygons of 6, 4, 3, 10, and 5 sides, it is shown that where the radius of the circumscribed circle $R =$ 100000,

$$\text{crd } 60^\circ = R = 100000, \qquad \text{crd } 90^\circ = \sqrt{2}\,R = 141422,$$

$$\text{crd } 120^\circ = \sqrt{3}\,R = 173205, \qquad \text{crd } 36^\circ = \tfrac{1}{2}(\sqrt{5} - 1)R = 61803,$$

$$\text{crd } 72^\circ = \tfrac{1}{2}(\sqrt{10 - 2\sqrt{5}})R = 117557.$$

---

[1] For detailed expositions one may consult, e.g., A. Birkenmajer in Copernicus (1959) and Copernicus (1975), Kuhn (1957), Koyré (1970), Rosen in Copernicus (1978).

[2] Copernicus's trigonometry has been treated in the standard histories of the subject, e.g. Braunmühl (1900), 139 ff., Tropfke (1923), by subject, and in more historical detail in Zeller (1946), 39–56, and in the notes to Copernicus (1975), 387–91. Our own treatment will be brief.

To this is added a porism. Let $\bar{\alpha} = 180° - \alpha$; then

$$\text{crd } \bar{\alpha} = \sqrt{(2R)^2 - \text{crd}^2\, \alpha}.$$

The porism is immediately applied to find

$$\text{crd } 144° = 190211, \qquad \text{crd } 108° = 161803.$$

2. Proof of Ptolemy's Theorem: If any quadrilateral be inscribed in a circle, the product of the diagonals is equal to the sum of the products of the opposite sides. Thus (Fig. 1), given inscribed quadrilateral $ABCD$,

$$AC \cdot BD = AB \cdot CD + AD \cdot CB.$$

3. Using Theorem 2, it is shown that if $\alpha > \beta$, given crd α and crd β,

$$\text{crd } (\alpha - \beta) = \text{crd } \alpha \text{ crd } \bar{\beta} - \text{crd } \bar{\alpha} \text{ crd } \beta.$$

4. The half-angle theorem is derived by a method differing from the *Almagest* and *Epitome*, although the resulting formula is the same, namely,

$$\text{crd}^2\, \tfrac{1}{2}\alpha = 2R(R - \tfrac{1}{2} \text{ crd } \bar{\alpha}).$$

Where $R = 1$, this is equivalent to

$$\text{crd } \tfrac{1}{2}\alpha = \sqrt{2 - \text{crd } \bar{\alpha}}.$$

5. Again using Theorem 2, given crd α and crd β,

$$\text{crd } [180° - (\alpha + \beta)] = \text{crd } \bar{\alpha} \text{ crd } \bar{\beta} - \text{crd } \alpha \text{ crd } \beta,$$

from which crd $(\alpha + \beta)$ may be found by the porism.
6. Proof that if $\alpha > \beta$,

$$\frac{\text{arc } \alpha}{\text{arc } \beta} > \frac{\text{crd } \alpha}{\text{crd } \beta}.$$

This is followed by the problem of showing that for small arcs, the ratios of the arcs and chords converge to equality, specifically, that to the accuracy of five-place tabulation, the chord of 1° may be found by linear interpolation between the chords of $1\tfrac{1}{2}°$ and $\tfrac{3}{4}°$, which are known through Theorem 4.

These procedures will suffice to compute a table of chords or of sines, and Copernicus then gives his "Table of Straight Lines Subtended in a Circle" which, as mentioned, is a table of sines for $R = 100{,}000$ for one quadrant at intervals of 0;10°, with a column of differences between each entry for interpolation. Whether Copernicus computed the table himself is unclear.[3]

---

[3] We have checked and eliminated the following sine tables, all at intervals of 0;1°, as the source of Copernicus's table (either by computation or rounding or simple copying): Regiomontanus, *Tabulae directionum*, $R = 60{,}000$; Regiomontanus, *Compositio tabularum sinuum rectorum* (Nuremberg, 1541 and earlier manuscripts) $R = 6{,}000{,}000$ and $R = 10{,}000{,}000$; P. Apianus, *Instrumentum primi mobilis* (Nuremberg, 1534) $R = 100{,}000$ (Copernicus owned this book, which contains also Jābir ibn Aflaḥ); Copernicus, *De lateribus*, $R = 10{,}000{,}000$ (this table, which Rheticus is thought to have added to Copernicus's little book, seems to agree with

Aside from some scribal errors in M and further typographical errors in N, just under 10% of the 540 entries are in error by 1 in the last place (47 cases of -1, 2 of $+1$). However, it is not altogether certain that Copernicus used precisely this table for his own work. More convenient tables at 0;1° intervals were available in manuscript, e.g. those of Regiomontanus, and he may well have had access to them. Although we have found a number of instances in which an error in a sine or chord given elsewhere in *De revolutionibus* corresponds to an error in the table, there are far too many errors that do not correspond to the table to permit an unambiguous decision.

Chapter 13 consists of seven theorems on the solution of plane triangles, and Chapter 14 of fifteen theorems on the solution of spherical triangles. Although some of the theorems occur in the *Almagest* and the *Epitome*, such a systematic exposition is not present in either work. Here Copernicus is instructing his reader and may, for all one knows, have been instructing himself. In the following, we will write sin α and cos α where Copernicus would write half-chord 2α and half-chord $(180° - 2\alpha)$. Whenever possible, however, Copernicus uses chords instead of half-chords. For example, the sine theorem for a triangle always appears as a ratio of chords in the circumscribed circle (Fig. 2):

$$a/b = \text{crd } 2\alpha/\text{crd } 2\beta.$$

Following Ptolemy, Copernicus's way of distinguishing circumferential and central angles in the circumscribed circle is to write

$$\text{angle } CAB = \alpha, \quad \text{where 4 right angles are } 360°,$$

$$\text{angle } CAB = 2\alpha, \quad \text{where 2 right angles are } 360°.$$

Then the chord to a given angle, or to the arc of the circumscribed circle,

$$a = 2 \sin \alpha = \text{crd } 2\alpha.$$

For the solutions to plane triangles, Copernicus gives, in 13,i, the law of sines

$$\frac{a}{\text{crd } 2\alpha} = \frac{b}{\text{crd } 2\beta} = \frac{c}{\text{crd } 2\gamma}, \quad \text{i.e.} \quad \frac{a}{\sin \alpha} = \frac{b}{\sin \beta} = \frac{c}{\sin \gamma}. \tag{A}$$

For the case of two sides and the included or exterior angle, that is, where the included angle is acute or obtuse, in 13,iv,v, he draws perpendicular *CD*,

---

Regiomontanus's table for the same radius, although Rheticus gives differences for minutes, while Regiomontanus gives them for seconds).

Among Copernicus's manuscript materials (Curtze (1875), 221, 243; Prowe (1883–84) II, 216) are a sexagesimal sine table for $R = 1;0,0,0$ and, as a handwritten addition to the *Tabula foecunda* (tangent table) in the 1490 Augsburg *Tabulae directionum*, a table of secants, i.e. $1/\cos \alpha$ for $R = 100{,}000$. Both are at 1° intervals, as is Regiomontanus's tangent table. Copernicus appears to make no use of the tangent or secant in *De rev.*

which is shown in Fig. 2 for an acute interior angle, and applies the law of cosines

$$a^2 = (c - b \cos \alpha)^2 + (b \sin \alpha)^2 = b^2 + c^2 - 2bc \cos \alpha. \qquad \text{(B)}$$

In the case of finding the angles of a triangle of given sides, in 13,vii, the law of cosines is used in the form

$$\cos \alpha = \frac{b^2 + c^2 - a^2}{2bc}. \qquad \text{(C)}$$

Somewhat more interesting is an alternate solution to the case of three given sides, likewise in 13,vii, that is directly applicable to the problem of finding the length of a planet's retrograde arc and the location of the stationary points.[4]

The solutions to spherical triangles, in 14,iii,iv, are based upon the following relations for a right triangle (Fig. 3):

$$\text{crd } 2\alpha = \text{crd } 2a/\text{crd } 2c, \quad \text{i.e.} \quad \sin \alpha = \sin a/\sin c, \qquad (1)$$

$$\text{crd } 2\bar{c} = \text{crd } 2\bar{a} \text{ crd } 2\bar{b}, \quad \text{i.e.} \quad \cos c = \cos a \cos b, \qquad (2)$$

$$\text{crd } 2\beta = \text{crd } 2\bar{\alpha}/\text{crd } 2\bar{a}, \quad \text{i.e.} \quad \sin \beta = \cos \alpha/\cos a. \qquad (3)$$

All three are consequences of the Menelaus Theorem, the first two known since antiquity and applied in the *Almagest* and the *Epitome*. The third, usually known as "Jābir's Theorem"—from Jābir ibn Aflaḥ's *De astronomia* I,14—follows immediately from applying (1) to right triangle *DEB*:

$$\sin \beta = \sin \bar{\alpha}/\sin \bar{a} = \cos \alpha/\cos a.^5 \qquad \text{(3a)}$$

The formula is derived and applied in *Epitome* II,34, which Copernicus in fact uses in II,3 for finding the angle of intersection of the ecliptic and meridian (see (8) below), and he received the 1534 edition of Jābir from Rheticus in 1539, although this is rather late for Copernicus to have made much use of it. As a consequence of (3) and (1) applied to β, Copernicus also gives the extended relation

$$\frac{\text{crd } 2b}{\text{crd } 2c} = \frac{\text{crd } 2\bar{\alpha}}{\text{crd } 2\bar{a}} = \text{crd } 2\beta, \quad \text{i.e.} \quad \frac{\sin b}{\sin c} = \frac{\cos \alpha}{\cos a} = \sin \beta. \qquad (4)$$

However, he does not use this formula when, in II,3, it would be appropriate to give the general relation between longitude, declination, right ascension, and the obliquity of the ecliptic (see (7) below).

---

[4] See below, p. 464. n. 37.

[5] On the Menelaus Theorem and the spherical trigonometry in the *Almagest*, see the standard histories in n. 2, Pedersen (1974), 69–78, *HAMA*, 26–30. Copernicus's statement of Jābir's Theorem has been noted in Copernicus (1975), 389. The theorem is also proved by Regiomontanus in *De triangulis* IV,18.

Ever since pointed out by Clavius, it has been well known that Copernicus's exposition of plane and spherical trigonometry contains a rather absent-minded error. If two triangles, plane or spherical, have two angles and one side (except for a special spherical case), or two sides and the included angle equal, they are congruent, and with this information they may be solved for their remaining parts. However (Fig. 4(a) and (b)), if two sides (*AB* and *BC*) and an opposite acute angle (*A*) are given, the case is ambiguous since the other angles (*C* and *B*) may be either acute or obtuse, so one can neither prove congruence nor solve the triangles without further information, specifically, whether a perpendicular (as from *B*) falls inside or outside the triangle. Copernicus misses this entirely three times over:

(1) for the solution of a plane triangle in 13,vi,
(2) for the congruence of spherical triangles in 14,viii,
(3) for the solution of a spherical triangle in 14,xi.[6]

It is difficult to evaluate the significance of such errors in considering whether Copernicus developed his exposition of trigonometry himself, borrowed it from a defective source, or was careless in adapting a correct source. It is also remarkable that Rheticus, who certainly knew something about trigonometry, did nothing to correct these mistakes when editing *De lateribus et angulis triangulorum.*

Although these trivialities have received excessive attention, since they are easy to spot, they are nevertheless of some interest for whatever information they may provide about Copernicus's sources. In 1539 Rheticus brought Copernicus a copy of Regiomontanus's *De triangulis omnimodis* (1533). Regiomontanus specifically considers the ambiguous case of two sides and an opposite acute angle, for a plane triangle in I,51 and for a spherical triangle in IV,29. If Copernicus gave any kind of serious attention to this book, it is hard to see how he could have avoided noticing these theorems and correcting his errors. This leads one to question the frequent assertion that Theorems xiii–xv of chapter 14, which are inserted out of place in M (ff. 24–25), were based upon Copernicus's study of Regiomontanus.[7] The three theorems, part of the spherical trigonometry, are:

xiii. Given the sides of a triangle, the angles are given.
xiv. If a given arc less than a semicircle be divided into two arcs, and the ratio of the sines of the arcs be given, the arcs are given.
xv. Given the angles of a triangle, the sides are given.

[6] Discussed in Zeller (1946), 50–51. Similar objections by Nunes are translated by Rosen in Copernicus (1978), 365–67. Nunes and Clavius object also to an ambiguity in 14,xii, the case of two angles and a side, but the problem here applies only to a special case of a triangle constructed on the extension of the base of a triangle whose vertical sides are equal to 180°, the criterion for which is in Regiomontanus, *De triangulis* III,48. Obvious examples are triangles formed by meridians standing on the equator.

[7] The question of Copernicus's use of *De triangulis omnimodis* is discussed in detail in Zeller (1946), 52–56. The paper on which Copernicus wrote the additional theorems in M can be dated to ca. 1541, after he received Regiomontanus's book.

Regiomontanus does treat each of these problems, but one must still be wary of concluding that Copernicus's exposition is based upon Regiomontanus's. Theorem xiv, corresponding to *De triangulis* IV,21, is also proved in *Almagest* I,13 and *Epitome* I,17. Theorem xv corresponds to *De triangulis* IV,33 and Theorem xiii to IV,34 (alternate constructions for the latter are given in V,3,4). Copernicus's proofs of these two theorems are less clear than Regiomontanus's proofs, and differ in a number of details. Although it is possible to reduce both expositions to basically the same procedures, it seems odd that Copernicus would show changes for the worse if his source were in fact *De triangulis*. Also, these theorems are simply too important to have been omitted from Copernicus's original exposition. All these reasons lead us to believe that Copernicus's trigonometry, mistakes and all, was entirely his own composition, and was based upon little more than what he could learn from the *Almagest* and the *Epitome*. If he did use Regiomontanus for a late revision, he did not use him well.

2. Spherical Astronomy

It is obviously indifferent to spherical astronomy whether the apparent rotation of the celestial sphere is due to its own motion or to the rotation of the earth. Copernicus can therefore take over his spherical astronomy unchanged from earlier sources, and in this particular field there had been some progress, at least in the computation of useful tables, since Ptolemy. Most of the more elaborate tables were in Arabic and thus unknown to Copernicus, but in the fifteenth century Giovanni Bianchini in Italy produced an extensive set of tables for spherical astronomy, and Regiomontanus likewise put together his very useful *Tabulae directionum* (which, according to Cardano, were plagiarized from Bianchini).[8] Copernicus owned the 1490 Augsburg edition of the *Tabulae directionum*, and this work, along with the *Almagest* and the *Epitome*, formed the basis of his exposition of spherical astronomy.[9]

Obliquity of the Ecliptic (II,2)

The only observational quantity necessary to spherical astronomy is the obliquity of the ecliptic ε which is found by measuring the altitude or zenith distance of the sun at summer and winter solstice, and then bisecting the resulting arc between the tropics. In *Almagest* I,12 Ptolemy describes two

[8] Cardano's remarks on Regiomontanus's horoscope in Cardano (1563) 5, 498. Bianchini's spherical tables survive in manuscript, but we have not checked to see whether Cardano tells the truth. According to G. Rosińska (*Math. Rev.* 83 m:01015), it appears that he does.

[9] The spherical astronomy in the *Almagest* is treated in *HAMA*, 30–52, Pedersen (1974), 94–121.

instruments for making these measurements. The first is a graduated metal meridian circle with a rotating inner circle containing two diametrically opposite prisms with two pointers to the outer circle. When the sun crosses the meridian, and the higher prism exactly shadows the lower, the zenith distances may be read directly on the graduated circle. The other is a wood or preferably stone plinth aligned in the meridian on which is drawn, from a center in the upper southern corner, a graduated quadrant. A horizontal gnomon is placed at the center point, and at meridian transit the gnomon will cast a shadow across the quadrant giving the zenith distance. Yet another instrument is described in *Epitome* I,22 consisting of a graduated quadrant, presumably of metal, with a rotating radius on which are mounted two pin-hole sights. When the rays of the sun in the meridian pass through both pin holes, the zenith distance may be read on the quadrant.

Copernicus describes only the plinth—a rather large one three or four cubits square—which he says should be made of wood, or better of stone or metal, although it is unclear whether he himself ever made or used such an instrument.[10] Nor is it easy to decide what value of ε he found "from 30 years of frequent observation" (III,6) since II,2, III,2, and III,6 contain a variety of values both in M and N. All that interests us here though is that his values fall close to 23;28°, the obliquity found by Peurbach and Regiomontanus (*Epitome* I,22), which is the value for which he computes his tables and which he conjectures to be the minimum value of ε. He explains that Ptolemy and, according to his reading of the *Almagest*, Eratosthenes and Hipparchus found 23;51,20°, so he believes that the maximum value is 23;52°.[11] Thus, all tables dependent upon ε are computed for 23;28°, and in some a correction column is provided to adjust for higher values.

Ecliptic and Equatorial Coordinates (II,3–4)

In II,3 Copernicus sets out the computation of three fundamental quantities which he then tabulates. These are shown in Fig. 5 in which E is the equinox, BE a quadrant of the equator, AE a quadrant of the ecliptic, and from the pole F of the equator descends the solstitial colure FAB. Take a point G on the ecliptic of longitude λ, and let a meridian FGH descend from F. Following *Epitome* I,23 Copernicus first finds the declination δ from the equivalent of

$$\sin \delta = \sin \varepsilon \sin \lambda \tag{5}$$

[10] In II,14 (N, 44v) Copernicus describes the use of, presumably, this instrument (he uses the plural *instrumentis* although he has described only one) to find the longitude of the sun from its meridian altitude and a declination table, and then, taking $\Delta\lambda \approx 0;2,30^{\circ/\mathrm{h}}$, to estimate its longitude at any time between meridian transits. It is just such meridian altitudes that are used to determine the equinoxes in III,13. See below, pp. 148–149.

[11] On Copernicus's theory of the variation of the obliquity, see below, pp. 133 ff.

which is an application of spherical formula (1). To find the right ascension α, Copernicus, using *Epitome* I,24, solves

$$\sin \bar{\alpha} = \sin \bar{\lambda}/\sin \bar{\delta}, \tag{6a}$$

another application of (1), and then takes $\alpha = 90° - \bar{\alpha}$. This is equivalent to a direct application of formula (2) in the form

$$\cos \alpha = \cos \lambda/\cos \delta. \tag{6b}$$

Epitome II,33 gives the useful relation of the four fundamental quantities

$$\sin \bar{\delta}/\sin \bar{\varepsilon} = \cos \delta/\cos \varepsilon = \sin \lambda/\sin \alpha, \tag{7}$$

which Copernicus does not mention. However, from (3), the relation leads, in II,34, to the angle γ of the intersection of the ecliptic and meridian,

$$\sin \gamma = \sin \bar{\varepsilon}/\sin \bar{\delta} = \cos \varepsilon/\cos \delta, \tag{8}$$

used by Copernicus. He calls γ the "meridian angle," and its principal application, for finding the angle made by the ecliptic with the horizon, is explained in II,10.

Next it is shown that each of these functions is symmetrical to the equinoxes and solstices, so that tabulation for a single quadrant will serve for any point of the ecliptic. The tables are computed from 1° to 90° at intervals of 1°, using the minimum value $\varepsilon = 23;28°$, and a correction column is provided giving the difference for the maximum $\varepsilon = 23;52°$, that is, for each function f, the correction d is found from

$$d = f(\varepsilon_{\max}) - f(\varepsilon_{\min}),$$

where d is positive for δ, negative for α and γ.[12] To find δ, α and γ for each quadrant, observe the following rules:

for	$0° \le \lambda \le 90°$	enter with	λ	read	$+\delta$,	α,	γ,
for	$90° \le \lambda \le 180°$	enter with	$180° - \lambda$	read	$+\delta$,	$180° - \alpha$,	$180° - \gamma$,
for	$180° \le \lambda \le 270°$	enter with	$\lambda - 180°$	read	$-\delta$,	$180° + \alpha$,	$180° - \gamma$,
for	$270° \le \lambda \le 360°$	enter with	$360° - \lambda$	read	$-\delta$,	$360° - \alpha$,	γ.

II,4 takes up the important problem of transformation of coordinates from ecliptic coordinates, longitude and latitude, to equatorial coordinates, right ascension and declination, and also shows with which degree of the ecliptic any point on the celestial sphere will culminate. The subject is treated in *Almagest* VIII,5 and *Epitome* VIII,6–7. In Fig. 6 let ELC be a quadrant of the equator, EKB a quadrant of the ecliptic, and let H be a star of known

---

[12] The use of the correction d is explained below, p. 145. Copernicus's instructions for the addition or subtraction of d are not altogether clear, and are mixed up in the translations. Both "the latter" (*his*), α and γ, are negative; "the former" (*illis*), δ, is positive (N, 30v31–34).

longitude $\lambda = EK$ and latitude $\beta = KH$. We wish to find $\delta = NH$ and $\alpha = EN$, and since $\alpha_M = \alpha_N$, M is the point of the ecliptic with which H will culminate, that is, cross the meridian, because all points of the same right ascension culminate simultaneously. Hence we must also find $\lambda_M = EM$. First, in triangle KEL, by formula (3),

$$\cos \kappa = \cos \lambda \sin \varepsilon, \qquad \cos KL = \cos \varepsilon / \sin \kappa,$$

and by formula (1),

$$\sin EL = \sin KL / \sin \varepsilon.$$

Thus,

$$HL = \beta + KL.$$

Then, in triangle HLN, from (1),

$$\sin \delta = \sin NH = \sin \kappa \sin HL$$

gives the declination, and from (2)

$$\cos NL = \cos HL / \cos \delta,$$

so that

$$\alpha = EN = EL - NL$$

gives the right ascension. Finally, entering the right ascension table inversely with α, one finds the corresponding longitude λ_M of the point M with which H culminates. To find λ_M trigonometrically, consider triangle EMN in which, from (3) and (1).

$$\cos \gamma = \cos \alpha \sin \varepsilon, \qquad \sin EM = \sin \alpha / \sin \gamma.$$

and $\lambda_M = EM$ is the required ecliptic longitude.

All of this may seem rather laborious compared with the modern formulas for transformation of coordinates, but Copernicus also gives a method of using the preceding tables to solve part of the problem. If KE be taken as the "right ascension" of LE, inverse use of the right ascension table will give LE. Then, entering the declination table with LE, LK will be given as the "declination," and likewise entering the meridian angle table with LE, angle κ will be given as the "meridian angle." The reasons are obvious if the various formulas are compared and it is noted that ε governs triangle KEL. From here one must proceed as before trigonometrically through triangle HLN. However, when α_N has been found, inverse use of the right ascension table will directly give λ_M.

Shadow Lengths (II,6)

In II,5 Copernicus explains right ascension, when the celestial equator and parallel circles rise perpendicular to the horizon at the earth's equator, and oblique ascension, when the equator and parallel circles rise at some other angle at any other latitudes. In II,6, following *Almagest* II,5 and *Epitome* II,9–10, he shows how to determine the length of shadows cast by a gnomon for a given solar altitude, and conversely. Thus in Fig. 7, given a known altitude α, gnomon AB will cast a shadow of length

$$BC = \frac{\cos \alpha}{\sin \alpha} AB.$$

For the converse problem, where the shadow length BC is given, Copernicus must solve

$$AC^2 = AB^2 + BC^2$$

and then find

$$\cos \alpha = BC/AC \qquad \text{or} \qquad \sin \alpha = AB/AC.$$

The lack of a tangent function in solving these problems is an obvious disadvantage.

Length of Daylight and Ascensional Corrections (II,7–8)

II,7 is on the problem of determining the length of daylight and the sun's rising point on the horizon for a given solar longitude and a given terrestrial latitude of the observer, and conversely, to determine the observer's latitude from the length of the longest day. The problems are treated in *Almagest* II,2,3,7 and *Epitome* II,1–7. In Fig. 8, $ABCD$ is the meridian, BED the horizon, and AEC the equator. Take a point G on the horizon, and let a circle of declination FGH fall from the pole F of the equator. Then G will rise with point E of the equator and culminate, cross the meridian, with H which has the same right ascension as G.

First let G be any point of known declination. Thus in triangle EGH we know $\delta = HG$, and we presume knowledge of the latitude φ of the observer, so the co-latitude $\bar{\varphi} = 90° - \varphi = GEH$. The arc $a = EG$ is called the "ortive amplitude" (*latitudo ortus*—amplitude of rising), and shows how far from E on the horizon G will rise or set. From (1)

$$\sin a = \sin \delta/\sin \bar{\varphi} = \sin \delta/\cos \varphi. \tag{9}$$

The time from the rising to the culmination of G, called the "semidiurnal arc," is shown by $s = AEH$. Now the semidiurnal arc of E is obviously 90°. Thus the "difference" or "correction" (*differentia*) for the semidiurnal arc of G is $n = EH$, that is, from (2)

$$\cos n = \cos a/\cos \delta. \tag{10}$$

The ascensional correction n is an important auxiliary quantity with many uses. Having found n, the semidiurnal arc of G,

$$s = 90° + n, \tag{11}$$

where n is positive for positive δ and negative for negative δ.

If G is the location of the sun, then s is the time from sunrise to noon, and $2s$ is the length of daylight. $\bar{s} = HC = 180° - s$ may be called the "semi-nocturnal arc," so $2\bar{s}$ is the length of night. In II,8 Copernicus explains the conversion of $2s$ from degrees to hours, namely,

$$2s°/15° = \text{equal hours of daylight,} \tag{12}$$

$$2s°/12 = \text{length of seasonal hour in degrees,} \tag{13}$$

from which

$$\frac{2s°/12}{15°} = s°/90° = \text{length of seasonal hour in equal hours.} \tag{14}$$

If further, as Copernicus assumes in his example, G is the summer solstice, then $\delta = \varepsilon$, n reaches its maximum value, and thus

$$\mathrm{M} = 2s_{\max} = 2(90° + n_{\max}) \tag{15}$$

is the length of the longest day and

$$\mathrm{m} = 2s_{\min} = 360° - 2s_{\max} \tag{16}$$

is the length of the shortest day at the winter solstice.

The converse problem, of finding φ from a given M or a, where ε is assumed known, is easily solved. Given M, find

$$n = \tfrac{1}{2}(\mathrm{M} - 180°), \tag{17}$$

and since at the solstice $\delta = \varepsilon$,

$$\cos a = \cos n \cos \varepsilon, \tag{18}$$

so that

$$\cos \varphi = \sin \varepsilon / \sin a. \tag{19}$$

If a is known, only the last equation need be used.

Adapting the proof in *Epitome* II,8—a proof also appears in *Almagest* II,3—Copernicus next shows that a, n, and thus s are symmetrical to the solstices and equinoxes. This is followed by an analemma construction allowing the same quantities to be found by plane triangles. In Fig. 9 let $ABCD$ be the meridian circle and, in the plane of the meridian, let AEC be the diameter of the equator and BED the diameter of the horizon. Take a point F of declination $\delta = AF$, and draw the diameter of its parallel FKG which will intersect the axis of the equator in K and the line of the meridian, i.e. the intersection of the planes of the meridian and horizon, in N. The observer's latitude $\varphi = LD$. Next, consider $r = FK$ the radius, called in Indian astronomy the "day-radius," of the circle described by a body of declination δ. Where $r = \sin 90° = 1$ corresponds to a semidiurnal arc of $90°$, $KN = \sin n$, called the "earth-sine" by the Indians, will give the ascensional difference n. Now, following Copernicus's procedures,

$$\frac{EK}{FK} = \frac{\sin \delta}{\cos \delta} \quad \text{and} \quad \frac{KN}{EK} = \frac{\sin \varphi}{\cos \varphi},$$

and thus

$$\frac{KN}{FK} = \frac{(\sin\varphi/\cos\varphi)EK}{(\cos\delta/\sin\delta)EK} = \frac{\sin\varphi\sin\delta}{\cos\varphi\cos\delta} = \frac{\sin n}{r}. \qquad (20)$$

Use of the tangent function would of course simplify this to

$$\sin n = \tan\varphi\tan\delta.$$

In order to find the ortive amplitude a, it is necessary to find EN, and thus KN, where the radius of the horizon $BE = \sin 90° = 1$, for then $EN = \sin a$. Since $FK/BE = \cos\delta/1$, from the above relation

$$\frac{KN}{BE} = \frac{\sin\varphi}{\cos\varphi}\cdot\sin\delta,$$

and since $EN = KN/\sin\varphi$,

$$\frac{EN}{BE} = \frac{\sin\varphi\sin\delta}{\cos\varphi\sin\varphi} = \frac{\sin\delta}{\cos\varphi} = \sin a. \qquad (21)$$

It should be noted that the analemma construction for finding a and n is not more efficient than the strictly spherical procedure of the *Almagest* and *Epitome*. We have gone through Copernicus's presentation of it at some length because it is quite remarkable that he presents it at all, for his source and purpose are hardly clear. The construction is very old, probably going back to Hipparchus who may have used it to solve similar problems in his *Commentary on Aratus*, and such analemmata have a long and increasingly diverse history in Greek—including of course Ptolemy's *Analemma*—Indian, and medieval Arabic and Latin astronomy. Copernicus comes at the end of a long tradition, and is something of an antique collector. It is just as interesting, and perhaps more difficult, to explain his attachment to antiquated and, by standards of the *Almagest* and *Epitome*, curious procedures than to explain his original inventions.[13]

Copernicus then gives a useful table of the ascensional differences or corrections n that fills fully five pages. It is a double-entry table arranged for values of φ from 31° to 60° (approximately from Alexandria to Oslo) and of δ between 1° and 36°, each at 1° intervals. To give an example of its use, at a latitude $\varphi = 54°$, the sun or any other body at a declination $\delta = 20°$ will have an ascensional correction $n = 30;4°$. Its semidiurnal arc will thus

[13] A reconstruction of Hipparchus's analemma solution to finding the semidiurnal arc of a star with $\delta = 27\frac{1}{3}°$ is given in *HAMA*, 301–02, and later analemmata, including Ptolemy's, are treated at 839–56. Indian applications, in the *Pañcasiddhāntikā*, are analyzed in Neugebauer and Pingree (1971) II, 40–46. The more serious literature on the history of sun dials contains much on analemma constructions. Interestingly, Pedro Nunes, in his *De crepusculis* (Nunes (1943) and below n. 24), published in 1542, gives a virtually complete, and very elegant, exposition of spherical astronomy using almost exclusively analemmata. Another analemma used by Copernicus in III,2 is discussed below, pp. 131–132.

be $s = 90° + 30;4° = 120;4°$, and since $2s = 240;8°$, the length of daylight or the time the body is above the horizon will be $240;8°/15° = 16;0,32$ equal hours. Thus a seasonal hour will be $240;8°/12 = 20;0,40°$, or about 1;20 equal hours. Ortive amplitudes a are not tabulated, but they may easily be derived from the table by applying

$$\cos a = \cos n \cos \delta. \tag{22}$$

It was not hard for Copernicus to provide this table for most of it was merely copied from Regiomontanus. In the *Tabulae directionum* Regiomontanus gives a table of n for φ (1° – 60°) and δ (1° – 32°) at 1° intervals, the arrangement being the same as Copernicus's table. Within the range φ (31° – 60°), of some 960 values common to the two tables, all but 12 in N, and only 4 in M, are identical, and these 4 are emendations by Copernicus of obvious typographical errors in the 1490 Augsburg edition of the *Tabulae directionum* that he owned. Thus all Copernicus had to do was compute the additional entries for δ (33° – 36°), and of the 102 entries, 31 are in error—21 by 0;1° which is merely the effect of rounding, but 10 by errors as high as −0;38° (for $\varphi = 33°$, $\delta = 33°$). If, as Cardano says, Regiomontanus plagiarized from Bianchini, Copernicus returned the compliment to Regiomontanus.

Oblique Ascension and Applications of Right and Oblique Ascension (II,9,11)

II,9 is concerned with oblique ascension, a quantity used to determine the time of rising and setting with many applications. It is treated in detail in *Almagest* II,7–9 and *Epitome* II,19–27. We define oblique ascension ρ as follows (Fig. 10): A point G of longitude λ measured from the vernal equinox ♈ rises at the horizon with point E of the equator. If G is on the ecliptic, it is called the "horoscopus" or "ascendant," the rising point of the ecliptic. The declination δ of G is shown by GH, its right ascension α by ♈H, and its ascensional correction n by EH. Then the oblique ascension of G

$$\rho = \alpha - n, \tag{23}$$

where n is positive for positive δ and negative for negative δ. ρ is dependent upon terrestrial latitude, and at the equator $\rho = \alpha$. We may thus define ρ as *the arc of the equator that rises simultaneously with an arc of the ecliptic*, and designate the oblique ascension of λ by $\rho(\lambda)$. The rising time of any arc of the ecliptic is thus the difference in the oblique ascensions of its end points, that is, for arc $\Delta\lambda = \lambda_2 - \lambda_1$,

$$\rho(\Delta\lambda) = \rho(\lambda_2) - \rho(\lambda_1). \tag{24}$$

Since at the time that λ is setting $\lambda + 180°$ is rising, the setting time of the arc $\Delta\lambda$ is

$$\rho(\lambda_2 + 180°) - \rho(\lambda_1 + 180°) = \rho(\lambda_2) - \rho(\lambda_1) = \rho(\Delta\lambda), \tag{24a}$$

exactly the same as the rising time. The points rising and setting need not be on the ecliptic since from λ and β, one may find α and δ, and then ρ.

Copernicus provides a table of ρ applicable only to points of the ecliptic which follows a second table of right ascension intended to be used with it. Both tables are arranged by zodiacal signs at intervals of $\Delta\lambda = 6°$, and the oblique ascension table has columns for φ (39° – 57°) at intervals of 3°.[14] In the *Tabulae directionum* Regiomontanus gives tables of ρ for φ (0° – 60°) with both λ and φ at 1° intervals. Copernicus does not prove the symmetries of ρ—they are demonstrated in the *Almagest* and *Epitome*—but they follow immediately from the symmetries of n and s demonstrated in II,7. This allows an easy check of Copernicus's table. For example, for $\varphi = 54°$, $\lambda_1 = 90°$ has $\rho_1 = 53;28°$, but $\lambda_2 = 270°$ has $\rho_2 = 306;42°$. Thus $\rho_1 + \rho_2 = 360;10°$, while it should be 360°, so at least one is wrong. For $\lambda = (90°, 270°)$, $\delta = \varepsilon$, and interpolating in the table of ascensional differences for $\varphi = 54°$, $\delta = 23;28°$, one finds $n = 36;42°$. Since $\alpha(90°) = 90°$ and $\alpha(270°) = 270°$,

$$\rho_1 = 90° - 36;42° = 53;18°, \qquad \rho_2 = 270° + 36;42° = 306;42°,$$

and ρ_1 is thus in error by 0;10° (as also in M).

The applications of right and oblique ascension are numerous, and thus the tables of oblique ascension are used constantly for standard problems of which Copernicus gives examples in II,11. We shall give seven, and for simplicity we shall take in all cases $\varphi = 48°$ (approximately the latitude of Vienna).

1. The point of the equator rising with a point of the ecliptic. In Fig. 11 ♈E rises with $\lambda =$ ♈G, and thus ♈$E = \rho(\lambda)$. Let $\lambda =$ ♌ 0°, so from the table $\rho = 98;5°$. This means that ♌ 0° rises 98;5° = 6;32^h after ♈.

2. Length of daylight. This may be found directly from ρ without using α, n or s. If the sun rises at G, then at sunset $G + 180°$ will be rising because the ecliptic and horizon, being great circles, bisect each other. Thus the length of daylight d will be the oblique ascension of a semicircle of the ecliptic counted forward from G, that is, by (24)

$$d = 2s = \rho(\lambda + 180°) - \rho(\lambda). \tag{25}$$

Now ♌ 0° + 180° = ♒ 0° for which $\rho = 326;19°$, so when the sun is at ♌ 0°

$$d = 326;19° - 98;5° = 228;14° = 15;13^h.$$

---

[14] In M, 38v, facing the oblique ascension table, Copernicus began writing a table, but stopped after completing four columns of the field. Rather interestingly, it is an inverse oblique ascension table, i.e. entering with ρ, which increases uniformly with time, one can find the arc of the ecliptic that has risen with a given arc of the equator. Thus, the entry column, which is blank, would have read 6°–180°, and the field of the completed columns runs from Aries 0° to Virgo 30° (only the sign ♈ appears over the columns). Such a table, unknown to us elsewhere, would have many applications, a few of which are mentioned in the following notes, since one often needs to know what arc of the ecliptic rises or sets during a certain interval of time. It seems unfortunate that Copernicus did not finish the table, explain its use, and publish it in N. Whether it represents his own contribution to spherical astronomy, we do not know.

Therefore

$$s = \tfrac{1}{2}d = 114;7^\circ = 7;36^{\text{h}},$$

and since the sun is taken to culminate at local noon, it will rise at 4;24 A.M. and set at 7;36 P.M. Call a seasonal hour of day h', and

$$h' = 228;14^\circ/12 = 19;1,10^\circ = 1;16^{\text{h}}.$$

3. Time of rising or setting. Given the longitudes of the sun and any other object, first determine the interval between their risings or settings, that is $\rho(\Delta\lambda)$ given by (24). For objects not in the ecliptic, α and δ must be found in order to find ρ. Example for time of moonset (see Fig. 12): $\lambda_\odot$ = ♌0°, $\lambda_☾$ = ♍0° and $\beta_☾ = 5^\circ$, for which $\alpha = 153;56^\circ$ and $\delta = 16;10^\circ$. From (1.) $\rho_\odot = 98;5^\circ$ and from (23)

$$\rho_☾ = \alpha - n = 153;56^\circ - 18;47^\circ = 135;9^\circ.$$

Thus

$$\Delta t = \rho_☾ - \rho_\odot = 135;9^\circ - 98;5^\circ = 37;4^\circ = 2;28^{\text{h}},$$

and if the sun sets at 7;36 P.M., the moon will set at 10;4 P.M.

4. For a given solar longitude and time of day, find the horoscopus, the rising point of the ecliptic. Call the longitude of the sun $\lambda_\odot$, and define the time as t equal hours after sunrise. Then $t \cdot 15^\circ$ have passed since sunrise when $\lambda_\odot$ rose. Thus

$$\rho_G = \rho(\lambda_\odot) + t \cdot 15^\circ, \qquad (26)$$

and the ecliptic longitude λ_G corresponding to ρ_G will be the horoscopus. If the time is given in seasonal hours, first convert to mean time using (14), that is, for t' seasonal hours h', $t = (s^\circ/90^\circ)t'$. Example (see Fig. 13): $\lambda_\odot$ = ♌0°, hence we may assume the results of (2.), at $3^{\text{h}'}$ after noon. Since from sunrise to noon there are by definition $6^{\text{h}'}$, so $t' = 9^{\text{h}'}$ and

$$t = (114;7^\circ/90^\circ)9^{\text{h}'} = 11;24,42^{\text{h}}.$$

Thus, from (26)

$$\rho_G = 98;5^\circ + 11;24,42 \cdot 15^\circ = 269;16^\circ,$$

and from inverse use of the oblique ascension table λ_G = ♐ 5;35°.[15]

5. Given the longitude of the sun and the time of day, find the midheaven, that is, the degree of the ecliptic in the meridian (see Fig. 14). Define the time as t equal hours after noon. Since the equator and therefore anything moving parallel to it in the daily rotation cross the meridian at right angles, $t \cdot 15^\circ$

[15] Here it would be useful to have the inverse oblique ascension table (above note 14), since we could enter with $\rho = 269;16^\circ$ and then read by an easier interpolation ♐ 5;35° (but Copernicus did not write the part of the table for $\rho > 180^\circ$). If we wished then to know the horoscopus 2^{h} later, we would add 30° to ρ and directly read the corresponding λ_G.

measures the degrees of right ascension elapsed since the sun was in the midheaven. Just as (24) gives the difference of oblique ascension with respect to the horizon, so

$$\alpha(\Delta\lambda) = \alpha(\lambda_2) - \alpha(\lambda_1) \tag{27}$$

gives the difference of right ascension with respect to the meridian. Hence, find the right ascension of the sun $\alpha(\lambda_\odot)$, and

$$\alpha_M = \alpha(\lambda_\odot) + t \cdot 15^\circ \tag{28}$$

gives the right ascension of the midheaven t hours after noon. From α_M find λ_M inversely in the right ascension table. If seasonal hours are given, convert to equal hours using (14). Example: $\lambda_\odot = ♌0^\circ$, $t' = 3^{h'}$, seasonal hours, after noon. From (2.) and (14)

$$t = (114;7^\circ/90^\circ)3^{h'} = 3;48,14^h,$$

and since $\alpha(♌0^\circ) = 122;12^\circ$, so from (28)

$$\alpha_M = 122;12^\circ + 3;48,14 \cdot 15^\circ = 179;16^\circ,$$

and from the right ascension table $\lambda_M = ♍29;12^\circ$.

6. Given the horoscopus, to find the midheaven. In Fig. 15, we wish to find λ_M when given λ_G. From (1.) we know $♈E = \rho_G$, and $AE = 90^\circ$. So the right ascension of the midheaven

$$\alpha_M = ♈A = ♈E - 90^\circ = \rho_G - 90^\circ, \tag{29}$$

taking $360^\circ + \rho_G$ if $\rho_G < 90^\circ$, and λ_M is found inversely in the right ascension table.[16] In II,9 Copernicus gives a method using the semidiurnal arc $s = AH$, from which

$$\alpha_M = ♈A = ♈H - AH = \alpha_G - s, \tag{30}$$

taking $\alpha_G + 360^\circ$ if $\alpha_G < s$. The equivalence of the two formulas is obvious because

$$\alpha_M = \rho_G - 90^\circ = (\alpha_G - n) - 90^\circ = \alpha_G - (90^\circ + n) = \alpha_G - s.$$

One then finds λ_M by inverse use of the right ascension table.

Example. From (4.) $\lambda_G = ♐5;35^\circ$ for which $\rho_G = 269;16^\circ$. Therefore, from (29)

$$\alpha_M = 269;16^\circ - 90^\circ = 179;16^\circ,$$

[16] In order to avoid negative values of α_M or the addition of 360°, it was usual, beginning with the *Handy Tables* and found commonly in medieval tables, e.g. Khwārizmī, Battānī, *Toledan Tables*, *Alfonsine Tables*, to tabulate $\alpha'(\lambda) = \alpha(\lambda) + 90^\circ$, which we call the "normed right ascension." Thus, adapting (29)

$$\alpha'_M = \alpha_M + 90^\circ = (\rho_G - 90^\circ) + 90^\circ = \rho_G, \tag{29a}$$

that is, we simply carry ρ_G from the oblique ascension table to the normed right ascension table and inversely read λ_M. See *HAMA* 42, 979.

and from the right ascension table $\lambda_M = ♍29;12°$ which, as one would expect, agrees with the result of (5.). Copernicus's method is somewhat more laborious since we should first have to find for λ_G the new quantities α_G and its corresponding s which differs from the s found in (2.) Since it was necessary to find ρ_G in order to find λ_G in the first place, it is obviously more convenient to find α_M directly from ρ_G. Taking $\lambda_G = ♐5;35°$, the reader may confirm for himself that both methods produce the same result (and if he is using N, he will discover, incidentally, that $\delta(66°)$ should read 21;20°, as in M).

7. Given the midheaven, to find the horoscopus. This is, of course, the inverse of (6.). From the right ascension table, for λ_M find α_M, then

$$\rho_G = \alpha_M + 90°, \tag{31}$$

and by inverse use of the oblique ascension table ρ_G leads to λ_G.[17] Example: For $\lambda_M = ♍29;12°$ the right ascension $\alpha_M = 179;16°$ so that

$$\rho_G = 179;16° + 90° = 269;16°,$$

and inversely from the oblique ascension table $\lambda_G = ♐5;35°$, in agreement with (4.). In II,9 Copernicus also gives a trigonometrical solution to this problem. In Fig. 15, for λ_M find the declination δ_M which, in the configuration of the figure is negative. Since arc $BA = \bar{\varphi}$, so arc $MB = \bar{\varphi} - \delta_M$. Now, in triangle BMG, angle BMG is the meridian angle γ, and may be found from the table in II,3, while MBG is a right angle. Thus, from (3)

$$\cos MGB = \cos MB \sin BMG,$$

and although Copernicus does not mention it here, MGB is the angle v between the ecliptic and horizon, treated specifically in II,10. Then from (1)

$$\sin MG = \sin MB/\sin MGB,$$

and finally

$$\lambda_M + MG = \lambda_G.^{18}$$

Intersection of Ecliptic with Horizon and Circles of Altitude (II,10,12)

II,10 is concerned with the angles between the ecliptic and the horizon, which are treated in *Almagest* II,11 and *Epitome* II,35–38. In Fig. 16 the ecliptic intersects the horizon at the rising point G, and we wish to find $v = MGB$.

---

[17] Obviously the normed right ascension $\alpha'(\lambda)$ described in note 16 would allow us to carry α'_M directly to the oblique ascension table to find ρ_G. The inverse oblique ascension table (note 14) would then be convenient for finding λ_G.

[18] This procedure is essentially the same as formulas (35) and (36) of II,10 described below.

First case: Assume the longitude of the horoscopus λ_G is given. Then in right triangle MGB

$$\sin v = \sin MB/\sin MG,$$

so we must first find MB and MG. Since we are given λ_G, we can find ρ_G, and from (6.)

$$\alpha_M = \rho_G - 90°, \tag{29}$$

so that finding the corresponding λ_M,

$$MG = \Delta\lambda = \lambda_G - \lambda_M.$$

Now MB is the altitude of the midheaven, as used earlier in Copernicus's solution to (7.), although here with δ_M positive,

$$MB = \bar{\zeta}_M = \bar{\varphi} + \delta_M. \tag{32}$$

Thus the solution for v may be written

$$\sin v = \sin(\bar{\varphi} + \delta_M)/\sin(\lambda_G - \lambda_M). \tag{33}$$

Second case: Assume we are given λ_M of the midheaven. Extend GB and GM to quadrants at L and N—the latter point N is called in II,11 the *nonagesimus*, the "nonagesimal" or 90th degree of the ecliptic from the ascendant—and complete triangle LBZ from the zenith Z. From λ_M we can find $\bar{\zeta}_M$ as before, or directly find the zenith distance ζ_M from

$$\zeta_M = 90° - \bar{\zeta}_M = \varphi - \delta_M. \tag{34}$$

Further, $ZMN = \gamma$ is the meridian angle known from (8) or the table in II,3, and thus from (1)

$$\sin \bar{v} = \sin \zeta_M \sin \gamma,$$

which may also be written

$$\cos v = \sin(\varphi - \delta_M) \sin \gamma. \tag{35}$$

A result of the second case pointed out by Copernicus is that v and λ_M may be used to find λ_G, that is, following from (33)

$$\sin(\lambda_G - \lambda_M) = \sin(\bar{\varphi} + \delta_M)/\sin v, \tag{36}$$

which is identical to the solution to (7.) given in II,9. Angle v is then tabulated, exactly as ρ, for λ_G, the ascendant, at intervals of 6° and for geographical latitudes φ (39° – 57°) at intervals of 3°.[19]

[19] This table contains a strange error in M (40r) in that the values of v for ♈0° and ♎0° are reversed (they differ greatly), and in general it appears from spot checking to be one of the least accurately computed tables in *De rev*. Since finding λ_G and λ_M both involve inverse interpolation in the oblique ascension and right ascension tables, it is easy to see how inaccuracies in computations depending upon them can accumulate.

The final topic, and computationally the most laborious topic, in spherical astronomy is the angles and zenith distances formed by the intersections of the ecliptic and any circle of altitude, that is, a great circle passing through the zenith. These are treated in II,12, following the expositions in *Almagest* II,12 and *Epitome* II,39–45. From the meridian angle (7) or (8) and from (34), we already know how to find these quantities for the midheaven, the intersection of the ecliptic and meridian. For the ascendant, the angle is simply $90° - v$, and the zenith distance obviously is 90°. For any other circle of altitude, consider Fig. 17 in which we wish to find for the intersection of the ecliptic at S with vertical circle ZST, the angle $\xi = ZSM = GST$, and the zenith distance $\zeta = ZS$. It is assumed that we know λ_S and the time t from the culmination of S in the midheaven at M. Thus, modifying (28)

$$\alpha_M = \alpha(\lambda_S) + t \cdot 15°, \tag{37}$$

where t is positive when counted after culmination and negative before culmination. Consequently from α_M we know λ_M, and thus $MS = \lambda_S - \lambda_M$. Further, the entire arc $MG = \lambda_G - \lambda_M$ can be found from (7.), and then $SG = \lambda_G - \lambda_S$. From λ_M one finds $\bar{\zeta}_M$ and ζ_M from (32) and (34).

The remainder of the solution Copernicus leaves to the reader with the remark that it follows from the *demonstrata sphericorum*, although this is not terribly helpful. What he doubtless has in mind can be found in the *Epitome*. From *Epitome* II,43

$$\sin \bar{\zeta} = \frac{\sin (\lambda_G - \lambda_S)}{\sin (\lambda_G - \lambda_M)} \sin \bar{\zeta}_M, \tag{38}$$

and $\zeta = 90° - \bar{\zeta}$ gives the zenith distance of S. Then, in triangle STG from (1)

$$\sin v = \sin \bar{\zeta} / \sin (\lambda_G - \lambda_S), \tag{39}$$

which is another, although rather roundabout way of finding v, and from (3)

$$\sin \xi = \cos v / \cos \bar{\zeta}. \tag{40}$$

An easier solution is given in *Epitome* II,44. Complete the quadrants GMN and GBL, and draw perpendicular ZNL. Then N is the nonagesimal, so from (33) or (35) $v = \bar{\zeta}_N$ are known, from which

$$\sin \bar{\zeta} = \sin v \sin (\lambda_G - \lambda_S) \tag{41}$$

and as before

$$\sin \xi = \cos v / \cos \bar{\zeta}. \tag{40}$$

Copernicus does not prove the symmetries of v, ζ, and ξ to the solstices and equinoxes—the proofs are given in the *Almagest* and *Epitome*—nor does he tabulate ζ and ξ, although Ptolemy does so in *Almagest* II,13. However, Ptolemy's table, one of the largest in the *Almagest*, is not really satisfactory. One must enter with three arguments: φ by maximum length of daylight M for seven climata M (13^h–16^h) at intervals of $0;30^h$, λ by zodiacal sign of 30°, and t by hours from noon. The steps between entries are quite large, and interpolation uncertain. Hence one can understand Copernicus's reluctance to carry out the labor of computing such a table, even for the restricted latitudes φ (39°–57°). The principal application of ζ and ξ is to determine the lunar parallax in a circle of altitude and resolve it into its components of longitude and latitude. Since this procedure is required almost exclusively for solar eclipses (and lunar occultations), one might as well compute only as necessary, or do the entire job by computing tables of the components of lunar parallax themselves. Ptolemy provides such tables in the *Handy Tables*, which therefore do not contain tables for ζ and ξ, and Ptolemy's practice was followed in most medieval tables, both Arabic and Latin. The application of ζ and ξ will therefore be taken up in the section on lunar parallax.[20]

3. Risings and Settings (II,13)

Since the risings and settings of stars and planets depend upon the diurnal rotation, Copernicus chooses to consider visibility phases in II,13 following the spherical astronomy. It is evident that these highly complex phenomena, so important in ancient astronomy, particularly for meteorological prognostication from phases of fixed stars, are of little interest to him, and his remarks on the subject—the only attention given to it in his book—are not of much value. Visibility phases of stars are treated in *Almagest* VIII,4–6 and *Epitome* VIII,3–15. Ptolemy also wrote a separate work on the subject, the *Phaseis*, of which only Book II survives. The more complex phases of the planets are treated in *Almagest* XIII,7–10 and in *Epitome* XIII,22–29.[21]

The "ancient mathematicians," says Copernicus, distinguished true and apparent risings and settings. His own description is commonplace and qualitative, but here we can be more specific. True phases depend only upon oblique ascension ρ since all points of the celestial sphere on the eastern horizon have the same oblique ascension ρ and all points on the western horizon $\rho - 180°$. Thus, two bodies of oblique ascension ρ rise and set together; two bodies of oblique ascensions ρ and $\rho - 180°$ rise and set opposite to each other. Let the oblique ascension of the sun be ρ. Then if a star or planet likewise has the oblique ascension ρ, it will rise at sunrise and set at sunset. But if it has the oblique ascension $\rho - 180°$, it will set at sunrise and rise at sunset. The same principles apply to the simultaneous or opposite risings and settings of any two bodies.

True phases, however, are of little interest since the star or planet is invisible when they occur. More important are the apparent phases, for which the sun must be below the horizon by some distance depending, at the very

[20] See below, pp. 264–265.

[21] On these subjects see *HAMA*, 230–61, 926–31, 1017–25.

least, upon the brightness of the star or planet. This distance, shown in Fig. 18 as h, is called the *arcus visionis* (*Epitome* VIII,12), although the term is not used by Copernicus. h is the perpendicular depression of the sun S below the horizon when a star $*$ or planet P is seen to rise after its period of invisibility. We are given the time t of its rising. From t, by our solar and planetary theory find λ_S, λ_P and β_P, and it is assumed that λ_* and β_* are known. From the preceding spherical astronomy, we can find the oblique ascension of S, P, and $*$. Then we must find the midheaven M and the ascendant G, and thus arc MG is known. Hence we can find $\bar{\zeta}_M$, v, and from $\sigma = \lambda_S - \lambda_G$, the arc of the ecliptic below the horizon. Thus in triangles MGB and SGK,

$$\frac{\sin h}{\sin \sigma} = \frac{\sin MB}{\sin MG} = \sin v.$$

so that the arcus visionis follows from

$$\sin h = \sin \sigma \sin v. \tag{42}$$

Note (1) that the only observation entering into this procedure is the rather simple one that the planet or star rose at a certain time, and (2) that atmospheric refraction is neglected. It is further assumed (incorrectly) that h remains the same for all values of φ and for all locations of the star or planet on the horizon.

Obviously in Book II Copernicus can do nothing about solving this problem—and indeed he says nothing about finding h—but he does give the following values for the planets from *Almagest* XIII,7 and *Epitome* XIII,23:

Planet	h
Saturn	11°
Jupiter	10°
Mars	$11\frac{1}{2}$°
Venus	5°
Mercury	10°

He also says that the arcus visionis is about 12° for the first-magnitude (*primariis*) stars, and 18° for dusk or dawn (*crepusculum vel diluculum*), that is, evening or morning astronomical twilight, so that when the sun has reached 18° below the horizon the "smaller" (*minores*) stars, presumably sixth magnitude, begin to appear. The sources of these numbers are interesting, but not a little puzzling.

The value $h = 12°$ for first-magnitude stars is, to our knowledge, not in Copernicus's sources, although it is not implausible for it has been found that for such stars in Ptolemy's *Phaseis* an arc of 11° or 12° underlies the heliacal risings and setting (arcs of 6° or 7° are used for the opposite acronychal risings and cosmical settings).[22] However, in the *Planetary Hypotheses* Ptolemy gives 15° for all first-magnitude stars near the ecliptic.[23] In any case, there is no way Copernicus could know anything of these values. On the other hand, the Latin version of al-Battānī (cap. 49) gives the rising and setting arc of first-magnitude stars as "12 time-degrees of the equator" (12 *graduum ex temporalibus circuli aequinoctialis*).[24] However, this refers, not to the arcus visionis, but to the difference of right ascension of the sun and the star, shown by arc $\Delta\alpha = \alpha_S - \alpha_*$ in the figure, from which the difference in oblique ascension for a given φ and the difference in longitude may be computed. One may wonder if Copernicus saw this number, either directly in Battānī or more likely in some derivative source, and either misinterpreted it or found it already misinterpreted as an arcus visionis. We can only suggest the possibility, fully aware that Copernicus appears to know little from Battānī beyond what is reported in the *Epitome*.

The value $h = 18°$ for twilight, which Copernicus defines precisely as one now defines astronomical twilight—when the faintest stars become visible—is fairly common in Arabic sources, as is 19°, and is still the conventional modern value. It appears, most notably, in Ibn Muᶜādh's *Treatise on Twilight* which, translated into Latin as *Liber de crepusculis matutino et uespertino*, circulated in manuscripts containing Ibn al-Haytham's *Optics*. Accompanied by Pedro Nunes's much larger work on the same subject, the *De crepusculis* was printed in 1542 under the name of Ibn al-Haytham, a confusion of authorship that has only recently been corrected.[25] Obviously the printing was too late to be used by Copernicus, but that the *De crepusculis* is the source of his value for twilight, either from a manuscript or through an intermediary work, cannot be doubted.

The application of the arcus visionis h is to find $\Delta\lambda = \lambda_S - \lambda_P$, the elongation between the sun and a planet at the various phases, thereby allowing, with some difficulty, the date of a phase to be predicted. The non-trivial procedure for finding $\Delta\lambda$ is explained in the *Almagest* and *Epitome*, and in the *Handy Tables* Ptolemy provides extensive tables of $\Delta\lambda$ applicable to the

---

[22] As determined in Vogt (1920); see also *HAMA*, 930–31.

[23] Goldstein (1967), 9.

[24] Al-Battānī (1537), 77; Nallino (1899–1907), I,120, where the number in the Arabic text is 15° of right ascension. Ulugh Beg gives $h = 12°$ for first-magnitude stars, and increases h at 2° intervals up to 22° for sixth magnitude (Nallino I,286, from Sédillot's text and translation of Ulugh Beg).

[25] Ibn Muᶜādh's treatise is translated from a Hebrew version (the Arabic does not survive) in Goldstein (1977) with an excellent commentary. The Latin text, from the 1542 edition under the name of Ibn al-Haytham, is most readily available in Nunes (1943), 129–43. The correct attribution of the *De crepusculis* was clarified in Sabra (1967). In fact, Ibn Muᶜādh gives both 18° and 19°, and prefers 19° for his own computations, but 18° seems to have become the canonical number—at least, it is still used today. Nunes, in his far more interesting treatise on twilight, does not consider h to be independent of φ.

seven climata, values of φ corresponding to M (13^h–16^h) at intervals of $0;30^h$. Phases were of course fundamental to Babylonian astronomy and celestial omina, but by the time of Ptolemy their role in horoscopic astrology was secondary, so it is interesting to see his devoting so much attention to what must have been a highly taxing but unrewarding exercise. Copernicus gives a one-sentence summary of the procedure for finding $\Delta\lambda$, and then drops the subject.

4. The Catalogue of Stars (II,14)

Chapter 14 of Book II is a curious introduction to Copernicus's catalogue of stars. Having considered the diurnal motion, he says, the demonstrations of the annual revolution ought to come next. However, he will instead follow "some of the ancient mathematicians," who believed that "the phenomena of the fixed stars" ought to take precedence. (If one wonders who these ancient mathematicians were, one can consult M, 42r, where Copernicus wrote and deleted *Solensis Aratus ac*, i.e. "Aratus of Soli and some of the ancient mathematicians.") He also says that he places the catalogue first because all planetary motions are referred to the fixed stars, that is, longitudes are sidereal. He acknowledges that this arrangement disagrees with Ptolemy, who places his star catalogue, in *Almagest* VII,5–VIII,1 after the solar and lunar theory. The reason for Ptolemy's arrangement is that he needed tropical longitudes of stars, which he measured with reference to certain fundamental zodiacal stars, e.g. Regulus, Spica, whose tropical longitudes were previously measured by taking their distance from the moon. Hence, it was first necessary to establish solar and lunar theory, including lunar parallax, before the tropical longitudes of the fundamental stars could be found.

Copernicus, however, decided to give his stellar longitudes sidereally because he believed that the equinoxes moved nonuniformly as a result of the nonuniform precession. Hence he could put his catalogue anywhere he saw fit even if he had measured the positions himself, which he had not, since all that mattered was the relative positions of the stars among themselves. He could select any point as his reference for sidereal longitude, and he chose the first star of Aries (γ Arietis) in Ptolemy's catalogue, for which $\lambda = 6;40°$, as sidereal longitude $\lambda^* 0°$. Thus all stars in Ptolemy's catalogue could be adapted to sidereal longitude by subtracting $6;40°$, and after some difficulty making up his mind, Copernicus decided to count the longitudes continuously from $0°$ to $360°$, rather than in zodiacal signs of $30°$ as in Ptolemy's catalogue.

Even though Copernicus did not determine the coordinates of a single star in the catalogue by observation, he goes through the exercise of describing in detail the construction and use of Ptolemy's armillary sphere, called an *astrolabe*, which is described in *Almagest* V,1 and *Epitome* V,1.[26] Ptolemy used this instrument for observations of the moon, stars, and planets since it allows a direct reading of longitude and latitude, and Copernicus says in V,4 (N, 143r21) that "astrolabic instruments" are used to find positions of planets near opposition. Since his description of the instrument is so elaborate, and since he offers no alternative for his own observations of the planets, one might surmise that he possessed and used such a device. But this is hardly certain. His only detailed account of the use of an armillary, which directly follows its description, is of Ptolemy's measurement of the longitude of Regulus.[27] For his own determinations of the longitude of Spica in 1515 and 1525, he measures only the meridian altitude, probably using the parallactic instrument, and then computes the longitude with an analemma.[28] While it is best to use an armillary after the solar theory has been developed, it is not absolutely necessary, and thus Copernicus's measurement of meridian altitudes of Spica suggests that he did not have such an instrument, at least in 1515 and 1525.

Copernicus's catalogue is, as we have mentioned, nothing more than an adaptation of Ptolemy's catalogue to arbitrarily defined sidereal coordinates, subtracting 6;40° from all longitudes; latitudes are assumed to have remained unchanged. Altogether too much has been made of this trivial adjustment. Although unknown to Copernicus, Ptolemy had done the same thing in the *Handy Tables*, using Regulus ($\lambda = 117;54°$ in -323) as $\lambda^*0°$, and measuring the longitudes of zodiacal stars and the sidereally fixed apogees of the planets accordingly. Since the choice of a zero point of sidereal longitude is arbitrary, and since some value of precession must always be added to use the catalogue, as Copernicus does for the occultation of Aldebaran in IV,27, it can hardly be said that any star catalogue is really independent of precession. It must be emphasized that the sidereal positions in Copernicus's catalogue are useless by themselves since all the spherical procedures of Book II necessary to determine the "phenomena of the fixed stars" depend upon distances from equinoxes and solstices, i.e. upon tropical coordinates. It is of course for this reason that Ptolemy's catalogue in the *Almagest*, and all its medieval adaptations, give tropical coordinates for the epoch at which the catalogue is to be used. With Copernicus's catalogue, one must always

---

[26] There is an excellent description of the instrument and its use in Rome (1927). For a brief description and discussion of whether Copernicus used an armillary, see above, pp. 68–69.

[27] Cf. *Almagest* VII,2 and *Epitome* VII,2. Copernicus's description, which in general follows the *Epitome*, contains a number of discrepancies with Ptolemy's, and the date is incorrectly converted to Olympiad 229,1 instead of 229,2 and to 24 Feb. 139 instead of 23 Feb. The following differences occur: 5^h after noon (for $5\frac{1}{2}^h$, probably just by omission) true sun in ♓ $3\frac{1}{24}°$ as in *Epitome* (for $3\frac{1}{20}°$); ♊ 4° in midheaven as in *Epitome* (for $\frac{1}{4}$ of ♊, i.e. ♊ $7\frac{1}{2}°$); distance of Regulus from the moon $57\frac{1}{10}°$ i.e. 57;6° as in *Epitome* (for $57;10° = 57\frac{1}{6}°$). Copernicus adds the information that according to his own determination of lunar parallax, the moon was between $5\frac{1}{3}°$ and $5\frac{2}{5}°$ of Gemini. In the "Letter against Werner" (Rosen (1971), 97) the date is given incorrectly as 22 Feb., but the time correctly as $5\frac{1}{2}^h$ after noon.

[28] Below, pp. 131–132.

add the precession, even to use it at his own time. The first really new star catalogue made in Europe, by Tycho, again gives tropical coordinates, for 1600, and provides a table of precession to adjust for later epochs.[29]

Perhaps the most curious thing about Copernicus's catalogue is the trouble he had making it, for he depended upon erroneous and discordant sources for Ptolemy's catalogue, that he had to choose between, and then he made a fair number of scribal errors in his transcription and computational errors when subtracting 6;40°. And he became particularly confused when attempting to rearrange the stars of Pegasus in order of increasing longitude, but did not succeed in doing so. These subjects have been treated in great detail in the literature,[30] and we shall note only the longitudes of two stars that are here worthy of attention. Copernicus's principal sources for the catalogue were the 1515 Venice edition of the Gerard of Cremona translation of the *Almagest* and the translation of the catalogue alone printed in Book XVII of Giorgio Valla's *De expetendis et fugiendis rebus* (Venice, 1501), and he also may have made some use of the catalogue in his 1492 Venice edition of the *Alfonsine Tables*. It appears that he did not yet have the George of Trebizond translation of the *Almagest*, first printed in Venice in 1528, but since exactly when Copernicus acquired this book is unknown, although eventually he did, one cannot use this information to date the catalogue, except that it is definitely among the earliest parts of M to be written. Initially, he seems to have relied upon Valla to his peril, and by all the bad luck, Valla's catalogue gives the longitude of γ Arietis itself as ♈ $6\frac{1}{2}°$ instead of $6\frac{2}{3}°$. While Copernicus adopted the correct 6;40° from the 1515 *Almagest* for his subtractions, the reading in Valla, which at first he must have taken seriously, caused him a fair amount of confusion in converting Ptolemy's apsidal longitudes of the planets, which are entered into the catalogue in M, where Copernicus is sometimes in error by 0;10°.[31] Then, by more bad luck, Valla's longitude of Spica, which was Copernicus's reference star for his determination of precession, is ♍ $26\frac{1}{2}°$ instead of $26\frac{2}{3}°$. Although he followed the correct reading from the 1515 *Almagest* for his catalogue, a relic of his earlier reliance upon Valla remains in III,2, where he first states, and later corrects, Ptolemy's longitude of Spica as given by Valla.[32]

[29] Tycho's catalogue of 1000 stars is printed from manuscript in Brahe (1913–29), 3, 331–77. The first printing was in the *Rudolphine Tables* (Kepler (1937–) 10, 105–29).

[30] A comparative table of Copernicus's catalogue and its sources is given in Dobrzycki (1963), and with admirable industry Rosen has annotated some eighty discrepancies (those in Pegasus are especially notable) in Copernicus (1978), 372–81.

[31] M, 61v, for Saturn 226;30° instead of 226;20°, one of Ptolemy's apsidal longitudes entered by Copernicus into his star catalogue in M, but (correctly) omitted in N. One wonders if these were entered at a time when he believed the apsidal lines to be sidereally fixed, as he did when he wrote the *Commentariolus*, so that Ptolemy's values, assuming that they were accurate, would be true for all times. Also in error by 0;10° is the apsidal longitude of Venus, 48;30° for 48;20°—which Copernicus did consider sidereally fixed in *De rev.*—in the Uppsala notes (see Swerdlow (1973), 427).

[32] Below, p. 132.

3

The Motions of the Earth

Introduction

Anyone who thinks that Copernican theory is "simpler" than Ptolemaic theory has never looked at Book III of *De revolutionibus*. In a geocentric system the earth is at rest—as indeed it appears to be—and any apparent motions in the heavens that *we* know to result from its motions are distributed among a number of objects, i.e. the sun, the individual planets, the sphere of the fixed stars, everything in its proper place as it actually appears. But when Copernicus worked through the consequences of his own theory, he had to attribute to the earth no less than three fundamental motions and a number of secondary motions. That all these compounded motions forced upon a single and, to all appearances, quiescent body seemed implausible to his contemporaries is not to be wondered at, especially because the end result was nothing other than reproducing the same apparent motions in the heavens that had been accounted for all along (and without making asumptions that contradicted contemporary natural philosophy, common sense, and the most casual or most meticulous observations then possible of the behavior of the earth and of objects on or near its surface).

The three fundamental motions of the earth, described qualitatively by Copernicus in I,11, are the rotation on its axis, completed in a day, the revolution about the sun, completed in a year, and the precession of the equinoxes, completed in some long period of uncertain duration. Today we describe precession as a slow motion of the earth's axis of rotation—which remains nearly parallel to itself as the earth moves about the sun—about a line parallel to the axis of the ecliptic (or, without invoking an "axis" of the ecliptic, about a line perpendicular to the plane of the ecliptic). Copernicus, however, just to explain the parallelism of the earth's axis in the course of moving about the sun, had to invoke an opposite annual rotation of the axis about the line parallel to the axis of the ecliptic, which he called the "motion of the inclination" (*motus declinationis*). Were these motions equal, the earth's axis would maintain a constant direction, but by assuming that the opposite motion of the axis slightly exceeds the annual motion of the earth, the axis will slowly change its orientation, causing the celestial equator to intersect the ecliptic in points continually moving in a westward direction, that is, opposite to the direction of the motion of the earth about the sun.

But these are only the fundamental motions described qualitatively. When it came to a detailed investigation of the periods of the motions, and of the other "constants" of the motion of the earth with respect to the sun and to the fixed stars, Copernicus found that he had to account in some way for a number of secondary motions of rather long period that seemed to be indicated by the observations of astronomers as far back as Timocharis

(3rd cen. B.C.) and by the variation of constants determined as far back as Hipparchus (2nd cen. B.C.). These were:

1. A variation in the length of the tropical year.
2. A variation in the rate of precession.
3. A diminution of the obliquity of the ecliptic.
4. A diminution in the maximum equation of the sun, that is, its departure from uniform angular motion.
5. A nonuniform motion of the apsidal line of the sun (or of the earth).

It is of course ironic that all these secondary motions that put Copernicus to so much trouble were illusory, the result of errors in observation and resulting theory by Copernicus and his predecessors. Not that some of the motions do not really occur, but in truth they are all essentially uniform within the historical period, and far slower than the erroneous observations used by Copernicus would appear to indicate. But Copernicus had good reason for wishing to account for these long-period motions in that most of them were already a standard part of medieval astronomy.

Ptolemy, having only about 450 years of earlier observations (other than lunar eclipses), and these often of questionable accuracy, could do little more than establish parameters applicable to his own time, or at best to the period of about 300 years between Hipparchus and himself. Agreement with Hipparchus's determinations, and a complicated interplay of observational errors of old and of his own, caused him to assume the constancy of parameters for his solar theory, precession, and obliquity of the ecliptic. Being the first to develop a complete and consistent theory of the planetary system, Ptolemy had to use the best parameters determined for his time, and leave it to future generations to decide whether they had remained constant.

But when, in the ninth century, Islamic astronomers began their own investigations of such fundamental quantities as the length of the tropical year, the precession, and the obliquity, it was found that marked changes had occurred since Ptolemy. Astronomers thus had two choices. They could base their theories entirely upon their own determinations, ignoring or considering erroneous the parameters in the *Almagest*, or they could attempt in some way to develop models producing long-period variations in fundamental parameters. The first alternative seems to have been the choice of, to name only an astronomer later known in the west, al-Battānī who, comparing Ptolemy's and his own observations, found a rather short tropical year and rapid rate of precession. But the second alternative was pursued by the author of a treatise called *De motu octavae sphaerae*, usually attributed to Thābit ibn Qurra, which contains an ingenious, if far-fetched, model for the simultaneous variation of the tropical year, precession, and obliquity.[1] It was also known by the time of Battānī that the eccentricity and apsidal

[1] References, above, p. 43, n. 3, and Goldstein (1964). In *De anno solis*, Thābit has only a uniform precession, although he derives two different values for it, and thus the tropical year is uniform.

direction of the sun were no longer the same as Ptolemy's determinations, and the variation of these quantities also required a theory to account for them, the most influential in the west being that of az-Zarqāl.[2]

Through the *Epitome* and Peurbach's *Theoricae novae planetarum*, Copernicus had at least some knowledge of the theories attributed to Thābit and az-Zarqāl, and through the *Alfonsine Tables*, Peurbach's *Theoricae*, and Johann Werner's *De motu octavae sphaerae* (1522), he could learn of yet other ways of accounting for a variable precession, tropical year, and obliquity.[3] Thus, in taking up these problems he was doing nothing fundamentally new, for he was simply attributing to the earth all the motions formally attributed to the sphere of the fixed stars. The principal distinction of Copernicus's theory was to subsume the inequalities of all these long-period motions under a single period of variation, a period he originally worked out for the precession and change of obliquity, and then forced upon the change of eccentricity and motion of the apsidal line. These are the principal subjects of Book III of *De revolutionibus*, and since they present problems not treated in the *Almagest* and *Epitome*, much of Copernicus's originality in the invention of models and the derivation of parameters appears in Book III, which is also where he has some of his worst difficulties.

1. Precession and Variation of Obliquity

Statement of the Problem (III,1)

Copernicus begins with a historical account. The "ancient mathematicians," he says, did not distinguish the tropical and sidereal year, for the "Olympic years," which they measured from the rising of *Canicula*, either Procyon or Sirius, were assumed to be the same as years measured from the summer solstice.[4] But Hipparchus of Rhodes, *vir mirae sagacitatis*, found the sidereal year to be longer than the tropical year, and hence that the fixed stars possessed a slow and not immediately perceptible eastward (*in consequentia*) motion. Through the passage of time, this motion has now

---

[2] The evidence for az-Zarqāl's solar theory is assembled and analyzed in Toomer (1969), along with a discussion of its relation to Copernicus. See also below, pp. 157–159.

[3] Descriptions of earlier theories pertinent to Copernicus can be found in Dobrzycki (1965) and briefly in Swerdlow (1973), 445–50. Previous studies of the precession theory in *De rev.* are Moesgaard (1968) and Swerdlow (1975), (1980).

[4] In III,11 (N 76v12–17) Copernicus says that the Olympic Game was celebrated at the time of the summer solstice *and* the rising of Canicula for the Greeks, "as Censorinus and other trustworthy authorities report." Censorinus 21,6 refers to the *diebus dumtaxat aestivis, quibus agon Olympicus celebratur*, which Copernicus could have interpreted as meaning (near?) the solstice. In fact, at Copernicus's Era of the Olympiad (−774 Jul 1 noon), according to his solar theory the sun is at 90;36° (below, p. 178), only about 15^h past the solstice, while Procyon rises some 10 days and Sirius some 20 days later. In Copernicus's star catalogue, Canicula refers only to Procyon. On the evidence there is for the time of the Olympic Games, see Samuel (1972), 191–94.

become very clear, and further, it is found to be nonuniform. And by another wonder of nature (*naturae miraculum*), the obliquity of the ecliptic does not appear as large to us as it did to Ptolemy. Many theories, including a back-and-forth libration within fixed limits, i.e. trepidation, and a ninth, tenth, and now even an eleventh sphere have been proposed to account for these motions, but without success. Nevertheless, he concludes, in so far as possible from the record of the observations, we shall demonstrate more certain things (*efficiemus certiora*) concerning these subjects.

In somewhat more technical terms, consider Figs. 1 and 2 in which Γ marks Copernicus's zero point of sidereal longitude $\lambda^*0°$ in the ecliptic and E is the pole of the ecliptic. "Precession" means that the vernal equinox $\overline{V}$ in the ecliptic, or the north pole $\overline{N}$ of the equator, rotate slowly in a westward direction, i.e. in the sense of decreasing λ^*. The angular velocity of this motion will be denoted by $\bar{\pi}$. The distance $\overline{N}E = \bar{\varepsilon}$ is the obliquity of the ecliptic since the angle made by the intersection of the ecliptic and equator is equal to arc $\overline{N}E$. The observational record available to Copernicus, however, indicated that the motion of the vernal equinox is not uniform and the obliquity of the ecliptic is not fixed. Hence the parameters $\bar{\pi}$ and $\bar{\varepsilon}$ must now be defined as mean values, as distinct from the true precession π and true obliquity ε which are subject to periodic variations. Copernicus's object is to construct a kinematic model—of course based on the principle of uniform circular motion—that will fit the observational record of the variation of π and ε as closely as possible.

In the same figures, we consider two orthogonal components in any motion of the north pole. If $\overline{N}$ changes its position in the plane of great circle SET, which is perpendicular to $\overline{V}O\overline{W}$ and the plane of the ecliptic, then the obliquity ε is altered while the equinoxes $\overline{V}$ and $\overline{W}$ remain in position, i.e. at a fixed distance from Γ. If, however, the north pole moves out of the plane SET, but such that ε remains constant, that is, if $\overline{N}$ moves on a small circle of radius $E\overline{N} = \bar{\varepsilon}$, then the equinox moves away from Γ. Since the motion of the equinox with respect to Γ measures precession, we see that the precession corresponds to the amount the north pole rotates in a circle with respect to SET. Any arbitrary motion of the north pole, which is always resolvable into components parallel and perpendicular to SET, will therefore produce a combination of a variation of the obliquity and of the rate of precession. The problem now is to determine how the true north pole is to be moved with respect to $\overline{N}$ in order to account for the observational variations of π and ε.

The Observational Record (III,2)

For the evidence of the variability of precession, Copernicus uses nine ancient observations of Spica (α Virginis), Regulus (α Leonis), and the northernmost of the three stars in the forehead of Scorpio (β Scorpii) recorded in *Almagest* VII,2–3 and the star catalogue, and in *Epitome* VII,2,5;

two observations made by al-Battānī recorded in *Epitome* VII,6; and two observations of his own. The observations are summarized in Table 1, in which Call. means Callippic Period and E.A. means Era of the Death of Alexander, in which the years are Egyptian years of 365 days.[5]

Copernicus gives some details of the 1525 observation of Spica. These are not all one could wish since he does not give a specific date, although in III,12 he computes according to his theory the position of Spica for 16 April 1525, which could possibly be the actual date of the observation. Nor does he explain what sort of instrument he used, although since the observation was of a meridian altitude, he could have used the parallactic instrument described in IV,15, which measures zenith distances and which in IV,16 he says he used to find the zenith distance of the moon.[6] Nevertheless, since it is the only account of the reduction of an observation in *De rev.*, it is of some interest.

He observed the meridian altitude of Spica to be very nearly (*proxime*) 27°. Since Copernicus takes the latitude of Frauenburg to be $\varphi = 54;19,30°$,[7] so $\bar{\varphi} = 35;40,30°$ and the declination of Spica

$$\delta = 27° - 35;40,30° = -8;40,30° \approx -8;40°.$$

The latitude is assumed to have remained unchanged from Ptolemy's catalogue at $\beta = -2°$. The obliquity ε is taken to be 23;28,30°. What Copernicus then presents at length is nothing more than a transformation of coordinates: given δ and β, and knowing ε, to find λ. He uses an analemma construction similar to one used by Johann Werner to solve the identical problem in *De motu octavae sphaerae*.[8]

In Fig. 3 *ABCD* is the meridian, *AEC* the diameter of the equator, *BED* the diameter and *GEF* the axis of the ecliptic, and *E* the autumnal equinox

---

[5] Notes to table: 5./6. one year too late A.D., although E.A. 422 is correct. 8./9. from Ptolemy's star catalogue, specifically for Thoth 1 Antoninus 1 (E.A. 461) = 137 Jul 20 (j.d. 1771,298), but of course applicable to Antoninus 2 as assumed by Copernicus. 8. 176;30° error in Valla (1501) for 176;40°. 10. 134;5°/*Epitome* ♌ 14;50°/*Epitome* (1496) ♌ 4;50°/Battānī (1537) 79v ♌ 14°/Nallino I,124 ♌ 14°. 11. 227;50°/*Epitome* ♏ 17;50°/Battānī (1537) ♏ 17;20°/Nallino ♏ 17;45°. 12. 197;14°/M ♎ 17;14° altered from 17;10°.

[6] Johann Werner used the parallactic instrument for measuring altitudes of fixed stars, see note 8. On the instrument, see below, p. 233.

[7] In 1584 Tycho sent one of his assistants to Frauenberg, among other reasons, to check the latitude with a sextant. The result was 54;22,15°, larger than Copernicus's figure by 0;2,45°, and the student brought back not only the sextant, "completely undamaged," but also Copernicus's parallactic instrument (Brahe (1946), 45–46). The correct latitude is 54;21.6° (Copernicus (1975), 401).

[8] Indeed, his whole procedure is the same as Werner's in *De motu octavae sphaerae* (1522), Prop. 1–2. In Prop. 1 Werner describes using the parallactic instrument, which he calls "Ptolemy's rulers" (*regulae Ptolemaei*) to take meridian altitudes of Regulus, Spica, and α Librae in 1514/15, and assuming the latitude of Nuremberg to be 49;23,30°, he finds their declinations. In Prop. 2 he takes $\varepsilon = 23;28,30°$, as does Copernicus, and in his example for Spica, assumes Ptolemy's latitude, which he presumably does also for the other stars. His method of finding λ is somewhat more direct than Copernicus's, but is essentially the same.

♎0°. Let arc $AB = -\varepsilon = 23;28,30°$, arc $AM = -\delta = 8;40°$, and arc $BH = -\beta = 2°$. Draw the parallels, of declination MN and of latitude HL, and the star will be at their intersection O. We wish to find the distance $\Delta\lambda$ of the star from the equinox. The procedure used by Copernicus is equivalent to the relation

$$\sin \Delta\lambda = \frac{\cos \beta - \sin(\varepsilon + \beta) - \sin \delta}{\sin \varepsilon \cos \beta}, \tag{1}$$

and he solves the problem numerically step by step where $BE = R = 100000$. Thus

$$HIK = \frac{\sin ABH}{\sin AB} \cdot BE = \frac{\sin(\varepsilon + \beta)}{\sin \varepsilon} R = \frac{43010}{39832} R = 107978,$$ [9]

and

$$OK = \frac{\sin AM}{\sin AB} \cdot BE = \frac{\sin \delta}{\sin \varepsilon} R = \frac{15069}{39832} R = 37831,$$

so that

$$HO = HIK - OK = 70147.$$

But

$$HOI = BE \cos BH = R \cos \beta = 99939,$$

and thus

$$OI = HOI - HO = 29792, \quad \text{(N 29892, M 7 altered to 8)}$$

so that in parallel HL, where $HOI = r = 100000$,

$$\sin \Delta\lambda = \frac{OI}{HOI} r = \frac{29792}{99939} r = 29810.$$

And since arc $\sin \Delta\lambda = 17;20,36\ldots° \approx 17;21°$, the longitude of Spica in 1525 is ♎17;21°. Although Copernicus gives no details, the longitude of ♎17;14° for 1515 was computed in the same way from $\delta = -8;36°$.[10]

Copernicus then corrects Ptolemy's longitude of Spica from ♍26;30°, the value from Giorgio Valla, to 26;40°, and with some trivial errors in the chronological intervals, computes the following rates of precession in degrees per Egyptian year:

Observers	Δt	$\Delta\lambda$	$\Delta\pi$
Tim.–Ptol.	432^y	4;20°	1°/100^y
Hip.–Ptol.	266	2;40	/100
Men.–Bat.	782	11;55	/66
Ptol.–Bat.	741	[11;30]	/65
Bat.–Cop.	645	9;11	/71

[9] Correctly sin 23;28,30° = .39835, but rather interestingly, $\sin^{-1}$ .39832 = 23;28,24°, Copernicus's computed value of ε for 1525.

[10] Correct computation gives 17;10°, which Copernicus wrote in M (72v36) and then altered to 17;14° to agree with the completed theory. Recomputation in Swerdlow (1975), 58–59.

The conclusion drawn from these figures is that the precession was slower in the 400 years before Ptolemy than between Ptolemy and Battānī, and the latter was in turn more rapid than the precession between Battānī and Copernicus.

Copernicus's records of the obliquity ε indicate a continual diminution:[11]

Observer	Date	ε
"Aristarchus"	[−279]	23;51,20°
Ptolemy	140	23;51,20
Battānī	880	23;36
az-Zarqāl	1070	23;34
Prophatius	1300	23;32
Copernicus	[1515]	23;28,30 ("not greater than")

About all one can conclude from this list is that the obliquity has been continuously decreasing, rather slowly at the beginning and more rapidly later, but Copernicus assumes instead that it is in fact periodic, as he had remarked in II,2 (N, 29r27), that he would show by a "probable enough conjecture" (*coniectura satis probabili*) that it oscillates over a range of 0;24° between the limits 23;28° and 23;52°. It is also apparent that while the precession has gone from slow to fast to slow, that is, through about one cycle of variation, the obliquity has gone in only one direction, through about half a cycle. Copernicus, however, makes the assumption that the period of the obliquity is *exactly* twice the period of the variation of the precession and that the extrema of each occur at *exactly* the same moment, specifically, that the slowest precession and the maximum obliquity coincide at some time between Timocharis and Ptolemy. He later assumes that this moment is *exactly* midway between Timocharis and Ptolemy, and after that corrects this assumption.

[11] "Aristarchus" is an error for Eratosthenes owing to Copernicus's guessing at the meaning of the corrupt form *Archusianus* in the 1515 *Almagest*. In II,2 which was written later than III,2 and III,6 he correctly names Eratosthenes, whose name is given correctly in George of Trebizond's translation (Venice, 1528) and of course in the Greek *Almagest* (Basel, 1538). Copernicus is unsure of Battānī's value, giving 23;35° or 23;36° with a variety of alterations and misprints in M and N. In the *Toledan Tables*, 23;33,30° is attributed to az-Zarqāl in the heading of the declination table; cf. Toomer (1968), 30–31. For Prophatius (Jacob b. Machir b. Tibbon), the year 1300 is indeed the year of his *Almanach*, but 23;32° is not used, and its origin is unknown; cf. Zinner (1943), 510–11. In cancelled lines of III,6 in M (79r26–32), Copernicus mentions a number of contemporary measurements of the obliquity: Peurbach 23;28°, Dominico Maria di Novara 23;29+°, Regiomontanus 23;28,30°. All three are found in Werner (1522), Prop. 25, although Novara's is simply 23;29°; cf. Swerdlow (1975), 63–65.

The Model (III,3–5)

Referring back to Figs. 1 and 2, we recall that the object of a model for the variable precession and obliquity is to cause the true north pole to move with respect to the mean pole $\bar{N}$ such that the obliquity and rate of precession are varied in accordance with the observational record. This can be attempted in various ways, e.g. in Fig. 4(a) a small circle, a polar epicycle, could be described about $\bar{N}$ and the north pole N made to move on it. Such a polar epicycle was in fact used by az-Zarqāl to vary the obliquity while the variation of precession was produced by circles around the equinoxes.[12] But here a circle would be too constraining since the same radius $\bar{N}N$ would determine both the departure from uniform precession, angle δ, and the change of obliquity $\varepsilon - \bar{\varepsilon}$, and the periods of both would necessarily be the same. Another, and rather attractive, possibility would be to let N move on an ellipse (b) about $\bar{N}$. By suitably choosing the ratio of the axes of the ellipse, δ and $\varepsilon - \bar{\varepsilon}$ could take on any desired value without constraining each other, and Copernicus even knows how to generate an ellipse from uniform circular motions. However, the periods of δ and ε would still be the same, while it is required that the latter be twice the former. But this suggests an interesting solution to our problem. What if we assume that N moves along a path formed of two adjacent ellipses (c), crossing from one to the other as it passes $\bar{N}$? Again, the axes of the ellipses may be chosen independently for δ and $\varepsilon - \bar{\varepsilon}$, but now, as required, the period of ε is exactly twice the period of δ. The problem now is how to generate such ellipses, or some figure with essentially the same properties. A simple procedure, at least in principle, would be to define a set of rectangular coordinates (d) such that N oscillates parallel to $E\bar{N}$ in twice the period that it oscillates on a line perpendicular to $E\bar{N}$. In this way, while we would get nothing so elegant as the adjacent ellipses, we would get a figure of the desired properties, that $\varepsilon - \bar{\varepsilon}$ and δ may be chosen independently with the period of ε twice the period of δ.[13]

Whether this was the course of Copernicus's analysis we do not know—logic is not necessarily history—but his solution was indeed to let the pole N move along the path generated by two perpendicular oscillations, "librations," as he called them, as shown in Fig. 5 in which the dimensions, which may be chosen independently, are drawn in the ratio 7/3 of Copernicus's final parameters. We can see that angle ϑ controls the motion of N parallel to $E\bar{N}$, and thus the change of obliquity, while 2ϑ controls the motion perpendicular to $E\bar{N}$ which determines the variation of the precession, so that the period of ε will be twice the period of δ, as required.

[12] Goldstein (1964), 238–44.

[13] It has been noted in Birkenmajer (1972), an interesting study of the curve produced by Copernicus's model, that Clavius compared the curve to two ellipses tangent at their minor axes, a clever insight into the purpose of the model.

The next problem—and for Copernicus an important one—is whether the rectilinear oscillations, the librations, may be generated by uniform circular motions. Fortunately, he knows how to do this, and had already done so for the latitude theory of the planets in the *Commentariolus*, using a device originally invented by Naṣīr ad-Dīn aṭ-Ṭūsī (1201–74).[14] In Fig. 6(a), let radius $AB = r_1$ rotate uniformly through ϑ, while equal radius $BC = r_2 = r_1$ rotates in the opposite direction through 2ϑ. Extend AB to D, and connect DC, which will be perpendicular to AC since ACD is an angle in a semicircle. Thus,

$$AC = AD \cos\vartheta = (r_1 + r_2)\cos\vartheta = 2r_1 \cos\vartheta, \tag{2}$$

and C will describe a straight line passing through A with limits $AE = AF = (r_1 + r_2) = 2r_1$.

Copernicus also points out, in a rather well-known, although often misinterpreted, passage crossed out in M (75r27–34) that if the radii of the circles are unequal, then "the circles will describe, not a straight line, but the conical or cylindrical section that mathematicians call an ellipse." This is illustrated in Fig. 6(b) in which the radii r_1 and r_2 are unequal. The proof that the path of C is an ellipse is trivial. Consider $AD = r_1 + r_2$ the diameter of a circle, and DCH will be perpendicular to AE. Thus

$$AH = (r_1 + r_2)\cos\vartheta$$

and

$$CH = DH - DC = (r_1 + r_2)\sin\vartheta - 2r_2 \sin\vartheta = (r_1 - r_2)\sin\vartheta.$$

And since $AE = r_1 + r_2$ and $AK = r_1 - r_2$, therefore

$$\frac{(r_1 - r_2)^2 \sin^2\vartheta}{(r_1 - r_2)^2} + \frac{(r_1 + r_2)^2 \cos^2\vartheta}{(r_1 + r_2)^2} = 1.$$

which is the equation of an ellipse with center at origin. The condition of unequal circles is just what is found in Copernicus's lunar model and in the double-epicycle model of Ibn ash-Shāṭir used by Copernicus for the planets in the *Commentariolus*. In both cases a small ellipse is generated—rotating in the case of the moon, fixed in direction in the case of the planets—and Copernicus appears by this remark to be aware of this property of the models.[15]

Finally, one must ask how a rectilinear motion of the earth's pole will be transformed to the equator and motion of the equinox. In the case of the motion of $\overline{N}$ along SET (see Fig. 2) changing ε, it is clear that the angle ε at $\overline{V}$ will be varied sinusoidally, but it is not immediately evident how a sinusoidal motion of $\overline{N}$ perpendicular to SET will shift $\overline{V}$ along the ecliptic. Therefore, consider Fig. 7 in which the two circles producing the rectilinear

[14] On the transmission of Ṭūsī's device to the west, see above, pp. 47–48.

[15] See below, p. 197 and Swerdlow (1973), 470.

motion have been moved down to the mean equinox $\bar{V}$, and since the period of the variation of the precession is one-half the period of the variation of the obliquity, the radii rotate through 2ϑ and 4ϑ. The motion of the two circles will cause N' to oscillate, exactly as the north pole does, along a great circle, $\bar{W}\bar{N}\bar{V}$ in Fig. 2, perpendicular to the equator, and will thus cause the true equator to oscillate with N' and thus intersect the ecliptic at the true equinox V which oscillates to either side of $\bar{V}$. We wish to find an expression for the motion of V along the ecliptic. Now, $\bar{V}N' = 2r \sin 2\vartheta$, and in triangle $\bar{V}N'V$

$$V\bar{V} = \bar{V}N'/\sin \varepsilon = 2r \sin 2\vartheta/\sin \varepsilon. \tag{3}$$

Were ε constant, we could thus define a maximum equation

$$\delta_{\max} = 2r/\sin \varepsilon,$$

and then consider the partial equation $\delta = V\bar{V}$ as simply $\delta_{\max} \sin 2\vartheta$. In fact, ε is not constant, but its variation is so small, $\pm 0;12°$ where $\bar{\varepsilon}$ is 23;40°, that it would produce a variation of only about 0;1° in $\delta_{\max}$. This can be neglected—Copernicus does so in III,7 (N 73r)—and thus the equation of precesion $\delta_P = V\bar{V}$ may be treated sinusoidally as

$$\delta_P = \delta_{P_{\max}} \sin 2\vartheta. \tag{4}$$

On the other hand, whether a sinusoidal variation is in fact a true representation of δ_P and ε is not a question Copernicus ever addresses, good medieval astronomer that he is.

Before turning to the determination of the parameters, one further point should be noted. In the preceding analysis we have been operating with plane triangles while in fact all the lines are actually on a spherical surface. When examined spherically, the whole mechanism breaks down. In Fig. 6(a), either C will not oscillate on a great circle, or ACD will not be a right angle because of the spherical excess in triangle ADC. The only way to compensate for this would be to adjust the motions through ϑ and 2ϑ to take account of the spherical excess, but this would violate the principle of uniform circular motion. Likewise, in Fig. 7 $\bar{V}N'$ will not be an arc of a great circle, so $V\bar{V} = \delta_P$ will vary in no simple way. However, the oscillations involved are all so small, $\pm 0;12°$ for the variation of ε and $\pm 0;28°$ for the motion of the equator, producing a maximum equation of $\pm 1;10°$ for the precession, that Copernicus asks that straight lines be substituted for great circles, and the request is only reasonable.

Derivation of Parameters (III,6–7,9–11)

We have mentioned that it is only in Book III that Copernicus had to derive parameters that were not considered in the *Almagest* and the *Epitome*. Unfortunately, his method was not to invent ingeneous ways of using strategically selected observations, or to devise sophisticated geometrical solutions to problems with several unknowns, but rather to ignore as many observations as he wished, and make as many prior assumptions as he needed, in order to solve a problem in the most elementary way. Then he goes back over his work, correcting his earlier assumptions, or some of them, by trial and error until the results are fairly consistent. And in the case of two fundamental parameters, the mean precession and the anomaly of precession, he really does not show how he found his values, but expects his word to be taken that they are correct.

(*a*) *The Anomaly of the Obliquity and Precession*: ϑ *and* 2ϑ (III,6)

The fundamental parameter for the variation of the obliquity is what Copernicus calls (N 66r29–35) the "anomaly of the inclination" (*anomalia declinationis*) ϑ, and for the inequality of precession, the "anomaly of the equinoxes" (*anomalia aequinoctium*) 2ϑ. Obviously if ϑ is found 2ϑ is known, and *vice versa*, but Copernicus does not explain how he found either. He says only that if one first conjectures that the 1819 years from Timocharis (E.A. 30) to his own 1525 observation of Spica (E.A. 1849) are equivalent to 360° of 2ϑ, then it turns out by a "more careful calculation" (*examinatiori calculo*) that 2ϑ has in fact exceeded 360° by 21;24°, and that the period of the anomaly is thus 1717 Egyptian years (N 69r32–69v8). There are various ways of reaching this number, but they all involve making some kind of prior assumption about the range of the variation of the obliquity, or the location of certain observations within the cycle of ϑ or 2ϑ, or the mean rate of precession. Now it is true that in II,2 Copernicus assumes by a "probable enough conjecture" that ε varies over a range of 0;24° from 23;28° to 23;52°, and in III,7 he assumes that Ptolemy and Timocharis are exactly symmetrical to the zero point of 2ϑ. And it also appears that he made prior assumptions about the mean precession.[15a] Our problem therefore is not an absence of information, but that there are too many procedures by which Copernicus could have made his "more careful calculation." We shall, however, set out one that seems the most economical in its prior assumptions, with the assurance that an attentive reader should be able to invent two or three more.

Consider the variation of the obliquity, which Copernicus conjectures to lie between 23;28° and 23;52°, so that its mean value $\bar{\varepsilon} = 23;40°$ which varies within $\pm 0;12°$. Thus, in Fig. 8 $E\bar{N} = \bar{\varepsilon} = 23;40°$ and $\rho = 0;12°$. At the time of Ptolemy $\varepsilon = EN_P = 23;51,20°$, and at Copernicus's observations of Spica $\varepsilon = EN_C = 23;28,30°$, so that $N_P\bar{N} = 0;11,20°$ and $N_C\bar{N} = 0;11,30°$. Thus, at the time of Ptolemy, the anomaly

$$\vartheta_P = \cos^{-1} N_P\bar{N}/\rho = \cos^{-1} 11,20/12,0 = 19;11°,$$

and at the time of Copernicus $\bar{\vartheta} = 180° - \vartheta$ is

$$\bar{\vartheta}_C = \cos^{-1} N_C\bar{N}/\rho = \cos^{-1} 11,30/12,0 = 16;36°,$$

[15a] See below, p. 139.

so that the change of anomaly between Ptolemy and Copernicus

$$\Delta\vartheta = 180^\circ - (\vartheta_P + \bar{\vartheta}_C) = 144;13^\circ.$$

Now we must find the time that has elapsed between the observations. Copernicus takes Ptolemy's obliquity to apply to his entire period, but the most likely guess for his choosing a specific year would be the second or third year of Antoninus, thus E.A. 462/63, containing Ptolemy's observation of Regulus and the equinoxes. Copernicus's own obliquity, used for reducing the observations of Spica in 1515 and 1525, he originally found in Johann Werner's *De motu octavae sphaerae*, where Werner says that it was originally determined by Regiomontanus (so prior to 1476), but he himself applies it to the end of 1514 and the beginning of 1515, thus E.A. 1839.[16] The interval between the observations is 1376 or 1377 years—in treating the obliquity and precession, Copernicus rounds to integral years—and simply because we know it works, we shall use 1376 years. The period of 180° of ϑ is thus

$$(180^\circ/144;13^\circ)\ 1376^y = 1717;24^y,$$

and if 144;13° is rounded to $144\frac{1}{4}^\circ$, the period is 1717;1 years. In any case, it is not hard to propose a plausible derivation of Copernicus's 1717-year period, although naturally it is hard to be certain of just what he did.[17]

The full period of ϑ is 3434 Egyptian years, in which a full cycle of the obliquity is completed, and of 2ϑ 1717 Egyptian years, the period of the anomaly of the precession. The intervals between the observations that will be of interest for further derivations are as follows:

Observers	Δt	$\Delta 2\vartheta$
Tim.–Ptol.	432^y	90;35°
Ptol.–Bat.	742	155;34
Bat. + 543^y	543	113;51
Total	1717	360;0

(5)

(*b*) *The Mean Precession*: $\bar{\pi}$ (III,6)

In the case of the period of the mean precession, Copernicus gives some information, but it is still difficult to determine how he found his result. Here is what he says: In the 1819 years between Timocharis's observation of Spica in E.A. 30 (2) and his own observation in E.A. 1849 (13), the tropical longitude of the star increased by 25;1°. But in the 102 years after Timocharis, the longitude must have changed by 1;4°, leaving 23;57° of both mean and true precession for the complete 1717-year cycle preceding E.A. 1849, so that the whole period of the precession is 25,816 Egyptian years.

[16] Werner (1522), Prop. 25; see Swerdlow (1975), 63–65.

[17] For another derivation, see Moesgaard (1968), 130–33. M (70), following the star catalogue, contains a cancelled table of $\vartheta = 0;6,17,29,36^{\circ/y}$, implying a period of about $3433;10^y$, which was originally accompanied by a corresponding table of the mean precession. The origin of this table is mysterious, and it must have preceded the present version of III,6 in M.

How Copernicus arrived at the difference of 1;4° has been plausibly explained by Moesgaard as follows:[18] In III,7 Copernicus assumes that the zero point of the anomaly falls exactly midway between Timocharis and Ptolemy. Thus any true precession for some number of years following Timocharis will be equal to the precession for the same number of years preceding Ptolemy, and if the rate of change of the precession is not too great, a linear proportionality will approximate the precession for different periods following Timocharis and preceding Ptolemy. Now, from observations 4/5 and 7/8, the precession in the 40 years between Menelaus and Ptolemy was 0;25°. Thus,

$$(102^y/40^y)0;25° = 1;4°$$

seems to explain the 1;4° in the 102 years following Timocharis.

However, let us not yet be too confident. For, Copernicus says, if the mean precession has advanced 23;57° in 1717 years, its complete period must be 25,816 years, and its annual motion is $0;0,50,12,5^{°/y}$. But this is simply untrue since

$$(360°/23;57°)\ 1717^y = 25808;46^y.$$[19]

But Copernicus has not miscomputed because $360°/25816^y = 0;0,50,12,5^{°/y}$ and $1717 \cdot 0;0,50,12,5° = 23;56,35\ldots° \approx 23;57°$, that is, these numbers are all consistent and indicate that 23;57° was derived from the complete period, and not *vice versa*. Hence it is hard to know what the above derivation of 1;4° really means, for at best it can be understood only as a confirmation that when the predetermined 23;57° has been subtracted from 25;1°, the remainder of 1;4° approximately represents the precession in the 102 years following Timocharis.[20] But this is only true if the zero point of the anomaly lies midway between Timocharis and Ptolemy, an assumption to be corrected in III,9. Thus we still do not know how Copernicus derived his mean rate of precession, but we shall return to the question in the section on solar theory and the length of the year.[21]

(c) *The Maximum Equation of Precession*: $\delta_{P_{max}}$ (III,7)[22]

In order to find the maximum equation of precession and the corresponding displacement of the pole of the equator, Copernicus uses the observed precession between Timocharis (E.A. 30) and Ptolemy (E.A. 462), and, as previously mentioned, assumes that the zero point of the anomaly fell exactly

---

[18] Moesgaard (1968), 133; cf. the remarkable misunderstanding in Copernicus (1978), 386 (n. 129:21).

[19] 25809 was entered as an emendation for 25816 in the 1854 Warsaw edition of *De rev.*, but Copernicus undoubtedly intended 25816.

[20] Further evidence for the priority of 23;57° is that M (78v–79r) contains earlier figures 25;3° and 1;6°, leaving as before 23;57°.

[21] See below, pp. 156–157.

[22] The order of presentation in this chapter was greatly altered in E, the correction sheet to N, and we follow the revised version.

midway between them (hence E.A. 246 = A.D. −77). In Fig. 9 the observed precession, from observations (2) and (8), is $\pi = 4;20°$, which gives the motion of the true equinox V. By computation for 432 years, the mean precession

$$\bar{\pi} = 7{,}12 \cdot 0;0{,}50{,}12{,}5° = 6;1{,}27° \approx 6°$$

showing the displacement of the mean equinox $\bar{V}$. The difference of 1;40° is twice the equation of precession δ_P at the time of each observation. So in Fig. 10, from (5) $\Delta 2\vartheta = 90;35°$ is the change in the anomaly between Timocharis and Ptolemy, and $V_T V_P = 2\delta_P = 1;40°$ is the change of the equation. The radius r measures the maximum equation $\delta_{P_{max}}$, which it is required to find. Now

$$\delta_P / r = \sin \tfrac{1}{2}\Delta 2\vartheta,$$

and since

$$\delta_P = \tfrac{1}{2} V_T V_P = 0;50° \qquad \text{and} \qquad \tfrac{1}{2}\Delta 2\vartheta = 45;17{,}30°,$$

$$\delta_{P_{max}} = r = \delta_P / \sin \tfrac{1}{2}\Delta 2\vartheta = 0;50°/.7107 = 1;10{,}21\ldots° \approx 1;10°.$$

For the displacement of the pole of the equator, consider Fig. 7 in which the motion that takes place at the pole is moved down to the equinox. In triangle $\bar{V}N'V$, from (3)

$$\bar{V}N' = V\bar{V} \sin \bar{\varepsilon},$$

or substituting the maximum values

$$\bar{V}N'_{max} = \delta_{P_{max}} \sin \bar{\varepsilon} = 1;10° \sin 23;40°$$

$$= 1;10° \cdot .4014 = 0;28{,}5° \approx 0;28°$$

gives the maximum displacement of the poles of the equator.

(d) *Correction of the Location of* $\vartheta = 0°$ (III,9)

It must next be investigated whether the assumption that the zero point of ϑ fell midway between Timocharis and Ptolemy was correct. Copernicus does this by checking whether the precession at the time of al-Battānī is compatible with this assumption. In Fig. 11 we have the position of the anomaly 2ϑ at the time of Timocharis T, Ptolemy P, and Battānī B, and the corresponding position of the equinox V_T, V_P, V_B. From the intervals of time and from observations (2/8) and (7/11), the differences are:

Observers	Δt	$\Delta\pi$	$\Delta\bar{\pi}$	$\Delta 2\vartheta$	$\Delta\pi - \Delta\bar{\pi}$
Tim.–Ptol.	432^y	4;20°	6;0°	90;35°	−1;40°
Ptol.–Bat.	742	11;30	10;21	155;34	+1;9

Now, in the figure, from the observations

$$V_T V_P = -1;40°, \qquad V_P V_B = +1;9°, \qquad V_B V_T = 1;40° - 1;9° = +0;31°,$$

and since

$$V_T \bar{V} = V_P \bar{V} = 0;50°, \qquad \text{so} \qquad V_B \bar{V} = 1;9° - 0;50° = +0;19°$$

which is the equation of precession δ_P at the time of Battānī. But by theory

$$2\vartheta_B = 45;17,30° + 155;34° = 200;51,30°,$$

from which

$$\delta_P = V_B \bar{V} = 1;10° \sin(200;51,30° - 180°)$$
$$= 1;10° \cdot .356 \approx 0;24°, \quad \text{(Corr. 0;24,55°)}$$

so there is an error of $0;24° - 0;19° = 0;5°$ in computing the equation of precession for the time of Battānī.

Hence something is wrong, and while it could be almost anything—the anomaly, the mean precession, the equation of precession—Copernicus considers only that the location of the zero point of 2ϑ is incorrect and must be changed. This he does by trial and error (and a last alteration even appears in M 85v7–19 and margin).[23] He changes $2\vartheta_P$ from 45;17,30° to 42;30°, and consequently $2\vartheta_T = 311;55°$ and $2\vartheta_B = 198;4°$. Thus (neglecting signs),

$$V_P \bar{V} = 1;10° \sin 42;30° = 0;47,17°, \quad \text{(Copernicus } 0;47\tfrac{1}{2}°\text{)}$$
$$V_T \bar{V} = 1;10° \sin 311;55° = 0;52°,$$
$$V_B \bar{V} = 1;10° \sin 198;4° = 0;21,42°, \quad \text{(Copernicus 0;21°)}$$

so that

$$V_P V_B = 0;47,30° + 0;21° = 1;8,30° \approx 1;9°,$$
$$V_B V_T = 0;52° - 0;21° = 0;31°,$$
$$V_T V_P = 0;47,30° + 0;52° = 1;39,30° \approx 1;40°,$$

in agreement with the observations. Thus, the following values of ϑ are required at each observation:

$$\vartheta_T = 155;57,30°, \qquad \vartheta_P = 21;15°, \qquad \vartheta_B = 99;2°. \tag{6}$$

(e) *Limits of the Obliquity*: ε_{min} *and* ε_{max} (III,10)

In II,2 Copernicus says that it is by a conjecture that the obliquity is taken to vary between 23;28° and 23;52°, and according to our reconstruction,

---

[23] In the second appendix of his *Apollonius Gallus* (1600), François Viète criticized Copernicus's procedure, and showed in Prob. 1 of the appendix how to derive by proper geometry the radius of the circle and the zero point of the anomaly from three observations. Viète's solution is given in Swerdlow (1975), 89–95.

this assumption was used in deriving the 3434-year cycle of the anomaly. Now in III,10, Copernicus "proves" what he had previously assumed. That the chapter was an afterthought is shown by the fact that it is added out of place in M (76v), and required renumbering the subsequent chapters of Book III. Further, that the chapter is indeed a late addition, after 1539, is shown by Rheticus's ignorance of its contents in the *Narratio prima.* And in connection with adding this chapter, Copernicus altered the obliquity for his own time to 23;28,24°, a value computed from the theory.[24]

Referring again to Fig. 8, at the time of Ptolemy in 139 (E.A. 462) $\varepsilon = EN_P = 23;51,20°$, but now at the time of Copernicus in 1525 (E.A. 1849) $\varepsilon = EN_C = 23;28,24°$. Thus $EN_P - EN_C = N_P N_C = 0;22,56°$, from which we must find the maximum variation 2ρ. Now at the time of Ptolemy $\vartheta_P = 21;15°$, and for the 1387 intervening years $\Delta\vartheta = 145;24,16° \approx 145;24°$, so that at the time of Copernicus $\bar{\vartheta}_C = 180° - (\vartheta_P + \Delta\vartheta) = 13;21°$. Thus, where $\rho = 1000$,

$$N_P\overline{N} = \rho \cos 21;15° = 932, \qquad N_C\overline{N} = \rho \cos 13;21° = 973,$$

so that $N_P N_C = 1905$. But where $N_P N_C = 0;22,56°$,

$$2\rho = (2000/1905)\ 0;22,56° = 0;24,1° \approx 0;24°.$$

Thus, $\varepsilon_{min} = 23;28°$, $\varepsilon_{max} = 23;52°$, $\bar{\varepsilon} = 23;40°$. Of course all of this is circular since a variation of 0;24° for ε was assumed in the first place, and possibly used to derive the period of ϑ.[25] About the most charitable interpretation is to say that Copernicus was "confirming" the accuracy of his prior assumption.

(*f*) *Epoch Positions of the Mean Precession and Anomaly*: $\bar{\pi}_0$ *and* ϑ_0 (III,11)

Copernicus must now give the positions of $\bar{\pi}$ and ϑ at some fundamental epoch for use in computation, but he chooses to give them for four epochs out of a list of no less than seven whose dates he establishes. His entire list is as follows:

A. Olympiad 1 I
B. Nabonassar 1
C. Alexander 1
D. Caesar's Calendar Reform 1
E. Augustus (Egyptian calendar) 1
F. Augustus (Julian calendar) 1
G. Christ (*Anno Domini*) 1

---

[24] Swerdlow (1975), 59–63, 97–98. In M the following demonstration contains some erroneous numbers that were corrected in N.

[25] In fact, since 2ρ was really assumed in advance, Copernicus could have solved a more interesting problem, namely, from $\Delta\vartheta$ and the ratio of $N_P N_C$ to 2ρ, he could find the correct locations of ϑ_P and ϑ_C, that is, test whether $\vartheta 0°$ and $2\vartheta 0°$, for the obliquity and the precession respectively, in fact coincide as he assumed. This must have been evident to Viète since in Prob. 2 of the second appendix to *Apollonius Gallus*, he shows a rigorous geometrical solution to this problem. Of course, if one uses three observed values of ε, then Viète's Prob. 1 allows one to find both 2ρ and the location of $\vartheta 0°$, that is, to investigate completely the range of the variation of ε and the location of each observation within the cycle.

Copernicus's epochs and his sources—principally Censorinus—are discussed in the following appendix.[26] Here we need consider only the four epochs actually used for computation, which are given with their intervals in Egyptian years and days:

Epoch	Δt	
A. Olympiad 1 I Hekatombaion 1 noon		
	$451^y\ 247^d$	
C. Alexander 1 Thoth 1 noon		
	278 $118\frac{1}{2}$	(7)
D. Caesar 1 Jan. 1 midnight		
	45 12	
G. Christ 1 Jan. 1 midnight		
Total	775 $12\frac{1}{2}$	

By giving a number of epochs, Copernicus was following in the tradition of the *Alfonsine Tables*, which gave epochs for everything from the Deluge (−3101 Feb 17, originally the Indian Era of the Kaliyuga) through Era Alfonso (1252 Jun 0), and since the tables of mean motions were arranged purely by days, they could be adapted to any calendar whatever. His choice, however, is peculiar, perhaps deliberately classical. He omits B. Era Nabonassar, the epoch of the *Almagest*, and E. the Egyptian Era Augustus, the epoch of Ptolemy's *Canobic Inscription*, which of course he did not know. On the other hand, he includes, with an error of one year, the rather uncertain Era of the Olympiad, and an even stranger choice is the year of Julius Caesar's calendar reform,[27] neither of the epochs having previously been used for astronomical purposes.

Now if we can find $\bar{\pi}$ and ϑ for any epoch, the intervals just given will easily allow computation of the remaining epochs. Copernicus takes π and ϑ for Ptolemy to apply specifically to Antoninus 2 (E.A. 462) Pharmuthi 9, which he says corresponds to 139 Feb 24 (corr. Feb 23).[28] This date, considered as referring to midnight rather than the actual time of the observation, follows A.D. 1 Jan. 1 by 138 julian years 55 days, or in Egyptian years $138^y\ 89^d$, and thus follows Olympiad 1 I by $913^y\ 101\frac{1}{2}^d$, in which time $\Delta\bar{\pi} = 12;44°$ and $\Delta\vartheta = 99;44°$ (corr. 99;44,33°). For Antoninus 2 it was shown that $\vartheta = 21;15°$ or $2\vartheta = 42;30°$, from which $\delta_P = -0;48°$ (which in III,9 was 0;47,30°). Since by observation $\pi = 6;40°$,

$$\bar{\pi} = \pi - \delta_P = 6;40° + 0;48° = 7;28°,$$

[26] Some literal quotations from Censorinus, whom Copernicus mentions as a source, are given by Rosen in Copernicus (1978), 390. For the appendix on Copernicus's chronology, see below, pp. 183–188.

[27] Copernicus's era of Caesar's calendar reform should not be confused with the *Era Caesaris* in the *Alfonsine Tables*, which is simply the Spanish Era −37 Jan 0.

[28] See appendix below, p. 188.

and at Olympiad 1 I

$$\bar{\pi}_0 = 7;28° - 12;44° = 354;44° = \lambda^*0° - 5;16°,$$

which tells us that the mean equinox preceded γ Arietis by 5;16°, and

$$\vartheta_0 = 21;15° - 95;45° = 285;30°,$$

noting that 95;44° has become 95;45°. By adding the appropriate $\Delta\bar{\pi}$ and $\Delta\vartheta$ for each interval in the order A. → C. → D. → G., Copernicus arrives at the following epoch values:

Epoch	$\bar{\pi}_0$	ϑ_0
A. Olympiad	354;44°	285;30°
C. Alexander	1;2	332;52
D. Caesar	4;55	2;2
G. Christ	5;32	6;45

(8)

The Tables and Their Use (III,6,8,12)

The tabulations of the mean precession $\bar{\pi}$ and anomaly ϑ follow III,6. Copernicus's tables of mean motions are purely sexagesimal—for 1–60 Egyptian years of 365 days, and for 1–60 days—and are very convenient to use simply by shifting place-values for powers of 60 = 1,0. The *Alfonsine Tables* tabulate mean motions sexagesimally, but for days only, and thus all intervals must be converted to days. Copernicus's tables offer the advantage that if one is reckoning in an era counted in Egyptian years, one may use the years directly. Naturally, if julian years are used, one must first convert to Egyptian years, which makes the use of A.D. dating somewhat inconvenient. Likewise, hours must be converted to sexagesimal fractions of a day. Tables for these conversions were available to Copernicus and his contemporaries in the *Alfonsine Tables*. The tabulated mean motions are based upon, for $\bar{\pi}$,

$$\bar{\pi}^{\mathrm{y}} = 6,0°/7,10,16^{\mathrm{y}} = 0;0,50,12,5^{°/\mathrm{y}},$$

$$\bar{\pi}^{\mathrm{d}} = 0;0,50,12,5°/6,5^{\mathrm{d}} = 0;0,0,0,8,15^{°/\mathrm{d}}, \qquad (9a)$$

and for ϑ,

$$\vartheta^{\mathrm{y}} = 6,0°/57,14^{\mathrm{y}} = 0;6,17,24,9^{°/\mathrm{y}},$$

$$\vartheta^{\mathrm{d}} = 0;6,17,24,9°/6,5^{\mathrm{d}} = 0;0,1,2,2^{°/\mathrm{d}}. \qquad (9b)$$

In order to find $\bar{\pi}$ and ϑ or 2ϑ for a given date, one uses the tables to find $\Delta\bar{\pi}$ and $\Delta\vartheta$ for Δt since epoch, and then forms

$$\bar{\pi} = \bar{\pi}_0 + \Delta\bar{\pi}, \qquad \vartheta = \vartheta_0 + \Delta\vartheta, \qquad 2\vartheta = 2\cdot\vartheta. \qquad (10)$$

III,8 is on the tabulation of the corrections or equations, which Copernicus likes to call by the Greek term *prosthaphaereses*, for finding π and ε from $\bar{\pi}$ and $\bar{\varepsilon}$ using arguments 2ϑ and ϑ. The first two columns of the table following III,8 give the argument 2ϑ or ϑ for 3°–180° and 180°–357° at 3° intervals. The third column gives the correction δ_P of the equinox, in which the "partial corrections" (*particulares differentiae*) are computed from the maximum correction by

$$\delta_P = \delta_{P_{max}} \sin 2\vartheta = 1;10^\circ \sin 2\vartheta. \qquad (11)$$

In using this column, one enters with 2ϑ, and directly reads δ_P, which is negative for $2\vartheta \leq 180^\circ$ and positive for $2\vartheta \geq 180^\circ$. Then

$$\pi = \bar{\pi} + \delta_P, \qquad (12)$$

and since π is the difference between sidereal and tropical longitude, a tropical longitude

$$\lambda = \lambda^* + \pi. \qquad (13)$$

Column 4 gives the coefficient c, called by Copernicus "proportional minutes," for finding the variation of the obliquity ε. The tabulated values of c are computed from

$$c = 0;30(1 + \cos \vartheta), \qquad (14)$$

and are used as a positive correction to $\varepsilon_{min} = 23;28^\circ$. One therefore enters with ϑ, reads c in column 4, and forms

$$\varepsilon = 23;28^\circ + c \cdot 0;24^\circ. \qquad (15)$$

The coefficient c is also used with the difference d in the tables for δ, α, and γ in II,3. Thus, to correct these quantities for the variability of ε, one regards the tabulated values as minima for δ, and as maxima for α and γ, and having found c for a given time, forms

$$\delta = \delta_{min} + c \cdot d(\delta), \qquad \alpha = \alpha_{max} - c \cdot d(\alpha), \qquad \gamma = \gamma_{max} - c \cdot d(\gamma).^{29} \quad (16)$$

The tables of oblique ascension and meridian angles following II,9 are based upon the minimum obliquity, and thus should also have a correction of the form d to be used with c. But none is provided.

Example. Compute the longitude of Spica and the obliquity of the ecliptic for A.D. 1525 Apr 16. This example is given by Copernicus in III,12, and since he computes for an integral number of days, the time is evidently

[29] In Copernicus's explanation (N 78v21–25), by a slip the meridian angle γ is described as the angle of intersection of the ecliptic and "equator" (*aequinoctialis*), when he means "meridian." This passes without comment in all the translations, although it has been correctly noted in Copernicus (1975), 407. It is also of interest that all the translations misinterpret the correction to α and γ, making the former negative and the latter positive. Both "the latter" (*his*) are negative; only δ "the former" (*illis*) is positive. Cf. also II,3 (N 30v31–34) and above, p. 106, n. 12.

taken to be midnight Apr. 16/17. Whether this is supposed to be the date, although not the time, of his measurement of the altitude of Spica in 1525 (III,2) is not clear. The epoch used in the computation is that of Christ, A.D. 1 Jan 0/1 midnight. Thus the elapsed time is 1524 julian years 106 days or, in Egyptian years $1525^y\ 122^d = 25{,}25^y + 2{,}2^d$. From the tables in III,6:

Δt	$\bar{\pi}$	ϑ	
$25{,}0^y$	20;55, 2°	2,37;15, 3°	
25^y	0;20,55, 2	2;37,15, 3	
$2{,}0^d$	0; 0,16	0; 2, 4	
2^d	0; 0, 0,16	0; 0, 2, 4	
	21;16,13,18	2,39;54,24, 7	
A.D. 1	5;32	6;45	
	26;48	2,46;39 ≈ 166;40°	$2\vartheta = 333;20°$

Thus from the table in III,8,

$$\delta_P(2\vartheta) = +0;32°, \qquad c(\vartheta) = 0;1,$$

so that

$$\pi = \bar{\pi} + \delta_P = 26;48° + 0;32° = 27;20°, \quad \text{(Copernicus 27;21°)}$$

and

$$\varepsilon = 23;28° + 0;1 \cdot 0;24° = 23;28,24°,$$

the obliquity inserted as an alteration in M and used as though observed in III,10. From the star catalogue, for Spica $\lambda^* = 170;0°$, so

$$\lambda = \lambda^* + \pi = 170;0° + 27;20° = 197;20° = \text{♎}17;20°.$$

Copernicus gives 17;21°, the value obtained from observation in III,2. Let us hope we have really recomputed an observed value.

In order to show the application of c to the tables in II,3, Copernicus takes the following example: For 880 Egyptian years after Christ, about the time of al-Battānī, $\vartheta = 98;59° \approx 99°$, so $c = 0;25$. Thus,

$$\varepsilon = 23;28° + 0;25 \cdot 0;24° = 23;38°$$

which does not compare too badly with Battānī's obliquity of 23;35° (which, however, Copernicus is unsure of). Taking an ecliptic longitude of ♉ 3° = 33°, from (16) the declination

$$\delta = 12;32° + 0;25 \cdot 0;12° = 12;37°,$$

the right ascension

$$\alpha = 30;50° - 0;25 \cdot 0;4° = 30;48°,$$

and the meridian angle

$$\gamma = 70;0^\circ - 0;25 \cdot 0;20^\circ = 69;52^\circ.$$

Verification of Precession and Obliquity

In order to compare the precession computed from Copernicus's theory with the precession implied by the observations of tropical longitudes of stars from Timocharis to Copernicus, we subtract from the longitudes of stars in Table 1 the sidereal longitudes in Copernicus's star catalogue, leaving a precession that is, for practical purposes, defined as $\pi = 6;40^\circ$ at the time of Ptolemy in 139. We may then compare the precession computed for the dates assumed by Copernicus—neglecting small discrepancies of a year or two in Copernicus's dates—and the results are as follows

Observer	Date	π Obs.	π Comp.
Timocharis	−293	2;20°	2;18°
Hipparchus	−128	4;0	4;0
Menelaus	99	6;15	6;15
Ptolemy	138	6;40	6;40
Battānī	880	18;10	18;10
Copernicus	1515	27;14	27;14
Copernicus	1525	27;21	27;20

The results could hardly be better, although it should be noted that there is really a discrepancy of 0;4° with the 1515 observation of Spica that Copernicus altered from ♎17;10° to ♎17;14° in M.

In the same way, we may take the values of the obliquity of the ecliptic that Copernicus believed, not necessarily correctly, to have been observed at each date, and compute the obliquity from his theory.

Observer	Date	ε Obs.	ε Comp.
"Aristarchus"	−279	23;51,20°	23;51,4°
Ptolemy	139	23;51,20	23;51,11
Battānī	880	23;36	23;38,8
az-Zarqāl	1070	23;34	23;34,12
Prophatius	1300	23;32	23;30,25
Werner – Copernicus	1515	23;28,30	23;28,23
Copernicus	1525	23;28,24	23;28,20

Here again the agreement is satisfactory, which is remarkable since only the values for 139 and 1515 were used to establish the parameters. Note, however, that the value 23;28,24° for 1525 was computed from Copernicus's tables, and was not observed, while computing directly from the theory gives 23;28,20°.

2. Solar Theory

The Inequality of the Tropical Year (III,13)

Again Copernicus begins with a historical introduction. Since the precession of the equinoxes and solstices is nonuniform, it follows that the period of the sun's return to any one of them, the tropical year, will also be nonuniform. And this can be proved by examining the different values found for the length of the tropical year over the centuries. For Callippus and, so Copernicus says, Aristarchus and Archimedes found the year measured with respect to the summer solstice to be $365\frac{1}{4}$ days. But Ptolemy, aware of the difficulty of determining the time of solstices, chose rather to use observations of equinoxes made by Hipparchus, who had found a slight deficit from $365\frac{1}{4}$ days. By comparing equinoxes he observed himself with those observed by Hipparchus, Ptolemy determined that this deficit was $1/300^{d}$, so that the length of the tropical year was $365\frac{1}{4} - 1/300^{d} = 365;14,48^{d}$.[30]

However, this value has not remained constant, as a number of later comparisons will show. Copernicus then gives a series of observations of equinoxes—two each by Hipparchus and Ptolemy, one by al-Battānī, and two by himself—which he uses to show the variation of the tropical year. His sources for the earlier observations were *Almagest* III,1 and *Epitome* III,2. Table 2 gives all the solar observations reported by Copernicus, and includes, besides the autumnal equinoxes A. and vernal equinoxes V., Copernicus's observation of the sun at Scorpio 15°. The dates are given as in the text, and converted to julian dates counted from midnight.

Copernicus says nothing of how he observed the equinoxes. In II,14 he explained how to use the instruments that he has described for finding the obliquity—although he had mentioned only one, the plinth—to find the longitude of the sun.[30a] The principle is very simple. Observe the altitude $\bar{\zeta}_M$ of the sun at meridian transit, i.e. when it is in the midheaven at local apparent noon. Then, knowing one's latitude φ, $\bar{\zeta}_M - \bar{\varphi} = \delta$, the declination, and by inverse use of the declination table, or trigonometrically by

[30] In fact, Hipparchus had already estimated the deficit as about $1/300^{d}$ (*Almagest* III,1; Manitius 1, 145:24–146:2). However, earlier in the same chapter (Manitius 1, 140:7–11) Ptolemy says simply that Hipparchus found the year shorter than $365\frac{1}{4}$ days, and that is probably what Copernicus has in mind. *Epitome* III,2 attributes to Hipparchus a year of $365\frac{1}{4}$ days exactly.
[30a] See above, p. 105.

$\sin\lambda = \sin\delta/\sin\varepsilon$, find the longitude λ. Obviously if $\bar{\zeta}_M = \bar{\varphi}$, $\delta = 0°$, and the sun is at an equinox. However, since it is most unlikely to be at an equinox at meridian transit, it will probably be necessary to interpolate between observations of transits before and after the equinox, assuming for interpolation that $\Delta\delta \approx 0;24^{\circ/d} = 0;1^{\circ/h}$, which is convenient, or that $\Delta\lambda \approx 1^{\circ/d} = 0;2,30^{\circ/h}$, as Copernicus recommended in II,14.

By comparing the intervals between the reported observations with the corresponding number of tropical revolutions of the sun, Copernicus can easily demonstrate the variability of the tropical year. However, since the observations were made at different terrestrial longitudes—Rhodes and Alexandria which, following Ptolemy, he assumes to lie on the same meridian, Raqqa in Syria, and Frauenburg in Warmia—he must adjust the intervals for the differences in longitude. He does this, assuming that

(a) Frauenburg–Alexandria $\approx 1^h = 15°$, (corr. $10;15° = 0;41^h$)

(b) Alexandria–Raqqa $= 10° = 0;40^h$, (corr. $9;8° \approx 0;37^h$) (17)

(c) Frauenburg–Raqqa $\approx 25° = 1;40^h$.[31] (corr. $19;23° \approx 1;18^h$)

The intervals between the observations and the resulting tropical years are given in Table 3.[32] Since a fair amount of rounding has entered into the last three comparisons, the numbers should not be pressed too closely. Nevertheless, the trend is clear that the tropical year was longest when the precession was slowest, between Hipparchus and Ptolemy, the year was shortest when the precession fastest, between Ptolemy and Battānī, and then somewhat longer as the precession becomes somewhat slower, between Battānī and Copernicus. Copernicus also shows that any one interval is incompatible with the substitution of the tropical year resulting from another interval, e.g. if the year from 3–7 be multiplied by the interval of 743^{ty} from 3–5, there will be an excess of about half a day over the correct interval 3–5.

Therefore, he concludes, it is more correct to measure the uniformity of the solar year with respect to the fixed stars as, so Copernicus thinks, Thābit ibn Qurra was the first to discover. His reasoning was that since the year is longer when the precession is slower, and shorter when faster, in a constant proportion, i.e. $\Delta ty = c/\Delta\pi$, the sidereal year must be uniform, and Thābit gave its length as $365;15,23^d = 365^d\ 6;9,12^h$.[33] Still, Copernicus

[31] On Copernicus's geography, see the appendix below, pp. 188–190.

[32] The deficit (4) of 5.–7., $5^d{-}1\frac{1}{4}^h$, is correctly $158^d\ 6^h - 153^d\ 6\frac{3}{4}^h = 4^d\ 23\frac{1}{4}^h = 5^d - \frac{3}{4}^h$. But Copernicus definitely used the wrong value $5^d - 1\frac{1}{4}^h$ since $(5^d - 1\frac{1}{4}^h)/633^y \approx 1^d/128^y$, as in the text, while $(5^d - \frac{3}{4}^h)/633^y \approx 1^d/127^y$.

[33] The value $365^d\ 6;9,12^h$ is from *Epitome* III,2. In *De anno solis* Thābit established various values for the sidereal year, but considered $365;15,23,34,43^d$ the most secure. He also derived a year of $365;15,22,47,30^d$, and either could be the source of the number in the *Epitome*. Cf. Neugebauer (1962a), 280–83, esp. para. [103, 108, 110]. Copernicus's sidereal year, first stated in III,14, will be taken up in the commentary to III,18, below, pp. 154–155.

continues, it is not easy to find the length of the sidereal year because the motion of the sun is affected by four causes of apparent nonuniformity:

(1) The nonuniform precession.
(2) The annual inequality, due to the eccentricity.
(3) The variation of the eccentricity.
(4) The nonuniform motion of the apsidal line.

The first cause has already been explained, and it is the task of the rest of Book III to explain the others. The annual inequality is also called the "first inequality," and the third and fourth causes together constitute the "second inequality."

The Model for the First Inequality (III,15)

Ptolemy's model for the motion of the sun, shown in Fig. 12(a) and lettered heliocentrically for the sun at S and the earth at E, is a simple eccentric, or a concentric carrying an epicycle that rotates with equal velocity in the direction opposite to its motion on the concentric. The parallelogram $SDEC$ shows the equivalence of the two models, in which $R = R'$ and consequently $e = r$, although one can also make the models of different radii R and R' provided that $e/R = r/R'$. Since neither model violates Copernicus's principle of uniform circular motion, he adopts them, preferring for computation the eccentric, and exhaustively explains their properties of equivalence, symmetry of equations to the apsidal line, location of maximum equation, etc.—all closely following *Almagest* III,3 and *Epitome* III,4–12.[34] He also shows (see Fig. 12(b)) that a rotation of the epicycle not equal to its motion on the concentric is equivalent to an eccentric with a moveable apsidal line, here following *Almagest* IV,5 and *Epitome* IV,7–8. In this case the apsidal line rotates through the difference between the motion on the concentric $\bar{\kappa}$ and the rotation of the epicycle $\bar{\alpha}$, i.e. $\Delta A = \Delta(\bar{\kappa} - \bar{\alpha})$. If $\bar{\kappa} > \bar{\alpha}$, A moves forward, if $\bar{\kappa} < \bar{\alpha}$, backward.

Derivation of Eccentricity and Direction of the Apsidal Line (III,16–17)

The problem here is to locate the center of the sun's (or the earth's) eccentric with respect to its distance and direction from the observer (or from the sun). In the most general case it is of the form: Given three points on the circumference of a circle, to find a point, within or without the circle,

[34] It is interesting that in the case of the lunar and planetary theory, where Copernicus cannot copy from the *Almagest* and *Epitome* because his models differ from Ptolemy's, he omits such demonstrations.

from which lines drawn to the three points will contain given angles. However, when possible, the problem is reduced to special cases that are easier to solve by specifically choosing the given angles. Copernicus first reviews Ptolemy's derivation, from *Almagest* III,4 and *Epitome* III,13, which uses the angles and intervals of time between the two equinoxes and the summer solstice. This has the advantage of making the given angles right angles, although the disadvantage of necessitating an observation of the time of the solstice, which is difficult.

From observation, the intervals are

$$♈0° - ♋0° = 94\tfrac{1}{2}^{d}, \qquad ♋0° - ♎0° = 92\tfrac{1}{2}^{d}.$$

Letting the mean motion of the sun be $360°/365\tfrac{1}{4}^{d} \approx 0;59,8^{°/d}$, in Fig. 13

$$\Delta\bar{\lambda}_1 = 93;9°, \qquad \Delta\bar{\lambda}_2 = 91;11°,$$

so that

$$FK = KH = \tfrac{1}{2}(\Delta\bar{\lambda}_1 + \Delta\bar{\lambda}_2) = \tfrac{1}{2}(184;20°) = 92;10°.$$

Thus,

$$\delta_1 = FK - 90° = 2;10°, \qquad \delta_2 = \Delta\bar{\lambda}_1 - FK = 0;59°,$$

and where $R = 10000$

$$CD = OE = R \sin \delta_1 = 378, \qquad CE = OD = R \sin \delta_2 = 172,$$

so that

$$e = OC = (OE^2 + OD^2)^{1/2} = 415$$

is the eccentricity.[35] In the *Almagest*, the result is given sexagesimally as $e = 2;29,30$ where $R = 60$, and rounded to $e/R \approx 2;30/60 = 1/24$. Then, since

$$\text{angle } EOC = \sin^{-1}(CE/OC) = \sin^{-1}(172/415) \approx 24;30°,$$

the longitude of the apogee

$$\lambda_A = ♋0° - 24;30° = ♊5;30°.$$

Since Hipparchus had found the same result nearly 300 years earlier, Ptolemy assumed that the solar eccentricity was invariable, and the apsidal line tropically fixed. But, Copernicus continues, it has since become evident that neither assumption is true. For (*Epitome* III,13) al-Battānī found

$$♈0° - ♋0° \approx 93^{d}\,14^{h} = 93;35^{d},$$

$$♈0° - ♎0° = 186^{d}\,14;45^{h} = 186;37^{d},$$

[35] In M (96v–97r), this derivation contains a series of demonstrably wrong numbers leading to $e = 414$, and although everything was corrected in N (87r), all the critical editions and all the translations, except Duncan's (Copernicus (1976), 172), which is fortunately based upon N, retain some or all of the wrong numbers, often with no particular consistency.

from which

$$e = 347, \qquad \lambda_A = \text{♋}0° - 7;43° = \text{♊}22;17°,$$

and, so Copernicus believes, az-Zarqāl found the same eccentricity as al-Battānī, but found that $\lambda_A = \text{♋}0° - 12;10° = \text{♊}17;50°$.[36] All of this indicates that the motion of the earth is affected by another inequality.

And so, "from ten and more years" of investigating these matters, and particularly from observations made during 1515/16, Copernicus finds that

$$\begin{aligned} \text{♈}0° - \text{♎}0° &= 186;5,30^d, \\ \text{♎}0° - \text{♏}15° &= 45;16^d, \qquad (18) \\ \text{♎}0° - \text{♈}0° &= 178;53,30^d. \end{aligned}$$

Note that there are some problems with these intervals. Observations 7/9 from III,13 give

$$\text{♎}0° - \text{♈}0° = 178^d\,21;50^h = 178;54,35^d,$$

compared with $178;53,30^d$ in III,16. But more remarkably, the sum

$$\text{♈}0° - \text{♈}0° = 186;5,30^d + 178;53,30^d = 364;59^d$$

falls about a quarter-day short of the length of the tropical year.[37]

Copernicus had observed Scorpio 15° because it avoids the difficulty of determining the time of a solstice, as is required in Hipparchus's and Ptolemy's method of deriving the elements of the solar orbit. The advantage is obvious since at the solstice, the meridian altitude of the sun changes by only $\Delta\delta \approx 0;0,14^{\circ/d}$, while at ♏15° $\Delta\delta \approx 0;17^{\circ/d}$, i.e. more than 70 times as much. Here Copernicus was using his own variation of a procedure described in the *Epitome*. In *Epitome* III,13 it is reported that az-Zarqāl determined the time of the four midpoints between the equinoxes and solstices, i.e. ♉15°, ♌15°, ♏15°, ♒15°, a method used earlier by Thābit ibn Qurra in *De anno solis*.[38] In this case $\Delta\delta \approx 0;17^{\circ/d}$, and there is the further benefit that the angles of apparent motion are right angles, so Ptolemy's simple method can still be used to find the elements. However, *Epitome* III,14 recommends using the equinoxes and the entry of the sun into a sign adjacent to an equinox, say ♉0° or ♏0°, and shows the trigonometric solution to finding the eccentricity and apsidal direction from three such points on a circle, which is a special case in that the equinoxes still lie on a straight

---

36 For Battānī's derivation, see Nallino 1, 43–44; Battānī (1537), cap. 27, ff. 27v–29r. M 97r28 gives, less accurately, $e = 346$. Copernicus's understanding of az-Zarqāl is distinctly incorrect; see below, pp. 157–159.

37 In III,13 in M (88), Copernicus made changes in the times of his observations, (7) originally being $1;30^h$ earlier and (9) $1;35^h$ earlier, but these changes do not account for the discrepancies in the intervals in III,16. The problem is that the implied time of observation (6) is more than $\frac{1}{4}^d$ late. Cf. also below, p. 172.

38 See Neugebauer (1962a), 274–75.

line passing through the earth. This is followed by a solution to the general problem of any three points on a circle with no special conditions.

Copernicus uses the first, and easier, solution, but combining the equinoxes with ♏15°, the midpoint between the autumnal equinox and the winter solstice. Thus, in Fig. 14, we know the intervals of time between observations of the sun at F when seen in (1) ♎0°, (2) ♏15°, and (3) ♈0°. Again assuming, as Copernicus recommends in III,13 (N 80v26–32) that $\bar{\lambda} = 360°/365\frac{1}{4}^{d} \approx 0;59,8^{°/d}$, from

$$\Delta t_{12} = 45;16^{d}, \qquad \Delta t_{13} = 178;53,30^{d},$$

it follows that

$$\Delta\bar{\lambda}_{12} = 44;37°, \qquad \Delta\bar{\lambda}_{13} = 176;19°.$$

Thus, since $\alpha' =$ ♏15° − ♎0° = 45°,

$$2\gamma = 2\alpha' - 2\alpha = 90° - 44;37° = 45;23°.$$

So

$$\text{arc } AC = \Delta\bar{\lambda}_{13} - \Delta\bar{\lambda}_{12} = 176;19° - 44;37° = 131;42°,$$
$$\text{arc } CAD = \text{arc } AC + 2\gamma = 177;5,30°, \quad (\text{corr. } 177;5°)$$

so that where $R = 100{,}000$

$$\text{crd } AC = 2 \sin \tfrac{1}{2}AC = 182494, \qquad (\text{corr. } 182495)$$
$$\text{crd } DC = 2 \sin \tfrac{1}{2}CAD = 199934. \quad (\text{corr. } 199936)$$

Now, in triangle AFC,

$$CF = \frac{\text{crd } 2\alpha}{\text{crd } 2\alpha'} \cdot AC = \frac{\text{crd } 44;37°}{\text{crd } 90°} \cdot AC = \frac{75918}{141421} \cdot 182494 = 97967,$$

so that

$$FK = \tfrac{1}{2}DC - CF = 99967 - 97967 = 2000,$$
$$EK = \tfrac{1}{2}\text{crd } (180° - CAD) = \tfrac{1}{2}\text{crd } 2;54,30° = 2534, \quad (\text{corr. } 2538)$$

and thus

$$FE = (FK^2 + EK^2)^{1/2} = 3228,$$

that is, $e \approx 323$ where $R = 10{,}000$, or $e/R \approx 1;56/60 \approx 1/31$. For the direction of the apsidal line, in triangle EKF

$$\sin^{-1} EFK = \sin^{-1} (EK/EF) = \sin^{-1} (2534/3230) = 51;40,34° \approx 51;40°,$$

and

$$\lambda_A = BFAp = EFK + \alpha' = 51;40° + 45° = 96;40° = ♋6;40°.$$[39]

[39] It should be noted that several steps in Copernicus's derivation contain computational errors, and M (97) shows a derivation starting from slightly different values of Δt and leading to $e = 322$, which was then altered to 323. All of this is characteristic of Copernicus's work. The alterations of the times of the equinoxes (7) and (9) in III,13 (above n. 37) are not sufficient to explain these changes.

Thus, as Copernicus concludes, since the time of Ptolemy, the eccentricity has diminished from $\frac{1}{24}$ to about $\frac{1}{31}$, while the apogee has moved from $24\frac{1}{2}°$ west to $6\frac{2}{3}°$ east of the summer solstice.

In III,17, Copernicus compares the maximum equations c_{max} produced by the various eccentricities, where $c_{max} = \sin^{-1}(e/R)$ and $R = 10{,}000$:

Hipparchus, Ptolemy	$e = 415$	$c_{max} = 2;23°$	
Battānī, az-Zarqāl	347	1;59	(19)
Copernicus	323	1;51	

Since the model for the first anomaly is a simple eccentric, the partial correction c, from the mean anomaly $\bar{\alpha}$, is given by

$$\sin c = \frac{e \sin \bar{\alpha}}{(R^2 + e^2 + 2Re \cos \bar{\alpha})^{1/2}}, \tag{20}$$

or, with a tangent function, more simply by

$$\tan c = e \sin \bar{\alpha}/(R + e \cos \bar{\alpha}). \tag{21}$$

Mean Motion, Length of Sidereal Year, Positions at Epoch (III,18–19)

In III,14 Copernicus had given the length of the sidereal year as $365;15,24,10^d = 365^d\ 6;9,40^h$, and computed the tables of mean motions accordingly. The derivation of this value is treated in III,18. In order to determine the mean sidereal motion of the sun, Copernicus must use two observations as far apart as he can find. He thus chooses Hipparchus's autumnal equinox (1) from −146, and his own autumnal equinox (7) for 1515. Since each is of a true tropical longitude, $\lambda = ♎0°$, he first converts to sidereal longitude λ^* by applying the precession for each date, and then finds the corresponding mean longitude $\bar{\lambda}^*$, using the appropriate e and λ_A^* for each observation.

1. Hipparchus's observation is for E.A. 177, for which the precession $\pi = 3;50°$. Hence

$$\lambda^* = ♎0° - 3;50° = 176;10°,$$

$$\lambda_A^* = ♊5;30° - 3;50° = 61;40°,$$

and the true anomaly

$$\alpha = \lambda^* - \lambda_A^* = 114;30°.$$

In Fig. 15(a), where $e = 415$ and $\alpha = 114;30°$,

$$c = \sin^{-1}(e \sin \alpha/R) \approx 2;10°,$$

$$\bar{\alpha} = \alpha + c = 116;40°,$$

$$\bar{\lambda}^* = \lambda^*_{\mathrm{A}} + \bar{\alpha} = 178;20°.$$

2. Copernicus's observation is for E.A. 1840, for which Copernicus takes $\pi = 27;15°$ (corr. 27;14°) so that

$$\lambda^* = ♎0° - 27;15° = 152;45°,$$

$$\lambda^*_{\mathrm{A}} = ♋6;40° - 27;15° = 69;25°,$$

$$\alpha = \lambda^* - \lambda^*_{\mathrm{A}} = 83;20°.$$

And in Fig. 15(b), where $e = 323$ and $\alpha = 83;20°$,

$$c = \sin^{-1}(e \sin \alpha/R) = 1;50°,$$

$$\bar{\alpha} = \alpha + c = 85;10°,$$

$$\lambda^* = \lambda^*_{\mathrm{A}} + \bar{\alpha} = 154;35°.$$

Between the two observations, counting complete revolutions (r),

$$\Delta\bar{\lambda}^* = 1661^{\mathrm{r}} + 154;35° - 178;20° = 1660^{\mathrm{r}} + 336;15°,$$

$$\Delta t = 1662^{\mathrm{y}}\ 37;18,45^{\mathrm{d}}.$$

Thus, the length of the sidereal year is

$$1^{\mathrm{sy}} = \frac{6,0° \cdot 1662^{\mathrm{y}}\ 37;18,45^{\mathrm{d}}}{6,0° \cdot 1660^{\mathrm{r}} + 336;15°} = 365;15,24,11,10\ldots^{\mathrm{d}}, \qquad (22)$$

which Copernicus rounds to $365;15,24,10^{\mathrm{d}} = 365^{\mathrm{d}}\ 6;9,40^{\mathrm{h}}$.[40] However, the annual motion in III,14, $5,59;44,49,7,4^{\circ/\mathrm{y}}$, was not computed from this value, which would give $\ldots7,55^{\circ/\mathrm{y}}$, but appears to be miscomputed from $365;15,24,11^{\mathrm{d}}$, which correctly gives $\ldots7,2^{\circ/\mathrm{y}}$. That so fundamental a parameter contains a computational error is not pleasant.

It is now a simple matter to establish the mean position of the sun at epoch, which Copernicus does in III,19. Hipparchus's observation (1) followed Era Alexander by $176^{\mathrm{y}}\ 362;30^{\mathrm{d}}$ at the meridian of Rhodes, but for

[40] In III,14 (M, 89v–90r) Copernicus originally rounded to $365;15,24^{\mathrm{d}}$, and tabulated a mean sidereal motion based upon

$$6,0° \cdot 6,5^{\mathrm{d}}/6,5;15,24^{\mathrm{d}} = 5,59;44,49,17,52^{\circ/\mathrm{y}},$$

but then rounded, first to $365;15,24,11^{\mathrm{d}}$, the basis of the revised tabulation from $5,59;44,49,7,4^{\circ/\mathrm{y}}$, and finally to $365;15,24,10^{\mathrm{d}}$. The tropical motion (M, 89v,93r) was correspondingly altered from

$$5,59;44,49,17,52^{\circ/\mathrm{y}} + 0;0,50,12,5^{\circ/\mathrm{y}} = 5,59;45,39,29,57^{\circ/\mathrm{y}}$$

to $5,59;45,39,19,9^{\circ/\mathrm{y}}$. On the final tabulations, see below, pp. 166–167.

the meridian of Frauenburg, by one hour less or 176^y $362;27,30^d$. For this period, from the table in III,14, $\Delta\bar{\lambda}^* = 312;42,37° \approx 312;43°$, and thus for E.A.1,

$$\bar{\lambda}_0^* = 178;20° - 312;43° = 225;37°.$$

Since for E.A. 1, $\bar{\pi}_0 = 1;2°$, the tropical longitude with respect to the mean equinox,

$$\bar{\lambda}_0 = \bar{\lambda}_0^* + \bar{\pi}_0 = 226;39°,$$

although Copernicus gives 226;38°. By adding the motions for the appropriate intervals, Copernicus then lists the positions for his four standard epochs.[41]

The Mean Tropical Year and Mean Rate of Precession

In our earlier examination of Copernicus's derivation of 25816^y as the period of the precession, we determined that, unless he computed very badly indeed, he could not have found the period as claimed in the text.[42] The question of how in fact he derived the period must now be considered in another way. Since the precession is not only the motion of the equinox, but also the arc measuring the difference between the sidereal and tropical motion of the sun in any period, it may also be derived from the difference between the sidereal and tropical year, that is,

$$\frac{360° \cdot 365^d}{ty} - \frac{360° \cdot 365^d}{sy} = \bar{\pi}^{°/y}. \tag{23}$$

If now we examine III,14 and the following tables in M (87v–93v), we find an original and corrected mean tropical motion of the sun which are each derived from adding $\bar{\pi} = 0;0,50,12,5^{°/y}$ to an original and corrected sidereal motion.[43] Each implies a slightly different tropical year, which we compare with the tropical motion and year of the *Alfonsine Tables*:

$$\begin{array}{llll} (1)\ \text{Original} & 5,59;45,39,29,57^{°/y} & ty \approx 365;14,33,2^d & \\ (2)\ \text{Corrected} & 5,59;45,39,19,9 & 365;14,33,13 & (24) \\ (3)\ \text{Alfonsine} & 5,59;45,39,22,2 & 365;14,33,10 & \end{array}$$

The point of the comparison is simply that all three are very close, the differences,

$$(1)\text{–}(3) \approx -0;1°/455^y \approx -0;4°/1800^y,$$

$$(2)\text{–}(3) \approx +0;1°/1250^y \approx +0;1,30°/1800^y,$$

[41] The epoch positions for the solar theory are tabulated together below, p. 165.

[42] See above, p. 139.

[43] See above, p. 155 and n. 40.

being insignificant during the 1800 years preceding Copernicus. It is not unreasonable to assume that Copernicus began with the Alfonsine tropical year, and derived a provisional rate of precession from the difference between the Alfonsine tropical year and some provisional sidereal year, say, that of Thābit, $365;15,23^d$, or Copernicus's own estimate in the *Commentariolus*, $365^d\ 6;10^h = 365;15,25^d$, or exactly midway between them, $365;15,24^d$, the value originally used in III,14.[44] If each is taken in combination with the Alfonsine tropical year rounded to $365;14,33^d$, the results are:

$$\begin{array}{llll} (1) & sy = 365;15,23^d & \bar{\pi} \approx 0;0,49,15^{\circ/y} & \\ (2) & 365;15,24 & 0;0,50,14 & (25) \\ (3) & 365;15,25 & 0;0,51,13 & \end{array}$$

None of these simple comparisons leads exactly to Copernicus precession of $0;0,50,12,5^{\circ/y}$, but all are sufficiently close that some sort of adjustment in either the precession or the Alfonsine tropical year could well be the origin of Copernicus's final value.

Exactly how such an adjustment was made is difficult to know. In principle Copernicus needs some estimate of the tropical motion of the sun during the 1717-year period of ϑ, or some estimate of the precession in the same period. Unfortunately, the estimate of $\pi = \bar{\pi} = 23;57°$ from III,6 will not work here, just as it did not work in Copernicus's account of his derivation. But it is not necessary to believe that he was so systematic, for the derivation could have been done in a less rigorous, and therefore less securely reconstructable, way. So while the exceedingly close similarity of Copernicus's and the Alfonsine tropical year may offer a valuable clue to the origin of his mean rate of the precession, a precise derivation remains unknown. At least, we have been unable to reconstruct one, although it is easy enough to come rather close.

The Model for the Second Inequality (III,20)

What Copernicus means by the second inequality, or anomaly, of the sun is that the earth's apsidal line moves nonuniformly, and its eccentricity is subject to a periodic variation. The model he invents to account for the inequality is motivated by a passage in *Epitome* III,13 to which we have already referred, and which we now give complete. After explaining that Hipparchus and Ptolemy derived an eccentricity of 2;30 where $R = 60$, and an apsidal direction of Gemini 5;30°, the *Epitome* continues:

> Al-Battani found the aforesaid eccentricity to be 2;4,45, and arc *BH* [between the summer solstice and the apogee] 7;43°. However, az-Zarqāl, although he altered the mean motion, nevertheless found the same eccentricity as al-Battānī, but

---

[44] Thābit's value, above, p. 149, n. 33; for the sidereal year in the *Commentariolus*, see Swerdlow (1973), 451–54; for the original value in M, see above, p. 155, n. 40.

found arc *BH* to be 12;10°, which certainly seems remarkable since az-Zarqāl was later than al-Battānī. Consider, therefore, whose observation you would trust. Al-Battānī found 93 days and about 14 hours from the vernal equinox to the summer solstice, but from the vernal equinox to the autumnal equinox 186 days 14;45 hours, and thus he set the maximum equation of the sun at 1;59,10°. Az-Zarqāl, 193 years after al-Battānī, made four observations near the four midpoints between the equinoctial and solstitial points, and found *BH* to be 12;10°. For that reason he was forced to say that the center of the sun's eccentric moves in a certain small circle just as is the case for Mercury.[45]

The passage contains a number of erroneous statements about az-Zarqāl, following from the belief that the *Toledan Tables* were the work of az-Zarqāl, and further, from failing to understand that the longitudes in the *Toledan Tables* are sidereal rather than tropical.[46] Thus, the longitude of the solar apogee in the *Toledan Tables*, 77;50°, which the *Epitome* gives incorrectly as ♋0° − 12;10°, is a sidereal longitude and applies, not to the time of az-Zarqāl, but to Era Hijra (622). And while az-Zarqāl did introduce a small circle into his solar model, its purpose was only to vary the size of the eccentricity, while the motion of the apsidal line was uniform and direct. Finally, az-Zarqāl himself did not find the same eccentricity as al-Battānī, although Battānī's eccentricity was used in the *Toledan Tables*.

Copernicus's own understanding of the motion of the apsidal line is given in a historical introduction to III,20. Ptolemy, he says, thought the apsidal line to be tropically fixed, "others," by which he means the *Alfonsine Tables* (and following the *Alfonsine Tables*, himself, when he wrote the *Commentariolus*) thought it was sidereally fixed. Az-Zarqāl thought it moved non-uniformly, and even moved (tropically) retrograde, the evidence being that in the 740 years between Ptolemy and al-Battānī it moved forward about 17°, but in the following 193 years it appeared to move retrograde by about $4\frac{1}{2}$°. And so, Copernicus concludes, az-Zarqāl thought that the center of the sun's eccentric moved on a small circle, causing the apogee to move back and forth and varying the eccentricity. But Copernicus rejects a retrograde tropical motion of the apsidal line as being incompatible with the rest of the history of the observations, although rather than suspect the accuracy of the observations of either al-Battānī or az-Zarqāl, he remarks that since both are equally careful, it is unclear who we should prefer to follow. (It is unfortunate that in all this speculation it did not occur to Copernicus that az-Zarqāl could be following his own convention of measuring longitudes sidereally, for then the problem of the retrograde apsidal line would simply disappear. On the other hand, making any kind of choice between

[45] Following the text in Venice lat. 328, f. 25v; cf. Toomer (1969), 307, Swerdlow (1980), 223.

[46] Az-Zarqāl's solar theory, and the misinterpretations of it in the *Epitome* and by Copernicus, have been thoroughly discussed in Toomer (1969), although more recent literature unfortunately continues to perpetuate the old errors. Some of the information, and misinformation, in the *Epitome* was taken from the *Almagesti minoris libri VI* (unpublished) III,11, an anonymous and, on the whole, quite intelligent summary of the first six books of the *Almagest* that appears to date from the late thirteenth century. Peurbach followed it quite closely in preparing the first six books of the *Epitome*.

al-Battānī and az-Zarqāl is irrelevant to Copernicus because in deriving the parameters of his model he ignores both of them.) Nevertheless, while in Copernicus's own model for the second inequality the tropical motion of the apsidal line is always direct, the sidereal motion is in fact subject to stations and is retrograde for more than one-third of the 3434-year cycle of the anomaly.

The model for the second anomaly is given in three forms, shown in Fig. 16, in which S is the sun, E the earth, R the radius of the earth's orbit, and $r_1 > r_2$ are two smaller radii.

1. Double epicycle. Let C_1 move in the positive sense, i.e. of increasing longitude, on a circle about S of radius R, completing a revolution in a sidereal year. Let C_2 move on an epicycle of radius r_1 in the negative sense, slightly slower than the motion of C_1, completing a revolution in an anomalistic year. The difference of these motions is the mean motion of the apsidal line, and it is direct. Finally, let E move very slowly in the negative sense on an epicycle of radius r_2, completing a revolution in the 3434-year period of the anomaly of the obliquity ϑ. It is assumed that C_1C_2E are extended in a straight line when $\vartheta = 0°$.

2. Eccentric epicycle. Completing parallelogram SC_1C_2F, we see that the effect of the first epicycle is equivalent to a slowly rotating eccentric of eccentricity $SF = e_1 = r_1$, that completes a rotation in the mean sidereal period of the apsidal line. C_2 then moves in a circle of radius R about F, and r_2 again has the period of ϑ such that r_2 is parallel to SF when $\vartheta = 0°$.

3. Double eccentric. Complete parallelogram EC_2FG, and the second epicycle is now seen to be equivalent to a rotating central eccentricity $FG = e_2 = r_2$ with the period of ϑ such that SFG are in a straight line when $\vartheta = 0°$. E now moves about G in a circle of radius R.

These equivalences can easily be derived from the properties of eccentric and epicyclic models already set out in III,15, so Copernicus explains them at length, and expresses his doubt about which is in fact correct, although, he says, the constant agreement (*perpetua consonantia*) of the calculations and appearances compels him to believe that some one of them must be. However, for all further purposes, he uses the third form, which he had described first, that of a double eccentric with the center of the earth's orbit moving in a small circle.

But what is more interesting is yet another alternative that Copernicus presents at the end of III,25:

> Nevertheless, we are also not unaware of the fact that if someone would suppose that the center of the annual revolution be fixed, as though it were the center of the world, but the sun be moveable by two motions similar and equal to those which we have demonstrated for the center of the eccentric, everything will appear just as before, the same numbers and the same demonstration, for nothing would be changed in them, other than the arrangement itself, especially in so far as applies to the sun. For then the motion of the center of the earth would be a perfect and simple motion about the center of the world, since the two other motions have been

> granted to the sun. And therefore an uncertainty will still remain, with regard to the center of the world, which one of the two [the sun or the center of the earth's eccentric] it is, just as at first we stated inconclusively (ἀμφιβολικῶς) that the center of the world is in the sun or near it. But we shall say more of this question in the explanation of the five wandering stars, where we shall even resolve [it] to the best of our ability, thinking it sufficient if at this time we have adopted true and the least erroneous numbers for the appearance of the sun.[47]

The alternate model is shown in Fig. 17. The center of the eccentric G is assumed at rest, "as though it were the center of the world," while as is made clear by completing parallelogram $GFSH$, the sun now moves in a small circle about H. The result is in every way equivalent to the models of Fig. 16, and of course yet more variations are possible.

Copernicus's remarks also provide some valuable, and often overlooked, evidence for what he believed his theory was about, what he knew he could prove, and what he knew he could resolve only "to the best of our ability." He has proposed three indistinguishable models for the motion of the earth with the sun at rest. He now describes a fourth model that "someone" might propose—Copernicus evidently does not consider it seriously—in which the sun moves in a small circle, and this is likewise indistinguishable from the others, at least as pertains to the sun. For the moment, he says, he can do nothing more than adopt correct parameters for the apparent motion of the sun. All of this, needless to say, is very much in the spirit of poor, abused Osiander, but these are Copernicus's own words. He is in fact far more reasonable and candid about his uncertainty than modern historians, who seem to think that he could really prove his theories, would have us believe.

But he also offers the possibility that in the planetary theory he will resolve the issue "to the best of our ability," specifically, that he will try to determine whether the sun or the center of the earth's eccentric is the center of the world. By the "center of the world," Copernicus means a point, or the point, that does not move, as when he says that someone might suppose the center of the earth's eccentric to be at rest "as though it were the center of the world" (*tanquam centrum mundi*). (We must remember that Copenicus envisions the planetary system as *filled* with contiguous spheres, all of which are in motion in one way or another.) In any case, the center of the world by definition cannot move, and the question is therefore whether the sun or the center of the earth's eccentric, the true sun or the mean sun, occupies such a position. That is not an easy question to answer.

---

[47] N 96r1–15, M 104r36–104v15. The passage is not without difficulty, and contains four textual differences between N and M. Three do not affect the meaning, and we have simply followed N, but for the fourth, in the last sentence, we read *qua* with M rather than *quas* with N. Copernicus (1873) emends to *quam*, which is possible, but not really necessary. Copernicus is referring to the consideration in Book V of the possibility that the motion of the center of the earth's orbit affects the eccentricities of the planets, under the assumption that the centers of their eccentrics maintain an invariable distance (although not direction) from the sun. It must be understood that Copernicus does not, and indeed cannot, answer the question, although he has an opinion. See the discussion immediately following, and below, pp. 359–361.

For all kinematical purposes, the mean sun is the "center" of Copernicus's system. The second anomaly of the planets is represented by the motion of the earth with respect to the mean sun, not the true sun. Likewise, the first anomaly is represented by models in which the eccentricities and apsidal lines are taken with respect to the mean sun. And the planes of the orbits in the latitude theory intersect in the mean sun. In fact, the exact location of the true sun is not pertinent to Copernicus's planetary theory, and both the models and tables of Books V and VI would be exactly the same if the sun were in China rather than near the center of the earth's orbit. This is of course a frightful exaggeration, but it is a fact that the true sun cannot be located, the center of the world cannot be determined, by appeal to Copernicus's planetary theory.

Nevertheless, although Copernicus cannot resolve the question, there is no doubt about what he considers the correct answer. In V,4,16,23 he considers a diminution of the eccentricities of Mars and Venus that has occurred since Ptolemy, and he assumes that it is caused by the center of the earth's eccentric approaching the centers of their eccentrics, "the sun in the intervening time remaining immovable" (V,16, N, 158r17). Evidently he assumes in addition that the distances, but not necessarily the directions, of the centers of the planets' eccentrics from the true sun have remained unchanged. Of course Copernicus cannot prove such an arrangement, and "someone" can equally well suppose that all of this is true, but that the sun still moves in a small circle, and carries with it the centers of the planet's eccentrics such that their distances from the immovable center of the earth's orbit are altered exactly as in Copernicus's theory. It would be naïve to think that Copernicus had not himself anticipated this argument, and understood that there was no answer to it. Hence, as he knows very well, he must remain confronted with a variety of models, and can only chose between them *pro posse nostro*.

Variation of the Eccentricity and Equation of the Apsidal Line (III,21)

Now comes the hard part. Copernicus assumes that the period of the motion of e_2 is that of ϑ, 3434 years. He has three values of the eccentricity at at least three different dates—say, that of Ptolemy, Battānī, and himself—for the intervals between which he knows $\Delta\vartheta$. Thus, he has the opportunity of solving a nontrivial problem: Given three points on the circumference of a circle, to find a point from which three straight lines drawn to the three points will be in a given ratio. A solution to this problem is given by Viète as Prob. V of the second appendix to *Apollonius Gallus*, but it is by no means easy, and there is no reason to believe that Copernicus could formulate the problem in this way, let alone find a solution. So again he must ignore much of the observational evidence, and make a number of questionable assumptions, in order to turn the problem into something that he can handle mathematically. In this case he assumes:

1. That the period of e_2 is exactly that of ϑ. This assumption would be required in any case. His reasoning is perhaps that since the obliquity has been diminishing, and since the eccentricity has been diminishing, both have the same period. And for the same reason

2. That the maximum eccentricity coincides with the maximum obliquity at $\vartheta = 0°$. In a correct solution from three given eccentricities, it would be possible to find the location of the various eccentricities with respect to $\vartheta = 0°$, and it would not necessarily turn out that the zero point for the obliquity and for the eccentricity were the same.

3. That the maximum eccentricity is exactly 417 where $R = 10{,}000$. One assumes that Copernicus reasoned that the maximum eccentricity, occurring between Hipparchus and Ptolemy, could be only a little larger than the 415 that they found, hence, by some sort of trial and error, 417. This assumption makes the whole problem trivial, for all Copernicus now has to do is use one eccentricity, his own, and solve for the minimum eccentricity.

So in Fig. 18, G shows the position of the center of the earth's eccentric at the time of Copernicus's equinox observation (7) of 14 Sep 1515, which is E.A. 1840 Phao 6 (j.d. 2274,668). Since $\vartheta = 0°$ on 28 Aug -64 (j.d. 1697,922), which Copernicus gives as "64 years before the birth of Christ" and equates with E.A. 259, so strictly

$$\Delta t = 576{,}746^{\text{d}} = 1580^{\text{y}}\ 46^{\text{d}}.$$

Copernicus rounds this to 1580^{y}, and gives correctly $\vartheta \approx 165;39°$. From the derivation in III,16 $e = SG = 323$, and by assumption $e_1 + e_2 = SB = 417$ where $R = 10{,}000$. Since in triangle BSG

$$2\alpha = 180° - \vartheta = 14;21°, \qquad \text{crd}\ 2\alpha = 2498,$$

thus

$$\text{crd}\ 2\gamma = (SB/SG)\ \text{crd}\ 2\alpha = (417/323)\ 2498 = 3225 = \text{crd}\ 18;34°,$$

$$2\gamma = 360° - 18;34° = 341;26°,$$

$$2\delta = 360° - (2\gamma + 2\alpha) = 4;13°.$$

And since crd $2\delta = 735$ (corr. 736), consequently

$$BG = SB\ (\text{crd}\ 2\delta/\text{crd}\ 2\gamma) = 417\ (735/3225) \approx 95,$$

$$BD = BG \cdot 2R/\text{crd}\ (180° - 2\alpha) = 190\ R/19843 \approx 96.$$

Therefore,

$$e_2 = \tfrac{1}{2}BD = 48, \qquad e_1 = 417 - 48 = 369,$$
$$e_{\text{max}} = e_1 + e_2 = 417, \qquad e_{\text{min}} = e_1 - e_2 = 321. \qquad (26)$$

It is probably not a coincidence that Ptolemy's eccentricity 415 differs from the maximum by 2, and Copernicus's eccentricity 323 differs from the minimum by 2. The maximum equation of the apogee is then

$$\delta_{\max} = \sin^{-1}(e_1/e_2) = \sin^{-1} 1300/R = 7;28,11^\circ \approx 7;28^\circ.^{48} \qquad (27)$$

For any value of ϑ, the eccentricity

$$e = (e_1^2 + e_2^2 + 2e_1e_2 \cos \vartheta)^{1/2} \qquad (28)$$

and the equation of the apogee

$$\delta = \sin^{-1}(e_2 \sin \vartheta/e). \qquad (29)$$

This is perhaps the weakest of Copernicus's demonstrations—really little more than an act of faith—and shows the difficulty he had when confronted with mathematical problems for which solutions had not previously been worked out.

Mean Motion of the Apogee (III,22)

At the time that $\vartheta = 0^\circ$, the following quantities are equal:

$$\lambda_A^* = \bar{\lambda}_A^*, \qquad \pi = \bar{\pi}, \quad \text{and thus} \quad \lambda_A = \bar{\lambda}_A.$$

Hence if Copernicus can find the true tropical longitude for -64, by subtracting π he will have the mean sidereal longitude, which he needs to find the mean motion. But how can he find λ_A for -64? He cannot, so he simply assumes that the value found by Hipparchus in -145, and found by Ptolemy in 139, $\lambda_A = 65;30^\circ$, applies also to -64. Since at this time, E.A. 259, $\pi = \bar{\pi} = 4;38,30^\circ$ (corr. 4;38,8 on 28 Aug -64), in Fig. 19

$$\lambda_A^* = \bar{\lambda}_A^* = 65;30^\circ - 4;38,30^\circ = 60;51,30^\circ \approx 60;52^\circ.$$

Note that in III,18, at the time of Hipparchus in E.A. 177, $\lambda_A^* = 61;40^\circ$, so in 82 years the apogee has *regressed* by 0;48,30°. This is a consequence of Copernicus's theory to which we shall return.[49] At the time of Copernicus's observation of the equinox (7) in E.A. 1840, $\vartheta = 165;39^\circ$ and $\pi = 27;15^\circ$, so that

$$\lambda_A^* = 96;40^\circ - 27;15^\circ = 69;25^\circ,$$

and, from III,21, where 2δ was 4;13°, $\delta = -2;6,30^\circ \approx -2;7^\circ$, and thus

$$\bar{\lambda}_A^* = 69;25^\circ + 2;7^\circ = 71;32^\circ.$$

[48] Originally in M (101v), Copernicus assumed $SG = 322$ (cf. above, p. 153, n. 39), $SB = 416$, and found $e_2 = 47\frac{1}{2}$, $e_1 = 368\frac{1}{2}$, so that $\delta_{\max} = \sin^{-1}(47\frac{1}{2}/368\frac{1}{2}) = 7;24^\circ$, and column 3 of the original solar correction table (M, 103) was computed with these values. On further problems in columns 4 and 6 of the correction table, due to taking $e_{\max} = 414$, see below, pp. 168–169.

[49] See below, pp. 165–166.

Therefore, in $\Delta t = 1580^y$,

$$\Delta\bar{\lambda}^*_A = 71;32° - 60;52° = 10;40°.$$

Copernicus gives 10;41° and then says that

$$\Delta\bar{\lambda}^*_A/\Delta t = 0;0,24,20,14^{°/y},$$

although

$$10;41°/1580^y = 0;0,24,20,30^{°/y},$$

or perhaps more strictly, using the interval from III,21

$$10;41°/1580^y\ 46^d = 0;0,24,20,23^{°/y},$$

so again it is hard to tell exactly how Copernicus computed his parameters. Calling the sun's mean anomalistic motion $\bar{\alpha} = \bar{\lambda}^* - \bar{\lambda}^*_A$,

$$\bar{\alpha} = 5,59;44,49,7,4^{°/y} - 0;0,24,20,14^{°/y} = 5,59;44,24,46,50^{°/y}.$$ [50]

Positions at Epoch (III,19,23)

From the equinox observation (7) on 14 Sep. 1515, which Copernicus says is Olympiad 573,2 (corr. 573,3), to the first Olympiad there have been $2290^y\ 281;46^d$, which is correct for his date of Olympiad 1,1. For this interval, the change of the solar anomaly (mod. 360°) $\Delta\bar{\alpha} = -42;49°$ (corr. $-42;48,9°$). At the time of the observation, from III,18,22,

$$\bar{\alpha} = \bar{\lambda}^* - \bar{\lambda}^*_A = 154;35° - 71;32° = 83;3°,$$

so that at Olympiad 1,1

$$\bar{\alpha}_0 = 83;3° - 42;49° = 40;14°.$$

By adding the appropriate $\Delta\bar{\alpha}$ for each Δt, Copernicus then gives the positions at each standard epoch. The following table summarizes the epoch positions of the mean motions of the solar theory:

$\bar{\lambda}^*_0$ mean sidereal longitude (III,19),

$\bar{\lambda}_0 = \bar{\lambda}^*_0 + \bar{\pi}_0$ mean tropical longitude (III,19),

$\bar{\alpha}_0$ mean anomaly (III,23),

$\bar{\lambda}^*_{A0} = \bar{\lambda}^*_0 - \bar{\alpha}_0$ mean sidereal longitude of apogee.

---

[50] In M (102r), Copernicus originally derived an apsidal motion of $0;0,24,43,26^{°/y}$. When subtracted from the original value of $\bar{\lambda}^* = 5,59;44,49,17,52^{°/y}$ (see above, n. 40), the resulting mean anomaly $\bar{\alpha} = 5,59;44,24,34,26^{°/y}$ was the basis of a cancelled table in M (94r).

The last is not given by Copernicus, but it seems informative to place it in the table.[51]

Epoch	$\bar{\lambda}_0^*$	$\bar{\lambda}_0$	$\bar{\alpha}_0$	$\bar{\lambda}_{A0}^*$	
A. Olympiad	96;16°	90;59°	40;14°	56;2°	(30)
C. Alexander	225;37	226;38	166;31	59;6	
D. Caesar	272;4	276;59	211;4	61;0	
G. Christ	272;31	278;2	211;14	61;17	

Remarks on the Second Inequality

The model for the second anomaly of the earth is something more than it appears, and perhaps something more than Copernicus understood. We had noted earlier that the true longitude of the apogee regressed by 0;48,30° between −145 and −64. In fact it regresses considerably more than that. By applying Apollonius's Theorem for finding the location of the stationary points in the small circle on which the center of the earth's orbit moves, we find that between −679 and +552, for over 1231 years, or more than one-third of the period of ϑ, the apsidal line moves retrograde a total of 4;22°. However, in this same period the precession increases by 15;22°, so the tropical longitude does not decrease, but increases by 15;22° − 4;22° = 11°.[52] Thus, while Copernicus rejected the *tropical* retrogradation that he believed az-Zarqāl to have assumed, his model still produces a sidereal retrogradation which, to Copernicus, is a real rather than merely an apparent motion.

Although it is not clear whether Copernicus knew this, it is hard to believe that he did not at least check his results against the apsidal longitudes of Hipparchus and Ptolemy. For once he decided to ignore the apparent retrogradation between al-Battānī and az-Zarqāl, the primary motivation of his model was to keep the tropical longitude of the apsidal line as close as

[51] There are several discrepancies when checked by recomputation from the intervals between the epochs. Correctly $\bar{\lambda}_{0A}^* = 96;17°$. Taking the sidereal longitudes as given, then $\bar{\lambda}_0$ for A., C., and G. are in error by −0;1°, which is hard to understand since all Copernicus has to do is add the precession from III,11. This means that tropical longitudes computed from these epochs using $\bar{\lambda}_0$ will be 0;1° lower than when using $\bar{\lambda}_0^*$. Correctly $\bar{\alpha}_{0D} = 211;5°$, $\bar{\alpha}_{0G} = 211;12°$. A number of earlier, and very inconsistent, values in M have been restored to the texts of the critical editions and the translations following them, doubtless without checking. All these epochs are said by Copernicus to apply to the meridian of Frauenburg and Cracow, which he believes to be the same. In fact, they are very close since Cracow is east of Frauenburg by only 0;15° = 0;1^h.

[52] There is a more extended discussion of the retrogradation in Swerdlow (1980), 231–35. On Apollonius's Theorem, see *HAMA*, 190–91, 267–70, and below, pp. 461–463.

possible to 65;30° between the time of Hipparchus and of Ptolemy, and then to bring it up to 96;40° at his own time. This in itself necessitates a sidereal retrogradation, even before the mean motion of the apsidal line is established in III,22. The following table shows the true and mean sidereal longitudes λ_A^* and $\bar{\lambda}_A^*$ required, or implied, by the apsidal directions of Hipparchus, Ptolemy, Copernicus, and also by Copernicus's assumption that $\lambda_A = 65;30°$ in −64, when the equation of the apsidal line δ is taken from Copernicus's model.

E.A.	Hipparchus 177	$\vartheta = 0°$ 259	Ptolemy 463	Copernicus 1840
λ_A	65;30°	65;30°	65;30°	96;40°
π	−3;50	−4;38	−6;40	−27;15
λ_A^*	61;40	60;52	58;50	69;25
δ	−1;0	0;0	+2;24	+2; 7
$\bar{\lambda}_A^*$	60;40	60;52	61;14	71;32

(31)

Note that between Hipparchus and Ptolemy the true sidereal longitude λ_A^* has regressed by 2;50°. This is simply a result of subtracting the precession at each date from 65;30°, and there is no way around it without rejecting one of the two determinations. However, when the equation of the apsidal line δ for each date is applied, the resulting mean longitude $\bar{\lambda}_A^*$ increases, but the resulting values are not consistent with the mean motion established in III,22, and thus the theory will not correctly give $\lambda_A = 65;30°$ at each date. In fact, computation from the theory gives 65;7° for Hipparchus and 66;29° for Ptolemy. The reason is that Copernicus's assumption that 65;30° applies exactly to −64 is simply incompatible with producing the same longitude in −145 and 139. Copernicus doubtless made the assumption because it appeared to him the best way of getting a reasonable approximation to the recorded values of Hipparchus and Ptolemy, and it is a fourth assumption in addition to the three that were required in III,21 to find the separate eccentricities and the equation of the apsidal line. As it turns out, the only apsidal longitude that Copernicus's theory reproduces correctly is his own tropical longitude in 1515 for which the theory gives 96;39° compared with 96;40° by observation. But this is not surprising since it was the only observationally derived longitude used in determining the mean motion.

The Tables and Their Use (III,14,24,25)

We begin with the tables of mean motions following III,14. The length of the sidereal year finally settled upon by Copernicus was $365;15,24,10^d$, although the mean sidereal motion of the sun, which Copernicus calls the "simple mean motion" (*motus aequalis simplex*) is slightly miscomputed from

$$\begin{aligned}\bar{\lambda}^{*y} &= 6{,}0^{\circ}\cdot 6{,}5^{d}/6{,}5;15{,}24{,}11^{d} = 5{,}59;44{,}49{,}7{,}4^{\circ/y}, \quad (\text{Corr.}\ldots 7{,}2)\\ \bar{\lambda}^{*d} &= \bar{\lambda}^{*y}/6{,}5^{d} = 0;59{,}8{,}11{,}22^{\circ/d}.\end{aligned} \tag{32}$$

If we add to each of these the mean precession

$$\bar{\pi}^{y} = 0;0{,}50{,}12{,}5^{\circ/y}, \qquad \bar{\pi}^{d} = 0;0{,}0{,}8{,}15^{\circ/d}, \tag{9a}$$

we get the mean tropical motion of the sun, which Copernicus calls the "mean composite motion" (*motus aequalis compositus*),

$$\begin{aligned}\bar{\lambda}^{y} &= 5{,}59;45{,}39{,}19{,}9^{\circ/y},\\ \bar{\lambda}^{d} &= 0;59{,}8{,}19{,}37^{\circ/d}.\end{aligned} \tag{33}$$

The mean sidereal motion of the apsidal line was

$$\bar{\lambda}_{A}^{*} = 0;0{,}24{,}20{,}14^{\circ/y}. \tag{34}$$

When this is subtracted from the mean sidereal motion of the sun, the result is the "motion of anomaly" (*motus anomaliae*),

$$\begin{aligned}\bar{\alpha}^{y} &= 5{,}59;44{,}24{,}46{,}50^{\circ/y},\\ \bar{\alpha}^{d} &= \bar{\alpha}^{y}/6{,}5^{d} = 0;59{,}8{,}7{,}22^{\circ/d}.\end{aligned} \tag{35}$$ [53]

All of these are tabulated for 1–60 years and for 1–60 days. In order to find $\bar{\lambda}^*$, $\bar{\lambda}$, and $\bar{\alpha}$ for a given date, one uses these tables to find $\Delta\bar{\lambda}^*$, $\Delta\bar{\lambda}$, and $\Delta\bar{\alpha}$ for Δt since epoch, and then forms

$$\bar{\lambda}^* = \bar{\lambda}_0^* + \Delta\bar{\lambda}^*, \qquad \bar{\lambda} = \bar{\lambda}_0 + \Delta\bar{\lambda}, \qquad \bar{\alpha} = \bar{\alpha}_0 + \Delta\bar{\alpha}. \tag{36}$$

The table of corrections follows a brief description of its arrangement in III,24. The first two columns are the argument for 3°–180° and 180°–357° at 3° intervals. They serve as the common entry for arguments of ϑ for the long-period corrections and α for the annual equation. The corrections are graphed in Fig. 20, and a representative configuration is shown in Fig. 21.

Column 3, called the "correction of center" (*prosthaphaeresis centri*) corresponds to the correction between the mean and true apogee, in both sidereal and tropical longitude,

$$c_{3(\vartheta)} = \lambda_{A}^{*} - \bar{\lambda}_{A}^{*} = \lambda_{A} - \bar{\lambda}_{A}. \tag{37}$$

c_3 is a function of ϑ, and thus has a period of 3434 years. Using the notation of Fig. 21, it is computed from

$$c_3 = \sin^{-1}(e_2 \sin\vartheta/e), \qquad \text{where} \quad e = (e_1^2 + e_2^2 + 2e_1e_2\cos\vartheta)^{1/2}. \tag{38}$$

[53] Cf. earlier values of all these quantities above, p. 155, n. 40, p. 164, n. 50.

The use of the tangent function would allow the simpler formula

$$c_3 = \tan^{-1}(e_2 \sin \vartheta/(e_1 + e_2 \cos \vartheta)). \tag{39}$$

c_3 is used to correct $\bar{\alpha}$, the mean anomaly, called by Copernicus the "first and simple anomaly" (*prima ac simplex anomalia*), which gives the distance of the sun from the mean apogee, i.e.

$$\bar{\alpha} = \bar{\lambda}^* - \bar{\lambda}_A^* = \bar{\lambda} - \bar{\lambda}_A, \tag{40}$$

to the "corrected anomaly" (*anomalia coaequata*), giving the distance of the sun from the true apogee,

$$\alpha = \bar{\lambda}^* - \lambda_A^* = \bar{\lambda} - \lambda_A. \tag{41}$$

Thus,

$$\alpha = \bar{\alpha} + c_{3(\vartheta)}, \tag{42}$$

where c_3 is positive for $\vartheta \leq 180°$, since in this semicircle of ϑ, $\lambda_A \leq \bar{\lambda}_A$, and negative for $\vartheta \geq 180°$, since here $\lambda_A \geq \bar{\lambda}_A$.

Column 5, $c_{5(\alpha)}$, is the annual solar equation, which Copernicus calls the "correction of the [annual] sphere" (*prosthaphaeresis orbis*), and here it is tabulated for the minimum eccentricity 321, except for a peculiar adjustment near the maximum equation, an alteration of originally correct values. The computation of c_5 has been explained already in the commentary to III,17.[54] In M c_5 was correctly computed for the eccentricity 321, which gives a maximum equation of 1;50°, but in N the maximum equation for $\alpha = (90° - 93°)$ has been changed to 1;51°, which would correspond to an eccentricity of 323, but the reason for the change only becomes clear after examining column 6.

Column 6, $c_{6(\alpha)}$, gives for each value of α the excess of the solar equation at maximum eccentricity over the equation at minimum eccentricity. Since $e_{max} = 417$ and $e_{min} = 321$, one would expect for the maximum value of c_6, at $\alpha \approx 93°$,

$$c_{6\,max} = \sin^{-1} .0417 - \sin^{-1} .0321 = 2;23° - 1;50° = 0;33°$$

but in fact c_6 gives 0;32°, and was computed with earlier, cancelled values of the eccentricity in M, that is, from

$$c_{6\,max} = \sin^{-1} .0414 - \sin^{-1} .0321 = 2;22° - 1;50° = 0;32°.$$[55]

The reason that c_5 was raised, incorrectly, to 1;51° was to bring $c_5 + c_6$ to 1;51° + 0;32° = 2;23°, the correct maximum equation for $e_{max} = 417$. Obviously, it was c_6, and not c_5, that required the adjustment, and in a clumsy way the entry for $\alpha = 96°$ was adjusted to 0;33°, as the graph in Fig. 20 makes fairly obvious, although correctly all of c_6 ought to have been recomputed.

[54] See above, p. 154.

[55] $e = 414$ was the eccentricity originally computed incorrectly in III,16 (M, 96v–97r) using Ptolemy's intervals between equinoxes and the summer solstice; see above, p. 151, n. 35.

Column 4, $c_{4(\vartheta)}$, is the coefficient, called "proportional minutes," to interpolate in c_6 for values of the eccentricity, and thus of the maximum solar equation, between minimum and maximum values. It is computed in the same way that Ptolemy computes coefficients of interpolation, for example, in the lunar correction tables in *Almagest* V,8, and the method of computation is explained in *Almagest* V,7 and *Epitome* V,11. Let $e_{(\vartheta)}$ be the eccentricity and $c_{\max(\vartheta)} = \sin^{-1}(e_{(\vartheta)}/R)$ the maximum equation for a given ϑ. Then take the extreme values for $\vartheta = 0°$ and $\vartheta = 180°$, that is,

$$c_M = c_{\max(0°)}, \qquad c_m = c_{\max(180°)},$$

and for any value of ϑ, $c_{4(\vartheta)}$ gives the ratio

$$c_{4(\vartheta)} = \frac{c_{\max(\vartheta)} - c_m}{c_M - c_m} = \frac{c_{\max(\vartheta)} - c_m}{c_{6\max}}. \tag{43}$$

c_4 is tabulated sexagesimally, and varies from $60' = 1;0$ to $0'$ as ϑ varies from $0°$ to $180°$, and conversely for $180°$ to $360°$. From $e_{\max} = 417$ and $e_{\min} = 321$, it would follow that

$$c_M - c_m = c_{6\max} = 2;23° - 1;50° = 0;33°,$$

so that c_4 should be computed from

$$c_{4(\vartheta)} = \frac{c_{\max(\vartheta)} - 1;50°}{0;33°}. \tag{44}$$

But spot checking shows that this is not so in the table, for $c_{6\max}$ was erroneously taken as $0;32°$ as explained previously. Thus, three of the four corrections in the solar tables have one thing or another wrong with them—and even c_3 is not computed very accurately—although the cumulative effect can change the solar equation by only about $\pm 0;2°$ compared to a correct computation.

Columns 4, 5, and 6 are always used together to find the annual solar correction $c_{(\alpha,\vartheta)}$ from

$$c_{(\alpha,\vartheta)} = c_{5(\alpha)} + c_{4(\vartheta)} \cdot c_{6(\alpha)}. \tag{45}$$

The true sidereal or tropical longitude of the sun then follows from

$$\lambda^* = \bar{\lambda}^* + c_{(\alpha,\vartheta)}, \tag{46}$$

or

$$\lambda = \bar{\lambda} + \delta_P + c_{(\alpha,\vartheta)}, \tag{47}$$

where $c_{(\alpha,\vartheta)}$ is negative for $\alpha \leq 180°$ and positive for $\alpha \geq 180°$. If the mean tropical longitude has not been determined, then the sidereal longitude must be corrected with the true precession, found as explained before, so that $\lambda = \lambda^* + \pi$.

Example. Find the longitude of the sun for A.D. 1525 Apr 16/17 midnight at the meridian of Frauenburg. This is the same date already used for the example of the computation of the precession and obliquity. Thus we know that $\Delta t = 25{,}25^{y} + 2{,}2^{d}$ and that

$$\vartheta = 166;40^{\circ}, \qquad \bar{\pi} = 26;48^{\circ}, \qquad \delta_{P} = 0;32^{\circ}, \qquad \pi = 27;20^{\circ}.$$[56]

Although it is not strictly necessary we shall compute both the mean tropical and mean sidereal longitudes $\bar{\lambda}$ and $\bar{\lambda}^*$, as well as the mean anomaly $\bar{\alpha}$. From the tables in III,14:

Δt	$\bar{\lambda}^*$	$\bar{\lambda}$	$\bar{\alpha}$
$25{,}0^{y}$	53,40;27,56°	54, 1;22,58°	53,30;19,30°
25^{y}	5,53;40,27,56	5,54; 1,22,58	5,53;30,19,30
$2{,}0^{d}$	1,58;16,22	1;58;16,39	1,58;16,14
2^{d}	1;58,16,22	1;58,16,39	1;58,16,14
	1,34;23, 2,18	1,55;39,16,37	1,24; 4,19,44
A.D. 1	4,32;31	4,38; 2	3,31;14
	6;54	33;41	4,55;18
	π 27;20	δ_P 0;32	
	$\bar{\lambda}'$ 34;14°	$\bar{\lambda}'$ 34;13°	

We call $\bar{\lambda}' = \bar{\lambda}^* + \pi = \bar{\lambda} + \delta_P$ the "adjusted mean longitude." The discrepancy of 0;1° is not unreasonable considering the small discrepancies of 0;1° in the various epoch positions. From the tables in III,24, for $\vartheta = 166;40^{\circ}$,

$$c_{3(\vartheta)} = +1;59^{\circ}, \qquad c_{4(\vartheta)} = 0;1,$$

so

$$\alpha = \bar{\alpha} + c_3 = 295;18^{\circ} + 1;59^{\circ} = 297;17^{\circ},$$

$$c_{5(\alpha)} = +1;36^{\circ}, \qquad c_{6(\alpha)} = 0;29^{\circ},$$

$$c_{(\alpha,\vartheta)} = c_5 + c_4 \cdot c_6 = 1;36^{\circ} + 0;1 \cdot 0;29^{\circ} \approx +1;36^{\circ},$$

and thus the tropical longitude of the sun

$$\lambda = \bar{\lambda}' + c_{(\alpha,\vartheta)} = 34;14^{\circ} + 1;36^{\circ} = 35;50^{\circ} = \text{♉} 5;50^{\circ}.$$

Comparing this longitude with Tuckerman (1964), we find for 1525 Apr 16, 4:00 P.M. G.M.T., $\lambda = 35.71^{\circ}$. Thus, for the following midnight Frauenburg ($l = 19;40^{\circ}$ E $\approx +1;20^{h}$), $\lambda = 35;59^{\circ}$, so the longitude from Copernicus's tables is 0;9° too low.

[56] See above, p. 146.

Application. We now have the information necessary to answer a question that may be of interest, or at least of curiosity. In III,2 Copernicus reported observing a meridian transit of Spica in 1525 in which he found $\delta = -8;40°$, and then determined that if $\beta = -2;0°$ and $\varepsilon = 23;28,30°$, $\lambda = ♎17;21°$. Then in III,12 he used his precession theory to compute the longitude of Spica for 16 Apr 1525, and found ♎17;21° (although strictly his theory leads to 17;20°).[57] If in fact this was the date of Copernicus observation, at what time did he observe Spica in the meridian? The time of Spica's meridian transit in relation to local apparent noon is given by the difference in right ascension of the star and the sun since, by definition, the sun is in the meridian at local apparent noon. In what follows, we shall use the methods and tables of Copernicus's spherical astronomy explained in Chapter 2.

For $\lambda_{\odot} = ♉ 5;50°$ and terrestrial latitude $\varphi = 54;20°$,

$$\delta_{\odot} = 13;29°, \qquad \alpha_{\odot} = 33;31°, \qquad n = 19;31°, \qquad \rho_{\odot} = \alpha_{\odot} - n = 14°.$$

The semidiurnal arc of the sun,

$$s_{\odot} = 90° + n = 109;31° = 7;18^{h},$$

so on that date the sun rose at 4;42 A.M. and set at 7;18 P.M., both in equal hours from local apparent noon. In the case of Spica, for $\lambda_{*} = ♎17;21°$, $\beta_{*} = -2;0$, and $\delta_{*} = -8;40°$, we find

$$\alpha_{*} = 195;13°, \qquad n = -12;16°, \qquad \rho_{*} = 195;13° + 12;16° = 207;29°.$$

Therefore

$$\Delta\alpha = \alpha_{*} - \alpha_{\odot} = 195;13° - 33;31° = 161;42° = 10;47^{h},$$

which means that Spica culminated $10;47^{h}$ after the sun, that is, at 10;47 P.M., so Copernicus did not have to stay up very late. Further, since

$$\Delta\rho = \rho_{*} - \rho_{\odot} = 207;29° - 14° = 193;29° = 12;54^{h},$$

Spica rose at 4;42 A.M. + $12;54^{h}$ = 5;36 P.M., i.e. $1;42^{h}$ before sunset, The period between rising and culmination is thus $10;47^{h} - 5;36^{h} = 5;11^{h}$, and indeed this can be confirmed, for the semidiurnal arc of the star

$$s_{*} = 90° + n = 90° - 12;16° = 77;44° = 5;11^{h},$$

so Spica will set at 10;47 P.M. + $5;11^{h}$ = 3;58 A.M.

Verification of the Solar Theory

In the preceding example we compared a position of the sun computed from Copernicus's tables with a position from modern theory, and found a discrepancy of $-0;9°$ which is really of little interest. It is more informative,

---

57 See above, pp. 131–132, 146.

however, to recompute the observations cited by Copernicus since the purpose of his theory was to reproduce such observations over a period of about 1660 years, particularly those of Hipparchus and Ptolemy. The results are shown in Table 4, in which the time is in each case reduced to the meridian of Frauenburg.

All in all, the theory is remarkably good, especially when one considers all the unconfirmed assumptions that went into establishing the variation of the eccentricity and the motion of the apsidal line. Since it was only the observations of Hipparchus and Ptolemy on one end, and of Copernicus on the other, that were used to establish the solar parameters, it is not surprising that they are reproduced the most accurately. Even so, one would not guess that the compromises in locating the apsidal line and the variation of the precession—essential in giving the tropical longitudes—would give such excellent agreement with the observations. The discrepancy in (5) is understandable since al-Battānī's equinox played no direct part in Copernicus's derivation of parameters. The same is true of the more serious error of $+0;14^{\circ}$ in (6), the vernal equinox of 1515, which Copernicus never uses and dates in III,16 only as $186;5,30^{d}$ before (7). Evidently, it should have been about $0;15^{d} = 6^{h}$ earlier, that is, the interval should be about $0;15^{d}$ longer, and this deficit would entirely account for the curious error of $\frac{1}{4}^{d}$ in the interval of $364;59^{d}$ between (6) and (9), which should be one year of about $365\frac{1}{4}^{d}$. How Copernicus could have overlooked such a discrepancy is hard to say. In fact the vernal equinox of 1515 occurred on 11 Mar at about 7;30 P.M. at Frauenburg. Perhaps Copernicus did observe it more-or-less correctly, and merely made an error in noting the interval between (7) and (6). We have here, in any case, a good example of something that is not at all uncommon, namely, that a theory is more accurate than a single observation. Had Copernicus recomputed (6) with his completed theory, he would doubtless have discovered the error in his observation or, as may be the case, in his recording of his observation.

3. The Equation of Time (III,26)

The last subject taken up in Book III, in III,26, is the equation of time, called by Copernicus the "correction of the natural day" (*diei naturalis differentia*) after the standard medieval term "equation of days" (*aequatio dierum*). His account follows *Almagest* III,9, and also draws upon the elegant presentation in *Epitome* III,22–30.[58] The equation of days is essentially equivalent to the modern equation of time, that is, the difference between the right ascension of the true sun and of a fictitious mean sun that moves in, or simply parallel to, the equator. But its application and computation are somewhat different in that it is concerned with finding the difference between

[58] The equation of time in the *Almagest* is discussed in *HAMA*, 61–68.

the apparent time and mean time elapsed during some interval. It is thus an interval measuring the difference of two intervals, so we designate it as

$$\Delta E = \Delta\bar{t} - \Delta t, \tag{48}$$

where $\Delta\bar{t}$ and Δt are the intervals of mean and apparent time respectively.

We define the true or apparent day as the period between successive meridian transits of the true sun. From Chapter 2 and from the preceding example of determining the time of the meridian transit of Spica,[59] we know that the interval of time between meridian transits depends upon the difference of right ascension,

$$\Delta t = \Delta\alpha = \alpha_2 - \alpha_1. \tag{49}$$

It is obvious that the period between successive meridian transits of the sun is nonuniform, for:

1. The sun moves nonuniformly in the ecliptic since

$$\Delta\lambda = \lambda_2 - \lambda_1 = (\bar{\lambda}_2 + c_2) - (\bar{\lambda}_1 + c_1) \tag{50}$$

and in general $c_2 \neq c_1$.

2. The right ascension of the sun increases nonuniformly, i.e.

$$\Delta\alpha = \alpha_2 - \alpha_1 \neq \Delta\lambda = \lambda_2 - \lambda_1, \tag{51}$$

because the ecliptic is inclined to the equator.

The differences on successive days are very small, but they accumulate periodically into a noticeable quantity of about $0;30^h$. Since tables of mean motion require the use of mean, not apparent, time from the epoch of the tables, it is necessary to apply the equation of time to any apparent time, e.g. equal hours since apparent noon, to find the mean time for the summation of the mean motions. This is particularly important in the case of the moon where $0;30^h$ corresponds to a mean motion in longitude of about $0;16°$, about one-half the lunar diameter, which is critical in computing an eclipse or occultation. For the sun and planets, the equation of time is not essential, and is usually neglected in early astronomy. We shall, following Copernicus's exposition, examine the two components of the equation of time separately.

1. Considering the nonuniform motion of the sun in the ecliptic, the difference during any interval amounts to

$$\Delta E_1 = (\lambda_2 - \lambda_1) - (\bar{\lambda}_2 - \bar{\lambda}_1) = c_2 - c_1, \tag{52}$$

that is, the difference in the solar equations. This can accumulate up to the difference between the maximum positive and negative equations when the motion of the sun is measured between the points where the maximum equations occur. Using Ptolemy's value $c_{max} = 2;23°$,

$$\Delta E_{1\,max} = +2;23° - (-2;23°) = 4;46° \approx 4\tfrac{3}{4}° = 0;19^h.$$

[59] See above, p. 171.

Consequently, in Fig. 22, taking the points of the maximum equations, which are symmetrical to the apsidal line,

$$(\lambda_2 - \lambda_1) - (\bar{\lambda}_2 - \bar{\lambda}_1) = c_{2\,\max} - c_{1\,\max} = -\Delta E_{1\,\max} = -4;45^\circ = -0;19^{\mathrm{d}},$$

$$(\lambda_1 - \lambda_2) - (\bar{\lambda}_1 - \bar{\lambda}_2) = c_{1\,\max} - c_{2\,\max} = +\Delta E_{1\,\max} = +4;45^\circ = +0;19^{\mathrm{h}},$$

so that

$$+\Delta E_{1\,\max} - (-\Delta E_{1\,\max}) = 2\Delta E_{1\,\max} = 9;30^\circ = 0;38^{\mathrm{h}}.$$

This means that if one begins timing meridian transits of the sun at λ_1, the transit at λ_2 will fall $0;19^{\mathrm{h}}$ early, and if one counts from λ_2, the transit at λ_1 will fall $0;19^{\mathrm{h}}$ late, compared to the uniform motion of the sun and the uniform measure of time given by $\Delta\bar{\lambda}$. And the two intervals will differ by $0;38^{\mathrm{h}}$. These differences ΔE_1 are symmetrical to the apsidal line.

2. The second component of the equation of time depends upon the differences of right ascension and true longitude, that is,

$$\Delta E_2 = (\alpha_2 - \alpha_1) - (\lambda_2 - \lambda_1). \tag{53}$$

An inspection of Copernicus's right ascension table for one quadrant in II,3 shows that α does not increase uniformly with λ, and that at $\lambda = 46^\circ$

$$(\lambda - \alpha)_{\max} \approx 46^\circ - 43;31^\circ = 2;29^\circ \approx 2;30^\circ.$$ [60]

Thus the maximum difference occurs approximately at (see Fig. 23)

$$♈0^\circ + 46^\circ = ♉16^\circ, \qquad ♎0^\circ - 46^\circ = ♌14^\circ,$$

$$♎0^\circ + 46^\circ = ♏16^\circ, \qquad ♈0^\circ - 46^\circ = ♒14^\circ.$$

Therefore, as Copernicus explains, following *Epitome* III,27, for

$$\Delta\lambda_{\mathrm{a}} = ♌14^\circ - ♉16^\circ = 88^\circ,$$

$$\Delta\alpha_{\mathrm{a}} = 136;30^\circ - 43;30^\circ = 93^\circ,$$

$$\Delta\alpha_{\mathrm{a}} - \Delta\lambda_{\mathrm{a}} = +5^\circ = +0;20^{\mathrm{h}} = +\Delta E_{2\,\max}.$$

Conversely, for

$$\Delta\lambda_{\mathrm{b}} = ♏16^\circ - ♌14^\circ = 92^\circ,$$

$$\Delta\alpha_{\mathrm{b}} = 223;30^\circ - 136;30^\circ = 87^\circ,$$

$$\Delta\alpha_{\mathrm{b}} - \Delta\lambda_{\mathrm{b}} = -5^\circ = -0;20^{\mathrm{h}} = -\Delta E_{2\,\max}.$$

Thus,

$$+\Delta E_{2\,\max} - (-\Delta E_{2\,\max}) = 2\Delta E_{2\,\max} = 10^\circ = 0;40^{\mathrm{h}}.$$

Again, this means that if one begins measuring meridian transits at ♉16°, the transit at ♌14° will be $0;20^{\mathrm{h}}$ late, and from ♌14° to ♏16°, $0;20^{\mathrm{h}}$ early,

[60] For the precise value and location of $(\lambda - \alpha)_{\max}$, see below, pp. 180–182.

compared to a right ascension increasing uniformly with the true longitude. The difference between the two intervals will thus be 0;40ʰ. Since the differences ΔE_2 are symmetrical to the lines through the equinoxes and solstices, the same relations hold for the arcs from ♏16° to ♒14° and from ♒14° to ♉16°.

Combining these two corrections, the equation of time, or the equation of days, ΔE, is their algebraic sum

$$\Delta E = \Delta E_1 + \Delta E_2. \tag{54}$$

Since the two components have no common points of symmetry, ΔE will vary in a complex way.[61] ΔE_2 is obviously a function of the tropical longitude of the sun, while ΔE_1 depends upon the anomaly, i.e., its distance from the apsidal line. Since the apsidal line is not tropically fixed, the relation of the two components will gradually change, and tables of the equation of time must frequently be recalculated.[62] The variation can be seen by examining maximum value ΔE_{max} at the time of Ptolemy and Copernicus.

Following *Almagest* III,9 and *Epitome* III,28, Copernicus says that according to Ptolemy $\Delta E_{max} = 8\frac{1}{3}°$, and occurs in the interval between ♏0° and ♒15°. Thus, $\lambda_1 = 210°$, $\lambda_2 = 315°$. Call the true anomaly of the sun $\kappa = \lambda - \lambda_A$. Then where $\lambda_A = 65;30°$

$$\kappa_1 = 210° - 65;30° = 144;30°, \qquad \kappa_2 = 315° - 65;30° = 249;30°.$$

Since $c_{max} = 2;23°$, observing the proper sign of c,

$$c_1 = c_{max} \sin 144;30° = -1;23°, \qquad c_2 = c_{max} \sin 249;30° = +2;14°,$$

and from $\bar{\lambda} = \lambda - c$,

$$\bar{\lambda}_1 = 210° + 1;23° = 211;23°, \qquad \bar{\lambda}_2 = 315° - 2;14° = 312;46°.$$

Therefore,

$$\Delta E_1 = \Delta\lambda_{21} - \Delta\bar{\lambda}_{21} = 105° - 101;23° = +3;37°.$$

---

[61] See the analysis and graph in *HAMA*, 66–68, 1222.

[62] The tabulation in the *Handy Tables*, which accompanies the table of normed right ascension (above, p. 114, n. 16), is discussed in *HAMA*, 984–86. Following the *Handy Tables*, tabulating the normed right ascension and equation of time together became the standard practice—e.g. in Battānī, the *Toledan Tables*, the *Alfonsine Tables*—the epoch being chosen or adjusted so that the correction is always positive or always negative. All such tables are useful only so long as the tropical longitude of the solar apsidal line has not changed by very much, and thus they cannot be used to determine the equation of time over a long interval. The 1492 Venice edition of the *Alfonsine Tables* contains two additional tables of the equation of time, one of which is computed for 1456, and in the *Prutenic Tables* (1551) Reinhold gives two tabulations, one with positive and negative, the other with only negative corrections, which he says can be used without sensible error until 1650. The computation of the latter table, in which the correction from apparent to mean time is always negative, and from mean to apparent time always positive, is explained in *Epitome* III,30.

For the second component, where $\varepsilon = 23;51,20°$, we have

$$\alpha_1 = \alpha(210°) = 207;50°, \qquad \alpha_2 = \alpha(315°) = 317;33°,$$

so that

$$\Delta E_2 = \Delta\alpha_{21} - \Delta\lambda_{21} = 109;43° - 105° = +4;43°,$$

and the equation of time

$$\Delta E = \Delta E_1 + \Delta E_2 = 3;37° + 4;43° = 8;20° = 0;33,20^{\mathrm{h}}.$$

Ptolemy rounds the components to

$$\Delta E = 3\tfrac{2}{3}° + 4\tfrac{2}{3}° = 8\tfrac{1}{3}° = (\tfrac{1}{2} + \tfrac{1}{18})^{\mathrm{h}} = 0;33,20^{\mathrm{h}}.$$

Note, however, that we have actually gone through more steps than necessary in finding ΔE since $\Delta\lambda_{21}$ was used to find both components. This is didactically informative, but for practical use it is only necessary to solve Ptolemy's formula

$$\Delta E = \Delta E_1 + \Delta E_2 = (\Delta\lambda_{21} - \Delta\bar{\lambda}_{21}) + (\Delta\alpha_{21} - \Delta\lambda_{21})$$
$$= \Delta\alpha_{21} - \Delta\bar{\lambda}_{21}. \tag{55}$$

We can use this formula to check Copernicus's estimate of $\Delta E_{\max}$ for his own time. He says that between ♏10° and ♒20° "or near by", $\Delta E_{\max} = 7;48°$. If now $\lambda_A = 96;40°$, for $\lambda_1 = 220°$ and $\lambda_2 = 320°$,

$$\kappa_1 = 220° - 96;40° = 123;20°, \qquad \kappa_2 = 320° - 96;40° = 223;20°,$$

and where $c_{\max} = 1;51°$,

$$c_1 = c_{\max} \sin 123;20° = -1;33°, \qquad c_2 = c_{\max} \sin 223;20° = +1;16°,$$
$$\bar{\lambda}_1 = 220° + 1;33° = 221;33°, \qquad \bar{\lambda}_2 = 320° - 1;16° = 318;44°.$$

Thus,

$$\Delta\bar{\lambda}_{21} = 318;44° - 221;33° = 97;11°.$$

And where $\varepsilon = 23;28,30°$, from the table in II,3,

$$\alpha_1 = \alpha(220°) = 217;37°, \qquad \alpha_2 = \alpha(320°) = 322;23°,$$

so that

$$\Delta\alpha_{21} = 322;23° - 217;37° = 104;46°,$$

and therefore

$$\Delta E = \Delta\alpha_{21} - \Delta\bar{\lambda}_{21} = 104;46° - 97;11° = 7;35° = 0;30,20^{\mathrm{h}},$$

compared with Copernicus's estimate of $7;48° = 0;31,12^{\mathrm{h}}$.[63]

---

[63] Correctly $\alpha(220°) = 217;35°$ and $\alpha(320°) = 322;25°$, and this raises ΔE to $7;39° = 0;30,36^{\mathrm{h}}$, which checking adjacent values shows to be about the maximum. How Copernicus reached 7;48° is not clear, but may be nothing more than a computational error. In *Epitome* III,30, $\Delta E = 0°$ at ♒21;15°.

As a consequence of Copernicus's theory of precession and motion of the solar apsidal line, the variation of the equation of time over long periods is quite complicated, and he says that the irregular precession can cause the equation to exceed $10^\circ = 0;40^h$, although how he reached this number is not clear.[64] Fortunately, in computing the equation of time for any interval, one need not be concerned with all its possible variations over centuries, but only with the mean longitude of the sun and the right ascension corresponding to the true longitude of the sun at the beginning and end of the interval. Note that all that is of interest here are tropical longitudes, so as was the case for all applications of spherical astronomy, Copernicus's use of sidereal longitudes is inconvenient and necessitates additional work.

In order to find the equation of time for any interval, one proceeds as follows: It is assumed that, either by observation or computation, the true longitudes of the sun, λ_1 and λ_2, are known for the beginning and end of the interval. Using the appropriate value of ε, from λ_1 and λ_2 find α_1 and α_2. Next, one must find the mean tropical motion of the sun with respect to the mean equinox, what Copernicus tabulates as the "mean composite motion of the sun." If λ_1 and λ_2 are known only by observation, then for the interval Δt, take $\Delta\bar{\lambda}_{21}$ from the table of the mean composite motion in III,14. If λ_1 and λ_2 have been computed, $\bar{\lambda}_1$ and $\bar{\lambda}_2$ may have appeared as steps in the computation, or may be found from

$$\bar{\lambda} = \lambda - c - \delta_P, \qquad (56)$$

where c is the annual solar equation and δ_P the equation of precession. Since δ_P changes rather slowly, by not more than $1'/8^y$, it may be neglected for shorter intervals. The equation of time is then

$$\Delta E = \Delta\alpha_{21} - \Delta\bar{\lambda}_{21} = (\alpha_2 - \alpha_1) - (\bar{\lambda}_2 - \bar{\lambda}_1), \qquad (55)$$

and the interval of mean time

$$\Delta\bar{t} = \Delta t + \Delta E. \qquad (57)$$

It is often required to determine the interval of mean time from one of the standard epochs. This is necessary when computing positions from tables since the summation of the mean motions assumes an interval of mean time since the epoch. Therefore, we must find for Copernicus's four standard epochs the mean longitude of the sun and the right ascension of the true longitude of the sun at epoch. The mean longitude $\bar{\lambda}_0$ has already been given in the table of the solar epoch positions,[65] so we must compute from the solar theory the true longitude at epoch λ_0 and its corresponding right

[64] Thus $20^\circ = 1;20^h$ for the difference of the maximum and minimum intervals of apparent time which, he goes on to say, can produce an error of $\frac{5}{6}^\circ$ in the motion of the moon. Indeed, since the maximum true velocity of the moon is, e.g. from the *Alfonsine Tables*, $0;36,4^{\circ/h}$, so $1;20^h \cdot 0;36,4^{\circ/h} = 0;48,5,20^\circ \approx 0;50^\circ = \frac{5}{6}^\circ$.

[65] See above, p. 165.

ascension α_0, all of which are given in the following table along with the quantity $\alpha_0 - \bar{\lambda}_0$, which will be useful.

Epoch	$\bar{\lambda}_0$	λ_0	α_0	$\alpha_0 - \bar{\lambda}_0$
A. Olympiad	90;59°	90;36°	90;39°	−0;20°
C. Alexander	226;38	227;4	224;31	−2;7
D. Caesar	276;59	278;11	278;56	+1;57
G. Christ	278;2	279;5	279;55	+1;53

(58)

Copernicus tries to determine the equation of time between the Olympiad and Christ, but makes rather a mess of it. He correctly computes λ_0 for the Olympiad as 90;36°, but when he computes λ_0 for Christ, he forgets to add the excess $c_4 \cdot c_6 \approx 0;18°$, so he gets ♑8;48° when he should get ♑9;5° or 9;6° (the uncertainty due to the 0;1° discrepancy in the epoch). From this point on, he finds α(♑8;48°) = 279;35° (corr. 279;36°), and thus

$$\Delta E = (279;35° - 90;39°) - (278;2° - 90;59°)$$
$$= 188;56° - 187;3° = 1;53° = 0;7,32^h \approx 0;7\tfrac{1}{2}^h.$$ [66]

Correctly, the equation of time is

$$\Delta E = (\alpha_{0G} - \alpha_{0A}) - (\bar{\lambda}_{0G} - \bar{\lambda}_{0A})$$
$$= (279;55° - 90;39°) - (278;2° - 90;59°)$$
$$= 189;16° - 187;3° = 2;13° = 0;8,52^h.$$

Exactly what the computation means is not immediately obvious. Copernicus established his epoch positions of the sun in III,19 by calculating backward from Hipparchus's autumnal equinox (1) of E.A. 177 to E.A. 1 with no correction for the equation of time, that is, from an apparent time in E.A. 177 he has computed back to "noon" of Thoth 1 E.A. 1. However, the noon is not an apparent noon, but simply noon with respect to an interval of apparent time from E.A. 177. The remaining epochs are calculated by intervals of apparent time from E.A.1 in the order (E.A.1) C. → D. → G. → A., so none of them is an apparent noon or midnight. In the case of the sun and planets, this makes little difference, but for the moon it is necessary to define

[66] Menzzer (Copernicus (1879), Anmerk. 35–36) checks this computation, but makes the same error as Copernicus, leaving out $c_4 \cdot c_6$, which is obviously wrong since in A.D. 1 e = 416, that is, near maximum. In N (97v20) $\Delta\alpha$ reads 188;54°, when it should be 188;56°. On the other hand, the critical editions and the translations following them put into the text, with no understanding of what they are doing, completely erroneous numbers from M (106r), including $\Delta\alpha$ = 178;54°, no less than 10° in error, which would give ΔE = 178;54° − 187;3° = −8;9° = −0;32,36ʰ—i.e. larger than Copernicus's ΔE_{max}—and hardly compatible with Copernicus's result of +1;53° ≈ +0;7½ʰ.

the epochs as apparent noon or midnight, and to know the interval of mean time between them. Therefore, the preceding computation of the equation of time between A. and G. is meant to be applied as a correction from apparent to mean time, that is

$$\Delta\bar{t} = \Delta t + \Delta E = 775^{y}\ 12\tfrac{1}{2}^{d} + 0;8,52^{h},$$

and indeed in IV,7 the interval of mean time, incorrectly computed as $775^{y}\ 12\frac{1}{2}^{d}\ 0;7\frac{1}{2}^{h}$, is used to determine the epoch positions of the lunar elongation and anomaly at the first Olympiad.[67]

In applying the equation of time to an interval since epoch for a given $\alpha(\lambda)$ and $\bar{\lambda}$, we have

$$\Delta E = (\alpha - \alpha_0) - (\bar{\lambda} - \bar{\lambda}_0) = (\alpha - \bar{\lambda}) - (\alpha_0 - \bar{\lambda}_0). \qquad (59)$$

Since $\alpha_0 - \bar{\lambda}_0$ is constant for a particular epoch, we can determine it once and for all, and then use it directly in calculations of ΔE. Thus, it has been included in the preceding table. Then the mean time

$$\Delta\bar{t} = \Delta t + \Delta E. \qquad (57)$$

Copernicus calls the interval Δt "apparent" (*apparens*) time or "simply" (*simpliciter*), and the interval $\Delta\bar{t}$ "correctly" (*examinatim*), "exactly" (*exacte*) or "uniform" (*aequale*) time, and he uses some further variations of these terms.

Example. Find the equation of time for 16/17 Apr 1525, apparent local midnight, since an assumed apparent local midnight of 1 Jan A.D. 1. From the previous example, we have the mean longitude of the sun $\bar{\lambda} = 33;41°$ and the true longitude $\lambda =$ ♉ 5;50°, for which the right ascension $\alpha = 33;31°$. Thus, with respect to the Epoch of Christ

$$\Delta E = (\alpha - \bar{\lambda}) - (\alpha_0 - \bar{\lambda}_0) = (33;31° - 33;41°) - (+1;53°)$$
$$= -0;10° - 1;53° = -2;3° = -0;8,12^{h}.$$

Since the interval was originally taken as apparent time, the interval of mean time

$$\Delta\bar{t} = \Delta t + \Delta E = 1525^{y}\ 242^{d} + (-0;8,12^{h}) = 1525^{y}\ 242^{d} - 0;8,12^{h},$$

which means that the mean time since the epoch, the uniform measure of elapsed time, is less than the apparent time by $0;8,12^{h}$. If one wished to be extremely accurate, one would recompute the position of the sun for $1525^{y}\ 242^{d} - 0;8,12^{h}$, but for the sun and likewise the planets such a difference is obviously insignificant since in $0;8^{h}$ the sun moves only about 0;0,20°. In the case of the moon, however, a difference of $0;8^{h}$ corresponds to a motion in longitude of about 0;4°, so should be taken into account.

---

[67] See below, p. 212.

Supplementary Remark: The Quantity and Location of $(\lambda - \alpha)_{max}$

From Copernicus's table of right ascension, we have estimated that $(\lambda - \alpha)_{max} \approx 2;29°$ occurs at $\lambda \approx 46°$. It is of interest to inquire into the precise quantity and location, for this also gives the maximum reduction of longitude to right ascension. A solution to the problem, and quite a clever one, is given in *Epitome* III,25, as part of the exposition of the equation of time, following Jābir ibn Aflaḥ's *De astronomia* I,18.

In Fig. 24, let AB be a quadrant of the equator, AC a quadrant of the ecliptic, say, from ♈ to ♋, and let fall DB, the solstitial colure, from the pole of the equator D. Thus, $BC = \varepsilon$ and $DC = \bar{\varepsilon} = 90° - \varepsilon$. Take a point E in DB such that

$$\frac{\sin DB}{\sin DE} = \frac{\sin DE}{\sin DC},$$

that is,

$$\sin^2 DE = \sin \bar{\varepsilon} = \cos \varepsilon = \cos^2 BE.$$

Draw a small circle from E parallel to the equator and meeting the ecliptic in F. Thus, the declination of F,

$$\cos \delta_F = \cos GF = \cos BE = \sqrt{\cos \varepsilon}, \tag{60}$$

and F is the required point at which $\lambda_F - \alpha_F$ is a maximum.

The proof in the *Epitome* is indirect, showing that for points on either side of F, $\lambda - \alpha < \lambda_F - \alpha_F$. Thus, on either side of F, take points H and K, draw arcs DHL and DKM, and perpendiculars FN and FO. Now, by assumption,

$$\frac{\sin DG}{\sin DF} = \frac{\sin DF}{\sin DC} = \frac{\sin DE}{\sin DC},$$

so that

$$\frac{\sin GL}{\sin FN} = \frac{\sin DG}{\sin DF} = \frac{\sin DF}{\sin DC} < \frac{\sin DH}{\sin DC} = \frac{\sin FH}{\sin FN},$$

the last equality because DC is perpendicular to HC, and FN perpendicular to DH, and the angle at H is common. Thus,

$$\sin GL < \sin FH \qquad \text{and} \qquad \text{arc } GL < \text{arc } FH.$$

But

$$FA - FH = HA, \qquad GA - GL = LA,$$

so that

$$FA - GA = FA - (LA + GL) = (FA - LA) - GL,$$
$$HA - LA = (FA - FH) - LA = (FA - LA) - FH,$$

and since $GL < FH$,

$$FA - GA > HA - LA.$$

Then on the other side of F,

$$\frac{\sin MG}{\sin FO} = \frac{\sin DG}{\sin DF} = \frac{\sin DF}{\sin \mathrm{DC}} > \frac{\sin DK}{\sin DC} = \frac{\sin KF}{\sin FO},$$

so that

$$\sin MG > \sin KF \qquad \text{and} \qquad \text{arc } MG > \text{arc } KF.$$

But

$$FA + KF = KA, \qquad GA + MG = MA,$$

so that

$$FA - GA = FA - (MA - MG) = (FA - MA) + MG,$$
$$KA - MA = (FA + KF) - MA = (FA - MA) + KF,$$

and since $MG > KF$,

$$FA - GA > KA - MA. \qquad \text{Q.E.D.}$$

Following the proof in the *Epitome*, it is stated that

$$DF = 90° - \delta_F = 73;13° \qquad \text{and} \qquad GF = \delta_F = 16;47°.$$

Thus,

$$\cos \delta_F = \sin 73;13° = \sqrt{\cos \varepsilon},$$
$$\varepsilon = \cos^{-1}(\sin^2 73;13°) = 23;33,47°.$$

But in fact, it is more likely that $DF = 73;13°$ was computed from

$$DF = \sin^{-1}(\cos 23;33,30°)^{1/2} = 73;13,12° \approx 73;13°.[68]$$

From $\delta_F = 16;47°$ and $\varepsilon = 23;33,30°$,

$$\lambda = \sin^{-1}(\sin \delta_F/\sin \varepsilon) = 46;15,28° \approx 46;15°,$$
$$\alpha = \cos^{-1}(\cos \lambda/\cos \delta_F) = 43;45,23° \approx 43;45°,$$
$$(\lambda - \alpha)_{max} = 46;15° - 43;45° = 2;30°,$$

---

[68] This is rather interesting since 23;33,30° is the obliquity attributed to az-Zarqāl in the *Toledan Tables* (above, p. 133, n. 11), and attributed by az-Zarqāl (if indeed he is the author of the precepts to the *Toledan Tables* known under his name) and by Ibn Ezra to Yahya ibn Abī Mansūr (if indeed that is the meaning of the corrupt forms Almeon Almansuris, etc.), one of the astronomers of the Caliph al-Maʾmūn, about 830 (Nallino 1, 157). One would guess that here the *Epitome* is copying from something dependent upon the *Toledan Tables* since in I,22 the obliquity was given as 23;28°.

all of which values are given in the *Epitome*. Using instead Copernicus's minimum obliquity, $\varepsilon = 23;28°$, which is also given in *Epitome* I,22, we have

$$\delta_F = \cos^{-1} (\cos \varepsilon)^{1/2} = 16;42,49,27°,$$

$$\lambda = \sin^{-1} (\sin \delta_F/\sin \varepsilon) = 46;14,10°,$$

$$\alpha = \cos^{-1} (\cos \lambda/\cos \delta_F) = 43;45,50°,$$

$$(\lambda - \alpha)_{max} = 46;14,10° - 43;45,50° = 2;28,20°.$$

Note that the sum $\lambda + \alpha = 90°$. *Epitome* III,26 gives two proofs that this is rigorously true, and thus λ and α are symmetrical to 45°. Indeed, from

$$\sin \lambda_F = \sin \delta_F/\sin \varepsilon = \sqrt{1 - \cos \varepsilon}/\sin \varepsilon,$$

we have

$$\sin \lambda_F = \frac{\sin \frac{1}{2}\varepsilon\sqrt{2}}{2 \sin \frac{1}{2}\varepsilon \cos \frac{1}{2}\varepsilon} = \frac{\frac{1}{2}\sqrt{2}}{\cos \frac{1}{2}\varepsilon} = \frac{\sin 45°}{\cos \frac{1}{2}\varepsilon}, \quad (61)$$

[69]

and

$$\sin \alpha_F = \frac{\cos \varepsilon \sin \lambda_F}{\cos \delta_F} = \frac{\cos \varepsilon}{\sqrt{\cos \varepsilon}} \cdot \frac{\sin 45°}{\cos \frac{1}{2}\varepsilon} = \frac{\sin 45°}{\cos \frac{1}{2}\varepsilon} \sqrt{\cos \varepsilon}. \quad (62)$$

Then since

$$\cos (\lambda - \alpha) = 2 \sin \lambda \sin \alpha + \cos (\lambda + \alpha),$$

and $\lambda_F + \alpha_F = 90°$,

$$\cos (\lambda_F - \alpha_F) = \frac{2 \sin^2 45°}{\cos^2 \frac{1}{2}\varepsilon} \sqrt{\cos \varepsilon} = \frac{2\sqrt{\cos \varepsilon}}{1 + \cos \varepsilon} \quad (63)$$

and

$$\sin (\lambda_F - \alpha_F) = \left[1 - \left(\frac{2\sqrt{\cos \varepsilon}}{1 + \cos \varepsilon}\right)^2\right]^{1/2} = \frac{1 - \cos \varepsilon}{1 + \cos \varepsilon} = \tan^2 \tfrac{1}{2}\varepsilon, \quad (64)$$

the standard formula for maximum reduction of longitude to right ascension.

Appendix: Copernicus's Chronology and Geography

The observations in the *Almagest* and the *Epitome* are dated principally in the Egyptian Calendar using the Era of Nabonassar or the Era of the

[69] Delambre (1819), 287–88, in commenting on Epitome III,25, derives this formula somewhat differently as a proof of the proposition—since he had proved the formula in his *Traité Complet d'Astronomie* (1814) I, 25)—and then remarks of the proof in the *Epitome*, "Sa solution n'est ni si complète ni si methodique que la mienne; mais elle est curieuse."

Death of Alexander (also called the Era of Philip Arrhidaeus or Era Philip). Copernicus uses the Julian Calendar and the Christian Era for his own observations, and thus had to convert dates from one calendar to the other in order to determine the time elapsed between the ancient and modern observations. But his interest in chronology went somewhat beyond this problem, for in III,11 he set out a series of calendrical epochs with the intervals between them, and he consistently gives the epoch positions of mean motions for four selected epochs. Most of the observations from the *Almagest* that Copernicus used were made in Alexandria—he uses none of the observations made in Babylon—and his own observations were made in Frauenburg and the surrounding region, in Bologna and Rome, and he also uses three observations made by Bernhard Walther in Nuremberg. In order to reduce all the observations to the same meridian—he selects a meridian that he considers to pass through both Frauenburg and Cracow—he must know the differences of longitude of the locations from which the observations were made. Thus, in addition to chronology, Copernicus required a certain amount of geography. In this appendix, we shall briefly examine the chronology and geography in *De revolutionibus*, and while the sources of the chronology are easily determined, the sources of the geography remain far less certain.

Chronology

In III,11, prior to determining the epoch positions of the mean precession and the anomaly of the obliquity and precession, Copernicus enumerates a series of epochs with the intervals between them. The sources available to him for establishing such a chronology were the *Almagest*, *Epitome*, *Alfonsine Tables*, Eusebius's *Chronicon*, and Censorinus's *De die natali*—Ptolemy's *Kanon Basileion*, generally included in manuscripts of the *Handy Tables*, was not printed until the early seventeenth century—and without some further check it was not easy for Copernicus to correlate the information from his sources. His fundamental list of epochs came from Censorinus, but he needed more information to connect Censorinus's epochs with dating in the Christian Era, and this was provided by the *Epitome*, although the *Alfonsine Tables* could also be used for this purpose. In addition, Copernicus gives intervals to Ptolemy's observation of Regulus (from II,14) since it is fundamental to establishing the epochs of the mean precession and the anomaly. By using the intervals given by Copernicus, which are expressed in Egyptian years and days in seven cases and in julian years in two, we can connect the epochs to A.D. 1 Jan 1 and give equivalent julian dates in the Christian Era and the corresponding julian day numbers, counted from midnight of Jan 1, −4712, used in modern chronological reckoning and astronomy. The results are shown in the following table in which the letter *e* or *j* shows whether the years of the era are counted in Egyptian or julian years, and the letter *m* or *n* shows whether the days are measured from midnight or noon.

Era	Year of Era		Month – Day		Julian Date	Julian Day
A.	Olympiad 1 I	j	Hekatombaion 1	n	−774 Jul 1	1438,536.5
B.	Nabonassar 1	e	Thoth 1	n	−746 Feb 26	1448,638.5
C.	Alexander 1	e	Thoth 1	n	−323 Nov 12	1603,398.5
D.	Caesar 1	j	Jan 1	m	−44 Jan 1	1704,987
E.	Augustus 1	e	Thoth 1	n	−29 Aug 31	1710,708.5
F.	Augustus 1	j	Jan 1	m	−26 Jan 1	1711,562
G.	Christ 1	j	Jan 1	m	1 Jan 1	1721,424
H.	Antoninus 2	e	Pharmouthi 9	m	139 Feb 24	1771,883

The intervals given by Copernicus in Egyptian years are:

A.B.	27^y	247^d	=	$10{,}102^d$
B.C.	424			154,760
C.D.	278	$118\frac{1}{2}$		101,588.5
D.E.	15	$246\frac{1}{2}$		5,721.5
E.G.	29	$130\frac{1}{2}$		10,715.5
A.G.	775	$12\frac{1}{2}$		282,887.5
G.H.	138	89		50,459
A.H.	913	$101\frac{1}{2}$		333,346.5

The following intervals are given in julian years:

D.F.	18^{jy}		=	$6{,}575^d$
F.G.	27			9,862
G.H.	138	55^d		50,459

These intervals form the basis of the preceding table.[1]

We may now attempt to reconstruct Copernicus's establishment of the dates of the epochs. The easiest are the three that can be dated independently of Censorinus, namely, B., Nabonassar, C., Alexander, and E., Augustus in the Egyptian calendar. The intervals between these are given in *Almagest*

[1] The interval A.G. is given by Copernicus in III,19 and elsewhere. In M and N interval A.H. is missing $\frac{1}{2}^d$, but this is supplied correctly in the errata sheet to N (at 77.9). Interval G.H. ends at midnight of Feb 24 following Ptolemy's observation of Regulus, which was made after sunset of Feb 23. On Copernicus's dating of this observation in II,14, see below, p. 188. Other sets of intervals written by Copernicus in his copy of the 1515 *Almagest* are given in Birkenmajer (1900), 244.

III,7 and *Epitome* III,21—B.C. = 424^y and C.E. = 294^y—and the *Epitome* also gives the important, and correct, information that between noon of Thoth 1 Nabonassar 1 and the beginning of the Christian Era, meaning here noon of Dec 31 preceding A.D. 1 Jan 1, there are $747^y\ 130^d$. It follows that with respect to midnight of A.D. 1 Jan 1:

$$\text{B. Nabonassar 1 Thoth 1} = -747^y\ 130\tfrac{1}{2}^d,$$
$$\text{C. Alexander 1 Thoth 1} = \text{Nabonassar 1} + 424^y = -323^y\ 130\tfrac{1}{2}^d,$$
$$\text{E. Augustus 1 Thoth 1} = \text{Alexander 1} + 294^y = -29^y\ 130\tfrac{1}{2}^d.$$

These agree with Copernicus's intervals, and B. and C. are all that are required to convert dates in the *Almagest* to julian dates. The only error in Copernicus's treatment of these is that he gives E. as III Kal. Sep. (Aug 30) while his intervals require Pri. Kal. Sep. (Aug 31), which is Thoth 1 of the moveable Egyptian calendar in −29. In fact, III Kal. Sep. (Aug 30) is Thoth 1 of Augustus 1 in the Alexandrian fixed calendar, that is, by julian intercalation, and corresponds to Epagomenal 5 in the moveable calendar.

Censorinus 21 gives seven concordances for the year of the consuls Pius (text: *Utpii*) and Pontianus, which correctly correspond to the following years:[2]

A. Olympiad 1014 (254 II)	Jul 1, 238–Jun 30, 239
A.U.C. 991	Jan 1, 238–Dec 31, 238
B. Nabonassar 986	Jun 25, 238–Jun 24, 239
C. Philip (Alexander) 562	Jun 25, 238–Jun 24, 239
D. Caesar 283	Jan 1, 238–Dec 31, 238
Augustus (Alexandrian) 267	Aug 29, 237–Aug 28, 238
F. Augustus (Julian) 265	Jan 1, 238–Dec 31, 238

In order to convert the epochs of eras A., D., and F. to the Christian Era, Copernicus had to begin with his knowledge that Thoth 1 Nabonassar 1 precedes the Dec 31 before A.D. 1 Jan 1 by $747^y\ 130^d$. Therefore, Thoth 1 Nabonassar 986, i.e. 985 completed Egyptian years, follows this date by

$$985^y - (747^y\ 130^d) = 237^y\ 235^d,$$

[2] The use of julian years for Olympiad dating and counting the Olympiad year from Jul 1 are conventions in chronology followed by Copernicus, but not necessarily by Censorinus. The exact day is uncertain, and is given by Censorinus as *diebus dumtaxat aestivis*, which Copernicus interprets as both the solstice (in III,1) and as Jul 1 (in III,11). The Era of the Foundation of Rome, A.U.C. *ab urbe condita*, is not given in *De revolutionibus*, but is listed by Copernicus in the notes in his 1515 *Almagest* transcribed in Birkenmajer (1900), 244. Censorinus's Era of Augustus 267 is in the fixed Alexandrian calendar, and is not the same as Copernicus's Era of Augustus E. in the moveable Egyptian calendar. In the moveable calendar Censorinus would have given Augustus 268, extending from Jun 25, 238 to Jun 24, 239, just as eras B. and C. Copernicus seems not to have realized this, and it is odd that his dating of Thoth 1 of E. corresponds to the Alexandrian calendar while his intervals D.E. and E.G. definitely intend the Egyptian calendar.

and converting to julian years by $237/4 = 59^d\ (+\frac{1}{4}^d)$,

$$237^y\,235^d = 237^{jy}\,235^d - 59^d = 237^{jy}\,176^d,$$

which corresponds to A.D. 238 June 25. And indeed, Censorinus says that in his reference year Thoth 1 of Nabonassar 986 and Philip 562 was on VII Kal. Jul., that is, June 25. Let us assume that Copernicus got this far, and knows that the year in the Christian Era containing Thoth 1 Nabonassar 986 is A.D. 238. Counting back from A.D. 238 Jan 1, he can find the epoch of Caesar's Calendar Reform D. and the julian Era of Augustus F.,

$$\text{D. } 238 \text{ Jan } 1 - 282^{jy} = -44 \text{ Jan } 1 = 1 \text{ Jan } 1 - 45^{jy},$$
$$\text{F. } 238 \text{ Jan } 1 - 264^{jy} = -26 \text{ Jan } 1 = 1 \text{ Jan } 1 - 27^{jy}.$$ [3]

Unfortunately, the Era of the Olympiad A. poses a problem. If it is assumed that Olympiad 1014, i.e. 254 II, begins on Jul 1, 238, then computing as before,

$$\text{A. } 238 \text{ Jul } 1 - 1013^{jy} = -775 \text{ Jul } 1 = 1 \text{ Jan } 1 - 776^y\,13^d,$$

which is the correct year, the days being counted from noon to noon. However, Copernicus's interval reaches only to −774 Jul 1, and this error requires an explanation. One possibility is that he took Olympiad 1014 to begin on Jul 1, 239, but this would be an odd way to read Censorinus. It seems more likely that he used Eusebius's *Chronicon*, which we shall see he did consult, and this led to the error. Eusebius gives the birth of Christ as falling in Olympiad 194 III, i.e. year 775, but A.D. 1 Jan 1 in fact is in Olympiad 194 IV, year 776. If Copernicus took Olympiad 194 III to begin at the Jul 1 preceding A.D. 1, he would compute for the year of the first Olympiad,

$$\text{A. } 0 \text{ Jul } 1 - 774^{jy} = -774 \text{ Jul } 1 = 1 \text{ Jan } 1 - 775^y\,13^d,$$

counting from noon to noon, or from 1 Jan 1 midnight, less by $\frac{1}{2}^d$, that is, $775^y\,12\frac{1}{2}^d$, which is the interval given by Copernicus, and thus accounts for his error. Of course, it is hard to understand why he would prefer Eusebius to Censorinus for establishing the date of the first Olympiad.[4]

Further evidence that Copernicus did use Eusebius can be seen in his distinction of Nabonassar from the later Babylonian king Nabuchodonassar, as he spells the name, the Nebuchadnezzar of the Bible, or Nabuchodonosor

[3] Note that we subtract 1 less than the calendar year to reach year 1 of the era; subtracting the calendar year would give year 0. Of course Copernicus would not have used negative A.D. years as we have, but he still would have found that D. and F. precede A.D. 1 Jan 1 by 45 and 27 julian years respectively.

[4] A confirmation that Copernicus has Olympiad 1 I late by one year is to be found in II,14 (N 46r11) where he gives the year of Ptolemy's observation of Regulus in Feb 139 as Olympiad 229 I, while correctly it is 229 II.

of the Vulgate. Instead, he seems to identify Nabonassar with the biblical Shalmaneser, king of Assyria, whom, following Eusebius, he calls Salmanassar, king of the *Chaldeans*. His remarks on this differ in M and N, M reading as follows:

> Ptolemy established the most remote limit of these epochs as the beginning of the reign of Nabonassar of the Chaldeans, whom many, misled by the similarity of the name, believe to be Nabuchodonassar, whom a consideration of the chronology and the computation of Ptolemy show to have been much later, which [computation] coincides with Salmanassar, king of the Chaldeans among the historical writers.[5]

What this means is that Ptolemy's Nabonassar is not Nabuchodonassar, but coincides with, and is presumably identical to, Salmanassar of the historical writers. In N (76v9–12), the middle of the sentence is omitted, apparently deliberately, and Nabuchodonassar is not mentioned:

> Ptolemy established the most remote limit of the epochs as the beginning of the reign of Nabonassar of the Chaldeans, which coincides with Salmanassar, king of the Chaldeans among the historical writers.

Copernicus's, presumably original, identification of Nabonassar with Salmanassar was followed by writers on chronology later in the sixteenth century, e.g. Osiander's son-in-law Johann Funck, until it was shown to be in error by Joseph Scaliger.[6]

The confusion of Nabonassar and the later Nabuchodonosor is due to the identical transcription of the names in the Arabic version of Ptolemy's *Kanon Basileion*, preserved for example in al-Battānī, in which Nabonassar, Nabopolassar, and Nabọkolassar appear as Buchtanaṣṣar I, II, and III, the last being Nabuchodonosor.[7] In the Gerard of Cremona translation of the *Almagest*, Buchtanaṣṣar is translated as Nabuchodonosor, and this reading is followed in the *Alfonsine Tables* and in most manuscripts of the *Epitome*.[8] Copernicus decided that the Nabonassar mentioned by Censorinus is not the same as Nabuchodonosor, but may instead be identical to Salmanassar, and he could do this using the Olympiad dating in Jerome's translation of Eusebius's *Chronicon* where the conquest of Samaria by *Salmanassar rege Chaldaeorum* is dated to Olympiad 8 III, that is, the 31st year of the Olympiads. Since Copernicus takes the Epoch of Nabonassar to fall in the 28th year of the Olympiads, Nabonassar "of the Chaldeans"

---

[5] M 85r24–28: Huius rei supremum scopum constituit Ptolemaeus principium regni Nabonassarii Caldeorum, quem plerique nominis affinitate decepti Nabuchodonassar esse putarunt, quem longe posteriorem fuisse ratio temporum ac supputatio Ptolemaei declarat, quae apud historiographos in Salmanassar Caldeorum regem cadit. (In the following passage from N, *quae apud* is changed to *quod apud*.)

[6] Personal communication from A. Grafton.

[7] Nallino 2, 1 and note at 192.

[8] In Venice lat. 328 the name is usually given correctly as Nabonassar; the 1496 Venice edition gives Nabuchodonosar.

must coincide with and possibly be identical to Salmanassar, the king of the Chaldeans in the *historiographos*, i.e. Eusebius and later chronological writers. The distinction of Nabonassar from Nabuchodonosor also follows from Eusebius, who dates the capture of Judea by *Nabuchodonosor rex Babyloniae* to Olympiad 42 III, about 140 years after Nabonassar or Salmanassar.[9]

A further point worthy of note in Copernicus's chronology is his dating of Ptolemy's observation of Regulus. This occurred on Antoninus 2 Pharmouthi 9 after sunset, and corresponds to 139 Feb 23. In III,11 Copernicus uses the observation as though it occurred at midnight of Feb 24, which is not necessarily an error, but in II,14 he gives the correct time of the observation and gives the date as Feb 24, which appears to be an error of one day. If this is an error rather than a rounding to the closest midnight, it could follow from an error in the text of Censorinus, who says that one hundred years before his time, in Antoninus 2, Thoth 1 fell on *XII* Kal. Aug. (Jul 21). The number is wrong, and was emended by Joseph Scaliger (*De emendatione temporum*, ed. 1629, 201D) to *XIII* Kal. Aug. (Jul 20), but Copernicus had only the uncorrected text. He could easily determine that Antoninus 2, which is Nabonassar 886 or Alexander 462, begins in A.D. 138, and thus following the text of Censorinus, Thoth 1 Antoninus 2 corresponds to 138 Jul 21. Since the interval to Pharmouthi 9 is $7 \cdot 30^d + 8^d = 218^d$, so Copernicus would find

$$\text{Antoninus 2 Pharmouthi 9} = \text{138 Jul 21} + 218^d = \text{139 Feb 24},$$

accounting for the error of one day.[10]

Geography

Determining the sources of Copernicus's geography turns out to be more difficult, and may depend partially upon unpublished information to which he had access concerning the differences in longitude of various cities. He says in IV,7 that observations of solar and lunar eclipses in Frauenburg and Cracow show that both lie on the same meridian, meaning that phases of the eclipses must have been observed at the same time in each location. This is essentially correct since Frauenburg is only $-0;15^\circ = -0;1^h$ west of Cracow, and a difference of $0;1^h$ in eclipse phases would not have been notice-

[9] There are several editions of Jerome's translation of Eusebius from the fifteenth and early sixteenth centuries—of which we have consulted Venice (1483) and Basel (1529)—all extended by later chronicles up to contemporary events.

[10] In the letter of 1524 on Werner's *De motu octavae sphaerae*, Copernicus gives the date as 22 Feb, and also gives Ptolemy's autumnal equinox of 26 Sep 139 as 25 Sep 140. Werner had misdated the observation of Regulus to 22 Feb 150, an error of eleven years, and while Copernicus corrected the year, he missed the error in the day, although his interval of $138^h\ 88^d\ 5\frac{1}{2}^h$ from noon of A.D. 1 Jan 1 is correct. Cf. translation in Rosen (1971), 94–97.

able. Further, Copernicus must also have assumed that the meridian of Frauenburg–Cracow is also applicable to other locations in Warmia, at least for observations of the slower planets, since observations of Mars in 1518 and of Saturn and Jupiter in 1520 when Copernicus was in Allenstein (Olsztyn), actually about $+0;3^h$ east of Cracow, are noted for the same meridian. Consequently, the reference of an observation to the meridian of Frauenburg–Cracow tells us only that Copernicus was somewhere reasonably close to this meridian, and not the exact location of the observation.[11] He also gives in IV,7 the curious information that Dyrrachium in Macedonia, formerly Epidamnus and now Durrës or Durrazo in Albania, is on the same meridian, which is also correct since Durrës is less than $-0;1^h$ west of Frauenburg and less than $-0;2^h$ west of Cracow.[12]

Thus far we have been giving modern differences in longitude, which are not necessarily the same as Copernicus's sources. Aside from his own observations in Warmia, he uses observations made in Alexandria, Rhodes, Raqqa (Syria), Bologna, Rome, and Nuremberg, and all of these locations must be reduced to the meridian of Cracow. We have found no single source containing longitudes of these cities that correspond to the differences in time used by Copernicus for his reductions, which are as follows:[13]

Nuremberg	$\Delta t = -1;0^h$	$\Delta l = -15°$
Bologna	$-0;36$	-9
Rome	$-0;20$	-5
Rhodes	$+0;50$	$+12;30$
Alexandria	$+1;0$	$+15$
Raqqa	$+1;40$	$+25$

In the following table, we give Δl and a computed Δt from Ptolemy's *Geography*, using the 1482 Ulm edition, Schoener's *Opusculum geographicum*

[11] From documents in Biskup (1973), it can be seen that Copernicus's administrative duties necessitated considerable travel in Warmia. Nevertheless, most of his observations were probably made in Frauenburg.

[12] In Ptolemy's *Geography*, using the 1482 Ulm edition, which contains the coordinates common to most of the early Latin printings, Cracow (Carrodunum) is at $l = 42;40°$ and Dyrrachium at $l = 45°$, $+2;20° = +0;9^h$ to the east. In the introduction to Map 10 of Europe, Ptolemy gives Dyrrachium as $-1;0^h$ west of Alexandria, exactly the difference of time used by Copernicus for Cracow. According to Schoener (1533), Dyrrachium is $+1;25° = +0;6^h$ east of Cracow.

[13] Copernicus gives (IV,27) the longitude of Bologna as about 9° west of Cracow and (IV,14) Rome as 5° west. For other locations we have computed Δl from Copernicus's Δt. In his copy of Regiomontanus's *Ephemerides* for 1491–1506 (Augsburg, 1492), Copernicus wrote the longitudes, for Bologna $33\frac{1}{2}°$ and for Rome $36\frac{1}{2}°$; cf. Birkenmajer (1900), 515, 521. If, as in Ptolemy's *Geography*, Cracow is taken as 42;40°, then for Bologna $\Delta l = -9{,}10°$ and for Rome $\Delta l = -6;10°$.

(1533), a contemporary geography, and *The Times Atlas of the World* (1967) for comparison with modern values.

	Ptolemy		Schoener		Times Atlas	
	Δl	Δt	Δl	Δt	Δl	Δt
Nuremberg	$-11;10°$	$-0;45^h$	$-10;0°$	$-0;40^h$	$-8;50°$	$-0;35^h$
Bologna	$-9;20$	$-0;37$	$-8;25$	$-0;34$	$-8;35$	$-0;34$
Rome	$-6;0$	$-0;24$	$-7;0$	$-0;28$	$-7;5$	$-0;28$
Rhodes	$+16;0$	$+1;4$	—	—	$+8;19$	$+0;33$
Alexandria	$+17;50$	$+1;11$	$+16;0$	$+1;4$	$+10;0$	$+0;40$
Raqqa	$+28;10$	$+1;53$	—	—	$+19;8$	$+1;17$

In no case is the agreement with Copernicus's intervals good, except perhaps that in the older sources Alexandria is about $+1^h$ east of Cracow and Rhodes slightly west of Alexandria. Copernicus's interval of $10° = 0;40^h$ between Alexandria and Raqqa, used in III,13 for reducing al-Battānī's equinox observations, is given in *Epitome* III,2, which is definitely Copernicus's source. The edition of the *Alfonsine Tables* (1492) used by Copernicus gives an interval of $-14° = -0;56^h$ between Nuremberg and Cracow, which could be the source of Copernicus's interval of -1^h, but it is hard to understand why he would use this source for Nuremberg since he does not use it for Bologna and Rome. It is possible that he determined the intervals for Nuremberg, Bologna, and Rome from reports of eclipse observations, and did not rely upon published sources.

4

Lunar Theory and Related Subjects

Introduction

Book IV of *De revolutionibus* is concerned with the subjects treated in Books IV–VI of the *Almagest* and *Epitome*: the first and second inequalities of the moon; the parallax, distance, and size of the sun and moon; the apparent diameters of the sun, moon, and shadow of the earth; the theory of eclipses. The brevity of Copernicus's treatment of eclipses accounts for the most significant reduction from the exposition in the *Almagest* and *Epitome*, the latter of which is especially important for the second half of Book IV. Since the recent discovery that Copernicus's lunar model was anticipated some two hundred years earlier by Ibn ash-Shāṭir of Damascus (1306–75), nothing of much originality is left to Copernicus's lunar theory.[1] Yet from the point of view of the historian who is concerned with Copernicus's sources and procedures, Book IV is very interesting as a highly synthetical, and highly medieval, exposition of methods and even numerical parameters from Greek, Indian, Arabic, and medieval Latin sources that found their way into the *Alfonsine Tables* and the *Epitome*, and then into the book usually said to mark a revolution in astronomy.

In his introduction to Book IV Copernicus remarks that the moon indicates nothing about the motion of the earth, "except perhaps for the daily motion" (a curious exception), and says that he does not differ from the opinions of the ancients that its motion takes place about the earth. But ironically, the one and only place in which Copernicus was able to correct an essential error of the kinematic theories of "the ancients" has nothing to do with heliocentricity and the motion of the earth, but is concerned with the properly geocentric motion of the moon. Ptolemy's lunar model for the second inequality had the serious problem that at quadrature the moon was moved to about one-half its distance at conjunction and opposition, which was completely irreconcilable with the absence of any appreciable variation in its apparent size. The model used by Ibn ash-Shāṭir and Copernicus avoids this difficulty, keeping the moon at a reasonable distance from the earth throughout its orbit.

Copernicus had described the model earlier in the *Commentariolus*, there merely adapting numerical parameters to fit the maximum lunar equations at syzygy and quadrature from the *Alfonsine Tables*.[2] What remained was the harder task of deriving a proper set of parameters for the lunar theory and for the parallax, distance, and apparent diameter of the moon and sun required for the computation of eclipses. To judge from the great number of alterations in M, and from the further changes in N, these derivations gave Copernicus much difficulty, and many small inconsistencies remain in the finished version of Book IV. Nevertheless, the lunar theory works quite

[1] On Ibn ash-Shāṭir's lunar model, see Roberts (1957), King (1975).

[2] Swerdlow (1973), 459–61.

well and, as was true of the solar theory, reproduces the various observations used by Copernicus with more than tolerable accuracy.

1. The Lunar Theory

The Problems of Ptolemy's Lunar Model (IV,1–2)

In Book IV of the *Almagest* Ptolemy explains the various periodic motions of the moon that must be accounted for in lunar theory—the motions completing revolutions in synodic, tropical, anomalistic, and draconitic months—and derives the parameters for the first lunar inequality, which depends upon the moon's distance from the apogee of its eccentric or equivalent epicyclic model (see Fig. 12(b) of Chapter 3). He had discovered, however, that the moon is subject to a second inequality, depending upon its distance from the mean sun. The effect of this inequality, as treated by Ptolemy in Book V, is two-fold. In the first place, the equation of the first inequality is varied such that, whatever value it may have, it is always increased as the moon moves from conjunction or opposition to quadrature, a correction that corresponds closely to the evection of modern lunar theory. The second effect, the *prosneusis* or inclination of the mean apogee of the lunar epicycle, is equivalent to a bimonthly variation in the direction of the apsidal line of the moon's orbit such that, counting from conjunction, it regresses in the first and third quadrants, and advances in the second and fourth. This is more remotely related to the modern description of the effect of the evection on the apsidal line, and likewise on the eccentricity, which depends upon twice the elongation of the sun from the lunar perigee, the period thus being about 206 days.

Copernicus begins his exposition of lunar theory with a description of Ptolemy's model and a series of objections. The model is shown in Fig. 1. With the earth at O and the mean sun in the direction $\bar{\odot}$, we describe a circle about O of radius $e = OM$, and let M move in the negative sense, that is, of decreasing longitude, through $\bar{\eta} = \bar{\lambda}_{☾} - \bar{\lambda}_{\odot}$, the mean elongation of the moon from the mean sun. M is in turn the center of a circle of radius $R - e$ with apogee A and perigee B, and the motion of M will cause the entire circle to rotate westward about O, completing a rotation in a mean synodic month. Then let a point C on the circle be the center of the lunar epicycle, which moves eastward with respect to the mean sun, uniformly about O through $\bar{\eta}$. Consequently arc $AC = 2\bar{\eta}$, and C will traverse circle ACB twice in a mean synodic month, twice reaching apogee and perigee. It is apparent that the distance OC will vary from $OA = (R - e) + e = R$ when $2\bar{\eta} = 0°$, to $OB = (R - e) - e = R - 2e$ when $2\bar{\eta} = 180°$. Next, we describe an epicycle of radius $r = CP$, and let the moon at P move in the negative sense through the mean anomaly $\bar{\alpha}$, which is measured, not with respect to the true apogee G lying on GCO, but with respect to a mean apogee

F having a *prosneusis*, an inclination along FCN toward a point N lying directly opposite M on the circle about O. The model has the following effects:

1. $c = POC$ is the equation of the anomaly, a function of α. The variation of the distance of the center of the epicycle C will cause c to increase as $2\bar{\eta}$ approaches 180°, that is as $\bar{\eta}$ approaches 90° at mean quadrature. The true elongation from the mean sun is then $\eta = \bar{\eta} + c$.
2. $c_3 = FCG = NCO$ is a correction to $\bar{\alpha}$ such that $c_3 = 0°$ for $2\bar{\eta} = (0°, 180°)$, at syzygy and quadrature, $c_3 \geq 0°$ for $0° \leq 2\bar{\eta} \leq 180°$ in the first and third quadrants from mean conjunction, and $c_3 \leq 0°$ for $180° \leq 2\bar{\eta} \leq 360°$, in the second and fourth quadrants. The true anomaly, measured from apogee G as seen from O, is $\alpha = \bar{\alpha} + c_3$.

The following parameters, derived by Ptolemy, will be of use for further reference:

$$
\begin{array}{lll}
R = 60, & r = 5;15, & c_{3\,\max} = 13;9° \text{ at } 2\bar{\eta} \approx 114°, \\
e = 10;19, & R + r = 65;15, & c_{\max\,(2\bar{\eta}=0°)} = 5;1°, \\
R - e = 49;41, & R - r = 54;45, & c_{\max\,(2\bar{\eta}=180°)} = 7;40°[3]. \\
R - 2e = 39;22, & R - 2e - r = 34;7, &
\end{array}
\quad (1)
$$

Copernicus raises four objections to the model:

1. The motion of the center of the epicycle C is uniform with respect to the center of the earth O, nonuniform with respect to the center of the eccentric M, and therefore nonuniform in the eccentric circle ACB. This violates the principle of uniform circular motion. As Copernicus remarks in a frequently quoted passage:

> But if this is so, what shall we reply to the axiom, that the motion of the heavenly bodies is uniform except that it seems nonuniform with respect to [its] appearance, if the apparent, uniform motion of the epicycle is in fact nonuniform, and the very opposite of the established and assumed principle will take place?[4]

2. The motion of the moon on the epicycle is uniform with respect to the apogee F lying on the line inclined towards N, rather than with respect to G lying on the line to the center of the earth O. Since the center of the epicycle moves uniformly about O, measuring the mean anomaly from F is regarded as causing a nonuniform motion of the moon on the epicycle, and thus a second violation of the principle of uniform circular motion.

---

[3] In principle; only 7;39° is recoverable from the table in V,8.

[4] N 99v9–15. Note that the objection refers specifically to the paradoxical situation in Ptolemy's lunar model, in which the apparent motion of the center of the epicycle is uniform while its motion in the eccentric is nonuniform, the very opposite of what Copernicus believes should happen.

3. The model causes the distance of the moon to vary between $R + r = 65;15$ and $R - 2e - r = 34;7$, a ratio of nearly 2 to 1. Using the distances in terrestrial radii (*tr*) from *Almagest* V,17, Copernicus says that "by the agreement of all mathematicians" the greatest distance is $64;10^{tr}$ and, because of the behavior of the model, the least distance $33;33^{tr}$. Therefore the parallax at least distance ought to be about twice the parallax at greatest distance. However, he continues, by observation it is seen that the parallax at quadrature, even when the moon is in the perigee of the epicycle, differs by little or nothing from the parallax at eclipses of the sun and moon, as will be shown later.[5]

4. If the distance of the moon varies in a ratio of nearly two to one, then its apparent diameter ought likewise to vary in the same ratio, and the apparent size of its body in a ratio of four to one. Thus, if the entire moon were illuminated at quadrature, it would appear four times as great as at opposition, and even the half-moon ought to give twice the light of the full moon at opposition. However, observation shows the contrary to be true, and use of a diopter shows the diameter of the moon to vary only by the amount produced by its motion on the epicycle. This last objection was already made in much the same form in *Epitome* V,22.

Therefore, Ptolemy's model must be replaced by one that preserves uniform circular motion, and does not produce a variation in the lunar distance incompatible with the observed variation in parallax and apparent diameter.

Copernicus's Model (IV,3)

Ptolemy's model worked on the principle of optically enlarging the effect of the lunar epicycle by moving it closer to the observer at quadrature. The model of Ibn ash-Shāṭir and Copernicus works on the alternate principle of truly enlarging the epicycle itself at quadrature. The device for doing so is illustrated in a simplified form in Fig. 2. Let the moon at conjunction be at the point B of a secondary epicycle, whose center C_2 moves on a primary epicycle in the negative sense through the mean anomaly (which is here assumed to be zero so that centers C_1 and C_2 lie on the line from the earth O to the mean sun $\bar{\odot}$). Now, however, the moon P moves in the positive sense from B toward D through $2\bar{\eta}$ while C_1 moves through $\bar{\eta}$, so that when $\bar{\eta} = 90°$ at quadrature, the moon will be at D. Thus the effective radius of the epicycle has increased from C_1B to C_1D, the difference being twice the radius of the secondary epicycle. The effective radius will then decrease toward opposition, and once again increase toward the next quadrature, completing two cycles in a mean synodic month.

The model becomes more interesting when the mean anomaly $\bar{\alpha}$ is included as shown in Fig. 3. Here we can see that the double epicycle in

[5] Below, pp. 236–238.

fact generates an ellipse around the center of the first epicycle of semimajor axis $r_1 + r_2$ and semiminor axis $r_1 - r_2$. Thus, for an arbitrary value of $\bar{\alpha}$, at mean conjunction the moon is at B on the minor axis of the ellipse. As C_2 moves in the negative sense through $\bar{\alpha}$, and P moves in the positive sense through $2\bar{\eta}$, the resulting path of P about C_1 is an ellipse, the semiminor axis of which, C_1B, rotates forward through $\bar{\eta}$ with respect to radius C_1C_2.[6] Or, in another way, C_1B and the ellipse may be considered to rotate backwards through $\bar{\alpha} - \bar{\eta}$ with respect to C_1A and the apogee A of the first epicycle. When $\bar{\eta} = 90°$ at mean quadrature, the moon has reached the major axis of the ellipse at D. The period of a complete traversal of the ellipse by P will be a mean synodic month, the period of $\bar{\eta}$, although P will not move uniformly on the ellipse. For the figure also shows that the motion of P on the second epicycle produces a correction c_3 to $\bar{\alpha}$ that is positive in the first and third quadrants of $\bar{\eta}$ and negative in the second and fourth, just as the corresponding correction in Ptolemy's model. In another way, measured from B, c_3 is positive in the first and third quadrants of the ellipse, and negative in the second and fourth. Then $\alpha = \bar{\alpha} + c_3$, and the moon may be considered to move nonuniformly on the ellipse through α, measured from the direction C_1A, or through $\bar{\eta} + c_3$ measured from the semiminor axis C_1B. Finally, the equation c, a function of α, corrects the mean elongation $\bar{\eta}$ to the true elongation $\eta = \bar{\eta} + c$.

Copernicus also mentions that an equivalent eccentric model is possible. although he does not describe it. The conversion of the double epicycle to the eccentric form is shown in Fig. 4. Completing parallelogram OC_1C_2F gives an eccentricity $e_1 = r_1$, so that if a circle of radius R is drawn about F, the moon at P will move on the second epicycle through $2\bar{\eta} - \bar{\alpha}$ with respect to FC_2. This illustrates clearly the term for the evection in Copernicus's model. In the epicyclic form, the return of P to line C_1A, or the return of the true anomaly $\alpha = AC_1P$ to the same value, likewise has the period of the evection $2\bar{\eta} - \bar{\alpha}$. Then, completing parallelogram FC_2PG, one has $e_2 = r_2$, and the fully eccentric form with P moving on a circle of radius R about G. Here it can be seen that the true apogee A will alternately advance and regress on either side of the mean apogee $\bar{A}$. While different characteristics of the lunar motion are clearer in one form than another, all three forms are fully equivalent, as was the case for the three forms of the solar model.

Preliminary Mean Motions (IV,4)

What remains is to find the parameters of the lunar theory—the radii of the epicycles, the mean motions of the elongation, anomaly, and the argument of latitude, and their positions at epoch. However, in order to begin

[6] The description of the rotation of the ellipse in Swerdlow (1973), 458 n. 2, is wrong (as was immediately remarked upon in a letter from D. T. Whiteside).

solving these problems, it is necessary to have some preliminary values of the mean motions. Any reasonable values will do, e.g. those of the *Almagest* or of the *Alfonsine Tables*, but attempting to follow Ptolemy precisely, Copernicus begins with an interesting, although common, misunderstanding of Hipparchus's lunar cycle of 126007 days that is described in *Almagest* IV,2 and already implicitly misunderstood in *Epitome* IV,3.

He first reviews, with some uncertainty and error, three of the Greek calendrical cycles of tropical years and synodic months mentioned in Censorinus 18, namely, the Metonic Cycle,

$$\text{M.C.} = 19^{\text{ty}} = 235^{\text{sm}} = 6940^{\text{d}},$$

the Callippic Cycle,

$$\text{C.C.} = 4\ \text{M.C.} - 1^{\text{d}} = 76^{\text{ty}} = 940^{\text{sm}} = 27759^{\text{d}},$$

and the Hipparchan Cycle

$$\text{H.C.} = 4\ \text{C.C.} - 1^{\text{d}} = 304^{\text{ty}} = 3760^{\text{sm}} = 111035^{\text{d}}.$$ [7]

However, these cycles do not take into account the moon's motion of anomaly and argument of latitude. Therefore, he says, Hipparchus, by comparing lunar eclipses he observed with earlier eclipses observed by the "Chaldeans", found that

$$\begin{aligned} 345^{\text{y}}\ 82^{\text{d}}\ 1^{\text{h}} &= 126007^{\text{d}}\ 1^{\text{h}} \\ &= 4267 \text{ synodic months} \\ &= 4573 \text{ anomalistic months} \\ (&= 4630\tfrac{1}{2} \text{ draconitic months} + \sim 11\tfrac{1}{3}^{\circ}) \qquad (2) \\ &= 4612 \text{ sidereal months} - \sim 7\tfrac{1}{2}^{\circ} \\ &= 345 \text{ sidereal years} - \sim 7\tfrac{1}{2}^{\circ}. \end{aligned}$$

Copernicus mentions only the terms for the synodic and anomalistic months. Hipparchus designed this cycle, which can fall between eclipses about 345 years apart, as a test of the Babylonian System B lunar periods

$$\begin{aligned} 1 \text{ syn. mo.} &= 29;31,50,8,20^{\text{d}}, \\ 251 \text{ syn. mo.} &= 269 \text{ anom. mo.}, \qquad (3) \\ 5458 \text{ syn. mo.} &= 5923 \text{ drac. mo.}, \end{aligned}$$

and confirmed that the Babylonian periods were quite accurate. He did not derive the mean motions of his lunar theory from the cycle, but, according to Ptolemy, used the confirmed Babylonian periods directly.[8]

---

[7] On such cycles, see *HAMA*, 619–24, 296–97.

[8] On the origin and function of the 126007^{d} cycle, see Toomer (1980), Aaboe (1955), *HAMA*, 309–12.

Copernicus, however, following the interpretation in the *Epitome* (it must be admitted that Ptolemy is not altogether clear on this point), takes the cycle as a source of Hipparchus's synodic month, and computes that

$$1 \text{ syn. mo.} = 126007^{d}\, 1^{h}/4267 = 29;31,50,8,\mathit{9,20}^{d},$$

a mistake that has been repeated in modern literature, entirely independent of Copernicus. He then claims that dividing this number into 360° gives a mean daily elongation of 12;11,26,41,20,18°, although the division would correctly give 12;11,26,41,*24,42*°. In fact, and fortunately, Copernicus has merely rounded from the value in the *Almagest*, based upon the correct synodic month used by Hipparchus, that is,

$$\bar{\eta}^{d} = 6,0°/29;31,50,8,20^{d} = 12;11,26,41,20,17,59^{°/d}.$$

Thus he saves himself from the error of computing what he believes to be Hipparchus's mean motions from the wrong synodic month.[9] Then, in agreement with the mean motion table in the *Almagest*, the annual elongation

$$\bar{\eta}^{y} = 6,5 \cdot \bar{\eta}^{d} = 12^{r} + 129;37,21,28,29^{°/y}. \quad \text{(N, 31 for 21)}$$

To find the mean motion of the anomaly, he notes that dividing by 17, 4267/4573 = 251/269. Therefore, the mean anomaly in one year will be

$$\bar{\alpha}^{y} = (269/251)\bar{\eta}^{y} = 13^{r} + 88;43,8,40,20^{°/y},$$

and in one day

$$\bar{\alpha}^{d} = \bar{\alpha}^{y}/6,5^{d} = 13;3,53,56,29^{°/d}.$$

Using the relation 5458 syn. mo. = 5923 drac. mo., he finds for the annual mean argument of latitude

$$\bar{\omega}^{y} = (5923/5458)\bar{\eta}^{y} = 13^{r} + 148;42,46,20,3^{°/y}, \quad \text{(M, . . . 19,3)}$$

and for the daily motion

$$\bar{\omega}^{d} = \bar{\omega}^{y}/6,5^{d} = 13;13,45,39,40^{°/d}.$$

Copernicus then gives small corrections to the annual motions, first according to Ptolemy by comparing them with the tables in the *Almagest*, and then according to his own derivations. As we shall see later, the corrections do not all follow from the observations.[10] And to make matters more confusing, M contains an earlier set of corrections and mean motions that are inconsistent in themselves, and unfortunately influence the final values in N.[11] All of this material, which will be needed for reference, is

---

[9] Unfortunately, Menzzer (Copernicus (1879), Anm., 36–37) laboriously does just that, and creates much unnecessary confusion.

[10] Below, pp. 222–223.

[11] The critical editions and the translations relying upon them have adopted contradictory readings from N and M apparently at random and without checking a single computation.

given in Table 1.[12] The provisional values, following directly from the Babylonian period relations, are designated Hip. for Hipparchus since Copernicus presumes that they were Hipparchus's mean motions. Since the provisional value of $\bar{\eta}$ used by Copernicus in fact followed from the correct synodic month of $29;31,50,8,20^d$, he could well be right.

Observations of the Moon

Copernicus gives a total of fourteen lunar observations, of which nine are lunar eclipses. Eight of the observations were his own, and six are from the *Almagest*. Most of the parameters of his lunar theory are derived from the eclipses, and of the five remaining observations, one is used as a test of the lunar theory, two are for the determination of lunar parallax, one is a test of the parallax, and one that is not used shows Ptolemy's lunar parallax. Table 2 gives: the number of the observation, first the nine eclipses, then the other five with the unused observation last; the date and apparent time from the text and as a julian date counted from midnight and reduced where necessary to the meridian of Frauenburg; the julian day number counted from midnight; and a description of the observation. In the description, E indicates an eclipse, followed by the magnitude in digits ($'$), where the apparent diameter of the moon is always $12'$, and the direction from which the shadow appears to move across the moon (in fact the direction towards which the moon moves through the shadow).

The reason for basing the parameters of the lunar theory, in so far as possible, on eclipses is that the moon's apparent position is affected by a substantial parallax—as much as one degree, about twice the lunar diameter—on account of the appreciable size of the earth in relation to the distance of the moon. And the parallax of the moon cannot be found, and observations adjusted accordingly, until the complete lunar theory has been developed. Consequently, Copernicus turns to the classical procedure, probably invented by Hipparchus and used constantly by Ptolemy, of finding the position of the moon by means of lunar eclipses. Since at the midpoint of a lunar eclipse the moon is for all practical purposes directly opposite the sun, by calculating the true longitude of the sun and adding to it 180°, the true geocentric longitude of the moon is known independently of any effects of parallax at the earth's surface. Hence the longitude of the sun must be computed for each lunar eclipse. For the remaining observations it is also necessary to know the longitude of the sun since Copernicus's lunar theory itself gives only the elongation of the moon from the mean longitude of the sun, specifically, from the adjusted mean longitude $\bar{\lambda}'_{\odot} = \bar{\lambda}_{\odot} + \delta_P$, where $\bar{\lambda}_{\odot}$ is the mean longitude from the mean equinox, Copernicus's "mean composite motion," and δ_P is the equation of the precession. The true longitude of the moon then follows from adding the elongation to $\bar{\lambda}'_{\odot}$.

[12] In M 110v33 the correction to $\bar{\varpi}$, $-0;0,1,1,44°$, appears to be altered from $0;0,1,47(?),57°$.

Table 3 contains all the required information about the sun for the apparent time of each lunar observation. Following the number and year of the observation, we tabulate:

(1) C. $\bar{\lambda}'_{\odot}$, adjusted mean longitude reported by Copernicus;
(2) C. $\lambda_{\odot}$, true longitude reported by Copernicus.

The remaining columns are recomputed from Copernicus's solar theory:

(3) $\bar{\lambda}_{\odot}$, mean longitude from mean equinox;
(4) $\bar{\lambda}'_{\odot}$, adjusted mean longitude;
(5) $\lambda_{\odot}$, true longitude;
(6) $\alpha_{\odot}$, right ascension of $\lambda_{\odot}$;
(7) $\alpha_{\odot} - \bar{\lambda}_{\odot}$, quantity required for the equation of time.

For observations 1/2/3 Copernicus has reported Ptolemy's computed positions of $\lambda_{\odot}$ and calculated $\bar{\lambda}'_{\odot}$ from

$$\bar{\lambda}'_{\odot} = \lambda_{\odot} - c_{\max} \sin(\lambda_{\odot} - \lambda_A),$$

where $c_{\max} = 2;23°$ and $\lambda_A = 65;30°$. The recomputations in column (5) show a maximum error of only 0;3°, for observation (1), which is further evidence that Copernicus's solar theory correctly reproduces Ptolemy's theory for Ptolemy's time. Comparison of columns (2) and (5) as a whole also shows that Copernicus has computed accurately.

The First Inequality (IV,5)

By the first inequality, Copernicus means the inequality in the motion of the moon depending only upon the mean anomaly $\bar{\alpha}$. The first inequality can be isolated from the effects of the second inequality only at mean syzygy when $2\bar{\eta} = 0°$ and, as shown in Fig. 5, the moon is at the point B of the second epicycle. Thus the moon may be considered to move on an epicycle of radius $r = r_1 - r_2 = C_1B$ in a model of the kind illustrated in Fig. 12(b) of Chapter 3.[13] The problem then is to find (1) the radius r of the epicycle, (2) the direction of its center C_1 as seen from O, and (3) the location of the moon at B on the epicycle at a given time. Since there are three unknowns, or in another way, since a circle is determined in magnitude and position by three points on its circumference, it is necessary to use three observations of the moon at or near syzygy, which is done using lunar eclipses when the true elongation η is 180°. If now $\eta = 180°$, then using Ptolemy's figures, the mean elongation will fall within the range

$$\bar{\eta} = 180° \pm (c_{☾} + c_{\odot}) = 180° \pm (5;1° + 2;23°) = 180° \pm 7;24°.$$

It is a nice question whether this is close enough to 180° that the second inequality may be ignored and the moon assumed to be at the point B of the

[13] Above, p. 150.

second epicycle. The equivalent problem arises for Ptolemy's model, and is investigated in *Almagest* V,10 and *Epitome* V,12, after the parameters of the model have been established. Two cases must be considered, for the moon at greatest equation of the anomaly where $2\bar{\eta} = 2 \cdot 7;24°$, and for the moon at zero equation where $2\bar{\eta} = 2c_{\odot} = 2 \cdot 2;23°$, and for both it is shown that the effect of the second anomaly is negligible. Although Copernicus does not examine the question, the same is not quite true of his lunar model.[14]

The method of finding the radius of the lunar epicycle, or the eccentricity in an eccentric model, was probably invented by Hipparchus, whose results, and problems, are discussed all too cryptically in *Almagest* IV,11 and, following the *Almagest*, in *Epitome* IV,19. The problem is of the form: given a circle with three points on its circumference, to find a point within (eccentric model) or without (epicyclic model) the circle, from which lines drawn to the three points will contain given angles. Since the three points may be anywhere on the eccentric or epicycle, none of the special conditions used to simplify the derivation of the solar eccentricity can be applied.[15]

Ptolemy's own derivation for an epicyclic model is in *Almagest* IV,6 and *Epitome* IV,9–11. He also explains briefly how the derivation is done in the case of an eccentric model, and his description is considerably amplified in *Epitome* IV,14. Copernicus uses only the epicyclic form of his lunar model. Ptolemy goes through the demonstration twice—first using a very old set of eclipses observed in Babylon in −720/−719 and then a set he had observed in Alexandria in 133/134/136. The reason is that he also needs to find the mean longitude and mean anomaly at two widely separated epochs in order to correct Hipparchus's mean motions. Copernicus does the same, using Ptolemy's second set and his own observations from 1511/1522/1523, thereby providing a base line of nearly 1400 years for checking the mean motions, compared with Ptolemy's interval of about 850 years. One may wonder why he chose not to use the older Babylonian eclipses, which would have given an interval of fully 2240 years.[16]

1. *Ptolemy's Derivation*

Copernicus reviews Ptolemy's derivation from eclipses 1/2/3 of Table 2. The pertinent information for the sun is contained in Table 3. It is first necessary to determine the intervals of mean time between the eclipses, and find the mean motions of the moon in longitude and anomaly. For all this, Copernicus simply adopts the numerical values given by Ptolemy, but we shall go through the procedures.

---

[14] See below, pp. 215–216.

[15] The general solution to the problem is treated in Herz (1887–94) 1, 62–63, Pedersen (1974), 172–77, *HAMA*, 73–76. Here we are concerned with the specific numerical solutions given by Copernicus.

[16] It is of interest that Copernicus uses none of the ten Babylonian eclipses reported in the *Almagest*.

Using the values of $\alpha_\odot$ and $\bar{\lambda}_\odot$ from Table 3, the equation of time is

$$\Delta E_{12} = (\alpha_\odot - \bar{\lambda}_\odot)_2 - (\alpha_\odot - \bar{\lambda}_\odot)_1 = -1;48^\circ = -0;7,12^{\mathrm{h}} \approx -\tfrac{1}{8}^{\mathrm{h}},$$
$$\Delta E_{23} = (\alpha_\odot - \bar{\lambda}_\odot)_3 - (\alpha_\odot - \bar{\lambda}_\odot)_2 = +7;8^\circ = +0;28,32^{\mathrm{h}} \approx +\tfrac{1}{2}^{\mathrm{h}},$$

so that from $\Delta\bar{t} = \Delta t + \Delta E$, the intervals of mean time are

$$\Delta\bar{t}_{12} = 1^{\mathrm{y}}\ 166^{\mathrm{d}}\ 23\tfrac{3}{4}^{\mathrm{h}} - \tfrac{1}{8}^{\mathrm{h}} = 1^{\mathrm{y}}\ 166^{\mathrm{d}}\ 23\tfrac{5}{8}^{\mathrm{h}},$$
$$\Delta\bar{t}_{23} = 1^{\mathrm{y}}\ 137^{\mathrm{d}}\ 5^{\mathrm{h}} + \tfrac{1}{2}^{\mathrm{h}} = 1^{\mathrm{y}}\ 137^{\mathrm{d}}\ 5\tfrac{1}{2}^{\mathrm{h}}.$$

During these intervals, the mean motions in longitude, $\Delta\bar{\lambda} = \Delta(\bar{\lambda}_\odot + \bar{\eta})$, and anomaly, $\Delta\bar{\alpha}$, are

$$\Delta\bar{\lambda}_{12} = 169;37^\circ, \quad \Delta\bar{\alpha}_{12} = 110;21^\circ,$$
$$\Delta\bar{\lambda}_{23} = 137;34^\circ, \quad \Delta\bar{\alpha}_{23} = 81;36^\circ.[17]$$

But since at each eclipse $\lambda_☾ = \lambda_\odot + 180^\circ$, the motion in true longitude, from the values in Table 3, column (2), is

$$\Delta\lambda_{12} = \Delta\lambda_{\odot 12} = 161;55^\circ, \qquad \Delta\lambda_{23} = \Delta\lambda_{\odot 23} = 138;55^\circ,$$

and thus the differences between the true and mean motion in longitude,

$$\delta_1 = \Delta\lambda_{12} - \Delta\bar{\lambda}_{12} = -7;42^\circ, \qquad \delta_2 = \Delta\lambda_{23} - \Delta\bar{\lambda}_{23} = +1;21^\circ.$$

Now in Fig. 6(a), let the earth be at D, draw the epicycle with center K, and let the moon be at A in the first eclipse, at B in the second, and at C in the third. We are given that

$$\Delta\bar{\alpha}_1 = 110;21^\circ, \qquad \delta_1 = -7;42^\circ,$$
$$\Delta\bar{\alpha}_2 = 81;36^\circ, \qquad \delta_2 = +1;21^\circ,$$
$$\Delta\bar{\alpha}_3 = 360^\circ - (\Delta\bar{\alpha}_1 + \Delta\bar{\alpha}_2) = 168;3^\circ, \qquad \delta_3 = -\delta_1 - \delta_2 = +6;21^\circ.$$

Considering only the absolute value of δ, in triangle BDE,

$$2\beta_1 = \Delta\bar{\alpha}_1 = 110;21^\circ, \qquad 2\delta_1 = 15;24^\circ,$$
$$2\gamma_1 = 2\beta_1 - 2\delta_1 = 94;57^\circ,$$
$$DE = \mathrm{crd}\ 2\gamma_1 = 2\sin\gamma_1 = 147396,$$
$$BE = \mathrm{crd}\ 2\delta_1 = 2\sin\delta_1 = 26798.$$

In triangle CDE,

$$2\beta_2 = \Delta\bar{\alpha}_1 + \Delta\bar{\alpha}_2 = 191;57^\circ, \qquad 2\delta_3 = 12;42^\circ,$$
$$2\gamma_2 = 2\beta_2 - 2\delta_3 = 179;15^\circ,$$
$$DE = \mathrm{crd}\ 2\gamma_2 = 2\sin\gamma_2 = 199996,$$
$$CE = \mathrm{crd}\ 2\delta_3 = 2\sin\delta_3 = 22120.$$

---

[17] These figures are directly from the *Almagest*, and are not quite consistent with the provisional mean motions.

Thus, in units of triangle BDE, where $DE = 147396$,

$$CE = \text{crd } 2\delta_3\left(\frac{\text{crd } 2\gamma_1}{\text{crd } 2\gamma_2}\right) = 22120\left(\frac{147396}{199996}\right) = 16302.$$

In triangle BEC,

$$\beta_3 = \beta_2 - \beta_1 = \tfrac{1}{2}\Delta\bar{\alpha}_2 = \tfrac{1}{2}(81;36^\circ) = 40;48^\circ,$$

$$CQ = CE \sin \beta_3 = 10652, \qquad EQ = CE \cos \beta_3 = 12341,$$

$$BQ = BE - EQ = 26798 - 12341 = 14457,$$

$$BC = (BQ^2 + CQ^2)^{1/2} = 17960. \quad \text{(Corr. 17957)}$$

But where the radius of the epicycle $r = 100{,}000$,

$$BC = \text{crd } \Delta\bar{\alpha}_2 = \text{crd } 81;36^\circ = 130684.$$

Consequently, now

$$DE = 147396\,(130684/17960) = 1072684, \quad \text{(Corr. 1072511)}$$

$$CE = 16302\,(130684/17960) = 118637, \quad \text{(Corr. 118620)}$$

and

$$\text{arc } CE = 2 \sin^{-1} \tfrac{1}{2}CE = 72;46,10^\circ.$$ [18]

Thus,

$$\text{arc } AE = \Delta\bar{\alpha}_3 - \text{arc } CE = 168;3^\circ - 72;46,10^\circ = 95;16,50^\circ,$$

$$AE = \text{crd } AE = 2 \sin \tfrac{1}{2}AE = 147786,$$

$$DEA = DE + EA = 1072684 + 147786 = 1220470.$$

Therefore, in Fig. 6(b), by Euclid III,36, (Text III,30)

$$DA \cdot DE = (R + r)(R - r),$$

and by Euclid II,6,

$$R^2 = (R + r)(R - r) + r^2 = DA \cdot DE + r^2 \approx 1.3191786 \times 10^{12},$$

so that

$$R = 1148556. \quad \text{(Corr. 1148555)}$$

Thus, where $R = 100{,}000$

$$r = 100000^2/1148556 = 8706, \quad \text{(Corr. 8706.58)}$$

---

[18] Correctly 72;46,0°, but Ptolemy gives 72;46,10° (Ptolemy (1912–13), 232:4). It appears as though the preceding numbers were not actually computed by Copernicus in the form shown here, but were simply converted from the sexagesimal values in the *Almagest*. The same explanation holds for small errors in the rest of the derivation.

is the radius of the epicycle. Where $R = 60$, this is equivalent to $r = 5;13,25$. Ptolemy gives 5;14, but this number is rounded from $60^2/689;8 = 5;13,26$, so Copernicus's result agrees very well. The maximum equation will be

$$c_{\max} = \sin^{-1} .08706 = 5;0°.$$

Ptolemy finally rounds r to 5;15, and so finds

$$c_{\max} = \sin^{-1} (5;15/60) = 5;1°.$$

Next, in Fig. 6(b), drawing KNO perpendicular to DEA,

$$NE = \tfrac{1}{2}AE = \tfrac{1}{2}(147786) = 73893,$$

$$DN = DE + NE = 1072684 + 73893 = 1146577,$$

$$DK = R = 1148556.$$

Thus,

$$\gamma_3 = \sin^{-1} (DN/DK) = 86;38,30°, \quad \text{(Corr. 86;38,10°)}$$

$$\text{arc } LO = 180° - \gamma_3 = 93;21,30°,$$

$$\text{arc } AO = \tfrac{1}{2} \text{ arc } AE = \tfrac{1}{2}(95;16,50°) = 47;38,25° \approx 47;38,30°,$$

so that

$$-\bar{\alpha}_1 = \text{arc } LO - \text{arc } AO = 45;43°,$$

is the distance of the moon at the first eclipse from the apogee of the epicycle. Checking over the entire derivation by an accurate and consistent calculation, we find that where $R = 1$

$$r = .0870656 \quad \text{and} \quad -\bar{\alpha}_1 = 45;43,26°,$$

so the various errors in Copernicus's demonstration hardly affect the final result.

For $-\bar{\alpha}_1 = -45;43°$,

$$\bar{\alpha}_2 = -\bar{\alpha}_1 + \Delta\bar{\alpha}_1 = 64;38°,$$

$$\bar{\alpha}_3 = \bar{\alpha}_2 + \Delta\bar{\alpha}_2 = 146;14°,$$

$$\bar{\alpha}_1 = \bar{\alpha}_3 + \Delta\bar{\alpha}_3 = 314;17°.$$

Then, since $\gamma_3 \approx 86;38°$,

$$c_1 = 90° - \gamma_3 = +3;22°,$$

$$c_2 = c_1 + \delta_1 = -4;20°,$$

$$c_3 = c_2 + \delta_2 = -2;59°. \quad \text{(M.N. 2;49°)}$$

Finally, from $\bar{\lambda} = \lambda - c$, and from $\bar{\eta} = \bar{\lambda} - \bar{\lambda}'_{\odot}$, we find the mean longitude and mean elongation, which we tabulate along with $\bar{\alpha}$:

No.	λ	$-\ c$	$=\ \bar{\lambda}$	$-\ \bar{\lambda}'_{\odot}$	$=\ \bar{\eta}$	$\bar{\alpha}$	
1.	223;15°	+3;22°	219;53°	42;21°	177;32°	314;17°	
2.	25;10	−4;20	29;30	206;43	182;47	64;38	(4)
3.	164;5	−2;59	167;4	341;44	185;20	146;14	

2. *Copernicus's Derivation*

Copernicus uses eclipses 4/5/6 that he observed himself between 1511 and 1523. Why they should span an interval of twelve years is not clear since they should be chosen as close together as possible in order that no error in the provisional mean motions can accumulate. From the information for the sun in Table 3, we find for the equation of time,

$$\Delta E_{45} = +2;18^\circ = +0;9,12^{\mathrm{h}}, \qquad \Delta E_{56} = +0;55^\circ = +0;3,40^{\mathrm{h}}.$$

From the difference between the apparent and mean time in the text, we can see that Copernicus uses

$$\Delta E_{45} = (1 - \tfrac{1}{15})^{\mathrm{h}} - \tfrac{3}{4}^{\mathrm{h}} = 0;56^{\mathrm{h}} - 0;45^{\mathrm{h}} = 0;11^{\mathrm{h}},$$ [19]

$$\Delta E_{56} = 0;4^{\mathrm{h}}.$$

Thus, for the intervals of mean time between the eclipses,

$$\Delta\bar{t}_{45} = 10^{\mathrm{y}}\ 337^{\mathrm{d}}\ 0;45^{\mathrm{h}} + 0;11^{\mathrm{h}} = 10^{\mathrm{y}}\ 337^{\mathrm{d}}\ 0;56^{\mathrm{h}},$$

$$\Delta\bar{t}_{56} = 354^{\mathrm{d}}\ 3;5^{\mathrm{h}} + 0;4^{\mathrm{h}} = 354^{\mathrm{d}}\ 3;9^{\mathrm{h}}.$$

For these intervals, from the provisional mean motions,

	4.–5.	5.–6.
$\Delta\lambda = \Delta\lambda_{\odot}$	329;47°	349;9°
$\Delta\bar{\lambda} = \Delta(\bar{\lambda}_{\odot} + \bar{\eta})$	334;47	346;10
$\delta = \Delta\lambda - \Delta\bar{\lambda}$	−5;0	+2;59
$\Delta\bar{\alpha}$	250;36	306;43

[19] A bad reading from M (116r31) of $(1 - \frac{1}{5})^{\mathrm{h}} = \frac{4}{5}^{\mathrm{h}} = 0;48^{\mathrm{h}}$ has been adopted, again without checking, by the critical editions and the translations that follow them. This would give $\Delta E = 0;48^{\mathrm{h}} - 0;45^{\mathrm{h}} = +0;3^{\mathrm{h}}$, very far from the correct $+0;9^{\mathrm{h}}$. Copernicus should have written $(1 - \frac{1}{10})^{\mathrm{h}} = 0;54^{\mathrm{h}}$, so that $0;54^{\mathrm{h}} - 0;45^{\mathrm{h}} = +0;9^{\mathrm{h}}$. In fact M once read this way, but *decima* was deleted, and replaced by *sexta* and then *quinta* in the margin.

Therefore, in Fig. 7(a),

$$\Delta\bar{\alpha}_4 = 250;36^\circ, \qquad \delta_4 = -5^\circ,$$

$$\Delta\bar{\alpha}_5 = 306;43^\circ \qquad \delta_5 = +2;59^\circ,$$

$$\Delta\bar{\alpha}_6 = (\Delta\bar{\alpha}_4 + \Delta\bar{\alpha}_5) - 360^\circ = 197;19^\circ, \qquad \delta_6 = -\delta_4 - \delta_5 = +2;1^\circ.$$

Now, considering only the absolute value of δ, in triangle DBE

$$2\beta_1 = 360^\circ - \Delta\bar{\alpha}_5 = 53;17^\circ, \qquad 2\delta_5 = 5;58^\circ,$$

$$2\gamma_1 = 2\beta_1 - 2\delta_5 = 47;18^\circ, \quad \text{(Corr. 47;19°)}$$

$$DE = \text{crd } 2\gamma_1 = 2 \sin \gamma_1 = 8024, \quad \text{(Corr. 8023)}$$

$$BE = \text{crd } 2\delta_5 = 2 \sin \delta_5 = 1042. \quad \text{(Corr. 1041)}$$

In triangle ADE,

$$2\beta_2 = \Delta\bar{\alpha}_6 = 197;19^\circ, \qquad 2\delta_6 = 4;2^\circ,$$

$$2\gamma_2 = 2\beta_2 - 2\delta_6 = 193;17^\circ,$$

$$DE = \text{crd } 2\gamma_2 = 2 \sin \gamma_2 = 19865, \quad \text{(Corr. 19866)}$$

$$AE = \text{crd } 2\delta_6 = 2 \sin \delta_6 = 702. \quad \text{(Corr. 704)}$$

Hence, in units of triangle DBE, where $DE = 8024$,

$$AE = \text{crd } 2\delta_6(\text{crd } 2\gamma_1/\text{crd } 2\gamma_2) = 702\ (8024/19865) = 283. \quad \text{(Corr. 284)}$$

In triangle BEA,

$$\beta_3 = \beta_1 + \beta_2 = \tfrac{1}{2}\Delta\bar{\alpha}_4 = \tfrac{1}{2}(250;36^\circ) = 125;18^\circ,$$

$$AQ = AE \sin \beta_3 = 231, \qquad EQ = AE \cos \beta_3 = 163\tfrac{1}{2},$$

$$BQ = BE + EQ = 1042 + 163\tfrac{1}{2} = 1205\tfrac{1}{2},$$

$$AB = (BQ^2 + AQ^2)^{1/2} = 1227.$$

But where the radius of the epicycle $r = 10{,}000$,

$$AB = \text{crd } \Delta\bar{\alpha}_4 = \text{crd } 250;36^\circ = 16323,$$

and thus now

$$DE = 8024\ (16323/1227) = 106751, \quad \text{(Corr. 106745)}$$

$$BE = 1042\ (16323/1227) = 13853, \quad \text{(Corr. 13862)}$$

so that

$$\text{arc } BE = 2 \sin^{-1} \tfrac{1}{2}BE = 87;41^\circ,$$

$$\text{arc } CBE = \text{arc } BE + (360^\circ - \Delta\bar{\alpha}_5) = 87;41^\circ + 53;17^\circ = 140;58^\circ.$$

Therefore,

$$CE = \text{crd } CBE = 2 \sin \tfrac{1}{2}CBE = 18851,$$

$$CED = CE + DE = 18851 + 106751 = 125602.$$

And in Fig. 7(b),

$$CD \cdot DE = (R + r)(R - r) = R^2 - r^2,$$

$$R^2 = CD \cdot DE + r^2 \approx 1.3508139 \times 10^{10},$$

$$R = 116226. \qquad \text{(Corr. 116224.5)}$$

Thus, where $R = 100{,}000$,

$$r = 10^8/116226 = 8604,$$

which, Copernicus remarks, "agrees with what we find reported by most of those who preceded us since the time of Ptolemy." Indeed,

$$\sin^{-1}(8604/100{,}000) = 4;56^\circ,$$

the maximum equation at syzygy in the *Alfonsine Tables*, and the same value Copernicus used in the *Commentariolus*.[20] Where $R = 60$, the equation in the *Alfonsine Tables* is based upon an epicyclic radius $r = 5;10$, while Copernicus's value leads to 5;9,45.

Next, in Fig. 7(b), drawing *FLM* perpendicular to *CED*,

$$LE = \tfrac{1}{2}CE = \tfrac{1}{2}(18851) = 9425\tfrac{1}{2} \approx 9426,$$

$$DEL = DE + LE = 106751 + 9426 = 116177,$$

$$DF = R = 116226.$$

Therefore,

$$\gamma_3 = \sin^{-1}(DL/DF) = 88;21^\circ, \qquad \text{(Corr. 88;20,10°)}$$

$$\text{arc } CM = \tfrac{1}{2} \text{ arc } CBE = \tfrac{1}{2}(140;58^\circ) = 70;29^\circ,$$

and thus

$$\bar{\alpha}_6 = 180^\circ - (CM + \gamma_3) = (180^\circ - 158;50^\circ) = 21;10^\circ,$$

---

[20] The original version of the demonstration in M (116r–117v) led to $r = 8614$ (117r18), which likewise gives a maximum equation of 4;56°. The parameter has an extensive earlier history, a maximum lunar equation of 4;56° being found in: Varāhamihira, *Pañcasiddhāntikā* (Indian, 6th cen.); Brahmagupta, *Khaṇḍakhādyaka* (Indian, 7th cen.); al-Khwārizmī, *Zij as-Sindhind* (Arabic, early 9th cen.); *Alfonsine Tables* (Spanish, late 13th cen.); Ibn ash-Shāṭir, *Nihāyat as-Sūl* (Arabic, mid. 14th cen.). These are only the most well-known sources. Note further that Copernicus does not believe that the maximum equation has actually diminished since Ptolemy, as was true for the solar theory, but that 4;56° is a more accurate value than 5;1° (which is not true).

is the distance of the moon at the third eclipse from the apogee of the epicycle. Checking the entire computation without errors or roundings, we find that where $R = 1$,

$$r = .086054 \quad \text{and} \quad \bar{\alpha}_6 = 21;12,27^\circ,$$

so once again Copernicus's demonstration is quite accurate. The remaining quantities, $\bar{\alpha}$, c, $\bar{\lambda}$, and $\bar{\eta}$ are all correctly computed, and we tabulate them as follows:

No.	λ	$-\ c$	$=\ \bar{\lambda}$	$-\ \bar{\lambda}'_{\odot}$	$=\ \bar{\eta}$	$\bar{\alpha}$	
4.	22;25°	+0;22°	22;3°	204;13°	177;50°	183;51°	
5.	352;12	−4;38	356;50	173;59	182;51	74;27	(5)
6.	341;21	−1;39	343;0	163;2	179;58	21;10	

Correction of Mean Elongation and Anomaly (IV,6)

In IV,4 Copernicus computed provisional values of the mean elongation and anomaly that he considered to be those of Hipparchus. They are found in lines 1 and 2 of Table 1, and we shall designate them as $\bar{\eta}_h$ and $\bar{\alpha}_h$. From the previous demonstrations, he has found values of $\bar{\eta}$ and $\bar{\alpha}$ at dates nearly 1400 years apart, and he now carefully selects two eclipses to correct the Hipparchan values, following the method explained in *Almagest* IV,7 and *Epitome* IV,12. He uses the second eclipse from each set, that is, (2) and (5), between which, from the preceding demonstrations,

$$\Delta\bar{\eta}_{25} = \bar{\eta}_5 - \bar{\eta}_2 = 182;51^\circ - 182;47^\circ = 0;4^\circ,$$[21]

$$\Delta\bar{\alpha}_{25} = \bar{\alpha}_5 - \bar{\alpha}_2 = 74;27^\circ - 64;38^\circ = 9;49^\circ.$$

The interval of apparent time between the eclipses, $\Delta t = 1388^y\ 302^d\ 3;20^h$, and from the values in Table 3, the equation of time,

$$\Delta E_{25} = (\alpha_{\odot} - \bar{\lambda}_{\odot})_5 - (\alpha_{\odot} - \bar{\lambda}_{\odot})_2 = +3;37^\circ = +0;14,28^h \approx +0;14^h.$$

Thus the interval of mean time,

$$\Delta\bar{t}_{25} = 1388^y\ 302^d\ 3;34^h = 23,8^y\ 5,2;8,55^d,$$

for which, using the provisional values,

$$\Delta\bar{\eta}_h = \Delta\bar{t} \cdot \bar{\eta}_h = 17165^r + 359;38^\circ, \quad (\text{Acc. } 359;37,48^\circ)$$

$$\Delta\bar{\alpha}_h = \Delta\bar{t} \cdot \bar{\alpha}_h = 18397^r + 9;39^\circ, \quad (\text{Acc. } 9;38,37^\circ)$$

[21] By selecting two eclipses in which $\bar{\eta}$ was most nearly the same, Copernicus insures that no possible effects of the second inequality introduce errors into the mean anomaly. See below, p. 216, n. 32.

and using the tables in *Almagest* IV,3, he also finds that from Ptolemy's corrected value of $\bar{\alpha}$,

$$\Delta\bar{\alpha}_p = \Delta\bar{t} \cdot \bar{\alpha}_p = 18397^r + 9;11^\circ. \quad (\text{Acc. } 9;10,59^\circ)$$

Thus, compared to the mean motions derived from the eclipses,

$$\begin{aligned} \delta(\bar{\eta}_h) &= \Delta\bar{\eta} - \Delta\bar{\eta}_h = 0;4^\circ - 359;38^\circ = +0;26^\circ, \\ \delta(\bar{\alpha}_h) &= \Delta\bar{\alpha} - \Delta\bar{\alpha}_h = 9;49^\circ - 9;39^\circ = +0;10^\circ,^{22} \\ \delta(\bar{\alpha}_p) &= \Delta\bar{\alpha} - \Delta\bar{\alpha}_p = 9;49^\circ - 9;11^\circ = +0;38^\circ. \end{aligned} \tag{6}$$

At this point Copernicus concludes the chapter by remarking that these corrections increase the provisional values, and agree with the numbers set out in IV,4.

However, he has not explained how the corrections to the provisional values (Table 1, line 5) were in fact derived. But a reconstruction is easy. Following the procedure described in the *Almagest* and *Epitome*, where δ is the difference between the correct and provisional mean motion for some period $\Delta\bar{t}$ stated in Egyptian years, the annual correction d given by Copernicus in IV,4 should follow from

$$d = \delta/\Delta\bar{t}.$$

Letting $\Delta\bar{t} = 1388^y\ 302^d\ 3;34^h \approx 23,8;49,40^y$ to sufficient accuracy,

$$\begin{aligned} d(\bar{\eta}_h) &= 0;26^\circ/\Delta\bar{t} = 0;0,1,7,24^\circ, \\ d(\bar{\alpha}_h) &= 0;10^\circ/\Delta\bar{t} = 0;0,0,25,55^\circ. \end{aligned}$$

Copernicus gives

$$d(\bar{\eta}_h) = 0;0,1,7,56^\circ, \qquad d(\bar{\alpha}_h) = 0;0,0,26,55^\circ,$$

which do not agree. In fact, he has computed with the values of δ that would result from carrying the provisional mean motions to seconds, i.e.,

$$\Delta\bar{\eta}_h = 359;37,48^\circ, \qquad \Delta\bar{\alpha}_h = 9;38,37^\circ.$$

In this way

$$\begin{aligned} d(\bar{\eta}_h) &= 0;26,12^\circ/\Delta\bar{t} = 0;0,1,7,55^\circ, \\ d(\bar{\alpha}_h) &= 0;10,23^\circ/\Delta\bar{t} = 0;0,0,26,55^\circ, \end{aligned} \tag{7}$$

which agree very well with Copernicus's values.

When these corrections are added to the provisional mean motions, line 1 of Table 1, the result is the corrected mean motions in N given in line 6. The daily mean motions should follow from dividing the annual motion by 365^d, which would give

$$\bar{\eta}^d = 12;11,26,41,31^{\circ/d}, \qquad \bar{\alpha}^d = 13;3,53,56,34^{\circ/d}.$$

---

[22] Omitted in N, but in M (118r margin).

However, Copernicus has merely carried over daily motions, line 9 of Table 1, that followed from the cancelled annual motions in M given in line 8. In the case of $\bar{\eta}$, this makes no difference, but $\bar{\alpha}$ is in error by -4 in the last place, which hardly matters unless the daily motions are multiplied by a large number of days.

Mean Elongation and Anomaly at Epoch (IV,7)

In order to establish the mean elongation and mean anomaly at the Epoch of Christ, Copernicus takes eclipse (2) observed by Ptolemy in 134 Oct 20, 22;0^{h}. The first step is to find the equation of time between the observation and epoch G. It was shown earlier that for this epoch $\alpha_0 - \bar{\lambda}_0 = +1;53°$, while Copernicus's incorrect computation in III,26 implies $\alpha_0 - \bar{\lambda}_0 = +1;33°$,[23] and this influences his result. Thus using the correct value for G,

$$\Delta E = (\alpha - \bar{\lambda})_2 - (\alpha_0 - \bar{\lambda}_0)_{\mathrm{G}} = -6;7° = -0;24,28^{\mathrm{h}},$$

while from Copernicus's value

$$\Delta E = -5;47° = -0;23,8^{\mathrm{h}} \approx -0;23^{\mathrm{h}},$$

which is what Copernicus uses. Thus, the mean time since epoch is

$$\begin{aligned}\Delta\bar{t} &= 133^{\mathrm{y}}\ 325^{\mathrm{d}}\ 22^{\mathrm{h}} - 0;23^{\mathrm{h}} = 133^{\mathrm{y}}\ 325^{\mathrm{d}}\ 21;37^{\mathrm{h}} \\ &= 2,13^{\mathrm{y}}\ 5,25;54,2,30^{\mathrm{d}},\end{aligned}$$

for which

$$\Delta\bar{\eta} = 332;49°, \quad (\text{Acc. } 332;49,28°)$$

$$\Delta\bar{\alpha} = 217;32°, \quad (\text{Corr. } 217;30,26°)$$

Taking the values of $\bar{\eta}$ and $\bar{\alpha}$ at eclipse (2),

$$\bar{\eta}_{0\mathrm{G}} = 182;47° - 332;49° = 209;58°,$$

$$\bar{\alpha}_{0\mathrm{G}} = 64;38° - 217;32° = 207;6°.$$

For the latter, Copernicus gives 207;7°, perhaps following from

$$\bar{\alpha}_{0\mathrm{G}} = 64;38° - 217;30,26° = 207;7,34°.$$

To determine the remaining epochs, it is first necessary to find the equation of time between the Epoch of Christ and the other three standard epochs. This can be done easily. Call the later epoch *b* and the earlier *a*. Then

$$\Delta E = (\alpha_{0\mathrm{b}} - \alpha_{0\mathrm{a}}) - (\bar{\lambda}_{0\mathrm{b}} - \bar{\lambda}_{0\mathrm{a}}) = (\alpha_0 - \bar{\lambda}_0)_{\mathrm{b}} - (\alpha_0 - \bar{\lambda}_0)_{\mathrm{a}}.$$

---

[23] From the wrong value $\lambda_{\odot} = ♑8;48°$, $\alpha - \bar{\lambda} = 279;35° - 278;2° = +1;33°$; correctly, from $\lambda_{\odot} = ♑9;5°$, $\alpha - \bar{\lambda} = 279;55° - 278;2° = +1;53°$. See above, p. 178. This error enters several of Copernicus's computations of the equation of time with respect to the Epoch of Christ.

Since we have already tabulated $\alpha_0 - \bar{\lambda}_0$ for each epoch,[24] we need only take their differences from epoch G to find ΔE. Thus, between the Olympiad and Christ,

$$\Delta E_{AG} = +1;53^\circ - (-0;20^\circ) = +2;13^\circ = +0;8,52^h,$$

in agreement with our earlier computation.[25] Between Alexander and Christ,

$$\Delta E_{CG} = +1;53^\circ - (-2;7^\circ) = +4^\circ = +0;16^\circ,$$

and between Caesar and Christ,

$$\Delta E_{DG} = +1;53^\circ - (+1;57^\circ) = -0;4^\circ = -0;0,16^h \approx 0^h.$$

In III,26 Copernicus had already computed ΔE_{AG} incorrectly as $+0;7\frac{1}{2}^h$, and he uses it here. ΔE_{CG} and ΔE_{DG} he gives correctly as $+0;16^h$ and 0^h respectively, which seems a bit puzzling with his incorrect value of $(\alpha_0 - \bar{\lambda}_0)_G$. Thus, the intervals of mean time are

$$\Delta\bar{t}_{AG} = 775^y\ 12^d\ 12;7,30^h = 12,55^y\ 12;30,18,45^d,$$

$$\Delta\bar{t}_{CG} = 323^y\ 130^d\ 12;16^h = 5,23^y\ 2,10;30,40^d,$$

$$\Delta\bar{t}_{DG} = 45^y\ 12^d.$$

For these intervals, we have correctly:

	$\Delta\bar{\eta}$	$\Delta\bar{\alpha}$
A.G.	170;13,54°	160;45,37°
C.G.	259;14,23	121;25,37
D.G.	219;19,17	189;8,37

(8)

From which the following epochs are established (including the mean argument of latitude $\bar{\omega}$ found in IV,14):[26]

Epoch	$\bar{\eta}_0$	$\bar{\alpha}_0$	$\bar{\omega}_0$
A. Olympiad	39;44°	46;20°	272;3°
C. Alexander	310;44	85;41	49;0
D. Caesar	350;39	17;58	118;56
G. Christ	209;58	207;7	129;45

(9)

[24] Above, p. 178.
[25] Above, p. 178.
[26] See below, pp. 223–225. $\bar{\alpha}_{0A}$ should be 46;21°.

Copernicus remarks that these epochs apply to Cracow, which lies on the same meridian as Frauenburg, as he had learned from simultaneous observations of lunar and solar eclipses, and also to "Dirrhachium," formerly Epidamnus, in Macedonia, here evidently defined as a province of the Roman Empire. Once again, he is quite correct, Dyrrachium, modern Durrës (Durazzo) in Albania being at 19;28° E, only 0;12° = $0;0,48^h$ west of Frauenburg. From Ptolemy's *Geography* III,13,3; IV,5,9, Dyrrachium or Epidamnus in Macedonia is at $l = 45°$ while Alexandria is at $l = 60\frac{1}{2}°$. Hence Dyrrachium is about 15° = 1^h west of Alexandria, just as Copernicus assumes for Frauenburg and Cracow. It seems reasonable to assume that the *Geography* was Copernicus's source.

The Second Inequality (IV,8–9)

Following the demonstration of the first inequality and the mean motions and epoch positions of longitude and anomaly in *Almagest* IV,6–8, Ptolemy goes on immediately to the determination of the mean motion and epoch position of the argument of latitude in IV,9–10, deferring consideration of the second inequality until Book V. He could do so because, as he later demonstrates in V,10, the greatest effect of the second inequality possible at true syzygy produces only negligible changes in the equation of the first inequality. Since the mean argument of latitude and its position at epoch are found by means of lunar eclipses, application of the second inequality is unnecessary, and may be deferred. Copernicus's model, however, behaves somewhat differently—the effect of the second inequality is not negligible at syzygy—and although he presents no demonstration, he appears to know it. Consequently, he develops the model for the second inequality, shows how to compute positions with it, and only then considers the argument of latitude, taking the second inequality into account.

Exactly how Ptolemy succeeded in isolating and correctly describing the second inequality is not clear from the *Almagest*, but it must have depended upon detailed comparisons between carefully selected observed and computed lunar positions at various values of α and η, in which the avoidance of parallax in longitude was an essential part. The behavior of the second inequality is described in *Almagest* V,2 and *Epitome* V,3, and at the beginning of this chapter.[27] In *Almagest* V,4 Ptolemy sets out two observations, one of his own and one by Hipparchus, chosen such that the second inequality produces its maximum effect, while the longitudinal parallax is minimal. In the first observation $\bar{\eta} \approx -90°$, $\bar{\alpha} \approx +90°$, and the observed value of the equation c was about $-7;40°$, compared to a computed $c \approx -5°$. In the second $\bar{\eta} \approx -90°$, $\bar{\alpha} \approx -90°$, the observed correction was $+7;40°$, while

[27] Above, p. 194.

by computation $c \approx +5^\circ$. In both cases the correction of the first inequality was increased by 2;40°, and from the maximum correction of 7;40°, Ptolemy derives the eccentricity $e = 10;19$.[28]

Copernicus's treatment of the second inequality is much briefer, and has no independent observations or analysis of observations underlying it. He takes as given the behavior of the inequality described by Ptolemy, and also takes as given a maximum correction of 7;40° when $\bar{\eta} = \pm 90^\circ$. The determination of the radii of the epicycles in his lunar model is thus trivial. Figure 8(a) shows the configuration for maximum equation c_{max} at syzygy and quadrature. It has previously been determined that at syzygy, $c_{max} = 4;56^\circ$, and where $R = 10{,}000$,

$$r_{min} = r_1 - r_2 = 860.4 \approx 860,$$

and now we are given that at quadrature, $c_{max} = 7;40^\circ$, so that

$$r_{max} = r_1 + r_2 = R \sin 7;40^\circ = 1334.$$

Therefore,

$$2r_2 = (r_1 + r_2) - (r_1 - r_2) = 1334 - 860 = 474,$$

$$r_2 = 237, \qquad r_1 = (r_1 - r_2) + r_2 = 860 + 237 = 1097.$$[29]

The second effect of the second inequality is the *prosneusis*, the inclination of the mean apogee of the epicycle that provides a correction to the mean anomaly $\bar{\alpha}$. Ptolemy had established the magnitude of the correction from two observations by Hipparchus in which $\bar{\eta} \approx \pm 45^\circ$.[30] Copernicus, however, uses no observations. The equivalent of the prosneusis in Copernicus's model is the angle subtended at the center of the first epicycle by the radius of the second epicycle. Thus, in Fig. 8(b) r_2 subtends the angle c_3 at C_1. But r_1 and r_2 have just been determined in order to produce the required maximum lunar equations at syzygy and quadrature. Thus c_3 is already given, and need not, indeed, cannot, be found independently from observation. From the previously found values of r_1 and r_2,

$$c_{3\,max} = \sin^{-1}(r_2/r_1) = \sin^{-1}(237/1097) = 12;28,36^\circ,$$

---

[28] See *HAMA*, 84–88. The corresponding section of the *Epitome* is V,5–6.

[29] Where $R = 60$, $r_1 = 6;35$, $r_2 = 1;25,20$. Ibn ash-Shāṭir used 6;35 and 1;25 respectively, producing the identical maximum equations. In the *Commentariolus*, Copernicus had used 7;36° for the maximum equation at quadrature, following the *Alfonsine Tables*. It appears as though the *Alfonsine Tables* took the equation at syzygy, 4;56°, from Khwārizmī, whose fundamentally Indian lunar theory does not include the second inequality. The difference of 2;40° between syzygy and quadrature was then taken from the *Toledan Tables*—indeed, the entire column for this correction appears to be copied from the *Toledan Tables*—which was in turn based upon Battānī, and through Battānī, upon the *Handy Tables*. It is notable that Copernicus is in a sense the end product and one of the last examples of this synthetical, medieval tradition.

[30] *Almagest* V,5, *Epitome* V,8; see *HAMA*, 88–91.

which will occur when

$$2\bar{\eta} = 90° - c_{3\,\max} = 77;31,24°, \qquad \bar{\eta} \approx 38;46°.$$

In Ptolemy's model, from the tables in *Almagest* V,8 $c_{3\,\max} \approx 13;9°$ at $2\bar{\eta} \approx 114°$ or $\bar{\eta} \approx 77°$, rather different from the correction in Copernicus's model.[31] c_3 is used to find the true anomaly α from $\alpha = \bar{\alpha} + c_3$, where c_3 is positive for values of $2\bar{\eta} \leq 180°$ and negative for $2\bar{\eta} \geq 180°$.

Effect of the Second Inequality at Syzygy

Copernicus does not investigate the question of whether the second inequality becomes negligible at all true syzygies, as Ptolemy does in *Almagest* V,10. That he employs the second inequality for computing the conditions of the eclipses used to find the mean argument of latitude and its position at epoch seems to indicate that he considered it essential. At any rate, the problem is worth examining. Two cases must be considered: (1) for maximum lunar equation, (2) for zero lunar equation.

1. When the lunar equation is maximum, mean and true syzygy can differ by the sum of the maximum solar and lunar equations which, in Copernicus's theory, can reach

$$c_{\odot\,\max} + c_{☾\,\max} = 2;23° + 4;56° = 7;19°.$$

Thus, under extreme conditions, at the time of true syzygy $2\bar{\eta}$ can reach 14;38°. Since the moon is at maximum equation, the equation changes slowly as a function of the true anomaly $\alpha = \bar{\alpha} + c_3$, and the effect of c_3 on α can be ignored. However, as illustrated in Fig. 9(a), the effective radius of the epicycle r is perpendicular to the line from the earth to the moon, so the equation $c_{\max}$ will be at its most sensitive to any changes in r. Now, when $2\bar{\eta} = 14;38°$, r is increased from 860 to about 870, for which $c_{\max} = 4;59°$, differing by only 0;3° from 4;56°. Hence, under these conditions, the effect of the second inequality is negligible. For the same circumstances, Ptolemy finds an increase of 0;2°.

2. When the lunar equation is zero, mean and true syzygy can differ by only the maximum solar equation $c_{\odot\,\max} = 2;23°$, meaning that $2\bar{\eta} = 4;46°$. The configuration is shown in Fig. 9(b). Since the moon is near the perigee of the epicycle, the effective radius of the epicycle is directed toward the observer, and a change in its length will have little effect on the lunar equation. Thus the effect on r may be neglected. However, any displacement

---

[31] Delambre (1819), 276–77, in commenting on Nunes's investigation of the precise values from Ptolemy's theory, finds $c_{3\,\max} = 13;8,6°$ at $2\bar{\eta} = 114;3,48°$. He remarks that Nunes finds 13;8,7° at 114;3,53°, which is close enough, although "La solution de Nonius occupe onze pages in-folio." Cf. Grafton (1973), 535 n. 17.

of the moon from line C_1C_2 to the second epicycle will have its greatest effect under these conditions, so we must find c_3 and examine its effect on the lunar equation. For $2\bar{\eta} = 4;46^\circ$, $c_3 = 1;19^\circ$, which means that $\alpha = \bar{\alpha} + c_3 = 180^\circ + 1;19^\circ = 181;19^\circ$. Using the least radius of the epicycle $r = 8604$, we find for the lunar equation $c = 0;7,26^\circ$, compared with $c = 0^\circ$ when c_3 is neglected. The same configuration in Ptolemy's model produces an equation of 0;4°, which he remarks corresponds to $\frac{1}{8}^{\text{h}}$ (since the moon moves about $\frac{1}{2}^\circ$ in one hour), and considers to be within the limits of observational error. The difference in Copernicus's model, however, amounts to nearly $\frac{1}{4}^{\text{h}}$, which is not negligible in determining the time of a lunar eclipse. Consequently, in computing the eclipses for finding the mean argument of latitude and its position at epoch—where the error can be doubled between two eclipses—he correctly includes the effect of the second inequality. Doubtless he carried out an analysis of this kind following Ptolemy, and determined the possible errors resulting from omitting the second inequality.[32]

Trigonometric Computation of a Lunar Position (IV,10)

In order to demonstrate the use and accuracy of his lunar theory, Copernicus recomputes an observation of the moon by Hipparchus reported in *Almagest* V,5 and *Epitome* V,8 that was used by Ptolemy for deriving the eccentricity producing the maximum effect of the prosneusis. The observation, number (10) in Table 2, was made in E.A. 197 Payni 17, $3\frac{1}{3}$ seasonal hours after noon in Rhodes, which Ptolemy converts to 4 equal hours. Hipparchus observed the sun at ♋11 − $\frac{1}{10}^\circ$ = ♋10;54° and the moon at about ♌29°, and thus the observed elongation was 48;6°. Copernicus assumes that the observation was made with a spherical astrolabe. Ptolemy says that at this time the moon showed no parallax in longitude, meaning that it was in the nonagesimal, the 90th degree of the ecliptic from the horoscopus. In the notation of Chapter 2, $\lambda_{\text{N}} = \lambda_{\text{G}} - 90^\circ$.[33]

Copernicus elaborates upon the description of the observation, and it is of interest to check his statements as an application of his spherical astronomy. For a solar longitude of ♋10;54° at 4^{h} after noon, by Chapter 2 (5.), the right ascension of the midheaven,

$$\alpha_{\text{M}} = \alpha(\lambda_{\odot}) + t \cdot 15^\circ = 101;53^\circ + 60^\circ = 161;53^\circ,$$

[32] It should also be noted that the effect of the second inequality on $\bar{\alpha}$ can amount to c_3 for $2\bar{\eta} = 14;38^\circ$, which is nearly 4°, a nontrivial quantity when correcting the mean anomaly as Copernicus does in IV,6. It is probably for this reason that he chooses eclipses (2) and (5), in which $\bar{\eta}$ is nearly the same in each, 182;47° and 182;51° respectively. Thus, c_3 has about the same effect in both, −0;46° and −0;47°, so that $\Delta\alpha \approx \Delta\bar{\alpha}$.

[33] Above, p. 116. See the discussion of the observation in *HAMA*, 91–93.

so that the longitude of the midheaven $\lambda_M = ♍10;18°$. Copernicus gives ♍10°. From Chapter 2 (7.), the oblique ascension of the horoscopus,

$$\rho_G = \alpha_M + 90° = 161;53° + 90° = 251;53°.$$

The latitude of Rhodes, $\varphi = 36°$, is not contained in Copernicus's oblique ascension tables, which begin at $\varphi = 39°$. One could solve for the corresponding longitude trigonometrically, but it is easier, and more likely what Copernicus did, to use the tables of oblique ascension in Regiomontanus's *Tabulae directionum* where, in the table for $\varphi = 36°$, for $\rho_G = 251;53°$, we find $\lambda_G = ♏28;50°$. Copernicus gives ♏29°. The longitude of the nonagesimal,

$$\lambda_N = \lambda_G - 90° = ♏29° - 90° = ♌29°,$$

and since the moon was observed in ♌29°, it may be concluded that it was without parallax in longitude.[34]

The time since noon in Rhodes was 4^h, which Copernicus says was $3\frac{1}{6}^h$ in Cracow, Rhodes now being taken as $\frac{1}{6}^h = 2;30°$ west of Alexandria. Note that throughout Book III Copernicus followed Ptolemy's assumption that Rhodes and Alexandria were on the same meridian, 1^h east of Frauenburg or Cracow.[35] In the case of the solar theory, a difference of $\frac{1}{6}^h$ hardly mattered, but in the case of the moon it corresponds to a lunar motion of about 0;5°, and thus Copernicus correctly takes it into account. Thus, from the Epoch of Alexander, the apparent time at the meridian of Cracow–Frauenburg was

$$\Delta t = 196^y\ 286^d\ 3;10^h,$$

for which

$$\bar{\lambda}'_\odot = ♋12;3°, \qquad \lambda_\odot = ♋10;40°.$$

Ptolemy gives 12;5° and 10;40° respectively. Note that the true longitude is 0;14° lower than Hipparchus's measurement. The equation of time, taken for the interval from Era Alexander, is

$$\Delta E = (\alpha - \bar{\lambda})_{10} - (\alpha_0 - \bar{\lambda}_0)_C = +1;56° = +0;7,44^h.$$

Copernicus takes $\Delta E = +0;10^h = +2;30°$, and gives the mean time,

$$\Delta t = 196^y\ 286^d\ 3;20^h = 3,16^y\ 4,46;8,20^d,$$

for which the mean elongation and anomaly are

$$\bar{\eta} = \bar{\eta}_0 + \Delta\bar{\eta} = 310;44° + 94;20,40° = 45;4,40° \approx 45;5°,$$

$$\bar{\alpha} = \bar{\alpha}_0 + \Delta\bar{\alpha} = 85;41° + 247;21,47° = 333;2,47° \approx 333°.$$

---

34 Using the *Tabulae directionum* for the entire computation, which Copernicus may well have done, gives the same results: $\lambda_M = ♍10°$, $\lambda_G = ♏29°$, $\lambda_N = ♌29°$.

35 Above, p. 149.

The configuration of the model is shown in Fig. 10 for

$$2\bar{\eta} = 90;10^\circ, \qquad \bar{\alpha} = 333^\circ.$$

First, in triangle C_1C_2P, where $r_1 = 1097$ and $r_2 = 237$,

$$C_1P = r = (r_1^2 + r_2^2 - 2r_1r_2 \cos 2\bar{\eta})^{1/2} = 1123,$$

$$c_3 = \sin^{-1}(r_2 \sin 2\bar{\eta}/r) = +12;11^\circ.$$

Therefore,

$$\alpha = \bar{\alpha} + c_3 = 333^\circ + 12;11^\circ = 345;11^\circ,$$

$$\gamma = \alpha - 180^\circ = 165;11^\circ,$$

and in triangle OC_1P, where $R = 10{,}000$ and $r = 1123$,

$$OP = (R^2 + r^2 - 2Rr \cos \gamma)^{1/2} = 11089,$$

$$c = \sin^{-1}(r \sin \gamma/OP) = 1;29^\circ,$$

which is positive since $\alpha > 180^\circ$. Therefore, the true elongation of the moon from the mean sun

$$\eta = \bar{\eta} + c = 45;5^\circ + 1;29^\circ = 46;34^\circ,$$

and since the mean sun was at ♋12;3°, the true longitude of the moon

$$\lambda_☾ = \bar{\lambda}'_☉ + \eta = ♋12;3^\circ + 46;34^\circ = ♌28;37^\circ.$$

If now Copernicus compared this with the position observed by Hipparchus, ♌29°, there would be an error of −0;23°. Instead, he takes the computed solar longitude, ♋10;40° in preference to Hipparchus's observed ♋10;54°, and finds the true elongation from the true sun

$$\eta' = \lambda_☾ - \lambda_☉ = ♌28;37^\circ - ♋10;40^\circ = 47;57^\circ.$$

But from Hipparchus's observation

$$\eta' = ♌29^\circ - ♋10;54^\circ = 48;6^\circ,$$

so the computed elongation disagrees with the observation by −0;9°. What Copernicus has done here might seem improper, but it is not. All he is interested in testing is the ability of his lunar model to reproduce an observed true elongation of the moon from the true sun. Thus, it does not matter where the sun is taken to be or what the longitude of the moon turns out to be—these both depend upon the accuracy of the solar theory, which is not here being tested.

One would think that an error of 0;9° was not worth worrying about, but Copernicus can even explain it. Since the orbit of the moon is inclined to the ecliptic, an arc along the inclined circle will not generally be equal to an arc of longitude on the ecliptic. It is exactly the same as the reduction from longitude to right ascension in the equator. If we call the distance of the moon from the northern limit of latitude ω, and the inclination of the orbit ι, then the distance $\Delta\lambda$ from the limit along the ecliptic is given by

$$\cot \Delta\lambda = \cot \omega \cos \iota. \tag{10}$$

Copernicus says that at the time of the observation, the moon was near the midpoint of the quadrant between the southern limit of latitude ($\omega = 180°$) and the ascending node ($\omega = 270°$), and that this location causes an increase of 0;7° when reduced to the ecliptic. Anticipating Copernicus's latititude theory, we find that at the time of the observation

$$\omega = \bar{\omega} + c = 222;6° + 1;29° = 223;35°,$$

very close to the midpoint of the quadrant at 225°. Therefore, letting $\iota = 5°$,

$$\Delta\lambda = \cot^{-1} (\cot 223;35° \cos 5°) = 223;42°,$$

and just as Copernicus says

$$\Delta\lambda - \omega = +0;7°.$$

This means that the longitude reduced to the ecliptic, which is presumably what Hipparchus measured—since the spherical astrolabe measures longitudes parallel to the ecliptic—must be increased from 47;57° to 48;4°, leaving a discrepancy between observation and computation of only −0;2°. Copernicus calls this a "remarkable agreement." We concur.

Correction of the Mean Argument of Latitude (V,13)

Ptolemy's method of finding the mean argument of latitude is treated in *Almagest* IV,9 and *Epitome* IV,15, and Copernicus follows the method with two modifications. Its principle is to find two lunar eclipses as far apart as possible in which the moon has completed an integral number of revolutions in its inclined orbit with respect to a node. For this to be the case, four conditions must be met:

(1) The eclipses must be near the same node, ascending or descending.
(2) The moon must be eclipsed from the same side, north or south.
(3) The eclipses must be partial, and of the same magnitude.
(4) The anomaly of the moon must be the same or symmetrical to the apogee of the epicycle.

The last condition insures that the distances of the moon, and thus the apparent diameters of the moon and shadow, are the same in both eclipses. Copernicus mentions a fifth condition that makes very little difference, and would be hard to fulfill along with the other four.

(5) The eclipses should be at the same longitude (correctly, the anomaly of the sun should be the same).

The diameter of the shadow also varies as a function of the solar anomaly on account of the changing distance between the earth and sun. Strictly, this implies that the solar anomaly should be the same in each eclipse, but rather less precisely, Copernicus says that the longitude (*locus*) should be the same, which would only be true if the apsidal line were fixed.[36]

As it happens, he cannot find two eclipses that fulfill the first four conditions, but only two near opposite nodes in which the moon is eclipsed on opposite sides by about the same magnitude, meaning that an integral number plus about one-half revolutions of the argument of latitude have taken place. Thus the modified conditions are (see Fig. 11):

(1a) The eclipses are at opposite nodes, indicating about $(n + \frac{1}{2})^r$ in the argument of latitude.
(2a) The moon is eclipsed from opposite sides, indicating that the moon is on opposite sides of the nodes, i.e. yet closer to $(n + \frac{1}{2})^r$.

Conditions (3), magnitude, and (4), anomaly, are the same, but Copernicus can find only an approximation to (3), so he must make a further adjustment.

A. The first eclipse (7), from E.A. 150, of 7′ from the north, is taken from *Almagest* VI,5 and *Epitome* VI,6 where it is used to find the maximum apparent diameter of the moon and shadow near least distance in syzygy. Copernicus says that the true sun was in the sixth degree of Taurus. By computation from his theory for $\Delta t = 149^y\ 206^d\ 13\frac{1}{3}^h$ from E.A. 1, we find

$$\bar{\lambda}'_{\odot} = \text{♉}\, 5;4^\circ, \qquad \lambda_{\odot} = \text{♉}\, 6;13^\circ,$$

which compares well with Ptolemy's reported $\lambda_{\odot} = \text{♉}\, 6\frac{1}{4}^\circ$, and even better with computation from Ptolemy's theory, which gives ♉ 6;13°. With respect to E.A. 1, the equation of time,

$$\Delta E = (\alpha - \bar{\lambda})_7 - (\alpha_0 - \bar{\lambda}_0)_C = -0;56^\circ + 2;7^\circ = +1;11^\circ = +0;4,44^h.$$

Copernicus somehow finds $\Delta E = +0;10^h$, and gives for the interval of mean time,

$$\Delta\bar{t} = 149^y\ 206^d\ 13;30^h = 2,29^y\ 3,26;33,45^d,$$

for which, by computation,

$$\bar{\eta} = 182;42^\circ \qquad \bar{\alpha} = 163;35^\circ.$$

Copernicus gives $\bar{\alpha} = 163;33^\circ$, which, he says, nearly agrees with Ptolemy, who gives 163;40°. For $2\bar{\eta} = 5;24^\circ$,

$$c_3 = +1;29^\circ, \qquad \alpha = \bar{\alpha} + c_3 = 165;2^\circ,$$
$$c = -1;23^\circ, \qquad \eta = \bar{\eta} + c = 181;19^\circ.$$

[36] On the variation of the shadow as a function of solar anomaly, see below, pp. 251–254. It amounts to less than 0;1°.

B. The second eclipse (8), from 1509, was about 8′ from the south. The apparent time of the eclipse from E.A. 1 is $1832^y\ 295^d\ 11;45^h$,[37] for which

$$\bar{\lambda}'_{\odot} = \text{♊}20;29°, \qquad \lambda_{\odot} \approx \text{♊}21°.$$

The equation of time,

$$\Delta E = (\alpha - \bar{\lambda})_8 - (\alpha_0 - \bar{\lambda}_0)_C = 0;18° + 2;7° = +2;25° = +0;9,40^h,$$

which Copernicus rounds to $0;10^h$, and gives as the mean interval

$$\Delta\bar{t} = 1832^y\ 295^d\ 11;55^h = 30,32^y\ 4,55;29,47,30^d.$$

For this time, by computation,

$$\bar{\eta} = 182;18°, \qquad \bar{\alpha} = 159;55°,$$

$$c_3 = +1;16°, \qquad \alpha = \bar{\alpha} + c_3 = 161;11°, \quad \text{(Copernicus 161;13°)}$$

$$c = -1;44°, \qquad \eta = \bar{\eta} + c = 180;34°.$$

For the first eclipse, Ptolemy had found that the true argument of latitude from the northern limit $\omega = 98;20°$, showing that the moon was *below* the descending node. In order to find that the second eclipse was in fact above the ascending node, Copernicus would need some provisional value for the argument of latitude. For example, using the *Alfonsine Tables*, for 1509, Jun 3, 0^h, the mean argument of latitude $\bar{\omega} = 275;29°$, which is sufficient to confirm that the moon was *above* the ascending node. Thus, conditions (1a) and (2a) are satisfied. The anomaly in the first eclipse was 165;2°, and in the second 161;13°, so condition (4) is met. Copernicus also points out that the sun was near apogee in both eclipses—it is in fact within 30° of the apogee—so even condition (5) is approximately satisfied. However, the magnitude of the first eclipse was 7′ and the second 8′, leaving condition (3) unfulfilled. Therefore, an adjustment must be made.

If the eclipses were of the same magnitude, the difference in the argument of latitude $\Delta\omega = 180°$, and in Fig. 12, they would lie on the straight line AB at equal distances from the nodes. But since eclipse A is 1′ less than eclipse B, it is farther from the node, say at A', and $\Delta\omega' = \text{arc } A'B < 180°$. To estimate the difference, Copernicus observes that the apparent diameter of the moon is about 0;30°, one-twelfth of which, corresponding to 1′, is 0;2,30°. Then since the inclination of the lunar orbit $\iota = 5°$, a difference d of 0;2,30° in the eclipsed part of the moon corresponds to a difference in the argument of latitude δ of

$$\delta = d/\sin\iota = 0;2,30°/\sin 5° = 0;28,41° \approx 0;30°.$$

Thus $AA' = \delta = 0;30°$ and $\Delta\omega' = \Delta\omega - \delta = 179;30°$.

What Copernicus has done here may appear innocuous, but it is not. The elegance of Ptolemys method was that it required no assumptions about

[37] N 115v27, M 123v22 by error *tribus quintis*, i.e. $0;36^h$; but correctly $0;45^h$ at N 116r3, M 124r1.

the inclination of the lunar orbit or the apparent diameter of the moon. By introducing estimates of the quantities, Copernicus could be introducing errors into $\Delta\omega$, since $\delta \approx 11\frac{1}{2}d$, that will seriously affect the correction to the provisional value of $\bar{\omega}$. But there is a further problem in that eclipse magnitudes are notoriously hard to estimate. In the present case, the magnitudes were not 7′ and 8′, but 7.4′ and 7.7′.[38] Hence the difference was not 1′ but $\frac{1}{3}'$, and δ should be 0;10° instead of 0;30°. Since Copernicus's correction to $\bar{\omega}$ turns on a difference of 0;33°, an error of 0;20° makes the whole operation rather meaningless.

In the first eclipse $c_A = -1;23°$, and in the second $c_B = -1;44°$. Thus the mean argument of latitude between the eclipses,

$$\Delta\bar{\omega} = \Delta\omega' - (c_B - c_A) = 179;30° - (-0;21°) = 179;51°.$$

The interval of apparent time,

$$\Delta t = 1683^y\ 88^d\ 22;25^h,$$

while Copernicus gives erroneously 22;35^h, and the equation of time

$$\Delta E = (\alpha - \bar{\lambda})_8 - (\alpha - \bar{\lambda})_7 = +1;14° = +0;4,56^h \approx +0;5^h,$$

although Copernicus says that the mean time and apparent time were the same. He then concludes the chapter by saying that during this time $\Delta\bar{\omega} = 22577^r + 179;51°$, which agrees with the values set out earlier. This, however is not true, since from Copernicus's tables in IV,4, for the correct Δt, $\Delta\bar{\omega} = 179;49°$, an error of $-0;2°$, and for Copernicus's incorrect Δt, $\Delta\bar{\omega} = 179;54°$, an error of $+0;3°$. Curiously M (124r25) originally read 179;54°, evidently computed with the incorrect Δt, and this was altered to 179;51° to agree with the motion derived from the observations.

But there is more trouble to come. Let us now try to find the annual correction given in IV,4 (see Table 1, line 5). For the correct mean and apparent time between the eclipses $\Delta t = 1683^y\ 88;56,2,30^d$, using the provisional value of $\bar{\omega}$,

$$\Delta\bar{\omega}_h = \Delta t \cdot \bar{\omega}_h = 22577^r + 180;18°,$$

differing from $\Delta\bar{\omega}$ between the eclipses by

$$\delta(\bar{\omega}_h) = \Delta\bar{\omega} - \Delta\bar{\omega}_h = 179;51° - 180;18° = -0;33°.$$

Letting $\Delta t = 1683^y$, the correction for each year,

$$d(\bar{\omega}_h) = -0;33°/\Delta t = -0;0,1,10,35°,$$

while in IV,4 Copernicus gives $-0;0,1,2,42°$, which, added to the provisional value, gives the value used for the tabulation. On the other hand, the incorrect Δt leads to a correction of $-0;0,0,57,45^{°/y}$, which is also not helpful,

[38] Oppolzer (1887). Copernicus seems aware of some imprecision since he describes eclipse (8) as "very nearly" (*proxime*) 8′.

while Copernicus's correction implies an interval of time close to 22;32^h. Then the daily motion should follow from dividing the annual motion by 365^d, and should be $\bar{\omega}^d = 13;13,45,39,29^{\circ/d}$. However, Copernicus gives 13;13,45,39,22$^{\circ/d}$, once again carrying over an earlier value from M, just as he did for $\bar{\alpha}$.

Mean Argument of Latitude at Epoch (IV,14)

The previous demonstration shows the mean motion of the moon with respect to a node or limit of its inclined orbit, but does not establish a value at a particular time. In order to find a specific position of the moon with respect to a node, Ptolemy uses two eclipses at the same distance, on the same side, of opposite nodes (see Fig. 13). Finding $\Delta\omega$ between the eclipses, it follows that $\frac{1}{2}(180^\circ - \Delta\omega)$ is the true distance of the moon from its node in each eclipse, from which its mean position and position at epoch may be determined. The method is explained in *Almagest* IV,9 and *Epitome* IV,16 and Copernicus follows it without modification.

Unlike the previous demonstration of the mean argument of latitude, it is not necessary that the eclipses be separated by a long interval, but it is sufficiently hard to find two eclipses meeting the required conditions that Ptolemy uses eclipses about 245 years apart, and Copernicus uses a pair 1366 years apart. The moon must be eclipsed:

(1) near opposite nodes, ascending and descending;
(2) from the same side, north or south;
(3) by the same magnitude;
(4) at equal distances from the apogee of the epicycle.

By checking eclipses in the *Almagest* with Ptolemy's tables, and by checking modern eclipses with the *Alfonsine Tables*, Copernicus can find two eclipses meeting these conditions.

A. The first eclipse is the second of the three observed by Ptolemy (2), and was used previously to find the radius of the epicycle at syzygy. Hence, it has already been determined that

$$\bar{\eta} = 182;47^\circ, \qquad \bar{\alpha} = 64;38^\circ, \qquad c = -4;20^\circ.$$ [39]

The moon was eclipsed by 10′ from the north, and from the tables in the *Almagest*, $\bar{\omega} \approx 100^\circ$, about 10° below the descending node.

B. The second eclipse (9) was observed by Copernicus at Rome on 6 Nov 1500. The eclipse was 10′ from the north, and according to the *Alfonsine Tables*, $\bar{\omega} \approx 261^\circ$, about 9° below the ascending node. Likewise, from the *Alfonsine Tables* $\bar{\alpha} \approx 295;43^\circ = -64;17^\circ$, which is nearly symmetrical to $\bar{\alpha} = 64;38^\circ$ in the first eclipse. The same estimate can of course be made with

[39] Above, p. 206.

Copernicus's theory. Thus, all four conditions are satisfied. The eclipse was observed at 2 hours after midnight in Rome. Copernicus takes Cracow–Frauenburg to be $5^\circ = 0;20^h$ east of Rome, so that the time in Cracow was $2;20^h$ after midnight, and the interval of apparent time from E.A. 1, $\Delta t = 1824^y\ 84^d\ 14;20^h$. From the solar theory,

$$\bar{\lambda}'_{\odot} = ♏24;34^\circ, \qquad \lambda_{\odot} = ♏23;16^\circ,$$

and thus the equation of time,

$$\Delta E = (\alpha - \bar{\lambda})_9 - (\alpha_0 - \bar{\lambda}_0)_C = -0;58^\circ = -0;3,52^h \approx -0;4^h,$$

which gives a mean time of

$$\Delta\bar{t} = 1824^y\ 84^d\ 14;16^h = 30,24^y\ 1,24;35,40^d.$$

From the lunar theory,

$$\bar{\eta} = 174;15,24^\circ \approx 174;16^\circ, \qquad \bar{\alpha} = 294;44^\circ, \quad (\text{N } 294;40^\circ)$$

$$c_3 = -3;9^\circ, \quad (\text{Corr. } -3;7^\circ), \qquad \alpha = \bar{\alpha} + c_3 = 291;35^\circ,$$

$$c = +4;28^\circ, \qquad \eta = 178;44^\circ.$$

The interval of apparent time between the eclipses $\Delta t = 1366^y\ 358^d\ 4;20^h$, and the equation of time

$$\Delta E = (\alpha - \bar{\lambda})_9 - (\alpha - \bar{\lambda})_2 = +1;9^\circ = +0;4,36^h.$$

Copernicus takes $\Delta E = +0;4^h = +1;0^\circ$, and gives the interval of mean time

$$\Delta\bar{t} = 1366^y\ 358^d\ 4;24^h = 22,46^y\ 5,58;11^d.$$

For this interval, the mean argument of latitude

$$\Delta\bar{\omega} = 159;54,59^\circ \approx 159;55^\circ.$$

Therefore, in Fig. 14, letting the first eclipse be at A and the second at B,

$$\text{arc } \bar{A}\bar{B} = \Delta\bar{\omega} = 159;55^\circ,$$

$$c_A = -4;20^\circ, \qquad c_B = +4;28^\circ,$$

and the true argument of latitude between the eclipses,

$$\Delta\omega = AOB = \Delta\bar{\omega} + (c_B - c_A) = 159;55^\circ + 8;48^\circ = 168;43^\circ.$$

Since the moon is taken to be equidistant from the node in each eclipse, the nodal distance

$$v = \tfrac{1}{2}(180^\circ - 168;43^\circ) = 5;38,30^\circ \approx 5;39^\circ,$$

and the mean argument of latitude measured from the northern limit N at the first eclipse

$$\bar{\omega}_A = 90^\circ + v - c_A = 90^\circ + 5;39^\circ + 4;20^\circ = 99;59^\circ.$$

Ptolemy's observation followed E.A. 1 by an apparent time of $\Delta t = 457^y\ 91^d\ 10^h$. For this interval, the equation of time,

$$\Delta E = (\alpha - \bar{\lambda})_2 - (\alpha_0 - \bar{\lambda}_0)_C = -2;7^\circ = -0;8,28^h.$$

Copernicus takes $\Delta E = -0;6^h = -1;30^\circ$, and gives for the mean time

$$\Delta\bar{t} = 457^y\ 91^d\ 9;54^h = 7,37^y\ 1,31;24,45^d,$$

for which

$$\Delta\bar{\varpi} = 50;58,33^\circ \approx 50;59^\circ.$$

Thus at E.A. 1,

$$\bar{\varpi}_{0C} = 99;59^\circ - 50;59^\circ = 49^\circ.$$

The remaining epochs are computed in the order C. → A. → D. → G. All the equations of time, intervals of mean time, mean motions between epochs, and positions at epoch are computed correctly. The results have already been tabulated along with the epochs of the mean elongation and mean anomaly.[40]

The Tables and Their Use (IV,4,11,12)

The tables of mean motions in IV,4 are based upon the following values for Egyptian years and days:

	Annual	Daily	
$\bar{\eta}$	$2,9;37,22,36,25^{\circ/y}$	$12;11,26,41,31^{\circ/d}$	
$\bar{\alpha}$	1,28;43,9,7,15	13;3,53,56,30	(11)
$\bar{\varpi}$	2,28;42,45,17,21	13;13,45,39,22	

As already mentioned, all three values of the daily motions were carried over from earlier versions in M (111–113) that resulted from dividing previous, rejected annual motions by 365^d. In the case of $\bar{\eta}$, there is no difference in the fourth fractional place, but the last place of $\bar{\alpha}$ is in error by -4 and of $\bar{\varpi}$ by -7. Copernicus does not tabulate the mean sidereal or tropical motion of the moon, but only its mean elongation. This means that to find the longitude of the moon it is always necessary first to compute the adjusted mean longitude of the sun, $\bar{\lambda}'_{\odot}$, and then add to it the corrected elongation η. Since the mean and true longitude of the sun are both required

40 Above, p. 212.

to find the equation of time, which must always be found before computing a lunar position, one normally computes the longitude of the sun along with the longitude of the moon. The lunar mean motions are tabulated for 1–60 years and for 1–60 days. To find their value at a given mean time since epoch—the use of mean time is essential—for $\Delta\bar{t}$ find $\Delta\bar{\eta}$, $\Delta\bar{\alpha}$, $\Delta\bar{\omega}$, and form

$$\bar{\eta} = \bar{\eta}_0 + \Delta\bar{\eta}, \qquad \bar{\alpha} = \bar{\alpha}_0 + \Delta\bar{\alpha}, \qquad \bar{\omega} = \bar{\omega}_0 + \Delta\bar{\omega}. \tag{12}$$

The table of corrections in IV,11 follows a brief description of its arrangement and calculation that in turn rests upon the example for Hipparchus's observation of -126 given in IV,10. The first two columns are the arguments, for 3°–180° and 180°–357° at intervals of 3°, and are used for entering the remaining columns with arguments $\bar{\eta}$, α, and ω. Columns 3–6 are corrections for finding the true elongation of the moon from the sun η. They are graphed in Fig. 15 (along with c_3 and c_6 from the *Almagest*), and shown in position on the lunar model in Fig. 16.

Column 3, $c_{3(2\bar{\eta})}$, called in the table "correction of epicycle b," is the correction of the mean anomaly corresponding to the prosneusis in Ptolemy's model (Alm. c_3 in Fig. 15). As previously applied in IV,10, it is computed from

$$c_3 = \sin^{-1}(r_2 \sin 2\bar{\eta}/r), \tag{13}$$

where the instantaneous radius of the epicycle,

$$r = (r_1^2 + r_2^2 - 2r_1 r_2 \cos 2\bar{\eta})^{1/2}. \tag{14}$$

Since $r_1 = 1097$ and $r_2 = 237$,

$$c_{3\max} = \sin^{-1}(r_2/r_1) = 12;28,36° \approx 12;29°,$$

but for $2\bar{\eta} = 78°$, Copernicus gives 12;28° (corr. 12;28,34°), and c_3 is frequently in error by $\pm 0;1°$. Using more accurate values $r_1 = 1097.25$ and $r_2 = 236.85$, which would follow from finding the radii to six places, gives $c_{3\max} = 12;27,57° \approx 12;28°$, but discrepancies follow when other values are checked. Note that c_3 and *Alm.* c_3 can differ by over 4°, but are about equal at octants where $2\bar{\eta} = 90°$. c_3 is used to correct the mean anomaly $\bar{\alpha}$ to the true or corrected anomaly

$$\alpha = \bar{\alpha} + c_{3(2\bar{\eta})}, \tag{15}$$

where c_3 is positive for $2\bar{\eta} \leq 180°$ and negative for $2\bar{\eta} \geq 180°$.

Column 5, $c_{5(\alpha)}$, called the "correction of epicycle a," is the equation of the first inequality, that is, in Fig. 17, for the least radius of the epicycle $r_{\min} = r_1 - r_2 = C_1B$, when $2\bar{\eta} = 0°$ at mean syzygy. In principle, it is computed from

$$\sin c_5 = r \sin\alpha/(R^2 + r^2 + 2Rr\cos\alpha)^{1/2}, \tag{16}$$

where $R = 100{,}000$ and $r = 8604$. However, Copernicus did not have to compute it at all. It was shown earlier that the maximum equation at syzygy,

$$c_{5\,\max} = \sin^{-1}(8604/100{,}000) = 4;56^\circ,$$

is identical to the corresponding maximum lunar equation in the *Alfonsine Tables*, as indirectly acknowledged by Copernicus in his remark that his epicycle radius "agrees with what we find reported by most of those who preceded us since the time of Ptolemy."[41] In the *Alfonsine Tables*, the correction is tabulated to seconds at intervals of 1° in column 6 of the lunar correction table. All Copernicus had to do to form his c_5 was round to minutes at 3° intervals, and every value in his table is compatible with this source.[42] c_5 is more accurately computed than c_3.

Column 6, $c_{6(\alpha)}$, called the "excess," is the difference to be added to c_5 for the maximum effect of the second inequality, in Fig. 17 for the greatest radius of the epicycle $r_{\max} = r_1 + r_2 = C_1D$, when $2\bar{\eta} = 180^\circ$ at mean quadrature. In principle, c_6 is computed by first finding a provisional c'_{5+6} from the formula used for c_5, but with $r = 1334$, and then subtracting c_5 to find $c_6 = c'_{5+6} - c_5$. The maximum value is therefore $c_{6\,\max} = 7;40^\circ - 4;56^\circ = 2;44^\circ$. Unfortunately, it is not possible to recover most of the values of c_6 in the table by this method, and even where the sum $c_5 + c_6$ should reach a maximum of 7;40°, the table gives only 7;38°. It appears as though Copernicus used a less laborious way of computing c_6—consider his method of finding c_5—but while it is easy enough to find ways of coming within 0;1° of his values, we have found no computation that always agrees precisely. In any case, his error from the correct method of computation does not appear to exceed 0;2°, although it is unfortunate that the maximum error occurs exactly at the maximum equation.

Column 4, $c_{4(2\bar{\eta})}$, called "proportional minutes," is the coefficient of interpolation for values of $2\bar{\eta}$ between 0° and 180°, and is computed in the same way as the corresponding column, $c_{4(\vartheta)}$, in the solar theory.[43] If $r_{(2\bar{\eta})}$ is the effective radius of the epicycle for a given value of $2\bar{\eta}$, which has been calculated previously from (14) for finding c_3, then

$$c_{\max(2\bar{\eta})} = \sin^{-1}(r_{(2\bar{\eta})}/R) \qquad (17)$$

will be the maximum equation for a given $2\bar{\eta}$. Let the extreme values for $2\bar{\eta} = 0^\circ$ and $2\bar{\eta} = 180^\circ$ be

$$c_m = c_{\max(0^\circ)} = c_{5\,\max} = 4;56^\circ, \qquad c_M = c_{\max(180^\circ)} = 7;40^\circ.$$

[41] Above, p. 208.

[42] For 48°, 3;26,30° is rounded to 3;26°, and for 57°, 3;56,30° is rounded to 3;56°.

[43] Above, p. 169.

Then for any value of $2\bar{\eta}$, $c_{4(2\bar{\eta})}$ is the ratio

$$c_{4(2\bar{\eta})} = \frac{c_{\max(2\bar{\eta})} - c_{\mathrm{m}}}{c_{\mathrm{M}} - c_{\mathrm{m}}} = \frac{c_{\max(2\bar{\eta})} - c_{5\max}}{c_{6\max}}$$

$$= \frac{c_{\max(2\bar{\eta})} - 4;56^{\circ}}{2;44^{\circ}}. \tag{18}$$

$c_{4(2\bar{\eta})}$ is tabulated sexagesimally from 0′ to 60′ as $2\bar{\eta}$ varies from 0° to 180°, and conversely for 180° to 360°.[44]

Columns 4, 5, and 6 are always used together to find the lunar equation $c_{(\alpha, 2\eta)}$ from

$$c_{(\alpha, 2\bar{\eta})} = c_{5(\alpha)} + c_{4(2\bar{\eta})} \cdot c_{6(\alpha)}. \tag{19}$$

The true elongation η of the moon from the *mean* sun then follows from

$$\eta = \bar{\eta} + c_{(\alpha, 2\bar{\eta})}, \tag{20}$$

where $c_{(\alpha, 2\bar{\eta})}$ is negative for $\alpha \leq 180°$ and positive for $\alpha \geq 180°$. The true tropical longitude of the moon is computed by adding to η the adjusted mean longitude of the sun, that is (see Fig. 18),

$$\lambda_{☾} = \bar{\lambda}'_{\odot} + \eta, \tag{21}$$

and the true sidereal longitude, which is seldom of interest, by adding the mean sidereal longitude of the sun,

$$\lambda^*_{☾} = \bar{\lambda}^*_{\odot} + \eta. \tag{22}$$

It is of interest to note that nowhere in the lunar theory does Copernicus use sidereal longitude, even when analyzing an occultation of a fixed star in IV,27.[45] Finally, the true elongation of the moon from the *true* sun is found from

$$\eta' = \lambda_{☾} - \lambda_{\odot} = \eta - c_{\odot}, \tag{23}$$

where $c_{\odot} = c_{(\alpha, \vartheta)}$ of the solar theory. It is η' that is of interest in the computation of eclipses, and we shall use it as a test of Copernicus's lunar theory.[46]

Column 7, $c_{7(\omega)}$, is the latitude of the moon, which Copernicus assumes to reach a maximum of 5°, the value used by Hipparchus, confirmed by Ptolemy in *Almagest* V,12, and ubiquitous in medieval Arabic and European

---

[44] There are frequent errors of ± 1. To insure accurate interpolation, c_4 should be tabulated to two places, as Ptolemy does in *Almagest* V,8 and as Copernicus does for the corresponding coefficient for the planets in V,33 (below, pp. 451–452).

[45] Below, pp. 266–269.

[46] Below, p. 231.

lunar theory. In order to use column 7, one first finds the true argument of latitude from the northern limit (see Fig. 19),

$$\omega = \bar{\omega} + c_{(\alpha,\, 2\bar{\eta})}, \tag{24}$$

where $c_{(\alpha,\, 2\bar{\eta})}$ is the same correction found for the elongation, and is negative for $\alpha \leq 180°$ and positive for $\alpha \geq 180°$. With ω as argument, one reads directly in column 7 the latitude

$$\beta = c_{7(\omega)}, \tag{25}$$

where β is northern, positive for $270° \leq \omega \leq 90°$ and southern, negative for $90° \leq \omega \leq 270°$. In principle, column 7 is computed from

$$\sin c_7 = \cos \omega \sin \iota, \tag{26}$$

where $\iota = 5°$. However, Copernicus had no need to compute the column. The *Alfonsine Tables* give lunar latitudes for $\iota = 5°$ to seconds of arc at 1° intervals counted from the ascending node at $\omega = -90°$. A comparison shows that Copernicus's column 7 was formed by rounding the corresponding latitudes from the *Alfonsine Tables* to minutes at intervals of 3°. Thus, two columns of the lunar correction tables, 5 and 7, are rounded from the *Alfonsine Tables*, saving Copernicus some superfluous work and the possibility of computational errors.

Example. In Bologna on 9 Mar 1497 at the end of the fifth equal hour of night, Copernicus observed the moon to occult the star Aldebaran (α Tauri). This is observation (13) in Table 2, and appears in IV,27. The apparent time elapsed from the Epoch of Christ is $1497^y\ 76^d\ 23^h$ at Bologna, but since Copernicus takes Cracow to be about $9° = 0;36^h$ to the east, the apparent time at the meridian of Cracow–Frauenburg is $1497^y\ 76^d\ 23;36^h$. For this time, compute the longitude and latitude of the moon.

Copernicus says that $\lambda_\odot = ♓ 28;30°$, and we find

$$\bar{\lambda}'_\odot = ♓ 26;39°, \qquad \lambda_\odot = ♓ 28;30°,$$

in perfect agreement. From the information in Table 3, the equation of time

$$\Delta E = (\alpha - \bar{\lambda})_{13} - (\alpha_0 - \bar{\lambda}_0)_G = +2;35° - 1;53° = +0;42° = +0;2,48^h,$$

although Copernicus gives $+0;4^h$, evidently following from his incorrect value of $(\alpha_0 - \bar{\lambda}_0)_G = +1;33°$, thus

$$\Delta E = +2;35° - 1;33° = +1;2° = +0;4,8^h \approx +0;4^h.$$

The interval of mean time is therefore

$$\Delta\bar{t} = 1497^y\ 76^d\ 23;40^h = 24,57^y\ 1,16;59,10^d.$$

For this time, from the tables in IV,4:

Δt	$\bar{\eta}$	$\bar{\alpha}$	$\bar{\omega}$
24,0^y	2,57; 2,34°	5,15;38,54°	5, 6; 6,56°
57^y	3, 8;30,28,35	16;59,39,53	3,16;37, 1,28
1,0^d	11;26,41,31	1, 3;53,56,30	1,13;45,39,22
16^d	3,15; 3, 7, 4	3,29; 2,23, 4	3,21;40,10,29
0;59^d	11;59,15,14	12;50,50, 2	13; 0,31,53
0;0,10^d	0; 2, 1,54	0; 2,10,38	0; 2,12,17
	3,44; 4, 8,18	4,18;27,54, 7	1,21;12,31,29
A.D.1	3,29;58	3,27; 7	2, 9;45
	1,14; 2	1,45;35	3,30;58

Copernicus gives $\bar{\eta} = 74°$. From the correction table in IV,11, for $2\bar{\eta} = 148;4°$,

$$c_{3(2\bar{\eta})} = +5;30°, \qquad c_{4(2\bar{\eta})} = 0;56,$$

so that

$$\alpha = \bar{\alpha} + c_3 = 105;35° + 5;30° = 111;5°. \quad \text{(Copernicus 111;10°)}$$

Thus

$$c_{5(\alpha)} = -4;44°, \qquad c_{6(\alpha)} = -2;43°,$$

$$c_{(\alpha,\, 2\bar{\eta})} = c_{5(\alpha)} + c_{4(2\bar{\eta})} \cdot c_{6(\alpha)}$$

$$= -(4;44° + 0;56 \cdot 2;43°) = -7;16°.$$

Consequently

$$\eta = \bar{\eta} + c_{(\alpha,\, 2\bar{\eta})} = 74;2° - 7;16° = 66;46°,$$

$$\lambda_{☾} = \bar{\lambda}'_{\odot} + \eta = \text{♓} 26;39° + 66;46° = \text{♊} 3;25°.$$

Copernicus gives ♊ 3;24°.

For the latitude,

$$\omega = \bar{\omega} + c_{(\alpha, 2\bar{\eta})} = 210;58° - 7;16° = 203;42°,$$

while Copernicus has 203;41°. Thus, from column 7

$$\beta = c_{7(\omega)} = -4;35°, \quad \text{i.e. to the south,}$$

in agreement with Copernicus. Recomputing λ and β directly from the theory gives exactly the same result.

Verification of the Lunar Theory

It is informative to test whether Copernicus's lunar theory correctly reproduces the observations used for its derivation. This will demonstrate its consistency, although not its accuracy compared to modern theory, a matter of at best secondary interest since Copernicus's object was to develop a theory consistent with the observations at his disposal. One should not think it an invalid test to recompute the very observations used to derive the parameters. The observations are all independent, and serve different functions in Copernicus's derivations. Thus, 7/8/9 were used for latitude theory, and are independent of the parameters for longitude. Only (2) and (5) were used directly to find $\bar{\eta}$ and $\bar{\alpha}$, while the remaining eclipses of 1–6 affect the mean motions only indirectly.

In order to carry out the test, we have used the theory directly to compute the true elongation of the moon from the mean sun, that is,

$$\eta = \lambda_{☾} - \bar{\lambda}'_{\odot} = \bar{\eta} + c_{☾}, \tag{20}$$

$c_{☾}$ being the lunar correction. Since we have earlier computed from the solar theory the adjusted mean longitude and true longitude of the sun, $\bar{\lambda}'_{\odot}$ and $\lambda_{\odot}$, the true elongation of the moon from the true sun η' can be found from

$$\eta' = \lambda_{☾} - \lambda_{\odot} = \eta - (\lambda_{\odot} - \bar{\lambda}'_{\odot}) = \eta - c_{\odot}, \tag{23}$$

as is clear from Fig. 18. We have made the examination for the nine lunar eclipses cited by Copernicus, for which η' should be exactly 180°. In each case we have converted the reported time to apparent time from Era Alexander, and with the recomputed $\alpha_{\odot}$ and $\bar{\lambda}_{\odot}$ in Table 3, computed the equation of time and found the interval of mean time. In Table 4, after the number and year of the eclipse, the next column gives the mean time $\Delta\bar{t}$ in days from E.A. 1. This is followed by the mean elongation $\bar{\eta}$, the lunar correction $+c_{☾}$, the true elongation from the mean sun $\eta = \bar{\eta} + c_{☾}$, the inverse solar correction $-c_{\odot} = \bar{\lambda}'_{\odot} - \lambda_{\odot}$ taken from Table 3, the true elongation from the true sun $\eta' = \eta - c_{\odot}$, and the error from true opposition $\eta' - 180°$.

The results, as was true of the solar theory, are quite good. To estimate the discrepancy from the time of true opposition, simply double the minutes of $\eta' - 180°$, e.g. (1) is short of true opposition by $0;4^{h}$, meaning that the computed opposition will be $0;4^{h}$ later than the observation. Because of roundings, the tabulated values represent the theory to $\pm 0;1°$, so the results should not be pressed too closely. In any case, it can be seen that the errors in the theory are small, and not systematic. The worst discrepancy, for (3) Ptolemy's eclipse of 136, is less than one-half hour, while all the rest are accurate to better than one-quarter hour. It would be pointless to attempt an explanation of errors so small since the observations themselves contain even greater inaccuracies.

2. The Parallax and Apparent Diameter of the Sun and Moon

Chapters 15–27 of Book IV, corresponding to *Almagest* V,11–19 and *Epitome* V,13–32, are devoted to subjects required primarily for the calculation of eclipses. The *Epitome* was here of particular importance to Copernicus because it contained additional methods, not in the *Almagest*, but developed by later Indian and Arabic astronomers who devoted attention to whatever could facilitate the routine of eclipse calculation. Copernicus had little of originality to add to such practical matters, yet his work is on the whole conscientious and, as we shall see in the following section on eclipses, successful.

There is, however, one subject of great importance in this part of Book IV, and that is the lunar parallax. Copernicus objected to Ptolemy's model because, among other reasons, it would greatly exaggerate the apparent diameter and parallax of the moon near quadrature. That the apparent diameter does not increase noticeably near quadrature is so obvious as to require no proof. But parallax is another matter, and it is now incumbent upon him to demonstrate that near quadrature the parallax is less than predicted by Ptolemy's model. This he does by two direct measurements of the parallax very near quadrature, and more strikingly by a computation of an occultation of Aldebaran by the moon that is without doubt the most remarkable demonstration in all of *De revolutionibus*. It is one of only three observational tests performed by Copernicus—the two others are the longitude of Spica in III,12, for whatever that is worth, and the lunar elongation in IV,10. The problem of the occultation is such that it constitutes a test of nearly everything that has preceded it—spherical astronomy, star catalogue, precession, solar theory, lunar theory, lunar parallax—and the results are by any standard extraordinary.

There is also a particular problem with this part of Book IV that deserves notice. It appears as though the revision from the earlier version in M was here done less carefully, or at least less thoroughly, than in the preceding sections. In fact Copernicus made drastic revisions in places, both in M and in the revision prepared for N, but many vestiges of earlier versions remain, and we shall have to look more often to M, and to cancelled readings in M, in order to find the basis of apparent inconsistencies in N. This is not to say that readings in M should be preferred to those in N—this is seldom the case—but that Copernicus's revision was not sufficiently extensive or consistent to eliminate and replace all remains of his earlier, and superseded, work.[47]

---

[47] This part of *De revolutionibus* has been treated in detail in Henderson (1973), which contains an examination of the alterations in M that elucidates much of what lies behind Copernicus's derivations of the various distances, parallaxes, and apparent diameters that are not altogether consistent. Although our own exposition is rather different—using M only in so far as it clarifies N—we have found her study very helpful. It is the only work yet done that considers the alterations in M and N with real understanding.

Parallax of the Moon (IV,15–16)

The lunar model gives geocentric positions of the moon that are independent of the location of the observer on the surface of the earth. Thus, the difference between the observed and computed zenith distance of the moon in a vertical circle, i.e. a great circle perpendicular to the horizon, is a direct measurement of the lunar parallax. In order to measure the moon's zenith distance, Copernicus uses an instrument described in *Almagest* V,12 and, in a somewhat modified form based upon al-Battānī's description, in *Epitome* V,13. The instrument is illustrated, at least in principle, in Fig. 20(a), and its use is shown in Fig. 20(b).

Take three *regulae*—finely planed, straight pieces of wood—two of them at least four cubits (about 2 m) in length, the third somewhat longer. Set upright and secure any post that has been very carefully *decussatus*—a term, apparently of technical meaning in this context, that we do not understand—and smoothed or polished (*palus aliquis optime decussatus et leuigatus*). Attach one of the pieces of equal length to it by appropriate hinges on which it can be turned around as would be suitable to a door (*quasi ianuam deceret, possit circumuolui*), i.e. so that it can rotate about a vertical axis. Near the top of this piece, at A attach the second equal piece AC, and near the bottom, at B, attach the longer piece BD, by fitting axles or pins into holes carefully drilled in the pieces so that they can rotate in the same vertical plane with the least possible wavering out of the plane. AC must be *exactly* equal to AB. On the upper surface of BD, draw a line down the middle, and taking $BE = AC$ *exactly*, divide BE into 1000 equal parts "or more if it is possible"(!). Then, letting $BD = BE\sqrt{2}$, so that BD will be the side of a square inscribed in a circle of radius AC, extend the division in the same parts, and thus $BD = 1000\sqrt{2} = 1414$ parts (each about 2 mm in length).[48] Draw a line on AC equal to BE so that its end at C can touch the divided line on BD, that is, the end of AC at C must be able to slide along the marked surface of BD—bevelling would be helpful—and some sort of device, not mentioned, should hold it in place. Finally, attach to the side of AC two sights (*specilla*), as in a diopter, so that the line of sight through them is parallel to the line drawn on the surface of AC.

Copernicus describes the use of this instrument for observing a star (*sidus*), a term which includes the moon. Although he does not say so, it appears plausible that he used it to take the altitude of Spica at meridian

[48] The fineness of the graduations seems excessive or fictitious. The instrument described in the *Epitome* is graduated for use with a sexagesimal chord table so that $BD = 60\sqrt{2} = 84;51$ parts, which means in principle a division into 85 parts or some small multiple of 85. Battānī explains the use of graduations of either $30\sqrt{2}$, which will give the sine of half the measured arc, or $60\sqrt{2}$, which will give the chord of the entire arc. Cf. Battānī (1537) cap. 57, ff. 89–90; Nallino 1, 143–44. Copernicus's instrument gives the chord decimally, as does a nearly identical instrument made by Regiomontanus and described in Schoener (1544), 23r ff.

transit, and it can also be used to find the oppositions of the superior planets.[49] In any case, it is one instrument that there is good evidence for his using, since Tycho later acquired a parallactic instrument, as it was called, that he was told belonged to Copernicus and was supposedly made with his own hands. Tycho writes that although it was made of wood and was unfit to use—among other problems, it appears to have warped—he was so pleased to have a relic of the "incomparable Copernicus," that he could not contain himself from writing a *Heroicum Carmen*, in hexameter, that very day.[50]

Ptolemy's parallactic instrument was limited to angles up to 60° since all three rods were of equal length. The instrument described by Battānī and in the *Epitome* could take angles up to 90° by the lengthening of *BE* to *BD*, and while Ptolemy's and Battānī's instruments were fixed in the plane of the meridian, the instrument in the *Epitome* has *AB* inserted into a hole in the base so that it could be rotated in azimuth. Copernicus made two further modifications. He graduated the scale decimally rather than sexagesimally, because his sine table was decimal, and he mounted *AB* on a post by hinges. The first change was harmless, but the second may have been a source of trouble.[51] The use of the instrument is shown in Fig. 20(b). Observe the star or the moon through the sights, and note where *C* meets *BD*. Since $AC = AB = r$, the radius of a circle with center *A*,

$$BC = r \text{ crd } \zeta' = 2r \sin \tfrac{1}{2}\zeta',$$

where ζ' is the apparent zenith distance, which can then be read inversely from a table of chords or sines. However reasonable its principle, the instrument does not appear to have been capable of much accuracy.

Copernicus reports Ptolemy's use of the parallactic instrument in two carefully selected observations. The first was made to determine the greatest lunar latitude and the inclination of the moon's orbit by observing the moon as close as possible to the zenith when its parallax is negligible. Under the best circumstances, shown in Fig. 21(a), which occur only once in about $18\frac{1}{2}$ years, the ascending node is at ♈0° while the moon is at ♋0° and thus at the northern limit of latitude. Hence the moon can culminate with its least possible zenith distance at exactly the time that the nodes and equinoxes

---

[49] Above, p. 131 and below, p. 311.

[50] Brahe (1913–29) 5, 45–47; 6, 265–67. Brahe (1946), 45–47. Cf. the partial translation by Rosen in Copernicus (1978), 412.

[51] See below, pp. 236–240. It would also appear that the long rods must have been difficult to manipulate properly, and that the instrument could scarcely have given better results than a graduated quadrant. Tycho is stronger in his criticism.

> In actual practice they do not yield results with the accuracy and certainty of the quadrants described before, however carefully they be constructed. For it is very difficult for the rulers to keep perfectly straight. If they are too long they bend by their own weight so that they will deviate from a straight line, and if they are too short they will not yield what is expected of them on account of their limited size. Brahe (1946), 47.

are at the horizon. The greatest latitude and inclination of the orbit will then be

$$\beta_{max} = \iota = \varphi - \varepsilon - \zeta. \tag{27}$$

Ptolemy took for Alexandria $\varphi = 30;58°$, $\varepsilon = 23;51°$, and measured $\zeta \approx 2\frac{1}{8}°$. Hence

$$\beta_{max} = \iota = 30;58° - 23;51° - 2;7,30° = 4;59,30° \approx 5°,$$

in agreement with Hipparchus.

The second observation was made for finding the lunar parallax when the moon's zenith distance was large, although not the largest possible. The conditions are shown in Fig. 21(b). The ascending node was near ♎0° and the moon was near ♑0° and the northern limit of latitude. Ptolemy's observation (14) was made on 1 Oct 135, and by computation

$$\lambda_{☾} = ♑3;10°,[52] \qquad \beta = +4;59°,$$

and since the declination of ♑3;10° is $-23;49°$, the true zenith distance

$$\zeta = \varphi + \delta - \beta = 30;58° + 23;49° - 4;59° = 49;48°. \tag{28}$$

However, the observed zenith distance ζ' was 50;55°, the difference being the parallax

$$p = \zeta' - \zeta = 50;55° - 49;48° = 1;7°.[53] \tag{29}$$

From the parallax, Ptolemy finds a geocentric lunar distance of $39;45^{tr}$ (terrestrial radii), and a corresponding relative distance in his lunar model of 40;25 where $R = 60$. It then follows that the greatest distance of the moon at syzygy

$$s_M = R + r = 59^{tr} + 5;10^{tr} = 64;10^{tr},$$

and the least distance at quadrature

$$s_m = R - 2e - r = 59^{tr} - 20;17^{tr} - 5;10^{tr} = 33;33^{tr}.$$

The corresponding horizontal parallaxes P, from $\sin P \approx \tan P = 1/s$, are

$$P_m = 0;53,34°, \qquad P_M = 1;43°. \quad \text{(Corr. 1;42)}$$

Copernicus reviews all this, although without giving the date of the observation, but he can make no use of it since the last thing he wants is a

---

[52] Copernicus, following *Epitome* V,15, gives ♑3;9°, as has been noted by Rosen as a mistake in Copernicus (1978), 413. Correctly, from Ptolemy's tables, the longitude is in fact ♑3;9°, so Ptolemy must have rounded while, more interestingly, Peurbach or, more likely, Regiomontanus, must have recomputed and emended the text accordingly. Such a check probably indicates some suspicion of the validity of Ptolemy's parallax measurement since *Epitome* V,22 remarks upon the problem of Ptolemy's lunar distance near quadrature. Cf. below, p. 251.

[53] The problems of Ptolemy's observation and its results are discussed in Delambre (1817) 2, 209–13 and in *HAMA*, 101–03.

lunar parallax of 1;7° at $\zeta \approx 50°$, leading to a lunar distance of about 40^{tr}. If one recomputes the position of the moon with Copernicus's theory, which he does not do, the results are

$$\lambda_☾ = ♑3;16°, \qquad \beta = +5;0°,$$

essentially in agreement with Ptolemy. Hence, according to Copernicus's lunar theory, there is nothing wrong with Ptolemy's computed position, so the error must lie in the values of φ or ε or in the observed zenith distance of 50;55°. Copernicus does not say this—he is rightly hesitant about criticizing older observations upon which he relies so much—but remarks that he has found on numerous occasions that the parallax is far different. Now he must prove that Ptolemy's parallax is incorrect, which he does by presenting two observations, in both of which the moon was very near mean quadrature and the parallax derived by observation was smaller than predicted by Ptolemy's model. It must be understood that the fundamental purpose of the following demonstrations is not to find the distance of the moon, for Copernicus has no doubt that Ptolemy's distance at syzygy is essentially correct, but to show that Ptolemy's large parallax near quadrature, even though apparently confirmed by an observation, cannot be right.

1. In the first observation (11) of 27 Sep 1522, $17;40^h$, the moon was observed in the meridian with an apparent zenith distance $\zeta' = 82;50°$. From the Epoch of Christ $\Delta t = 1522^y\ 284^d\ 17;40^h$, and from the solar position in Table 3, the equation of time

$$\Delta E = -2;25° - 1;53° = -4;18° = -0;17,12^h.$$

From his faulty value of $(\alpha_0 - \bar{\lambda}_0)_G$, Copernicus computes

$$\Delta E = -2;25° - 1;33° = -3;58° = -0;15,52^h \approx -0;16^h,$$

and gives for the mean time

$$\Delta\bar{t} = 1522^y\ 284^d\ 17;24^h = 25,22^y\ 4,44;43,30^d.$$

Therefore, from the lunar theory,

$$\bar{\eta} = 87;6°, \qquad \bar{\alpha} = 357;39°, \qquad \text{(Corr. 357;40)}$$

$$c_3 = +1;1°, \quad \text{(Corr. 1;2)} \qquad \alpha = \bar{\alpha} + c_3 = 358;40°,$$

$$c = +0;7°, \quad \text{(Corr. 0;9)} \qquad \eta = \bar{\eta} + c = 87;13°,$$

$$\bar{\varpi} = 197;1°, \qquad \varpi = \bar{\varpi} + c = 197;8°,$$

$$\lambda_☾ = \bar{\lambda}'_\odot + \eta = 195;20° + 87;13° = 282;33°, \qquad \beta = -4;47°.$$

Copernicus determines the declination of the moon simply by adding

$$\delta = \delta(282;33°) + \beta = -22;54° - 4;47° = -27;41°,$$

which is not strictly correct, although since the moon is near ♑0° it is close to an accurate computation which gives $\delta = -27;39°$. While Copernicus does not do so, it is useful to check whether the moon was in the meridian at the time of the observation. The right ascension of the moon $\alpha_☾ = 284;9°$, and the right ascension of the true sun $\alpha(193;29°) = 192;24°$. Thus at $5;40^h$ of apparent time after noon, the right ascension of the mid-heaven,

$$\alpha_M = 192;24° + 5;40 \cdot 15° = 277;24°, \qquad (2.28)$$

which means that the moon is out of the meridian by

$$\Delta\alpha = \alpha_M - \alpha_☾ = -6;45° = 6;45° \text{ east} = -0;27^h, \qquad (2.27)$$

that is, the moon is not yet in the meridian, and will not be there for nearly $\frac{1}{2}^h$. This is a nontrivial error, and the following determination of the zenith distance is not valid, for $\varphi - \delta = \zeta$ only in the meridian. But Copernicus simply takes $\varphi = 54;19°$ for Frauenburg and, as shown in Fig. 22, computes

$$\zeta = \varphi - \delta = 54;19° + 27;41° = 82°. \qquad (30)$$

Hence from the observed $\zeta' = 82;50°$, the parallax

$$p = \zeta' - \zeta = 0;50°. \qquad (29)$$

Copernicus remarks that according to Ptolemy's table (*Almagest* V,18), the parallax would be 1;17° (corr. 1;17,35°), that is, 0;27° greater.

2. The second observation (12) of 7 Aug 1524, $18;0^h$, showed an apparent zenith distance $\zeta' = 81;55°$. From the Epoch of Christ $\Delta t = 1524^y\, 234^d\, 18;0^h$, and from the solar position in Table 3, the equation of time

$$\Delta E = +1;29° - 1;53° = -0;24° = -0;1,36^h,$$

although Copernicus finds

$$\Delta E = +1;29° - 1;33° = -0;4° = -0;0,16^h \approx 0^h,$$

and says that the mean time is equal to the apparent time. Thus, from $\Delta\bar{t} = 25,24^y\, 3,54;45^d$,

$$\bar{\eta} = 97;6°, \quad (\text{Corr. } 97;7) \qquad \bar{\alpha} = 242;10°, \quad (\text{Corr. } 242;11)$$

$$c_3 = -2;30°, \qquad \alpha = \bar{\alpha} + c_3 = 239;40°,$$

$$c = +7;0°, \qquad \eta = \bar{\eta} + c = 104;6°,$$

$$\bar{\varpi} = 193;19°, \qquad \varpi = \bar{\varpi} + c = 200;17°, \quad (\text{Corr. } 200;19)$$

$$\lambda_☾ = \bar{\lambda}'_\odot + \eta = 145;33° + 104;6° = 249;39°, \qquad \beta = -4;41°,$$

$$\delta = \delta(249;39°) + \beta = -21;55° - 4;41° = -26;36°.$$

Correctly computed $\delta = -26;34°$. Again, we check whether the moon is in the meridian. The right ascension $\alpha_☾ = 247;12°$, and the right ascension of the true sun $\alpha(144;14°) = 146;33°$. For $6;0^h$ apparent time after noon, the right ascension of the midheaven

$$\alpha_M = 146;33° + 6 \cdot 15° = 236;33°, \tag{2.28}$$

and thus the distance of the moon from the meridian

$$\Delta\alpha = \alpha_M - \alpha_{☾} = -10;39° = 10;39° \text{ east} = -0;42,36^h, \tag{2.27}$$

that is, the moon will not be in the meridian for nearly $\frac{3}{4}^h$, which is even worse than the last observation. Again, what follows is not valid, for Copernicus computes the geocentric zenith distance from

$$\zeta = \varphi - \delta = 54;19° + 26;36° = 80;55°, \tag{30}$$

and thus the parallax is

$$p = \zeta' - \zeta = 81;55° - 80;55° = 1°, \tag{29}$$

while, he says, Ptolemy's table gives 1;38° (corr. 1;35).

Comment. Copernicus has indeed proved his point—the parallax at quadrature is far less than required by Ptolemy's model—but it can hardly be said that the demonstration is satisfactory. There are three fundamental problems. At the end of IV,27 Copernicus remarks that the high latitude of Frauenburg allowed him to observe the moon much closer to the horizon than Ptolemy could in Alexandria ($\varphi \approx 31°$). He considers this an advantage because the observable parallax can be much larger, and therefore more easily detected. This, however, is untrue. Aside from the fact that Ptolemy's measurement of $\zeta' = 50;55°$ is mistaken, the observable parallax at $\zeta \approx 50°$ is about 0;45°, which is quite large enough to be measured. On the other hand, by observing the moon so close to the horizon, Copernicus was inviting a rather large error due to refraction, of which he appears to have no knowledge. For $\zeta' = 83°$ the mean refraction is about 0;7°, and correction for refraction will increase the parallax at each of Copernicus's observations, and therefore bring the moon closer, although not nearly so close as implied by Ptolemy's lunar model.

A second problem concerns, not the observation, but the computed latitude of the moon. There is nothing wrong with Copernicus's argument of latitude, but errors in the latitude itself will be at their greatest when the moon is at quadrature. Correctly, the lunar latitude is

$$\beta = 5;9° \sin \nu + 0;8,48° \sin (2\eta - \nu) + \cdots \tag{31}$$

where $\nu = \omega + 90°$ is the true argument measured from the ascending node. The errors in Copernicus's latitudes are +0;16° in the first observation and +0;18° in the second, but since the correct latitudes would increase the zenith distances, they are partially compensated by the neglect of refraction in the observation. Adjusting the computed latitude and the observed zenith distance accordingly, we reach the following results which we compare with Copernicus's figures.

	1. Cop.	1. Adj.	2. Cop.	2. Adj.	
β	$-4;47°$	$-5;3°$	$-4;41°$	$-4;59°$	
δ	$-27;41$	$-27;54$	$-26;36$	$-26;51$	
ζ	82;0	82;13	80;55	81;10	(32)
ζ'	82;50	82;57	81;55	82;1	
p	0;50	0;44	1;0	0;51	
s	$68;13^{tr}$	$77;33^{tr}$	$56;44^{tr}$	$66;45^{tr}$	

The last figure is the resulting geocentric distance, which is included because it gives the parallax more immediate significance. Copernicus was to some extent lucky. Even his errors in δ worked to compensate the errors in β, so somehow his results were tolerable. But the adjusted results are obviously not, for the parallax in 1. is far too low, and the general effect is to increase the lunar distance by about 10^{tr}.

But the most serious problem of all is that, by the position that follows from Copernicus's own computation, the moon is not in the meridian at either observation—in the first it is 6;45° east and in the second 10;39° east, and these are very large errors. The correct zenith distance outside the meridian is not $\zeta = \varphi - \delta$, but

$$\zeta = \cos^{-1}(\sin\varphi \sin\delta + \cos\varphi \cos\delta \cos h),$$

where $h = \alpha_M - \alpha_☾$ is the hour angle of the moon. Solving correctly for ζ with Copernicus's figures and the adjusted figures, the results are as follows:

	1. Cop.	1. Adj.	2. Cop.	2. Adj.	
ζ	82;12°	82;25°	81;26°	81;41°	
ζ'	82;50	82;57	81;55	82;1	(33)
p	0;38	0;32	0;29	0;20	
s	90^{tr}	107^{tr}	117^{tr}	170^{tr}	

Hence the results are in all cases disastrous, and it must be concluded that Copernicus's observations are simply without meaning for establishing the lunar parallax. And the error must lie in the observations themselves since it is not possible for Copernicus's solar and lunar theory to produce such large errors in hour angle. Either the recorded times of the observations are in error by $\frac{1}{2}^h$ and $\frac{3}{4}^h$ respectively, which seems excessive, or Copernicus really did observe the moon nearly 7° and 11° east of the meridian. It is hard to believe that he could make such errors in locating his meridian. At this point one begins to wonder whether it was such a good idea to mount the parallactic instrument on hinges.

Distance of the Moon (IV,17)

Copernicus has managed, one way or another, to find parallaxes at quadrature smaller than those required by Ptolemy's model and apparently confirmed by Ptolemy's observation (14) of 1 Oct 135, so at least to his own satisfaction, he has proved his point. It is now a simple matter to find the geocentric distances, and in the preceding analysis we have already done so. In Fig. 23 the moon is at P, and is seen by the observer at O to have an apparent zenith distance ζ', while by computation its true or geocentric zenith distance with respect to E is ζ. Their difference

$$p = \zeta' - \zeta \qquad (29)$$

is the parallax in the circle of altitude. Now,

$$EP = \text{crd } 2(\zeta + p) = \text{crd } 2\zeta',$$

$$EO = \text{crd } 2p,$$

and thus in units of $EO = r_e = 1^{\text{tr}}$, the geocentric distance

$$s = \text{crd } 2\zeta'/\text{crd } 2p = \sin \zeta'/\sin p. \qquad (34)$$

For each observation:

1.	$\zeta = 82°$	$\zeta' = 82;50°$	$p = 0;50°$	$s \approx 68^{\text{tr}}$	(35)
2.	80;55	81;55	1;0	56;41[54]	

In order to determine the greatest and least geocentric distances produced by the model, Copernicus must next find the relative distance in the lunar model corresponding to the absolute distances at one of these observations. As luck would have it, he chooses the second observation, in which the moon was farthest from the meridian, so everything that follows is quite invalid. The configuration of the model at the second observation is shown in Fig. 24. From the previous computation,

$$2\bar{\eta} = 194;12°, \qquad \bar{\alpha} = 242;10°,$$

$$c_3 = -2;30°, \qquad \alpha = \bar{\alpha} + c_3 = 239;40°,$$

$$c = +7;0°, \qquad EC_1P = \alpha - 180° = 59;40°,$$

$$\gamma = 180° - (EC_1P + c) = 113;20°.$$

[54] Correctly, for 1. $s = 68;13^{\text{tr}}$ and for 2. $s = 56;44^{\text{tr}}$. M gives for 2. $s = 56;42^{\text{tr}}$. Ptolemy's theory would give distances of about 44^{tr} and 36^{tr}.

And where $R = 100{,}000$,

$$EP = R(\sin EC_1P/\sin\gamma) = R(86310/91821) = 93998$$

is the relative distance in the model corresponding to the absolute distance of $56;41^{\text{tr}}$ in the second observation. Hence, the mean distance in terrestrial radii

$$R^{\text{tr}} = (R/EP)\,56;41^{\text{tr}} = 60;18^{\text{tr}}.$$[55]

For the radii of the epicycles

$$(r_1 - r_2)^{\text{tr}} = (8600/R)\,60;18^{\text{tr}} = 5;11^{\text{tr}},$$

$$(r_1 + r_2)^{\text{tr}} = (13340/R)\,60;18^{\text{tr}} = 8;2^{\text{tr}}, \quad \text{(Corr. 8;3)}$$

so that

$$r_1 = 6;36,30^{\text{tr}}, \qquad r_2 = 1;25,30^{\text{tr}}.$$

And thus for the four limits (*limites*) of lunar distance which will be utilized for the calculation of the parallax table in IV,24, the following distances are established:

I	quadrature	$2\bar{\eta} = 180°$	$\bar{\alpha} = 0°$	$R + (r_1 + r_2) = 68;20^{\text{tr}}$ [56]
II	syzygy	$2\bar{\eta} = 0$	$\bar{\alpha} = 0$	$R + (r_1 - r_2) = 65;30$
III	syzygy	$2\bar{\eta} = 0$	$\bar{\alpha} = 180$	$R - (r_1 - r_2) = 55;8$
IV	quadrature	$2\bar{\eta} = 180$	$\bar{\alpha} = 180$	$R - (r_1 + r_2) = 52;17$

(36)

From Ptolemy's model and parallax, the corresponding distances are

I $39;53^{\text{tr}}$, II $64;10^{\text{tr}}$, III $53;50^{\text{tr}}$, IV $33;33^{\text{tr}}$.

Hence the distances II and III at syzygy are much the same, while the distances I and IV at quadrature are more reasonable in Copernicus's model. If we use instead Copernicus's first observation, and assume the moon to be in the meridian, then for the distance $68;13^{\text{tr}}$, following strictly from the observation, $EP = 109587$. Consequently, $R = 62;15^{\text{tr}}$, $r_1 - r_2 = 5;21^{\text{tr}}$, $r_1 + r_2 = 8;18^{\text{tr}}$, and at the same four limits,

I $70;33^{\text{tr}}$, II $67;36^{\text{tr}}$, III $56;54^{\text{tr}}$, IV $53;57^{\text{tr}}$.

Both sets are within reason—although in fact the variation of distance is more than twice the true variation of lunar distance—but it must be remembered that the underlying measurements of parallax, especially for the second observation, were only fortuitously plausible.

[55] M likewise gives 60;18, although 56;42 in M should give 60;19, which Copernicus in fact uses. Cf. n. 56.

[56] Limit I follows from $R = 60;18$, but the rest from $R = 60;19$.

Hypothetical Determination of the Apparent Diameter of the Moon and Shadow (IV,18)

Almagest V,14 and *Epitome* V,18 present a derivation of the apparent diameters of the moon and shadow from two partial lunar eclipses when the moon is near greatest distance at syzygy. The same demonstration is repeated for the diameters at least distance in *Almagest* VI,5 and *Epitome* VI,6.[57] Copernicus does not use this method, and never really explains how he found his values of the apparent diameters, but he goes through the motions of setting out an example that will produce exactly Ptolemy's result for the moon at greatest distance, although not for the shadow.

The eclipses are shown in Fig. 25, in which it is assumed that the moon is north of the ecliptic. The first eclipse is of 3′ with the moon at a latitude β of 0;47,54°. The second eclipse is of 10′ and the latitude of the moon is 0;29,37°. For what follows, one should correctly use the arc BC perpendicular to the lunar orbit, as Ptolemy does in his demonstration. However, al-Battānī goes through the same procedure assuming that $BC = \beta$—it makes but a small difference—and he is followed in this approximation by the *Epitome* and consequently by Copernicus.[58] The difference of the eclipsed part of the moon in the two eclipses is

$$10' - 3' = 7' \qquad \text{or} \qquad 0;47,54^\circ - 0;29,37^\circ = 0;18,17^\circ,$$

and thus the diameter of the moon, which corresponds to 12′, is

$$(12/7)\ 0;18,17^\circ = 0;31,20,34^\circ \approx 0;31,20^\circ.$$

In the first eclipse, of 3′, the center of the moon was 3′ or $\frac{1}{4}$ of the lunar diameter, that is, 0;7,50°, above the upper rim of the shadow. Hence, the radius of the shadow is

$$0;47,54^\circ - 0;7,50^\circ = 0;40,4^\circ.$$

On the other hand, in the second eclipse, of 10′, the center of the moon was 4′ or $\frac{1}{3}$ the lunar diameter, 0;10,27°, below the upper edge of the shadow. Thus, the radius of the shadow is again

$$0;29,37^\circ + 0;10,27^\circ = 0;40,4^\circ.$$

The diameter of the shadow is therefore 1;20,8°, and the ratio of the diameter of the shadow to the diameter of the moon

$$1;20,8/0;31,20 = 2;33,26,$$

more-or-less in agreement with Ptolemy, who found

$$1;21,20/0;31,20 = 2;35,44 \approx 2;36 = 13/5.$$

---

[57] Description and criticism of Ptolemy's method in *HAMA*, 103–08.

[58] Battānī (1537), cap. 30, f. 37r; Nallino 1, 58. The substitution of latitude for the perpendicular to the lunar orbit is likewise standard in computing magnitudes of eclipses. Cf. below, pp. 282–283.

Solar Distance and Related Topics (IV,19–20)

The only reason for finding the distance of the sun in pre-modern astronomy is to show that its parallax is essentially negligible within the accepted limits of observational error, and so may be ignored or routinely deducted from the lunar parallax in the computation of a solar eclipse. The canonical value of the solar parallax, following Ptolemy, is 0;2,51°, often rounded to 0;3°, which is slightly over 1′ in a solar eclipse. One would think that to Copernicus the solar distance would have more significance, since the mean distance from the earth to the sun, the radius of the earth's orbit, is the common unit by which the distances of the planets are measured in a heliocentric theory. Thus, an absolute value of the solar distance allows the determination of the absolute distances of the planets, something impossible in geocentric theory without additional assumptions such as nested spheres—which Copernicus still appears to subscribe to—or some kind of proportionality of velocity or period to distance. But it turns out that the solar distance has no special importance for Copernicus since he examines it only for the usual reasons, and never applies its absolute value to the planetary distances. His determination follows the classical method invented by Hipparchus for finding the lunar distance from an assumed solar distance, and adapted by Ptolemy for finding the solar distance from a known lunar distance.[59]

Copernicus begins with a review of Ptolemy's procedure from *Almagest* V,15 and *Epitome* V,19–20. The demonstration depends upon the fact that at some distance, the apparent diameter of the moon is equal to the apparent diameter of the sun. In Fig. 26, we draw in the same plane, great circles of the sun S, moon M, and earth E, and let line TE tangent to the sun and moon show the apparent radius r' of each when equal, and line TFV show the shadow cone of the earth, with the moon's transit through the shadow at U where $EU = EM$. It is assumed that the lines r_s, r_m, r_e to the points of tangency do not differ from radii of the bodies. Where $s_m = EM$ is the distance of the moon,

$$r_m = s_m \tan r' \approx s_m \sin r'. \tag{37}$$

Call the ratio of the apparent diameter of the shadow to the apparent diameter of the moon k, and since $EU = EM$,

$$r_u = kr_m. \tag{38}$$

Again, since $EU = EM$, r_e is the arithmetic mean between r_u and $r_m + i$, so that

$$r_m + i = 2r_e - r_u,$$
$$i = 2r_e - (r_u + r_m).$$

[59] Cf. *HAMA*, 109–11, 325–29.

Now, letting $SE = s_s$,

$$\frac{i}{r_e} = \frac{TN}{TF} = \frac{SM}{SE} = \frac{SE - EM}{SE} = \frac{s_s - s_m}{s_s},$$

so that

$$s_m = s_s - s_s \cdot \frac{i}{r_e} = s_s \cdot \frac{r_e - i}{r_e},$$

and the distance of the sun

$$s_s = \frac{s_m r_e}{r_e - i} = \frac{s_m r_e}{r_u + r_m - r_e}. \tag{39}$$

Or, letting $r_e = 1$, the whole procedure amounts to

$$s_s = s_m/[(k + 1)s_m \sin r' - 1]. \tag{40}$$

To find the distance $s_v = EV$ to the vertex of the shadow, we have

$$\frac{r_u}{r_e} = \frac{VE - UE}{VE} = \frac{s_v - s_u}{s_v},$$

$$s_u = s_v - s_v \cdot \frac{r_u}{r_e} = s_v \cdot \frac{r_e - r_u}{r_e},$$

and the distance to the vertex is then

$$s_v = \frac{s_u r_e}{r_e - r_u}. \tag{41}$$

And since $s_u = s_m$, letting $r_e = 1$,

$$s_v = s_m/(1 - k s_m \sin r'). \tag{42}$$

1. *Ptolemy's Demonstration*

Ptolemy assumes that the apparent diameters of the sun and moon are equal at greatest distance at syzygy, and has established that

$$s_m = 64;10 r_e, \qquad r' = 0;15,40°, \qquad k = 2;36.$$

Therefore,

$$r_m = 0;17,33 r_e, \qquad r_u = 0;45,38 r_e,$$

$$i = 0;56,49 r_e, \qquad r_e - i = 0;3,11 r_e,$$

$$s_s = 64;10 r_e/0;3,11 = 1209;25 r_e \approx 1210 r_e.$$

Since the apparent diameter of the sun is assumed invariable, it is not certain which solar distance this applies to, and Ptolemy offers no information in the *Almagest*. However, in the *Planetary Hypotheses* he takes 1210^{tr} as a mean distance, and gives the greatest and least distance as 1260^{tr} and 1160^{tr} respectively.[60] The distance to the vertex of the shadow is then

$$s_{\mathrm{v}} = 64;10r_{\mathrm{e}}/(1 - 0;45,38) = 267;58r_{\mathrm{e}} \approx 268r_{\mathrm{e}}.$$

Ptolemy's demonstration has the virtue of being consistent and of working at all, which is more than can be said for what follows. Note that $r_{\mathrm{e}} - i = r_{\mathrm{u}} + r_{\mathrm{m}} - r_{\mathrm{e}}$ is a very small quantity, precariously close to zero.

2. *Al-Battānī's Demonstration*

Copernicus next reviews some of the parameters and the result of al-Battānī's demonstration as reported in *Epitome* V,21, although he cannot believe that Battānī could be responsible, and says that it is by "others" and only attributed to Battānī. The account in the *Epitome* is not quite faithful to either the Arabic text of al-Battānī or the Latin translation by Plato of Tivoli, and the problems, which are bad enough in the original, are even worse in the *Epitome*. Now Copernicus says that according to "others," the greatest distance of the moon is $64;10^{\mathrm{tr}}$, the apparent diameter of the sun at apogee is 0;31,20°, and the ratio of the diameter of the shadow to the diameter of the moon is $2\frac{3}{5}$ to 1, all in agreement with Ptolemy. However, they deny that the apparent diameter of the moon at greatest distance is greater than 0;29,30°, and therefore set the diameter of the shadow at about 1;16,45°.[61] He continues, "They think that it follows from these that the distance of the apogee of the sun from the earth is 1146 parts, and the axis of the shadow 254 parts, where the radius of the earth is one part, attributing these to the discovery of the noted philosopher of Raqqa, although they can be put together by no computation."

Some further analysis will show what Copernicus means. Since the apparent diameter of the moon is 0;29,30° at $64;10^{\mathrm{tr}}$, one must find the distance at which its angular diameter is equal to that of the sun, 0;31,20°. The *Epitome* gives this as $60;55^{\mathrm{tr}}$, a number Copernicus does not mention, although in both the Arabic and Latin texts of Battānī it is $60;58^{\mathrm{tr}}$.[62] In any case, we, and Copernicus when he was studying the problem, have for the demonstration of the solar distance

$$s_{\mathrm{m}} = 60;55r_{\mathrm{e}}, \qquad r' = 0;15,40°, \qquad k = 2;36.$$

With these values,

$$r_{\mathrm{m}} = 0;16,39r_{\mathrm{e}}, \qquad r_{\mathrm{u}} = 0;43,17r_{\mathrm{e}},$$

$$r_{\mathrm{e}} - i = r_{\mathrm{u}} + r_{\mathrm{m}} - r_{\mathrm{e}} = -0;0,4r_{\mathrm{e}},$$

---

[60] Goldstein (1967), 7.

[61] $2;36 \cdot 0;29,30° = 1;16,42°$, which Copernicus evidently rounds to $1;16\frac{3}{4}°$. The *Epitome* gives the radius of the shadow as 0;38,20°—which is probably correct even though the text of Battānī, both Arabic and Latin, usually reads 0;38,30°—implying a diameter of 1;16,40°. These details do not really matter since only $k = 2;36$ is used in the demonstration.

[62] Battānī (1537), cap. 30, f. 39r; Nallino 1, 61. Cf. Swerdlow (1973a).

which means that (39) would have a negative denominator, giving a negative solar distance, of about $-54825r_e$, which is nonsense. Hence Copernicus's remark that these numbers cannot be put together by any computation. In fact, what Battānī did was simply to take the ratio $s_s/s_m \approx 18\frac{4}{5}$ from Ptolemy's solar and lunar distance, and then compute

$$s_s = 18;48 \cdot 60;58^{tr} = 1146;10^{tr} \approx 1146^{tr}.$$

3. *Copernicus's Demonstration*

The first step is to arrange the parameters. Copernicus makes no independent determination of these, but adjusts the values of Ptolemy and al-Battānī to fit his own solar and lunar theory. He says that the apparent diameter of the sun at apogee is 0;31,40° since it should now be a little greater than before Ptolemy, evidently because the solar eccentricity has decreased from 415 to 323.[63] Thus, the sun is now somewhat closer at apogee and the apparent diameter is increased to

$$\frac{10415}{10323} \cdot 0;31,20° = 0;31,37° \approx 0;31,40°.$$

It should, however, be noted that Ptolemy considered 0;31,20° to apply to all solar distances, not just apogee, as Copernicus himself mentions (N 121r16–17). The apparent diameter of the moon, taken by Copernicus as 0;30° at greatest distance at syzygy, poses more problems. It does not follow from the reduction of the radius of the effective lunar epicycle at syzygy from 8706 to 8604 since Copernicus regards the latter as an improved determination rather than a real reduction. He has increased the greatest distance at syzygy from $64;10^{tr}$ to $65;30^{tr}$, but this too he considers an improvement rather than a real increase. Hence it would seem that the diameter of 0;30° should be independent of any earlier parameters. However, if one takes al-Battānī's diameter of 0;29,30°, and assumes that it has increased with the increase of distance, in other words, that both the distance and apparent diameter had been underestimated in the same proportion, then

$$\frac{65;30}{64;10} \cdot 0;29,30° = 0;30,6° \approx 0;30°.$$

But the reasoning behind this seems so contrived, that one may hope that it is only a coincidence, although it probably is not.

The diameter of the shadow at greatest distance is given as 1;20,36°, for, Copernicus says, the ratio of the shadow to the moon should be a little greater than $13/5 = 2;36$, namely $403/150 = 2;41,12$. The reasoning behind this adjustment appears to be as follows: If the diameter of the moon is

[63] Above, p. 154. On all that follows in this section, cf. Henderson (1973), 156–65, 199–201.

0;30° at a distance of 65;30^{tr}, then it will be equal to the diameter of the sun, 0;31,40°, as is required for the demonstration of solar distance, at

$$\frac{0;30}{0;31,40} \cdot 65;30^{tr} = 62;3^{tr} \approx 62^{tr},$$

which is exactly the distance Copernicus gives for equal lunar and solar apparent diameters. Now, Ptolemy took the ratio 13/5 to apply in general but derived it for the distance 64;10^{tr} where he took the diameters of the sun and moon to be equal. If, however (see Fig. 27), the diameters are equal at a lesser distance, 62^{tr}, then at that distance the shadow must be somewhat larger, since its diameter is inversely as its distance from the earth. But 64;10/62 ≈ 31/30, and thus when adjusted in proportion to the change of distance, the ratio becomes

$$\frac{31}{30} \cdot \frac{13}{5} = \frac{403}{150}.$$

Hence the diameter of the shadow at greatest distance,

$$(403/150)\ 0;30° = 1;20,36°.$$

It can also be seen from Fig. 27 that the true diameter of the moon will be smaller in Copernicus's demonstration than in Ptolemy's, as indeed turns out to be the case. Whether Copernicus really derived the apparent diameter of the moon and the ratio of the shadow to the moon by manipulations of this kind, which are not entirely free of contradiction, is difficult to know, but the certain derivation of the solar diameter makes them more than plausible.

Copernicus now has the following parameters:

$$s_m = 62r_e, \qquad r' = 0;15,50°, \qquad k = 2;41,12.$$

Thus,

$$r_m = 0;17,8r_e, \qquad r_u = 0;46,1r_e, \quad \text{(Corr. 0;46,1,53)}$$

$$i = 0;56,51r_e, \qquad r_e - i = 0;3,9r_e,$$

$$s_s = 62r_e/0;3,9 = 1180;57r_e.$$

But Copernicus gives $s_s = 1179r_e$, which in fact did not come from this computation, but from an earlier version in M (130r). In this cancelled version, Copernicus mistakenly computed for greatest distance in the syzygy without realizing that the apparent diameters of the sun and moon were not equal at that distance, or at least without realizing that the derivation requires equality of apparent diameters. At this stage, he took

$$s_m = 65;30r_e, \qquad r' = 0;15°, \qquad k = 79/30 = 2;38.$$ [64]

---

[64] Perhaps from Battānī's diameter of the shadow 1;16,45° ≈ 1;17°, adjusted by (65;30/64;10) 1;17° = 1;18,36°, so that $k \approx$ 1;19/0;30 = 79/30.

First

$$r_m = 0;17,9r_e, \qquad r_u = 0;45,10r_e,$$

However, he also computed, from $r' = 0;15,50°$,

$$r_m = 65;30r_e \sin 0;15,50° = 0;18,11r_e, \quad \text{(Corr. 0;18,6)}$$

and then put his results together to find

$$i = 2r_e - (r_u + r_m) = 2r_e - (0;45,10r_e + 0;18,11r_e)$$
$$= 0;56,39r_e,$$

which is not very consistent. Then

$$r_e - i = 0;3,21r_e \approx 0;3,20r_e,$$

and the solar distance

$$s_s = 65;30r_e/0;3,20 = 1179r_e \text{ (exactly)}.$$

Of course the derivation is invalid, but it is remarkable that in the revision Copernicus could rearrange all of his parameters and still come close to 1179 since the calculation is so sensitive to change in $r_e - i$ that, e.g. 65;30/0;3,*21* = 117*3*.[65] His reason for preserving 1179 seems to be that by the time he caught his error, consequences of the number were already embedded in IV,21,23,24, and he did not want to do a lot of tedious revision for what he must have considered a trivial alteration.

For the vertex of the shadow, correctly

$$r_e - r_u = r_e - 0;46,1r_e = 0;13,59r_e,$$

so that in principle,

$$s_v = 62r_e/0;13,59 = 266;1r_e.$$

Copernicus gives $265r_e$, which really follows from $s_m = 65;30r_e$ and $r_u = 0;45,10r_e$ in M. Thus $r_e - r_u = 0;14,50r_e$, and

$$s_v = 65;30r_e/0;14,50 = 264;56r_e \approx 265r_e.$$

Having determined the distance of the sun, it is now an easy matter to find its size (in IV,20). Copernicus first says that $s_s/s_m = 18$. In fact 1179/62 = 19, but from the earlier version in M, 1179/65;30 = 18 exactly (since 0;3,20 = 1/18). He then says that the diameter of the sun $d_s = 18d_m = 5;27d_e$. Now here

$$18 \cdot 0;17,8 = 5;8,24, \qquad \text{but} \qquad 18 \cdot 0;18,11 = 5;27,18,$$

so he is still operating with discarded values from M, which is doubly in error since the computation supposed d_m for a distance at which the apparent

[65] See the interesting analysis of the sensitivity of the method in the commentary by E. S. Abers and C. F. Kennel to Henderson (1975), 130–36.

diameters of the sun and moon are equal. He gives another way of finding the same result. Referring to Fig. 26,

$$\frac{EF}{EV} = \frac{ST}{SV}, \qquad \text{that is,} \quad \frac{r_e}{s_v} = \frac{r_s}{s_s + s_v},$$

and thus

$$d_s = \frac{s_s + s_v}{s_v} d_e = \frac{1179 + 265}{265} d_e = 5;26,55 d_e \approx 5;27 d_e. \tag{43}$$

Since spheres are in the ratio of the cubes of their diameters, the volume of the sun v_s is

$$v_s = 5;27 d_e^3 = 161;52,43 v_e \approx (162 - \tfrac{1}{8}) v_e.$$

For the moon, he remarks that

$$\frac{d_m}{d_e} = \frac{0;17,9}{1} \approx \frac{2}{7}, \tag{44}$$

(0;17,*9* is a cancelled value of r_m in M) and thus

$$\frac{v_m}{v_e} = \left(\frac{2}{7}\right)^3 = \frac{1}{42;52,30} = \frac{1}{43 - \frac{1}{8}},$$

$$\frac{v_s}{v_m} = (162 - \tfrac{1}{8})(43 - \tfrac{1}{8}) = 6940;23,$$

but Copernicus gives 7000 − 62 = 6938. In the *Almagest* these quantities are

$$d_e = 3\tfrac{2}{5} d_m, \qquad d_s = 5\tfrac{1}{2} d_e, \qquad v_s \approx 6944\tfrac{1}{2} v_m \approx 170 v_e.$$

Parallax and Apparent Diameter of the Sun (IV,21)

In the preceding demonstration of the solar distance, the apparent diameter of 0;31,40° was assumed to be the minimum, so the distance of 1179^{tr} applies to the apogee of the sun. Since the solar eccentricity is taken to be 323, the greatest relative distance is 10323 and the least 9677.[66] Hence the least distance is (9677/10323) $1179^{tr} = 1105^{tr}$, and the mean distance is 1142^{tr}. The corresponding horizontal parallaxes, computed from $P \approx \sin^{-1}(1/s)$, are as follows:

s	1179^{tr}	1142^{tr}	1105^{tr}
P	0;2,55°	0;3°	0;3,7°

(45)

[66] Copernicus gives 9678, which follows from the cancelled eccentricity of 322 in M (above, p. 153, n. 39), but it makes no difference in the following computation.

For all purposes, including the parallax table, Copernicus uses the mean value of 0;3°.

Copernicus now rejects the apparent diameter of 0;31,40° for greatest distance, and computes a new value from

$$d'_s = \text{crd}^{-1}(5;27/1179) = 0;31,48°. \quad \text{(Corr. 0;31,47)}$$

Then for least distance

$$d'_s = \text{crd}^{-1}(5;27/1105) = 0;33,54°,$$

the difference being 0;2,6°. For computing the intermediate apparent diameters, he says one may use the intermediate distances, which is what he does in his table of the semidiameter of the sun, or, as "some" do, let

$$d'_s = (66/5)v_s^{\circ/\text{h}}, \tag{46}$$

where v_s is the true velocity of the sun measured in degrees per hour. Copernicus knows this rule through *Epitome* V,21 and Peurbach's *Theoricae novae planetarum*, and it has a long history, going back through al-Battānī and al-Khwārizmī to Indian astronomy.[67] However, the rule is not compatible with Copernicus's extrema, and he does not use it. The method of finding $v_s^{\circ/\text{h}}$ is explained in *Epitome* VI,4. Let the mean velocity be $\bar{v}_s$, and let $\Delta c = c_{(\alpha+1°)} - c_{(\alpha)}$ be the difference of the solar equation for α and $\alpha + 1°$. Then the true velocity will be

$$v_s = \bar{v}_s + \bar{v}_s \cdot \Delta c.$$

From Copernicus's tables $\bar{v}_s = 0;2,28^{\circ/\text{h}}$, at apogee $\Delta c = -0;2°$, and at perigee $\Delta c = +0;2°$. Consequently

$$v_{s\,\min} = 0;2,23^{\circ/\text{h}}, \qquad v_{s\,\max} = 0;2,33^{\circ/\text{h}}.$$

Hence from (46)

$$d'_s\min = 0;31,28°, \qquad d'_{s\,\max} = 0;33,40°,$$

which do not agree with Copernicus's values. It is hard to know why he bothers to give this antiquated rule, since there is no way to reconcile it with his values of the apparent diameter. There is a corresponding rule for the moon also mentioned in *Epitome* V,21, namely,

$$d'_m = (47/48)v_m^{\circ/\text{h}}, \tag{47}$$

but Copernicus says nothing of it.

Parallax and Apparent Diameter of the Moon (IV,22)

The distances of the moon at the four limits have already been determined in IV,17, and now the corresponding horizontal parallaxes are found from $P \approx \sin^{-1}(1/s)$, where s is the distance at each limit. For finding the appar-

[67] Analysis and further references in Swerdlow (1973a).

ent diameter, Copernicus makes up a simple rule. It is clear from Fig. 28 that the parallax and apparent diameter are inversely proportional to the distance of the moon. It was shown in IV,20 that $r_m/r_e \approx 2/7$, hence the apparent radius r' and diameter d' will be

$$r'_m = (2/7)P, \qquad d'_m = (4/7)P. \tag{48}$$

The following table contains the lunar distances s_m at each limit, and the corresponding parallax P and apparent diameter d'_m, as given in the text. A single error is that at limit II, P should be 0;52,29°.

Limit	s_m	P	d'_m
I	68;21[tr]	0;50,18°	0;28,45°
II	65;30	0;52,24	0;29,57 ≈ 0;30°
III	55;8	1;2,21	0;35,38
IV	52;17	1;5,45	0;37,34

(49)

In concluding, Copernicus remarks that according to the model of Ptolemy and "others," the apparent diameter at the fourth limit would be nearly 1°, and consequently the half moon would give as much light as the full moon. Essentially the same point is made in *Epitome* V,22, where it is found from al-Battānī's figures that at the fourth limit $r'_m = 0;28,11°$, so that $d'_m = 0;56,22°$, indeed, nearly 1°. The *Epitome* likewise finds this remarkable.

Variation of the Shadow (IV,23)

Copernicus first says that from the ratio $k = 403/150 = 2;41,12$, we find the diameter of the shadow as a function of the diameter of the moon,

$$d'_u = kd'_m. \tag{50}$$

Thus, for greatest and least distance

$$d'_{u\,min} = k\ 0;30° = 1;20,36°,$$

$$d'_{u\,max} = k\ 0;35,38° = 1;35,44°,$$

and the difference $\Delta d'_u = 0;15,8°$. But most of IV,23 is devoted to a rather fine point, the variation of the apparent radius of the shadow depending upon the distance of the earth from the sun, which is thus a function of the solar anomaly. The variation amounts to less than 0;1°, and considering the imprecision of the apparent diameters of the moon and shadow, it seems a waste of time to worry over something so minute. Ptolemy does not mention the subject, but it is briefly referred to by Battānī, who sets its maximum value at $\sim$0;0,50°,[68] and its calculation is explained in *Epitome* V,22, which Copernicus follows. It is used for the calculation of magnitudes and durations of lunar eclipses—it can make a difference of about 0;4^h—and is tabulated in the *Alfonsine Tables* and Peurbach's *Eclipse Tables*.

In Fig. 29, the radii of the sun, earth, and shadow stand at S, E, and U, and the vertex of the shadow cone is V. From the similarity of triangles FYT and VEF, $YT/YF = EF/EV$, that is,

$$\frac{r_s - r_e}{s_s} = \frac{r_e}{s_v} \quad \text{or} \quad s_v = \frac{s_s}{r_s - r_e} r_e \tag{51}$$

gives the distance to the vertex of the shadow cone. Then since $EF/EV = UW/UV$,

$$\frac{r_e}{s_v} = \frac{r_u}{s_v - s_u} \quad \text{or} \quad r_u = \frac{s_v - s_u}{s_v} r_e \tag{52}$$

is the true radius of the shadow for a given solar distance. Finally, the apparent radius

$$r_u' \approx \sin^{-1}(r_u/s_u). \tag{53}$$

Note that the distance of the shadow s_u is the distance at which the moon crosses the shadow, which can vary from 65;30^{tr} at limit II to 55;8^{tr} at limit III. In the *Epitome* r_u' is computed for greatest and least solar distance with the moon at both limits, giving four extreme values, but Copernicus computes only for the distance 62^{tr} at which the apparent diameters of the sun and moon are assumed to be equal. Thus for the sun at apogee

$$s_v = 1179r_e/4;27 = 264;56r_e \approx 265r_e,$$

as in IV,19, and

$$r_u = 203r_e/265 = 0;45,58r_e,$$

but Copernicus simply gives $r_u = 0;46,1r_e$, which he had found in IV,19. Then

$$r_u' = \sin^{-1}(0;46,1/62) = 0;42,32°,$$

which of course agrees with $r_u' = 2;41,12 \cdot 0;15,50° = 0;42,32°$ implicit in IV,19. For the sun at perigee,

$$s_v = 1105r_e/4;27 = 248;19r_e,$$
$$r_u = 186;19r_e/248;19 = 0;45,1r_e,$$
$$r_u' = \sin^{-1}(0;45,\mathit{0}/62) = 0;41,35°.$$

---

[68] Battānī (1537), cap. 30, f. 37v; Nallino 1, 58:21–23.

The variation of the shadow between solar apogee and solar perigee is thus

$$\Delta r'_u = 0;42,32° - 0;41,35° = 0;0,57°,$$

and if Copernicus had done the entire computation consistently, the result would be 0;0,51°. But a more serious problem is that all of this is incompatible with the radii of the shadow for the moon at limits II and III. From the apparent diameters given at the beginning of the chapter,

$$\text{II} \quad r'_u = 0;40,18°, \qquad \text{III} \quad r'_u = 0;47,52°.$$

Applying the procedure just explained gives, for the sun at apogee,

$$\text{II} \quad r'_u = 0;39,31°, \qquad \text{III} \quad r'_u = 0;49,23°,$$

while placing the sun at perigee gives

$$\text{II} \quad r'_u = 0;38,38°, \qquad \text{III} \quad r'_u = 0;48,31°.$$

The variation due to the descent of the sun is now, for limit II 0;0,53°, and for limit III 0;0,52°, more in keeping with the accurate computation for $s_m = 62^{tr}$. But since the values for the sun at apogee differ considerably from Copernicus's figures, by 0;0,47° at II and by 0;1,31° at III, it seems pointless to worry about the variation of the shadow at all.

In the original, cancelled version in M (131v–132r), Copernicus had computed the variation of the shadow more accurately, although using the greatest lunar distance at syzygy $65;30^{tr}$ instead of 62^{tr}. Evidently the revision accompanied his change of the demonstration of the solar distance in IV,19, but he could have been more careful in his changes. In the earlier version s_v, from (51), is the same as in the revision, but with $s_u = 65;30r_e$, from (52), for the sun at apogee

$$r_u = 199;30r_e/265 = 0;45,10r_e,$$

in agreement with the original demonstration in IV,19, and for the sun at perigee,

$$r_u = 182;49r_e/248;19 = 0;44,14r_e. \quad \text{(Corr. 0;44,10)}$$

Then from (53), the apparent diameter with the sun at apogee,

$$r'_u = \sin^{-1}(0;45,10/65;30) = 0;39,30°,$$

(note that 0;39,30°/0;15° = 79/30 = k from the original version of IV,19) and for the sun at perigee,

$$r'_u = \sin^{-1}(0;44,14/65;30) = 0;38,41°,$$

leading to the variation

$$\Delta r'_u = 0;39,30° - 0;38,41° = 0;0,49°,$$

which is closer to an accurate computation than the result of the revision.

The original version also shows the basis of an otherwise incorrect statement that Copernicus makes at the end of the chapter. From the demonstration of the variation of the shadow just given, he claims that when the sun is at apogee $r'_u/r'_m > 13/5$, while when the sun is at perigee $r'_u/r'_m <$ 13/5. In the revision this is not so, for

$$\frac{0;42,32^\circ}{0;15,50^\circ} = 2;41 > \frac{0;41,35^\circ}{0;15,50^\circ} = 2;38 > \frac{13}{5} = 2;36,$$

but from the original version, where at the distance $65;30^{\mathrm{tr}}$ $r'_m = 0;15^\circ$,

$$\frac{0;39,30^\circ}{0;15^\circ} = 2;38 > \frac{13}{5} = 2;36 > \frac{0;38,41^\circ}{0;15^\circ} = 2;34,44.$$

He then says that if one follows the opinion of the "ancients," and everywhere uses 13/5, the error will be small. At least for his original version this was true, but it cannot be held to apply correctly to the revision. It should be noted that these internal contradictions follow from Copernicus's flawed revision of the solar distance in IV,19 and from his failure to correct all of its consequences accordingly.

Table of Apparent Semidiameters

Copernicus provides a table of the apparent semidiameters of the sun, moon, and shadow, and of the variation of the shadow. He gives no explanation of its construction and, curiously, no explanation of its use. He must have assumed that his readers were familiar with such tables, since exactly the same sort are found in the *Alfonsine Tables* and Peurbach's *Eclipse Tables*. The table is in six columns.

Columns 1 and 2 contain the arguments in 6° intervals for 6°–180° and for 180°–354°. They are entered with the true anomaly of the sun α_s and the true anomaly of the moon α_m.

Column 3, $c_{3(\alpha_s)}$, gives the apparent semidiameter of the sun as a function of the true solar anomaly α_s over the range 0;15,50° to 0;16,57°. Note that Copernicus has discarded the minimum diameter of 0;31,48° established in IV,21, and returned to 0;31,40° used in IV,19, while the maximum does follow from 0;33,54° of IV,21. Both $c_{3(\alpha_s)}$ for the sun and the following $c_{4(\alpha_m)}$ for the moon were computed in the same way, by the method described in *Epitome* VI,6, which in turn is based upon Jābir ibn Aflaḥ.[69]

[69] Cf. the computation of the coefficient $c_{9(\alpha)}$ in the parallax table, below, p. 260. The following description is slightly shortened from the one in the *Epitome*, but is essentially the same. *Epitome* V,22 shows how to find the apparent diameters of the sun and moon directly from the true diameters and absolute distances, but it is more trouble than it is worth, and Copernicus does not use it. For the source of the *Epitome*, cf. Jābir (1534), 77.

Call the distance of the sun or of the moon at syzygy ρ, which is computed from

$$\rho = (R^2 + r^2 + 2Rr \cos \alpha)^{1/2}, \tag{54}$$

where for the sun r is the radius of the epicycle or the equivalent eccentricity, and for the moon the radius of the epicycle at mean syzygy, $r = r_1 - r_2$. ρ will vary between a greatest distance of $R + r$ and a least distance of $R - r$, with a total variation of $2r$. Take $\rho - (R - r)$, and find the ratio

$$\frac{\rho - (R - r)}{2r} = g. \tag{55}$$

Corresponding to the variation in distance is a variation in apparent radius from $r'_{\min}$ at greatest distance to $r'_{\max} = r'_{\min} + \delta$ at least distance, and since the angles are small, it is assumed that the variation of r' is proportional to g. Hence, for any value of α,

$$r'_{(\alpha)} = r'_{\min} + g\delta. \tag{56}$$

In the case of the sun, $\delta = 1;7°$, and c_3 is computed from

$$c_{3(\alpha_s)} = 0;15,50° + g1;7°. \tag{57}$$

We have tested the column, finding ρ and g from $r = e = 323$, and the greatest difference from the tabulated values is $\pm 0;0,1°$, which is probably the result of rounding.

Column 4, $c_{4(\alpha_m)}$, gives the apparent semidiameter of the moon at syzygy as a function of the true anomaly over the range 0;15° to 0;17,49° established in IV,22 for limits II and III. The computation is the same as for the sun, but with the epicyclic radius $r = 860$ and $\delta = 2;49°$. Thus,

$$c_{4(\alpha_m)} = 0;15° + g2;49°, \tag{58}$$

and again checking shows errors of only $\pm 0;0,1°$.

Column 5, $c_{5(\alpha_m)}$, is the apparent semidiameter of the shadow as a function of the true lunar anomaly, since the apparent size of the shadow varies inversely with the distance of the moon where it crosses the shadow. c_5 is computed from

$$c_{5(\alpha_m)} = (403/150)c_{4(\alpha_m)} = 2;41,12c_{4(\alpha_m)}, \tag{59}$$

where 403/150 is the ratio k established in IV,19 and used in IV,23.[70]

Column 6, $c_{6(\alpha_s)}$, gives the variation in the radius of the shadow due to the changing distance of the sun. In IV,23 Copernicus established the range of this variation as 0;0,57° for a lunar distance of 62^{tr}, although originally in M he had found a variation of 0;0,49° for a distance of $65;30^{\text{tr}}$. He now

[70] In M (134v) c_5 was computed from $79/30 = 2;38$, a cancelled value from IV,19 (above, p. 247, n. 64) likewise cancelled in IV,23.

assumes that the same variation is valid for all lunar distances. It would be extremely tedious to compute the variation of the shadow by the method used in IV,23 to find the extrema, and that the extrema are incorrect shows that Copernicus must have computed c_6 in some other way, which must be the following: In computing the apparent semidiameter of the sun, $c_{3(\alpha_s)}$, he found a coefficient g from (55) above that expressed the variation of solar distance as a function of the true anomaly α_s. Applying the coefficient to the variation of the shadow gives

$$c_{6(\alpha_s)} = g \cdot 0;0,57^\circ. \tag{60}$$

This appears to reproduce c_6 quite well; occasional errors of $\pm 0;0,1^\circ$ must again be due to rounding. In M c_6 is computed for a variation of $0;0,50^\circ$.

Columns 3, 4, and 5 are graphed in Fig. 30, with c_5 at one-half the vertical scale of c_3 and c_4. One can easily use the graph or the table to estimate the possibility of total and annular central solar eclipses. According to Ptolemy, the minimum apparent diameter of the moon is equal to the diameter of the sun, so annular eclipses are entirely excluded. Here it is evident that all central eclipses for which $290^\circ \leq \alpha_m \leq 70^\circ$ will be annular, and all for which $115^\circ \leq \alpha_m \leq 245^\circ$ will be total. This covers 270° of the circle. In the two 45° arcs from 70° to 115° and from 245° to 290°, central eclipses may be annular or total depending upon the specific values of α_m and α_s.

Use of the table. The table is very simple to use. For the apparent radius of the sun, enter column 1 or 2 with α_s, and directly read $r'_s = c_{3(\alpha_s)}$, interpolating as necessary. For the radius of the moon, enter with α_m and read $r'_m = c_{4(\alpha_m)}$. Then for the shadow, again enter with α_m and find $r'_u = c_{5(\alpha_m)}$. Finally, correct the radius of the shadow for the change of solar distance by entering with α_s and finding the correction $c_u = c_{6(\alpha_s)}$, which is always negative, and forming

$$r'_u - c_u = c_{5(\alpha_m)} - c_{6(\alpha_s)}. \tag{61}$$

Example. Assume that $\alpha_s = 156^\circ$ and $\alpha_m = 288^\circ$. Thus,

$$r'_s = c_{3(\alpha_s)} = 0;16,54^\circ, \qquad r'_m = c_{4(\alpha_m)} = 0;15,56^\circ,$$

$$r'_u - c_u = c_{5(\alpha_m)} - c_{6(\alpha_s)} = 0;42,40^\circ - 0;0,54^\circ = 0;41,56^\circ.$$

Table of Parallax and Its Use (IV,24–25)

For computing parallaxes in the circle of altitude, Copernicus provides a table much like the one in *Almagest* V,18, the computation of which is explained in *Almagest* V,17 and *Epitome* V,23–24. But because the lunar distance varies much less between syzygy and quadrature in Copernicus's model than in Ptolemy's, he can simplify the columns for interpolation.

The parallax table is arranged such that one enters with a true zenith distance that has been found previously by computation from the lunar theory and any necessary spherical astronomy. In the course of computing the lunar position, the mean elongation and true anomaly necessary for interpolation have already been found. The table contains nine columns.

Columns 1 and 2 contain the arguments for 6°–180° and 180°–354° at intervals of 6°. For columns 3–7, only column 1 is entered with 2ζ, twice the geocentric zenith distance; thus column 1 functions as a tabulation from 3°–90° at 3° intervals. Columns 8 and 9 are entered with $2\bar{\eta}$ and α respectively, for which columns 1 and 2 are used as written.

Columns 3–7 are all in principle based upon the same formula for finding the parallax p at a given geocentric zenith distance ζ when the geocentric distance s is known. Thus, in Fig. 31, the horizontal parallax P appears when the object is at P_0 where $\zeta = 90°$, and approximately

$$P = \sin^{-1}(r_e/s), \tag{62}$$

the approximation because of the absence of a tangent function and because P is small. The values of s and P have already been determined for the sun at greatest, least, and mean distance, and for the moon at the four limits of distance. For any other zenith distance, *Almagest* V,17 and *Epitome* V,23 recommend an approximation. In the figure, letting $r_e = 1$,

$$OQ = \sin\zeta, \qquad EQ = \cos\zeta, \qquad PQ = s - \cos\zeta.$$

Assume that $OP \approx PQ$, and solve

$$p = \sin^{-1}[\sin\zeta/(s - \cos\zeta)], \tag{63}$$

another substitution of sine for tangent. Unfortunately, many values in Copernicus's table cannot be reproduced by this formula, the errors frequently amounting to 3 or 4 seconds in the case of the moon, which are beyond the imprecision of rounding. In particular, its use would expose the error of taking P_{ii} as 0;52,24° instead of 0;52,29°, which Copernicus did not seem to detect. Errors of this size are of no real significance in the application of the table, but they leave doubt about exactly how it was computed. In any case, Copernicus computed p at all zenith distances for the sun at mean distance and for the moon at the four limits, and then tabulated the following quantities:

Column 3, $c_{3(2\zeta)}$, the solar parallax for $s = 1142r_e$.
Column 4, $c_{4(2\zeta)}$, the deficit of the lunar parallax at the first limit below the parallax at the second limit, i.e. $p_i - p_{ii} = -c_4$.
Column 5, $c_{5(2\zeta)}$, the parallax at the second limit, p_{ii}.
Column 6, $c_{6(2\zeta)}$, the parallax at the third limit, p_{iii}.
Column 7, $c_{7(2\zeta)}$, the excess of the parallax at the fourth limit over the parallax at the third limit, $p_{iv} - p_{iii} = c_7$.

Column 8, $c_{8(2\bar{\eta})}$. Much of the chapter is devoted to the explanation of $c_{8(2\bar{\eta})}$, the coefficient of interpolation for the position of the moon on the smaller epicycle. What Copernicus must show is that the same coefficient is applicable when the smaller epicycle is at both its greatest and least geocentric distance on the larger epicycle. Otherwise, he would have to provide two distinct columns, as Ptolemy must for the motion of the moon on the epicycle because the variation of the distance of the epicycle due to its motion on the eccentric is so large.

In Fig. 32, the smaller epicycle is shown at the apogee of the larger epicycle when $\alpha = 0°$, the condition for c_4, and at the perigee when $\alpha = 180°$, the condition for c_7, with the moon at the identical value of $2\bar{\eta}$ at each location. In order to show that the same coefficient of interpolation, $c_{8(2\bar{\eta})}$, is applicable to both locations, it must be shown that

$$\frac{EP - EB}{2r_2} \approx \frac{EB' - EP'}{2r_2}, \tag{64}$$

and this ratio will also define $c_{8(2\bar{\eta})}$. For each situation, the effective radius of the epicycle is

$$C_1P = C_1P' = r = (r_1^2 + r_2^2 + 2r_1r_2 \cos 2\bar{\eta})^{1/2}. \tag{14}$$

Correctly, one next finds

$$c_{3(2\eta)} = \sin^{-1}(r_2 \sin 2\bar{\eta}/r), \tag{13}$$

and then

$$\begin{aligned} EP &= (R^2 + r^2 + 2Rr \cos c_3)^{1/2}, \\ EP' &= (R^2 + r^2 - 2Rr \cos c_3)^{1/2}. \end{aligned} \tag{65}$$

One can then find the ratios

$$\frac{EP - EB}{2r_2}, \qquad \frac{EB' - EP'}{2r_2},$$

and see if they are close enough to equality to be represented by the same coefficient of interpolation.

In fact they are, but Copernicus's demonstration is a hopeless mess that ends up proving nothing. He does not proceed as described above, but after finding $C_1P = C_1P'$, he says they are to be measured along EC_1 to form

$$EQ = EC_1 + C_1P, \qquad EQ' = EC_1 - C_1P'. \tag{66}$$

He then takes

$$QB = EQ - EB, \qquad Q'B' = EB' - EQ', \tag{67}$$

and since the total variation in distance is $2C_2P = 2C_2'P'$, his final, and rather redundant, steps are to show that

$$\frac{QB(60/EB)}{2C_2P(60/EB)} \approx \frac{Q'B'(60/EB')}{2C_2'P'(60/EB')}. \tag{68}$$

This, however, is not only cumbersome, but simply a truism, since of necessity $QB = Q'B'$, and all distances from E have been cancelled out. However, all of this is masked in the text by Copernicus's round-about procedure and extraordinary number of computational errors.

In his example, Copernicus lets $2\bar{\eta} = 60°$, and says that $r = 6;7$ where $r_1 = 6;36$ and $r_2 = 1;25$, which is not true, for correctly $r = 6;1$. Anyway, let $C_1P = C_1P' = r = 6;7$, and

$$EQ = 60;18 + 6;7 = 66;25, \qquad EQ' = 60;18 - 6;7 = 54;11,$$

$$QB = 66;25 - 65;30 = 0;55, \qquad Q'B' = 55;8 - 54;11 = 0;57.$$

Correctly and consistently $EB = 65;29$ and $EB' = 55;7$, and with these values $QB = QB' = 0;56$. Copernicus says that $QB = Q'B' = 0;55\frac{1}{2}$. He then computes, one painful step at a time,

$$\frac{0;55\frac{1}{2}(60/65;30)}{2;51(60/65;30)} = \frac{0;46}{2;37} \approx 0;18,$$

although correctly he should have found $0;51/2;37 = 0;19,29$, and

$$\frac{0;55\frac{1}{2}(60/55;8)}{2;51(60/55;8)} = \frac{0;55}{3;7} \approx 0;18,$$

but here correctly one has $1;0/3;7 = 0;19,15$. (Copernicus remarks, "Yet they differ a little by some seconds.") So even with all the preceding errors, the result should be about 0;19. And of course nothing has been proved since the different distances from E have been cancelled out of the computation.

At the end of all this trouble, Copernicus says that if instead we use the coefficient of interpolation $c_{4(2\bar{\eta})}$ from the lunar correction table, we will in no way commit an error. And that is exactly what he does. Hence c_8 has nothing to do with the method just described at such length. It is perhaps notable that for $2\bar{\eta} = 60°$, $c_{4(2\bar{\eta})} = 0;18$, which is perhaps why Copernicus makes so many errors and ends up with 0;18 in this demonstration.

If by chance one wishes to do the comparison correctly rather than in Copernicus's way, then where $2\bar{\eta} = 60°$,

$$C_1P = C_1P' = 6;1, \qquad c_3 = 11;50°,$$

$$EP = 66;12,44, \qquad EP' = 54;24,48,$$

$$EP - EB = 0;42,44, \qquad EB' - EP' = 0;43,16.$$

Thus it is already clear that the change of distance is about the same at both positions. The coefficient of interpolation is in each case

$$c = 0;42,44/2;51 = 0;15,0,$$

$$c = 0;43,16/2;51 = 0;15,10 \approx 0;15.$$

Column 9, $c_{9(\alpha)}$. After the wasted effort of column 8, few problems are posed by column 9, the coefficient of interpolation for motion on the larger epicycle or, more precisely, the effective epicycle at syzygy. The method of computation is explained in *Almagest* V,17 and *Epitome* V,24. Figure 33 shows the epicycle of radius

$$r = r_1 - r_2 = 6;36,30^{\text{tr}} - 1;25,30^{\text{tr}} = 5;11^{\text{tr}}.$$

Where, Copernicus says, $R = 60;19^{\text{tr}}$, the distance of the moon varies between $R + r = 65;30^{\text{tr}}$ and $R - r = 55;8^{\text{tr}}$, over a range of $2r = 10;22^{\text{tr}}$. Compute ρ from

$$\rho = (R^2 + r^2 + 2Rr \cos \alpha)^{1/2}, \qquad (54)$$

and $c_{9(\alpha)}$ gives the ratio

$$c_{9(\alpha)} = \frac{R + r - \rho}{2r_2} = \frac{65;30 - \rho}{10;22}, \qquad (69)$$

rounded to minutes from 0 to 60′. c_9 is exactly the coefficient g computed for the semidiameter of the moon in the table of semidiameters, although g must have been computed to at least two places.[71]

Use of the table. The application of the columns is summarized as follows:

$c_{3(2\zeta)}$	p_s	mean distance			
$c_{4(2\zeta)}$	$p_{ii} - p_i$	$2\bar{\eta} = 180°$	$\alpha = 0°$	quadrature	
$c_{5(2\zeta)}$	p_{ii}	$2\bar{\eta} = 0$	$\alpha = 0$	syzygy	
$c_{6(2\zeta)}$	p_{iii}	$2\bar{\eta} = 0$	$\alpha = 180$	syzygy	(70)
$c_{7(2\zeta)}$	$p_{iv} - p_{iii}$	$2\bar{\eta} = 180$	$\alpha = 180$	quadrature	
$c_{8(2\bar{\eta})}$	interpolation	$2\bar{\eta}$ between syzygy and quadrature			
$c_{9(\alpha)}$	interpolation	α between apogee and perigee			

The resulting parallaxes at each limit are graphed in Fig. 34.

Let $\alpha = 0°$, the moon at apogee, and $2\bar{\eta}$ be some value. Then for the zenith distance ζ, compute the minimum parallax

$$p_m = c_{5(2\zeta)} - c_{8(2\bar{\eta})} \cdot c_{4(2\zeta)}. \qquad (71)$$

[71] Cf. above, p. 255.

Next, let $\alpha = 180°$, the moon at perigee, and for the same values of $2\bar{\eta}$ and ζ, compute the maximum parallax

$$p_{\mathrm{M}} = c_{6(2\zeta)} + c_{8(2\bar{\eta})} \cdot c_{7(2\zeta)}. \tag{72}$$

Now, interpolate between the difference $p_{\mathrm{M}} - p_{\mathrm{m}}$ for some value of α by

$$\Delta p = c_{9(\alpha)}(p_{\mathrm{M}} - p_{\mathrm{m}}). \tag{73}$$

Finally, the difference is added to the minimum parallax p_{m} to form the true parallax

$$p = p_{\mathrm{m}} + \Delta p. \tag{74}$$

Example. Copernicus gives an example in which $\zeta = 54°$, about the zenith distance of the culminating point of the equator in Frauenburg, the mean elongation $\bar{\eta} = 15°$, and the true anomaly $\alpha = 100°$. Take $2\zeta = 108°$, and find

$$c_4 = 0;1,48°, \qquad c_5 = 0;42,50°, \qquad c_6 = 0;50,59°, \qquad c_7 = 0;2,46°.$$

For $2\bar{\eta} = 30°$ and $\alpha = 100°$, find

$$c_{8(2\bar{\eta})} = 0;5, \qquad c_{9(\alpha)} = 0;34.$$

Hence,

$$p_{\mathrm{m}} = 0;42,50° - 0;5 \cdot 0;1,48° = 0;42,50° - 0;0,9° = 0;42,41°,$$

$$p_{\mathrm{M}} = 0;50,59° + 0;5 \cdot 0;2,46° = 0;50,59° + 0;0,14° = 0;51,13°,$$

$$\Delta p = 0;34\ (0;51,13° - 0;42,41°) = 0;34\ (0;8,32°) = 0;4,50°,$$

$$p = 0;42,41° + 0;4,50° = 0;47,31°.$$

And if the sun likewise has a zenith distance $\zeta = 54°$, take $2\zeta = 108°$, and find in column 3

$$p_{\mathrm{s}} = 0;2,26°.$$

The difference between the lunar and solar parallax, the adjusted parallax $p' = p - p_{\mathrm{s}}$, which is used in computing solar eclipses, is

$$p' = 0;47,31° - 0;2,26° = 0;45,5°.$$

Since the lunar and solar positions are computed with the correction tables only to minutes, it is generally superfluous to retain the seconds in a fully computed parallax—the tabulation of c_8 and c_9 only to one place in itself makes the seconds only approximate—and the components of the parallax in longitude and latitude will surely be rounded to minutes. Copernicus remarks correctly that the differences c_4 and c_7 make so little difference that it is usually adequate to use only the terms for syzygy, c_5 and c_6. In this way

$$p = p_{\mathrm{m}} + \Delta p = c_5 + c_{9(\alpha)}(c_6 - c_5), \tag{75}$$

a very simple computation. In the previous example, this would give

$$p = 0;42,50^\circ + 0;34\,(0;50,59^\circ - 0;42,50^\circ)$$
$$= 0;42,50^\circ + 0;4,37^\circ = 0;47,27^\circ,$$

differing by only 0;0,4° from the full computation. Inspection of the table shows that this approximation is more than sufficient since, letting $c_9 = 1;0$, the greatest difference that can occur between the abbreviated and full computation is $c_8 \cdot c_7$ which, for $2\bar{\eta} = 60^\circ$, amounts to about 0;1°. Hence for eclipses in particular, it is never necessary to use the fully computed parallax.

Resolution of the Components of Parallax in Longitude and Latitude (IV,26)

The parallax in the circle of altitude is seldom of use by itself, for what is usually required is the effect of the parallax on the location of the moon in a different coordinate system; in the case of early astronomy, in longitude and latitude. The problems involved are nothing more than transformations of coordinates—in modern terms, from zenith distance and hour angle to longitude and latitude, or from parallax and hour angle to change of longitude and latitude—but the solutions can be cumbersome, and in *Almagest* V,19 Ptolemy makes some approximations to simplify the solutions. Ptolemy's exposition is systematically enlarged with specific trigonometric formulas in *Epitome* V,26–32. The approximations, shown in Fig. 35, are

(1) to substitute plane for spherical triangles,
(2) to take as the parallax of the moon p the parallax of the point of corresponding longitude in the ecliptic λ_m,
(3) to consider the circle of altitude through the moon parallel to the circle of altitude through λ_m,

(2) and (3) amount to ignoring the latitude of the moon in finding and resolving the parallax.

Under these simplifying assumptions, the resolution of p into its components in longitude p_λ and latitude p_β is straightforward. In order to find p from the table of parallax, we must first find the zenith distance ζ corresponding to λ_m. This is a lengthy enough procedure, using much of the spherical astronomy of Book II, and finally, in II,12, the determination of ζ and the angle of intersection ξ of the ecliptic and circle of altitude. Having found $p = MM'$ from the parallax table, we resolve it into its components p_λ and p_β parallel and perpendicular to the ecliptic. Since from approximation (3), the angle at M' is taken to be ξ,

$$p_\lambda = p \cos \xi, \qquad p_\beta = p \sin \xi. \tag{76}$$

The *Epitome* shows how to do this spherically, and gives as corollaries to V,26 the correct spherical relations

$$\sin p_\beta = \sin p \sin \xi, \qquad \sin p_\lambda = \sin p \sin \kappa, \qquad \cos p = \cos p_\lambda \cos p_\beta, \tag{77}$$

where κ, called the "angle of longitude," is the angle opposite p_λ and is given by $\sin \kappa = \cos \xi / \cos p_\beta$. Then the apparent longitude,

$$\lambda' = \lambda + p_\lambda, \tag{78}$$

where p_λ is positive for λ between the ascendent and the nonagesimal, i.e. $\lambda_G \geq \lambda \geq \lambda_N = \lambda_G - 90°$, and negative for λ between the nonagesimal and the descendent, i.e. $\lambda_N \geq \lambda \geq \lambda_D = \lambda_G - 180°$. The apparent latitude

$$\beta' = \beta + p_\beta, \tag{79}$$

where p_β is positive for the midheaven north of the zenith, which never happens above $\varphi \approx 23\frac{1}{2}°$N, and negative for the midheaven south of the zenith, which never happens below $\varphi \approx 23\frac{1}{2}°$S.[72] Obviously, for all of Europe and the Mediterranean, p_β is always negative.

Ptolemy also shows how to find the components of parallax without assumption (3) while retaining (1) and (2), and this leads to much complication, particularly the treatment of very large spherical triangles as though they were plane, that is not satisfactorily resolved.[73] He explains the method with a numerical example, and is followed by the *Epitome* which also gives a spherical solution in V,31–32. Copernicus shows how to find the components without assumption (3), but he does not include an example, and it is unclear whether he intends the solution to be plane or spherical. His figure is spherical, but this was also true of the simplified solution where he specifically said to use plane triangles. At any rate, he concludes, as did Ptolemy, that taking the latitude of the moon into account when finding its parallax is more trouble than it is worth, and makes very little difference, particularly in northern regions where the moon does not reach the zenith. Where the moon can reach the zenith, the maximum difference will be the maximum parallax for $\zeta = \beta = 5°$, which is $c_6 + c_7 = 0;5,50° \approx 0;6°$. And within the solar eclipse limits, where $\zeta = \beta < 1\frac{1}{2}°$, $c_6 + c_7 = 0;1,45°$.

Example. At the end of the lunar theory, we computed the position of the moon for a mean time in Cracow of 9 Mar 1497, 23;40^h, and found

$$\lambda_☾ = ♊\,3;25°, \qquad \beta = -4;35°, \qquad \lambda_☉ = ♓\,28;30°.^{74}$$

[72] If the lunar latitude were included in finding ζ, φ would be increased to about $\pm 28\frac{1}{2}°$, but approximation (2) means that only the relation of the midheaven to the zenith need be considered.

[73] See *HAMA*, 115–16.

[74] Above, pp. 229–230.

For $\lambda_{☾}$ Copernicus found ♊3;24°. The purpose of the computation is to examine Copernicus's test of the lunar parallax in IV,27 by recomputing an occultation of Aldebaran. We shall now continue the examination by finding the parallax and its components in longitude and latitude as observed in Bologna ($\varphi = 44;30°$, a necessary element not explicitly given by Copernicus) at a local time of 11^h after noon.

From the longitude of the sun at 11^h after noon, the right ascension of the midheaven,

$$\alpha_M = \alpha(358;30°) + 11 \cdot 15° = 358;37° + 165° = 163;37°, \qquad (2.28)$$

for which $\lambda_M =$ ♍12;14°. The oblique ascension of the horoscopus

$$\rho_G = \alpha_M + 90° = 253;37°, \qquad (2.31)$$

for which, by inverse use of the oblique ascension table, $\lambda_G =$ ♏25;53°. Copernicus gives ♏26°. Using the table of angles of the ecliptic and horizon, for $\lambda_G =$ ♏26°, $\nu = 55;42°$, but Copernicus gives $\nu = 59\frac{1}{2}°$. We can check the result by computation from (2.33). For $\lambda_M =$ ♍12;14°, the declination $\delta = +6;59°$, and since $\bar{\varphi} = 45;30°$, the altitude of the midheaven

$$\bar{\zeta}_M = \bar{\varphi} + \delta_M = 52;29°. \qquad (2.32)$$

Since

$$\lambda_G - \lambda_M = \text{♏}26° - \text{♍}12;14° = 73;46°,$$

$$\nu = \sin^{-1}(\sin \bar{\zeta}_M / \sin(\lambda_G - \lambda_M)) = 55;42°, \qquad (2.33)$$

which agrees exactly with Copernicus's table. Hence his error of nearly 4° is specific to this computation, not to the table.

We next consider the moon at $\lambda_{☾} =$ ♊3;24°, Copernicus's value. Since $\lambda_G =$ ♏26°, the descendent $\lambda_D = \lambda_G - 180° =$ ♉26°, and

$$\lambda_{☾} - \lambda_D = \text{♊}3;24° - \text{♉}26° = 7;24°.$$

The arc

$$\lambda_M - \lambda_D = \text{♍}12;14° - \text{♉}26° = 106;14°,$$

and we have established that $\bar{\zeta}_M = 52;29°$. Thus, the altitude of the moon, from (2.38) arranged for the descendent is

$$\bar{\zeta}_{☾} = \sin^{-1}\left(\frac{\sin(\lambda_{☾} - \lambda_D)}{\sin(\lambda_M - \lambda_D)} \sin \bar{\zeta}_M\right) = 6;6°, \qquad \zeta_{☾} = 83;54°.$$

Copernicus gives 84°. We can check the result in another way. From (2.41)

$$\sin \bar{\zeta}_{☾} = \sin \nu \sin(\lambda_{☾} - \lambda_D),$$

which gives correctly

$$\bar{\zeta}_{☾} = \sin^{-1}(\sin 55;42° \sin 7;24°) = 6;6°, \qquad \zeta_{☾} = 83;54°,$$

in agreement with (2.38), or with Copernicus's incorrect v,

$$\bar{\zeta}_{☾} = \sin^{-1}(\sin 59;30° \sin 7;24°) = 6;22°, \qquad \zeta_{☾} = 83;38°,$$

which can still be rounded to 84°. All that remains is to find the angle ξ between the ecliptic and the circle of altitude. Using (2.40) with the correct v,

$$\xi = \sin^{-1}(\cos v/\cos \bar{\zeta}) = \sin^{-1}(\cos 55;42°/\cos 6°) = 34;31°,$$

and with the incorrect v,

$$\xi = \sin^{-1}(\cos 59;30°/\cos 6°) = 30;41°.$$

Copernicus, however, gives $\xi \approx 29°$, which is very wrong, implying $v > 61°$, since correctly $\xi + v > 90°$ by the spherical excess.

In the computation of the position of the moon, Copernicus found $\bar{\eta} = 74°$ and $\alpha = 111;10°$, which we use along with $\zeta = 84°$ to find the parallax in the circle of altitude from the parallax table, which gives to full accuracy,

$$p = p_{\mathrm{m}} + \Delta p = 0;50,17,30° + 0;9,59,49° = 1;0,17,19°, \qquad (74)$$

while Copernicus gives $p = 1°$. Hence, with the correct value $\xi = 34;31°$,

$$p_{\lambda} = 1° \cos 34;31° = 0;49,26°, \qquad p_{\beta} = 1° \sin 34;31° = 0;34,0°.$$

With Copernicus's value $\xi = 29°$

$$p_{\lambda} = 0;52,29°, \qquad p_{\beta} = 0;29,5°,$$

but Copernicus gives

$$p_{\lambda} = 0;51°, \qquad p_{\beta} = 0;30°,$$ [75]

which is strange since the computation is so trivial. Note that almost every step of Copernicus's computation of the parallax and its components has contained some kind of error, although his result is not far from the correct parallax since the arcs are so small. Finally, using Copernicus's numbers, the apparent position of the moon is

$$\lambda'_{☾} = ♊\,3;24° - 0;51° = ♊\,2;33°, \qquad \beta'_{☾} = -4;35° - 0;30° = -5;5°.$$

With the fully accurate computation of $\lambda_{☾}$, p_{λ} and p_{β},

$$\lambda'_{☾} = ♊\,3;25° - 0;49° = ♊\,2;36°, \qquad \beta'_{☾} = -4;35° - 0;34° = -5;9°.$$

---

[75] The text of N, 129r22–23, reads *parallaxis lunae pars una, longitudinis scrup. LI. latitudinis scru. XXX*, which means "the parallax of the moon [in the circle of altitude] was 1°, in longitude 0;51°, in latitude 0;30°." However, the critical editions have worked wonders with the comma after *una*—it was omitted in M, 137v3—and the translations have done even worse, e.g. reading a parallax in longitude of 1;51°, which simply cannot happen, or dropping the 1° parallax in the circle of altitude entirely. One might suggest that some computation ought to precede emendation.

Test of the Lunar Parallax (IV,27)

Copernicus performs a test of the lunar parallax, and also of the lunar theory, by showing that together they correctly reproduce the occultation of Aldebaran that he observed in Bologna on 9 Mar 1497 at 1^h before midnight. The test is particularly interesting since the moon was at 74° of mean elongation, where Ptolemy's theory would have it close to the earth and displaying a large parallax. The report of the observation is as follows:

> For we observed that the moon was about to occult (*occultura*) the bright star of the Hyades, which the Romans call *Palilicium*,[76] and having waited, we saw the star touched (*applicatam*) by the dark part of the lunar body, and immediately (*iamque*) hidden between the horns of the moon at the end[77] of the fifth hour of night, but nearer to the southern horn by nearly a third of the width or diameter of the moon. And since by computation the star was in ♊2;52° with a southern latitude of $5\frac{1}{6}$°, it was evident that by observation (*secundum visum*) the center of the moon preceded [i.e. was west of] the star by half a diameter, and therefore its apparent position was in longitude ♊2;36°, in latitude about 5;2° (M 137r19–20, 5;6°).

The report reads authentically enough, particularly the description of the sudden extinction of the star on contact with the moon, but it does present a problem, namely, whether to accept the estimated lunar latitude in N, −5;2°, or in M, −5;6°. The star is supposed to be at

$$\lambda_* = ♊2;52^\circ, \qquad \beta_* = -5;10^\circ,$$

and was $\frac{1}{2}d'_m = 2;52^\circ - 2;36^\circ = 0;16^\circ$ from the center of the moon, but was nearly $\frac{1}{3}d'_m$ nearer to the southern horn than to what?—the center, the northern horn? This is important for choosing between the latitudes in M and N. Both possibilities are shown in Fig. 36. Neither value is really satisfactory, although $\beta = -5;6^\circ$, giving $Ca' = 0;4^\circ$, corresponding to an occultation at a, $\frac{1}{3}d'_m$ nearer S than N, is close to the result of Copernicus's computation, $\beta = -5;5^\circ$, just given. Yet an occultation at b, closer to the southern horn at $\frac{1}{3}d'_m$ from C, is what really happened, that is, what Copernicus must have observed. For this reason, the reading $\beta = -5;2^\circ$, raising the center of the moon such that $Cb' = 0;8^\circ$ above the star, is closer to the true observation. We shall have to consider both possibilities.

The components of parallax implicit in the observation are

$$p_\lambda = ♊3;24^\circ - ♊2;36^\circ = +0;48^\circ,$$

$$p_\beta = -5;2^\circ + 4;35^\circ = -0;27^\circ$$

or

$$p_\beta = -5;6^\circ + 4;35^\circ = -0;31^\circ,$$

[76] See Menzzer's note in Copernicus (1879), Anm. 47, n. 335.

[77] The word *fine* is not in N, but in M, 137r13. In fact, he observed the occultation forty minutes earlier at $10\frac{1}{3}^h$ after noon. See below, p. 269.

compared with Copernicus's computed

$$p_\lambda = 0;51°, \qquad p_\beta = -0;30°,$$

or with the correctly computed

$$p_\lambda = 0;49,26°, \qquad p_\beta = -0;34°.$$

Interestingly, p_λ from the observation agrees best with the accurate computation. Taking the reading $\beta = -5;6°$ produces the best agreement with either computation of p_β.

Now let us examine the occultation more closely, beginning with the location of the star. According to Copernicus's star catalogue, Aldebaran is at

$$\lambda^* = 36;0°, \qquad \beta = -5;10°.$$

From the precession theory, for midnight 9/10 Mar 1497,

$$\pi = \bar{\pi} + \delta_P = 26;24,42° + 0;37,42° = 27;2,24°, \tag{3.12}$$

and thus the tropical longitude of Aldebaran,

$$\lambda = \lambda^* + \pi = \text{♊}\,3;2°. \tag{3.13}$$

But Copernicus gave ♊ 2;52°, no less than $0;10° = \frac{1}{3}d'_m$ too low. Hence, from the correctly precessed position of the star, the occultation would occur about $0;20^h$ later than the time reported by Copernicus. It is hard to see how he could make such an error in computing the longitude since it is crucial to the success of his test of the parallax.

The situation is illustrated in Fig. 37, and the following table shows the location of the observed and computed positions.

		λ	β	Location	
Aldebaran	Comp. Cop.	♊ 2;52°	−5;10°	A	
Aldebaran	Recomp.	3;2	−5;10	B	
Aldebaran	Correct	2;46	−5;30	C	
True moon	Comp. Cop.	3;24	−4;35	M	(80)
True moon	Recomp.	3;25	−4;35	N	
App. moon	Observed	2;36	−5;2 or −5;6	M_0	
App. moon	Comp. Cop.	2;33	−5;5	M'	
App. moon	Recomp.	2;36	−5;9	N'	

Since the moon was not at syzygy, from the rule (48) in IV,22, $r'_m = (2/7)P \approx 0;17°$, and the drawing is done accordingly. The observed position of the moon M_0 is shown for $\beta = -5;2°$. In any case, with Aldebaran at A, either Copernicus's computation, placing the moon at M', or the recomputation, placing the moon at N', adequately reproduces the occultation of A. Were M_0 placed at $\beta = -5;6°$, the agreement with the two computations would be even better. But unfortunately, by recomputation from Copernicus precession theory, the star was not at A, but at B, and so the computed position does not reproduce the occultation. Now it is true that since B is 0;10° east of A, an occultation will occur about 20 minutes later, but that hardly confirms the computed true and apparent positions of the moon. It might charitably be suggested that Copernicus made an error of $-0;20^h$ in recording the time of the observation, and by coincidence made an exactly compensating error of $-0;10°$ in computing the precession, but we shall soon see that the occultation actually occurred $0;40^h$ earlier, so mere error is out of the question.

If this is not bad enough, there is worse to come, for Aldebaran was in fact neither at A nor B, but at C, and by Copernicus's lunar theory and parallax would not be occulted at all since it was fully 0;10° south of the lower limb of the moon. Thus it was a combination of errors, not only in Copernicus's computation of the precession, but also in the coordinates of Aldebaran in Ptolemy's catalogue, that allowed Copernicus to recompute the occultation of the star. Indeed, Copernicus was very lucky not to have determined the position of Aldebaran independently, but to have chosen to represent Ptolemy rather than nature, as Kepler later said. It is, however, even more remarkable that if we compute the parallax from Ptolemy's tables for the conditions of the observation, we find that $p = 1;31,47°$, and that

$$p_\lambda = -1;16°, \qquad p_\beta = -0;52°,$$

which applied to the true position M, gives an apparent position P of

$$\lambda'_{☾} = \text{♊} 2;8°, \qquad \beta'_{☾} = -5;27°,$$

and the moon at P will very nicely occult the star a little below its central line. The difference in longitude corresponds to about $1;16^h$, so the occultation will occur after midnight, but at least with Ptolemy's parallax the star will be occulted, while with Copernicus's it will not. Thus, if the correct position of Aldebaran were used, actual observation would have confirmed Ptolemy's very large lunar parallaxes and great variation in lunar distance—since the parallax implies a distance of about $37;15^{tr}$—and flatly contradicted Copernicus's theory.

This seemingly perverse result can be explained when the position and parallax of the moon are computed with modern tables.[78] In this way, for 10:00 P.M. and 11:00 P.M. at Bologna:

[78] P. V. Neugebauer (1914) for λ and β; P. V. Neugebauer (1929) for p_λ and p_β. Coordinates of Aldebaran from P. V. Neugebauer (1912), checked against *Yale Catalogue of Bright Stars* (1964).

	10:00 P.M.	11:00 P.M.
$\lambda_{☾}$	3;13°	3;47°
p_{λ}	−0;47	−0;47
$\lambda'_{☾}$	2;26	3;0
$\beta_{☾}$	−4;47,24	−4;46
p_{β}	−0;29,24	−0;32
$\beta'_{☾}$	−5;17	−5;18

(81)

The true positions are indicated in the figure by M_{10} and M_{11}, and the apparent by M'_{10} and M'_{11}. The true latitudes are about 0;12° below Copernicus's value, and it is for this reason that his theory does not produce the correct occultation. On the other hand, the parallaxes themselves agree very well with Copernicus's computation, and show that the parallax from Ptolemy's table is much too large. Note finally that the apparent positions M'_{10} and M'_{11} bring the star into the path of the moon such that at about 10:16 P.M. an occultation takes place near the southern limb of the moon, in good agreement with the description of the observation—if the apparent latitude is interpreted as −5;2°—although not with the time. However, and this is of considerable interest, in Copernicus's volume of the 1492 Augsburg editions of Regiomontanus's *Calendarium* and *Ephemerides* for 1491–1506, the time is correctly recorded as $10\frac{1}{3}^{h}$,[79] which shows that he actually knew when the occultation occurred.

Comment. The preceding analysis is of interest for a number of reasons, not the least of them being its complexity. The test of lunar parallax in IV,27 requires almost every topic treated thus far in *De revolutionibus*:

Book I: plane and spherical trigonometry;
Book II: obliquity of the ecliptic, equatorial and ecliptic coordinates, transformation of coordinates, oblique ascension and its various applications, horizon angle, zenith distance, angle of ecliptic with circle of altitude, star catalogue;
Book III: precession, variation of obliquity and its effect on equatorial coordinates, solar theory, equation of time;
Book IV: lunar theory for both inequalities, lunar latitude, lunar parallax in circle of altitude and resolved into components in longitude and latitude, apparent diameter of moon.

Each time an apparent position of the moon is computed, almost all of this is necessary, just as it is necessary in even more complexity in modern lunar theory.

---

[79] Birkenmajer (1900), 515; Zinner (1943), 409. Copernicus received this book after the death of Hildebrand Ferber in 1530, but the record can only refer to his own observation in Bologna.

A practical application of Copernicus's lunar theory, such as the very demanding computation of an occultation, shows that it is not a simple matter to determine its accuracy. And it makes no sense at all to speak blithely of the "saving the phenomena" so fashionable among philosophers and historians who do not waste time on the tedious exercise of computation. It is not such a simple matter to decide whether Copernicus's theory is accurate, and it is neither simple nor necessarily informative to compare it with modern theory. Here we have seen a long series of computations, containing numerous errors along the way, that reproduces the occultation of a star at the very moment the occultation was supposed to have been observed, "which," Copernicus remarks at the end of IV,27, "agrees completely with the observation, so that all the less will anyone doubt that our hypotheses and what follow from them are correct." This is perhaps Copernicus's strongest statement of confidence in the entire book, but he can only make it because he changed the time of the observation by $+0;40^h$ and the longitude of the star by $-0;10°$, which is not what is usually meant by "saving the phenomena." In fact, were the star in its correct location by modern theory, Copernicus's lunar and parallax theories would have given the moon the correct apparent longitude for the occultation about $\frac{1}{2}^h$ too late. But in no way could Copernicus's theory give the moon the correct latitude to occult the star, while by all the strange ironies, Ptolemy's parallax could do just that, although with a time about 2^h too late. Thus, one must be rather careful about comparing older theories with each other, and with modern computations, and then making judgments about "saving the phenomena." And it is probably more effort than it is worth.

It is not hard to find the numerical values of Copernicus's errors by comparison with modern theory, and they are fairly small. Copernicus's true longitude at 11:00 P.M. is about 0;23° too low—less than the diameter of the moon—and his latitude about 0;11° too close to the ecliptic, a small but crucial distance in an occultation. The latitude in Copernicus's theory is computed from $\beta = 5° \sin v$, where $v = \omega + 90°$. But correctly, and taking into account the principal inequality of the lunar latitude, discovered by Tycho,

$$\beta = 5;9° \sin v + 0;8,48° \sin (2\eta - v) + \cdots. \tag{31}$$

Thus, using the results of the lunar position computed for the occultation,

$$\eta = 66;46°, \qquad v = 293;42°, \qquad \beta = -4;43° - 0;3° = -4;46°,$$

and adding Copernicus's parallax in latitude

$$\beta'_{☾} = -4;46° - 0;30° = -5;16°,$$

which will nicely produce an occultation of Aldebaran near the southern limb at $\beta = -5;30°$, in excellent agreement both with modern theory and with Copernicus's report of the observation.

The problems of the occultation are a good example of the difficult situation of astronomy brought about by the continuous use of elements and methods that originated anywhere during the preceding 1400 years, and Book IV of *De revolutionibus* is nothing if not a collection of the relics of antiquity and the middle ages—from Alexandria, from Ujjain, from Raqqa, Damascus, and Toledo. Only against this background can one understand why Copernicus would uncritically locate a star with a catalogue nearly 1400 years old, why he would accept an estimate of the maximum lunar latitude nearly 1700 years old. For the latter, he gives as a reason that he is too far north to find the maximum lunar latitude, meaning implicitly by exactly the method explained by Ptolemy. And only against this background can one fully appreciate the revolutionary character of Tycho Brahe's new methods of observation, of the reduction and analysis of the results of observation, of his new cataloguing of the fixed stars, and of his new investigation of the latitude of the moon and its inequalities. And Hveen is north of Frauenburg.

3. The Theory of Eclipses

Copernicus concludes Book IV with a highly abbreviated treatment of eclipses in Chapter 28–32. He remarks at the end that what he has done will suffice concerning the moon since eclipses have been treated in greater detail by "others," and he wishes to hasten on to the planets. The reference is to *Almagest* VI, *Epitome* VI, and perhaps to Peurbach's *Eclipse Tables*, compared to all of which, Copernicus has provided little more than an outline. Nevertheless, he has some methods of his own, particularly for finding the time between mean and true conjunction or opposition, that differ from the earlier sources, and as condensed as his exposition is, it presents everything of importance in the strategy of eclipse calculation, and with a few additions, can be applied directly to the calculation of an eclipse. To illustrate this clearly, we shall expand it somewhat and provide numerical examples.

Mean Conjunction and Opposition (IV,28)

The first step in computing an eclipse is to find the time of mean conjunction, $\bar{\eta} = 0°$, for a solar eclipse, or mean opposition, $\bar{\eta} = 180°$, for a lunar eclipse, and the value of the mean anomaly of the moon, mean argument of latitude, and mean anomaly of the sun for that time. This is usually done with special tables, but Copernicus gives a direct method that is the basis of the computation of such tables of mean syzygies. Take a time t_0 prior to the required syzygy, and compute for it the mean elongation $\bar{\eta}$. Then where $\bar{v}_\eta^{\circ/\mathrm{d}}$ is the mean daily elongation,

$$\bar{\eta}/\bar{v}_\eta^{\circ/\mathrm{d}} = \Delta t, \qquad (\bar{\eta} + 180°)/\bar{v}_\eta^{\circ/\mathrm{d}} = \Delta t', \tag{82}$$

give the time Δt elapsed since the preceding conjunction, or the time $\Delta t'$ since the preceding opposition. Let the length of the mean synodic month be $\overline{M}$, and

$$\overline{M} - \Delta t = \Delta t_1, \qquad \overline{M} - \Delta t' = \Delta t'_1, \tag{83}$$

is the time from t_0 to the following conjunction or opposition, which will therefore be at

$$t_1 = t_0 + \Delta t_1, \qquad t'_1 = t_0 + \Delta t'_1. \tag{84}$$

The procedure can be shortened by computing directly

$$\Delta t_1 = (360° - \bar{\eta})\bar{v}_{\eta}^{\circ/\mathrm{d}}, \qquad \Delta t'_1 = (180° - \bar{\eta})/\bar{v}_{\eta}^{\circ/\mathrm{d}}, \tag{85}$$

finding t_1 or t'_1 as before, and observing that

$$t'_1 = t_1 + \tfrac{1}{2}\overline{M}. \tag{86}$$

Next, for t_1 or t'_1 compute the mean anomaly of the moon $\bar{\alpha}_{☾}$, the mean argument of latitude $\bar{\omega}$, and the mean anomaly of the sun $\bar{\alpha}_{\odot}$. Note that the time t_1 or t'_1 is a mean time with respect to the epoch.

Tables allowing these operations to be done rapidly are provided by Ptolemy in *Almagest* VI,3 and by Peurbach in his *Eclipse Tables*, to name sources known to Copernicus,[80] and the computation and use of such tables is described in *Almagest* VI,2,4 and *Epitome* VI,1–3. In using them, one enters with t_0, and finds by summation $\bar{\lambda}_{\odot}$ or $\bar{\alpha}_{\odot}$, $\bar{\alpha}_{☾}$, $\bar{\omega}$, at t_1 or t'_1. Copernicus, on the other hand, in giving instructions for the direct computation of the various quantities, beginning with $\bar{\eta}$ at t_0, is following the method of the *Alfonsine Tables*. However, after determining t_1, $\bar{\alpha}_{☾}$, $\bar{\omega}$, and $\bar{\alpha}_{\odot}$ for one mean syzygy, say the first one after Jan 0 of a given year, it is often of interest to investigate a series of syzygies for the remainder of the year. To assist in this Copernicus does provide a table of multiples of $\overline{M}$ and of $\Delta\bar{\alpha}_{☾}$, $\Delta\bar{\omega}$, and $\Delta\bar{\alpha}_{\odot}$ for $1 - 12\overline{M}$ and for $\frac{1}{2}\overline{M}$. Beginning with the quantities for the syzygy at t_1, one finds from the table, for syzygy n

$$\begin{aligned} t_n &= t_1 + n\overline{M}, & \bar{\alpha}_{☾n} &= \bar{\alpha}_{☾1} + n\Delta\bar{\alpha}_{☾}, \\ \bar{\omega}_n &= \bar{\omega}_1 + n\Delta\bar{\omega}, & \bar{\alpha}_{\odot n} &= \bar{\alpha}_{\odot 1} + n\Delta\bar{\alpha}_{\odot}. \end{aligned} \tag{87}$$

The table, however, poses problems. As accurately as it can be extracted from Copernicus's table in IV,4, the length of the mean synodic month $\overline{M}$ is

$$\overline{M} = (365^{\mathrm{d}} \cdot 360°)/(12^{\mathrm{r}} + \bar{\eta}^{\mathrm{y}}) = 29;31,50,7,53,45\ldots^{\mathrm{d}}. \tag{88}$$

[80] Copernicus copied part of the table of mean syzygies from Peurbach's *Eclipse Tables* on sheets bound in with his copies of the *Alfonsine Tables* and the *Tabulae directionum* (Uppsala University Library 34.VII.65). These were published without identification in Curtze (1875), 225–26 and Prowe (1883–84) 2, 224–25.

If we use the cancelled value of $\bar{\eta}^y$ in M (Table 1, line 8) the result is 29;31,50,7,54 . . .^{d}. But the table of syzygies in N gives multiples of 29;31,50,9^d, and the table in M (138r) multiples of 29;31,50,8,5 . . .^{d}, neither of which follows from Copernicus's values of $\bar{\eta}$. The synodic month in N could be rounded from Copernicus's incorrect computation of Hipparchus's value,[81] but using this would make little sense. Nor is either one a convenient rounding of a correct computation, and we are at a loss to explain their origin. The values of $\Delta\bar{\alpha}_☾$, $\Delta\bar{\omega}$, and $\Delta\bar{\alpha}_\odot$, follow in principle from multiplying the daily motion of each by $\bar{M}$, and then summing the results up to $12\bar{M}$, but here too the last place is sometimes inconsistent with the mean motions tabulated earlier. The following table lists the values for $\bar{M}$ and for $12\bar{M}$ from both N and M.

	N, $\bar{M}$	N, $12\bar{M}$	M, $\bar{M}$	M, $12\bar{M}$	
$\bar{M}$	29;31,50,9^d	354;22,1,48^d	29;31,50,8^d	354;22,1,37^d	
$\Delta\bar{\alpha}_☾$	25;49,0°	309;48,4°	25;49,0°	309;48,0°	(89)
$\Delta\bar{\omega}$	30;40,14	8;2,48	30;40,13	8;2,46	
$\Delta\bar{\alpha}_\odot$	29;6,18	349;15,38	29;6,18	349;15,37	

The table also gives values for $\frac{1}{2}\bar{M}$ in order to find mean opposition from a known mean conjunction.

Example. Find the first mean conjunction after Jan 0, 1500 along with its mean motions, and then find the first mean opposition in November with its corresponding mean motions. The interval of mean time from the Epoch of Christ to Jan 0, 1500 at a mean time of midnight is 1500^y 8^d = 25,0^y 8^d, for which from the tables of mean motions in IV,4 $\bar{\eta}$ = 341;54,43,32°. Hence 360° − $\bar{\eta}$ = 18;5,16,28°, and thus

$$\Delta t_1 = (360° - \bar{\eta})/\bar{v}_\eta^{\circ/d} = 1;29,1,28^d,$$

$$t_1 = 1500 \text{ Jan } 1;29,1,28^d = 1500 \text{ Jan } 1, 11;36,35^h.$$

For this time, from the tables in IV,4 and III,14,

$$\bar{\alpha}_{☾1} = 209;49,18°, \qquad \bar{\omega}_1 = 124;5,2°, \qquad \bar{\alpha}_{\odot 1} = 190;54,19°.$$

Hence these are the motions for the mean conjunction of 1 Jan 1500. In a leap year, Nov 1 follows Jan 1 by 305^d, so we look in the table of mean conjunctions and oppositions for the closest multiple of $\bar{M} < 305^d$, and

[81] Above, p. 199.

it is $\overline{M}_{10}$. Then since we want the next opposition, we add $\frac{1}{2}\overline{M}$ and its corresponding motions as follows:

	t	$\bar{\alpha}_{☾}$	$\bar{\omega}$	$\bar{\alpha}_{\odot}$
t_1	1;29, 1,28^{d}	3,29;49,18°	2, 4; 5, 2°	3,10;54,19°
$\overline{M}_{10}$	295;18,21,30	4,18;10, 3	5, 6;42,20	4,51; 3, 1
$\overline{M}_{1/2}$	14;45,55, 4,30	3,12;54,30	3,15;20, 7	14;33, 9
t'_{10}	311;33,18, 2,30	5, 0;53,51	4,26; 7,29	2,16;30,29

t'_{10} corresponds to 6 Nov 1500, 13;19,13^{h}, which is the mean time of the mean opposition.

True Conjunction and Opposition (IV,29)

Having obtained the time of mean syzygy, the next step is to find the time of true syzygy when $\lambda_{☾} = \lambda_{\odot}$ (mod. 180°). Copernicus has his own way of doing this, which he says is preferable to the standard procedure that uses what is called the *superatio*, the "excess" of the true velocity of the moon over the true velocity of the sun. He is referring to the method described in *Almagest* VI,4—and, with some refinements, in *Epitome* VI,4–5—that is the basis of tables for finding true syzygy in the *Alfonsine Tables*. Copernicus's description of his method is very concise, and we shall expand as necessary for intelligibility.

1. For the time of mean syzygy, compute the true longitudes of the sun and moon, $\lambda_{\odot}$ and $\lambda_{☾}$, or at least their corrections $c_{\odot}$ and $c_{☾}$. Then, as shown in Fig. 38(a),

$$\Delta c = c_{\odot} - c_{☾} = \Delta\lambda \quad (\text{mod. } 180°) \tag{90}$$

is the distance between the sun and moon. Observing the proper signs of the corrections, if $c_{\odot} \geq c_{☾}$, $\Delta c \geq 0°$, and true syzygy will follow mean; if $c_{\odot} \leq c_{☾}$, $\Delta c \leq 0°$, and true syzygy will precede mean.

2. Estimate the elapsed time between mean and true syzygy by

$$\Delta t^{h} = 2\Delta c, \tag{91}$$

since the mean velocity of the elongation $\bar{v}_{\eta} \approx 1;1^{\circ/2h}$. The sign of Δt is the same as that of Δc.

3. The hourly velocity of the true anomaly v_{α} at mean syzygy is given by

$$v_{\alpha}^{\circ/h} = \bar{v}_{\alpha}^{\circ/h} + \bar{v}_{2\eta}^{\circ/h} \cdot c_{3(2\bar{\eta})}, \tag{92}$$

because for any interval Δt,

$$\Delta\alpha = \Delta\bar{\alpha} + \Delta c_{3(2\bar{\eta})}. \tag{93}$$

This nice refinement for correcting $\bar{v}_\alpha$ to v_α at syzygy is from *Epitome* VI,4, and does not appear in the *Almagest*. Its effect on $\bar{v}_\alpha$ is significant, especially in Copernicus's lunar theory where, from his tables,

$$v_\alpha^{\circ/\mathrm{h}} = 0;32,40^\circ + 1;1^\circ \cdot 0;17^\circ = 0;49,57^\circ \approx 0;50^{\circ/\mathrm{h}}.$$

From Ptolemy's tables, since c_3 is lower near $2\bar{\eta} = 0^\circ$,

$$v_\alpha^{\circ/\mathrm{h}} = 0;32,40^\circ + 1;1^\circ \cdot 0;\,9^\circ = 0;41,49^{\circ/\mathrm{h}},$$

which is given in the *Epitome*.

4. Call the anomaly at mean syzygy α_1. Then at the end of the interval Δt at true syzygy, the anomaly will be

$$\alpha_2 = \alpha_1 + \Delta\alpha = \alpha_1 + \Delta t v_\alpha, \tag{94}$$

and we find from the lunar correction table $c_{(\alpha_2)}$. We shall call the correction at mean syzygy c_1 and the correction at true syzygy c_2, and as shown in Fig. 38(b), the corrected distance between mean and true syzygy

$$\Delta c' = \Delta c + (c_2 - c_1), \tag{95}$$

carefully observing the signs of c_2 and c_1.

5. Since the true velocity of the elongation during the interval Δt is

$$v_\eta^{\circ/\mathrm{h}} = \Delta c'/\Delta t, \tag{96}$$

the corrected interval of time from mean to true syzygy is

$$\Delta t' = \Delta c/v_\eta^{\circ/\mathrm{h}} = (\Delta c/\Delta c')\Delta t, \tag{97}$$

and the time of true syzygy

$$t' = t + \Delta t', \tag{98}$$

where the sign of $\Delta t'$ is that of Δt and Δc, i.e. positive for $c_\odot \geq c_☾$, and negative for $c_\odot \leq c_☾$.

Comment. Copernicus says that the preceding method has an advantage over the standard procedure in that it does not require iteration. He is for all practical purposes quite right, for the residual error in his method must be small. In the standard way, one finds for the time of mean syzygy the "excess" (*superatio*) of the true velocities

$$v_\eta = v_☾ - v_\odot, \tag{99}$$

and then finds the interval of time from

$$\Delta t = \Delta c/v_\eta. \tag{100}$$

This method is given in *Epitome* VI,5 as an improvement over Ptolemy's method from *Almagest* VI,4 which is also described in *Epitome* VI,5. In Ptolemy's procedure one uses only the true velocity of the moon. To estimate the motion of the sun during the interval between mean and true conjunction, Δc is replaced by $(13/12)\Delta c$, and the interval of time follows from

$$\Delta t = (13/12)\Delta c/v_{☾}.^{82} \qquad (101)$$

The problem with both methods is that $v_{☾}$, and thus v_η, is not constant over Δt, and thus an iteration may be required when, on computation, it is found that at the supposed time of true syzygy $\lambda_{☾} \neq \lambda_{\odot}$ (mod. 180°). On the other hand, Copernicus's value of v_η is integrated over Δt by being derived from $v_\eta = \Delta c'/\Delta t$, where the initial estimate of Δt from $\Delta t^{\mathrm{h}} = 2\Delta c = \Delta c/\bar{v}_\eta$, cannot but give a far better value of v_η than merely computing the velocity for mean syzygy. If this method of finding true syzygy is indeed his own, it may be Copernicus's most valuable contribution to practical astronomy. It is true that even Copernicus's method does not take into account the change in the velocity of the sun due to the increase of $\alpha_{\odot}$ during Δt, but he points out that the corresponding change in $c_{\odot}$ is less than 0;1°, so may safely be ignored.

6. For the time of true syzygy, compute $\lambda_{\odot}$ and find $\lambda_{☾} = \lambda_{\odot}$ (mod. 180°). This is Copernicus's instruction. However, it is easier to take $\bar{\lambda}_{☾}$ for mean syzygy, and compute for true syzygy

$$\lambda_{☾} = \bar{\lambda}_{☾} + \Delta\bar{\eta} + c_2 = \bar{\lambda}_{☾} + \Delta t'\bar{v}_\eta^{\circ/\mathrm{h}} + c_2. \qquad (102)$$

In the same way, from $\bar{\omega}$ at mean syzygy, the true argument of latitude at true syzygy

$$\omega = \bar{\omega} + \Delta\bar{\omega} + c_2 = \bar{\omega} + \Delta t'\bar{v}_\omega^{\circ/\mathrm{h}} + c_2. \qquad (103)$$

Finally, from ω, find the latitude β.

7. The time that results for true syzygy is a mean time with respect to the epoch used for finding mean syzygy. To convert it to an apparent time, compute the equation of time ΔE, and take

$$\Delta t = \Delta\bar{t} - \Delta E, \qquad (3.48)$$

where $\Delta\bar{t}$ is the mean time and Δt the apparent time since epoch.

8. As a final rule, Copernicus says that to adjust the time t to a meridian other than Cracow, for the difference in longitude Δl, find the time

$$t_l = t + 0;4^{\mathrm{h}}\Delta l, \qquad (104)$$

where the correction is positive to the east of Cracow and negative to the west.

---

[82] See the more detailed analyses of Ptolemy's method in Pedersen (1974), 223–26 and *HAMA*, 122–24.

Example. Find the true opposition corresponding to the mean opposition of 6 Nov 1500, 13;19^{h}. From the solar theory, we have

$$\bar{\lambda}_{\odot} = 234;24^{\circ}, \qquad \bar{\lambda}'_{\odot} = 235;1^{\circ}, \qquad c_{\odot} = -1;15^{\circ}, \qquad \lambda_{\odot} = 233;46^{\circ}.$$

From the lunar anomaly $\alpha_1 = 300;54^{\circ}$, $c_{☾} = +4;2^{\circ}$, and thus

$$\Delta c = c_{\odot} - c_{☾} = -1;15^{\circ} - 4;2^{\circ} = -5;17^{\circ}, \tag{90}$$

which means that true opposition precedes mean opposition. The estimate of the time between mean and true opposition is

$$\Delta t = 2\Delta c = -10;34^{\text{h}}. \tag{91}$$

Taking $v_{\alpha} = 0;50^{\circ/\text{h}}$,

$$\Delta\alpha = \Delta t v_{\alpha} = -10;34 \cdot 0;50^{\circ} = -8;48^{\circ},$$

$$\alpha_2 = \alpha_1 + \Delta\alpha = 300;54^{\circ} - 8;48^{\circ} = 292;6^{\circ}, \tag{94}$$

for which $c_2 = +4;24^{\circ}$. Consequently,

$$\Delta c' = \Delta c - (c_2 - c_1) = -5;17^{\circ} - (4;24^{\circ} - 4;2^{\circ}) = -4;55^{\circ}. \tag{95}$$

The true velocity of elongation is therefore

$$v_{\eta}^{\circ/\text{h}} = \Delta c'/\Delta t = 4;55^{\circ}/10;34^{\text{h}} = 0;27,55^{\circ/\text{h}}, \tag{96}$$

and thus the corrected time between mean and true syzygy

$$\Delta t' = \Delta c/v_{\eta}^{\circ/\text{h}} = -5;17^{\circ}/0;27,55^{\circ/\text{h}} = -11;21^{\text{h}}, \tag{97}$$

or equivalently,

$$\Delta t' = (\Delta c/\Delta c')\Delta t = (-5;17^{\circ}/-4;55^{\circ}) \cdot -10;34^{\text{h}} = -11;21^{\text{h}}, \tag{97}$$

so that the mean time of true syzygy

$$t' = t + \Delta t' = 13;19^{\text{h}} - 11;21^{\text{h}} = 1;58^{\text{h}}. \tag{98}$$

Now, for Nov 6 1500, 1;58^{h}, we find for the sun

$$\bar{\lambda}_{\odot} = 233;56^{\circ}, \qquad \bar{\lambda}'_{\odot} = 234;33^{\circ}, \qquad c_{\odot} = -1;16^{\circ}, \qquad \lambda_{\odot} = 233;17^{\circ}.$$

Hence the true longitude of the moon is

$$\lambda_{☾} = 233;17^{\circ} + 180^{\circ} = 53;17^{\circ},$$

or, without going through the trouble of finding $\lambda_{\odot}$,

$$\lambda_{☾} = 55;1^{\circ} - \tfrac{1}{2}(11;21 \cdot 1;1^{\circ}) + 4;2^{\circ} = 53;17^{\circ}. \tag{102}$$

In IV,14 Copernicus computed $\lambda_{\odot} = 233;16^{\circ}$ so that $\lambda_{☾} = 53;16^{\circ}$. Compared with modern computation, from Goldstine (1973) $\lambda_{☾} = 53;16^{\circ}$, so the agreement is perfect. The true argument of latitude,

$$\omega = 266;7^{\circ} - (11;21 \cdot 0;33,4^{\circ}) + 4;2^{\circ} = 263;54^{\circ}, \tag{103}$$

for which

$$\beta = -0;31^\circ.$$

From the computation of the sun at $1;58^h$, the equation of time since the Epoch of Christ,

$$\Delta E = -3;3^\circ - 1;53^\circ = -4;56^\circ = -0;19,44^h \approx -0;20^h,$$

and thus the apparent time

$$\begin{aligned}\Delta t = \Delta\bar{t} - \Delta E &= 6 \text{ Nov } 1500,\ 1;58^h + 0;20^h\\ &= 6 \text{ Nov } 1500,\ 2;18^h.\end{aligned}$$

Copernicus reported observing this lunar eclipse on 6 Nov 1500 at $2;0^h$ in Rome, but reduced to the meridian of Cracow, at $2;20^h$, so the agreement could hardly be better. To check the accuracy of the true syzygy against modern theory, we reduce the mean time to G.M.T. by

$$\text{G.M.T.} = 1;58^h - 0;4^h \cdot 19;40 \approx 0;39^h, \tag{104}$$

and again from Goldstine (1973) we find that reduced to G.M.T. the syzygy occurred at $0;42^h$. Again, the agreement is excellent.

Distinction of Ecliptic Syzygies (IV,30)

Almagest VI,5–6 and *Epitome* VI,7–14 present detailed analyses of the criteria for establishing eclipse limits and the intervals between ecliptic syzygies.[83] The limits of mean nodal distance are found so that the possibility of an eclipse at a given syzygy may be determined simply by finding the mean argument of latitude. The intervals between eclipses—generally of six, sometimes of five or seven months—allow the exclusion of other syzygies from investigation. Copernicus does none of this, but instead gives rules for evaluating each syzygy separately after the conditions of true opposition, in the case of a lunar eclipse, or apparent conjunction, in the case of a solar eclipse, have been determined.

A lunar eclipse can occur so long as the moon touches the edge of the shadow which, as shown in Fig. 39, happens when the distance of the moon from the ecliptic on the arc perpendicular to the lunar orbit $UM < (r'_u + r'_m)$. Copernicus gives the conventional, but slightly less accurate rule that an eclipse is possible if the latitude $\beta < (r'_u + r'_m)$.[84] From the maximum values of r'_u and r'_m,

$$\beta_{max} = 0;47,52^\circ + 0;17,49^\circ = 1;5,41^\circ, \tag{105}$$

[83] See *HAMA*, 125–34.

[84] The approximate rule is given, e.g. in the *Alfonsine Tables* (1492) Prop. 32, following Battānī (ed. 1537, f. 61v; Nallino 1, 97). The same rule is applied to solar eclipses except that the apparent latitude is used.

and the true nodal distance

$$\nu = \sin^{-1}(\sin \beta_{\max}/\sin 5°) = 12;40°. \qquad (106)$$

Solar eclipses are more complicated since the latitudinal component of the parallax can cause the apparent latitude of the moon to differ from the true latitude by as much as 1°, i.e. twice the diameter of the moon. This has a pronounced effect on the possibility of an eclipse, and greatly enlarges the eclipse limits. Copernicus also considers in this chapter the parallax in longitude, which affects only the time of the eclipse. This too can reach about a degree, and hence the time of true and apparent conjunction can differ by as much as two hours. However, these are extrema that can occur only at certain terrestrial latitudes—maximum parallax in longitude near the equator, maximum parallax in latitude near the poles—and cannot occur simultaneously. Copernicus says that finding apparent conjunction for a solar eclipse *plus satis habet negotii*, which is a nice understatement for an expression requiring an idiomatic rendering not usually found in scientific books. He does not explain the fully accurate, and very lengthy, iterative procedure from *Almagest* VI,10 and *Epitome* VI,22, but gives an alternate method described in the *Epitome* as *facilius*, although it too is more than enough trouble. Neither the *Epitome* nor Copernicus says so, but the "easier" method is meant to be used with tables of the longitudinal and latitudinal components of parallax—otherwise, it would be as much work as the strict iterative procedure. Such tables, found originally in the *Handy Tables*, became standard in medieval Arabic tables, and were known to Copernicus through the *Alfonsine Tables* and Peurbach's *Eclipse Tables*.[85]

There exists in Copernicus's hand a copy of a table of the components of parallax computed (not by Copernicus) for the latitude of Cracow, 51°, a part of which, for the sign of Taurus, is given in an expanded form in Table 5.[86] Tables of this kind are meant to be used principally for solar eclipses,

---

[85] The parallax tables in the *Handy Tables* are discussed in *HAMA*, 990–97. Battānī's tables (Nallino 2, 95–101) are identical. Various manuscripts of the *Toledan Tables* introduce different parallax tables along with those of Battānī and the *Handy Tables*. Cf. Toomer (1968), 97–112. The *Alfonsine Tables* are in turn based upon the collection in the *Toledan Tables*, and are completely derivative. All of the above tables are computed only for the moon at apogee of the epicycle, $\alpha_{☾} = 0°$. Peurbach's tables are considerably more ambitious. They are restricted to what he calls the sixth and seventh climates, $\varphi \approx (46° - 49°)$, but the tabulation is for $\alpha_{☾}$ at 30° intervals, and instead of giving p_λ, he works out the difference in time between true and apparent conjunction. In the *Prutenic Tables*, Reinhold computes parallax tables for no less than fourteen different latitudes, from 16° to 70°. It seems reasonable that they are based upon his own recomputation of Copernicus's lunar distance and parallax.

[86] On f. 113v of Copernicus's copy of the 1492 *Alfonsine Tables* with the heading *Tabula diversitatis aspectus ad 51 gradum*, which is the latitude of Cracow given in the *Alfonsine Tables*. The table is printed in Curtze (1875), 223 and Prowe (1883–84) 2, 221, although with *ad 51 g*[*radum*] misread as *ad minuta*. Curtze described the table as "astrologische Spielereien," and Prowe called this a "sachvertändiges Urteil." The table was correctly identified as concerning parallax in Birkenmajer (1900), 56–58, and is discussed in Neugebauer (1968a), 97–99.

and assume the moon in true conjunction with the sun. The parallaxes are computed by the approximate method described in IV,26, and for the moon at greatest distance at syzygy. A correction table is provided for adjusting to other distances as a function of the lunar anomaly. In using the table, one enters column 1 of the sign of $\lambda_{\odot}$ with the time of true conjunction on either side of apparent noon. Column 2 contains the adjusted parallax in longitude $p'_{\lambda} = p_{\lambda☾} - p_{\lambda\odot}$, and column 3 the adjusted parallax in latitude $p'_{\beta} = p_{\beta☾} - p_{\beta\odot}$. It is then necessary to find the same quantities for the following sign, and interpolate for the exact location of $\lambda_{\odot}$ since the parallaxes under each sign are computed for the beginning of the sign. Likewise, it may be necessary to interpolate between tables for different terrestrial latitudes if the observer's latitude does not fit one of the tables exactly. We shall explain Copernicus's method of finding apparent conjunction, the *facilius* method from the *Epitome*, with the assistance of the table, although without worrying about the various interpolations for lunar distance, solar longitude, and terrestrial latitude.

1. For the time t of true conjunction, find the adjusted parallax in longitude p'_{λ}. When the true conjunction is in the nonagesimal, $\lambda_{\odot} = \lambda_{N}$, $p'_{\lambda} = 0°$. As shown in Fig. 40(a), when the sun is east of the nonagesimal, that is, between the nonagesimal and the ascendant,

$$\lambda_{\odot} \geq \lambda_{N}, \qquad p'_{\lambda} \geq 0°.$$

As shown in Fig. 40(b), when the sun is west of the nonagesimal, between the nonagesimal and the descendant,

$$\lambda_{\odot} \leq \lambda_{N}, \qquad p'_{\lambda} \leq 0°.$$

The appropriate signs have been added to column 2 of Table 5.

2. If the sun is east of the nonagesimal, find the parallax at $t - 1^{h}$, and if the sun is west of the nonagesimal, find the parallax at $t + 1^{h}$. Call the parallax at the time of true conjunction $p'_{\lambda 1}$, and the parallax differing by 1^{h} $p'_{\lambda 2}$, and find

$$\Delta p'_{\lambda} = p'_{\lambda 2} - p'_{\lambda 1}, \tag{107}$$

carefully observing the signs.

3. The true velocity of elongation v_{η} at the time of true conjunction

$$v_{\eta}^{\circ/\text{h}} = \bar{v}_{\eta}^{\circ/\text{h}} + v_{\alpha}^{\circ/\text{h}} \cdot \Delta c, \tag{108}$$

where $v_{\alpha} = 0;50^{\circ/\text{h}}$ in accordance with (92), and $\Delta c = c_{(\alpha + 1°)} - c_{\alpha}$ for $t + 1^{h}$, or $\Delta c = c_{\alpha} - c_{(\alpha - 1°)}$ for $t - 1^{h}$. The apparent velocity of elongation $v_{\eta}'^{\circ/\text{h}}$ then follows from

$$v_{\eta}'^{\circ/\text{h}} = v_{\eta}^{\circ/\text{h}} \begin{cases} -\Delta p'_{\lambda} & \text{for } t - 1^{h}, \\ +\Delta p'_{\lambda} & \text{for } t + 1^{h}, \end{cases} \tag{109}$$

again carefully observing the sign of $\Delta p'_\lambda$. In general, $v'_\eta < v_\eta$ on both sides of the nonagesimal, since the decreasing positive parallax as the moon rises and the increasing negative parallax as it sets both act to decrease the moon's apparent motion in longitude. However, near the horizon this is not always so, for the parallax in longitude can decrease, particularly at extreme terrestrial latitudes where the inclination of the ecliptic to the horizon is small.

4. The parallax at true conjunction is $p'_{\lambda 1}$, originally p'_λ shown in Fig. 40. The time required for the moon to cross p'_λ is the time between true and apparent conjunction, which is given by

$$\Delta t = p'_\lambda / v'^{\circ/\mathrm{h}}_\eta, \tag{110}$$

and where t is the time of true conjunction, the time of apparent conjunction

$$t' = t \begin{cases} -\Delta t & \text{for } t - 1^\mathrm{h}, \\ +\Delta t & \text{for } t + 1^\mathrm{h}, \end{cases} \tag{111}$$

since, as shown in Fig. 40, east of the nonagesimal the parallax increases the apparent longitude, and west of the nonagesimal the parallax decreases the apparent longitude. The result is that east of the nonagesimal, apparent conjunction precedes true conjunction, and west of the nonagesimal, apparent conjunction follows true conjunction. The true longitude of apparent conjunction is the true longitude of the sun at t'. Where $\lambda_\odot$ is the longitude at t,

$$\lambda'_\odot = \lambda_\odot + \Delta t v_\odot, \tag{112}$$

where the sign of Δt is the same as in finding t'.

5. The true argument of latitude changes in the interval Δt by

$$\Delta\omega = \Delta t v_\omega^{\circ/\mathrm{h}} = \Delta t(v_\eta + 0;2,36)^{\circ/\mathrm{h}}, \tag{113}$$

so that at the time of apparent conjunction, the true argument of latitude

$$\omega' = \omega + \Delta\omega, \tag{114}$$

where the sign of $\Delta\omega$ is that of Δt. For ω' find the latitude β, and then for t' find from column 3 of the parallax table the adjusted parallax in latitude p'_β. Anywhere north of $\varphi \approx 23\frac{1}{2}^\circ$, p'_β is always negative, so the apparent latitude at the time of apparent conjunction is

$$\beta' = \beta - p'_\beta. \tag{79}$$

In Table 5 we have added minus signs to column 3.

Note that in finding the time of apparent conjunction and the apparent latitude in this way, we have had to determine the parallax in longitude twice and the parallax in latitude once for three different times. For each the moon had a different zenith distance and angle of intersection of the ecliptic and circle of altitude, both of which would have to be found three times in order to then compute the parallax and its components *ab initio* three times. Even with the interpolations that are still necessary in using the table, one can well imagine that doing all this without the table *plus satis habet negotii.*

Finally, Copernicus establishes the condition for a solar eclipse, that $\beta' < (r'_s + r'_m)$. Hence,

$$\beta'_{max} = 0;16,57^\circ + 0;17,49^\circ = 0;34,46^\circ, \tag{115}$$

to which corresponds a true nodal distance $\nu = 6;40^\circ$. Again it would have been preferable to take $r'_s + r'_m$ on the arc SM perpendicular to the lunar orbit, as shown in Fig. 41. Because of the parallax and the correction between mean and true conjunction, the mean nodal distance can reach about 20°.

Eclipse Magnitudes (IV,31)

Again letting the latitude approximate the distance on the arc perpendicular to the lunar orbit, in the case of a solar eclipse, shown in Fig. 42(a) the eclipsed part of the diameter in degrees, m°, is

$$m^\circ = r'_s + r'_m - \beta', \tag{116}$$

where $\beta' = SM'$ is the apparent latitude. If the diameter of the sun is taken as 12′, digits or *puncta*, then the conventional measure of magnitude

$$m' = 12(r'_s + r'_m - \beta')/d'_s = 6(r'_s + r'_m - \beta')/r'_s. \tag{117}$$

The greatest magnitude of a solar eclipse occurs when the sun is at apogee and the moon is at perigee with $\beta' = 0^\circ$ at mid-eclipse, in which case

$$m'_{max} = 6(0;15,50^\circ + 0;17,49^\circ)/0;15,50^\circ = 12;45'. \tag{118}$$

Since a total eclipse is 12′, this means that there will be a delay (*mora*) of the sun in the eclipse of $0;45r'_s/6 \approx 0;2^\circ$ from the beginning of totality to mid-eclipse. No special consideration is given to annular eclipses, which are of course possible with Copernicus's values of r'_s and r'_m. In the extreme case of the sun at perigee and the moon at apogee

$$m' = 6(0;16,57^\circ + 0;15^\circ)/0;16,57^\circ \approx 11;19', \tag{119}$$

that is, a central eclipse of less than 12′, which means that there will be a delay in the annular phase of $0;41r'_s/6 \approx 0;2^\circ$ until mid-eclipse.

For a lunar eclipse, shown in Fig. 42(b), the true latitude $\beta = UM$ approximates the distance perpendicular to the lunar orbit, and the part of the moon in the shadow is

$$m^\circ = r'_u + r'_m - \beta. \tag{120}$$

Converting to digits,

$$m' = 12(r'_u + r'_m - \beta)/d'_m = 6(r'_u + r'_m - \beta)/r'_m. \tag{121}$$

Because in Copernicus's theory, provided that the sun is at apogee, $r'_u = kr'_m$, where $k = 403/150 = 2;41,12$, all central lunar eclipses with the sun at apogee will be of the same magnitude regardless of the distance of the moon. The maximum value follows from

$$\begin{aligned} m'_{max} &= 6(r'_u + r'_m)/r'_m = 6(kr'_m + r'_m)/r'_m \\ &= 6(k + 1) = 6 \cdot 3;41,12 = 22;7,12'. \end{aligned} \tag{122}$$

Hence the moon will be delayed in a central eclipse for $10;7,12r'_m/6$ degrees of its diameter. Taking the maximum value of r'_m, this amounts to $10;7,12 \cdot 0;17,49°/6 = 0;30,3°$ as the maximum delay from the beginning of totality until mid-eclipse.

Example. In the true syzygy of 6 Nov 1500, we found that $\beta = -0;31°$ and $\alpha_☾ = 292;6°$. From the table of semidiameters, for $\alpha_☾ = 292;6°$,

$$r'_m = 0;15,50°, \qquad r'_u = 0;42,29°,$$

and just to be very careful, for $\alpha_\odot \approx 137°$, the variation of the shadow $c_u = 0;0,49°$, so that $r'_u - c_u = 0;41,40°$. Hence, the eclipsed part of the moon is

$$m° = 0;41,40° + 0;15,50° - 0;31° = 0;26,30°,$$

and the magnitude in digits

$$m' = 6 \cdot 0;26,30°/0;15,50° = 10;2'.$$

Copernicus reports the eclipse as 10′ from the north, so the agreement is excellent. According to modern computation, from Oppolzer (1887), the magnitude is 10.7′.

Phases and Duration of Eclipses (IV,32)

Solar eclipses and lunar eclipses $\leq 12'$ have three phases—first contact, mid-eclipse, last contact—usually referred to simply as beginning, middle, and end. The arc that the moon crosses from first contact to mid-eclipse is called the "minutes of immersion" (*minuta casus*), a term not used by Copernicus. Lunar eclipses $> 12'$ have five phases—first contact, beginning of totality, mid-eclipse, end of totality, last contact—essentially these terms being used in *Epitome* VI,18. The arc from first contact to the beginning of totality is called the "minutes of immersion" (*minuta casus*), and the arc from the beginning of totality to mid-eclipse, the "minutes of delay" (*minuta morae*). While it is recognized that the sun can have a small delay in a central eclipse, only three phases are considered.

The duration of an eclipse depends upon its magnitude, a function of the lunar latitude and the solar and lunar anomalies—which determine the apparent diameters of the sun, moon, and shadow—and upon the difference between the velocities of the sun and moon, which again depends upon their respective anomalies. And since the duration of a central solar eclipse can exceed two hours, the change of parallax in longitude can also affect the duration by changing the apparent velocity of the moon. Phases and durations are treated in *Almagest* VI,7,9–10 and *Epitome* VI,17–18,24–26 by first finding the length of the arcs traversed by the moon between each phase, and then dividing each arc by the velocity of elongation—true velocity for lunar eclipses, apparent velocity for solar—in order to find the durations of the phases. Two simplifying assumptions are made: that plane triangles may be substituted for spherical, and that the motion of the moon is parallel to the ecliptic during an eclipse, although the *Epitome* also shows how to solve the problem rigorously. Following Ptolemy, Copernicus points out that the assumption of parallelism, while not strictly true, can produce a difference in the arc of the moon's transit through the eclipse of only 0;2° or about $0;4^{h}$. *Epitome* VI,15 gives a more detailed investigation of the validity of the approximation.

The configuration for a solar eclipse is shown in Fig. 43(a) with the sun at S and the moon at first contact A, mid-eclipse C, and last contact E. The distance $SA = SE = r_s' + r_m'$, and at mid-eclipse $SC = \beta'$, the apparent latitude at apparent conjunction, which is related to the magnitude of the eclipse by

$$\beta' = r_m' + r_s'\left(1 - \frac{m'}{6}\right). \tag{123}$$

Then

$$AC = (SA^2 - SC^2)^{1/2} = [(r_s' + r_m')^2 - \beta'^2]^{1/2} = \Delta\eta_i \tag{124}$$

is the minutes of immersion, and dividing by the apparent velocity of elongation $v_\eta'^{\circ/h}$,

$$\Delta t_i = \Delta\eta_i/v_\eta'^{\circ/h} \tag{125}$$

is half the duration of the eclipse. Copernicus appears to assume that the duration of the second half, from mid-eclipse to last contact, is the same, although this can be far from true because the change of the moon's parallax during the eclipse can affect v_η'. The accurate, and rather tedious, procedure is explained in *Almagest* VI,10 and *Epitome* VI,26.

In a partial or minimally total lunar eclipse without a delay, much the same method applies. In Fig. 43(b), $UA = UE = r_u' + r_m'$ and $UC = \beta$ is the true latitude at true conjunction, related to the magnitude by

$$\beta = r_u' + r'_m\left(1 - \frac{m'}{6}\right). \tag{126}$$

The minutes of immersion follow from

$$AC = (UA^2 - UC^2)^{1/2} = [(r_u' + r_m')^2 - \beta^2]^{1/2} = \Delta\eta_i, \tag{127}$$

and half the duration from

$$\Delta t_i = \Delta\eta_i / v_\eta^{\circ/h}, \tag{128}$$

where v_η is the true velocity of elongation. Again it is assumed that the duration of the second half of the eclipse is the same, and in the case of a lunar eclipse the assumption is approximately correct since parallax does not affect the true velocity of elongation.

The five phases of a total lunar eclipse with a delay are shown in Fig. 43(c). At the beginning of the total phase, the moon at B is internally tangent to the shadow so that $UB = r'_u - r'_m$. Therefore

$$BC = (UB^2 - UC^2)^{1/2} = [(r'_u - r'_m)^2 - \beta^2]^{1/2} = \Delta\eta_d \tag{129}$$

gives the minutes of the delay, i.e. of the total phase, and

$$\Delta t_d = \Delta\eta_d / v_\eta^{\circ/h} \tag{130}$$

is half the duration of the delay. It is assumed that the second half of the delay is equal. Finally, the length of the partial phase of immersion is

$$AB = AC - BC = \Delta\eta_i - \Delta\eta_d, \tag{131}$$

and the duration of the partial phase is $\Delta t_i - \Delta t_d$.

Example. In the eclipse of 6 Nov 1500, we found that

$$r'_m = 0;15,50^\circ, \qquad r'_u = 0;41,40^\circ, \qquad \beta = -0;31^\circ.$$

Hence the minutes of immersion

$$\Delta\eta_i = [(0;15,50^\circ + 0;41,40^\circ)^2 - 0;31^2]^{1/2} = 0;48,26^\circ.$$

For $\alpha_{☾} = 292;6^\circ$, the true velocity of elongation

$$v_\eta = \bar{v}_\eta + v_\alpha \cdot \Delta c = 0;30,29^{\circ/h} + 0;50^{\circ/h} \cdot -0;2 = 0;28,49^{\circ/h},$$

and thus half the duration of the eclipse

$$\Delta t_i = 0;48,26^\circ / 0;28,49^{\circ/h} = 1;41^h.$$

According to Oppolzer (1887), $\Delta t_i = 1;35^h$.

Copernicus concludes with a curious, and archaic, part of eclipse calculation. In *Almagest* VI,7 Ptolemy says that most of those who observe eclipse prognostications, i.e., for weather prediction, estimate magnitudes by areas of the eclipsed body rather than by diameters. Ptolemy correctly considers this a crude method since it is hard enough to estimate linear digits of the lunar or solar diameter, let alone fractions of a circular area cut off by another circle. Nevertheless, he shows how to compute area digits, twelfths of the area of the eclipsed sun or moon, and provides a table for converting linear digits to area digits that is commonly reproduced, e.g. in the *Alfonsine Tables*.[87] In so doing, he gives a subsequently well-known

[87] See *HAMA*, 140–41.

approximation $\pi = 3;8,30$, remarking that it is nearly a mean between Archimedes's limits, i.e.

$$3\tfrac{1}{7} \approx 3;8,34 > 3;8,30 > 3\tfrac{10}{71} \approx 3;8,27.$$

Accurately, $\pi = 3;8,29,44,0 \ldots$, so the approximation is quite good. Ptolemy gives numerical examples of finding area digits for both a solar and lunar eclipse—they differ slightly since the large shadow covers a larger area of the moon for a given immersion than the moon covers of the sun—and *Epitome* VI,27 explains the method with the same examples. Copernicus paraphrases Ptolemy's remark that many people—although surely not in the sixteenth century—measure eclipses by areas, and gives a qualitative summary of the procedure for computing the areas.

5

Planetary Theory of Longitude

Introduction

It was mentioned in the introduction to the previous chapter that Book IV of *De revolutionibus* is of interest, not only for its great variety of subjects, but also because it is a synthetical treatise on lunar and eclipse theory, probably the last of its kind, that brings together techniques and numerical parameters accumulated over more than a millenium from diverse civilizations of antiquity and the middle ages. Lunar and eclipse theory were, in a sense, practical matters, used both for the prediction of eclipses, which are quite obvious and provide easily observable (although not necessarily easily usable) tests of theory and parameters, and also for the routine determination of first visibility of the moon in astronomical, rather than schematic, lunar calendars. Thus it is not surprising that what innovations there were during the middle ages took place principally in these subjects and in the equally practical solar theory and spherical astronomy, some of which found their way into Copernicus's sources and then into his book.

But planetary theory was something else. The only common application of planetary theory was for the casting of horoscopes, and astrologers, whether in the second century or the twentieth, have for the most part been content to use whatever methods of computing positions were available—and the easier the better—without much thought about accuracy. And unlike eclipses or first visibilities of the moon, it is by no means easy to perform the observations necessary to test the accuracy of planetary theory or parameters, and this made it difficult for even the handful of astronomers that may have had a serious interest in the subject to isolate the problems in earlier theory and make improvements. At the time of Copernicus, planetary theory was much the same as Ptolemy left it in the *Almagest*, and even the parameters were for the most part unchanged, except for the necessary adjustments of mean motions required by alterations of the mean motion of the sun. In the *Epitome*, although the presentation is up to its expected standard of clarity, Regiomontanus adds little by way of theory or method, and nothing by way of parameters, to what was already established in the *Almagest*. Hence, Book V of *De revolutionibus* is technically more thoroughly Ptolemaic, and less influenced by later astronomy in its methods and numerical parameters, than the preceding two books on the motions of the earth and moon. More than ever before, Copernicus could only follow Ptolemy's procedures to confirm or modify slightly Ptolemy's results, and only in the sidereal motions of the planetary apsidal lines that, mostly through a combination of errors in his own and in Ptolemy's observations, appeared considerably larger than their correct values, did he add a single element to those specified by Ptolemy.

Although Copernicus adds nothing to Ptolemy's methods, and makes only small modifications of Ptolemy's parameters, there are of course two important departures from Ptolemy's theory in Book V. The first, and better known, is the representation of the second anomaly of the superior planets by the motion of the earth around the mean sun, and of the inferior planets by the motion of the planet around the mean sun, in place of the epicycle used in Ptolemy's geocentric models. At the time of Copernicus, and to Copernicus himself, the importance of the innovation was that it allowed the unambiguous determination of the order and distances of the planets from the mean sun, which was the principal advantage of the heliocentric theory, whether Tychonic or Copernican in form, over the geocentric theory. The other departure from Ptolemaic theory is the use of models for the planets in which each separate motion is uniform and circular, or more physically, each component sphere rotates uniformly about an axis passing through its center, a principle violated by Ptolemy's models whether considered geometrically or physically. This physical principle about motions in the heavens was of the greatest importance to Copernicus, and its violation by Ptolemy was to him a great source of annoyance. It formed the first of his objections to Ptolemy's lunar theory, and in his earlier work, the *Commentariolus*, he says that it was concern over the violation of uniform circular motion in Ptolemy's planetary theory that provided the motivation for his own investigations.

Copernicus's models in Book V do succeed in preserving uniform circular motion just as he wished, but it is extremely improbable that they were his own inventions. During the last twenty-five years, principally through the publications of E. S. Kennedy, it has become apparent that a number of Arabic astronomers of the thirteenth and fourteenth centuries devised various models that would meet the same criterion, among them models identical in every respect except the heliocentric form and the specific numerical parameters to those used by Copernicus in the *Commentariolus* and *De revolutionibus*.[1] The recovery of the planetary theory of the astronomers of the "Marāgha School," as they have been called because of the association of some of them with the observatory of Marāgha in north-western Iran, is not only of great interest in itself, but has also demonstrated that much of what had been taken for Copernicus's own planetary theory is actually of medieval Arabic origin, and was transmitted to western Europe by an unknown route, perhaps by way of late Byzantine sources, to Italy at some time in the fifteenth century. A more detailed account of the transmission of this astronomical material, and evidence for how extensive it was, must await further research.[2]

None of this, however, is to say that Book V of *De revolutionibus* is not an impressive accomplishment, even though it may not be so impressive for diversity or historical interest as Book IV. Copernicus set out to do in Book V exactly what Ptolemy did in Books IX–XI of the *Almagest*, that is, to derive a complete and consistent set of elements of the orbits of all five planets, directly from observation and, in so far as possible, independently of

[1] The principal literature on the subject of Marāgha planetary theory is listed above, p. 46, n. 7.
[2] The current state of the evidence for the transmission is discussed above, pp. 47–48.

earlier elements. This was a task of great difficulty—it is not known whether anyone since Ptolemy even attempted it—and Copernicus, in spite of his isolation and unfavorable location for observation, was at least partially successful in attaining his goal. For the superior planets, he did derive new elements, using Ptolemy's methods for the most part correctly, and in so doing he found that the eccentricities of Saturn and Jupiter had remained constant since Ptolemy, the eccentricity of Mars had diminished slightly, and the apsidal lines all showed an obvious displacement in the intervening 1400 years. For the inferior planets, however, he could not derive new elements on his own, if for no other reason because he could not obtain the observations required to apply Ptolemy's methods. Hence, the best he could do was to use Ptolemy's observations directly in order to adapt Ptolemy's elements to his own models, and then make adjustments, of less than clear origin, in the eccentricity of Venus and the direction of the apsidal line of Mercury. The difficulty and tedium, in computation alone, of these derivations, and of the preparation of the text and tables of Book V, was immense, and since only the accumulated observations of later years could show whether Copernicus's elements were accurate, he could never really know if his long, solitary labors were worthwhile, a judgment that even now, and after a detailed study of Book V, is difficult to make.[3]

1. General Considerations

Model for the Second Anomaly (V,3)

Copernicus's most well-known contribution to astronomy is his model for the second or solar anomaly of the planets. Figure 1(a) shows the standard geocentric model for a superior planet with the earth at O and the planet P moving on an epicycle with center C, the motion of C and P both being in the positive sense. The period of P is the planet's mean synodic period, return

[3] While there are a number of special studies of various aspects of Copernicus's planetary theory, particularly his theory of Mercury (see below, p. 404, n. 2), there are no comprehensive analyses of Book V. The more general, and technically competent, studies of Copernicus's planetary theory are: Michael Maestlin, his *Appendix* to Kepler's *Mysterium Cosmographicum* (1596) in Kepler (1937–) 1, 132–45, trans in Grafton (1973); Delambre (1821) 1, 121–36; Herz (1887–94) 2, 74–94; Moesgaard (1974); Riddell (1980), 114–39. Each of these is written for a different purpose, and with a different emphasis. Maestlin, whose particular concern was distances, is purely descriptive, although quite thorough, and the parameters are taken from the *Prutenic Tables*. Delambre and Herz modernize considerably, and are not always numerically reliable, although Herz remains the best treatment of the subject and is still worth careful study. Moesgaard is concerned principally with the relation of Copernicus's observations and elements to modern theory, without considering the details of Copernicus's own derivations. There is some rather imaginative speculation about what could account for discrepancies (for example, see below, pp. 359–360). Riddell's quite penetrating analysis of the planetary theory, considered formally without examining the observations or derivations, is part of a comparative study of the planetary theory of Ptolemy, Copernicus, and Kepler.

to conjunction with the mean sun $\bar{\odot}$ at which time the planet is at the apogee *A* of the epicycle, and thus the radius of the epicycle *CP* is always parallel to the direction $O\bar{\odot}$ from the earth to the mean sun. It follows also that when at opposition to the mean sun, the planet will be at the perigee of the epicycle, and that the apparent retrograde motion of the planet will take place in the vicinity of the perigee. Complete parallelogram *OCPS*, and assume that the mean sun coincides with *S*. Then there are two historically interesting transformations. If *O* is taken to be at rest, as in Fig. 1(b), the result is that *S* moves in the positive sense about *O* while, with respect to the apogee *A*, *P* moves in the negative sense about *S*. This is the Tychonic model. If as in Fig. 1(c) *S* is taken to be at rest, then both *O* and *P* move about *S* in the positive sense. This is the Copernican model. In both cases *OS* may be taken as the common measure of the radii *SP* of the planetary orbits, thereby fixing the order and distances of the planets with respect to the mean sun.[4]

The standard geocentric model for the second anomaly of an inferior planet is shown in Fig. 2(a), in which the center of the epicycle *S* always lies in the direction $O\bar{\odot}$ from the earth to the mean sun while the planet *P* completes a rotation on the epicycle in its mean synodic period. At superior conjunction the planet is at the apogee *A* of the epicycle. If the mean sun is assumed to be at *S*, as in Fig. 2(b), the result is the Tychonic model, although such an arrangement for the inferior planets was already known in antiquity.[5] If in addition *S* is assumed at rest and *O* in motion, the result is the Copernican model shown in Fig. 2(c). Again in both models the distance *SP* may be measured in units of *OS*, thus determining the distances of the planets from the mean sun.

The Problem of the First Anomaly (V,2)

A more difficult, and more interesting, problem is posed by the first or zodiacal anomaly. The correct representation of the first anomaly is given by Kepler's first two laws:

(1) that the orbit of the planet is an ellipse with the sun at one focus,
(2) that the line joining the planet and the sun describes an area proportional to time.

The best earlier approximation to Kepler's laws was Ptolemy's model, which is shown heliocentrically and superimposed on Kepler's model in Fig. 3.[6]

[4] The preceding is merely an explanation of the transformation of the models, not of how Copernicus arrived at the heliocentric model for the second anomaly, concerning which see above, pp. 55–60.

[5] The literature on this particular subject is excessive in more than one sense. For a critical evaluation, cf. *HAMA*, 693–98.

[6] The eccentricity in the figure, and in all subsequent figures, is greatly exaggerated, as is the departure of the ellipse from its major auxiliary circle. For planetary eccentricities and ellipses drawn to scale, see *HAMA*, 1102–03.

According to Kepler, the planet moves on ellipse *APB*, and the line *SP* from the sun to the planet describes an area *ASP* that increases uniformly with time. In Ptolemy's model, the planet at *P′* moves on the major auxiliary circle of the ellipse *AP′B*, and the line *EP′* joining the empty focus to the planet describes an angle *AEP′* that increases uniformly with time. Kepler found that in the case of Mars, which has a relatively large eccentricity $e \approx .1$, the greatest difference between the two models, angle *PSP′* $\approx$ 0;8°.[7]

Copernicus's objections to Ptolemy's model were not on the grounds of accuracy—indeed, no one before Tycho and Kepler was in a position to evaluate its accuracy—nor on the grounds of a serious defect as in the case of the lunar distance at quadrature, but only that it violates the principle of uniform circular motion. The complete model is shown in Fig. 4. The earth is at O and the center of the epicycle *C* moves on the circle about *M* such that angle *AEC* increases uniformly. Point *E*, corresponding to the empty focus of Kepler's ellipse, was called in medieval astronomy the "equant point" (*punctum aequans*), and the circle *FC′G* described about *E* the "equant circle" or "equant eccentric" (cf. *Epitome* IX,7). In this way, the uniform motion of *C* can be said to take place either about *E* or on the circle *FC′G*. Copernicus objects that the motion of *C* ought to be uniform with respect to the center *M* of the circle on which it actually moves, that is, angle *AMC* ought to increase uniformly so that *C* moves uniformly on circle *ACB*. Extend *EC* to meet the far side of the epicycle at *H*, and the planet *P* now moves on the epicycle such that *HCP* increases uniformly, that is, the distance of the planet from apogee *H* increases uniformly, a necessary consequence of the uniform motion of *C* about *E*. Here Copernicus objects that the motion of *P* ought to be uniform with respect to the apogee *I* lying on *MC* extended, that is, angle *ICP* ought to increase uniformly, which would indeed be true if *C* moved uniformly about *M*.[8]

With the wisdom of hindsight, we know that Copernicus' objections were ill-founded and that Ptolemy's model was the single most important step towards a correct understanding of planetary motion prior to Kepler. Kepler himself already showed his intuitive understanding of its value in Chapter 22 of the *Mysterium Cosmographicum* where he pointed out that it was precisely because it did violate uniform circular motion—making the true orbital motion of the planet faster at perihelion and slower at aphelion—that in some way it must correspond to the true physical causes of planetary motion.[9] But here we are looking to the future, and to understand Copernicus's motivation we must look to the past. Considered historically, his

---

[7] *Astronomia nova* 19, Kepler (1937–) 3, 177–78, where strictly the comparison is between Ptolemy's model and Kepler's "vicarious hypothesis," which produces directions, although not distances, nearly indistinguishable from the ellipse and area law.

[8] The same objections are raised, quite succinctly, in the *Commentariolus* (cf. Swerdlow (1973), 434–35), and appear to have provided the original motivation for Copernicus's investigation of an alternative planetary theory.

[9] Kepler (1937–) 1, 75–77, and with his notes in the second edition of 1621 in *ibid.* 8, 120–23.

objections are of considerable interest because they were likewise the motivation of the principal innovations in late medieval planetary theory.

The objections can be reduced to the principle that as long as the motions of heavenly bodies are believed to be controlled by the rotation of material spheres, only uniform rotations about diametral axes are physically possible. The objections are thus on physical or mechanical rather than on merely philosophical grounds. That Ptolemy's model violates this principle is shown in Fig. 5, the sort of physical model found originally in Ptolemy's *Planetary Hypotheses* and known to Copernicus principally through Peurbach's *Theoricae novae planetarum.* The problem is first that the spherical body 3 with an axis passing through center *M* is expected to rotate so that it carries the epicycle uniformly about *E*, a point off the axis of the sphere. Second, the epicycle 4, another spherical body, is expected to regulate its rotation such that the planet is moved uniformly away from *H*, which has a *prosneusis*, an inclination towards the extraneous point *E*.

By the early eleventh century, Ptolemy's model was subject to criticism of this kind, and perhaps one could say that he brought it on himself by writing Book II of the *Planetary Hypotheses.* The point was raised by Ibn al-Haytham in his *Al-Shukūkᶜalā Baṭlamyūs* (Doubts concerning Ptolemy), a vitriolic and often obtuse, but historically important critique of what appeared to him to be problems and inconsistencies in the *Almagest* and the *Planetary Hypotheses.*[10] How common such criticisms were is not known, but it is clear that ultimately they led to attempts at devising physically unobjectionable planetary models. More than two hundred years after Ibn al-Haytham, a number of astronomers associated with the observatory at Marāgha developed various models that would effectively reproduce the apparent motion of Ptolemy's model, critically the uniform motion of the center of the epicycle with respect to the equant, while preserving uniform circular rotations of all the component spheres, or at least, so the inventors seemed to think. The first generation of these astronomers included Muʾayyad ad-Dīn al-ᶜUrḍī (d. 1266) and Naṣīr ad-Dīn aṭ-Ṭūsī (d. 1274), and they were followed in the course of the next century by Quṭb ad-Dīn ash-Shīrāzī (d. 1311) and Ibn ash-Shāṭir (d. 1375).[11]

We have already remarked upon Copernicus's use in the theory of the precession and variation of obliquity of Ṭūsī's device for the generation of rectilinear motion from two circular motions, and we have likewise considered the relation of Copernicus's lunar model to that of Ibn ash-Shāṭir.[12] In the *Commentariolus*, aside from the heliocentric representation

[10] Text edited by A. Sabra and N. Shehaby, Cairo, 1971. An English translation by D. Voss is in preparation. On the general contents, see Sabra (1972), 198–99 and Sabra (1978).

[11] See the general reference to the secondary literature above, p. 46, n. 7. The studies of most importance here are Kennedy and Roberts (1959), Kennedy (1966), King (1975), 358–60, Saliba (1979a), containing the important discovery that Shīrāzī's models, which are closest in form to those in *De revolutionibus*, were fundamentally the invention of ᶜUrḍī, Ragep (1982), the most detailed discussion of the physical basis of the planetary models.

[12] Above, pp. 135, 196.

of the second anomaly, all of Copernicus's planetary models are identical to those of Ibn ash-Shāṭir.[13] In *De revolutionibus* the models are somewhat altered, but are still based upon the principles of the Marāgha astronomers. The alterations are the result of discoveries and further analyses that Copernicus made in the course of working out the numerical parameters in *De revolutionibus*—in the *Commentariolus* parameters were merely adapted from the *Alfonsine Tables*—but nowhere do the alterations depend upon new principles.[14] In a very real sense, Copernicus can be looked upon as, if not the last, surely the most noted follower of the "Marāgha School."

Model for the First Anomaly (V,4)

In the following analysis, we shall consider the model for the superior planets, taking up the modifications for Venus and Mercury when treating those planets separately.[15] The fundamental model for the first anomaly is shown in Fig. 6 in the form used by al-ʿUrḍī and ash-Shīrāzī superimposed upon Ptolemy's model. On the apsidal line AB, we take an eccentricity $e_1 = OM'$ measured from the observer at O such that $e_1 = \frac{3}{2}e$, where e is the eccentricity of Ptolemy's model. M' is now the center of a circle of radius R, on which moves an epicycle of radius $r' = e_1/3 = e/2$. We call the center of the epicycle C_1 and the endpoint of r' C_2, which corresponds to the center of the epicycle C in Ptolemy's model. The motion of C_1 and C_2 are both in the positive sense through the "eccentric anomaly" $\bar{\kappa}$ such that C_2 lies on C_1M' at C_{2a} when $\bar{\kappa} = 0°$. Hence, when C_1 is at apogee A,

$$OC_{2a} = R + e_1 - r' = R + 3e/2 - e/2 = R + \mathrm{e},$$

and at perigee B,

$$OC_{2b} = R - e_1 + r' = R - 3e/2 + e/2 = R - e,$$

in agreement with Ptolemy's model. It is these extremal conditions that determine the relation of e_1 and r' to e. Now, we draw QQ' parallel to the apsidal line, and Q' must move on a circle of radius R and center M, while Q moves on a circle of parallel radius EQ and center E. The center of the epicycle in Ptolemy's model lies at C, the intersection of EQ and the circle of radius R about M. Since $EM' = C_1C_2 = r'$, while $EM'C_1 = M'C_1C_2 = \bar{\kappa}$, C_2 must also lie on EQ and thus move uniformly about E through $\bar{\kappa}$. This means that the motion of the center of the epicycle C_2 is uniform with respect to the same equant point E as in Ptolemy's model at a distance from O of $OE = e_1 + r' = 2e$.

Copernicus's model for the first anomaly is shown in Fig. 7, likewise superimposed upon Ptolemy's model, although with somewhat less detail.

[13] Cf. Swerdlow (1973), 467–69, 500–505. The parameters, of course, differ.

[14] Below, pp. 299 f., 356 ff.

[15] Below, pp. 372 ff., 405 ff.

The only difference from the model of Fig. 6 is the replacement of the earth O by the mean sun $\bar{S}$ and of the center of the epicycle C_2 by the planet P. All fundamental properties of the models are identical. It is curious that Copernicus never demonstrates that his model preserves uniform motion with respect to the equant, although he uses this most important property repeatedly in the derivations of the eccentricities of the superior planets.[16] He does, however, explain another important property of the model, the small departure of the path of the planet from circularity. In Ptolemy's models, with the exception of Mercury, the path of the center of the epicycle is circular. It can be seen in Figs. 6 and 7 that C_2 and P lie outside the circle passing through C, the center of the epicycle in Ptolemy's model, although because the eccentricities are small, the departure from the circle is very slight. Copernicus goes through some pains to prove that "by this composite motion, the star does not describe a perfect circle, in accordance with the theory of the ancient mathematicians, although the difference is insensible."[17] He does this by considering only the situation for $\bar{\kappa} = 90°$ shown in Fig. 8. Since when in the apsidal line P is at P_a and P_b, if the orbit of P were circular, it would move on a circle with center M, midway between P_a and P_b, and the radius of the circle would be MP. But for $\bar{\kappa} = 90°$, $PE = R$, and thus P lies momentarily on a circle FG about E (the circle about the equant in fact). Consequently $MP > R$, and P lies outside the circle about M, which is the circular path of the center of the epicycle in Ptolemy's model. And as shown in Fig. 9, P is mostly outside circle P_aP_b, coinciding with the circle only when in the apsidal line.

Kepler's remark in the *Astronomia nova* on this departure from circularity is well known:

> Ptolemy would rightfully raise this deviation (*exorbitationem*) of the path of the planet from the perfection of a circle as an objection to Copernicus; I do not object. For below in the Fourth Part, it will be demonstrated that by two physical forces, each simple in effect, combining to move the planet, it is necessarily brought about that the planet turns aside a little from a circle, not by moving outward as in this Copernican model, but in the opposite direction, that is, by entering towards the center.[18]

---

[16] Below, pp. 313–319. See also the correspondence between Maestlin and Kepler on equant motion in Copernicus's models in Grafton (1973), 526–29. Although clearly demonstrated by Herz (1887–94) 2, 96–97, and several times since, the quantity of modern literature asserting that equant motion is absent from Copernicus's models is not small.

[17] ... *quod sidus hoc motu composito, non describit circulum perfectum*[: M] *iuxta priscorum sententiam mathematicorum, differentia tamen insensibili* (N, 142v10–12, M, 151r5–8). Cf. the remarkable translation in Copernicus (1978), 242:30–33 and the note at 417 to 242:32, according to which ancient astronomers employed non-circular orbits. Rather like Copernicus's demonstration, a similar departure from circularity is pointed out by Ṭūsī concerning his lunar model in Carra de Vaux (1893), 351–52 and Hartner (1969).

[18] *Astronomia nova* 4, Kepler (1937–) 3, 75. Kepler's remarks would be incorrect and pointless according to the interpretation referred to in n. 17.

Kepler's point, of course, is that the departure from circularity in Copernicus's model, while not objectionable in itself, is in exactly the wrong direction. More important than the departure itself is the fact that "the difference is insensible," as Copernicus says although he provides no proof. Therefore, consider again Fig. 8 in which, since $ME = 2r' = e$, so $MP = (R^2 + e^2 \sin^2 \bar{\kappa})^{1/2}$, while $MC = R$. Hence the difference $MP - MC$, here $PC \approx e^2 \sin^2 \bar{\kappa}/2R$, or at most, for $\bar{\kappa} = 90°$ as illustrated, $e^2/2R$. In the case of Mars with the large eccentricity $e \approx .1$, $PC = .005$, which can change the apparent position of the planet by about 0;3°.

We may now summarize the relation of Ptolemy's and Copernicus's models for the first anomaly by stating their fundamental properties in a form analogous to Kepler's first two laws. For Ptolemy's model:

(1) The center of the epicycle describes a circle with its center removed from the earth by an eccentricity e.
(2) The line joining the center of the epicycle to a point on the apsidal line removed from the earth by a double eccentricity $2e$ describes an angle proportional to time.

For Copernicus's model:

(1) The planet describes a figure differing insensibly from a circle with its center removed from the mean sun by an eccentricity e.
(2) The line joining the planet to a point on the apsidal line removed from the mean sun by a double eccentricity $2e$ describes an angle proportional to time.

To complete the comparison with Kepler's laws, we need only point out that:

(1) the circle in Ptolemy's model and the nearly circular figure in Copernicus's model correspond to the major auxiliary circle to the ellipse,
(2) the point on the apsidal line at a distance $2e$ from the earth or mean sun respectively corresponds to the empty focus of the ellipse.

The Equation of Center

The correction produced by the model for the first anomaly is called the "equation of center," from the medieval term "center" (*centrum*) applied to the eccentric anomaly. Figure 10 shows Ptolemy's and Copernicus's models for the first anomaly superimposed heliocentrically with the mean sun at $\bar{S}$. In the former, the planet P moves on a circle about M while remaining always on a line EP that describes angle $\bar{\kappa}$, the mean eccentric anomaly, which is proportional to time. In Copernicus's model, by the combined rotations of $M'C$ and CP', the planet P' moves on a figure nearly indistinguishable from a circle, lying just outside of the circle about M, while likewise remaining on the line EPP' that describes the same mean eccentric anomaly $\bar{\kappa}$. With respect to $\bar{S}$, on the other hand, P and P' move nonuniformly through the true eccentric anomaly κ, that is angles $A\bar{S}P$ and $A\bar{S}P'$ respectively. The difference, $\kappa - \bar{\kappa} = c_3$, is the equation of center, and as would now be expected, c_3 is almost identical in the two models. In a simple modern formulation, using a tangent function, in Ptolemy's model

$$\tan c_{3P} = \frac{2e \sin \bar{\kappa}}{\sqrt{R^2 - e^2 \sin^2 \bar{\kappa}} + e \cos \bar{\kappa}}, \tag{1}$$

and in Copernicus's model

$$\tan c_{3C} = \frac{(e_1 + r') \sin \bar{\kappa}}{R + (e_1 - r') \cos \bar{\kappa}}. \tag{2}$$

If, according to the standard relation

$$e_1 = \tfrac{3}{2}e, \qquad r' = \tfrac{1}{2}e, \tag{3}$$

the equation in Copernicus's model is equivalent to

$$\tan c_{3C} = \frac{2e \sin \bar{\kappa}}{R + e \cos \bar{\kappa}}. \tag{4}$$

In the case of Mars, letting $r \approx .1$, the greatest difference between the two equations is about 0;3,21°, which occurs at $\bar{\kappa} \approx 93;46°$, and for the other planets the difference is always less.

Copernicus does not solve the equation of center in the form given above. Using a sine rather than a tangent, the equation for his model may be written

$$\sin c_3 = \frac{(e_1 + r') \sin \bar{\kappa}}{\sqrt{(R + (e_1 - r') \cos \bar{\kappa})^2 + (e_1 + r')^2 \sin^2 \bar{\kappa}}} \tag{5}$$

but neither does he solve for c_3 in this form. Instead, he divides c_3 into two components, a_1 and a_2, as shown in Fig. 11, and takes c_3 as their sum. The procedure is rather cumbersome. He first solves triangle $M'C\bar{S}$, finding

$$\bar{S}C = (R^2 + e_1^2 + 2Re_1 \cos \bar{\kappa})^{1/2}, \tag{6}$$

and then takes

$$a_1 = \sin^{-1} (e_1 \sin \bar{\kappa}/\bar{S}C). \tag{7}$$

Next, he forms the sum $\bar{\kappa} + a_1$, and in triangle $C\bar{S}P$, finds

$$\bar{S}P = (\bar{S}C^2 + r'^2 - 2\bar{S}Cr' \cos(\bar{\kappa} + a_1))^{1/2}, \tag{8}$$

which leads to

$$a_2 = \sin^{-1} (r' \sin (\bar{\kappa} + a_1)/\bar{S}P). \tag{9}$$

Finally, the equation of center itself,

$$c_3 = a_1 + a_2, \tag{10a}$$

and the corrected or true eccentric anomaly,

$$\kappa = \bar{\kappa} + c_3. \tag{10b}$$

Note that c_3 may be measured both at the planet P and at the intersection of $M'C$ and $\bar{S}P$ since EP and $M'C$ are parallel. The latter position of the angle brings out clearly the relation of c_3 to a_1 and a_2, but is not strictly the equation of center, which is properly the correction between the uniform and nonuniform motion of the planet, for the uniform motion of P takes place with respect to the equant E, not the center of the eccentric M'. As we have seen, in fact, the fundamental purpose of the model is to produce a uniform motion of the planet about the equant through uniform circular motions. Hence, we shall generally show the equant in figures of the model, and shall generally measure c_3 as angle $EP\bar{S}$. In the derivations of the elements of the planetary orbits, Copernicus applies the equation of center dozens of times, and the tables of corrections require nearly three hundred further solutions. In our examination of the derivations, unless the distances $\bar{S}C$ and $\bar{S}P$ are specifically needed, as they are for finding the radius of the orbit, we shall give only the final addition $c_3 = a_1 + a_2$ along with any disagreement with a correct solution.

Transformations of the Complete Model and Technical Terms

The complete geocentric model of al-ᶜUrḍī and ash-Shīrāzī is shown superimposed upon Ptolemy's model in Fig. 12. Since C_2 nearly coincides with the center of the epicycle C in Ptolemy's model, the position of the planet on the epicycle, which accounts for the second anomaly, is likewise nearly indistinguishable in the two models. The transformation of the Marāgha model to its heliocentric form, by a series of parallelograms after the manner of Fig. 1 for the second anomaly, is illustrated in Fig. 13. Parallelogram $O\bar{S}MM'$ establishes an apsidal line $\bar{S}A$ passing through the mean sun, $M'C_1CM$ locates the center of the small epicycle C, and finally C_1C_2PC locates the planet P. Note further that taking $EM = E'M'$ and drawing EP parallel to $E'C_2$ shows the uniform motion of P through the mean eccentric anomaly $\bar{\kappa}$ with respect to the equant E.

The resulting form of the model that is used by Copernicus in *De revolutionibus* is shown in Fig. 14. He also mentions in V,4 that it is possible to construct a double epicycle form of the model, as in fact he had used in the *Commentariolus*, and this equivalent form is shown in Fig. 15. Why he prefers the eccentric form—or as we know, why he changed from the double epicycle to the eccentric—is explained in an interesting remark about the change of eccentricities of Mars and Venus due to the motion of the center of the earth's orbit since the time of Ptolemy.

> But here we have selected an epicycle with an eccentric because, while always remaining between the sun and the center C [of the planet's eccentric (M in Fig. 14), the center of the annual sphere of the earth] D ($\bar{S}$ in Fig. 14) is found in the meantime to have changed position, as was shown in the solar appearances. But since the other planets have not likewise accomodated themselves to this change of position, it is necessary that some sort of correction follows in them which, although it is very small, is nevertheless detected in the case of Mars and Venus.[19]

We shall take up the significance of this statement when considering the specific planets.[20]

A more detailed comparison of the Ptolemaic and Copernican models for both anomalies is shown in Figs. 16(a) and 16(b), which are in exactly the same configuration, the former showing Ptolemy's model in solid lines and Copernicus's model in broken lines, and the latter the other way around. Note that in the Ptolemaic form, longitudes are measured tropically from ♈0°, the vernal equinox, while in the Copernican form longitudes are measured sidereally from γ Arietis, here denoted by Γ. Finally, the Copernican model is shown alone and complete in Fig. 17. Copernicus's technical vocabulary for the parts of the model is diverse, and not altogether constant. Some of the standard terms are as follows:

- $\bar{\lambda}^*_{\odot}$ *locus solis medius, medius motus solis*: mean position or longitude of the sun, mean motion of the sun, always understood to be the "simple" or sidereal mean longitude.
- $\bar{\lambda}^*$ *medius motus, motus proprius, locus medius, locus aequalis*: mean sidereal motion or longitude of the planet.
- $\bar{S}$ *centrum orbis terrae annui*: center of the annual sphere of the earth, i.e. what we usually call the "mean sun."
- A *summa apsis, apogaeus eccentri, apogaeus*: the "apogee" of the eccentric, meaning the point farthest from $\bar{S}$. Copernicus does not use the term "aphelion."[21]
- B *infima apsis, perigaeus*: the "perigee" of the eccentric, the point closest to $\bar{S}$.
- λ^*_A *locus summae apsidis*: the sidereal longitude of the "apogee."
- $\bar{\kappa}$ *anomalia eccentri*: mean eccentric anomaly $\bar{\kappa} = \bar{\lambda}^* - \lambda^*_A$.
- c_3 *prosthaphaeresis eccentri, aequatio eccentri*: equation of center.
- κ *anomalia eccentri aequata*: corrected eccentric anomaly $\kappa = \bar{\kappa} + c_3$.
- M *centrum circuli eccentri, centrum eccentri*: center of the eccentric.

---

[19] N, 143r5–10. Copernicus is thinking only of distance, not direction, and he is not strictly correct since the center of Venus's eccentric can be, and with his eccentricity is, closer than the center of the earth's eccentric to the true sun.

[20] Below, pp. 356 ff., 384 ff.

[21] The term ἀφήλιον was used in 1596 by Kepler in *Mysterium Cosmographicum* 15, Kepler (1937–) 1, 51, 53, for the greatest distance from the true sun, and in a note to the second edition of 1621, *ibid.* 8, 90, he claims to have invented *Aphelia* and *Perihelia*, which may well be true.

E Copernicus has no name for the equant point, but refers to a circle of radius R described about E as the "eccentric circle in which the planet is considered to move simply and uniformly" (N 151v6–7).[22]

T *apogaeus commutationis, apogaeus orbis*: the point on the earth's orbit, lying on a line through $\bar{S}$ parallel to EP and MC, that corresponds to the mean apogee of the epicycle in Ptolemy's model (point H in Figs. 4 and 16(a)).

U *perigaeus orbis*: the point on the earth's orbit corresponding to the mean perigree of the epicycle in Ptolemy's model.

V *verus apogaeus orbis*: the point on the earth's orbit, lying on $P\bar{S}$ extended to meet the orbit, corresponding to the true apogee of the epicycle in Ptolemy's model.

$\bar{\alpha}$ *anomalia commutationis, motus commutationis*: anomaly or motion of "commutation," a term that Copernicus uses as the equivalent of the Greek παράλλαξις; here it is the mean anomaly $\bar{\alpha} = \bar{\lambda}^*_{\odot} - \lambda^*$ or arc TO.[23]

α *anomalia commutationis aequata* or *coaequata*: corrected anomaly $\alpha = \bar{\alpha} + c_3$ or arc VO.

c *prosthaphaeresis commutationis, commutatio, parallaxis orbis*: the equation of the anomaly, i.e. the angle subtended at the planet by the radius of the *orbis terrae annuus*, hence the term *parallaxis orbis*.

λ^* *locus verus, locus apparens*: true or apparent sidereal longitude of the planet $\lambda^* = \bar{\lambda}^* + c$.

The orbit of the earth O about $\bar{S}$ is called the *orbis terrae annuus* or *orbis magnus terrae*, the "annual sphere" or "great sphere" of the earth, the word *orbis* referring to the rigid spherical body, the solid sphere, that carries the earth (and the moon) about the mean sun. The orbit of the center of the epicycle C is called the *circulus eccentrus, eccentrus*, or more physically the *orbis eccentrus* or *orbis*. For the epicycle, Copernicus usually uses the diminutive *epicyclius*, the "small epicycle." The term "deferent" is very seldom, if ever, used for the sphere that carries the epicycle.

Mean Motions (V,1)

In considering the mean motions, we dispense with the effects of the first anomaly and return to the simple representations of the second anomaly of Figs. 1 and 2. We define any zero point of longitude as λ_0, call the planet's

[22] In the *Commentariolus* (Prowe (1883–84) 2, 185:15), Copernicus refers to *aequantes quosdam circulos*, and in V,25 (N, 164v9) to Mercury's circle *quem recentiores appellant aequantem*, but both uses apply to Ptolemy's models.

[23] In IV,21, 22, 24, *commutatio* is used for the solar and lunar parallax.

mean motion in longitude $\bar{\lambda}$, the mean anomaly $\bar{\alpha}$, and the mean motion in longitude of the sun $\bar{\lambda}_{\odot}$. Then for a superior planet, as shown in Fig. 18(a),

$$\bar{\lambda} + \bar{\alpha} = \bar{\lambda}_{\odot}, \tag{11a}$$

and for an inferior planet in Fig. 18(b),

$$\bar{\lambda} - \bar{\alpha} = \bar{\lambda}_{\odot}. \tag{11b}$$

Note that $\bar{\alpha}$ is unaffected by the location of the zero-point of $\bar{\lambda}$ and $\bar{\lambda}_{\odot}$. Thus $\bar{\alpha}$ will be the same whether longitudes are measured tropically or sidereally, i.e.

$$\bar{\alpha} = \bar{\lambda}_{\odot} - \bar{\lambda} = (\bar{\lambda}^*_{\odot} + \pi) - (\bar{\lambda}^* + \pi) = \bar{\lambda}^*_{\odot} - \bar{\lambda}^*. \tag{12}$$

This identity is important for the form in which Copernicus arranges his tables of mean motions, tabulating only $\bar{\alpha}$, which may then be combined with $\bar{\lambda}^*_{\odot}$ or $\bar{\lambda}_{\odot}$ to yield mean sidereal or tropical longitudes. It also allows him to take over provisional values of $\bar{\alpha}$ in his demonstrations from tables arranged tropically, e.g. those of the *Almagest* or the *Alfonsine Tables.*

The relations given above hold instantaneously and for any interval, i.e. for a superior planet,

$$\Delta\bar{\lambda} + \Delta\bar{\alpha} = \Delta\bar{\lambda}_{\odot}, \tag{13a}$$

and for an inferior,

$$\Delta\bar{\lambda} - \Delta\bar{\alpha} = \Delta\bar{\lambda}_{\odot}. \tag{13b}$$

Consequently, we may define some integral numbers of years N^y in which $\bar{\lambda}_{\odot}$ completes an integral number of annual rotations N^r, $\bar{\lambda}$ completes an integral number of heliocentric rotations in longitude R^r, and $\bar{\alpha}$ an integral number of rotations in anomaly A^r so that for a superior planet,

$$R^r + A^r = N^r = N^y, \tag{14a}$$

and for an inferior planet

$$R^r - A^r = N^r = N^y. \tag{14b}$$

The periods required to fulfill these conditions exactly would be enormously long and not of much use. However, the relations are approximately satisfied by a set of periods, originally from Babylonian sources, given by Ptolemy in *Almagest* IX,3 along with small corrections that he found by comparing observations of planets about 400 years apart. Ptolemy appears not to know of the Babylonian origin of these periods, for he attributes them to Hipparchus, who probably reported them in some kind of test of their accuracy. The corrected periods are then the basis of Ptolemy's tabulations of the mean motions of $\bar{\lambda}$ and $\bar{\alpha}$.

Copernicus, who knows of the periods and their application from *Almagest* IX,3 and *Epitome* IX,4, as well perhaps as from almanachs such as that of Abraham Zacuto that are arranged for the same periods, follows Ptolemy's practice, at least in principle. Just as for the instantaneous locations, the periods may be either sidereal or tropical depending upon whether N is defined as sidereal or tropical years. Copernicus gives them sidereally, and they are as follows:[24]

Planet	A^r	R^r	$N^r = N^y$
Saturn	57	2	59
Jupiter	65	6	71
Mars	37	42	79
Venus	5	13	8
Mercury	145	191	46

(15)

The corrections to the periods take the form

$$N^y + c^d \quad \text{contain} \quad (R^r + c'^{\circ}) \pm A^r, \tag{16a}$$

A^r being added for superior and subtracted for inferior planets respectively. Since in N^y years the sun completes N^r rotations, we may say that

$$N^r + c'^{\circ} = N^y + c^d \cdot \bar{\lambda}_{\odot}^{*\circ/d}. \tag{17}$$

Hence, relation (16a) may also be written

$$N^r + c'^{\circ} = (R^r + c'^{\circ}) \pm A^r, \tag{16b}$$

and since A^r is an integer, necessarily both the sun and the planet have alike exceeded or fallen short of integral revolutions by c'°. This gives a simple test of the consistency of Copernicus's corrections since it is required that

$$c'^{\circ} = c^d \cdot \bar{\lambda}_{\odot}^{*\circ/d} \qquad \text{or} \qquad c'^{\circ}/c^d = \bar{\lambda}_{\odot}^{*\circ/d}. \tag{18}$$

As it turns out, the application of this test to the corrections shows that the exposition of the mean motions in V,1 leaves something to be desired. In fact, the chapter presents the most serious textual and numerical problems in *De revolutionibus*, many of the numbers being uncertain in origin although

[24] Ptolemy does not specify an R differing from N for Venus and Mercury since considered geocentrically the centers of their epicycles complete N^r about the earth in N^y while the planet, its motion measured from the apogee of the epicycle, completes A^r. If, however, as was the practice in Indian astronomy, the motion of the planet on the śīghra epicycle is measured from a fixed direction, then in N^y an inferior planet will complete $R^r = N^r + A^r$ (14b), and a superior planet will complete $N^r = R^r + A^r$, corresponding respectively to the motion of an inferior planet and the earth in the heliocentric arrangement. There is, of course, nothing heliocentric in Indian astronomy. See Pingree (1978), 556 ff. This method of stating periods of mean motions appears to have pre-Ptolemaic, Greek prototypes, e.g. the Keskinto Inscription, on which cf. *HAMA*, 698–705.

demonstrably incorrect. To begin, there are different sets of numbers in N and M that are inconsistent both with each other and among themselves. Further problems arise when it is attempted to reconstruct the derivations of the numbers from the observations that are supposed to be—and doubtless were in some form or other—their origin.

Altogether, Copernicus gives seven quantities for each planet:

(1) A^r
(2) $N^y + c^d$
(3) $R^r + c'^\circ$
(4) P_α^d, the period of the anomaly
(5) $\bar{\alpha}^{\circ/y}$
(6) $\bar{\alpha}^{\circ/d}$
(7) $\bar{\lambda}^{*\circ/d}$

The values of these quantities from N are shown in Table 1(a), and those from M in Table 1(b).[25] The possible methods for their derivation from observation will be explained later,[26] while here we shall consider the internal consistency of the numbers in V,1. The most curious problem is posed by the related corrections c^d and c'°. Since they are given to two fractional places, it should follow, even with possible inaccuracies due to rounding, that

$$c'^\circ/c^d = \bar{\lambda}_\odot^{*\circ/d} = 0;59,8 \pm 0;0,1.$$

However, as the following table shows, the requirement is only seldom fulfilled.

Planet	N c'°/c^d	M c'°/c^d
Saturn	0;58,41	0;59,22
Jupiter	0;59,30	0;59,14
Mars	0;59,11	0;59,8
Venus	0;58,40	0;59,8
Mercury	0;52,31	1;0,0

Obviously something is wrong with either c^d or c'°, but we do not yet know which. If we take as given the synodic period P_α, then c^d may be found from

$$c^d = A^r \cdot P_\alpha - N^y, \tag{19a}$$

[25] A few scribal or typographical errors have been corrected on the basis of the following analysis. N: Jupiter, c^d 5;*45*,*23* for 5;*54*,13, P_α 398;53,*2*,58 for . . . *3*,58. M: Saturn, $\bar{\alpha}^{\circ/d}$ add . . . *2*, Jupiter P_α 398;53 . . . for *23*, Venus $\bar{\alpha}^{\circ/d}$ 0;36,59 . . . for *49*. There are many other obviously wrong, even impossible, numbers, but one cannot simply correct what are probably Copernicus's errors of computation. It is essential to note that the numbers from N and M are independent, and cannot be used to emend each other or form some kind of combination—pastiche would be a better word—with fewest errors.

[26] Below, pp. 320–321.

where the year is the sidereal year of $365;15,24,10^d$. From the values of P_α in N and M respectively, we find c^d and compare it with the text.

Planet	N c^d	Comp. c^d	M c^d	Comp. c^d
Saturn	$+1;7,18^d$	$+1;7,18^d$	$+1;6,48^d$	$+1;6,48^d$
Jupiter	$-5;45,23$	$-5;45,23$	$-5;45,27$	$-5;45,25$
Mars	$+2;23,45$	$+2;23,45$	$+2;27,3$	$+2;27,8$
Venus	$-2;26,44$	$-2;26,44$	$-2;26,46$	$-2;26,46$
Mercury	$+0;25$	$+0;25,26$	$+0;34,23$	$+0;33,27$

The results are close enough to show that P_α and c^d are more-or-less consistent—although we cannot account for the discrepancies—and that it is c'° that must be faulty. The correct values, computed from c^d in N and M by

$$c'^\circ = 0;59,8,11\ c^d,$$

are as follows:

Planet	N c'°	M c'°
Saturn	$+1;6,20^\circ$	$+1;5,50^\circ$
Jupiter	$-5;40,25$	$-5;40,29$
Mars	$+2;21,41$	$+2;24,56$
Venus	$-2;24,37$	$-2;24,39$
Mercury	$+0;24,38$	$+0;33,53$

Since c'° appears in no further computations, its accuracy is really of little interest.

Copernicus writes as though he found P_α from the previously established values of c^d, which is done through

$$P_\alpha = (N^y + c^d)/A^r, \tag{19b}$$

again taking the sidereal year of $365;15,24,10^d$. The following table compares the fractional places resulting from this computation—the integers are secure—with the corresponding values in N and M.

Planet	N P_α	Comp. P_α	M P_α	Comp. P_α
Saturn	$\ldots;5,32,42^d$	$\ldots;5,32,42^d$	$\ldots;5,32,11^d$	$\ldots;5,32,11^d$
Jupiter	;53,2,58	;53,2,58	;53,2,56	;53,2,54
Mars	;56,13,55	;56,13,53	;56,19,7	;56,19,15
Venus	;55,17,50	;55,17,48	;55,17,24	;55,17,24
Mercury	;52,38,53	;52,38,42	;52,42,12	;52,42,35

In only a few cases is the agreement exact, but only for Mercury do the differences exceed $0;0,0,10^d$. Assuming that Copernicus did compute P_α from c^d, and not *vice versa*, the discrepancies could be due to his taking c^d to three places. However, we do not know how he found c^d, nor can we offer any other explanation of the discrepancies in P_α.

Next, Copernicus explains that the annual motion of the anomaly, $\bar{\alpha}^y$, is found from

$$\bar{\alpha}^y = (360^\circ \cdot 365^d)/P_\alpha. \tag{20}$$

As the following comparisons show, the errors here—if that is what they are—are rather great, few of the numbers agreeing with the recomputations beyond the seconds. Only the fractional places are shown.

Planet	N $\bar{\alpha}^y$	Comp. $\bar{\alpha}^y$	M $\bar{\alpha}^y$	Comp. $\bar{\alpha}^y$
Saturn	...;32,3,9,40°	...;32,2,41,10°	...;32,2,54,12°	...;32,3,9,41°
Jupiter	;25,8,15,6	;25,8,13,1	;25,8,15,6	;25,8,14,41
Mars	;28,30,36,4	;28,30,22,16	;28,29,13,12	;28,29,14,52
Venus	;1,45,3,40	;1,48,43,34	;1,48,54,30	;1,48,53,35
Mercury	;57,23,6,30	;57,23,9,57	;56,46,54,40	;56,50,42,30

Again we have no explanation. The daily motion of the mean anomaly is computed from $\bar{\alpha}^d = \bar{\alpha}^y/365^d$. All the values of $\bar{\alpha}^d$ are correct except that for Venus in N Copernicus merely carried over the value from M, while correctly, from $\bar{\alpha}^y$ it should be $0;36,59,27,57^{\circ/d}$. Finally, for the superior planets Copernicus computes the annual mean sidereal motion in longitude from

$$\bar{\lambda}^{*y} = \bar{\lambda}^{*y}_{\odot} - \bar{\alpha}^y = 359;44,49,7,4^\circ - \bar{\alpha}^y.$$

Assuming that the value in N for $\bar{\alpha}^y$ of Mars is correct, then $\bar{\lambda}^{*y}$ should be 191;16,18,31,0°. All other values are correct.

V,1 is followed by tables of $\bar{\alpha}^y$ and $\bar{\alpha}^d$ for 1–60 years and 1–60 days in the usual form. Those in N are based upon the values of $\bar{\alpha}^y$ and $\bar{\alpha}^d$ set out in the chapter. On the other hand, those in M have been altered in various ways to agree with $\bar{\alpha}^y$ and $\bar{\alpha}^d$ in the exposition in N, although they originally followed the values set out in M. Thus, Copernicus must have revised his mean motions—how is not clear—and revised his tables accordingly before taking up the revision of V,1. He altered the tables as follows: For Saturn (M, 144), he erased and wrote over the original numbers. Jupiter (145) was not altered. For Mars (146), he cut out and replaced the page. Venus (147) was originally based upon $\bar{\alpha}^y = 225;1,50,11,15^\circ$ and $\bar{\alpha}^d = 0;36,59,28,48^\circ$, and Copernicus merely noted the revised values for 30^y and 60^y and for 60^d at the bottom of the columns. The tables for Mercury (148) agree with V,1 in N, and were not altered.

The more serious problem of the mean motions, of what relation, if any, they have to the observations, will be taken up in the analyses of the individual planets.[27]

2. The Derivation of the Elements of the Orbits of the Superior Planets

We have mentioned that the bisection of the eccentricity in Ptolemy's planetary model resulted in the best earlier approximation to Kepler's laws. However, the model creates great difficulties in finding the eccentricity and the direction of the apsidal line since the motion of the center of the epicycle on the eccentric is in fact nonuniform, that is, it does not describe arcs of the eccentric proportional to time, and thus its motion with respect to the center of the eccentric does not correspond to its mean motion. In the case of the sun and moon, it was possible to solve restricted and general forms of the problem: Given a circle with three points on its circumference, to find a point from which lines drawn to the three points will contain given angles. However, as shown in Fig. 19, the situation for the planets is not so straightforward, for although the center of the epicycle C lies on a circle about M, its mean motion $\Delta\bar{\lambda}$ is known only with respect to the equant point E. In another way, the locations of C' on a circle about E are given, but the locations of C on the circle about M are not known. Consequently, we are not really "given a circle with three points on its circumference," i.e. given arcs and central angles. Instead, the problem is of the form: Three points being located on a circle, to find two internal points symmetrical to the center of the circle from which lines drawn to the points on the circle will contain given angles.

Ptolemy knows of no direct solution to this problem, and indeed, no direct geometrical solution has been found.[28] However, in a tour de force of possibly the most complex and extended calculations in all of ancient mathematics, he developed a method of successive approximation that allows the numerical values of the eccentricity and the direction of the apsidal line to be found to any desired degree of accuracy. Both the problem and the solution are remarkable. While Greek mathematics contains applications of the principle of exhaustion to converge upon an area or volume, it is far less obvious in principle how even to attack Ptolemy's problem, and his solution shows a very high order of mathematical intuition. Ptolemy explains the method

[27] Below, pp. 333 f., 346 f., 361 f., 396, 401, 438 ff.

[28] There is, however, an algebraic solution, employing an equation of the eighth degree by G. W. Hill in Hill (1900).

only by example, setting out the most detailed solution for Mars in *Almagest* X,7. *Epitome* X,13–20, on the other hand, uses the derivation for Mars to explain the method itself in great detail.[29] This is a matter of no small importance since the number of astronomers after Ptolemy who understood and could apply the method must have been very small. By giving so thorough an exposition, Regiomontanus was in a sense rediscovering the most sophisticated achievement of ancient astronomy. And Copernicus could well be the first astronomer to benefit from Regiomontanus's teaching by actually applying Ptolemy's procedure to the superior planets.

The method is directly applicable to Copernicus's model, even though the orbit of the planet itself departs slightly from circularity, and the reason is that the planet still moves uniformly with respect to the equant. It is this property upon which Copernicus's derivation of the eccentricities by Ptolemy's method depends because the principle underlying the method is to assume first that the center of the epicycle (or planet) moves on a circle with the equant point as center, that is, that C coincides with C' in Fig. 19, and then correct for the fact that the center of the epicycle (or planet) is really located on a circle (or a nearly circular figure) whose center M bisects the distance OE from the earth (or mean sun) to the equant point. Thus, far from abandoning the equant point, as is so often said, its preservation by Copernicus in his model is crucial to his derivation of the orbital elements of the planets. Step-by-step Copernicus's derivations follow the method of the *Almagest* and the *Epitome*, with only the slight difference in the equation of center that appears repeatedly in the corrections of the location of the planet from the circle about the equant to its proper position. In what follows, after considering in a general way Copernicus's observations of the superior planets, we shall explain the entire procedure for the derivation of the parameters, and in the sections on the individual planets go through the particular numerical work for each.

---

[29] More recent expositions of Ptolemy's procedure can be found in Pedersen (1974), 273–83, *HAMA*, 172–79, and earlier in Delambre (1817) 2, 352–57, which is rather modern in form and contains errors, and Herz (1887–94) 1, 113–16 with solutions for the individual planets at 118–26. Herz, *ibid.* 2, 77–82, also worked through Copernicus's solutions, but his procedure is modernized and there appear to be computational errors underlying some of his results.

Epitome X, 21–22 contains an alternate, and apparently simpler, method invented by Jābir ibn Aflaḥ, who presents it in Book VII of his *De astronomia* (1534), 121–22. However, while the method is mathematically easier than Ptolemy's, it seems nearly impossible to apply since it requires two pairs of oppositions with equal intervals of time between the members of each pair. If such pairs can be found, they will be symmetrical to the apsidal line, so that by bisecting the distance between corresponding members of each pair, the direction of the apsidal line may be found. The analogy with Ptolemy's method of finding the direction of the apsidal lines of the inferior planets is fairly obvious. But there is a great difficulty since, depending upon how closely one wants to press the *equal* intervals of time, such pairs may require several lifetimes to accumulate, during which time the apsidal line will have moved. Whether anyone ever tried to apply Jābir's method, we do not know. Copernicus certainly does not.

Observations of the Superior Planets

Copernicus presents seven observations of each of the superior planets, which can be grouped as follows:

(1) Three oppositions observed by Ptolemy;
(2) Three oppositions observed by Copernicus;
(3) One position outside of opposition observed by Copernicus.

The individual observations are tabulated and discussed in the sections on each planet. Here we wish to make some general remarks on Copernicus's use of Ptolemy's observations and on how he may have made his own.

The observations of oppositions by Ptolemy were made with the armillary, or spherical astrolabe, and were reported as tropical longitudes, which is what the instrument itself gives directly. Copernicus reduces the tropical to sidereal longitudes, which he can do in two different ways. The first, and strictly speaking the more correct since it corresponds to what Copernicus does for his own oppositions, is to compute for the time of Ptolemy's observation the mean sidereal longitude of the sun $\bar{\lambda}^*_{\odot}$. Then necessarily the sidereal longitude of the planet $\lambda^* = \bar{\lambda}^*_{\odot} + 180°$. This is what Copernicus does in the case of Saturn, but it produces a problem since the differences in sidereal longitude between the oppositions that result are not the same as Ptolemys differences in tropical longitude because of the precession, i.e. $\Delta\lambda - \Delta\lambda^* = \Delta\pi$. And over the period of nine years covered by Ptolemy's observations of Saturn, $\Delta\pi \approx 0;6°$ according to Copernicus's variable—and at the time of Ptolemy, rather slow—precession. Consequently the values of the apparent motion $\Delta\lambda^*$ between the oppositions, and likewise the mean motions $\Delta\bar{\lambda}^* = \Delta\bar{\lambda}^*_{\odot} - \Delta\bar{\alpha}$, found by Copernicus differ slightly from Ptolemy's $\Delta\lambda$ and $\Delta\bar{\lambda}$, and thus Copernicus is forced to alter Ptolemy's final values of the eccentric anomaly at each opposition in order to produce agreement with his own values of $\Delta\lambda^*$. The whole demonstration becomes rather messy, and it is clear that this is not a satisfactory way of reviewing Ptolemy's derivation.[30]

An alternative would be to subtract from Ptolemy's tropical longitudes the *correct* value of precession at the time of each according to Copernicus's theory. This will not produce exactly the same values of $\Delta\lambda^*$ as the previous method since tropical longitudes computed with Copernicus's theory by $\lambda = \lambda^* + \pi$ can differ slightly from tropical longitudes computed by Ptolemy. But the problem will still remain that $\Delta\lambda^* = \Delta\lambda - \Delta\pi$ cannot be the same as Ptolemy's $\Delta\lambda$. Hence, after getting into a fair amount of difficulty over Saturn, and since the second method would produce the same problems, Copernicus adopts an expediency for Jupiter and Mars that is not strictly correct. He assumes that the precession is the same for all three oppositions of each

[30] Below, pp. 324–327.

planet, i.e. that for over four years in the case of Jupiter and for over eight years in the case of Mars $\Delta\pi = 0°$, which is equivalent to assuming that the precession has stopped, and is obviously untrue. However, it allows him to use Ptolemy's values of $\Delta\lambda$ and $\Delta\bar{\lambda}$ directly and without any alteration since if $\Delta\pi = 0°$, then trivially $\Delta\lambda^* = \Delta\lambda$.

In the case of Copernicus's observations, it is of interest to inquire how they were made and reduced since he gives very little information. Two of the observations outside of opposition, of Saturn and Mars, are simply estimates of configurations with respect to fixed stars that were made without any instrument.[31] In the observation of Jupiter outside of opposition, he reports that he saw *per instrumentum* that Jupiter was west of β Scorpii by 4;31°. Since the *instrument* gives a direct measurement of difference of longitude, it would appear to be an armillary. Likewise, when describing in general terms the observations of oppositions, Copernicus says, "Such positions are taken from observations with astrolabic instruments (*per instrumenta astrolabica*), as was explained above."[32] When the graduated ecliptic ring of an armillary is set to coincide with the ecliptic on the celestial sphere, the instrument will give direct readings of tropical longitude. However, having properly aligned the ecliptic ring, the instrument may also be used to find the difference in longitude between a planet and a zodiacal star. Since the sidereal longitude of the star is known from the catalogue, the difference of longitude will give the sidereal longitude of the planet. This is what Copernicus appears to have done for the observation of Jupiter outside of opposition.

To apply the same technique to finding the time and longitude of an opposition, it is necessary to find the sidereal longitude of the planet λ^* in this way on successive nights near the anticipated time of opposition, and then:

A. (1) Compute for the time of each observation the mean sidereal longitude of the sun $\bar{\lambda}^*_{\odot}$ from the table in III,14.
(2) Find the difference in longitude $\Delta\lambda = \Delta\lambda^* = \lambda^* - \bar{\lambda}^*_{\odot}$ for each observation, and interpolate for the time t that $\Delta\lambda = 180°$.
(3) For the time t, again compute the mean sidereal longitude of the sun, and find the sidereal longitude of the planet from $\lambda^* = \bar{\lambda}^*_{\odot} + 180°$.

If, on the other hand, the armillary is used to find a tropical longitude λ of the planet by direct measurement, then:

B. (1) Compute for the time of each observation the adjusted mean longitude of the sun $\bar{\lambda}'_{\odot} = \bar{\lambda}^*_{\odot} + \pi$ from the table in III,14 and the precession theory.

[31] Below, pp. 335, 363 f.

[32] V,4, N, 143r20–22, referring to II,14, cf. above, pp. 121–122. Given the statements in the text, it appears that Copernicus actually used an armillary, yet considering the complexity of the instrument and the great difficulty, and doubtless expense, of making or procuring one, we still wonder whether his account is not more didactic than a record of his own practice.

(2) Find the difference in longitude $\Delta\lambda = \lambda - \bar{\lambda}'_{\odot}$ for each observation, and interpolate for the time t that $\Delta\lambda = 180°$.
(3) For the time t, compute the mean sidereal longitude of the sun, and find the sidereal longitude of the planet from $\lambda^* = \bar{\lambda}^*_{\odot} + 180°$.

Copernicus, however, need not have used an armillary to find the longitudes of the oppositions, for any device that can measure the zenith distance of the planet at meridian transit, such as the parallactic instrument, would be adequate to find its longitude.[33] In order to do this, measure the zenith distance of the planet ζ_p at meridian transit, and for the time of the observation compute the true tropical longitude of the sun $\lambda_{\odot}$. The right ascension of the midheaven α_M then follow from (2.28), and this is also the right ascension of the planet α_p since all points that cross the meridian together have the same right ascension. An obvious difficulty here is determining the time of meridian transit accurately since an error of $0;4^h$ will produce an error of 1° in right ascension. Then since the zenith distance of the point of the equator crossing the meridian is equal to the observer's latitude φ, the declination of the planet δ_p is $\delta_p = \varphi - \zeta_p$. Since we now have the right ascension and declination of the planet, α_p and δ_p, its longitude λ_p may be found by the transformation of coordinates explained in II,4.[34] Having found the tropical longitude of the planet from night to night near opposition in this way, one interpolates by method B. for the time and sidereal longitude of the opposition.

1. *Apparent and Mean Motion Between Oppositions*

The derivation of the eccentricity and the direction of the apsidal line depends upon oppositions since in order to isolate the effects of the first anomaly, it is necessary to find three locations of the planet unaffected by the second anomaly, that is, to eliminate the motion of the earth and observe the planet as it would appear from the mean sun. In Ptolemy's model that is equivalent to finding three positions of the center of the epicycle. From the rule, shown in Fig. 1(a), that the radius of the epicycle is always parallel to the direction from the earth to the mean sun, it follows that when the planet is directly opposite the mean sun

$$\bar{\lambda}_{\odot} - \lambda_p = \alpha = 180°,$$

and the planet will be exactly at the perigee of the epicycle. Under these conditions, shown in Fig. 20(a), the mean sun $\bar{S}$, the earth O, the planet P, and the center of the epicycle C all lie in a straight line, so that the direction OP gives the direction from the earth (or mean sun) to the center of the epicycle. Three such observations allow the determination of three locations

[33] Above, pp. 68, 233 f.
[34] Above, pp. 106–107.

of C on the eccentric circle about M in Fig. 20(b), and define two angles δ_1 and δ_2 of apparent motion between the oppositions. The intervals of time between the oppositions, along with some provisional value of the mean motion in longitude, then define two angles of mean motion $\bar{\delta}_1$ and $\bar{\delta}_2$ with respect to the equant point E.

The situation in Copernicus's model is much the same. As is clear from Fig. 1(c), when the planet is observed in opposition to the mean sun, again

$$\bar{\lambda}^*_{\odot} - \lambda^*_{\mathrm{p}} = \alpha = 180°,$$

and in Fig. 21(a) the direction OP from the earth to the planet coincides with the direction $\bar{S}P$ from the mean sun to the planet. Just as in Ptolemy's model an observation of the planet at opposition eliminates the effect of the epicycle, so in Copernicus's model an opposition eliminates the effect of the earth's orbit. Three such observations will establish three locations of P which lie, as shown in Fig. 21(b), on a nearly circular figure about M, the center of the eccentric in Ptolemy's model, and define the angles of apparent motion δ_1 and δ_2. From the interval of time between the oppositions, one then finds the angles of mean motion $\bar{\delta}_1$ and $\bar{\delta}_2$ with respect to the equant.

To determine values of the mean motion in longitude $\bar{\delta}$, Ptolemy uses provisional values of $\Delta\bar{\lambda}$ defined tropically that are derived from the period relations for mean motions. Copernicus instead first takes provisional values of $\Delta\bar{\alpha}$ from the tables in the *Almagest* or the *Alfonsine Tables*—since $\Delta\bar{\alpha}$ is the same whether $\Delta\bar{\lambda}$ is tropical or sidereal—and then finds sidereal values of $\bar{\delta}$ from

$$\bar{\delta} = \Delta\bar{\lambda}^* = \Delta\bar{\lambda}^*_{\odot} - \Delta\bar{\alpha}, \tag{21}$$

where $\Delta\bar{\lambda}^*_{\odot}$ is given by the table in III,14. Thus, the initial step of Copernicus's derivation is as follows: From three observations of the planet at opposition to the mean sun, find

$$\lambda^*_{(123)} = \bar{\lambda}^*_{\odot(123)} + 180°. \tag{22}$$

The angles of apparent motion are then

$$\delta_1 = \lambda^*_2 - \lambda^*_1, \qquad \delta_2 = \lambda^*_3 - \lambda^*_2. \tag{23a}$$

In the event that δ_1 or δ_2 exceeds 180°, he also finds

$$\delta_3 = \lambda^*_1 - \lambda^*_3 = 360° - (\delta_1 + \delta_2). \tag{23b}$$

From the interval of time between the observations find $\Delta\bar{\lambda}^*_{\odot}$ and a provisional value of $\Delta\bar{\alpha}$, and form

$$\bar{\delta}_1 = \Delta\bar{\lambda}^*_{\odot 12} - \Delta\bar{\alpha}_{12}, \qquad \bar{\delta}_2 = \Delta\bar{\lambda}^*_{\odot 23} - \Delta\bar{\alpha}_{23}, \tag{24a}$$

and if necessary also

$$\bar{\delta}_3 = 360° - (\bar{\delta}_1 + \bar{\delta}_2). \tag{24b}$$

2. *Solution for the Double Eccentricity and the Mean Eccentric Anomaly*

The situation that is now established is shown in Fig. 22. It is known by observation that as seen from $\bar{S}$, the planet P has moved through the apparent motion δ_1 and δ_2, and it is known by computation that with respect to the equant point E, it has moved through the mean motion $\bar{\delta}_1$ and $\bar{\delta}_2$. According to Copernicus's model, however, the path of the planet is a nearly circular figure about M, the point in Ptolemy's model that bisects the distance $\bar{S}E = e_1 + r' = 2e$ between the mean sun and the equant point. The angular motion of the planet with respect to M is not known, so there is no way of finding the distance $\bar{S}M = e_1 - r' = e$, the simple eccentricity of Ptolemy's model. Nor of course is the angular motion of P known with respect to M'; hence we cannot find $\bar{S}M' = e_1$ of Copernicus's model. If, however, it is assumed that the planet moves in the circle about the equant, that is, is located at P' rather than P, then on the circle about E

$$\text{arc } P'_1P'_2 = \bar{\delta}_1, \qquad \text{arc } P'_2P'_3 = \bar{\delta}_2,$$

and we are given three points on a circle. We next assume, falsely, that arc $P'_1P'_2$ is seen from $\bar{S}$ under the angle δ_1 and that arc $P'_2P'_3$ is seen under the angle δ_2. That this assumption is false is apparent from the figure since in fact

$$P'_1\bar{S}P'_2 = \varepsilon_1 + \delta_1 + \varepsilon_2, \qquad P'_2\bar{S}P'_3 = -\varepsilon_2 + \delta_2 - \varepsilon_3.$$[35]

In a later step we shall correct this false assumption by finding values of ε and modifying the angles $P'_1\bar{S}P'_2$ and $P'_2\bar{S}P'_3$ accordingly. But under the assumption that the motion of P' is seen from $\bar{S}$ under the angles of apparent motion δ, it is possible to solve the unrestricted problem of a circle with three given points seen from an interior point $\bar{S}$ in order to find the distance $\bar{S}E = 2e$ from the mean sun to the equant. This is Copernicus's next step.

The trigonometric solution is shown in Fig. 23(a), labelled in accordance with the diagram used by Copernicus for Saturn and Jupiter in V,6,11. A circle ABC is described about the equant point F, and the mean sun is located at D. Corresponding to P' in Fig. 22, let A be the position of the planet—really the projection of the planet on the circle about the equant—at the first opposition, B the position at the second, and C at the third. Take any location of the planet, say C, draw CDE, and complete triangles BDE and ADE. Now, in triangle BDE,

$$BE = \text{crd } 2\beta_1 = \text{crd } 2(180° - \delta_2),$$

$$DE = \text{crd } 2\gamma_1 = \text{crd } (2\delta_2 - \bar{\delta}_2),$$

[35] Here we are not taking the signs of ε into account. For the proper formulation, see below, p. 316.

and in triangle ADE,

$$AE = \text{crd } 2\beta_2 = \text{crd } 2(180^\circ - (\delta_1 + \delta_2)),$$

$$DE = \text{crd } 2\gamma_2 = \text{crd}(2(\delta_1 + \delta_2) - (\bar{\delta}_1 + \bar{\delta}_2)).$$

Hence, in units of triangle BDE,

$$AE = \text{crd } 2\beta_2(\text{crd } 2\gamma_1/\text{crd } 2\gamma_2).$$

In triangle ABE,

$$AB = (BE^2 + AE^2 - 2BE \cdot AE \cos \tfrac{1}{2}\bar{\delta}_1)^{1/2},$$

although Copernicus merely refers this step to the theory of plane triangles. He probably solves it in the more extended form

$$AQ = AE \sin \tfrac{1}{2}\bar{\delta}_1, \qquad EQ = AE \cos \tfrac{1}{2}\bar{\delta}_1,$$

$$BQ = BE - EQ, \qquad AB = (AQ^2 + BQ^2)^{1/2}.$$

But taking the radius of the circle as the unit,

$$\overline{AB} = \text{crd } \bar{\delta}_1,$$

and in the same units

$$BE = \text{crd } 2\beta_1(\text{crd } \bar{\delta}_1/AB), \qquad DE = \text{crd } 2\gamma_1(\text{crd } \bar{\delta}_1/AB).$$

Therefore,

$$\text{arc } BAE = 2 \sin^{-1} \tfrac{1}{2}BE, \qquad \text{arc } EABC = \text{arc } BAE + \bar{\delta}_2,$$

$$\text{arc } CE = 360^\circ - \text{arc } EABC,$$

and thus

$$CDE = \text{crd } CE, \qquad CD = CDE - DE.$$

Consequently, in Fig. 23(b), by Euclid III,35 and II,5,

$$CD \cdot DE = (R - 2e)(R + 2e) = R^2 - 4e^2,$$

and thus

$$2e = (R^2 - CD \cdot DE)^{1/2}$$

gives the double eccentricity from the mean sun to the equant.

Now, drawing FKL perpendicular to CDE,

$$DK = DE - KE = DE - \tfrac{1}{2}CDE, \qquad \gamma_3 = \sin^{-1}(DK/2e),$$

and

$$\text{arc } CH = \text{arc } CHL - \text{arc } HL = \tfrac{1}{2}\text{arc } CHE - \gamma_3$$

gives the distance of C from the "perigee" H. Thus, the mean eccentric anomaly at each opposition

$$\bar{\kappa}_3 = 180^\circ - \text{arc } CH,$$

$$\bar{\kappa}_2 = \bar{\kappa}_3 - \bar{\delta}_2, \qquad -\bar{\kappa}_1 = \bar{\kappa}_2 - \bar{\delta}_1.$$

This completes the first derivation of $2e$ and $\bar{\kappa}$.

3. *Test of the Derived Elements*

The values of the double eccentricity and the mean eccentric anomaly at each opposition have been found under the false assumption that the planet moves in the circle about the equant. It is next necessary to test whether the results of the derivation are correct by determining whether they will reproduce the angles δ_1 and δ_2 of apparent motion as seen from $\bar{S}$. This step is omitted by Ptolemy in his presentation until after he has found the correct elements. However, he had to perform it after each derivation of $2e$ and $\bar{\kappa}$ since it provides the criteria for terminating or continuing the derivations. Copernicus, although he shows even fewer steps than Ptolemy, must have done the same. The test is performed by placing the three oppositions on Copernicus's model as in Fig. 24. From the previous derivation, and from the standard division of $2e$, we have the quantities

$$2e, \qquad e_1 = \tfrac{3}{4} \cdot 2e, \qquad r' = \tfrac{1}{4} \cdot 2e,$$

and the mean eccentric anomalies $-\bar{\kappa}_1$, $\bar{\kappa}_2$, and $\bar{\kappa}_3$. We next find for each the equation of center c_3, and form the true eccentric anomalies

$$-\kappa_1 = -\bar{\kappa}_1 + c_{3(1)}, \qquad \kappa_2 = \bar{\kappa}_2 + c_{3(2)}, \qquad \kappa_3 = \bar{\kappa}_3 + c_{3(3)}. \quad (25)$$

Then the resulting angles of apparent motion at $\bar{S}$ are

$$\delta_1 = \kappa_2 - \kappa_1, \qquad \delta_2 = \kappa_3 - \kappa_2. \quad (26)$$

If by chance these agree with the observed values of δ_1 and δ_2, then the derived values of $2e$ and $\bar{\kappa}$ are correct, and the derivation is terminated. However, since the previous demonstration depended upon the incorrect assumption that δ applied to the motion of the projection of the planet P' in the circle around the equant, it is unlikely that the result will fit the observations since the planet was in fact in its proper orbit at P. Therefore, it is necessary to modify the initial assumption by correcting δ to a new angle δ' that will more accurately represent the motion of the projection of the planet on the circle about the equant, that is, more accurately show the angular motion of P' as seen from $\bar{S}$.

4. *Correction of δ to δ'*

We have already touched upon this correction in stating the problem of step 2 where we showed that in Fig. 22 the direction $\bar{S}P$ from the mean sun to the planet differs from the direction $\bar{S}P'$ from the mean sun to the projection of the planet on the equant circle by a small angle ε.[36] Our false assumption that P is located at P', or that the direction $\bar{S}P$ coincides with $\bar{S}P'$, was equivalent to assuming that $\varepsilon = 0°$, which is untrue except in the apsidal line and at $\bar{\kappa} = \pm 90°$. At the time we had no way of knowing ε since we did not yet know $2e$ and $\bar{\kappa}$, but now we can at least estimate ε and correct the angular motion of P' as seen from $\bar{S}$. This correction is the most important step in the derivation. The method of finding ε is shown in Fig. 25. In step 3, using the derived $\bar{\kappa}$, we found the equation of center for the planet at P, and it was c_3. The equation of center of the projection of the planet P' on the equant circle is $c'_3 = EP'\bar{S}$, which is given by

$$\tan c'_3 = \frac{(e_1 + r') \sin \bar{\kappa}}{R + (e_1 + r') \cos \bar{\kappa}} = \frac{2e \sin \bar{\kappa}}{R + 2e \cos \bar{\kappa}}. \tag{27}$$

Although Copernicus never sets out this part of the procedure, he would first find

$$\begin{aligned} \bar{S}P' &= (R^2 + (e_1 + r')^2 + 2R(e_1 + r') \cos \bar{\kappa})^{1/2} \\ &= (R^2 + 4e^2 + 4Re \cos \bar{\kappa})^{1/2}, \end{aligned} \tag{28}$$

and next

$$c'_3 = \sin^{-1}((e_1 + r') \sin \bar{\kappa}/\bar{S}P') = \sin^{-1}(2e \sin \bar{\kappa}/\bar{S}P'). \tag{29}$$

Then, the correction of the direction $\bar{S}P$ to $\bar{S}P'$, of κ to κ', is

$$\varepsilon = c_3 - c'_3. \tag{30}$$

The situation for all three positions is shown in Fig. 26. If we observe the correct signs of c_3 and c'_3, that is

$$(c_3, c'_3) \leq 0° \quad \text{for} \quad 0° \leq \bar{\kappa} \leq 180°,$$

$$(c_3, c'_3) \geq 0° \quad \text{for} \quad 180° \leq \bar{\kappa} \leq 360°,$$

then in all cases the signs of ε will be such that the corrected angles of P' on the equant circle,

$$\begin{aligned} P'_1\bar{S}P'_2 &= \delta'_1 = +\varepsilon_1 + \delta_1 - \varepsilon_2, \\ P'_2\bar{S}P'_3 &= \delta'_2 = +\varepsilon_2 + \delta_2 - \varepsilon_3. \end{aligned} \tag{31}$$

It is also apparent from Figs. 25 and 26 that δ' may be found in another way. Let $\bar{\kappa} + c'_3 = \kappa'$, and then

$$\delta'_1 = \kappa'_2 - \kappa'_1, \qquad \delta'_2 = \kappa'_3 - \kappa'_2.$$

While this method is faster, it is didactically less informative of the adjustment of the direction $\bar{S}P$ to $\bar{S}P'$, and Ptolemy does not use it.

[36] Above, p. 313.

Note that since c_3 and c'_3, and thus ε, depend upon values of $2e$ and $\bar{\kappa}$ that are known to be incorrect, they are only estimates, and thus δ' is not necessarily a completely accurate measure of the angular motion of P' as seen from $\bar{S}$. Nevertheless, it is more accurate than δ, and that is what makes the correction so significant. The original assumption that δ represents the angular motion of P', that $\varepsilon = 0°$, was false. But it has now to some degree been corrected since δ' must give a more accurate approximation to the motion of P' than did δ. The first solution for $2e$ used the apparent motion δ only because nothing better was yet at hand, although the solution really depends, not upon the motion of the planet P in its orbit, but upon the motion of its projection P' in the equant circle. Since δ' gives a better approximation of the motion of P' than δ, a new solution for $2e$ and $\bar{\kappa}$ substituting δ' for δ must give a better result than the previous solution. This is the linchpin of the iterative method, and Ptolemy's invention of the method must rank as one of the most remarkable examples of applied mathematics in antiquity. It is true, as remarked earlier, that methods of convergence had been used earlier, by Archimedes in particular,[37] but Ptolemy's application is so indirect, so far from obvious, that one cannot but recognize a mathematician of profound insight. He gives no proof that the successive adjustments will give increasingly better values of δ' and then eventually values of δ that agree with the observations, but a careful examination of the method shows intuitively that this must be so. And this Ptolemy understood. To what extent Copernicus understood the procedure is a subject for later discussion.[38]

5. *Iteration*

The values of $2e$ and $\bar{\kappa}$ found in step 2 were not adequate to reproduce the observed values of δ in the test of step 3. How inaccurate $2e$ and $\bar{\kappa}$ are, whether they are larger or smaller than the values that would correctly give the observed δ is fortuitous, depending upon the differences in the directions of P and P' at each opposition given by ε, which in turn depend upon the distances of the planet from the apsidal line and so far can only be estimated. However, the new value δ' will give better positions of P' on the equant circle, and these may be used to find $2e$ and $\bar{\kappa}$ again. Thus, for an iteration of step 2, we now use

$$\delta'_1, \qquad \delta'_2, \qquad \text{if necessary} \quad \delta'_3 = 360° - (\delta'_1 + \delta'_2),$$

and since the mean motion of P and P' with respect to the equant point remains the same, we use the original values of

$$\bar{\delta}_1, \qquad \bar{\delta}_2, \qquad \text{if necessary} \quad \bar{\delta}_3 = 360° - (\bar{\delta}_1 + \bar{\delta}_2).$$

---

[37] In M, 153r3–4, Copernicus noted the similarity of the procedure to approximating the area of a circle and other figures (*sicut accidit in circuli quadratura et aliis plerisque*), but deleted the remark.

[38] Below, pp. 341 ff., 355, 366 ff.

Having repeated step 2 and found new values of

$$2e, \quad e_1, \quad r', \quad \bar{\kappa}_1, \quad \bar{\kappa}_2, \quad \bar{\kappa}_3,$$

we repeat step 3 to test for δ_1 and δ_2. If these agree with the observed δ_1 and δ_2, the derivation is complete for $2e$ and $\bar{\kappa}$ are now correct. If not, we repeat step 4, find new values of ε, and correct the *original* values of δ (not δ'!) a second time to

$$\delta_1'' = +\varepsilon_1 + \delta_1 - \varepsilon_2, \qquad \delta_2'' = +\varepsilon_2 + \delta_2 - \varepsilon_3. \tag{31}$$

Once again we repeat step 2 to find new values of $2e$ and $\bar{\kappa}$. The procedure continues as before until values of δ computed in the test of step 3 agree with the values derived from the observations in step 1.

The results of successive iterations will give values of $2e$ and $\bar{\kappa}$ alternately above and below the final results that produce the observed values of δ, the changes from solution to solution, and the errors found in step 3, growing smaller as $2e$ and $\bar{\kappa}$ converge on the correct results. The convergence is quite rapid. How many iterations are necessary depend upon the required accuracy of the agreement between the observed and computed δ. Ptolemy takes δ to minutes, and finds agreement in two solutions each for Saturn and Jupiter, and three for Mars. Copernicus likewise takes δ to minutes, and while he does not say how many solutions were required, we shall see that satisfactory agreement is reached after three for Saturn and Jupiter, and four for Mars, although Copernicus appears to have stopped after three. Erasmus Reinhold, in his *Commentary on De revolutionibus*, begins with Copernicus's observations and shows the complete solutions for all three planets. He computes to seconds of arc, goes through four solutions for Saturn, six for Mars, the extraordinary number of nine for Jupiter, and finds agreement to less than one minute, although not to the second.[39] The possibility of computational error in such an *ingentem numerorum multitudinem*, as Copernicus's calls Ptolemy's exposition, is not small, and Reinhold's failure to find agreement for Jupiter in fewer than nine solutions could be the result of mistakes.

Yet, it is important to note that errors in the calculations, if they are not too serious, will not make much difference since the iterative method is to some extent self-correcting. Each time one finds new values of δ' from δ and ε, it is rather like starting over again to find $2e$ and $\bar{\kappa}$, but with a slight advantage in knowing a better estimate of δ'. A really serious error might produce a computed δ in step 3 worse than the previous test, but minor errors will still move the demonstration forward, although perhaps not rapidly. All that is retained from steps 2–4 when beginning step 2 again is the correction ε applied to the observed δ. If one computes long enough, and does not make continual errors, one must be put back on the right track. This is of some

[39] Berlin, Deutsche Staatsbibliothek MS Lat. fol. 391, ff. 192–202 (Saturn), 208–29 (Jupiter), 236–56 (Mars).

interest for Copernicus's derivations since in the case of Jupiter he makes an initial mistake in the first solution for $2e$ and $\bar{\kappa}$, resulting in a value of $2e$ that he finds suspiciously large. Whether he gave up the strict method at that point—he then introduces Ptolemy's value of $2e$—and turned to trial and error is a matter to be investigated.[40]

6. *Mean Anomaly, Mean Longitude, and Longitude of Apogee*

We next take any one of the oppositions, and consider its relation to the orbit of the earth in Fig. 27. In the final test of $2e$ and $\bar{\kappa}$, we determined the equation of center c_3 and the true eccentric anomaly at each location of P. Since necessarily at the time of an opposition the true anomaly

$$\alpha = \bar{\lambda}^*_{\odot} - \lambda^* = 180°,$$

so the mean anomaly

$$\bar{\alpha} = \alpha - c_3 = 180° - c_3, \tag{32}$$

noting that as the difference between α and $\bar{\alpha}$, c_3 is positive for $\bar{\kappa} \leq 180°$ and negative for $\bar{\kappa} \geq 180°$. The true longitude of the planet λ^* is known from observation, and thus the mean longitude, the angle at $\bar{S}$ measured from Γ to a direction $\bar{S}U$ parallel to EP and MC, is

$$\bar{\lambda}^* = \lambda^* - c_3, \tag{33}$$

where in this case $c_3 = \lambda^* - \bar{\lambda}^* = \kappa - \bar{\kappa}$ is negative for $\bar{\kappa} \leq 180°$ and positive for $\bar{\kappa} \geq 180°$. Then from the true longitude and the true eccentric anomaly, or from the mean longitude and the mean eccentric anomaly, the sidereal longitude of the apogee follows from

$$\lambda^*_A = \lambda^* - \kappa = \bar{\lambda}^* - \bar{\kappa}, \tag{34}$$

and the tropical longitude is found by adding the true precession, i.e.

$$\lambda_A = \lambda^*_A + \pi. \tag{35}$$

7. *Correction of the Mean Anomaly and Positions at Epoch*

Ptolemy corrects the mean anomaly, given provisionally by the Babylonian period relations, using observations made about 400 years previously of distances of the planets from fixed stars, including reported occulations for Mars and Jupiter. He evidently has no earlier observations of oppositions, even from Hipparchus, and it is probable that Ptolemy was the first one to make a practice of observing the time of oppositions. Since the positions of the planets were at arbitrary values of the true anomaly α, he had first to establish the radius of the epicycle in order to find the mean anomaly at each

---

40 Below, pp. 339–346.

observation. Copernicus, on the other hand, has the good fortune of having Ptolemy's oppositions at his disposal, for which the true anomaly is necessarily 180° and the mean anomaly has already been determined by $\bar{\alpha} = 180^\circ - c_3$. Hence, he can defer consideration of the ratio of the radius of the planet's orbit to the radius of the earth's orbit until after correcting the mean anomaly, and this likewise gives him more freedom in his choice of observations for the later demonstration.

The correction of the mean anomaly thus depends only upon its value at two widely spaced oppositions, and is straightforward, at least in principle. From $\bar{\alpha}$ at one of Ptolemy's oppositions and at one of his own—he always uses the latest in each set—Copernicus knows the mean motion of the anomaly $\Delta\bar{\alpha}$ during the interval Δt. $\Delta\bar{\alpha}$ is given in the form $p^{\mathrm{r}} + q^\circ$, where p^{r} can be found from any provisional value of $\bar{\alpha}^{\circ/\mathrm{y}}$ or $\bar{\alpha}^{\circ/\mathrm{d}}$, e.g. from the tables in the *Almagest*, by

$$p^{\mathrm{r}} = (\Delta t^{\mathrm{y}} \cdot \bar{\alpha}^{\circ/\mathrm{y}})/360^\circ = (\Delta t^{\mathrm{d}} \cdot \bar{\alpha}^{\circ/\mathrm{d}})/360^\circ, \tag{36}$$

taking only the integer part of the quotient, while $q^\circ = \Delta\bar{\alpha}$ (mod. 360°). He can then proceed in various ways, and unfortunately the text is often unclear about what he has done. The annual motion of the mean anomaly $\bar{\alpha}^{\circ/\mathrm{y}}$ can be computed directly from

$$\bar{\alpha}^{\circ/\mathrm{y}} = \Delta\bar{\alpha}^\circ/\Delta t^{\mathrm{y}} = (\Delta\bar{\alpha}^\circ \cdot 365^{\mathrm{d}})/\Delta t^{\mathrm{d}}. \tag{37}$$

Copernicus actually shows this procedure for Venus, but in no case do its results agree with the mean anomaly in V,1.[41] In another way, the planet's mean synodic period in days P^{d}_{α} can be found by

$$P^{\mathrm{d}}_{\alpha} = \Delta t^{\mathrm{d}}/\Delta\bar{\alpha}^{\mathrm{r}} = (\Delta t^{\mathrm{d}} \cdot 360^\circ)/\Delta\bar{\alpha}^\circ. \tag{38}$$

For Jupiter alone, the result agrees with (an emended) P_{α} in V,1 of N.[42] By yet another way, used by Ptolemy in the *Almagest*, the correction c^{d} to the number of years N^{y} in the period relation may be found from

$$\mathrm{N}^{\mathrm{y}} + c^{\mathrm{d}} = (\mathrm{A}^{\mathrm{r}} \cdot \Delta t^{\mathrm{d}})/\Delta\bar{\alpha}^{\mathrm{r}} = (\mathrm{A}^{\mathrm{r}} \cdot \Delta t^{\mathrm{d}} \cdot 360^\circ)/\Delta\bar{\alpha}^\circ = \mathrm{A}^{\mathrm{r}} \cdot \mathrm{P}^{\mathrm{d}}_{\alpha}, \tag{39}$$

where A^{r} is the number of anomalistic rotations in the period and c^{d} depends upon the length of the sidereal or tropical year used to reduce the result in days to $\mathrm{N}^{\mathrm{y}} + c^{\mathrm{d}}$. Ptolemy used the tropical year of $365;14,48^{\mathrm{d}}$, while Copernicus would necessarily use the sidereal year of $365;15,24,10^{\mathrm{d}}$. That this sidereal year underlies the mean motions in V,1, both in M and N, is certain from our earlier examination of the internal consistency of the numbers.[43] However, as it turns out, when c^{d} is computed directly from the observations, it only once agrees with any of the values from V,1 in either M or N.

Whatever procedure Copernicus may have adopted, that is, whatever quantity he chose to derive directly from the oppositions, it is then simple

---

[41] Below, p. 396.

[42] Below, p. 347.

[43] Above, pp. 304–305.

arithmetic to compute the remaining quantities, as has been shown previously in the discussion of V,1. However, only a single value in V,1, and that with an emendation, can be derived from the observations, and the chapter itself is filled with inconsistencies both small and large that are evidence of many computational errors. Given the numbers in the text of M and N for V,1, which it is important to note are completely independent in this chapter and cannot be used to emend each other, and given that it would be unwarranted to emend the text as a whole to produce internal consistency, a secure reconstruction of Copernicus's procedures appears unobtainable. It also cannot be ruled out that the mean motions were found by a completely different method. For example, Copernicus could take a provisional value of $\Delta\bar{\alpha}$ during the interval Δt between the two oppositions, e.g. from the *Almagest* or the *Alfonsine Tables*, and then correct the annual motion by the difference between the provisional and the true $\Delta\bar{\alpha}$ from the observations divided by Δt. This was exactly the procedure used for the moon, where Copernicus found small corrections to the Hipparchan mean motions just as Ptolemy did in the *Almagest*.[44] The procedure is certainly legitimate, but since Δt would necessarily be rounded in some way, it is difficult to know exactly what Copernicus might have done. Although it is not hard to come close to his results by trial, this would still not consistitute a conclusive reconstruction.

Having found the mean motion of the anomaly, the position at some epoch —Copernicus always uses the Epoch of Christ—follows from taking $\bar{\alpha}$ at one of Ptolemy's oppositions, finding $\Delta\bar{\alpha}$ for the interval Δt since the epoch, so that at epoch

$$\bar{\alpha}_0 = \bar{\alpha} - \Delta\bar{\alpha}. \tag{40}$$

The remaining epochs then follow from the intervals of time between them. Copernicus does not give epoch values of the mean sidereal longitude $\bar{\lambda}^*$, but they can be found easily from

$$\bar{\lambda}^*_0 = \bar{\lambda}^*_{\odot 0} - \bar{\alpha}_0, \tag{41a}$$

and the mean tropical longitude from the mean equinox is

$$\bar{\lambda}_0 = \bar{\lambda}_{\odot 0} - \bar{\alpha}_0, \tag{41b}$$

using the values of $\bar{\lambda}^*_{\odot 0}$ and $\bar{\lambda}_{\odot 0}$ from III,19.[45]

Finally, since the sidereal longitude of the apogee λ^*_A is known at the time of the oppositions observed by Ptolemy and Copernicus, the motion of the apogee $\Delta\lambda^*_A$ is known for the interval of time between them, and can be converted into an assumed uniform motion either in degrees of motion per unit of time or, as Copernicus prefers, in time per degree of motion. Copernicus does this for Jupiter and Saturn, but does not do so for Mars since in the case of Mars he believes that much of the motion of the apsidal

---

[44] Above, pp. 210, 222.

[45] Above, p. 165. Consequently the errors of $-0;1°$ in $\bar{\lambda}^*_{\odot 0}$ for A., and in $\bar{\lambda}_{\odot 0}$ for A., C., and G. will be carried over into $\bar{\lambda}^*_0$ and $\bar{\lambda}_0$. See above, p. 165, n. 51.

line is really due to the motion of the center of the earth's orbit.[46] Consequently, the motion of Mars' apogee will not be uniform and thus cannot be treated in this simple way.

8. *Distance of the Planet and Equation of the Anomaly*

The last subject treated for each planet is its distance from the mean sun, that is, the radius of its orbit with the orbit of the earth as the unit, which determines the equation of the second anomaly. This is equivalent to the ratio of the radius of the eccentric to the radius of the epicycle in Ptolemy's model, although with this difference: that in Copernicus's theory the radius of the earth's orbit is the common measure of the radii of all the planetary orbits. It is this assumption, shown earlier in Figs. 1 and 2, that allows the distances of the planets from the mean sun to be determined in what is often called the Copernican System. In Ptolemy's theory, on the other hand, there could be no necessary relation between the dimensions of the models for the individual planets. It is of course true, as also shown in Figs. 1 and 2, that in the Tychonic theory the radius of the mean sun's geocentric orbit serves the same function as the radius of the earth's orbit in the Copernican.[47] Copernicus's and Tycho's assumptions were, as far as could be known in the sixteenth century, essentially equivalent, and were of importance principally because the distances from a common center, the mean sun, provide a single explanation for the magnitude and behavior of the second anomaly of all the planets.

For finding the radius of the planet's orbit, Copernicus uses the same procedure used by Ptolemy for finding the radius of the epicycle. All that is required is a single observation of the planet when not at opposition. Since the effect of the first anomaly is now under control, the effect of the second anomaly on the observation can be isolated. One can then find in turn the equation of the anomaly and the ratio of the radii of the planet's orbit and the earth's orbit. From observation the sidereal longitude of the planet λ^* is known, and from the time of the observation one finds

$$\bar{\lambda}^*_{\odot}, \qquad \bar{\alpha}, \qquad \bar{\lambda}^* = \bar{\lambda}^*_{\odot} - \bar{\alpha}.$$

A representative situation is shown in Fig. 28. From the previous demonstration, we know e_1, r', and λ^*_A, and thus the mean eccentric anomaly $\bar{\kappa} = \bar{\lambda}^* - \lambda^*_A$. Following his procedure for finding the equation of center, Copernicus computes

$$\bar{S}C = (R^2 + e_1^2 + 2Re_1 \cos \bar{\kappa})^{1/2}, \tag{6}$$

$$a_1 = \sin^{-1} (e_1 \sin \bar{\kappa}/\bar{S}C), \tag{7}$$

$$\bar{S}P = (\bar{S}C^2 + r'^2 - 2\bar{S}Cr' \cos (\bar{\kappa} + a_1))^{1/2} \tag{8}$$

$$a_2 = \sin^{-1}(r' \sin (\bar{\kappa} + a_1)/\bar{S}P), \tag{9}$$

$$c_3 = a_1 + a_2, \qquad \kappa = \bar{\kappa} + c_3. \tag{10}$$

[46] Below, pp. 362–363.

[47] Above, pp. 291–292.

Therefore, the true heliocentric longitude $\lambda^{*\prime}$ and the equation of the anomaly c follow from

$$\lambda^{*\prime} = \bar{\lambda}^{*} + c_3 = \lambda_A^{*} + \kappa, \tag{42}$$

$$c = \lambda^{*} - \lambda^{*\prime}. \tag{43}$$

This is, incidentally, the only place in which Copernicus uses a true heliocentric longitude that is not the same as the true geocentric longitude, as it is at opposition. Now, in triangle $\bar{S}PO$

$$\gamma = |\bar{\lambda}_{\odot}^{*} - \lambda^{*}| \qquad (\text{mod. } 180°), \tag{44}$$

and thus, where the radius R of the planet's eccentric is the unit of distance, the radius of the earth's orbit r is

$$r = \bar{S}P\,(\sin c/\sin\gamma). \tag{45}$$

Let the relative greatest and least distances of the planet in units of R be

$$M' = R + (e_1 - r'), \qquad m' = R - (e_1 - r'). \tag{46}$$

Then taking r as the unit, the distances of the planet in terms of the radius of the earth's orbit are

$$M = M'/r, \qquad m = m'/r, \qquad \mu = R/r, \tag{47}$$

where μ is the mean distance. The corresponding maximum equations of the anomaly at each distance, which are inversely proportional to the distances, are

$$c_{\mathrm{m}} = \sin^{-1}(r/M), \qquad c_{\mathrm{M}} = \sin^{-1}(r/m), \qquad c_{\mu} = \sin^{-1}(r/\mu), \tag{48}$$

where r is now taken as the common unit of distance.

3. The Individual Planets

Copernicus's exposition for each planet is far from complete. He first reviews Ptolemy's derivation, giving the observations, setting up the mean motions between the oppositions (step 1), stating Ptolemy's final results for $2e$ and $\bar{\kappa}$ (from step 5), converting $2e$ to e_1 and r', confirming that the original oppositions are reproduced (step 3), and finding the mean anomaly and the position of the apogee at the time of one opposition (step 6). For his own derivations he is more detailed, although the iterations of steps 2–5 are omitted and in the case of Mars he does not even show the initial solution of step 2. Hence, from Copernicus account, it is not obvious how strictly he has derived the elements of the orbits. We have therefore performed the omitted iterations to check Copernicus's results which turn out to be accurate within the range of discrepancies expected in a very long computation that must contain some number of roundings and small errors. In our exposition, we shall set out the observations used by Copernicus, and then go through his presentation of Ptolemy's demonstration and his own, showing our check of the iterations only for Copernicus's elements. (Ptolemy's derivations are essentially as he reports them, containing only the smallest numerical discrepancies when checked through all their iterations.)

A. Saturn

Observations

The seven reported observations of Saturn are listed in Table 2, the form of which will also be followed for Jupiter and Mars. Numbers 1/2/3 are Ptolemy's oppositions, 4/5/6 are Copernicus's oppositions, and 7 is Copernicus's observation of the planet outside of opposition for determining the radius of its orbit. For observations 1–6 we give the following information:

(1) Date and Time as reported by Copernicus for the meridian of Frauenberg–Cracow.[48]
(2) Julian Day Number.
(3) $\bar{\lambda}^*_{\odot}$, sidereal longitude of the mean sun by computation.
(4) $\lambda^*_{\rm p} = \bar{\lambda}^*_{\odot} + 180°$, sidereal longitude of planet.
(5) Cop $\lambda^*_{\rm p}$, Copernicus's report.
(6) π, precession by computation.
(7) $\lambda_{\rm p} = \lambda^*_{\rm p} + \pi$, tropical longitude.
(8) Ptol $\lambda_{\rm p}$, tropical longitude in *Almagest* for 1/2/3.

For observation 7, column 4 is blank since the planet is not at opposition. Note that in all cases columns 4 and 5 are in agreement, which is not true for Jupiter and Mars. Observations 4/5/6 present a particular difficulty in that their times have been altered in M (154r28–35), not once, but twice, and the alterations for (6) are particularly puzzling. These seem to have affected the first solutions for $2e$, but curiously not for $\bar{\kappa}$, in M—at least, there are many visible alterations—but we have confined our examination to the final version, which is the same in M and N.

Review of Ptolemy's Derivation (V,5)

Ptolemy's derivation of the eccentricity and the direction of the apsidal line is in *Almagest* XI,5 and *Epitome* XI,10–15. From observations 1/2/3, we have

$$\Delta t_{12} = 6^{\rm y}\ 70;55^{\rm d}, \qquad \Delta t_{23} = 3^{\rm y}\ 35;50^{\rm d},$$

[48] For (1) Copernicus gives the month as *Mechyr*, evidently following the 1515 *Almagest*, f. 123v (as noted by Rosen, Copernicus (1978), 418 n. to 244:15) or *Epitome* XI, 10, both of which read *machur*. Yet the conversion to 26 Mar 127 corresponds to the correct Pachon 7, which Copernicus could easily find from Ptolemy's Δt_{12}.

and applying (23a) to the true sidereal longitudes,

$$\delta_1 = 68;23°, \qquad \delta_2 = 34;34°,$$

compared with Ptolemy's

$$\delta_1 = 68;27°, \qquad \delta_2 = 34;34°,$$

which are differences of tropical longitudes. Provisionally taking $\bar{\alpha}$ from the *Alfonsine Tables* for intervals Δt,

$$\Delta\bar{\alpha}_1 = 352;44°, \qquad \Delta\bar{\alpha}_2 = 356;43°,$$

exactly Copernicus's figures. The tables in the *Almagest* give $\Delta\bar{\alpha}$ 0;1° less. Hence, from (24a),

$$\bar{\delta}_1 = 75;39°, \qquad \bar{\delta}_2 = 37;51°,$$

while Ptolemy's differences of mean tropical longitudes were

$$\bar{\delta}_1 = 75;43°, \qquad \bar{\delta}_2 = 37;52°.$$

Copernicus now has the differences of mean and apparent longitude required for finding the elements of Saturn's orbit. But he does nothing with them. Nor does he go through Ptolemy's derivation. Instead, he says that the problem is very difficult and has no direct, *a priori* solution, so that it is necessary to use roundabout (*ambages*), *a posteriori* methods. He concludes, rather petulantly, "So Ptolemy, in examining these problems by a lengthy discourse, poured himself forth in an immense multitude of calculations (*ingentem numerorum multitudinem*) which I consider tedious and superfluous to review, especially because likewise in our own demonstrations that follow, we shall copy nearly the same method." This is a surprising acknowledgment of a predecessor without whose *ingentem numerorum multitudinem* Copernicus would have little to do.[49]

Copernicus then places the oppositions in his own model as shown in Fig. 29. In his first solution, Ptolemy finds

$$2e = E\bar{S} = 7;8/60 = 1189/10{,}000,$$

$$\bar{\kappa}_1 = -55;52°, \qquad \bar{\kappa}_2 = 19;51°, \qquad \bar{\kappa}_3 = 57;43°,$$

leading to corrected values

$$\delta'_1 = 68;42° \quad \text{in place of} \quad \delta_1 = 68;27°,$$

$$\delta'_2 = 34;38° \quad \text{in place of} \quad \delta_2 = 34;34°.$$

[49] N, 144r26–30. Professor Rosen, in Copernicus (1978), 419 n. to 245:22, finds Copernicus's remark commendable, and appears to think Copernicus intends to do something different, for he writes, "Copernicus' rejection of this overelaborate treatment by Ptolemy should be carefully considered by those who are fond of accusing Copernicus of merely parroting Ptolemy." In fact, Copernicus does not reject, but uses, the same "overelaborate treatment."

Ptolemy does not show the iteration with the new values δ', but gives his results

$$2e = E\bar{S} = 6;50/60 = 1139/10{,}000,$$

$$\bar{\kappa}_1 = -57;5^\circ \qquad \bar{\kappa}_2 = 18;38^\circ, \qquad \bar{\kappa}_3 = 56;30^\circ.$$

Copernicus, however, gives, as though they were Ptolemy's results

$$\bar{\kappa}_1 = -57;1^\circ, \qquad \bar{\kappa}_2 = 18;37^\circ, \qquad \bar{\kappa}_3 = 56;30^\circ,$$

and in the following test of the derivation he also uses $\bar{\kappa}_2 = 18;36^\circ$, $\bar{\kappa}_2 = 18;38^\circ$, and $\bar{\kappa}_3 = 56;29^\circ$. Now these are not copying errors, but deliberate, and rather inconsistent, alterations, evidently by some sort of trial and error, that Copernicus has made in order to produce his own differences of sidereal longitude δ_1 and δ_2 since they differ from Ptolemy's differences of tropical longitude. Copernicus also gives $2e = 1016$, but this seems to be an odd scribal error since he immediately gives correctly

$$e_1 = \tfrac{3}{4} \cdot 1139 = 854, \qquad r' = 1139 - 854 = 285.$$

In order to show that these values of e_1, r', and $\bar{\kappa}$ correctly produce his sidereal values of δ_1 and δ_2, Copernicus finds the equation of center for each value of $\bar{\kappa}$. From (6–10),

$$c_{3(1)} = 3;55^\circ + 1;22^\circ = 5;17^\circ, \qquad \text{(Accur. 5;17,40°)}$$

$$c_{3(2)} = -1;27^\circ - 0;32^\circ = -1;59^\circ,$$

$$c_{3(3)} = -3;53^\circ - 1;22^\circ = -5;15^\circ. \quad \text{(Accur. 5;15,37°)}$$

Hence, using Copernicus's rather inconsistent values of $\bar{\kappa}$, the corrected eccentric anomaly at each opposition, from (25),

$$\kappa_1 = -57;1^\circ + 5;17^\circ = -51;44^\circ,$$

$$\kappa_2 = 18;38^\circ - 1;59^\circ = 16;39^\circ,$$

$$\kappa_3 = 56;29^\circ - 5;15^\circ = 51;14^\circ,$$

and the differences of apparent longitude, by (26),

$$\delta_1 = 68;23^\circ \quad \text{compared with} \quad 68;23^\circ, \qquad \Delta = 0;0^\circ,$$

$$\delta_2 = 34;35^\circ \quad \text{compared with} \quad 34;34^\circ, \qquad \Delta = 0;1^\circ.$$

So by altering Ptolemy's values of $\bar{\kappa}$, Copernicus has reached very close agreement with the differences of sidereal longitude. Note that had he used Ptolemy's original values, the results would be

$$\kappa_1 = -51;48^\circ, \qquad \kappa_2 = 16;39^\circ, \qquad \kappa_3 = 51;15^\circ,$$

from which

$$\delta_1 = 68;27^\circ, \qquad \delta_2 = 34;36^\circ,$$

compared with Ptolemy's observed tropical differences of 68;27° and 34;34° respectively.

Next, consider only the third opposition, which is shown along with the orbit of the earth in Fig. 30.[50] Since

$$\lambda_3^* = 277;37°, \qquad \kappa_3 = 51;14°,$$

so from (34)

$$\lambda_A^* = \lambda^* - \kappa = 226;23° \approx 226;20°,$$

and since at the third opposition $\pi = 6;39° \approx 6;40°$, from (35)

$$\lambda_A = 226;20° + 6;40° = 233° = \text{♏}23°,$$

the tropical longitude of the apogee in A.D. 136, in agreement with Ptolemy. Since at opposition necessarily $\alpha = 180°$, so now taking $c_{3(3)} = 5;16°$, from (32)

$$\bar{\alpha}_3 = 180° - 5;16° = 174;44°,$$

the mean anomaly on 8 July 136, 11^h at the meridian of Cracow. This is precisely the value found by Ptolemy, and follows from the unaltered $\bar{\kappa}_3 = 56;30°$. Note that Copernicus's review of Ptolemy's derivation was filled with small errors and inconsistencies produced by the necessity of forcing out an agreement with the sidereal values of δ_1 and δ_2. The difficulty of doing this, and the unsatisfactory results, must have given Copernicus enough trouble that for Jupiter and Mars he finds a (strictly incorrect) way of using Ptolemy's differences of tropical longitudes as though they were sidereal.[51]

Copernicus's Derivation (V,6)

From observations 4/5/6, by (23a)

$$\delta_1 = 68;1°, \qquad \delta_2 = 86;42°,$$

and since

$$\Delta t_{45} = 6^y\ 70;33^d, \qquad \Delta t_{56} = 7^y\ 89;46^d,$$

from both the *Almagest* and the *Alfonsine Tables*, provisionally

$$\Delta\bar{\alpha}_{45} = 352;23°, \qquad \Delta\bar{\alpha}_{56} = 358;13°.$$

---

[50] Copernicus's figures showing the model for Saturn with the orbit of the earth (N, 145r, 147v, 149r; M, 153v, 156v, 158rv) are all incorrect in that they show the center of Saturn's eccentric, marked *D*, falling outside the orbit of the earth. Since Copernicus finds in V,9 that, where the radius of Saturn's eccentric $R = 10{,}000$, the radius of the earth's orbit $r = 1090$, and since $e_1 = 854$ and $2e = 1139$, so *M* (*D* in Copernicus's figures) must fall inside and *E* just outside the earth's orbit, as shown in Figs. 30, 33, and 35. The critical editions and translations retain Copernicus's figures without comment.

[51] Above, pp. 309–310, below, pp. 337, 349.

Thus, from (24a)

$$\bar{\delta}_1 = 75;38°, \qquad \bar{\delta}_2 = 88;29°,$$

while Copernicus gives $\bar{\delta}_1 = 75;39°$ which we use in the following demonstration.[52]

It is now assumed that, as shown in Fig. 31(a), the planet lies on a circle about F, the equant point in Ptolemy's, and so also Copernicus's, model. Let the mean sun be D, and let the projection of the planet on the circle about the equant be A at the first opposition, B at the second, and C at the third. We are given

$$\delta_1 = 68;1°, \qquad \delta_2 = 86;42°,$$

$$\bar{\delta}_1 = 75;39°, \qquad \bar{\delta}_2 = 88;29°.$$

In triangle BDE,

$$2\beta_1 = 186;36°, \qquad 2\gamma_1 = 84;55°,$$

$$BE = \text{crd } 2\beta_1 = 19968,^{53} \qquad DE = \text{crd } 2\gamma_1 = 13501.$$

In triangle ADE,

$$2\beta_2 = 50;34°, \qquad 2\gamma_2 = 145;18°,$$

$$AE = \text{crd } 2\beta_2 = 8542, \qquad DE = \text{crd } 2\gamma_2 = 19090.$$

Thus, in units of triangle BDE, where $DE = 13501$,

$$AE = \text{crd } 2\beta_2(\text{crd } 2\gamma_1/\text{crd } 2\gamma_2)$$

$$= 8542(13501/19090) = 6043. \quad (\text{Corr. } 6041)^{54}$$

In triangle ABE,

$$AQ = AE \sin \tfrac{1}{2}\bar{\delta}_1 = 3706, \qquad EQ = AE \cos \tfrac{1}{2}\bar{\delta}_1 = 4773,$$

$$BQ = BE - EQ = 19968 - 4773 = 15195,$$

$$AB = (AQ^2 + BQ^2)^{1/2} = 15647. \quad (\text{Corr. } 15640)$$

Now, letting the radius of the eccentric $R = 10{,}000$,

$$AB = \text{crd } \bar{\delta}_1 = \text{crd } 75;39° = 12266, \quad (\text{Corr. } 12265)$$

so that

$$BE = 19968(12266/15647) = 15664, \quad (\text{Corr. } 15653)$$

$$DE = 13501(12266/15647) = 10599. \quad (\text{Corr. } 10583)$$

---

[52] This difference of +0;1° may have something to do with the interval between the original times of 4/5 (M, 154), although neither Δt nor $\Delta\bar{\alpha}$ has been altered. See above, p. 324.

[53] Corr. 19967. Copernicus twice writes 19953, but always uses, and later writes, 19968.

[54] Cf. the explanation of the error by Rosen in Copernicus (1978), 420 n. to 248:28.

Therefore,

$$\text{arc } BAE = 2\sin^{-1}\tfrac{1}{2}BE = 103;7°,$$
$$\text{arc } EABC = \text{arc } BAE + \bar{\delta}_2 = 191;36°,$$
$$\text{arc } CE = 360° - \text{arc } EABC = 168;24°.$$

Hence,

$$CDE = 2\sin\tfrac{1}{2}CE = 19898,$$
$$CD = CDE - DE = 19898 - 10599 = 9299.$$

Consequently, in Fig. 31(b),

$$CD \cdot DE = R^2 - 4e^2 = 9299 \cdot 10599 = 98560101,$$
$$4e^2 = R^2 - CD \cdot DE = 1439899,$$
$$2e = 1200, \qquad \text{(Accur. 1199.95)}$$

or, where $R = 60$, $2e = 7;12$.

Next, drawing FKL perpendicular to CDE,

$$DK = DE - \tfrac{1}{2}CDE = 10599 - \tfrac{1}{2}(19898) = 650,$$

and where $2e = 10{,}000$,

$$DK = 650(10{,}000/1200) = 5411. \quad \text{(Corr. 5417)}$$

Therefore,

$$\gamma_3 = \sin^{-1}(5411/10{,}000) = 32;45°, \quad \text{(Corr. 32;45,30°)}$$
$$\text{arc } CHL = \tfrac{1}{2}\text{arc } CHE = \tfrac{1}{2}(168;24°) = 84;13°, \quad \text{(Corr. 84;12°)}$$
$$\text{arc } CH = \text{arc } CHL - \gamma_3 = 51;28°.$$

Thus, the mean eccentric anomaly at each opposition

$$\bar{\kappa}_6 = 180° - \text{arc } CH = 128;32°,$$
$$\kappa_5 = \bar{\kappa}_6 - \bar{\delta}_2 = 40;3°, \qquad \bar{\kappa}_4 = \bar{\kappa}_5 - \bar{\delta}_1 = -35;36°.$$

Copernicus's derivation contains a number of computational errors. Recomputing accurately, we find

$$2e = 1191 \qquad \text{or} \qquad 7;9/60,$$
$$\bar{\kappa}_4 = -35;57°, \qquad \bar{\kappa}_5 = 39;42°, \qquad \bar{\kappa}_6 = 128;11°.$$

Note that the error in $2e$ is very small, only 9/1191, while the error in $\bar{\kappa}$ is $+0;21°$. This is more-or-less as expected since the eccentricity is insensitive, while the direction of the apsidal line, or equivalently, the mean eccentric anomaly at each opposition, is very sensitive to small changes in the initial parameters or any irregularities in the calculation.

Copernicus now places the oppositions on his own model, shown in Fig. 32, in order to test whether the derived values of $2e$ and $\bar{\kappa}$ will correctly reproduce the observed values of δ_1 and δ_2. From the previous demonstration, for $2e = 1200$,

$$e_1 = \tfrac{3}{4} \cdot 1200 = 900, \qquad r' = \tfrac{1}{4} \cdot 1200 = 300,$$

and the mean eccentric anomaly,

$$\bar{\kappa}_4 = -35;36^\circ, \qquad \kappa_5 = 40;3^\circ, \qquad \bar{\kappa}_6 = 128;32^\circ.$$

He says that these will not reproduce the observations, but not wishing to bother the reader with all the details, he merely gives the result that

$$\delta_1 = 67;35^\circ \quad \text{compared with} \quad 68;1^\circ, \qquad \Delta = -0;26^\circ,$$
$$\delta_2 = 87;12^\circ \quad \text{compared with} \quad 86;42^\circ, \qquad \Delta = +0;30^\circ.$$

Recomputing, we find from (25) the true eccentric anomaly

$$\kappa_4 = -35;36^\circ + 3;49^\circ = -31;47^\circ,$$
$$\kappa_5 = 40;3^\circ - 4;13^\circ = 35;50^\circ,$$
$$\kappa_6 = 128;32^\circ - 5;34^\circ = 122;58^\circ,$$

and from (26) the differences in longitude

$$\delta_1 = 67;37^\circ \quad \text{compared with} \quad 68;1^\circ, \qquad \Delta = -0;24^\circ,$$
$$\delta_2 = 87;8^\circ \quad \text{compared with} \quad 86;42^\circ, \qquad \Delta = +0;26,$$

in slight disagreement with Copernicus's values.

The discrepancies between the observed and computed values of δ show that the derivation of $2e$ and $\bar{\kappa}$ is not yet correct, and it is necessary to iterate the entire procedure in order to find more accurate results. In order to do so, we must first correct the locations of the projection P' of the planet on the circle about the equant point E in Fig. 32. Thus, the equations of center c_3 of P must be found by (4) or by (6–10), and the equations of center c'_3 of P' by (27) or by (28–29), to then find the corrections ε by (30). In the present case, for each value of $\bar{\kappa}$:

	$\bar{\kappa}$	c_3	$-$ c'_3	$=$ ε
$\bar{\kappa}_4$	$-35;36^\circ$	$+3;48,38^\circ$	$+3;38,30^\circ$	$+0;10,8^\circ \approx +0;10^\circ$
$\bar{\kappa}_5$	40;3	$-4;13,19$	$-4;2,42$	$-0;10,37 \approx -0;11$
$\bar{\kappa}_6$	128;32	$-5;34,10$	$-5;47,34$	$+0;13,24 \approx +0;13$

The resulting value of δ' are then, from (31),

$$\delta'_1 = +0;10^\circ + 68;1^\circ + 0;11^\circ = 68;22^\circ,$$
$$\delta'_2 = -0;11^\circ + 86;42^\circ - 0;13^\circ = 86;18^\circ.$$

We may now iterate the derivation of $2e$ and $\bar{\kappa}$ using

$$\delta_1' = 68;22^\circ, \qquad \delta_2' = 86;18^\circ,$$

and as before

$$\bar{\delta}_1 = 75;39^\circ, \qquad \bar{\delta}_2 = 88;29^\circ.$$

The results are

$$2e = 1130, \qquad e_1 = 848, \qquad r' = 282,$$
$$\bar{\kappa}_4 = -39;31^\circ, \qquad \bar{\kappa}_5 = 36;8^\circ, \qquad \bar{\kappa}_6 = 124;37^\circ.$$

We next test the new values of $2e$ and $\bar{\kappa}$. From (6–10, 25),

$$\kappa_4 = -35;35^\circ, \qquad \kappa_5 = 32;29^\circ, \qquad \kappa_6 = 119;8^\circ,$$

and thus from (26),

$$\delta_1 = 68;4^\circ \quad \text{compared with} \quad 68;1^\circ, \qquad \Delta = +0;3^\circ,$$
$$\delta_2 = 86;39^\circ \quad \text{compared with} \quad 86;42^\circ, \qquad \Delta = -0;3^\circ.$$

So another iteration is required. Solving for the correction ε, we find

$$\varepsilon_4 = +0;9,26^\circ, \qquad \varepsilon_5 = -0;9,6^\circ, \qquad \varepsilon_6 = +0;11,12^\circ,$$

so that by (31)

$$\delta_1'' = +0;9^\circ + 68;1^\circ + 0;9^\circ = 68;19^\circ,$$
$$\delta_2'' = -0;9^\circ + 86;42^\circ - 0;11^\circ = 86;22^\circ.$$

So we again solve for $2e$ and $\bar{\kappa}$ using

$$\delta_1'' = 68;19^\circ, \qquad \delta_2'' = 86;22^\circ,$$
$$\bar{\delta}_1 = 75;39^\circ, \qquad \bar{\delta}_2 = 88;29^\circ,$$

and find

$$2e = 1139, \qquad e_1 = 854, \qquad r' = 285,$$
$$\bar{\kappa}_4 = -38;55^\circ, \qquad \bar{\kappa}_5 = 36;44^\circ, \qquad \bar{\kappa}_6 = 125;13^\circ.$$

Testing these results, we find from (6–10, 25),

$$\kappa_4 = -35;0^\circ, \qquad \kappa_5 = 33;0^\circ, \qquad \kappa_6 = 119;43^\circ,$$

so that from (26)

$$\delta_1 = 68;0^\circ \quad \text{compared with} \quad 68;1^\circ, \qquad \Delta = -0;1^\circ,$$
$$\delta_2 = 86;43^\circ \quad \text{compared with} \quad 86;42^\circ, \qquad \Delta = +0;1^\circ.$$

Strictly, another iteration is required for there still remain discrepancies of 0;1°. However, we have essentially reached Copernicus's results. Without showing his work, he says that he finds

$$2e = 1139, \qquad e_1 = 854, \qquad r' = 285,$$

in agreement with Ptolemy and with our third iteration, and that "by moving the apogee forward a little bit" (actually by $+3;14°$ from his first solution),

$$\bar{\kappa}_4 = -38;50°, \qquad \bar{\kappa}_5 = 36;49°, \qquad \bar{\kappa}_6 = 125;18°,$$

which differs by $+0;5°$ from our result. This is not surprising in view of the sensitivity of $\bar{\kappa}$ to roundings or errors in the computation. The question that remains is whether Copernicus reached these results by correct iteration of the solution for $2e$ and $\bar{\kappa}$, or simply took over Ptolemy's $2e$ and used some kind of trial and error to locate the apsidal line and thus $\bar{\kappa}$.[55]

He next shows that the derived parameters will produce the observed values of δ. For the equation of center, from (6–10),

$$c_{3(4)} = 2;52° + 1;3° = 3;55°,$$

$$c_{3(5)} = -2;45° - 0;59° = -3;44°,$$

$$c_{3(6)} = -4;13° - 1;18 = -5;31°, \quad (\text{Corr. } -5;29°)$$

which give, from (25), the corrected eccentric anomaly,

$$\kappa_4 = -34;55°, \qquad \kappa_5 = 33;5°, \qquad \kappa_6 = 119;47°.$$

Consequently, the computed differences in longitude, from (26),

$$\delta_1 = 68;0° \quad \text{compared with} \quad 68;1°, \qquad \Delta = -0;1°$$

$$\delta_2 = 86;42° \quad \text{compared with} \quad 86;42°, \qquad \Delta = 0;0°,$$

although the correct value of $c_{3(6)}$ would give $\delta_2 = 86;44°$ or $\Delta = +0;2°$. Hence, the agreement is not yet perfect, although Copernicus evidently considered it satisfactory or at least not worth the effort to improve.

There remains only the determination of the longitude of the apogee λ_A^* and the value of the mean anomaly $\bar{\alpha}$ at one opposition. Copernicus chooses the last of the three, with the planet at P_6 as illustrated in Fig. 33. The planet was observed at a true longitude of $\lambda^* = 0;7°$, although here he gives its position as $\lambda^* = 0;8°$. It was found that $\kappa_6 = 119;47°$, and thus by (34)

$$\lambda_A^* = 240;21° \approx 240;20°.$$

Likewise, at the third opposition the equation of center $c_{3(6)} = -5;31°$, so that from (32)

$$\bar{\alpha}_6 = 180° - 5;31° = 174;29°,$$

the mean anomaly at $6;24^h$ of 10 Oct 1527 at the meridian of Cracow.

[55] See below, pp. 366–367.

Correction of the Mean Anomaly and Positions at Epoch (V,7–8)

At the third oppositions of Ptolemy (3) and Copernicus (6) respectively,

$$\bar{\alpha}_3 = 174;44^\circ, \qquad \bar{\alpha}_6 = 174;29^\circ \qquad \Delta\bar{\alpha} = -0;15^\circ \quad (\text{mod. } 360^\circ).$$

Ptolemy's observation, reduced to the meridian of Cracow, was on 8 Jul 136, 11^h, and Copernicus's observation on 10 Oct 1527, $6;24^h$, although here Copernicus's reports $6;0^h$. Hence,

$$\Delta t = 1392^y\, 75^d\, 19;0^h = 1392^y\, 75;47,30^d,$$

while Copernicus gives $75;48^d$, a better approximation of the correct interval of $75^d\, 19;24^h = 75;48,30^d$. Taking any provisional value of the mean anomaly, e.g. from the *Almagest*, by (36),

$$\Delta\bar{\alpha} = \Delta t \cdot \bar{\alpha}^{\circ/y} = 1343;59\ldots^r \approx 1344^r,$$

which shows that between oppositions (3) and (6) the mean anomaly has changed by

$$\Delta\bar{\alpha} = 1344^r - 0;15^\circ = 1343^r + 359;45^\circ.$$

Copernicus then says that from the tables of mean motion, for Δt one finds $\Delta\bar{\alpha} = 1343^r + 359;48^\circ$, which is correct (accur. $359;47,41^\circ$), but he does not show how the value underlying the tables was derived. He remarks that what had been set out earlier concerning the mean motion of Saturn is therefore correct, but this is hardly true in light of the discrepancy of $0;3^\circ$. Originally, in M (157r14) he had said that the table would give $359;45^\circ$, in exact agreement with the observations, but this is simply untrue for both the revised and the original tabulation of Saturn's mean anomaly.[56]

The difference of $0;3^\circ$ between the mean motion computed from the tables in V,1 and derived from the observations indicates that the observations were not the direct source of the tables (unless one assumes computational error). Nor were the observations the direct source of any of the values in V,1. Correctly, if $\Delta\bar{\alpha} = 1344^r - 0;15^\circ$ during an interval $\Delta t = 1392^y\, 75;48^d$, then from (37) the annual motion is

$$\bar{\alpha}^y = 347;32,3,2,39,39\ldots^\circ,$$

which agrees with neither the original value in M nor the revised value in N (see Tables 1(a) and 1(b)). Likewise, from (38), the period of the anomaly

$$P_\alpha = 378;5,32,18,37,34\ldots^d,$$

which also disagrees with Copernicus's original and revised values. Finally, letting the sidereal year be $365;15,24,10^d$, from (39), the cycle

$$57P_\alpha = 21551;15,41,41^d = 59^y + 1;6,55,51^d,$$

---

[56] The alteration of xlv to xlviii in M, 157r14, has been noted by Rosen in Copernicus (1978), 421 n. to 252:8.

that is, the correction $c^d \approx +1;6,56^d$ agrees with neither M nor N. Thus, it is difficult to know exactly how Copernicus reached the values given in V,1, although one may reasonably assume that he made some use of oppositions (3) and (6).

The mean sidereal motion of the sun between the oppositions

$$\Delta\bar{\lambda}^*_{\odot} = 180;7^\circ - 97;37^\circ = 82;30^\circ,$$

and thus, using the correct $\Delta\bar{\alpha} = 359;45^\circ$ from the observations (not the erroneous 359;48° computed from the tables), the mean sidereal motion of Saturn

$$\Delta\lambda^* = \Delta\bar{\lambda}^*_{\odot} - \Delta\bar{\alpha} = 82;30^\circ - 359;45^\circ = 82;45^\circ.$$

Finally, the sidereal motion of the apogee

$$\Delta\lambda^*_A = 240;21^\circ - 226;23^\circ = 13;58^\circ,$$

and Copernicus estimates that

$$\Delta\lambda^*_A = 1;0,1/100^y \approx 1^\circ/100^y.$$

Ptolemy's third opposition, on Hadrian 20 Mesore 24, or 136 Jul 8, 11^h reduced to the meridian of Cracow, followed the Epoch of Christ by

$$\Delta t = 135^y\, 222;27,30^d \approx 135^y\, 222;27^d,$$

during which time the mean anomaly of Saturn changed by

$$\Delta\bar{\alpha} = 328;55,26^\circ \approx 328;55^\circ.$$

At the time of Ptolemy's observation $\bar{\alpha}_3 = 174;44^\circ$, and thus at the Epoch of Christ

$$\bar{\alpha}_{0G} = 174;44^\circ - 328;55^\circ = 205;49^\circ.$$

Copernicus then gives $\bar{\alpha}_0$ for the four standard epochs computed in the order G. → A. → C. → D. All the computations are correct, and we give the positions at epoch below along with the mean sidereal longitude of the planet computed from (41a) and the mean tropical longitude from (41b). We have also computed for each epoch the sidereal longitude of the apogee λ^*_A and the mean tropical longitude $\bar{\lambda}_A = \lambda^*_A + \bar{\pi}$, taking $\Delta\lambda^*_A = 0;0,36^{\circ/y}$. These additional quantities will be of use in the explanation of the use of the tables to find planetary positions.

Epoch	$\bar{\alpha}_0$	$\bar{\lambda}^*_0$	$\bar{\lambda}_0$	λ^*_{A0}	$\bar{\lambda}_{A0}$	
A. Olympiad	134;54°	321;22°	316;9°	217;17°	212;1°	
C Alexander	148;1	77;36	75;35	221;48	222;50	(49)
D. Caesar	35;21	236;43	241;38	224;35	229;30	
G. Christ	205;49	66;42	72;13	225;2	230;34	

Distance and Equation of the Anomaly (V,9)

Ptolemy's demonstration of the radius of Saturn's epicycle, using an observation of the planet outside of opposition, is in *Almagest* XI,6 and *Epitome* XI,16. Copernicus, for his equivalent demonstration of the ratio of the radius of Saturn's orbit to the orbit of the earth, uses an observation (7) made on 24 Feb 1514 at 5^h after midnight in which he observed Saturn in a straight line with the second and third stars in the forehead of Scorpio, δ and π Scorpii. The observation is of interest principally because it seems to have been made with no instrument, for Copernicus must simply have noticed that the planet was in line with the stars. Since by luck both stars have the same longitude $\lambda^* = 209°$ in his star catalogue, the sidereal longitude of Saturn is immediately determined to also be 209° without the necessity of instrumental measurement. It should be noted, however, that because Saturn was then nearly stationary, it is unclear how he can specify an exact time for the alignment. Further, a comparison of correct positions of Saturn and the stars by modern computation at the time of Copernicus's observation shows that they were not in a straight line anyway.

The configuration is shown in Fig. 34 in which Copernicus's sidereal longitudes have been converted to tropical to allow comparison with the correct positions. From Copernicus's precession theory, for 24 Feb 1514 $\pi = 27;13°$, and thus the tropical longitude of the stars and Saturn $\lambda = 209° + 27;13° = 236;13°$. We thus have the following locations:[57]

	λ Cop.	β	λ Mod.	β
δ Sco	236;13°	−1;40°	235;47°	−1;56°
π Sco	236;13	−5;0	236;10	−5;25
Saturn	236;13	[+2;36]	235;41	+2;14

Copernicus does not specify a latitude for Saturn. According to his latitude theory, which of course has nothing to do with the observation, the latitude would be +2;36°. In the figure we show both this point and the point at +2;14° corresponding to the correct latitude. It is immediately obvious from the figure that Saturn is not in line with the stars and that the stars are not at the same longitude. The error in Copernicus's longitude of Saturn is +0;32°, about the diameter of the moon, and the error from the point where Saturn would be in a straight line with the stars is more than 0;50°. And according to the correct position of Saturn, the planet is more than 0;20° ahead of the point that would be in line with the stars, or about 20 days in the motion of the planet. The most charitable explanation of Copernicus's strange observation is that it was made and noted quite casually in 1514,

[57] Modern computed positions from P. V. Neugebauer (1912) and Tuckerman (1964).

and then taken much too seriously some 20 or so years later when he worked out the contents of this chapter.

The positions in the model are shown in Fig. 35. Since epoch

$$\Delta t = 1514^{y}\,67^{d}\,5^{h} = 25,14^{y}\,1,7;12,30^{d} \approx 25,14^{y}\,1,7;13^{d},^{58}$$

for which, from the tables in III,14 and V,1,

$$\bar{\lambda}^*_{\odot} = 315;41^{\circ}, \qquad \bar{\alpha} = 116;31^{\circ}, \quad (\text{Corr. } 116;37^{\circ})$$

$$\bar{\lambda}^* = \bar{\lambda}^*_{\odot} - \bar{\alpha} = 199;10^{\circ}, \qquad \lambda^*_A \approx 240;20^{\circ},$$

$$\bar{\kappa} = \bar{\lambda}^* - \lambda^*_A = -41;10^{\circ},$$

and thus the equation of center, from (6-10),

$$c_3 = 3;1^{\circ} + 1;5^{\circ} = 4;6^{\circ}, \quad (\text{Accur. } 4;6,43^{\circ})$$

which gives, from (42), the true heliocentric longitude,

$$\lambda^{*\prime} = 199;10^{\circ} + 4;6^{\circ} = 203;16^{\circ}.$$

But since by observation the apparent geocentric longitude $\lambda^* = 209;0^{\circ}$, the equation of the anomaly is, from (43),

$$c = \lambda^* - \lambda^{*\prime} = 5;44^{\circ},$$

the angle subtended by the radius of the earth's orbit at the planet.

Now, where $R = 10{,}000$, it has been established that $e_1 = 854$ and $r' = 285$. Hence, by (6–8),

$$\bar{S}C = 10657, \quad (\text{Text } 10667, \text{ corr. } 10658)$$

$$\bar{S}P = 10455, \quad (\text{Text } 10465)$$

and in triangle $P\bar{S}O$, from (44),

$$\gamma = \bar{\lambda}^*_{\odot} - \lambda^* = 106;41^{\circ}.$$

Consequently, from (45),

$$r = \bar{S}P\,(\sin c/\sin\gamma) = 1090,$$

or, where $R = 60$, $r = 6;32$, differing only a little from Ptolemy's $r = 6;30$, as Copernicus remarks. The greatest and least relative distances of Saturn where $R = 10{,}000$ are, by (46),

$$M' = 10569, \qquad m' = 9431,$$

where also $r = 1090$. Thus, from (47), where $r = 1$,

$$\text{M} = 9;41,46 \approx 9;42, \qquad m = 8;39,8 \approx 8;39,$$

$$\mu = 9;10,27 \approx 9;10.$$

---

[58] N and M by error 77;13^{d}.

The maximum equations of the anomaly corresponding to these distances are, from (48), respectively

$$c_{\mathrm{m}} = 5;55^\circ, \qquad c_{\mathrm{M}} = 6;39^\circ, \quad (\text{Corr. } 6;38^\circ)$$

$$c_\mu = 6;15^\circ.$$

This completes the derivation of the elements of the orbit of Saturn.

B. Jupiter

Observations

The seven observations of Jupiter are set out in Table 3 in the same form used earlier for the observations of Saturn. Note now that the sidereal longitude of the planet computed from $\lambda_{\mathrm{p}}^* = \bar{\lambda}_\odot^* + 180^\circ$ and Copernicus's reported sidereal longitude often do not agree. In the case of observations 1/2/3 from Ptolemy, this can be explained. We have seen that Copernicus created a good deal of trouble for himself in reviewing Ptolemy's derivation of $2e$ and $\bar{\kappa}$ for Saturn by taking the correct differences of sidereal longitude at opposition δ_1 and δ_2, which naturally were not equal to Ptolemy's differences of tropical longitude. Hence he had to manipulate Ptolemy's resulting values of $\bar{\kappa}$ at each opposition in order to produce his own values of δ_1 and δ_2. For Jupiter he simply takes over Ptolemy's δ_1 and δ_2, and this enables him to use directly Ptolemy's $\bar{\delta}_1$ and $\bar{\delta}_2$ and resulting $\bar{\kappa}$. However, in order to do so, he must alter the sidereal longitudes of the oppositions so that their differences are equal to Ptolemy's differences of tropical longitude. He does this by subtracting a precession of 6;38° from Ptolemy's tropical longitudes, that is, for observations 1/2/3 Copernicus lets

$$\lambda_{\mathrm{p}}^* = \lambda_{\mathrm{p}} - \pi = \lambda_{\mathrm{p}} - 6;38^\circ.$$

One assumes that he chose 6;38° because it was in between the correct 6;37° or 6;39° at each opposition according to his precession theory. However, as noted earlier, this is equivalent to assuming that the precession has stopped for over four years, something that cannot happen. Nevertheless, it makes Copernicus's review of Ptolemy's demonstration quite straightforward.

Copernicus's own oppositions 4 and 5 show differences of +0;1° and +0;2° respectively from the correct $\lambda_{\mathrm{p}}^* = \bar{\lambda}_\odot^* + 180^\circ$, but for these discrepancies we have no explanation.

Review of Ptolemy's Derivation (V,10)

Copernicus again presents only the initial conditions of the three oppositions and the final demonstration that the derived values of $2e$ and $\bar{\kappa}$ are correct. He then finds the values of $\bar{\alpha}$ and $\bar{\lambda}_A^*$ for the third opposition.

Ptolemy's derivation appears in *Almagest* XI,1 and *Epitome* XI,1–6. From observations 1/2/3, by (23a),

$$\delta_1 = 104;43^\circ, \qquad \delta_2 = 36;29^\circ,$$

in agreement with Ptolemy. The intervals of time between the oppositions are

$$\Delta t_{12} = 3^y\ 106^d\ 23^h, \qquad \Delta t_{23} = 1^y\ 37^d\ 7^h,$$

for which the mean motions in longitude

$$\bar{\delta}_1 = 99;55^\circ, \qquad \bar{\delta}_2 = 33;26^\circ,$$

are taken directly from Ptolemy. In this way, as mentioned, Copernicus avoids the problems that resulted from his independent calculations of δ and $\bar{\delta}$ for Saturn.

In his first solution, Ptolemy found

$$2e = 5;23/60 = 897/10{,}000,$$

$$\bar{\kappa}_1 = 79;30^\circ, \qquad \bar{\kappa}_2 = 179;25^\circ, \qquad \bar{\kappa}_3 = 212;51^\circ,$$

leading to the corrected values

$$\delta_1' = 104;39^\circ \quad \text{in place of} \quad \delta_1 = 104;43^\circ,$$

$$\delta_2' = 36;37^\circ \quad \text{in place of} \quad \delta_2 = 36;29^\circ.$$

Iterating the derivation of $2e$ and $\bar{\kappa}$, Ptolemy finds that where $R = 60$, $2e = 5;30$ or, as Copernicus gives the complete results for $R = 10{,}000$,

$$2e = 917, \qquad e_1 = 687, \qquad r' = 229,$$

which are not quite consistent, and

$$\bar{\kappa}_1 = 77;15^\circ, \qquad \bar{\kappa}_2 = 177;10^\circ, \qquad \bar{\kappa}_3 = 210;36^\circ.$$

The demonstration that these values will produce the correct differences of longitude between the observations is shown in Fig. 36. For each value of $\bar{\kappa}$, the equation of center is, from (6–10),

$$c_{3(1)} = -3;48^\circ - 1;17^\circ = -5;5^\circ, \quad (\text{Corr. } -5;3)$$

$$c_{3(2)} = -0;12^\circ - 0;4^\circ = -0;16^\circ,$$

$$c_{3(3)} = +2;8^\circ + 0;39^\circ = +2;47^\circ,$$

so that, from (25), the true eccentric anomaly,

$$\kappa_1 = 72;10^\circ, \qquad \kappa_2 = 176;54^\circ, \qquad \kappa_3 = 213;23^\circ.$$

The apparent motion between the oppositions is then, from (26),

$$\delta_1 = 104;44^\circ \quad \text{compared with} \quad 104;43^\circ, \qquad \Delta = +0;1^\circ,$$

$$\delta_2 = 36;29^\circ \quad \text{compared with} \quad 36;29^\circ, \qquad \Delta = 0;0,$$

which proves the accuracy of the derivation.

Next, in Fig. 37 we consider the third opposition with the planet at $\kappa_3 = 213;23°$. The equation of center $c_{3(3)} = +2;47°$, and thus from (32) the mean anomaly

$$\bar{\alpha}_3 = 180° + 2;47° = 182;47°.$$

Then since the true sidereal longitude $\lambda_3^* = 7;45°$, the mean sidereal longitude, by (33),

$$\bar{\lambda}_3^* = 7;45° - 2;47° = 4;58°,$$

and the sidereal longitude of the apogee, from (34),

$$\lambda_A^* = 7;45° - 213;23° = 154;22°.$$

Finally, from (35), the tropical longitude of the apogee,

$$\lambda_A = 154;22° + 6;39° = 161;1° = ♍11;1°,$$

in excellent agreement with Ptolemy's $\lambda_A =$ ♍$11°$. Copernicus probably added the 6;38° used to reduce Ptolemy's tropical longitudes.

Copernicus's Derivation (V,11)

In Fig. 38(a), with the mean sun at D, we let the planet move on a circle around the equant F, and be at A at the first opposition, B at the second, C at the third. From observations 4/5/6 of λ^*, by (23a)–(23b),

$$\delta_1 = 208;6°, \qquad \delta_2 = 65;10°, \qquad \delta_3 = 86;44°,$$

for which the intervals of time are

$$\Delta t_{45} = 6^y\ 212;40^d, \qquad \Delta t_{56} = 2^y\ 66;40^d.$$ [59]

From Δt, we estimate $\Delta\bar{\alpha}$ using either the *Alfonsine Tables* or the *Almagest*, and find

$$\Delta\bar{\alpha}_{45} = 8;26°, \qquad \Delta\bar{\alpha}_{56} = 359;0°,$$

and thus, from (24a)–(24b), the differences of mean longitude,

$$\bar{\delta}_1 = 199;40°, \qquad \bar{\delta}_2 = 66;10°, \qquad \bar{\delta}_3 = 94;10°,$$

in agreement with Copernicus.

Now, in triangle CDE,

$$2\beta_1 = 2(180° - \delta_2) = 229;40°,$$

$$2\gamma_1 = 2\delta_2 - \bar{\delta}_2 = 64;10°,$$

$$CE = \text{crd } 2\beta_1 = 18150, \qquad \text{(Accur. 18150.6)}$$

$$DE = \text{crd } 2\gamma_1 = 10623.$$

[59] N and M by error 66;*39*^{d}.

However, Copernicus by error takes

$$DE = \text{crd}\,\bar{\delta}_2 = \text{crd } 66;10^\circ = 10918, \quad \text{(Corr. 10917)}$$

and this error of $+295/10{,}000$ in crd $2\gamma_1$ makes the rest of the demonstration invalid.[60] Nevertheless, we shall continue our exposition of Copernicus's derivation, and then examine the consequences of the error. In triangle ADE,

$$2\beta_2 = 2(180^\circ - (\delta_2 + \delta_3)) = 56;12^\circ,$$

$$2\gamma_2 = 2(\delta_2 + \delta_3) - (\bar{\delta}_2 + \bar{\delta}_3) = 143;28^\circ,$$

$$AE = \text{crd}\,2\beta_2 = 9420, \qquad DE = \text{crd}\,2\gamma_2 = 18992.$$

Thus, in units of triangle CDE, where $DE = 10918$,

$$AE = \text{crd}\,2\beta_2(\text{crd}\,2\gamma_1/\text{crd}\,2\gamma_2)$$
$$= 9420(10918/18992) = 5415.$$

In triangle ACE,

$$AQ = AE \sin \tfrac{1}{2}\bar{\delta}_3 = 3966, \qquad EQ = AE \cos \tfrac{1}{2}\bar{\delta}_3 = 3687,$$

$$CQ = CE - EQ = 18150 - 3687 = 14463,$$

$$AC = (AQ^2 + CQ^2)^{1/2} = 14997,$$

$$2\gamma_3 = 2 \sin^{-1}(AQ/AC) = 2 \sin^{-1}(3966/14997) = 30;40^\circ,$$

$$\text{arc } CAE = \bar{\delta}_3 + 2\gamma_3 = 94;10^\circ + 30;40^\circ = 124;50^\circ.$$

But letting the radius of the eccentric $R = 10{,}000$,

$$CE = \text{crd } CAE = 17727,$$

and thus in the same units,

$$DE = 10918(17727/18150) = 10665. \quad \text{(Corr. 10664)}$$

Since

$$\text{arc } BCAE = \bar{\delta}_2 + \text{arc } CAE = 66;10^\circ + 124;50^\circ = 191^\circ,$$

$$\text{arc } EB = 360^\circ - \text{arc } BCAE = 169^\circ,$$

thus

$$EB = \text{crd } 169^\circ = 19908,$$

$$BD = EB - DE = 19908 - 10665 = 9243.$$

Consequently, in Fig. 38(b),

$$DE \cdot BD = R^2 - 4e^2 = 10665 \cdot 9243 = 98576595,$$

$$4e^2 = R^2 - DE \cdot BD = 1423405,$$

$$2e = 1193,$$

or where $R = 60$, $2e = 7;9$.

[60] This has been noted in Pannekoek (1945).

Next, drawing FKL perpendicular to EDB,

$$DK = \tfrac{1}{2}EDB - BD = \tfrac{1}{2}(19908) - 9243 = 711,$$

$$\gamma_4 = \sin^{-1}(DK/FD) = \sin^{-1}(711/1193) = 36;35^\circ,$$

$$\text{arc } LHB = \tfrac{1}{2} \text{ arc } EHB = \tfrac{1}{2}(169^\circ) = 84;30^\circ,$$

$$\text{arc } HB = \text{arc } LHB - \gamma_4 = 47;55^\circ.$$

Therefore, in the form Copernicus uses,

$$\bar{\kappa}_5 = 180^\circ - \text{arc } HB = -132;5^\circ,$$

$$\bar{\kappa}_6 = \bar{\kappa}_5 + \bar{\delta}_2 = -65;55^\circ,$$

$$\bar{\kappa}_4 = \bar{\kappa}_6 + \bar{\delta}_3 = 28;15^\circ.$$

Now, it is not expected that the results of this first solution will agree precisely with the observed values of δ_1 and δ_2, but they are worth testing, even with Copernicus's error in crd $2\gamma_1$. Note first that $2e = 7;9$, while Ptolemy found in his first solution 5;23 and, after an iteration, a final value of 5;30. So Copernicus's value is quite distant. In Fig. 39, from Copernicus's derived $2e = 1193$, we set

$$e_1 = 895, \qquad r' = 298,$$

and arrange the mean eccentric anomalies as

$$\bar{\kappa}_4 = 28;15^\circ, \qquad \bar{\kappa}_5 = 277;55^\circ, \qquad \bar{\kappa}_6 = 294;5^\circ.$$

We then find, from (6–10, 25), the true eccentric anomalies,

$$\kappa_4 = 25;11^\circ, \qquad \kappa_5 = 233;11^\circ, \qquad \kappa_6 = 300;9^\circ,$$

and thus, from (26), the differences of longitude,

$$\delta_1 = 208^\circ \quad \text{compared with} \quad 208;6^\circ, \qquad \Delta = -0;6^\circ,$$

$$\delta_2 = 66;58^\circ \quad \text{compared with} \quad 65;10^\circ, \qquad \Delta = +1;48^\circ.$$

Hence the results are not very good, although, as we shall show, they can be improved considerably by an iteration. But Copernicus seems to lose all patience, and, rather like his remarks in his review of Ptolemy's derivation for Saturn, agains turns petulant.

> Doubtless the results agree insufficiently with the appearances, since the planet is not moving along the imagined eccentric, so that this method of demonstration, depending upon an uncertain principle, cannot yield anything certain, the evidence for which, among much else, is this, that for Ptolemy the demonstration produced a distance between the centers, greater in the case of Saturn, smaller in the case of Jupiter, than the proper distance, although for us it produced a distance for Jupiter greater than required, so that it is evidently clear that when different arcs of the circle are taken for the same planet, what is sought does not turn out in the same way.[61]

[61] N, 152r29–v2. Pannekoek (1945), 69, remarks, "This is a most astonishing utterance, since there is nothing to be objected in the method, which in all other computations has been used with good results."

What Copernicus is saying here is that because Ptolemy's first eccentricity for Saturn was larger than the final result, and for Jupiter smaller, while his own first eccentricity is too large, that *something must be wrong with the method*. He seems to believe that the problem lies in the admittedly false assumption that the planet is moving in the circle about the equant, and that the problem is shown by the fact that different arcs of this circle between different oppositions of the same planet will, upon first solution, produce eccentricities either larger or smaller than the true value. His analysis, however, if we understand it properly, is not correct, for there is simply no predicting whether the first solution will be larger or smaller than the final result—it depends upon the location of the three oppositions with respect to the apsidal line, which is initially unknown—while the essence of Ptolemy's method is that it must successively approximate a truer and truer fit to the three observations. One may now wonder whether Copernicus really understood Ptolemy's iterative method. That he made a computational error and found a very large value of $2e$ is not relevant to the successive convergence of further iterations, for the method is, within limits, self-correcting in later iterations.

We can show this by repeating the derivation using Copernicus's results, erroneous though they may be. For each value of $\bar{\kappa}$, from (6–10, 28–30), the correction

$$\varepsilon_4 = -0;8,44^\circ, \qquad \varepsilon_5 = -0;13,38^\circ, \qquad \varepsilon_6 = +0;8,23^\circ.$$

Thus, by (31)

$$\delta_1' = 208;11^\circ, \qquad \delta_2' = 64;48^\circ,$$

$$\delta_3' = 360^\circ - (\delta_1' + \delta_2') = 87;1^\circ,$$

and as before

$$\bar{\delta}_1 = 199;40^\circ, \qquad \bar{\delta}_2 = 66;10^\circ, \qquad \bar{\delta}_3 = 94;10^\circ.$$

Iterating the solution for $2e$ and $\bar{\kappa}$, we find

$$2e = 906, \qquad e_1 = 680, \qquad r' = 226,$$

$$\bar{\kappa}_4 = 47;11^\circ, \qquad \bar{\kappa}_5 = 246;51^\circ, \qquad \bar{\kappa}_6 = 313;1^\circ.$$

Note that each $\bar{\kappa}$ has increased by fully 18;56°, corresponding to a decrease of 18;56° in the longitude of the apogee, and that $2e = 906$ is reasonably close to Ptolemy's $2e = 5;30/60 = 917/10{,}000$. The derivation is obviously back on the right track. This becomes even clearer when we test the results. Computing the true eccentric anomalies, from (6–10, 25),

$$\kappa_4 = 43;30^\circ, \qquad \kappa_5 = 251;42^\circ, \qquad \kappa_6 = 316;42^\circ,$$

it follows, from (26), that

$$\delta_1 = 208;12^\circ \quad \text{compared with} \quad 208;6^\circ, \qquad \Delta = +0;6^\circ,$$

$$\delta_2 = 65;0^\circ \quad \text{compared with} \quad 65;10^\circ, \qquad \Delta = -0;10^\circ.$$

Hence, we now have much smaller residual errors, which show, however, that a further iteration is still required. The correction ε is, from (6–10, 28–30),

$$\varepsilon_4 = -0;6,23^\circ, \qquad \varepsilon_5 = -0;5,20^\circ, \qquad \varepsilon_6 = +0;6,23^\circ,$$

and therefore, from (31),

$$\delta_1'' = 208;5^\circ, \qquad \delta_2'' = 64;58^\circ, \qquad \delta_3'' = 86;57^\circ.$$

Solving for $2e$ and $\bar{\kappa}$ with these values of δ'' and the original values of $\bar{\delta}$, we find

$$2e = 915, \qquad e_1 = 686, \qquad r' = 229,$$

$$\bar{\kappa}_4 = 45;11^\circ, \qquad \bar{\kappa}_5 = 244;51^\circ, \qquad \bar{\kappa}_6 = 311;1^\circ.$$

The approach to Ptolemy's $2e = 917$ is now closer still, and $\bar{\kappa}$ has decreased by 2;0°. Finding the true eccentric anomaly, from (6–10, 25),

$$\kappa_4 = 41;35^\circ, \qquad \kappa_5 = 249;41^\circ, \qquad \kappa_6 = 314;51^\circ,$$

it follows, from (26), that

$$\delta_1 = 208;6^\circ \quad \text{compared with} \quad 208;6^\circ, \qquad \Delta = 0;0^\circ,$$

$$\delta_2 = 65;10^\circ \quad \text{compared with} \quad 65;10^\circ, \qquad \Delta = 0;0^\circ.$$

Thus, with the third solution, the agreement with the observations is perfect, and this terminates the derivation.

At the third opposition (6), the sidereal longitude of the planet $\lambda_6^* = 113;44^\circ$ and the true eccentric anomaly $\kappa_6 = 314;51^\circ$, and thus the longitude of the apogee

$$\lambda_A^* = \lambda_6^* - \kappa_6 = 158;53^\circ.$$

The equation of center $c_{3(6)} = +3;50^\circ$, so that the mean longitude of the planet

$$\bar{\lambda}_6^* = \lambda_6^* - c_{3(6)} - 109;54^\circ,$$

and since $\bar{\kappa} > 180^\circ$, the mean anomaly

$$\bar{\alpha}_6 = 180^\circ + c_{3(6)} = 183;50^\circ.$$

The preceding demonstration shows that even with the error made by Copernicus in the first derivation of $2e$ and $\bar{\kappa}$, the method itself is self-correcting and can still yield an accurate result—assuming here, as Copernicus does, that the approach of $2e$ towards Ptolemy's $2e = 917$ constitutes an accurate result. However, in light of Copernicus's apparent condemnation of the method as faulty following his initial derivation of $2e = 1193$, it is not certain how he reached his own final results, for he continues:

> In no other way was it possible to combine the uniform and apparent motion of Jupiter in the three stated oppositions, and after that in all oppositions, unless we take over the entire separation of the centers of the eccentricity reported by Ptolemy of 5;30 where the semidiameter of the eccentric is 60, or of 917 where the semidiameter is 10,000.[62]

Hence, Copernicus takes

$$2e = 917, \qquad e_1 = 687, \qquad r' = 229,$$

and further sets

$$\bar{\kappa}_4 = 45;2^\circ, \qquad \bar{\kappa}_5 = 244;42^\circ, \qquad \bar{\kappa}_6 = 310;52^\circ,$$

which differ by $-0;9^\circ$ from our third iteration. Had Copernicus not made his earlier remarks, one would naturally presume that he carried out the derivation properly, and found these values of $\bar{\kappa}$ and a value of $2e$ sufficiently close to Ptolemy's that it seemed reasonable to adopt 917, as our own derivation of 915 would seem to support. However, considering that he may well have given up on a correct solution after his first defective attempt, the possibility also remains that he simply adopted Ptolemy's $2e$ and then found $\bar{\kappa}$ by trial and error, as indeed his language suggests.

He next shows that the results will in fact produce the observed values of δ. Finding the equation of center, from (6–10),

$$c_{3(4)} = -2;39^\circ - 0;57^\circ = -3;36^\circ,$$
$$c_{3(5)} = +3;40^\circ + 1;10^\circ = +4;50^\circ,$$
$$c_{3(6)} = +2;51^\circ + 1;0^\circ = +3;51^\circ,$$

the true eccentric anomaly, from (25),

$$\kappa_4 = 41;26^\circ, \qquad \kappa_5 = 249;32^\circ, \qquad \kappa_6 = 314;43^\circ.$$

Therefore, the differences of longitude, from (26),

$$\delta_1 = 208;6^\circ \quad \text{compared with} \quad 208;6^\circ, \qquad \Delta = 0;0^\circ,$$
$$\delta_2 = 65;11^\circ \quad \text{compared with} \quad 65;10^\circ, \qquad \Delta = +0;1^\circ,$$

although Copernicus gives δ_2 as 65;10°, thereby producing exact agreement with the observed difference in longitude. In any case, even with the difference of $+0;1^\circ$, the agreement is within rounding error, and thus satisfactory.

The configuration for the third opposition is now shown in Fig. 40.[63] It has been established that

$$\lambda_6^* = 113;44^\circ, \qquad \kappa_6 = 314;43^\circ, \qquad c_{3(6)} = +3;51^\circ.$$

[62] N, 152v2–7.

[63] By a curious error, the figure in N, 153r, really belongs on 158r in V,16 for Mars, while the figure on 158r should be on 153r for Jupiter. The corresponding figures in M, 162v, 167v, are correct, so the printer must have exchanged woodcuts.

Consequently, the mean anomaly

$$\bar{\alpha}_6 = 180° + c_{3(6)} = 183;51°,$$

the mean longitude of the planet

$$\bar{\lambda}^*_6 = \lambda^*{}_6 - c_{3(6)} = 109;53°,$$

although Copernicus gives 109;52; and mistakenly calls it the *true* motion (*vero motu*) instead of the *mean* motion, and the longitude of the apogee

$$\lambda^*_A = \bar{\lambda}^*_6 - \kappa_6 = 159;1° \approx 159°,$$

all positions for 1 Feb. 1529 at 19^h after midnight at the meridian of Cracow.

It is of interest to examine what would happen if Copernicus's first derivation of $2e$ and $\bar{\kappa}$ had been solved correctly, and then iterated until the resulting values of δ agreed wth the observations. Correctly, the result of the first solution should be

$$2e = 947, \qquad e_1 = 710, \qquad r' = 237,$$

$$\bar{\kappa}_4 = 42;34°, \qquad \bar{\kappa}_5 = 242;14°, \qquad \bar{\kappa}_6 = 308;24°.$$ [64]

Testing for δ, we find

$$\kappa_4 = 39;1°, \qquad \kappa_5 = 247;7°, \qquad \kappa_6 = 312;31°,$$

$$\delta_1 = 208;6° \quad \text{compared with} \quad 208;6°, \qquad \Delta = 0;0°,$$

$$\delta_2 = 65;24° \quad \text{compared with} \quad 65;10°, \qquad \Delta = +0;14°.$$

Hence, we must repeat the derivation. From

$$\varepsilon_4 = -0;6,55°, \qquad \varepsilon_5 = -0;6,45°, \qquad \varepsilon_6 = +0;6,51°,$$

we have

$$\delta'_1 = 208;6°, \qquad \delta'_2 = 64;56°, \qquad \delta'_3 = 86;58°.$$

Again solving for $2e$ and $\bar{\kappa}$,

$$2e = 913, \qquad e_1 = 685, \qquad r' = 228,$$

$$\bar{\kappa}_4 = 45;36°, \qquad \bar{\kappa}_5 = 245;16°, \qquad \bar{\kappa}_6 = 311;26°,$$

which lead to

$$\kappa_4 = 41;59°, \qquad \kappa_5 = 250;6°, \qquad \kappa_6 = 315;14°,$$

$$\delta_1 = 208;7° \quad \text{compared with} \quad 208;6°, \qquad \Delta = +0;1°,$$

$$\delta_2 = 65;8° \quad \text{compared with} \quad 65;10°, \qquad \Delta = -0;2°.$$

Hence, we try a third solution. First, we find

$$\varepsilon_4 = -0;6,29°, \quad \varepsilon_5 = -0;5,43°, \quad \varepsilon_6 = +0;6,28°,$$

[64] Pannekoek (1945), 69, finds $2e = 945$ and $\gamma_4 = 22;46°$, which would lead to $\kappa_5 = 270° - \gamma_4 = 247;14°$, and must result from some error.

leading to

$$\delta_1'' = 208;5^\circ, \quad \delta_2'' = 64;57^\circ, \quad \delta_3'' = 86;58^\circ.$$

Solving for $2e$ and $\bar{\kappa}$,

$$2e = 913, \quad e_1 = 685, \quad r' = 228,$$

exactly as before, but now

$$\bar{\kappa}_4 = 45;25^\circ, \quad \bar{\kappa}_5 = 245;5^\circ, \quad \bar{\kappa}_6 = 311;15^\circ.$$

Again, we test for δ, finding

$$\kappa_4 = 41;49^\circ, \qquad \kappa_5 = 249;55^\circ, \qquad \kappa_6 = 315;4^\circ,$$

$$\delta_1 = 208;6^\circ \quad \text{compared with} \quad 208;6^\circ, \qquad \Delta = 0;0^\circ,$$

$$\delta_2 = 65;9^\circ \quad \text{compared with} \quad 65;10^\circ, \qquad \Delta = -0;1^\circ.$$

Since we are now within the rounding error, i.e. within 0;1°, we may at this point terminate the solution. Note that $2e$ has reached 913 while, iterating from Copernicus's erroneous first solution, $2e = 915$, and Ptolemy's value, which Copernicus adopts, is 917. All are sufficiently close to show that Ptolemy's double eccentricity is satisfactory. Once again, Copernicus was lucky to have had Ptolemy's *ingentem numerorum multitudinem.*

Correction of the Mean Anomaly and Positions at Epoch (V,12–13)

Ptolemy's third opposition (3) occurred on 8 Oct 136 at 4^h after midnight reduced to the meridian of Cracow and Frauenburg, and Copernicus's third opposition (6) on 1 Feb 1529 at 19^h. The interval of time is

$$\Delta t = 1392^y\, 99^d\, 15^h = 1392^y\, 99;37,30^d,$$

which Copernicus rounds to $1392^y\, 99;37^d$. It has been determined that at each opposition

$$\bar{\alpha}_3 = 182;47^\circ, \qquad \bar{\alpha}_6 = 183;51^\circ, \qquad \Delta\bar{\alpha} = 1;4^\circ$$

$$\bar{\lambda}_3^* = 4;58^\circ, \qquad \bar{\lambda}_6^* = 109;52^\circ, \qquad \Delta\bar{\lambda}^* = 104;54^\circ,$$

$$\lambda_{A3}^* = 154;22^\circ, \qquad \lambda_{A6}^* = 159;1^\circ, \qquad \Delta\lambda_A^* = 4;39^\circ.$$

Copernicus, however, gives

$$\Delta\bar{\alpha} = 1;5^\circ, \qquad \Delta\bar{\lambda}^* = 104;54^\circ, \qquad \Delta\lambda_A^* = 4\tfrac{1}{2}^\circ.$$

He says that for the interval Δt, "according to the calculation set out above," presumably referring to the mean motions in V,1, there corresponds $\Delta\bar{\alpha} = 1274^r + 1;5^\circ$, and indeed the tables give $1274^r + 1;5,46\ldots^\circ$. But as was true for Saturn, this does not explain how the value of $\bar{\alpha}$ underlying the tables in V,1 was derived since even the difference of $+0;0,46^\circ$ shows that the tables do not come directly from the observations.

Correctly, the observations lead to the following values. For $\Delta\bar{\alpha} = 1274^{\mathrm{r}} + 1;5°$ in $\Delta t = 1392^{\mathrm{y}}\, 99;37^{\mathrm{d}}$, the annual motion, from (37), is

$$\bar{\alpha}^{\mathrm{y}} = 329;25,8,13,3,54\ldots°,$$

and the period of the anomaly, from (38),

$$\mathrm{P}_{\alpha} = 398;53,2,57,57\ldots^{\mathrm{d}}.$$

Note that $\bar{\alpha}^{\mathrm{y}}$ agrees with neither N nor M, while P_{α} rounds to 398;53,2,58, the value in N (see Tables 1(a) and 1(b)). Taking the sidereal year as $365;15,24,10^{\mathrm{d}}$, the correction to Jupiter's period relation, from (39), is given by

$$65\mathrm{P}_{\alpha} = 25927;28,12,46^{\mathrm{d}} = 71^{\mathrm{y}} - 5;45,23,4^{\mathrm{d}},$$

or $c^{\mathrm{d}} \approx -5;45,23^{\mathrm{d}}$, which agrees with the emended $-5;45,23^{\mathrm{d}}$ in N (and further confirms the emendation).

Copernicus rounds the sidereal motion of the apogee to $4\frac{1}{2}°$ in 1392^{y}, which he says is equivalent to very nearly $1°/300^{\mathrm{y}}$. In fact, $4;30°(300/1392) = 0;58,11°$, but the correct motion of 4;39° would give $4;39°(300/1392) = 1;0,7°$, which is more likely what he has in mind.

Ptolemy's third opposition (3) on Antoninus 1 Athyr 20, 5^{h} after midnight in Alexandria, but 4^{h} after midnight at the meridian of Cracow, follows the Epoch of Christ by

$$\Delta t = 136^{\mathrm{y}}\, 314^{\mathrm{d}}\, 4^{\mathrm{h}} = 136^{\mathrm{y}}\, 314;10^{\mathrm{d}},$$

during which time, from the table in V,1,

$$\Delta\bar{\alpha} = 84;31,9° \approx 84;31°.$$

At the time of Ptolemy's observation $\bar{\alpha}_3 = 182;47°$, and thus at the Epoch of Christ

$$\bar{\alpha}_{0\mathrm{G}} = 182;47° - 84;31° = 98;16°.$$

Copernicus then gives the epochs of $\bar{\alpha}$ computed correctly in the order G. → A. → C. For some reason, epoch D. is omitted. In the following table we give the values of $\bar{\alpha}$, $\bar{\lambda}^*$, $\bar{\lambda}$, λ_{A}^*, and $\bar{\lambda}_{\mathrm{A}}$ computed for the four standard epochs just as was done for Saturn. For the motion of λ_{A}^* we have used Copernicus's estimate of $1°/300^{\mathrm{y}} = 0;0,12^{°/\mathrm{y}}$.

Epoch	$\bar{\alpha}_0$	$\bar{\lambda}_0^*$	$\bar{\lambda}_0$	$\lambda_{\mathrm{A}0}^*$	$\bar{\lambda}_{\mathrm{A}0}$
A. Olympiad	27;18°	68;58°	63;41°	151;20°	145;56°
C. Alexander	138;10	87;27	88;28	152;50	153;52
D. Caesar	23;35	248;29	253;24	153;46	158;41
G. Christ	98;16	174;15	179;46	153;55	159;27

(50)

Distance and Equation of the Anomaly (V,14)

In order to determine the radius of Jupiter's orbit, Copernicus follows the procedure used by Ptolemy in *Almagest* XI,2 and *Epitome* XI,7 for finding the radius of Jupiter's epicycle, and takes one observation of the planet outside of opposition. He says that on 18 Feb. 1520 at 6^h after midnight he saw "with an instrument" (*per instrumentum*) that Jupiter preceded, i.e. was west of, the first, more bright star in the forehead of Scorpio (β Scorpii) by 4;31°. Since the "instrument" gives a direct measurement of the difference in longitude, that is, it measures parallel to the ecliptic, one would suppose that Copernicus is referring to the armillary. The measurement of 4;31° rather than, say, $4\frac{1}{2}°$ seems to be asking a lot of the precision of any instrument Copernicus may have used. Nevertheless, the measurement is excellent. Modern computation gives for β Scorpii $\lambda = 236;29°$, $\beta = +1;5°$, and for Jupiter $\lambda = 231;58°$, $\beta = +1;9°$.[65] Hence $\Delta\lambda = 4;31°$, in exact agreement with Copernicus's report of his observation. Since from the star catalogue the sidereal longitude of β Scorpii is 209;40°, the observed sidereal longitude of Jupiter

$$\lambda^* = 209;40° - 4;31° = 205;9°.$$

The time from the Epoch of Christ

$$\Delta t = 1520^y\, 62^d\, 6^h = 1520^y\, 62;15^d,$$

for which, from the tables in III,14 and V,1,

$$\bar{\lambda}^*_{\odot} = 309;16°, \quad (\text{Accur. } 309;16,33°)$$
$$\bar{\alpha} = 111;15°, \quad (\text{Accur. } 111;15,56°)$$
$$\bar{\lambda}^* = \bar{\lambda}^*_{\odot} - \bar{\alpha} = 198;1°, \qquad \lambda^*_A = 159°,$$
$$\bar{\kappa} = \lambda^* - \lambda^*_A = 39;1°.$$

The configuration for the demonstration is shown in Fig. 41. Letting $e_1 = 687$ and $r' = 229$ where $R = 10{,}000$, the equation of center, from (6–10),

$$c_3 = -2;21° - 0;50° = -3;11°,$$

and thus the heliocentric longitude of Jupiter,

$$\lambda^{*\prime} = \bar{\lambda}^* + c_3 = 194;50°.$$

But by observation, the geocentric longitude $\lambda^* = 205;9°$, and thus the correction due to "parallax," the equation of the anomaly,

$$c = \lambda^* - \lambda^{*\prime} = 10;19°.$$

Now, where $R = 10{,}000$, by (6–8),

$$\bar{S}C = 10543, \qquad \bar{S}P = 10373, \quad (\text{Corr. } 10353)$$

[65] From P. V. Neugebauer (1912) and Tuckerman (1964). Jupiter was approaching first station.

and in triangle $P\bar{S}O$, from (44),

$$\gamma = \bar{\lambda}^*_{\odot} - \lambda^* = 104;7^\circ.$$

Consequently, from (45),

$$r = \bar{S}P\,(\sin c/\sin \gamma) = 1916, \quad \text{(Accur. 1915.53)}$$

where $R = 10{,}000$, or where $R = 60$, $r = 11;29,45 \approx 11;30$, which is Ptolemy's value for the radius of Jupiter's epicycle, as Copernicus points out. Rather than computing the relative distances of Jupiter in its model, Copernicus then says that where $r = 1$,

$$R = 10{,}000/1916 = 5;13,9,$$
$$e_1 = 5;13,9(687/10{,}000) = 0;21,30, \quad \text{(N \& M 0;21,29)}$$
$$r' = 5;13,9(229/10{,}000) = 0;7,10.$$

Therefore, the distances of Jupiter at apogee and perigee are respectively

$$M = R + (e_1 - r') = 5;27,29, \qquad m = R - (e_1 - r') = 4;58,49,$$

and the mean distance $\mu = 5;13,9$. The maximum parallax, or equation of the anomaly corresponding to these distances is, from (48),

$$c_{\mathrm{m}} = 10;35, \quad \text{(Corr. 10;33)}$$
$$c_{\mathrm{M}} = 11;35^\circ, \qquad c_{\mu} = 11;3^\circ.$$

C. Mars

Observations

The observations of Mars are shown in Table 4. In order to find sidereal longitudes for Ptolemy's oppositions 1/2/3, Copernicus subtracts a precession of 6;40°, i.e.

$$\lambda^*_{\mathrm{p}} = \lambda_{\mathrm{p}} - \pi = \lambda_{\mathrm{p}} - 6;40^\circ.$$

This means that he has made the precession stop for nearly nine years, which produces errors of as much as 0;4° compared to the strictly correct $\lambda^*_{\mathrm{p}} = \bar{\lambda}^*_{\odot} + 180^\circ$. Copernicus's longitudes 4/5/6 also show discrepancies of up to 0;4°, but these are harder to explain. In the case of (4), he computes the sun to be at 55;33° instead of 55;34°, but for (5) and (6) he does not report the solar longitude so one cannot be sure where he went wrong.

Review of Ptolemy's Derivation (V,15)

Mars is the first superior planet treated by Ptolemy, in *Almagest* X,7–10, and in the *Epitome* it is Mars that is used for the most detailed exposition of the method of deriving the eccentricity and the direction of the apsidal line, in X,13–20, and the remaining elements in X,23–25.[66] The demonstration of the eccentricity uses observations 1/2/3, which Copernicus reduces, as noted, by $\lambda_p^* = \lambda_p - 6;40°$, from which, by (23a)

$$\delta_1 = 67;50°, \qquad \delta_2 = 93;44°,$$

as in the *Almagest*. The intervening time

$$\Delta t_{12} = 4^y\, 69^d\, 20^h, \qquad \Delta t_{23} = 4^y\, 96^d\, 1^h,$$

during which, directly from the *Almagest*,

$$\bar{\delta}_1 = 81;44°, \qquad \bar{\delta}_2 = 95;28°.$$

Mars required three solutions for $2e$ and $\bar{\kappa}$ to produce agreement with the observed values of δ. In the first Ptolemy found that, where $R = 60$, $2e = 13;7$, in the second $2e = 11;50$, and in the third $2e = 12$.

Copernicus then shows that in his own model, Ptolemy's results are satisfactory. In agreement with Ptolemy, in Fig. 42 he sets

$$2e = 2000, \qquad e_1 = 1500, \qquad r' = 500,$$
$$\bar{\kappa}_1 = -41;33°, \qquad \bar{\kappa}_2 = 40;11°, \qquad \bar{\kappa}_3 = 180° - 44;21° = 135;39°.$$

The equation of center, from (6–10), is

$$c_{3(1)} = +5;7° + 1;56° = +7;3°, \qquad \text{(Corr. } +7;2°)$$
$$c_{3(2)} = -4;58° - 1;53° = -6;51°, \qquad \text{(Corr. } -6;50°)$$
$$c_{3(3)} = -6;42° - 1;52° = -8;34°,$$

and thus the true eccentric anomaly, from (25),

$$\kappa_1 = -34;30°, \qquad \kappa_2 = 33;20°, \qquad \kappa_3 = 127;5°.$$

The intervals of longitude between the oppositions are therefore

$$\delta_1 = 67;50° \quad \text{compared with} \quad 67;50°, \qquad \Delta = 0;0°,$$
$$\delta_2 = 93;45° \quad \text{compared with} \quad 93;44°, \qquad \Delta = +0;1°,$$

which is satisfactory agreement.

Then, in Fig. 43, we consider the third opposition, in which

$$\lambda_3^* = 235;54°, \qquad \bar{\kappa}_3 = 135;39°, \qquad c_{3(3)} = -8;34°, \qquad \kappa_3 = 127;5°.$$

Since $\bar{\kappa} < 180°$, the mean anomaly, from (32),

$$\bar{\alpha}_3 = 180° - 8;34° = 171;26°,$$

the mean sidereal longitude, from (33),

$$\bar{\lambda}_3^* = 235;54° + 8;34° = 244;28° \approx 244;30°,$$

66 X,21–22 give Jābir's method of finding the eccentricity and direction of the apsidal line. See above, p. 308, n. 29.

and the longitude of the apogee, from (34),

$$\lambda_A^* = 235;54^\circ - 127;5^\circ = 108;49^\circ \approx 108;50^\circ.$$

Adding the precession of 6;40° used to reduce Ptolemy's observations, we find the tropical longitude, from (35)

$$\lambda_A = 108;50^\circ + 6;40^\circ = 115;30^\circ = ♋ 25;30^\circ,$$

as in the *Almagest*.

Copernicus's Derivation (V,16)

From oppositions 4/5/6, by (23a),

$$\delta_1 = 187;29^\circ, \qquad \delta_2 = 70;18^\circ,$$

and the intervals of time

$$\Delta t_{45} = 6^y\ 191^d\ 19^h = 6^y\ 191;47,30^d \approx 6^y\ 191;45^d,$$

$$\Delta t_{56} = 4^y\ 72^d\ 9^h = 4^y\ 72;22,30^d \approx 4^y\ 72;23^d.$$

The rounding in Δt_{45} is equal to 1^h, or about 0;1,9° in $\Delta\bar{\alpha}$, but using the accurate Δt to estimate $\Delta\bar{\alpha}$ from the tables in the *Almagest*,

$$\Delta\bar{\alpha}_{45} = 19;22,36^\circ \approx 19;22^\circ, \qquad \Delta\bar{\alpha}_{56} = 347;18,24^\circ \approx 347;18^\circ,$$

so that, from (24a),

$$\bar{\delta}_1 = 168;7^\circ, \qquad \bar{\delta}_2 = 83^\circ,$$

which are Copernicus's values. Hence he did not use the rounded values of Δt in the text. Now Copernicus does not show his derivation of $2e$ and $\bar{\kappa}$, "that we may," he says, "pass over in silence the multitude, complexity, and tedium of the calculations." He reports simply that by the same method used for Saturn and Jupiter, he has found that where $R = 10{,}000$,

$$2e = 1960, \qquad e_1 = 1460, \qquad r' = 500,$$

$$\bar{\kappa}_4 = 125;29^\circ, \qquad \bar{\kappa}_5 = -66;24^\circ,[67] \qquad \bar{\kappa}_6 = 16;36^\circ.$$

Note that $2e$ is less than Ptolemy's value of 2000, and that $e_1/r' < 3/1$, so the division is not quite equivalent to a bisected eccentricity, a matter to be investigated.[68] Copernicus then shows that these values will reproduce the observations. But before examining his demonstration, let us take upon ourselves the "multitude, complexity, and tedium of the calculations" to see whether in fact Copernicus's results were derived strictly from the iterative procedure.

---

[67] N and M read 66;*18*°, but $\bar{\kappa}_4 + \bar{\delta}_1 - 360^\circ = \bar{\kappa}_6 - \bar{\delta}_2 = -66;24^\circ$, while $\kappa_5 - c_{3(5)} = -66;25^\circ$.

[68] Below, pp. 356–359.

In Fig. 44(a), we let the mean sun be at D, and drawing a circle around the equant F, we let the first opposition be at A, the second at B, and the third at C. We are given, by observation and (23a)–(23b),

$$\delta_1 = 187;29^\circ, \qquad \delta_2 = 70;18^\circ, \qquad \delta_3 = 102;13^\circ,$$

and by computation, from (24a)–(24b),

$$\bar{\delta}_1 = 168;7^\circ, \qquad \bar{\delta}_2 = 83;0^\circ, \qquad \bar{\delta}_3 = 108;53^\circ.$$

In triangle CDE,

$$2\beta_1 = 2(180^\circ - \delta_3) = 155;34^\circ,$$

$$2\gamma_1 = 2\delta_3 - \bar{\delta}_3 = 95;33^\circ,$$

$$CE = \text{crd } 2\beta_1 = 19547, \qquad DE = \text{crd } 2\gamma_1 = 14810.$$

In triangle BDE,

$$2\beta_2 = 2(180^\circ - (\delta_2 + \delta_3)) = 14;58^\circ,$$

$$2\gamma_2 = 2(\delta_2 + \delta_3) - (\bar{\delta}_2 + \bar{\delta}_3) = 153;9^\circ,$$

$$BE = \text{crd } 2\beta_2 = 2605 \qquad DE = \text{crd } 2\gamma_2 = 19453.$$

Thus, in units of triangle CDE, where $DE = 14810$,

$$BE = \text{crd } 2\beta_2(\text{crd } 2\gamma_1/\text{crd } 2\gamma_2)$$

$$= 2605(14810/19453) = 1983.$$

In triangle CBE,

$$BC = (CE^2 + BE^2 - 2\,CE \cdot BE \cos \tfrac{1}{2}\bar{\delta}_2)^{1/2} = 18110.$$

Now, letting the radius of the eccentric $R = 10{,}000$,

$$BC = \text{crd } \bar{\delta}_2 = 13252,$$

and thus

$$CE = 19547(13252/18110) = 14304,$$

$$DE = 14810(13252/18110) = 10837.$$

Therefore,

$$\text{arc } CBE = 2 \sin^{-1} \tfrac{1}{2}\,CE = 91;19^\circ,$$

$$\text{arc } EBCA = \text{arc } CBE + \bar{\delta}_3 = 91;19^\circ + 108;53^\circ = 200;12^\circ,$$

$$\text{arc } AE = 360^\circ - \text{arc } EBCA = 159;48^\circ.$$

Hence,

$$ADE = \text{crd } 159;48^\circ = 19690,$$

$$AD = ADE - DE = 19690 - 10837 = 8853.$$

So, in Fig. 44(b),

$$AD \cdot DE = R^2 - 4e^2 = 8853 \cdot 10837 = 95939961,$$

$$4e^2 = R^2 - AD \cdot DE = 4060039,$$

$$2e = 2015,$$

or, where $R = 60$, $2e = 12;5$. It is already apparent that the double eccentricity will be close to Ptolemy's $2e = 12$.

Drawing FKL perpendicular to ADE,

$$DK = DE - \tfrac{1}{2}ADE = 10837 - \tfrac{1}{2}(19690) = 992,$$

$$\gamma_3 = \sin^{-1}(DK/2e) = 29;30^\circ,$$

$$\text{arc } AHL = \tfrac{1}{2} \text{ arc } ALE = \tfrac{1}{2}(159;48^\circ) = 79;54^\circ,$$

$$\text{arc } AH = \text{arc } AHL - \gamma_3 = 50;24^\circ.$$

Thus, the mean eccentric anomaly at each opposition,

$$\bar{\kappa}_4 = 180^\circ - \text{arc } AH = 129;36^\circ,$$

$$\bar{\kappa}_6 = \bar{\kappa}_4 - \bar{\delta}_3 = 20;43^\circ, \qquad \bar{\kappa}_5 = \bar{\kappa}_6 - \bar{\delta}_2 = -62;17^\circ.$$

We now test for δ, letting

$$2e = 2015, \qquad e_1 = 1511, \qquad r' = 504,$$

$$\bar{\kappa}_4 = 129;36^\circ, \qquad \bar{\kappa}_5 = -62;17^\circ, \qquad \bar{\kappa}_6 = 20;43^\circ.$$

From (6–10, 25), the true eccentric anomaly,

$$\kappa_4 = 120;11^\circ, \qquad \kappa_5 = -52;37^\circ, \qquad \kappa_6 = 16;59^\circ,$$

and by (26),

$$\delta_1 = 187;12^\circ \quad \text{compared with} \quad 187;29^\circ, \qquad \Delta = -0;17^\circ,$$

$$\delta_2 = 69;36^\circ \quad \text{compared with} \quad 70;18^\circ, \qquad \Delta = -0;42^\circ,$$

so the agreement is not satisfactory and another solution is required. From (6–10, 28–30), we find the correction

$$\varepsilon_4 = +0;40,50^\circ, \qquad \varepsilon_5 = +0;24,25^\circ, \qquad \varepsilon_6 = -0;17,41^\circ,$$

leading to, by (31),

$$\delta_1' = 187;45^\circ, \qquad \delta_2' = 71;0^\circ, \qquad \delta_3' = 101;15^\circ.$$

Retaining the previous values of $\bar{\delta}$, we repeat the derivation of $2e$ and $\bar{\kappa}$, and find

$$2e = 1942, \qquad e_1 = 1456, \qquad r' = 486,$$

$$\bar{\kappa}_4 = 124;39^\circ, \qquad \bar{\kappa}_5 = -67;14^\circ, \qquad \bar{\kappa}_6 = 15;46^\circ.$$

Again we test for δ, finding from (6–10, 25),

$$\kappa_4 = 115;3^\circ, \qquad \kappa_5 = -57;26^\circ, \qquad \kappa_6 = 13;0^\circ,$$

resulting in, from (26),

$$\delta_1 = 187;30^\circ \quad \text{compared with} \quad 187;29^\circ, \qquad \Delta = +0;1^\circ,$$

$$\delta_2 = 70;26^\circ \quad \text{compared with} \quad 70;18^\circ, \qquad \Delta = +0;8^\circ.$$

Hence, another iteration is necessary.

From (6–10, 28–30), the correction

$$\varepsilon_4 = +0;35,2^\circ, \qquad \varepsilon_5 = +0;20,10^\circ, \quad \varepsilon_6 = -0;13,2^\circ,$$

so that, by (31),

$$\delta_1'' = 187;44^\circ, \qquad \delta_2'' = 70;51^\circ, \qquad \delta_3'' = 101;25^\circ.$$

Again, the original values of $\bar{\delta}$ are used to find

$$2e = 1958, \qquad e_1 = 1468, \qquad r' = 490,$$

$$\bar{\kappa}_4 = 125;37^\circ, \qquad \bar{\kappa}_5 = -66;16^\circ, \qquad \bar{\kappa}_6 = 16;44^\circ.$$

Note that $2e$ is now very close to Copernicus's value $2e = 1960$, although we are dividing e_1 and r' in the standard ratio. To test for δ, from (6–10, 25),

$$\kappa_4 = 116;2^\circ, \qquad \kappa_5 = -56;29^\circ, \qquad \kappa_6 = 13;47^\circ,$$

which give, by (26),

$$\delta_1 = 187;29^\circ \quad \text{compared with} \quad 187;29^\circ, \qquad \Delta = 0;0^\circ,$$

$$\delta_2 = 70;16^\circ \quad \text{compared with} \quad 70;18^\circ, \qquad \Delta = -0;2^\circ.$$

If $2e$ be divided into $e_1 = 1458$ and $r' = 500$, after the manner used by Copernicus, we find

$$\delta_1 = 187;29^\circ, \qquad \Delta = 0;0^\circ, \qquad \delta_2 = 70;15^\circ, \qquad \Delta = -0;3^\circ,$$

so the unusual division produces no improvement, and another iteration is in order. Using the standard division, $e_1 = 1468$ and $r' = 490$, we find the correction from (6–10, 28–30),

$$\varepsilon_4 = +0;36,16^\circ, \qquad \varepsilon_5 = +0;21,4^\circ, \qquad \varepsilon_6 = -0;13,58^\circ,$$

leading to, by (31),

$$\delta_1''' = 187;44^\circ, \qquad \delta_2''' = 70;53^\circ, \qquad \delta_3''' = 101;23^\circ.$$

Then for a fourth time with the original values of $\bar{\delta}$, we solve for $2e$ and $\bar{\kappa}$, finding

$$2e = 1954, \qquad e_1 = 1466, \qquad r' = 488,$$

$$\bar{\kappa}_4 = 125;25^\circ, \qquad \bar{\kappa}_5 = -66;28^\circ, \qquad \bar{\kappa}_6 = 16;32^\circ.$$

Now we test for δ by finding, from (6–10, 25)

$$\kappa_4 = 115;50^\circ, \qquad \kappa_5 = -56;41^\circ, \qquad \kappa_6 = 13;37^\circ,$$

which give, from (26),

$$\delta_1 = 187;29^\circ \quad \text{compared with} \quad 187;29^\circ, \qquad \Delta = 0;0^\circ$$

$$\delta_2 = 70;18^\circ \quad \text{compared with} \quad 70;18^\circ, \qquad \Delta = 0;0,$$

so at last the agreement is perfect, and the fourth solution terminates the derivation.

We may now examine Copernicus's results, namely,

$$2e = 1960, \qquad e_1 = 1460, \qquad r' = 500,$$

$$\bar{\kappa}_4 = 125;29^\circ, \qquad \bar{\kappa}_5 = -66;24^\circ, \qquad \bar{\kappa}_6 = 16;36^\circ.$$

$2e$ is almost identical to the result of our third solution, $2e = 1958$, and $\bar{\kappa}$ differs by $-0;8^\circ$, which is also very close. By good fortune $\bar{\kappa}$ is only $+0;4^\circ$ from the result of the last iteration, but this is not pertinent to the question of how Copernicus reached his results. We must conclude that he actually carried out three iterations, and did them quite accurately, since the differences in $2e$ and $\bar{\kappa}$ are easily within the range that would be produced by small errors and roundings in the course of the "multitude, complexity, and tedium of the calculation." More significantly, while for Saturn and Jupiter Copernicus might have taken over Ptolemy's $2e$ and then used trial and error to find suitable values of $\bar{\kappa}$, in the case of Mars there is no motivation for reducing Ptolemy's $2e$ from 2000 to 1960, and at the same time finding appropriate values of $\bar{\kappa}$, by pure trial and error. With two independent variables involved, it is difficult to imagine what kind of trial and error could be used, for it appears far easier to solve the problem strictly. Indeed, only by performing the derivation could Copernicus reach the surprising result—particularly in light of his treatment of Saturn and Jupiter—that Ptolemy's $2e$ was larger than required by oppositions 4/5/6. Finally, that his results are so close to our third solution provides further evidence that he also carried out three solutions. Hence, if we consider only Mars, and ignore Copernicus's earlier, intemperate remarks about his difficulties with Jupiter, it appears that he understood Ptolemy's method thoroughly and could apply it effectively. This was no inconsiderable accomplishment, for Copernicus may have been the first person—certainly the first modern European—to do so.

The confirmation of the elements is shown in Fig. 45. Copernicus computes the equation of center, from (6–10), and finds

$$c_{3(4)} = -7;24^\circ - 2;12^\circ = -9;36^\circ,$$

$$c_{3(5)} = +7;11^\circ + 2;36^\circ = +9;47^\circ, \quad \text{(Corr. 9;49)}$$

$$c_{3(6)} = -2;6^\circ - 0;50^\circ = -2;56^\circ.$$

The true eccentric anomaly, from (25), is then

$$\kappa_4 = 115;53°, \qquad \kappa_5 = -66;25 + 9;47° = -56;38°,$$
$$\kappa_6 = 13;40°.$$[69]

Therefore, from (26), the differences in longitude,

$$\delta_1 = 187;29° \quad \text{compared with} \quad 187;29°, \qquad \Delta = 0;0°,$$

$$\delta_2 = \ 70;18° \quad \text{compared with} \quad 70;18°, \qquad \Delta = 0;0°,$$

and the agreement is perfect. However, with the correct $\kappa_5 = -66;24° + 9;49° = -56;35°$, $\delta_1 = 187;32°$ and $\delta_2 = 70;15°$, leaving residual errors of $+0;3°$ and $-0;3°$, which is hardly perfect.

To find the mean anomaly, Copernicus considers the third opposition as shown in Fig. 46,[70] for which

$$\lambda_6^* = 133;20°, \qquad \bar{\kappa}_6 = 16;36°, \qquad c_{3(6)} = -2;56°, \qquad \kappa_6 = 13;40°.$$

The mean anomaly, from (32), is

$$\bar{\alpha}_6 = 180° - 2;56° = 177;4°,$$

the mean sidereal longitude, from (33),

$$\bar{\lambda}_6^* = 133;20° + 2;56° = 136;16°,$$

and the sidereal longitude of the apogee, from (34),

$$\lambda_A^* = 133;20° - 13;40° = 119;40°,$$

all for 22 Feb 1523, 5^h after midnight at the meridian of Frauenburg–Cracow.

Reduction and Division of the Eccentricity

For Jupiter and Saturn Copernicus retained Ptolemy's double eccentricity of 917 and 1139 respectively, and divided them in the standard way, $e_1 = \frac{3}{4} \cdot 2e$, $r' = \frac{1}{4} \cdot 2e$, thereby reproducing the equivalent of the bisected eccentricity in Ptolemy's model. However, for Mars Ptolemy found $2e = 2000$ and Copernicus found $2e = 1960$. The standard division of 2000 gives $e_1 = 1500$ and $r' = 500$, which Copernicus uses, and the standard division of 1960 would give $e_1 = 1470$ and $r' = 490$. But Copernicus instead divides 1960 into $e_1 = 1460$ and $r' = 500$, thus keeping the epicycle the same size it was following the division of Ptolemy's 2000, and takes the entire reduction of 40 away from e_1. The model is thus no longer equivalent to a bisected eccentricity, but now implies a division in the ratio

$$(e_1 + r')/(e_1 - r') = 1960/960 = 2;2,30/1 = 49/24.$$

---

[69] Note that now $\bar{\kappa}_5 = -66;25°$. Copernicus says that κ_6 is approximate, although it is exact.
[70] The diagram in N is incorrect; see above, p. 344, n. 63.

The difference in the division compared with the equivalent of a strict bisection is so slight as to have only a negligible effect, less than 0;1°, on the equation of center. However, the reduction of $2e$ from 2000 to 1960 reduces the maximum equation of center from 11;22° to 11;8°, which is not negligible.

As we have seen, there is no question that Copernicus's reduction of $2e$ to 1960 is entirely legitimate, following strictly from the observations on the third solution. However, why he should choose to give the entire reduction to e_1, while leaving r' the same as it was for $2e = 2000$, is not so obvious. The motivation is found in the passage from V,4 cited earlier concerning Copernicus's reason for selecting a model consisting of an eccentric plus epicycle rather than some equivalent, such as a double epicycle.[71] There he pointed out that the motion of the center of the earth's orbit with respect to the true sun has not been compensated by a similar motion of the center of the planetary orbits, that is, that the mean sun has shifted position with respect to the true sun and the centers of the planetary orbits. Hence in Fig. 14, $\bar{S}$ can move with respect to M, changing both the eccentricity e_1 and the direction of the apsidal line. Since the sum $e_1 + r' = 2e$, and the motion of $\bar{S}$ does not affect the radius r', any change in $2e$, as Copernicus has found for Mars, must be attributed entirely to a change in e_1. Hence the division of $2e = 1960$ into $e_1 = 1460$ and $r' = 500$, all the reduction from the earlier $2e = 2000$ being taken by e_1.

Copernicus says something more on this subject in V,16, specifically about the effect of the motion of the center of the earth's orbit on the eccentricity of Mars, following his confirmation of $2e$ and $\bar{\kappa}$. He reports that he finds the sidereal longitude of Mars's apogee at 119;40°, and then continues:

> In the time of Antoninus, Ptolemy found the apogee at 108;50°, and therefore up to our time it has moved forward by 10;50°. We also find the distance of the centers smaller by 40 parts, where the radius of the eccentric is given 10,000 parts, not because Ptolemy or we were in error, but, by clear evidence, because the center of the great sphere of the earth has approached the center of the sphere of Mars, while in the meantime the sun has remained unmoved. For these things correspond approximately to each other, as will appear more clearly elucidated below. (N, 158r10–18).

Here Copernicus says that the reduction in Mars's eccentricity has occurred because the center of the earth's orbit, the mean sun, has approached the center of Mars's orbit while the true sun, and presumably also the distance between the true sun and the center of Mars's orbit, has remained fixed. Unfortunately, he does not explain anything more of this interesting subject, although we can to some extent investigate his claim. For if the reduction of e_1 for Mars from 1500 to 1460 is entirely due to the motion of the mean sun with respect to the true sun and the center of Mars's orbit, then the distance between the true sun and the center of Mars's orbit ought to remain

[71] Above, p. 300.

unchanged. We can find the distance at the time of Ptolemy and Copernicus to see whether Copernicus's claim is true.

In Fig. 47, S is the true sun, $\bar{S}$ the mean sun, the center of the earth's eccentric, and M the center of Mars's eccentric. At the time of Ptolemy's third opposition (3), 8 Oct 137, by Copernicus's solar theory of Book III, the sun's eccentricity is 415 and the sidereal longitude of its apogee 59;49°. But in order to compare this eccentricity with that of Mars, the solar eccentricity must be measured in terms of the radius of Mars's eccentric. Anticipating Copernicus's demonstration in V,19, where the radius of Mars's eccentric $R = 10{,}000$, the radius of the earth's orbit $r = 6580$. Thus, we shall have for the eccentricities and apsidal longitudes of the sun and Mars measured from the mean sun at the time of Ptolemy,

$$e_S = .658 \cdot 415 = 273, \qquad \lambda^*_{AS} = 59;49°,$$

$$\bar{e}_M = 1500, \qquad \lambda^*_{AM} = 108;50°.$$

We establish the same quantities for the time of Copernicus's third opposition (6), 22 Feb 1523, finding

$$e_S = .658 \cdot 323 = 212.5, \qquad \lambda^*_{AS} = 69;35°$$

$$\bar{e}_M = 1460, \qquad \lambda^*_{AM} = 119;40°.$$

Now, in triangle $SM\bar{S}$,

$$\gamma_1 = \lambda^*_{AM} - \lambda^*_{AS},$$

and the eccentricity SM of Mars with respect to the true sun,

$$e_M = (\bar{e}_M^2 + e_S^2 - 2\bar{e}_M e_S \cos \gamma_1)^{1/2}.$$

To find the sidereal longitude of Mars's aphelion referred to the true sun $\lambda^{*\prime}_{AM}$, we find the difference d from

$$d = \sin^{-1}\left(\frac{e_S}{e_M} \sin \gamma_1\right),$$

and then

$$\lambda^{*\prime}_{AM} = \lambda^*_{AM} + d.$$

From the values established for Ptolemy's time, we find

$$\gamma_1 = 49;1°, \qquad e_M = 1337,$$

$$d = 8;52°, \qquad \lambda^{*\prime}_{AM} = 117;42°,$$

and from the values for Copernicus's time,

$$\gamma_1 = 50;5°, \qquad e_M = 1334,$$

$$d = 7;1° \qquad \lambda^{*\prime}_{AM} = 126;41°.$$

These are highly interesting results. For the time of Ptolemy and Copernicus respectively, the eccentricity of Mars is 1337 and 1334, that is, the values are practically identical. Thus, the investigation, which Copernicus himself must likewise have carried out, confirms his statement that the motion of the center of the earth's orbit accounts for the diminution of Mars's eccentricity from 1500 to 1460. Note further that the longitude of the aphelion of Mars with respect to the true sun has increased by 8;59°, or about 9°. Hence, while the eccentricity has remained, for all practical purposes, constant, the apsidal line has advanced 1;50° less than the 10;50° advance found for the motion of the apsidal line passing through the mean sun.

We can further examine the problem in the opposite way by taking as given a single value of Mars's eccentricity with respect to the true sun, and the values of $\lambda^{*\prime}_{AM}$ just found, and solving for the eccentricity referred to the mean sun at the time of Ptolemy and Copernicus. This time, we use

$$\gamma_2 = \lambda^{*\prime}_{AM} - \lambda^{*}_{AS},$$

leading to

$$\bar{e}_M = (e_M^2 + e_S^2 + 2\, e_M e_S \cos \gamma_2)^{1/2}.$$

Let us assume that $e_M = 1335$. Then we find for the time of Ptolemy,

$$\gamma_2 = 57;53°, \qquad \bar{e}_M = 1498,$$

and for the time of Copernicus,

$$\gamma_2 = 57;6°, \qquad \bar{e}_M = 1461.$$

Hence we see even more clearly that the motion of the center of the earth's orbit with respect to a constant, although advancing, eccentricity of Mars of about 1335 referred to the true sun will nicely account for the decrease from 1500 to 1460 in Mars's eccentricity referred to the mean sun. "For these things," as Copernicus remarks, "correspond approximately to each other."

Since the eccentricity of Mars referred to the true sun has turned out to be essentially of constant length, although not direction, between the time of Ptolemy and Copernicus, and since in V,4 Copernicus suggests that the distances from the true sun to the centers of all the planetary orbits—except significantly the earth's—are of constant length,[72] one might wonder whether eccentricities and apsidal lines passing through the true sun had any further significance in his planetary theory. This question has been affirmatively, although cautiously, answered by Moesgaard, who has found that, while the evidence is admittedly ambiguous, some of Copernicus's eccentricities and apsidal longitudes for the superior planets agree better with modern values determined for the true sun rather than the mean sun.[73]

---

[72] Above, p. 300.

[73] Moesgaard (1974), 248–49, 273–74, 299 ff.

He first hypothesizes, and then concludes, although with awareness of the difficulties, that Copernicus's apsidal longitudes are in fact directions from the true sun, and that Copernicus was advancing from a mean-sun to a true-sun system for the superior planets, although not for Venus and Mercury.

These conclusions were drawn on the basis of comparisons between Copernicus's elements and modern elements, without attempting to analyze Copernicus's derivations of the elements of the orbits of the superior planets. However, the derivations we have examined here show that Copernicus did not derive eccentricities and apsidal longitudes referred to the true sun. For we have found that, beginning from Copernicus's observations of oppositions to the mean sun, three solutions of Ptolemy's iterative procedure have sufficed to yield almost exactly Copernicus's values of $2e$ and λ_A^*. Specifically, taking the results of our third solution for each planet, and computing the longitude of the apogee from

$$\lambda_A^* = \lambda_6^* - c_{3(6)} - \bar{\kappa}_6,$$

we reach the following comparisons:[74]

	Copernicus		Recomp.	
Planet	$2e$	λ_A^*	$2e$	λ_A^*
Saturn	1139	240;21°	1139	240;24°
Jupiter	917	159;1	915	158;53
Mars	1960	119;40	1958	119;33

Since these values of $2e$ and λ_A^* were based upon oppositions to the mean sun, they can only be eccentricities and apsidal lines passing through the mean sun, and cannot properly be interpreted as referring to the true sun, however they may compare with modern elements. It may therefore be concluded that while Copernicus certainly speculated that eccentricities referred to the true sun may well be constant and unaffected by the motion of the center of the earth's orbit—and indeed he could prove this to be the case for Mars—he derived the elements of the orbits, whether strictly by Ptolemy's procedure or partially by trial and error, only with respect to the mean sun, the true sun playing no significant role in his planetary theory,

Nevertheless, the question that Copernicus raised in III,25 about the location of the "center of the world" has to some extent been answered,[75] for it

[74] Above, pp. 331 f., 344 f., 355 f. In the case of Jupiter, we have used the values that follow from continuing Copernicus's erroneous first solution. The correct solution gives $2e = 913$ and $\lambda_A^* = 158;40°$, which is also close to Copernicus's results.

[75] Above, pp. 159–161.

is the true sun that he considers to be fixed, and the center of the earth's orbit to be moving. And if there is any doubt about Copernicus's opinion, it is removed by the well-known passage in the *Narratio prima* in which Rheticus, presumably reflecting his teacher's beliefs, discusses at some length the effect of the motion of the center of the "great sphere" on the eccentricities of the five planets.[76] However, as treated by Copernicus and Rheticus, these are still only small modifications of a planetary theory referred to the mean sun—how, for example, they intended to reconcile the orbital planes of the latitude theory that intersect at the mean sun with the orbital centers at a constant distance from the true sun is difficult to know—and there is no reason to suppose that either Copernicus or Rheticus appreciated the significance of the change of Mars's eccentricity, as obvious or momentous as it may appear in retrospect. It was still a very long step, not only to Kepler's demonstrations in the *Astronomia nova*, but even to his speculations in the *Mysterium Cosmographicum*, and Copernicus was to remain quite unaware of his own riches.

Correction of the Mean Anomaly and Positions at Epoch (V,17–18)

From the third oppositions of Ptolemy (3) and Copernicus (6),

$$\bar{\alpha}_3 = 171;26^\circ, \qquad \bar{\alpha}_6 = 177;4^\circ, \qquad \Delta\bar{\alpha} = 5;38^\circ,$$

$$\bar{\lambda}_3^* = 244;30^\circ, \qquad \bar{\lambda}_6^* = 136;16^\circ, \qquad \Delta\bar{\lambda}^* = 251;46^\circ,$$

$$\lambda_{A3}^* = 108;50^\circ, \qquad \lambda_{A6}^* = 119;40^\circ, \qquad \Delta\lambda_A^* = 10;50^\circ.$$

The intervening time from 27 May 139, 21^h to 22 Feb 1523, 5^h,

$$\Delta t = 1384^y\,251^d\,8^h = 1384^y\,251;20^d,$$

although Copernicus writes, and actually uses, $251;19^d$, a curious error since the subtraction is trivial. He then says that from the earlier exposition of the mean motions, for this time $\Delta\bar{\alpha} = 648^r + 5;38^\circ$, and indeed from the tables in V,1 we find $\Delta\bar{\alpha} = 5;37,56^\circ \approx 5;38^\circ$. (The correct Δt gives $5;38,24^\circ \approx 5;38^\circ$.) But as was true for Jupiter and Saturn, this close agreement does not tell us exactly how the value in V,1 was derived. Directly from the observations, the annual motion for $\Delta\bar{\alpha} = 648^r + 5;38^\circ$ and $\Delta t = 1384^y\,251;19^d$, is, by (37),

$$\bar{\alpha}^y = 168;28,30,36,9,7\ldots^\circ,$$

[76] Text in Kepler (1937–) 1, 113; translation in Rosen (1971), 160–61; analysis in Moesgaard (1974), 300–02. While there are a number of problems in Rheticus explanation, for which the fault could lie with either teacher or student, the general meaning is unmistakable and clarifies any possible ambiguities in *De revolutionibus*.

which differs by +5 in the fourth fractional place from the value in N. This is at least better than Jupiter and Saturn where the differences occurred in the third place. The period of the anomaly, from (38),

$$P_\alpha = 779;56,12,50,41\ldots^{\mathrm{d}}$$

agrees with neither N nor M. From the sidereal year of 365;15,24,10^{d}, the correction to Mars's period relation, from (39), is given by

$$37P_\alpha = 28857;39,55,15^{\mathrm{d}} = 79^{\mathrm{y}} + 2;23,6,5^{\mathrm{d}},$$

or $c \approx +2;23,6^{\mathrm{d}}$, agreeing with neither N nor M. (The correct Δt does not produce agreement either.)

Copernicus now says that during the time between the observations the mean motion of the sun $\Delta\bar{\lambda}^*_\odot = 257;30^\circ$, although from the values in Table 4 it should be 257;32°. His error, however, can be explained. Since for opposition (3) Copernicus reports $\lambda^*_{\mathrm{p}} = 235;54^\circ$, he must have taken $\bar{\lambda}^*_{\odot 3} = 55;54^\circ$ instead of the correct 55;52°. And correctly for opposition (6) $\bar{\lambda}^*_{\odot 6} = 313;24^\circ$. Hence,

$$\Delta\bar{\lambda}^*_\odot = 313;24^\circ - 55;54^\circ = 257;30^\circ,$$

or so it would seem. Copernicus then says that the mean motion of Mars for Δt is 251;52°, which follows from

$$\Delta\bar{\lambda}^* = \Delta\bar{\lambda}^*_\odot - \Delta\bar{\alpha} = 257;30^\circ - 5;38^\circ = 251;52^\circ,$$

although from the mean longitudes of the planet found in V,15–16 it would be 251;46°, and from the correct $\Delta\bar{\lambda}^*_\odot = 257;32^\circ$ it would be 251;54°. That the number is not correct hardly matters since nothing is derived from it.

We gave above the sidereal longitude of Mars's apogee at the time of observations (3) and (6) along with its change of direction, but Copernicus does not mention Mars's apogee in the chapter on mean motions. The omission is deliberate. If indeed as Copernicus believes, and as we have shown, the reduction of the eccentricity of Mars is due to the motion of the center of the earth's orbit such that it approaches the center of Mars's orbit, then this motion must also cause a change in the direction of Mars's apsidal line.[77] However, the change of direction will be nonuniform, since the motion of the center of the earth's orbit has two components, and as was shown in the previous demonstration, the apsidal line of Mars passing through the true sun must itself rotate. Since Copernicus apparently worked out something like the same demonstration, he was aware of this. Evidently, he did not want to go through the trouble of developing a theory of the motion of Mars's apsidal line with respect to the mean sun, and in any case he probably considered the results of his demonstration tentative at best. Hence, the subject of the apsidal line is not raised in considering the mean motions, and

---

[77] See above, pp. 358–359.

indeed, Copernicus never explains how one is to find the longitude of Mars's apogee for a given time.

Ptolemy's observation (3) reduced to the meridian of Cracow occurred at 139 May 27, 21^h, which follows the Epoch of Christ by

$$\Delta t = 138^y\ 180;52,30^d,$$

although Copernicus rounds to $138^y\ 180;52^d$. However, he uses the difference in mean anomaly following from the unrounded interval, that is, from the tables in V,1,

$$\Delta\bar{\alpha} = 293;3,37^\circ \approx 293;4^\circ.$$ [78]

At observation (3) $\bar{\alpha} = 171;26^\circ$, and thus at the Epoch of Christ

$$\bar{\alpha}_{0G} = 171;26^\circ - 293;4^\circ = 238;22^\circ.$$

The remaining epochs are computed in the order G. → A. → C. → D., and contain both scribal and computational errors for D. We give the values in the text for $\bar{\alpha}_0$ followed by recomputations, and the following values of $\bar{\lambda}_0^*$ and $\bar{\lambda}_0$ are computed from the correct values of $\bar{\alpha}_0$. Since Copernicus does not suggest a mean motion for λ_A^*, its position at epoch is omitted from this table.

	Cop. $\bar{\alpha}_0$	Recomp. $\bar{\alpha}_0$	$\bar{\lambda}_0^*$	$\bar{\lambda}_0$
A. Olympiad	344;21°	344;20,35° ≈ 344;21°	111;55°	106;38°
C. Alexander	120;39	120;39,8 ≈ 120;39	104;58	105;59
D. Caesar	111;25	211;26,43 ≈ 211;27[79]	60;37	65;32
G. Christ	238;22	238;22	34;9	39;40

(51)

If, however, it is assumed, not legimately, that the apsidal line moves uniformly, its annual motion between the observations is about $0;0,28,10^{\circ/y}$, from which its positions at epoch or at any time can be estimated. In the *Prutenic Tables*, Reinhold uses $0;0,28,44,37^{\circ/y}$, a difference of $0;13,27^\circ$ in 1400 years.

Distance and Equation of the Anomaly (V,19)

In order to determine the radius of Mars's orbit, Copernicus uses an observation (7) of a near conjunction of the planet and α Librae that appears to have been made without the aid of any instrument. It is the earliest of

[78] N and M by error 293;*22*°, but never used. Nicely explained by Rosen in Copernicus (1978), 425 n. to 268:9.

[79] The scribal error of 100° has been dutifully retained in all editions and translations.

Copernicus's planetary observations used in *De revolutionibus*, and provides a curious lesson in the difficulty of estimating distance in an unaided observation.

> For this purpose we also observed a conjunction of Mars with the first, bright star of the Claws [of the Scorpion], called the "Southern Claw," that took place in the Year of Christ 1512 exactly on the Kalends of January. For in the morning at six equal hours before noon of that day, we saw Mars separated from the fixed star by $\frac{1}{4}°$, but inclined toward the solstial rising,[80] by which it was indicated that Mars had already drawn away from the star in longitude by $\frac{1}{8}°$ to the east (*in consequentia*), but in northern latitude by $\frac{1}{5}°$. It is known that the position of the star measured from the first star of Aries is 191;20° with a northern latitude of 0;40°. And so it was clear that the position of Mars was at 191;28° with a northern latitude of 0;51°.[81]

It is not clear how Copernicus decided to resolve the distance of $\frac{1}{4}° = 0;15°$ between the star and Mars into a longitudinal component of $\frac{1}{8}° = 0;7,30° \approx 0;8°$ and a latitudinal component of $\frac{1}{5}° = 0;12°$ which for some reason he converts to 0;11°.[82] In order to analyze the observation, we convert to tropical longitude by adding the precession according to Copernicus's tables for 1512 Jan 1, and find that for the star *s*,

$$\lambda_s = \lambda_s^* + \pi = 191;20° + 27;12° = 218;32°, \qquad \beta_s = +0;40°.$$

Hence, by Copernicus's estimate, the position of Mars *m* is

$$\lambda_m = 218;32° + 0;8° = 218;40°, \qquad \beta_m = +0;40° + 0;11° = +0;51°.$$

These locations are shown in Fig. 48 along with positions of α Librae *s*′ and Mars's *m*′ from modern computation, that is,

$$\lambda_{s'} = 218;17°, \qquad \beta_{s'} = +0;23°,$$

$$\lambda_{m'} = 218;38°, \qquad \beta_{m'} = +1;13°.$$[83]

The comparison is, to say the least, remarkable. The longitude of Mars is not 0;8° in advance of the star, but fully 0;21°, and the latitudes differ, not by 0;11°, but by 0;50°. And strangest of all, the distance between the two that

---

[80] *in ortu solstitialem deflexu*. About all this phrase can mean is toward the south-east part of the horizon. For the time of Copernicus's observation, from Tuckerman (1964) the sun is at 289;44°, and by Copernicus's solar theory at 289;38° which agrees quite well. The sun is thus about 20° past the winter solstice, and will rise at $8;17^h$, about $2\frac{1}{4}$ hours after the observation, with an ortive amplitude of 40° to the south, differing little from the ortive amplitude of the rising point of the solstice at 43° to the south. Either is in the south-east part of the horizon. Next, using the correct modern positions of the sun, α Librae, and Mars (below, n. 83), the midheaven is 203;6°, the ascendant 257;5°, and thus α Librae is 15;11° and Mars 15;32° east of the midheaven toward the ascendant. The altitude of α Librae is 21;14° and of Mars 21;58°. Hence, even though Mars has a slightly greater altitude than the star, an arc drawn from the star through Mars will appear to verge toward the south-east part of the horizon, in conformity with Copernicus's description.

[81] N, 159r7–18.

[82] In any case, from $\Delta\lambda = \frac{1}{8}°$ and $\Delta\beta = \frac{1}{5}°$, the separation would be only 0;14°.

[83] From P. V. Neugebauer (1912) and Tuckerman (1964).

Copernicus estimated as 0;15° is actually in excees of 0;54°, more than $3\frac{1}{2}$ times greater. It is difficult to know how he could underestimate the distance so severely.[84] If we compute the position of Mars from the tables in V,33 and VI,8, we find

$$\lambda^*_{m''} = 191;34°, \qquad \lambda_{m''} = 218;46°, \qquad \beta_{m''} = +0;57°.$$

Since both the longitudes and the latitude are 0;6° in excess of the estimates from the observation, it does not seem as though Copernicus computed the position and then adjusted the observation to fit. Fig. 48 shows that, by luck, the tropical longitude of Mars is only +0;2° from its correct position, but the tropical longitude is not of interest for Copernicus's demonstration.

The interval of time from the Epoch of Christ, which Copernicus fails to give, is

$$\Delta t = 1512^y\ 12;15^d.$$

By observation $\lambda^* = 191;28°$, and by computation

$$\bar{\lambda}^*_\odot = 262;1° \approx 262°, \qquad \bar{\alpha} = 98;28°, \qquad \lambda^*_A = 119;40°,$$

$$\bar{\lambda}^* = \bar{\lambda}^*_\odot - \bar{\alpha} = 163;32°, \qquad \bar{\kappa} = \bar{\lambda}^* - \lambda^*_A = 43;52°.$$

The positions are now shown in Fig. 49 in which, where $R = 10{,}000$, $e_1 = 1460$ and $r' = 500$. The equation of center, from (6–10) is

$$c_3 = -5;13° - 2° = -7;13°, \quad (\text{Corr. } -7;14°)$$

and thus the true heliocentric longitude of Mars,

$$\lambda^{*\prime} = \bar{\lambda}^* + c_3 = 156;19°.$$

By observation, the true geocentric longitude $\lambda^* = 191;28°$, so that the equation of the anomaly, or parallax,

$$c = \lambda^* - \lambda^{*\prime} = 35;9°.$$

Where $R = 10{,}000$, from (6–8),

$$\bar{S}C = 11097, \quad (\text{Corr. } 11099), \qquad \bar{S}P = 10776,$$

and since in triangle $P\bar{S}O$, from (44),

$$\gamma = \bar{\lambda}^*_\odot - \lambda^* = 70;32°,$$

[84] There is a possible, although curious explanation. At the time of the observation, the moon, at $\lambda \approx 81°$, $\beta \approx -4°$ about two days before full, was just setting on the western horizon (since the descendant was 77°), and because of the well-known illusion, the moon would appear very large. If Copernicus used the apparent diameter of the moon for comparison to estimate the distance between the star and Mars, it is possible—although how possible we do not know—that he could have mistakenly estimated the separation as about one-half the diameter of the setting, and illusively large, moon, which would give about $\frac{1}{4}°$. This admittedly speculative explanation cannot be proved.

the radius of the earth's orbit, from (45),

$$r = \bar{S}P(\sin c/\sin \gamma) = 10776(5757/9428) = 6580,$$

or, where $R = 60$, $r = 39;28,48$. Ptolemy found 39;30 for the radius of Mars's epicycle, and Copernicus remarks that his result is about the same. The greatest and least relative distances are, by (46),

$$M' = 10960, \qquad m' = 9040,$$

where $r = 6580$, and thus where $r = 1$, from (47),

$$M = 1;39,57, \qquad m = 1;22,26, \qquad \mu = 1;31,11.$$

The corresponding maximum equations of the anomaly, which in the case of Mars Copernicus does not give, are, from (48),

$$c_{\mathrm{m}} = 36;53^{\circ}, \qquad c_{\mathrm{M}} = 46;42^{\circ}, \qquad c_{\mu} = 41;9^{\circ}.$$

Conclusion to the Superior Planets

Having now subjected the reader to an *ingentem numerorum multitudinem*, it seems useful to summarize the results of Copernicus's demonstrations. The derivations of the radii of the planet's spheres, or their distances from the mean sun, are straightforward, aside from various problems in the observations and some computational errors, and in close agreement with Ptolemy, as indeed they should be since Ptolemy's ratios of the radii of the eccentrics and epicycles are essentially correct. On the other hand, the derivations of precisely the mean motions given in V,1 are uncertain, and there is little more we can say about them. However, the most interesting, and most difficult, derivations are of the double eccentricity and direction of the apsidal line of each planet, and it is these that are worthy of further discussion and summary. Consider first the values of $2e$ and λ_{A}^{*} found by Ptolemy, according to Copernicus's report, and by Copernicus:

	Ptolemy		Copernicus	
Planet	$2e$	λ_{A}^{*}	$2e$	λ_{A}^{*}
Saturn	1139	226;20°	1139	240;21°
Jupiter	917	154;22	917	159;0
Mars	2000	108;50	1960	119;40

The eccentricities of Saturn and Jupiter have remained constant while their apsidal lines have advanced about 14° and 4;40° respectively, but in the case of Mars, the eccentricity has decreased slightly and the apsidal line advanced about 11°. Looked at superficially, the results suggest that Copernicus could simply have adopted Ptolemy's eccentricities of Saturn and Jupiter, reduced the eccentricity of Mars—perhaps because he could show that the center of the earth's orbit was approaching the center of Mars's orbit—and used some kind of trial-and-error method to advance the apsidal lines to positions that could be confirmed for his own time, at least roughly, by a few observations, as he later attempted for Mercury.

However, our recomputation of the complete derivations shows this interpretation to be very unlikely, if not entirely false. Comparing Copernicus's solutions for $2e$ and $\bar{\kappa}_6$ for Saturn with our own recomputations, we find:

	Copernicus		Recomp.	
Solution	$2e$	$\bar{\kappa}_6$	$2e$	$\bar{\kappa}_6$
1st	1200	128;32°	1191	128;11°
2nd	?	?	1130	124;37
3rd	1139	125;18	1139	125;13

Considering the inevitable errors of computation and inaccuracies of rounding in the several hundred separate calculations that Copernicus had to carry out to derive the final results of the third solution, the agreement with the recomputation is excellent, and it is hard to imagine how Copernicus could have reached such results without actually going through the entire iterative procedure, and for the most part accurately.

Jupiter is more informative because Copernicus's error in the first solution does affect his final result. In the following comparison, recomputation (a) follows from Copernicus's erroneous first solution, while recomputation (b) is correct throughout:

	Copernicus		Recomp. (a)		Recomp. (b)	
Solution	$2e$	$\bar{\kappa}_6$	$2e$	$\bar{\kappa}_6$	$2e$	$\bar{\kappa}_6$
1st	1193	294;5°	[1193	294;5°]	947	308;24°
2nd	?	?	906	313;1	913	311;26
3rd	917	310;52	915	311;1	913	311;15

Two points are of interest here. First, the general agreement of the third solution in all three columns, even with Copernicus's error in the first solution, shows just how conveniently self-correcting Ptolemy's method is.[85] Second, the close agreement of recomputation (a) with Copernicus's results shows that, despite his exasperation at the improbably large eccentricity of his first solution, he probably did continue through two further iterations more-or-less accurately. Even if he did take over the precise $2e = 917$ from Ptolemy, it was probably because he had reached a result sufficiently close to warrant accepting Ptolemy's eccentricity as correct for his own time.

But the most interesting comparison is for Mars, for here Copernicus finds an eccentricity smaller than Ptolemy's:

	Copernicus		Recomp.	
Solution	$2e$	$\bar{\kappa}_6$	$2e$	$\bar{\kappa}_6$
1st	?	?	2015	16;59°
2nd	?	?	1942	15;46
3rd	1960	16;36°	1958	16;44
4th	—	—	1954	16;32

Copernicus gives only his final results for Mars, but the recomputation shows that on the third solution he must have reached a result very close to $2e = 1960$, and then rounded to that number. As we observed earlier, without actually carrying out the derivation, it is difficult to imagine why Copernicus would reduce Ptolemy's eccentricity when the eccentricities of Saturn and Jupiter had remained unchanged. (Of course we cannot be certain that Copernicus carried out the derivations in the order presented in his book.) It is possible that a prior assumption that the motion of the center of the earth's orbit was bringing it closer to the center of Mars's orbit—which could be demonstrated using Ptolemy's eccentricity and apsidal direction of Mars—could lead him to suspect that the eccentricity would diminish. But how much it would diminish also depends upon the motion of Mars's apsidal line, and this he could find only from carrying out the derivation. Surely trial-and-error methods would be even more trouble, and it is unlikely that their results would come so close to our third solution. Hence, for Mars also, Copernicus must have done the entire iterative derivation as far as the third solution, and done it accurately.

[85] Just to puzzle someone who might be reading this a generation, or just a decade, from now, one might say that the procedure, while not exactly user-friendly, is at least forgiving. In fact, the self-correction of the method makes it likely that one will reach an approximately correct result, that is, convergence will occur eventually and the original oppositions reproduced, but will leave one uncertain whether small inaccuracies remain due to earlier errors in the computation. This can be seen by comparing the third solutions of recomputations (a) and (b), although a fourth solution of (a) will decrease $2e$ and increase $\bar{\kappa}_6$ to produce a closer agreement with (b).

All in all, Copernicus's accomplishment in these derivations was considerable, whether his observations and final results are accurate by modern standards or not. He applied a very demanding method successfully, and derived directly from observation a new and, as far as he could know, verified set of elements for the superior planets, something that may not have been done by anyone since Ptolemy. The labor involved was immense—well over a thousand separate calculations for the derivations presented or implicit in *De revolutionibus*, and perhaps thousands more in unsuccessful trials with other sets of oppositions that produced obviously erroneous results, or serious computational errors with the sets actually used.[86] The tedium of so many calculations—interpolation in sine tables, multiplication and division of five- and six-digit numbers, extraction of roots from much larger numbers, and all this without even logarithms—is now hard to imagine, but one can at least understand why hardly any astronomers after Ptolemy had attempted and perhaps spent years on such an "immense multitude of calculations." If the orbital elements of the superior planets were not Copernicus's most important accomplishment, they were surely his most difficult.

4. The Inferior Planets

The transformation from a geocentric to a heliocentric representation of the motion of an inferior planet, illustrated earlier in Fig. 2, appears straightforward.[1] Since the direction from the earth to the center of the epicycle coincides with the direction from the earth to the mean sun, it is only necessary to assume that the center of the epicycle is in fact the mean sun, and that the earth is likewise in motion about the mean sun, to produce Copernicus's heliocentric model. This simple transformation applies to a uniform motion of the center of the epicycle. However, when the first anomaly is taken into consideration, the situation is no longer straightforward, but is instead not a little paradoxical. In a geocentric model, the first anomaly for both the superior and inferior planets is represented by the motion of the center of the epicycle on an eccentric about the earth, and the equation of center is a function of the distance of the center of the epicycle, measured by the mean eccentric anomaly, from an apsidal line passing through the earth. When transformed into a heliocentric model for a superior planet, the first anomaly now corresponds to the motion of the planet on an eccentric about

[86] One could, however, propose that the fact that Copernicus did not catch his error in the first solution for Jupiter shows that he carried out the derivations only once—not going through anything like Kepler's seventy trials with the vicarious hypothesis—and using only the oppositions presented in the text. It would be very interesting to know how much preliminary work stands behind the demonstrations in Copernicus's text, but like Ptolemy, he dismantles the scaffolding when the building is completed.

[1] Above, p. 292.

the mean sun, the equation of center becoming a function of the distance of the planet, measured by the same mean eccentric anomaly, from an apsidal line passing through the mean sun.[2] On the other hand, when the same transformation is made for an inferior planet, in which the motion of the center of the epicycle corresponds to the motion of the earth, the first anomaly and the equation of center have nothing to do with the motion of the planet, but depend only upon the motion of the earth, likewise measured by the mean eccentric anomaly, with respect to the planet's apsidal line passing through the sun.

Consider Fig. 50 which shows Ptolemy's model for Venus with the epicycle drawn to scale although with the eccentricity greatly exaggerated. The model is exactly like that for the superior planets except that the mean motion of the center of the epicycle, shown by the direction EC, is always parallel to the direction $O\bar{\odot}$ from the earth to the mean sun. The equation of center c_3 is a function of the motion of C through the mean eccentric anomaly $\bar{\kappa}$. The direct conversion to a heliocentric arrangement is shown in Fig. 51. Complete parallelogram $OEC\bar{S}$, and let the mean sun be at $\bar{S}$. Bisect $\bar{S}C$ at M', and with radius $M'O$, which will necessarily be equal to R, describe a circle about M'. The result is that the earth moves in the circle about M' with its motion uniform with respect to $\bar{S}$, the mean sun, which now functions as an equant point for the motion of the earth. The equation of center c_3 is thus a function of the motion of the earth through the mean eccentric anomaly $\bar{\kappa}$ measured from the "apogee" A' lying on the planet's apsidal line extended to meet the earth's orbit. Note that the term *apogee* has changed meaning, and is now defined as the point of the earth's orbit at which the earth is at its greatest distance from the center C of the planet's orbit. It is a rather geocentric usage since it means conversely the point at which the center of the orbit C is at its greatest distance from the earth. This is the way Copernicus uses the term *apogee* for the inferior planets, i.e. when the earth is at A' it is at apogee.[3] Note further that the direction $A'C$ lies in the same direction as OA in the geocentric model, and it is the (really geocentric) direction $A'C$ that Copernicus specifies as the direction of the apogee, the direction of the greatest distance of the center of the planet's orbit from the earth.

Now, it is often said that Copernicus's theory explains the "coincidence" that the centers of the epicycles of the inferior planets always lie in the direction from the earth to the mean sun by identifying the center of the epicycle—here we do not consider the first anomaly—with the mean sun. However, one coincidence is removed at the expense of another, for when the first anomaly is taken into account, it now follows that the equations of center of the inferior planets depend upon the motion of the earth with respect to their apsidal lines, and it is not obvious, except when understood as a direct transformation of Ptolemy's model, why this should be so.[4]

---

[2] Above, p. 299.

[3] Below, p. 381.

[4] The models for latitude of all the planets have an even more remarkable dependency upon the motion of the earth, concerning which see below, pp. 483 ff.

But the more serious problem, the true paradox, of the transformation becomes clear when it is noted that since the earth moves in a circle about M', the distance $O\bar{S}$ is not constant. In the heliocentric transformation of the solar theory, the earth moves in a circle of fixed radius about the mean sun—the motion of the mean sun with respect to the true sun is not now of concern—but here we see that the distance of the earth from the mean sun is necessarily variable. The situation becomes worse if we consider the heliocentric transformations of the motions of the sun, Venus, and Mercury together. In Fig. 52 we show the true sun at S, the mean sun at $\bar{S}$, and the orbit of the earth about the mean sun with the "apogee" at A_s' and the earth at O_s. Transforming Ptolemy's model as before, the point M' will be M_v' for Venus and M_m' for Mercury, the directions corresponding to directions of the apsidal lines.[5] Now the earth must lie in the direction $\bar{S}O_s$, but it does not lie in the same position in all three models. If we describe a circle of radius R about M_v', the earth will be at O_v; if the same circle is described about M_m', the earth will be at O_m. It is now evident that to fulfill simultaneously the conditions of the heliocentric transformation for the sun, Venus, and Mercury, the earth must be in three different positions at the same time, which is impossible. Consequently, a direct heliocentric transformation of Ptolemy's models for these three bodies leads to an absurdity.[6]

It is evident that the inferior planets present a peculiar problem for Copernicus in that he must both account for their first anomaly in a manner as close as possible to Ptolemy's model, although preserving as always uniform circular motion, and at the same time keep the distance from the mean sun to the earth constant, as it is in the solar theory. The latter can be done if, in Fig. 51, the displacement of the earth along $O\bar{S}$ can be transformed into a displacement of C, the center of the planet's orbit, such that $O\bar{S}$ can remain constant while OC takes on (approximately) the same length it would have if O moved in the circle about M'. Fortunately, the Marāgha models that Copernicus in some way had at his disposal already contain all the mechanism necessary for doing this. However, Copernicus had some difficulty in adapting them properly, so to best understand the form they finally take in *De revolutionibus* we shall have to investigate the earlier models in the *Commentariolus* and their peculiarities. This was not necessary in considering the superior planets since the models in the *Commentariolus* present no difficulties, and their transformation to the models in *De revolutionibus*, the motivation for which is explained by Copernicus, is straightforward.[7]

---

[5] Here we are neglecting the special features of the model for Mercury, which would introduce still greater difficulties.

[6] The same paradox arises in the Tychonic transformation, that is, assuming the earth at rest and the mean sun in motion, for the distance $O\bar{S}$ would still be variable, but now compelling a different location of the mean sun for each planet. Whether a consideration of this sort caused Ptolemy to reject the possibility that the mean sun was the center of the epicycles of Venus and Mercury—it is already sufficient that their eccentricities and apsidal directions differ—is a subject about which we will not speculate.

[7] Above, pp. 299–300.

A. Venus

Development of the Model

Ibn ash-Shāṭir's model for Venus is shown in Fig. 53. It differs from his model for the superior planets in that, as in Ptolemy's model, the direction EC, from the equant to the center of the principal epicycle, is parallel to the direction $O\bar{\odot}$ from the earth to the mean sun. The equation of center c_3 is produced by two epicycles of radii $r_1 = \frac{3}{2}e$ and $r_2 = e/2$, where e is the eccentricity in Ptolemy's model. The resultant motion of C is a uniform motion about E where $OE = 2e = r_1 + r_2$. When Copernicus adopted a model of this kind for the superior planets to a heliocentric arrangement in the *Commentariolus*, he placed the planet at C, and replaced the principal epicycle with the earth's orbit about the mean sun, as shown in Fig. 15.[8] Hence the planet came at the end of the radius of the smaller epicycle, of radius r_2, and thus moved uniformly about E, as indeed it should. He tried to do the same thing for Venus and Mercury, but the result was rather peculiar, and not exactly what he had in mind. The model for Venus is shown in Fig. 54. What is peculiar is that the equation of center can no longer be measured at the earth, for in order to find it at all, it is necessary to construct parallelogram $O\bar{S}C_1N$, draw an instantaneous apsidal line NEA at N, and extend NC_1 to meet the larger epicycle at G. It then follows that r_1 describes $\bar{\kappa}$ with respect to NC_1G, the planet moves uniformly through $\bar{\kappa}$ with respect to E, the equant on NEA, the equation of center c_3 is given by PNC_1 or EPN, and the equation of the anomaly c by NPO. Now, in a sense the model works well enough in that the planet is where it should be in relation to O, but the order of the radii and the necessity of constructing the point N is rather cumbersome, and the last thing this contorted model can be called is simple.

In order to understand the peculiar form of this model and the necessity of constructing point N, we must move back a step earlier than the *Commentariolus* to Copernicus's original derivation of the heliocentric theory. There is rather good evidence, a page of notes in Copernicus's hand setting out the parameters of the models in the *Commentariolus*, that shows that Copernicus reached the heliocentric theory through an examination of alternative eccentric models for the second anomaly of the superior and inferior planets respectively described at some length by Regiomontanus in *Epitome* XII,1–2.[9] The description in the *Epitome* applies only to the second anomaly, but when the double epicycle of Ibn ash-Shāṭir's model is added to account for the first anomaly, the result for an inferior planet is the model shown in Fig. 55. With the earth at O a circle of radius $e' = r$, the radius of the principal epicycle in the geocentric model, is described about O, and a point N moves on the circle through $\bar{\lambda}^{*\prime} = \bar{\lambda}^*_{\odot} + \bar{\alpha}$, that is, the sum of the mean motion of

[8] Above, p. 299.

[9] Swerdlow (1973), 471–78. Above, pp. 56–57.

the sun and the mean anomaly of the planet. N is then the center of an eccentric of radius R—which is why the distance ON is referred to in Copernicus's notes as an eccentricity—and the center C_1 of the larger epicycle moves on the eccentric in the negative sense through the mean anomaly $\bar{\alpha}$ measured from the direction ONF. Constructing an instantaneous apsidal line NEA upon N, it can be seen that the planet P moves uniformly through $\bar{\kappa}$ with respect to an equant E where $NE = 2e$. The radius of the first epicycle r_1 rotates through $\bar{\kappa}$ with respect to the direction NC_1G, the equation of center c_3 is given by PNC_1 or EPN, and the equation of the anomaly by NPO. All of this is exactly the same as the model in the *Commentariolus* (Fig. 54) in which the angles and motions can only be measured by constructing the apparently unmotivated point N. But N is motivated, and its necessity is understood, when seen as a remnant of an earlier, and not yet heliocentric, stage of Copernicus's planetary theory.[10]

By the time of *De revolutionibus*, Copernicus had realized the problems of the model in the *Commentariolus* and rearranged it properly. Note that in Ibn ash-Shāṭir's model in Fig. 53, the center C_1 always lies in the direction $O\bar{\odot}$ from the earth to the mean sun, while the smaller epicycle carries the center C of the principal epicycle. Evidently, in properly transforming the model to a heliocentric arrangement C_1 should be identified with the mean sun and C with the center of Venus's orbit, that is, the two epicycles must be placed so that they carry the center of Venus's orbit, and should not be placed on the orbit itself to carry the planet. The model in *De revolutionibus* is shown in Fig. 56(a), and its transformation from the *Commentariolus* model in Fig. 56(b). The epicycle r_1 has become the eccentricity e_1 and r_2 has become e_2. We now have the exact heliocentric equivalent of Ibn ash-Shāṭir's model, the center of Venus's orbit C always appearing to move uniformly with respect to an equant point E lying on an instantaneous apsidal line OEA passing through the earth. The equation of center c_3 is a function of the mean eccentric anomaly $\bar{\kappa}$, which is determined by the distance of the earth O from Venus's apsidal line passing through the mean sun $\bar{S}$, and c_3 is now measured from O, as it should be.[11]

The relation to Ptolemy's model is brought out more clearly in Fig. 57 showing only the first anomaly, that is, the relative motion of the earth and

---

[10] The same peculiarity is contained in the model for Mercury in the *Commentariolus*. Below, p. 409.

[11] It is also clear why for the inferior planets Copernicus cannot separate the components of the first anomaly into an eccentricity from the mean sun and a small epicycle on the planet's eccentric, for this arrangement would likewise make it impossible to measure c_3 as a single equation from O. However, the arrangement adopted for the inferior planets could be applied to the superior planets with no ill effects, and it is not obvious why Copernicus did not do so. Perhaps the decrease of Mars's eccentricity gave him a reason for wishing to keep the two components as distinct as possible, but there is nothing in the model for the inferior planets that prevents the same change in e_1, and in fact Copernicus decreases Venus's eccentricity considerably more than Mars's. See below, pp. 385–386.

the center C of Venus's orbit. Just as in the model for the superior planets, C in Copernicus's model falls just outside a circle about M', the center of the eccentric in Ptolemy's model. A comparison with Fig. 51 will show that in this way the distances OC are about the same in Copernicus's model and in the direct heliocentric conversion of Ptolemy's model, while in Copernicus's model the distance $O\bar{S}$ remains fixed. This eliminates the problem of the direct heliocentric conversion of Ptolemy's model. In principle $e_1 + e_2 = 2e$, but Copernicus modifies this relation for his own time so that his model no longer has the equivalent of a bisected eccentricity.[12] The equation of center for Venus is exactly like that of the superior planets (formulas 2 and 5, but substituting e_2 for r').[13] In solving it in steps, Copernicus draws OM dividing c_3 into components a_1 and a_2. He then applies (6–10), but substituting OM for $\bar{S}C$ in (6) and (8), and OC for $\bar{S}P$ in (8).

The complete model is shown in Fig. 58. Measured from the fixed sidereal direction Γ, the planet moves through its mean heliocentric motion in longitude $\bar{\lambda}^{*\prime} = \bar{\lambda}^*_{\odot} + \bar{\alpha}$, although Copernicus neither uses nor tabulates $\bar{\lambda}^{*\prime}$, but considers the mean motion of the planet to be the mean geocentric motion of the sun $\bar{\lambda}^*_{\odot}$. The mean eccentric anomaly $\bar{\kappa}$ is determined by the distance of the earth from the direction $\bar{S}A'$, where A' is the "apogee," meaning the location of the earth when it is at the greatest distance from the center of Venus's orbit, and is given by $\bar{\kappa} = \bar{\lambda}^*_{\odot} - \lambda^*_{\mathrm{A}}$. In order to find the mean anomaly $\bar{\alpha}$, we construct the instantaneous apsidal line OA at the earth, and designate the equant point at the distance $OE = e_1 + e_2 = 2e$. Then EC extended to meet the orbit at T divides $\bar{\lambda}^{*\prime}$ into $\bar{\lambda}^*_{\odot}$ and the mean anomaly $\bar{\alpha}$. Once again the equant is very much present in Copernicus's model, for the mean anomaly cannot be found without it. Correcting the mean anomaly to the true anomaly $\alpha = \bar{\alpha} + c_3$, the equation of the anomaly c can be found trigonometrically, and then the elongation from the mean sun $\Delta = c + c_3$ and the true longitude of the planet $\lambda^* = \bar{\lambda}^*_{\odot} + \Delta$, all additions being done algebraically with the proper sign.

Observations

Copernicus uses eleven observations of Venus (including one that is cancelled), all but one of which are from the *Almagest*.[14] The one exception is his own observation (11) of 12 Mar 1529, the latest observation cited in *De revolutionibus*, which he uses to correct the mean motion by comparing the position of Venus with a very early observation (9) by Timocharis on 12 Oct −271. This same observation had been used by Ptolemy in connection with an observation (10) of 18 Feb 134 for correcting the mean motion,

---

[12] Below, p. 385.

[13] Above, p. 298.

[14] Ptolemy's observations and his derivation of the elements of Venus have been discused in Czwalina (1959), 1–18, Wilson (1972), 209–23, Moesgaard (1974), 98–101, 255, who also considers Copernicus's use of the observations, Pedersen (1974), 298–308, *HAMA*, 152–58.

and in an early, cancelled version of V,23 in M Copernicus had used this later observation (10) rather than (9) for the same purpose.[15] Since Copernicus relies so completely on Ptolemy's observations of Venus, he not surprisingly establishes essentially Ptolemy's elements of the orbit, and he even concludes, with Ptolemy, that the apsidal line of Venus is sidereally fixed, unlike the independent motions that he finds for the other planets. Presenting only a single observation from his own time, there seems to be little else he can do. Nevertheless, at the end of V,22 he suddenly says that the eccentricity must be reduced, "as many observations teach us," although one can only wonder what these observations could be and why Copernicus does not use them in the text to derive modern elements.[16]

Copernicus's reports of the observations from the *Almagest* present a number of problems, principally with regard to the times of the observations, but these are on the whole not Copernicus's fault. Ptolemy's own reports of (1 to 8) in *Almagest* X,1–3 and *Epitome* X,1–4 are rather vague in that the time is given only as "morning" or "evening." Copernicus gives the times just as vaguely as *illuscente* and *diluculo* or *crepusculo* and *prima hora noctis*, by which he seems to intend only morning or evening, before sunrise or after sunset, rather than the more specific dawn or dusk, that is, the beginning of morning and the end of evening twilight. However, some more specific time is required in order to compute the mean position of the sun, which is necessary for the reduction and application of each observation. At the latitude of Alexandria, 31°, the sun's semidiurnal arc can vary by up to $\pm 1^h$ from 6^h, and the times of the observations can fall somewhat outside these limits. In checking these observations, we have computed the solar longitude by choosing a plausible time on the hour or half-hour before sunrise or after sunset in Alexandria, which depends upon the sun's tropical longitude, and then subtracting 1^h to adjust to the meridian of Cracow–Frauenburg.[17]

The observations are summarized in Tables 5 and 6. Table 5 gives the following information concerning date and time:

(1) Number of observation.
(2) Observer: Ptolemy (Ptol), Theon,[18] Timocharis (Tim), Copernicus (Cop).
(3) Egyptian date and time in text: M, morning; E, evening.
(4) Year of Era Alexander (E.A.).
(5) Time from midnight by our estimate.
(6) Julian date corresponding to Egyptian date.
(7) Julian day.

[15] Below, pp. 397–401.

[16] Below, pp. 384–389.

[17] Cf. the somewhat different times used by Moesgaard (1974), 100, 255.

[18] Called by Ptolemy "Theon the Mathematician," he is cited for three observations of Venus (1/3/5) and one of Mercury (5) made between 127 and 132 in Alexandria, for which reason, presumably, Copernicus calls him Theon of Alexandria. He is a, perhaps older, contemporary of Ptolemy, whose dated observations extend from 127 to 141. Nothing else is known of him, and there is no reason to identify him with any other person named Theon in antiquity (a very common name, there are 36 Theons in Pauly–Wissowa for a start).

For all the observations except (9), Copernicus gives Julian dates corresponding to the Egyptian dates, but those for (2), (5), and (6) are in error.[19] Since the solar longitudes are correct for the Egyptian dates, Copernicus must have computed from them, presumably from Era Alexander, and we have done the same in Table 6. Observations (10) and (11) were, however, computed from their Julian dates. A more serious problem is that the date of observation (3) is in error by eight years, a matter of some interest.[20]

Ptolemy's descriptions of the observations are fairly detailed, the position of Venus being given in relation to zodiacal stars, and the longitude deduced from the particular configuration. Copernicus abbreviates these considerably, giving for (1–4) only Venus's elongation from the mean sun and for (5–8) Venus's elongation and sidereal longitude, although the longitude is merely derived from the elongation. For (9–11) Copernicus's description is more detailed.

Table 6 gives the following information concerning solar longitude and the longitude and elongation of the planet:

(1) Number of observation.
(2) Cop $\bar{\lambda}^*_\odot$, Copernicus's sidereal longitude of the mean sun.
(3) $\bar{\lambda}^*_\odot$, by computation.
(4) π, precession by computation.
(5) $\bar{\lambda}'_\odot = \bar{\lambda}^*_\odot + \pi$, adjusted mean tropical longitude.
(6) Ptol $\bar{\lambda}_\odot$, tropical longitude in the *Almagest*.
(7) Obs λ_p, observed tropical longitude of Venus.
(8) Cop λ^*_p, Copernicus's reduction to sidereal longitude.
(9) Δ, elongation of Venus to the east (+) or west (−) of the mean sun.

Discrepancies will be discussed along with the applications of the observations.[21]

In deriving the elements of the orbits of the superior planets, Ptolemy, and following him Copernicus, used a rigorous and completely general method depending upon observations of any three oppositions for determining the eccentricity and the direction of the apsidal line, and upon any one observation of the planet outside opposition for determining the radius of the epicycle, or equivalently, the ratio of the radius of the earth's orbit to the radius of the planet's orbit. Such observations, which isolate the effects of the first and second anomalies, were not possible for the inferior planets since the analogue of an opposition, in which the planet, the earth, and the mean sun are collinear, would be an observation of the planet at inferior conjunc-

[19] (2) 142 for 140, (5) N XII, M *Duodecimo*, *Calend. Junii* for XIII, (6) V *Calend. Januarii* for XIV *Kal. Dec.*, but these Julian dates are not used. There is also an error in the Egyptian date of (10) in M (172r8), the text giving Tybi 20 (*vigesimi*) for 30, although here Copernicus uses the correct interval from the Epoch of Christ.

[20] Below, pp. 378–379.

[21] Textual corrections: Cop $\bar{\lambda}^*_{\odot 4}$ 265;27° for 265°; Cop λ^*_{p8} 7;10° for 7;50° (*sextante* for *dextante*, as frequently noted).

tion with the mean sun when unfortunately the planet is too close to the true sun to be visible. Consequently, Ptolemy used a less satisfactory method depending upon observations of the planet rather near greatest elongation from the mean sun while the mean sun was almost exactly at carefully selected positions with respect to the planet's apsidal line. It has been noted that Ptolemy's observations of what he calls greatest elongation are frequently in error by several days—in the case of Venus by even more than two weeks—from the correct time of greatest elongation.[22] There is no doubt that Ptolemy was himself aware of these discrepancies since, for example, observations (4) and (6), supposedly of greatest evening elongation, are only 37 days apart, while Ptolemy knew perfectly well that Venus can have but one greatest evening elongation in its synodic period of about 584 days or 1.6 years.[23] However, within a reasonable period of time there was no way of making observations of the planet exactly at greatest elongation and likewise with the mean sun exactly at the required locations with respect to the apsidal line. What is most crucial to the demonstrations is the location of the mean sun, and thus Ptolemy had no choice but to compromise on the proximity to greatest elongation, which is less critical. Ptolemy considers greatest elongation to occur when the equation of the anomaly is at its maximum and the line from the observer to the planet is tangent to the epicycle,[24] and near maximum the equation of the anomaly, and thus the elongation, change rather slowly, but not so slowly that Ptolemy could not determine the time of greatest elongation with more accuracy than the reports in the *Almagest*. Hence the reported observations appear to represent the best approximations that Ptolemy could find to the proper location of the mean sun and a fairly close approach to greatest elongation.

It is obvious that these observations require persistence, prior knowledge, or a bit of luck to obtain, and it is probably for this reason that Copernicus presents no observations of his own of Venus near greatest elongation, but relies entirely on Ptolemy's observations to establish the elements of Venus's orbit by Ptolemy's methods. In fact, during the period that Copernicus observed, it was simply not possible to assemble a suitable set of observations of Venus near greatest elongation—assuming that he understood that the planet had only to be *near* greatest elongation—and with the mean sun in the correct locations with respect to the apsidal line. The reason is simple. In an eight-year synodic cycle Venus's greatest elongations occur with the mean sun at longitudes distributed at intervals of about 72°, while the required locations are 90° apart, and the longitudes of the mean sun shift only about −2° in each cycle. Hence, Copernicus's reference to "many

[22] Wilson (1972), 213–19, Moesgaard (1974), 100–01, *HAMA*, 153 n. 1.

[23] In fact, the correct time of the greatest elongation falls in between the two observations on about Dec 9. Cf. references n. 22.

[24] This is not strictly the greatest elongation unless there is no equation of center, which can increase the elongation when the planet is not exactly at the point of tangency, as noted by Sawyer in Goldstein and Sawyer (1977), 169.

observations" must be limited to something less than the observations required to establish independently the elements of Venus's orbit.[25]

Supplementary Remark: The Date of Observation (3)

Correctly, observation (3) was made on Hadrian 12 Athyr 21/22, $18\frac{1}{2}^h$ after noon, which is 127 Oct 12, $6\frac{1}{2}^h$ after midnight (j.d. 1767,729) at Alexandria, and Ptolemy computed the tropical longitude of the mean sun to be 197;52°. In the Gerard of Cremona translation of the *Almagest*, the date appears as Hadrian 2, the night which Athyr 22 follows, i.e. Athyr 21/22,[26] corresponding to 117 Oct 14 (j.d. 1764,079), which is fully ten years prior to the earliest observations in the *Almagest* definitely attributable to Ptolemy. If one computes $\bar{\lambda}_\odot$ with Ptolemy's tables for this date, Era Nabonassar 865, Athyr 21, $18\frac{1}{2}^h$, one finds that $\bar{\lambda}_\odot = 200;18°$ or 2;26° in excess of the number in the text. Hence, either the date or the solar longitude or both are in error. One must therefore check Venus, which Ptolemy reports at $\lambda =$ 150;20°, or an elongation of $-47;32°$. For the same time the mean anomaly $\bar{\alpha} = 143;6°$, and already it is obvious that something is wrong since $\alpha < 180°$ puts Venus on the east or positive side of the epicycle, while the reported longitude and morning elongation show that it must be on the west or negative side where $\alpha > 180°$. Hence, the date is incorrect, but exactly how to fix it is not obvious.

This problem was evidently confronted by Regiomontanus in writing the *Epitome*, for in X,1 the date is given as Hadrian *4* Athyr 19/20, which Copernicus follows, and this date is Regiomontanus's own emendation of the faulty number in the Gerard translation. Now it is easy enough to suggest that Hadrian *2* ought to be changed to Hadrian *12*, and $\bar{\lambda}_\odot$ and $\bar{\alpha}$ checked for this date to see whether a plausible location results. But this was not what Regiomontanus did, since he emends to Hadrian *4* and also changes the day from Athyr 21/22 to Athyr 19/20. He must have asked, what is the date closest to Hadrian 2 Athyr 21/22 that will put the sun and Venus into the correct configuration? Since the period of the anomaly of Venus is related to the solar year by $5P_\alpha = 8^y - {\sim}2^d$, there is a possibility of being in error modulo $8^y - 2^d$ in the date of the observation. One would guess that Regiomontanus proceeded as follows: Beginning from Hadrian 2, for Athyr 21, $18\frac{1}{2}^h$, we have from the tables in the *Almagest*:

Year	$\bar{\lambda}_\odot$	$\bar{\alpha}$
Hadrian 2	200;18°	143;6°
Hadrian 3	200;4	8;8
Hadrian 4	199;49	233;9
⋮	⋮	⋮
Hadrian 12	197;52	233;22

[25] Discussed in more detail below, pp. 387–388.

[26] Likewise in the George of Trebizond translation, *Almagest* (1551), 234.

Now, in Hadrian 4 $\bar{\alpha}$ is in about the proper location for maximum western elongation, but the sun is 1;57° ahead of Ptolemy's 197;52°. Evidently, without computing forward to Hadrian 12, Regiomontanus decided to regress the date by 2 days, which reduces the solar longitude to 197;51°, and ended up with Hadrian 4 Athyr 19/20, in error by 8^y–2^d from the correct date. Copernicus, even though he must have been aware of the contradiction between the *Epitome* and the *Almagest*, followed the *Epitome* without further checking.[27]

Longitude of the Apsidal Line (V,20)

Observations (1–4) fall into two pairs, in each of which the elongations of Venus from the mean sun are equal and opposite, that is, equal morning and evening elongations. Ptolemy concludes that in each pair the center of the epicycle must be at the same distance from the earth, but at symmetrical points on either side of the apsidal line, and Copernicus concludes correspondingly that the earth must be at the same distance from the center of Venus's orbit at points symmetrical to the apsidal line.[28] The point midway between the longitudes of the mean sun in each pair will therefore give the longitude of one end of the apsidal line. In Fig. 59, the earth is shown at the four locations for Copernicus's demonstration. In the first pair

$$\bar{\lambda}^*_{\odot 1} = 337;41°, \qquad \bar{\lambda}^*_{\odot 2} = 119;0°, \qquad \Delta_1 = -\Delta_2 = 47;15°,$$

and one end of the apsidal line is directed toward

$$\lambda^*_A = \tfrac{1}{2}(337;41° + 119° - 360°) = 48;20,30° \approx 48;20°,$$

and the other end opposite at $\lambda^*_{A'} = 228;20°$. Likewise, in the second pair, using the correct $\bar{\lambda}^*_{\odot 3}$,

$$\bar{\lambda}^*_{\odot 3} = 191;23°, \qquad \bar{\lambda}^*_{\odot 4} = 265;27°,^{29} \qquad -\Delta_3 = \Delta_4 = 47;32°,$$

[27] Regiomontanus's failure to make correctly what now seems a trivial emendation, and Copernicus's (apparently unquestioning) acceptance of the dating in the *Epitome*, are of some interest for it shows something of the state of evaluating textual problems at their time. It appears to be only in the late sixteenth and early seventeenth centuries, with Tycho, Scaliger, and Kepler, that one finds a truly modern sense of emendation of astronomical texts (that unfortunately is not present in all modern editions).

[28] This is true for the four observations used in the demonstration, but the equality of greatest elongations does not by itself guarantee symmetry to the apsidal line. Because of the effect of the equation of center, except at the minimum and maximum values of greatest elongation, about 60° from apogee and perigee respectively, there are four points at which the greatest elongations will be equal, two evening and two morning elongations. One of each kind will form a pair symmetrical to the apsidal line, but members of different pairs will not be symmetrical. This has been pointed out by Sawyer in Goldstein and Sawyer (1977), 169 ff. Nevertheless, Ptolemy already knew a provisional direction of the apsidal line from taking the minimum and maximum sums of elongations (below, p. 381), and within the period of his observations, greatest elongations of Venus could occur in only five regions of the zodiac (below, pp. 387–388).

[29] N 160v25 and M 170r27 read 265°, but this is only a textual error for 265;27°.

and one end of the apsidal line is directed toward

$$\lambda^*_{A'} = \tfrac{1}{2}(191;23^\circ + 265;27^\circ) = 228;25^\circ \approx 228;20^\circ,$$

and the other end toward $\lambda^*_A = 48;20^\circ$. Note that if $\bar{\lambda}^*_{\odot 3} = 191;\mathit{13}^\circ$, the incorrect value in the text, the result is

$$\lambda^*_{A'} = \tfrac{1}{2}(191;13^\circ + 265;27^\circ) = 228;20^\circ \text{ exactly,}$$

so Copernicus probably altered $\bar{\lambda}^*_{\odot 3}$ in order to produce the precisely desired location of the apsidal line. Adding a precession of 6;40° for the time of Ptolemy, the tropical longitudes are $\lambda_A = \text{♉}\,25^\circ$ and $\lambda_{A'} = \text{♏}\,25^\circ$, as stated by Ptolemy.

Since the elongations Δ_3 and Δ_4 are larger than Δ_1 and Δ_2, one might conclude that the orbit of Venus is closer to O_3 and O_4 than to O_1 and O_2, and thus that the apogee is at 48;20° or ♉ 25° and the perigee at 228;20° or ♏25°. This is in fact the case, and because of the precise configurations chosen by Ptolemy, the distinction is already warranted. However, if, for example, observations (3) and (4) were reversed in direction such that Venus were on the other side of the mean sun, that is, if morning and evening elongations were reversed at P'_3 (evening) and P'_4 (morning) respectively, then $\Delta_3 = -\Delta_4 < \Delta_1 = -\Delta_2$, and one would conclude wrongly that O_3 and O_4 were near perigee and O_1 and O_2 near apogee. Therefore, in order to confirm the distinction, Ptolemy, in *Almagest* X,2 and *Epitome* X,2, and following him Copernicus, take observations (5) and (6) in which the mean sun is computed to be so close to the opposite ends of Venus's apsidal line that the direction to the mean sun, and thus that the earth, may be considered to lie in the apsidal line. Hence, in Fig. 60,

$$\bar{\lambda}^*_{\odot 5} = 48;50^\circ \approx 48;20^\circ, \qquad \Delta_5 = -44;48^\circ,$$

$$\bar{\lambda}^*_{\odot 6} = 228;54^\circ \approx 228;20^\circ, \qquad \Delta_6 = +47;16^\circ.$$[30]

Since $|\Delta_5| < |\Delta_6|$, consequently $O_5C > O_6C$, and when the earth is at O_5 or A', which according to Copernicus's definition is the "apogee" or the location of the earth when the center of Venus's orbit is most distant, the center C appears in the direction $\lambda^*_A = 48;20^\circ$, the geocentric direction of the apogee. Conversely, when the earth is at O_6 or A, the center C appears

---

[30] Ptolemy finds $\Delta = 47;20^\circ$ from

$$\Delta = \lambda - \bar{\lambda}_\odot = \text{♑}\,12;50^\circ - \text{♏}\,25;30^\circ = 47;20^\circ.$$

Copernicus adjusts the planet to sidereal longitude by $\lambda^* = \text{♑}\,12;50^\circ - 6;40^\circ = 276;10^\circ$, and then combines it with his computed $\bar{\lambda}^*_\odot$ to give

$$\Delta = \lambda^* - \bar{\lambda}^*_\odot = 276;10^\circ - 228;54^\circ = 47;16^\circ.$$

He does not comment on the difference, but he was aware of it since in an earlier cancelled version in M (170v10–12) he had set $\bar{\lambda}^*_\odot = 228;50^\circ$, i.e. ♏25;30° − 6;40°, and $\Delta = 47;20^\circ$ in agreement with Ptolemy. Further, in V,21, when finding the eccentricity and radius of the orbit, Copernicus alternates in a confusing way between 47;16° and 47;20°. Below, p. 382.

at 228;20°, the geocentric direction of the perigee, although Copernicus does not use this term for Venus, but instead refers to the "lowest apsis" as he usually calls the apogee the "highest apsis."

Ptolemy remarks further that when the mean sun is in Taurus, the sum of Venus's greatest elongations is a minimum, and in Scorpio the sum is a maximum, although he does not present an additional pair of observations to confirm the opposite elongations in or near the apsidal line. Such pairs showing minimum and maximum sums of elongations would constitute the best proof of the direction of the apsidal line because, owing to the contribution of the equation of center, the minimum and maximum single elongations do not occur exactly in the apsidal line, but at about 60° from the apsidal line, although the minimum and maximum sums of opposite, and also equal, elongations do occur in the apsidal line. In fact, Ptolemy did not have observations of equal and opposite elongations of Venus in the apsidal line—he does have such observations of Mercury at perigee, and they are highly significant to the development of his model[31]—but he evidently did have sufficient observations of opposite elongations of Venus with the mean sun in or near Taurus and Scorpio to determine that their sums were a minimum in the former and a maximum in the latter. This quite significant point about the sums of the elongations is not mentioned in the *Epitome*, and is likewise omitted by Copernicus.[32]

Radius of Orbit and Eccentricities (V,21–22)

Ptolemy determines the radius of Venus's epicycle in *Almagest* X,2 and *Epitome* X,3 using observations (5) and (6) under the assumption that the center of the epicycle is exactly in the apsidal line, and Copernicus finds the radius of the orbit from the same observations, assuming correspondingly that the earth is exactly in the apsidal line, as in the preceding demonstration. Under this assumption, the equation of center is zero, and the demonstration is straightforward. It is here that Copernicus introduces his definition of "apogee." His words are: "Let the position of the apogee be A, for when the earth is there the center of the sphere of Venus is most distant, at which time AB itself is the line of the mean motion of the sun directed towards 48;20°, but [when the earth is] in B directed towards 228;20°."[33] Thus, in Fig. 61, at observation (5) the earth is at A, which Copernicus calls the position of the apogee since when the earth is at A, the center C of Venus's orbit is most distant, lying in the geocentric direction 48;20°, and at observation (6) the earth is at B, which Copernicus does not call the perigee, C lies in the geocentric direction 228;20°. We shall henceforth use Copernicus's lettering,

[31] Below, pp. 417–418.

[32] The minimum and maximum sums of greatest elongations probably allowed Ptolemy a preliminary determination of the direction of the apsidal line, which could then be sharpened by observations (1–4) selected for their symmetries to the preliminary direction.

[33] N 161r16–20. Some translations of the passage are strange.

designating the "apogee" or "highest apsis" by A, and the perigee or "lowest apsis" by B, as in his diagrams.

When the planet is at greatest elongation at P_5 and P_6 respectively, the lines AP_5 and BP_6 are assumed tangent to the orbit forming right triangles ACP_5 and BCP_6. The following demonstration has a remarkable number of inconsistencies. From observations (5) and (6) it is known that

$$|\Delta_5| = 44;48°, \qquad |\Delta_6| = 47;20°.$$

Note that for Δ_6, Copernicus has resumed Ptolemy's value in place of his own 47;16°. In triangle ACP_5, where $AC = 10{,}000$,

$$CP_5 = AC \sin \Delta_5 = 7046,$$

and in triangle BCP_6, where $BC = 10{,}000$,

$$CP_6 = BC \sin \Delta_6 = 7346.$$

Correctly $\sin 47;20° = 7353$, but $\sin 47;16° = 7345$, so Copernicus has evidently used his own 47;16° after all. Now, in units of $AC = 10{,}000$,

$$BC = (\sin \Delta_5/\sin \Delta_6)\, AC,$$

and Copernicus claims that

$$BC = (7046/7346)\, 10{,}000 = 9582,$$

although correctly this would be 9592. Note however, that using $\sin \Delta_6 = \sin 47;20° = 7353$,

$$BC = (7046/7353)\, 10{,}000 = 9582,$$

so Copernicus is again using Ptolemy's $\Delta_6 = 47;20°$. These changes are a bit disconcerting, but, as we have learned from much experience, not uncharacteristic. Since

$$ACB = AC + BC = 19582,$$

so

$$A\bar{S} = \tfrac{1}{2}ACB = 9791, \qquad \bar{S}C = AC - A\bar{S} = 209,$$

but Copernicus gives $\bar{S}C = 205$, and he then proceeds to use it.[34] For he next says that where $R = A\bar{S} = 1;0$, the radius of the orbit

$$r = 7046/9791 = 0;43,10, \quad \text{(Accur. 0;43,10,42)}$$

and the eccentricity

$$e = 205/9791 = 0;1,15,22 \approx 0;1,15,$$

---

[34] If $BC = 9592$, then $ACB = 19592$, $AS = 9796$, and $\bar{S}C = 204$, rather close to Copernicus's 205.

all in agreement with Ptolemy. Then, letting $R = 10{,}000$,

$$r = 0;43,10\,R = 7193,^{35} \qquad e = 0;1,15\,R \approx 208.$$

We may check the results of Copernicus's derivation by noting that the entire procedure may be reduced to

$$e = R\frac{\sin \Delta_6 - \sin \Delta_5}{\sin \Delta_5 + \sin \Delta_6}, \tag{52}$$

$$r = (R + e)\sin \Delta_5 = (R - e)\sin \Delta_6.^{36} \tag{53}$$

With Copernicus's original values

$$\Delta_5 = 44;48^\circ, \qquad \Delta_6 = 47;16^\circ,$$

$$e = 208, \qquad r = 7193,$$

so the various errors have no effect on the final results. On the other hand, from Ptolemy's

$$\Delta_5 = 44;48^\circ, \qquad \Delta_6 = 47;20^\circ,$$

$$e = 213, \qquad r = 7196.$$

The eccentricity just found corresponds to Ptolemy's eccentricity e from the earth to the center of Venus's eccentric, that is, to OM' in Fig. 57. Hence, in Copernicus's model it is the least distance from the mean sun to the center of Venus's orbit, which is $e_1 - e_2 = e$. Note that $e = 208$ is about half the solar eccentricity of 415 found for Ptolemy's time in III,16, and what is more pertinent is that it is exactly half the maximum solar eccentricity of 416 assumed by Copernicus in the original version of III,21.[37] Ptolemy next finds the distance from the earth to the equant point, which turns out to be exactly $2e$, and Copernicus finds the corresponding greatest distance $e_1 + e_2 = 2e$ from the mean sun to the center of Venus's orbit. The derivation is based upon observations (7) and (8), in both of which the earth is almost exactly 90° from Venus's apsidal line, while the planet displays opposite, and very unequal, elongations in each. In the demonstration of the simple eccentricity, the earth was in the apsidal line, the equation of center was zero, and the elongations were affected only by the change of distance between the earth and the center of Venus's orbit. However, when the earth is 90° from the apsidal line, the equation of center is very near maximum, and the difference in the opposite elongations is due entirely to the difference in direction between the mean sun and the center of Venus's orbit as seen from the earth. It is thus possible to investigate the isolated effect of the equation

[35] Corr. 7194, which Copernicus originally wrote (M 171r9, 173r11) and used in the original version of V,23 (below, p. 400, n. 75).

[36] Cf. *HAMA*, 153–54.

[37] Above, p. 163, n. 48.

of center on the elongations, and find the eccentricity producing the maximum equation of center.

In observations (7) and (8), $\bar{\lambda}^*_{\odot} = 318;50°$, which is correct for (8), but was probably found for both observations by subtracting $\pi = 6;40°$ from Ptolemy's $\bar{\lambda}_{\odot} = 325;30°$, and the elongations,

$$|\Delta_7| = 43;35°, \qquad |\Delta_8| = 48;20°.$$

The configuration is shown in Fig. 62. If the earth were at A, the longitude of the mean sun would be the direction $A\bar{S}B$ or $\lambda^*_{\mathrm{A}} = 48;20°$, and thus at the time of the observations,

$$\text{angle } A\bar{S}O = 48;20° - 318;50° = 89;30° \approx 90°,$$

meaning that $\bar{S}O$ is about perpendicular to the apsidal line, as required. Now directly,

$$e_1 + e_2 = \bar{S}C = R \tan c_3 = R \tan \tfrac{1}{2}(\Delta_8 - \Delta_7), \tag{54}$$

but Copernicus solves in steps. Since OC bisects $\Delta_8 + \Delta_7 = 2c$,

$$2c = \Delta_8 + \Delta_7 = 91;55°, \qquad c = 45;57,30°,$$

and the equation of center

$$c_3 = c - \Delta_7 = \Delta_8 - c = 2;22,30° \approx 2;23°.$$

Thus, where $R = 10,000$, the corresponding eccentricity

$$e_1 + e_2 = \bar{S}C = R \tan c_3 = 416,$$

which Copernicus can find less directly or approximate adequately by $R \sin c_3 = 416$. The result is exactly twice the previous $e = e_1 - e_2 = 208$, and $e_1 + e_2$ is Ptolemy's double eccentricity $2e = 2;30/60$, which is the eccentricity of the solar model. Consequently, Copernicus has established from Ptolemy's observations

$$2e = 416, \qquad e_1 = \tfrac{3}{2}e = 312, \qquad e_2 = e/2 = 104,$$

applicable to the time of Ptolemy and in agreement with Ptolemy's own derivation. Note that the eccentricities e_1 and e_2 have no relation to the radius of the planet's orbit, but must be stated as a fraction of the radius of the earth's orbit since the equation of center depends upon the location of the earth and has no relation to the position of the planet.

Reduction of the Eccentricity from "Many Observations"

Copernicus then describes the structure of his model for Venus—he had not yet done so—and explains the relation of the motion of the center of Venus's sphere in the small circle to the motion of the earth, i.e. that C will make two revolutions in the circle about M for one revolution of the earth with respect to Venus's apsidal line. He then concludes the chapter:

> Through such a model (*hypothesim*) for Venus the uniformity and appearance [of its motions] agree with specific cases (*exemplis*) of every kind, as will soon be clear. Now all the things demonstrated thus far are likewise suitable for our times, [M: except that the eccentricity has decreased by about one-sixth part] so that what was all of 416 parts is now 350, as many observations teach us.[38]

The double eccentricity has been reduced to 350, which in V,23 Copernicus distributes as

$$2e = 350, \qquad e_1 = 246, \qquad e_2 = 104.$$

Hence, the new division is rather far from a bisected eccentricity, which would give $e_1 = 262.5$ and $e_2 = 87.5$, but is equivalent to division in the ratio

$$(e_1 + e_2)/(e_1 - e_2) = 350/142 \approx 2;28/1,$$

a much greater change than for Mars where the ratio was as close as 2;2,30/1.[39] Although the unusual division has only the smallest effect on the equation of center—less than 0;1°—the reduction of $2e$ from 416 to 350 reduces the maximum equation from 2;23° to 2;0°, and the reduction of e from 208 to 142 constricts the range of the maximum equation of the anomaly. For $e = 208$, c_{max} varied from 44;48° at apogee to 47;16° at perigee, but with $e = 142$, the range is only from 45;10° to 46;51°, a decrease of 0;47°.

The question arises of why Copernicus has reduced the eccentricity. His answer is "as many observations teach us," but he gives only one observation of his own (11), which unfortunately does not teach us anything about the reduction. If that is the entire explanation, there is nothing more that we can say about the subject. We have seen, however, that the reduction for Mars could be derived from the observations, and that it even made sense if the distance, although not the direction, from the true sun to the center of Mars's eccentric remained constant while the center of the earth's orbit approached the true sun and the center of Mars's eccentric.[40] In the case of

[38] N 162r25–30, M 172r margin. In M the reduction was originally "about one-fifth part" and $2e$ was reduced from 414 to 333, which was written over 332. The fraction thus applies to $(414 - 332)/414 = 1/5.05 \approx 1/5$ and $(416 - 333)/416 = 1/5.01 \approx 1/5$. $2e = 414$ corresponds to the cancelled solar eccentricity derived incorrectly for Ptolemy's time in the original version of III,16 (above, p. 151, n. 35), and thus this section was written prior to the revision of III,16 and III,21. Since in the *Narratio prima* (above, p. 361, n. 76) Rheticus also reports a reduction of one-fifth part, Copernicus's alteration to $2e = 350$ and "one-sixth part" was made after Sep 1539. With the revision, $(416 - 350)/416 = 1/6.3 \approx 1/6$, but the fraction was dropped in N. In V, 23 (M 173v26) Copernicus originally divided $2e = 333$ into $e_1 = 229$ and $e_2 = 104$, and these cancelled eccentricities account for his incorrect computation of the equation of center (below, p. 395, n. 61). Some of these cancelled numbers have been read differently in Copernicus (1975), 283, 285 apparatus.

[39] Above, p. 356.

[40] Above, pp. 358–359.

Venus, Copernicus reduces the eccentricity more seriously, but keeps the apsidal line sidereally fixed, meaning that with respect to the mean sun it maintains a constant direction, which seems strange since the mean sun, the center of the earth's orbit, is moving. One would like to know if there is a reasonable, and reconstructable, explanation for these changes.

Note first that the entire reduction is applied to e_1, which decreases from 312 to 246, while e_2 remains constant at 104. This agrees with the alteration for Mars, where likewise the reduction was taken by e_1 for the reason that the mean sun was supposed to be approaching the center of the planet's eccentric. In the case of Venus, in Fig. 62 $\bar{S}$ can be thought of as approaching M, while MC remains constant. Further, since the direction of the apsidal line has not changed, either the reduction has taken place directly along line $\bar{S}M$ or compensating motions of $\bar{S}$ and M have produced the same effect, which seems somewhat implausible. It thus appears that Copernicus was lacking observations to confirm or modify Ptolemy's direction of Venus's apsidal line, which he therefore had to accept on trust to be sidereally fixed, but did have some kind of evidence for reducing the eccentricity.

There is nothing unusual about reducing Venus's eccentricity and keeping its apsidal line sidereally fixed. Ptolemy had found that $2e$ for Venus was equal to the sun's eccentricity, although Venus's apogee was 10;30° west of the sun's. While he considered the solar apsidal line tropically fixed, since its tropical longitude had remained constant since the time of Hipparchus, he presumed, on rather less evidence, that the apsidal lines of Venus and the other planets were sidereally fixed.[41] The normal convention in medieval astronomy was to make the double eccentricity of Venus equal to the solar eccentricity—which is always reduced to something less than Ptolemy's—and likewise to have the apsidal lines of the sun and Venus coincident and sidereally fixed. This was already true in Indian astronomy, where Venus has only the epicyclic equivalent of an undivided eccentricity equal to that of the sun, and perhaps originally under Indian influence, the convention was followed by, to give only examples, al-Battānī, the *Toledan Tables*, the *Alfonsine Tables*, and Ibn ash-Shāṭir.[42] Copernicus himself adopts the same practice in the *Commentariolus* where he was relying on the *Alfonsine Tables*, but in *De revolutionibus* he instead follows Ptolemy in distinguishing the apsidal lines of Venus and the sun, and while he does reduce the double eccentricity of Venus, its reduction is somewhat less than that of the sun.[43]

Since there was an observational basis for the reduction of Mars's eccentricity, and since Copernicus claims that the reduction for Venus

---

[41] It is proved only for Mercury, *Almagest* IX,7 (Manitius 2, 132–36).

[42] See the discussion and references in Goldstein and Sawyer (1977), 166–68, Swerdlow (1973), 490.

[43] Note that the cancelled $2e = 333$ or 332 in M (above, p. 385, n. 38) is only slightly larger than the solar $e = 323$ (or 322, above, p. 153, n. 39), and will give a maximum equation of center of 1;54° compared with the sun's 1;51°.

is based upon "many observations," one can take him at his word and look for observations from Copernicus's time that demonstrate a reduction of Venus's eccentricity. It is most natural to expect that he found the reduction by repeating the previous derivations of e and $2e$ by Ptolemy's methods, thus requiring observations of the planet near greatest elongation when the mean sun is respectively in the apsidal line and 90° from the apsidal line. That Copernicus retains the same sidereal direction of the apsidal line found by Ptolemy is good evidence that he did not independently determine its direction by the method of symmetrical elongations. In any case, all that is necessary to investigate the eccentricities is the direction used by Copernicus, however he may have reached it.

Greatest elongations of Venus do not occur frequently; there is one evening and one morning greatest elongation in a synodic period of about 584 days or 1.6 years. Taking the implausibly wide limits for Copernicus's observations of 1500 to 1540, we can examine the dates of the greatest elongations—there are 52 of them—to see whether the mean sun is close enough to the required locations for the elongations to be of any use.[44] Taking 1520 as a mean date, with sufficient accuracy the tropical longitude of Venus's apogee during this period is

$$\lambda_A = \lambda_A^* + \pi = 48;20° + 27;16° = 75;36°.$$

From Copernicus's solar theory, the dates that the mean sun is in the apsidal line or 90° from the apsidal line during the period 1500 to 1540 are as follows:

$\bar{\kappa}$	$\bar{\lambda}_\odot^*$	$\bar{\lambda}_\odot'$	Date
0°	48;20°	75;36°	May 28–29
180	228;20	255;36	Nov 27–28
90	138;20	165;36	Aug 27–28
270	318;20	345;36	Feb 26–27

(55)

In five synodic periods of $8^y - \sim 2^d$, Venus will have five evening and five morning elongations, following which, in the next cycle the elongations will repeat with dates shifted about -2^d and the solar longitude about

[44] In fact, since Copernicus's solar theory depends upon his precession theory, which in turn uses a 1525 observation of Spica (above, pp. 131–132), it could not be applied until after 1525, perhaps several years after, and the precession is also required for assigning a tropical longitude to Venus's apsidal line. Hence, the period that Copernicus could have used his own solar theory to determine when to observe Venus was much shorter than the wide limits we are using in our analysis, perhaps even too short to obtain any useful observations at all. Yet, at an earlier period he could have used the *Alfonsine Tables* to locate the mean sun with sufficient accuracy, and some estimate of the precession since Ptolemy to find a tropical longitude of Venus's apsidal line.

-2°. Hence, in 40 years the dates will shift only about -10^d and the longitudes about -10°, which does not allow much opportunity for the mean sun to shift into the required positions if it is not already rather near them to begin with. Although there are ten greatest elongations in an eight-year cycle, they occur in only five regions of the ecliptic since an evening elongation is always followed two years later by a morning elongation in about the same location, the evening elongation recurring after another six years. The reason is that in 2^y and 6^y respectively Venus's mean anomaly increases by about

$$2(\tfrac{5}{8}\cdot 360^\circ) = 1^r + 90^\circ, \qquad 6(\tfrac{5}{8}\cdot 360^\circ) = 3^r + 270^\circ,$$

each corresponding closely to the increase of mean anomaly modulo 1^r between greatest elongations, 90° E to M and 270° M to E, and their sum is the 5^r cycles of anomaly completed in eight years. It is this relation that makes it possible to find alternate greatest elongations of Venus with the mean sun at the same longitude during a relatively brief period. But it likewise limits the elongations to five locations in eight years.

The dates of greatest elongations of Venus from 1500 to 1540 can be estimated to an accuracy of about $\pm 2^d$ by considering only mean synodic intervals of 583.92^d from single accurately determined E and M greatest elongations. These dates are shown below with the locations of $\bar{\lambda}'_\odot$ and the errors from the required $\bar{\lambda}'_\odot$, everything rounded to degrees.

Dates	$\bar{\lambda}'_\odot$	Req $\bar{\lambda}'_\odot$	Errors	
May 29–18	76°–67°	76°	0°–9°	
Oct 24–11	221–211	256	35–45	
Mar 19–6	5–355	346	19–9	(56)
Aug 12–Jul 31	149–140	166	17–26	
Jan 3–Dec 22	292–282	256	36–26	

The results should not be pressed too closely, but they do indicate whether the locations of the mean sun at the greatest elongations are favorable. In most cases they are not, as indeed they cannot be since the elongations occur at points separated by about $360^\circ/5 = 72^\circ$, while the required locations are 90° apart. The errors also show the number of days between the greatest elongations and the time the mean sun is in the required locations. For apogee at 76° conditions are excellent, but for perigee at 256° they are worthless. Hence Copernicus could not have made suitable observations for finding or confirming e and r by Ptolemy's method. For observations at 90° from the apsidal line the errors are smaller, no greater than the discrepancies that Ptolemy had to accept, and these could perhaps have been used to find a reduced value of $2e$. We have checked the twelve configurations near greatest elongation that occur between 1510 and 1532 on Feb 26–27 and Aug 27–28, taking the longitudes of Venus from Tuckerman (1964), and the resulting $2e$ turns out as small as 272 and as large as 368. It is in any case always less than Ptolemy's 416, and it is possible that one or another pair of elongations as used by Copernicus could give something close to 350, but there is no way of knowing which observations he used or if he in fact attempted to apply Ptolemy's method at all.[45] It is thus difficult to know what Copernicus did with his "many observations," or even what it was that he observed.

Correction of the Mean Anomaly and Positions at Epoch (V,23–24)

V,23–24 are of some interest as having undergone the most extensive revision of any part of *De revolutionibus* for which the entire earlier version survives in M. Although our general principle in this exposition has been to consider changes in M only when they are necessary to understand the final version in N, we shall here consider both versions. Copernicus originally corrected the mean anomaly using Ptolemy's observation (10) from 138 and his own (11) from 1529, an interval of 1391 years. In the revision he substituted for (10) Ptolemy's report of an observation (9) by Timocharis from −271, thereby gaining an interval of fully 1800 years, the longest of any interval used for correcting mean motions. The reason for Copernicus's revision will be considered following the exposition of first the revised and then the original version.

1. *Revised Version*[46]

Ptolemy reports that in the thirteenth year of Ptolemy Philadelphus, which is Era Nabonassar 476 or Era Alexander 52, on Mesore 17/18 at the twelfth seasonal hour of night, apparently about $6;30^h$ after midnight in Alexandria (−271 Oct, 12, $5;30^h$ in Cracow), Timocharis observed that Venus appeared to have exactly overtaken the star opposite *Protrygeter* (*Vindemiatrix*, ε Virginis).[47] Ptolemy identifies the star overtaken by Venus with the most western of the four stars in the left wing (η Virginis). Note that the report can be read as an exact contact conjunction, and so it was understood in the Gerard of Cremona translation, the *Epitome*, and by Copernicus, who says that Venus was seen to have "occupied the position of" or "covered" (*occupasse*) the star.[48]

---

[45] That he originally reduced $2e$ to 333 or 332 (above, p. 385, n. 38) makes any secure reconstruction of what he did even more unlikely.

[46] N 162r–164v, M 173r27–174r25, 174v11–175v10.

[47] Heiberg (1903) 2, 310:23–35 ὁ τῆς Ἀφροδίτης ἐφαίνετο κατειληφὼς τὸν ἀντικείμενον τῷ Προτρυγητῆρι ἀκριβῶς. καταλαμβάνω can also, although not in this case, mean "to cover."

[48] *Almagest* (1515), 111r *vidit stellam veneris iam comprehendisse secundum veritatem super stellam fixam. Epitome* X,6 *vidit eam coniunctam stellae. Almagest* (1551), 238 *Veneris, stella cernebatur exacte obscurasse oppositum praevindemiatrici.* N 162v2 *Venus uisa fuit occupasse stellam fixam.*

In fact this was not true. At the time of the observation, by modern computation:[49]

η Vir	$\lambda = 153;18°$	$\beta = +1;25°$
Venus	153;25	+1;13

Since $\Delta\beta = -0;12°$, the bodies were nowhere near touching, and since $\Delta\lambda = +0;7°$, the conjunction in longitude occured about $2;20^h$ earlier. Nor is the conjunction an exact contact according to Ptolemy's tables. Ptolemy reduces the longitude of the star in his catalogue, ♍8;15°, by a precession of −4;5° to ♍4;10°. If we now compute the position of Venus with the tables in the *Almagest*, we find:

η Vir	$\lambda = 154;10°$	$\beta = +1;30°$ or $+1;10°$[50]
Venus	154;9	+1;40

Hence, while the longitude is nicely reproduced, Venus is either 0;10° or 0;30° above the star (depending upon which textual reading of the star's latitude is accepted), so it is curious that the conjunction has been treated as a contact.

Copernicus's reduction of the observation is itself curious. According to his precession theory, for the time of the observation $\pi = 2;34°$. In the 1515 *Almagest*, the longitude of η Virginis is given as ♍ 8;10° instead of 8;15°, and thus Copernicus's catalogue gives $\lambda^* = 158;10° - 6;40° = 151;30°$.[51] Consequently, at the time of the observation one would have $\lambda = 151;30° + 2;34° = 154;4°$, which is 0;6° less than Ptolemy's 154;10°. However, Copernicus in fact reached the sidereal longitudes of both the star and the mean sun by subtracting from Ptolemy's tropical longitudes an incorrect precession of 2;40°, i.e.

$$\bar{\lambda}^*_{\odot} = 197;3° - 2;40° = 194;23°,$$

$$\lambda^* = 154;10° - 2;40° = 151;30°,$$

[49] P. V. Neugebauer (1912) and Tuckerman (1964).

[50] Manuscripts of the *Almagest* disagree on the latitude. Most give 1;10°, as do the 1515 *Almagest* and Copernicus, but 1;30°, given in Heiberg's D, is more nearly correct (although this is no argument for it). If the third component of Venus's latitude is omitted, the result is $\beta = +1;32°$, very close to the star indeed, but probably of no significance and, in any case still not a contact.

[51] It has been noted by Rosen (Copernicus (1978), 428 n. to 275:5) that in the star catalogue in M (60v31) the longitude has been altered, rather crudely, to 151;35°, which would be correct. Rosen believes the alteration to be in a foreign hand, concerning which we have no opinion, but the correction did not affect V,23 or the star catalogue in N.

and this accounts for the error of $-0;4^\circ$ in Copernicus's solar longitude (see Table 6).[52] Since the longitude of Venus is taken to be the same as that of the star, by this (strictly incorrect) procedure, Copernicus retains Ptolemy's elongation

$$\Delta = 154;10^\circ - 197;3^\circ = 151;30^\circ - 194;23^\circ = -42;53^\circ.^{53}$$

The configuration of the model is shown in Fig. 63. It has been established that where $R = 10{,}000$,

$$e_1 = 312, \qquad e_2 = 104, \qquad r = 7193,$$

and at the time of the observation

$$\bar{\lambda}^*_\odot = 194;23^\circ, \qquad \lambda^* = 151;30^\circ, \qquad |\Delta| = 42;53^\circ, \qquad \lambda^*_{\mathrm{A}} = 48;20^\circ.$$

Ptolemy says that the observation took place after greatest elongation, and thus Venus is located at P rather than at the first intersection of OP with the orbit, which would have the same elongation but would be before greatest elongation. The mean eccentric anomaly $\bar{\kappa} = 146;3^\circ$, which Copernicus gives equivalently as $B\bar{S}O = 180^\circ - \bar{\kappa} = 33;57^\circ$. From (6–10) adapted to Venus, the equation of center

$$c_3 = -1;1^\circ - 0;20^\circ = -1;21^\circ,$$

and the distances

$$OM = 9743, \qquad OC = 9831. \quad \text{(Corr. 9828)}$$

Thus, in triangle OCP,

$$c = \Delta + c_3 = 42;53^\circ - 1;21^\circ = 41;32^\circ,$$

$$\gamma_1 = \sin^{-1}(OC \sin c/r) = 64;59^\circ,$$

$$\gamma_2 = 180^\circ - (c + \gamma_1) = 73;29^\circ,$$

so that

$$\gamma_2 + c_3 = \bar{\alpha} - 180^\circ = 73;29^\circ - 1;21^\circ = 72;8^\circ,^{54}$$

although Copernicus gives $72;5^\circ$, and then finds

$$\bar{\alpha} = 252;5^\circ,$$

the mean anomaly at the time of the observation. Ptolemy had found $252;7^\circ$, and correctly Copernicus should have found $252;8^\circ$, in better agreement.

---

[52] Copernicus may also have taken a time for this observation about two hours too early. See below, p. 396, n. 63.

[53] *Almagest* (1515), (1551), and *Epitome* X,6 by error $\bar{\lambda}_\odot = ♎ 17\frac{1}{3}^\circ$, leading to $\Delta = ♎ 17;20^\circ - ♍ 4;10^\circ = 43;10^\circ$. Copernicus must have recomputed $\bar{\lambda}_\odot$ to get the correct elongation.

[54] N 162v37, M 175r9, for *EFG* read *LFG*. Perhaps Copernicus skipped a line in writing out this section. Disregard notes in Copernicus (1879), Anm., 59 n. 428 and Copernicus (1978), 428 n. to 275:25.

Copernicus's only reported observation of Venus (11) is of an occultation by the moon on 12 Mar 1529 at $19;30^h$ in Frauenburg. The report itself reads as follows:

> We ourselves observed another position of Venus in the Year of Christ 1529, on the 4th day before the Ides of March, one hour after sunset, and thus in the beginning of the eighth hour after noon. And we saw the moon begin to occult Venus on the dark side at a distance midway between both horns, and the occultation lasted until the end of that very hour [M: or a little more] until the planet was seen to emerge toward the west from the other side in the middle of the curved back (*gibbositatis*) of the horns. It is therefore clear that in the middle of this hour, or thereabout, there was a conjunction of the centers of the moon and Venus, and we happened upon this wonderful sight in Frauenburg. But Venus was still increasing its evening elongation, and not yet in the point of tangency of its sphere.[55]

The only problem with the report is its specification that the conjunction was central, that the planet entered into the middle of the space between the horns of the crescent and emerged from the middle of the back.[56]

The apparent time elapsed since the Epoch of Christ,

$$\Delta t = 1529^y\ 87^d\ 19;30^h = 25,29^y\ 1,27;48,45^d,$$ [57]

for which, from the solar and precession theory of Book III,

$$\bar{\lambda}^*_\odot = 332;12^\circ, \qquad \bar{\pi} = 26;51^\circ, \qquad \bar{\lambda}_\odot = 359;3^\circ, \qquad \pi = 27;22^\circ,$$

$$\bar{\lambda}'_\odot = 359;34^\circ, \qquad c_\odot = +1;50^\circ, \qquad \lambda_\odot = 1;24^\circ, \qquad \alpha_\odot = 1;17^\circ.$$

The equation of time

$$\Delta E = (\alpha - \bar{\lambda}) - (\alpha - \bar{\lambda})_{0G} = +2;14^\circ - 1;53^\circ = +0;21^\circ = +0;1,24^h,$$

and thus correctly, the interval of mean time

$$\Delta\bar{t} = \Delta t + \Delta E = 1529^y\ 87^d\ 19;31,24^h.$$

Copernicus, however, reports

$$\bar{\lambda}^*_\odot = 332;11^\circ, \qquad \pi = 27;24^\circ,$$

and while he does not report the remaining quantities, his equation of time was about $+0;4^\circ$, for he says that

$$\Delta\bar{t} = 1529^y\ 87^d\ 19;34^h.$$ [58]

---

[55] N 163r5–15, M 173r27–173v5.

[56] Below, pp. 394–395.

[57] The text gives $7\frac{1}{2}^h$, i.e. counted from noon, as in the description of the observation.

[58] Again, the text by error $7;34^h$ counted from noon.

Considering that the time of occultation was only roughly estimated, the inclusion of so small an equation of time is in any case not significant. For Copernicus's mean time, from the lunar theory of Book IV,

$$\bar{\eta} = 33;59^\circ, \qquad \bar{\alpha} = 206;5^\circ,$$

$$c_3 = +11;57^\circ, \qquad \alpha = \bar{\alpha} + c_3 = 218;2^\circ,$$

$$c = +3;58^\circ, \qquad \eta = \bar{\eta} + c = 37;57^\circ,$$

$$\bar{\omega} = 73;1^\circ, \qquad \omega = \bar{\omega} + c = 76;59^\circ,$$

$$\lambda^*_☾ = \bar{\lambda}^*_\odot + \eta = 332;11^\circ + 37;57^\circ = 10;8^\circ, \qquad \beta_☾ = +1;7^\circ,$$

$$\lambda_☾ = \lambda^*_☾ + \pi = 10;8^\circ + 27;24^\circ = 37;32^\circ = ♉7;32^\circ,$$

Copernicus reports only some of these quantities, and with a number of discrepancies:

$$\bar{\eta} = 33;57^\circ, \qquad \bar{\alpha} = 205;1^\circ, \qquad \bar{\omega} = 71;59^\circ,$$

$$\lambda^*_☾ = 10^\circ, \qquad \beta_☾ = +1;13^\circ, \qquad \lambda_☾ = ♉7;24^\circ.$$

The errors in $\bar{\alpha}$ and $\bar{\omega}$ suggest computation for a time about two hours earlier, and the results are that λ is in error by $-0;8^\circ$ and β by $+0;6^\circ$.

One must next compute the lunar parallax, and Copernicus gives almost no details of his procedure. He does not give the true longitude of the sun, but his value cannot differ by more than a minute or so from our $\lambda_\odot = 1;24^\circ$ at $19;30^h$, which is accurate enough for what follows. Thus, for $\lambda_\odot = 1;24^\circ$ at $7;30^h$ after noon at a latitude $\varphi = 54;20^\circ$, from the spherical astronomy of Book II:

$$\lambda_M = 112;1^\circ, \qquad \lambda_G = 196;11^\circ, \qquad v = 57;48^\circ,$$

and since the moon is considered to be in the ecliptic for the simple method of computing its parallax, its zenith distance and the angle of intersection of the ecliptic and circle of altitude,

$$\zeta = 72;10^\circ, \qquad \xi = 34;2^\circ.$$

Using Copernicus's values of $\bar{\eta}$ and $\bar{\alpha}$,

$$\bar{\eta} = 33;57^\circ \approx 34^\circ, \qquad \alpha = \bar{\alpha} + c_3 = 205;1^\circ + 11;57^\circ \approx 217^\circ,$$

we find the parallax from the table in IV,24 and resolve it into its components

$$p = 0;59,45^\circ, \qquad p_\lambda = -0;49,31^\circ, \qquad p_\beta = -0;33,26^\circ.$$ [59]

Copernicus reports only that

$$\lambda_G = ♎15^\circ, \qquad p_\lambda = -0;48^\circ, \qquad p_\beta = -0;32^\circ,$$

[59] If ζ and ξ are rounded to degrees, the abbreviated parallax computation gives $p = 0;58,42^\circ$, $p_\lambda = -0;49^\circ$, $p_\beta = -0;33^\circ$.

all rather close to our computation. A comparison of the resulting true and apparent positions follows:

	Cop.	Recomp.
$\lambda_{☾}$	37;24°	37;32°
p_λ	−0;48	−0;50
$\lambda'_{☾}$	36;36	36;42
$\beta_{☾}$	+1;13	+1;7
p_β	−0;32	−0;33
$\beta'_{☾}$	+0;41	+0;34

(57)

Copernicus concludes that at the midpoint of the occultation, the tropical position of Venus is

$$\lambda = ♉\, 6;36°, \qquad \beta = +0;41°,$$

and subtracting the precession, the sidereal longitude

$$\lambda^* = 36;36° - 27;24° = 9;12°,$$

as in M (173v18), although N (163r26) gives 9;11°. From the recomputed longitude and precession, the result is 9;20°, so Copernicus's error is −0;8°.

In the case of an occultation, it is of interest to compare Copernicus's observation and results with a modern computation. For $19;10^h$ and $20;10^h$ in Frauenburg (corresponding to 5:30 P.M. and 6:30 P.M. G.M.T.), we have for the moon and Venus:[60]

	$19;10^h$	$20;10^h$
Moon		
λ	37;21°	37;58°
p_λ	−0;43	−0;36
λ'	36;38	37;22
β	+1;1	+0;58
p_β	−0;30	−0;30
β'	+0;31	+0;28
Venus		
λ	36;58	37;1
β	+0;37	+0;37

(58)

The conjunction will thus occur at

$$t \approx 19;40^h, \qquad \lambda \approx 37;0°,$$

[60] Using P. V. Neugebauer (1914), (1929), and Tuckerman (1964).

so Copernicus's error is $-0;10^h$ in time, which is quite small, but $-0;24°$ in longitude. The correct latitude of the moon at the time of conjunction is about $+0;30°$ and the latitude of Venus $+0;37°$. Hence Venus was $+0;7°$, about one-half the lunar radius, north of the center of the moon, and could not have entered and emerged in line with the center, as Copernicus reports the observation.

The occultation is shown in Fig. 64 with the correct relative positions of the moon M and Venus P at the beginning P_I, middle P_M, and end P_E of the occultation. The direction $M\odot$ that determines the orientation of the crescent is within 1° of parallelism to the ecliptic. While it may not be easy to estimate exactly where in relation to the ends of the crescent the dark part of the moon occults the planet at P_I, it is clear that the emergence of the planet at P_E is not "in the middle of the curved back of the horns," as Copernicus reports.

The configuration of the model at the time of the observation is shown in Fig. 65. Copernicus uses the following values:

$$\bar{\lambda}^*_\odot = 332;11°, \qquad \lambda^* = 9;12°, \qquad \Delta = +37;1°, \qquad \lambda^*_A = 48;20°,$$

and it is now assumed that where $R = 10{,}000$,

$$e_1 = 246, \qquad e_2 = 104, \qquad r = 7193.$$

Since greatest elongation is yet to occur, the planet is at P rather than at the first intersection of OP with the orbit. Copernicus gives the mean eccentric anomaly correctly as $\bar{\kappa} = -76;9°$, but the equation of center is computed incorrectly as

$$c_3 = +1;15° + 0;35° = 1;50°,$$

when it should be

$$c_3 = +1;21,36° + 0;34,46° = +1;56,22° \approx +1;56°,$$

and the distances,

$$OM = 10056, \quad \text{(Corr. 10062)} \qquad OC = 10034.^{61} \quad \text{(Corr. 10040)}$$

Thus, in triangle OCP,

$$c = \Delta - c_3 = 37;1° - 1;50° = 35;11°,$$

$$\gamma_1 = \sin^{-1}(OC \sin c/r) = 53;30°,$$

$$\gamma_2 = 180° - (c + \gamma_1) = 91;19°,$$

---

[61] Copernicus in fact computed from the cancelled $e_1 = 229$ and $e_2 = 104$, from $2e = 333$ (above, p. 385, n. 38), which give strictly

$$c_3 = +1;16° + 0;35° = 1;51°, \qquad OM = 10057, \qquad OC = 10035,$$

very close to his results. That he did not change this computation when he changed the eccentricity is surprising.

so that

$$\gamma_2 - c_3 = 180° - \bar{\alpha} = 89;29°,$$

and

$$\bar{\alpha} = 90;31°$$

is the mean anomaly at the time of the observation.[62]

The correction of the mean anomaly is now straightforward, at least in principle. At the times of the observations of Timocharis (9) and Copernicus (11),

$$\bar{\alpha}_9 = 252;5°, \qquad \bar{\alpha}_{11} = 90;31°, \qquad \Delta\bar{\alpha} = 198;26°.$$

The interval between the two observations is

$$\Delta t = 1800^y\, 236^d\, 14;0^h = 30,0^y\, 3,56;35^d,$$

although Copernicus gives the days as "about" $236;40^d$, an inaccuracy of $+2;0^h$.[63] Estimating the complete rotations of $\Delta\bar{\alpha}$ from the tables in the *Almagest* by (36) as 1125^r, we have

$$\Delta\bar{\alpha} = 1125^r + 198;26°.$$[64]

Copernicus then claims that $\bar{\alpha}^y = 225;1,45,3,40°$, although correctly if $\Delta\bar{\alpha} = 1125^r + 198;26°$ and $\Delta t = 1800^y\, 236;40^d$, from (37–39)

$$\bar{\alpha}^y = 225;1,45,2,52,48\ldots°,$$

$$P_\alpha = 583;55,27,22,38\ldots^d,$$

$$5P_\alpha = 2919;37,16,53^d = 8^y - 2;25,56,27^d,$$

or $c \approx -2;25,56^d$, none of which agree with either N or M.[65] He also says that $\bar{\alpha}^d = 0;36,59,28°$, which could be a rounding of the correct $0;36,59,27,57^{°/d}$, but is not the value given in V,1 in M and N.[66]

The interval from Timocharis's observation (9) to the first Olympiad,

$$\Delta t = 503^y\, 228;40^d,$$

the observation now considered to be at 4^h after midnight. During this time $\Delta\bar{\alpha} = 290;39°$, and thus at the first Olympiad,

$$\bar{\alpha}_{A0} = 252;5° - 290;39° = 321;26°.$$

---

[62] With the correct c_3 and OC the result would be

$$c = 35;5°, \qquad \gamma_1 = 53;21°, \qquad \bar{\alpha} = c + \gamma_1 + c_3 = 90;22°,$$

so Copernicus's error amounts to $+0;9°$.

[63] Note that Copernicus's $\lambda^*_{\odot 9} = 194;23°$ corresponds to a time about two hours early for Timocharis's observation, i.e. about $3;30^h$ or 4^h after midnight at the meridian of Cracow, and in finding the epoch, he takes the time as 4^h after midnight.

[64] N and M give 1115^r, which is retained in the critical editions and translations, but Copernicus definitely computed correctly and uses 1125^r.

[65] From the correct $\Delta t = 1800^y\, 236;35^d$, $\bar{\alpha}^y = 225;1,45,9,2,35\ldots°$.

[66] See Table 1.

The remaining epochs are computed in the order A. → C. → D. → G. Copernicus does not give the mean heliocentric longitude $\bar{\lambda}^{*\prime} = \bar{\lambda}^{*}_{\odot} + \bar{\alpha}$ of Venus (or Mercury) since it is never used in computing the position of an inferior planet. The mean geocentric sidereal and tropical longitudes are those of the sun given in III,19.[67] Since the apsidal line is sidereally fixed at $\lambda^{*}_{A} = 48;20°$, its sidereal longitude is the same at all epochs while its mean tropical longitude is $\bar{\lambda}_{A0} = 48;20° + \bar{\pi}_0$ for each epoch. The mean anomaly and the mean tropical longitude of the apsidal line, which is not given by Copernicus, are as follows for each epoch:

Epoch	$\bar{\alpha}_0$	$\bar{\lambda}_{A0}$
A. Olympiad	321;26°	43;4°
C. Alexander	81;52	49;22
D. Caesar	73;2[68]	53;15
G. Christ	126;45	53;52

(59)

2. *Original Version*[69]

Ptolemy corrected the mean anomaly of Venus using Timocharis's observation (9) and an observation (10) of his own from Antoninus 2, the interval between the two being about 409 years. Copernicus originally corrected the mean anomaly using Ptolemy's observation rather than Timocharis's, but he must then have found flaws in his use of (10) and turned instead to (9), which also had the benefit of increasing the interval for correcting the mean anomaly from 1391^y to 1800^y. To anticipate our later analysis, the problem with (10) was that Copernicus derived a mean anomaly for the time of the observation about $\frac{1}{2}°$ less than Ptolemy had derived, and then he may have noted that if he tried to compute back to the mean anomaly at the time of observation (9), the result was in error by more than $2\frac{1}{2}°$ compared to Ptolemy's value. Clearly, something was wrong with the mean anomaly derived from (10) and (11).[70]

On Antoninus 2 Tybi 29/30[71] at $4\frac{3}{4}^h$ after midnight in Alexandria (138 Dec 16, $3;45^h$ in Cracow), Ptolemy observed Venus after its greatest elongation to fall in a straight line between the moon and the northern star in the

[67] Above, p. 165.

[68] By a strange scribal error in M and N, $\bar{\alpha}_{0D}$ is given as 70;26°, but Copernicus must have computed correctly 73;2° and used it to find $\bar{\alpha}_{0G} = \bar{\alpha}_{0D} + \Delta\bar{\alpha} = 73;2° + 53;43° = 126;45°$.

[69] M 172r6–173v27, 174r25–174v10. Of course, not in N.

[70] Discussed in more detail below, pp. 402–403.

[71] Copernicus writes (M 172r8, 174r29) Tybi 20, but the intervals he gives from the Epoch of Christ and between observations (10) and (11) show that he had the correct date, from which all his computations follow.

forehead of Scorpio (β Scorpii), one-and-one-half times farther from the moon than from the star. For the time of the observation, Ptolemy finds:

$$\bar{\lambda}_{\odot} = ♐ 22;9°, \qquad \lambda_{\odot} = ♐ 22;52° \approx ♐ 23°,$$

$$\bar{\lambda}_{☾} = ♏ 11;24°, \qquad \lambda_{☾} = ♏ 5;45°, \qquad \beta_{☾} = +5;0°,$$

and applying the correction for parallax, the apparent position of the moon,

$$\lambda'_{☾} = ♏ 5;45° + 1° = ♏ 6;45°, \qquad \beta'_{☾} = +5° - 0;20° = +4;40°.$$

The coordinates of β Scorpii,

$$\lambda_{*} = ♏ 6;20°, \qquad \beta_{*} = +1;20°,$$

and since Venus is one-and-one-half times farther from the moon than from the star, or $\frac{2}{5}$ the distance from the star to the moon, its coordinates are

$$\lambda = \lambda_{*} + \tfrac{2}{5}(\lambda'_{☾} - \lambda_{*}) = ♏ 6;20° + 0;10° = ♏ 6;30°,$$

$$\beta = \beta_{*} + \tfrac{2}{5}(\beta'_{☾} - \beta_{*}) = +1;20° + 1;20° = +2;40°.$$

Hence, the elongation of Venus from the mean sun,

$$\Delta = 216;30° - 262;9° = -45;39°.$$

Copernicus cannot accept Ptolemy's position, and reduce it to a sidereal longitude by subtracting the precession, since it depends upon an apparent position of the moon containing a larger correction for parallax than Copernicus's lunar model and lunar distance will permit. Consequently, he must recompute the true and apparent position of the moon. The apparent time elapsed from the Epoch of Christ at the meridian of Cracow,

$$\Delta t = 138^{y}\ 18^{d}\ 3;45^{h},$$

for which, from the solar and precession theory of Book III,

$$\bar{\lambda}^{*}_{\odot} = 255;30°, \qquad \bar{\pi} = 7;27°, \qquad \bar{\lambda}_{\odot} = 262;57°, \qquad \pi = 6;40°,$$

$$\bar{\lambda}'_{\odot} = 262;10°, \qquad c_{\odot} = +0;40°, \qquad \lambda_{\odot} = 262;50°, \qquad \alpha_{\odot} = 262;10°.$$

The equation of time,

$$\Delta E = (\alpha - \bar{\lambda}) - (\alpha - \bar{\lambda})_{0G} = -0;47° - 1;53° = -2;40° = -0;10,40^{h},$$

and thus correctly the interval of mean time,

$$\Delta\bar{t} = \Delta t + \Delta E = 138^{y}\ 18^{d}\ 3;34,20^{h}.$$

Copernicus reports

$$\bar{\lambda}^{*}_{\odot} = 255;30°, \qquad \lambda_{\odot} = ♐ 23°,$$

the latter probably just taken from Ptolemy. Then, evidently miscomputing ΔE as about $-0;4°$, he says that the mean time

$$\Delta\bar{t} = 138^{y}\ 18^{d}\ 3;41^{h},$$

and further, that 3;41^{h} are 0;9,32^{d}, although correctly 3;41^{h} = 0;9,22,30^{d}.[72] Using Copernicus's $\Delta\bar{t} = 138^{y}\ 18;9,32^{d}$,

$$\bar{\eta} = 319;18^\circ, \qquad \bar{\alpha} = 87;37^\circ,$$

$$c_3 = -12;27^\circ, \qquad \alpha = \bar{\alpha} + c_3 = 75;10^\circ,$$

$$c = -5;50^\circ, \qquad \eta = \bar{\eta} + c = 313;28^\circ,$$

$$\bar{\omega} = 12;19^\circ, \qquad \omega = \bar{\omega} + c = 6;29^\circ,$$

$$\lambda^*_{☾} = \bar{\lambda}^*_{\odot} + \eta = 255;30^\circ + 313;28^\circ = 208;58^\circ, \qquad \beta_{☾} = +4;58^\circ,$$

$$\lambda_{☾} = \lambda^*_{☾} + \pi = 208;58^\circ + 6;40^\circ = 215;38^\circ = ♏5;38^\circ.$$

Copernicus gives $\bar{\eta}$, $\bar{\alpha}$, $\bar{\omega}$, $\lambda^*_{☾}$, $\lambda_{☾}$, and $\beta_{☾}$, and all agree except for

$$\lambda^*_{☾} = 209;4^\circ, \qquad \lambda_{☾} = 209;4^\circ + 6;41^\circ = 215;45^\circ = ♏5;45^\circ,$$

the latter in exact agreement with Ptolemy. Perhaps Copernicus increased $\lambda^*_{☾}$ by 0;6° for this purpose.

Now, taking the solar longitude to be $\lambda_{\odot} = ♐23^\circ$ at an apparent time of 16;45^{h} after noon in Alexandria ($\varphi \approx 31$)°, the longitudes of the midheaven and horoscopus, and the horizon angle, are

$$\lambda_{M} = 151;30^\circ, \qquad \lambda_{G} = 244;33^\circ, \qquad v = 71;32^\circ,$$

and since the latitude of the moon is ignored in computing its parallax, the zenith distance and angle of the ecliptic and circle of altitude,

$$\zeta = 72;12^\circ, \qquad \xi = 19;26^\circ.$$

Finding the parallax from the table in IV,24, and resolving it into its components,

$$p = 0;54,38^\circ, \qquad p_{\lambda} = +0;51,31^\circ, \qquad p_{\beta} = -0;18,10^\circ.$$ [73]

Copernicus gives only the following:

$$\lambda_{M} = ♍2^\circ, \qquad \lambda_{G} = ♏25^\circ, \qquad p_{\lambda} = +0;51^\circ, \qquad p_{\beta} = -0;16^\circ,$$

and the apparent sidereal coordinates of the moon,

$$\lambda^{*\prime}_{☾} = 209;4^\circ + 0;51^\circ = 209;55^\circ, \qquad \beta'_{☾} = +4;58^\circ - 0;16^\circ = +4;42^\circ.$$ [74]

The sidereal coordinates of β Scorpii,

$$\lambda^*_{*} = 209;40^\circ, \qquad \beta_{*} = +1;20^\circ,$$

---

[72] Rosen (Copernicus (1978), 427 n. to 273:16) suggests, plausibly, that 32 is a scribal error for 23.

[73] The abbreviated parallax computation gives $p = 0;53,42^\circ$, $p_{\lambda} = +0;51^\circ$, $p_{\beta} = -0;18^\circ$.

[74] λ_{M} is exactly the value given by Ptolemy, while p_{λ} was altered from 0;52° and p_{β} from 0;18° (M 172v1–2), the last in better agreement with our computation. Combined with $\beta_{☾}$, it would give $\beta'_{☾} = +4;40^\circ$, identical with Ptolemy, so the alteration is hard to understand.

and thus, the sidereal coordinates of Venus

$$\lambda^* = \lambda^*_* + \tfrac{2}{5}(\lambda^{*\prime}_{☾} - \lambda^*_*) = 209;40° + 0;6° = 209;46°,$$

$$\beta = \beta_* + \tfrac{2}{5}(\beta'_{☾} - \beta_*) = +1;20° + 1;21° = +2;41° \approx +2;40°.$$

The elongation of Venus from the mean sun is thus

$$\Delta = 209;46° - 255;30° = -45;44°,$$

compared with Ptolemy's −45;39°, a difference of 0;5° that turns out to have a significant effect on the mean anomaly.

The reduction of the observation by Ptolemy and Copernicus is shown in Fig. 66 in which the scale is $\Delta\lambda = 4\Delta\beta$ and, where necessary, Copernicus's sidereal longitudes have been converted to tropical by the addition of 6;40°. The true moon is M, the apparent M', the star *, Venus P, and the subscripts p and c show the locations according to Ptolemy and Copernicus. Note the difference of 0;4° in the longitude of P.

To find the mean anomaly at the time of the observation, we now have the following positions in the model shown in Fig. 67, in which the planet is slightly past greatest elongation:

$$\bar{\lambda}^*_{\odot} = 255;30°, \qquad \lambda^* = 209;46°, \qquad |\Delta| = 45;44°, \qquad \lambda^*_{A} = 48;20°.$$

At the time of Ptolemy, where $R = 10{,}000$,

$$e_1 = 312, \qquad e_2 = 104, \qquad r = 7193.$$

The mean eccentric anomaly $\bar{\kappa} = 207;10°$, which Copernicus gives equivalently as $B\bar{S}O = \bar{\kappa} - 180° = 27;10°$, and the equation of center

$$c_3 = +0;50° + 0;16° = +1;6°. \quad \text{(Corr. 1;4°)}$$

The distances

$$OM = 9724, \ \text{(Corr. 9723)} \qquad OC = 9817, \ \text{(Corr. 9815)}$$

and thus in triangle OCP,

$$c = \Delta + c_3 = 46;50°,$$

$$\gamma_1 = \sin^{-1}(OC \sin c/r) = 84;19°,\text{[75]} \quad \text{(Corr. 84;32°)}$$

$$\gamma_2 = 180° - (c + \gamma_1) = 48;51°.$$

The error in γ_1 is rather serious, but following Copernicus's steps,

$$180° - \gamma_2 = 360° - \alpha = 131;6°, \quad \text{(Corr. 131;9°)}$$

$$360° - \bar{\alpha} = (360° - \alpha) - c_3 = 131;6° - 1;6° = 130;0°,$$

$$\bar{\alpha} = 230;0°,$$

differing by −0;32° from Ptolemy's 230;32°, a considerable disagreement.[76]

[75] Incorrectly labelled *EFG* (M 173r13–14) when it should be *FGE*. More important, the angle seems to have been computed from cancelled values,

$$\gamma_1 = \sin^{-1}(9817 \sin 46;49°/7194) = 84;17°.$$

[76] Copernicus originally wrote 229;58° (M 173r22).

Although Copernicus's derivation contains a number of errors, the most serious problem arises from the difference of 0;5° between his elongation of 45;44° and Ptolemy's 45;39°. Since the planet is rather close to greatest elongation, the anomaly is very sensitive to changes in the elongation, that is, small changes in the elongation correspond to large changes in the anomaly. In this case, letting $\Delta = 45;39°$, and keeping Copernicus's incorrect c_3 and OC,

$$c = 46;45°, \qquad \gamma_1 = 83;46°, \qquad \gamma_2 = 49;29°,$$
$$\gamma_2 + c_3 = \bar{\alpha} - 180° = 50;35°, \qquad \bar{\alpha} = 230;35°,$$

which is much closer to Ptolemy's 230;32°. However, it should be noted that the computation is so sensitive that with the same value of Δ but with the correct c_3 and OC, the result is $\bar{\alpha} = 230;58°$, which agrees no better than Copernicus's value although the error is in the opposite direction. Clearly, for correcting the mean anomaly, the planet should be observed as far as possible from greatest elongation, preferably close to inferior conjunction where large changes in the elongation correspond to small changes in the anomaly so the anomaly can be found quite accurately.

Copernicus next derives the mean anomaly at his own observation (11) of 1529 Mar 12, and this is the same as in the revised version since M is unaltered. The result is that at observations (10) and (11)

$$\bar{\alpha}_{10} = 230;0°,^{77} \qquad \bar{\alpha}_{11} = 90;31°, \qquad \Delta\bar{\alpha} = 220;31°.$$

The interval of time is

$$\Delta t = 1391^{y}\,69^{d}\,15;45^{h} = 23,11^{y}\,1,9;39,22,30^{d},$$

although Copernicus rounds the days to $69;39,23^{d}$. Estimating the complete rotations of $\Delta\bar{\alpha}$ from the tables in the *Almagest* by (36),

$$\Delta\bar{\alpha} = 869^{r} + 220;31°.$$

Copernicus says that the table will give the same value, and indeed the original table in M (147)[78] gives 220;31,11°, while the revised table in N(138v–139r) gives 218;32,20°, a difference of nearly 2° in 1391^{y} which is not inconsiderable. This shows that the unrevised table in M is more-or-less based upon observations (10) and (11), and it has already been seen that the revised table in N is more-or-less derived from observations (9) and (11).

The computation of the positions at epoch then follows from observation (10). The interval from the Epoch of Christ,

$$\Delta t = 138^{y}\,18;9,30^{d},$$

the days evidently rounded from the incorrect $18;9,32^{d}$ used in the previous computation of the position of the moon.[79] In this time, from the table in M (147) $\Delta\bar{\alpha} = 105;25°$, so that at epoch

$$\bar{\alpha}_{0G} = 230;0° - 105;25° = 124;35°,$$

[77] Copernicus first wrote 230;32° (M 174r26), Ptolemy's value, but then deleted the 32.

[78] Above, p. 306.

[79] Above, p. 399.

the mean anomaly at the Epoch of Christ. The remaining epochs are all correctly computed in the order G. → A. → C. → D. using the table in M or the value of $\bar{\alpha}$ underlying the table, and they are as follows:

$$\text{A. } 318;9^\circ, \quad \text{C. } 79;14^\circ, \quad \text{D. } 70;48^\circ, \quad \text{G. } 124;35^\circ. \tag{60}$$

Note that G. differs from the revised version by $-2;10^\circ$, while A., which admittedly is of little interest, differs by fully $-3;17^\circ$.

The question now arises of why Copernicus cancelled the derivation of the mean anomaly from observations (10) and (11) and substituted the revised derivation from (9) and (11). One possible explanation is that he was concerned about the difference of $0;32^\circ$ in $\bar{\alpha}_{10}$ in his and Ptolemy's demonstrations, but it is not clear whether he would consider his derivation in error or would instead believe that Ptolemy's value was incorrect since it followed from an erroneously large lunar parallax. Let us suppose, however, that using the $\bar{\alpha}$ of the original derivation, that is, of the original table in M, Copernicus tried to compute the mean anomaly at the time of observations (10) and (9) by computing back from observation (11), where $\bar{\alpha} = 90;31^\circ$, comparing his results with Ptolemy. He would find:

Observation	Δt	Comp. $\bar{\alpha}$	Ptol. $\bar{\alpha}$	$\Delta\bar{\alpha}$
(10)	$-1391^y\ 69;39,23^d$	$230;0^\circ$	$230;32^\circ$	$-0;32^\circ$
(9)	$-1800^y\ 236;40^d$	249;30	252;7	$-2;37$

The difference of $-0;32^\circ$ in $\bar{\alpha}_{10}$ was already in Copernicus's derivation, but the difference of $-2;37^\circ$ in recomputing $\bar{\alpha}_9$ is clearly intolerable since it is hard to believe that Ptolemy's derivation could show so large an error—no lunar parallax is involved—and it is always Copernicus's goal to reproduce, so far as possible, a close approximation to the observations and theory of the *Almagest* for the period of the observations reported by Ptolemy. Hence, this discrepancy would give Copernicus a reason for questioning the accuracy of the mean anomaly derived from (10) and (11).

Further, by carrying out a new derivation based upon (9) and (11), the interval between the observations is increased from under 1400^y to 1800^y, which should provide a more accurate value unless—and this Copernicus cannot know—observation (9) is considerably less accurate than (10). However, in substituting the revised mean anomaly from (9) and (11), a problem still remains since now $\bar{\alpha}_{10}$ shows a serious error:

Observation	Δt	Comp. $\bar{\alpha}$	Ptol. $\bar{\alpha}$	$\Delta\bar{\alpha}$
(10)	$-1391^y\ 69;39,23^d$	$231;59^\circ$	$230;32^\circ$	$+1;27^\circ$
(9)	$-1800^y\ 236;40^d$	252;5	252;7	$-0;2$

Obviously there is no way to reconcile observations 9/10/11 with a single value of $\bar{\alpha}$. The reason becomes clear when we compare the values of $\bar{\alpha}$ for each observation derived by Ptolemy and Copernicus with a modern computation.[80]

Observation	Ptol. $\bar{\alpha}$	Cop. $\bar{\alpha}$	Comp. $\bar{\alpha}$
(9)	252;7°	252;5°	251;2°
(10)	230;32	230;0	231;8
(11)	—	90;31	90;25

The errors in $\bar{\alpha}_9$ are about $+1°$ for both Ptolemy and Copernicus, while the error in $\bar{\alpha}_{10}$ is about $-0;30°$ for Ptolemy but $-1°$ for Copernicus, and the errors in the two observations are in opposite directions. Hence the mean anomaly derived from (10) and (11) will be too large, the mean anomaly derived from (9) and (11) will be too small, and there is no way a single value of $\bar{\alpha}$ computed back from $\bar{\alpha}_{11}$ can reproduce both $\bar{\alpha}_9$ and $\bar{\alpha}_{10}$ as derived by either Ptolemy or Copernicus. While Copernicus has no way of knowing the correct value of $\bar{\alpha}$ for each observation, he was surely aware of the inconsistencies previously shown, and thus it seems reasonable to believe that he chose to use the value derived from the earlier observation in the hope that the interval of 1800^y would at least distribute the cumulative error, which the two tests show to be smaller, over a longer period.

B. Mercury

Of the five planets visible to the unaided eye Mercury is the least often seen and is never visible very far from the horizon. Owing to the high eccentricity of its orbit, its greatest elongations from the true sun vary considerably, lying between about 17° and 29°, but because the plane of its orbit is inclined by some 7° to the ecliptic, even at greatest elongation it can be under the horizon and invisible for many locations on the earth. Since the inclination of the ecliptic to the horizon decreases with increasing terrestrial latitude, the periods of Mercury's visibility likewise decrease with the increasing latitude of the observer.[1] Copernicus says in V,30 (N 169r17–29) that for these reasons the planet could seldom be seen in Frauenburg. He mentions some of the zodiacal signs in which Mercury omits a morning or evening appearance because the elongation from the true sun required for visibility exceeds

---

[80] Using P. V. Neugebauer (1914), and taking the difference between the mean heliocentric longitude of Venus and the mean longitude of the sun. Compare (10) with 230;58° on p. 401.

[1] See the visibility tables in P. V. Neugebauer (1938), 315–16, which however, extend only to $\varphi = 40°$, while visibility would be even more restricted in Frauenburg ($\varphi \approx 54;20°$).

its greatest elongations in those signs—his description is rather vague and not complete—and in a well-known passage he remarks that his observing conditions were unfavorable because of fogs off the Vistula and bad weather that the "ancients" near the Nile were spared. Hence for Mercury he cites no observations of his own, but relies entirely on Ptolemy's for deriving the elements of its orbit, and uses three modern observations by Bernhard Walther, who observed in Nuremberg ($\varphi \approx 49;27°$, $l = 11;5°$ E), in order to establish the location of the apsidal line and the position of the mean anomaly in his time. As was the case for Venus, the result is an adaptation rather than a verification of Ptolemy's elements, the few modifications being imposed by the different structure of Copernicus's model and the requirement that it produce the same greatest elongations as Ptolemy's model, which it does very well.[2]

Ptolemy established the elements of Mercury, just as he did for Venus, by taking observations of the planet very near equal and opposite greatest elongations in order to locate the apsidal line, and then using greatest elongations with the direction of the mean sun in the apsidal line and perpendicular to the apsidal line to find the radius of the epicycle and the two eccentricities, to the center of the eccentric and equant respectively. In this way he discovered that, unlike every other planet, the equant is located at half rather than twice the distance from the earth to the center of the eccentric. Still more remarkable were observations—among them the same ones used to locate the apsidal line—that showed that the sum of Mercury's greatest elongations were larger when the center of the epicycle was $\pm 120°$ from the apogee than at 180°. Copernicus briefly describes this fundamental peculiarity of the planet at the beginning of his treatment of Mercury in V,25, and although he fails to report a crucial part of Ptolemy's observational evidence—the very observations that show the anomalous elongations—he accepts without question that Mercury's curious behavior is authentic.[3]

In order to account for Mercury's observed elongations, Ptolemy developed a model in which the center of the epicycle is brought closest to the earth at about $\pm 120°$ of eccentric anomaly, just as to account for the second inequality of the moon he had varied the distance between the earth and the center of the lunar epicycle so that the closest approach occurred at $\pm 90°$ of elongation from the mean sun. It is notable that for both the moon and Mercury, the very structure, not only the parameters, of Ptolemy's models were determined by specific observations, a method of investigation quite foreign to Copernicus, but to be applied once again with profound understanding by Kepler in his analysis of the motion of Mars. Ptolemy's principle

[2] The theories of Ptolemy and Copernicus for Mercury have received particular attention in addition to studies of their planetary theory in general. Among recent papers are Hartner (1955), (1974a), (1974b), Gingerich (1971), Wilson (1972), 224–44, Swerdlow (1975b). There are detailed expositions of Ptolemy's treatment of the planet in the *Almagest* in Pedersen (1974), 309–28, and *HAMA*, 158–69.

[3] Below, p. 421.

in both models was to vary the radius of the epicycle optically by changing its distance from the observer. For the moon Ibn ash-Shāṭir and Copernicus chose the alternate method of varying the effective radius of the epicycle itself by means of a second epicycle while keeping the distance from the observer to the center of the epicycle constant.[4] Something similar is done in the case of Mercury where Ibn ash-Shāṭir makes the radius of the epicycle variable in the form of a rectilinear oscillation of the planet along the radius. Copernicus considered no less than four models for Mercury, two of them equivalent to Ibn ash-Shāṭir's, two of them depending upon original, although unwarranted, modifications. In the following section we shall examine Ptolemy's model, and then show the various transformations used by Ibn ash-Shāṭir and Copernicus.

Development of the Models (V,25,32)

Ptolemy's model is shown in Fig. 68. The center of the epicycle C moves uniformly through the mean eccentric anomaly $\bar{\kappa}$ about the equant point E such that the direction EC always points toward the mean sun $\bar{\odot}$. The center N of the eccentric upon which the epicycle moves is carried in the opposite direction through $-\bar{\kappa}$ on a small circle about M, so placed that N coincides with E when $\bar{\kappa} = 180°$. We call the eccentricity OE of the equant e and the radius $MN = ME$ of the small circle e', and thus $OM = e + e'$. Now e and e' are in fact independent parameters that may or may not be equal, and in the *Almagest* IX,8–9 Ptolemy treats OE, EM, and MN as three independent parameters and derives them separately, but in the following analysis we shall assume that $EM = MN = e'$. The parameters that Ptolemy establishes in the *Almagest* are that where $R = 60$, $e = e' = 3;0$ and the radius of the epicycle $r = 22;30$.[5]

The equation of center in this model is rather complicated, although the complication was a secondary result of the introduction of the motion of the center of the eccentric in the small circle. The distance EC is given by

$$EC = (R^2 - e'^2(\sin \bar{\kappa} + \sin 2\bar{\kappa})^2)^{1/2} + e'(\cos \bar{\kappa} + \cos 2\bar{\kappa}), \qquad (61)$$

---

[4] Above, pp. 196–197.

[5] The tables for Mercury in the *Handy Tables* are based upon the parameters in the *Almagest*, but in the *Planetary Hypotheses*, which is a later work still, Ptolemy sets $e = 3;0$, $e' = 2;30$, and $r = 22;15$, leading to greatest elongations of 19;1° at apogee and 22;59° at $\bar{\kappa} = 180°$. In the *Canobic Inscription* of A.D. 146, which, through a remarkable discovery of N. T. Hamilton is now known to be earlier than the *Almagest*, Ptolemy gives the eccentricity of Mercury as 2;30, exactly that of the sun and of $2e$ for Venus, without any indication that the model for Mercury differs from that of the other planets (Heiberg (1907), 150). If in this earlier stage of his work Mercury had a normal model with $e = 2;30$ and $r = 22;30$, the resulting greatest elongations at apogee and $\bar{\kappa} = 180°$ are 21;6° and 23;2°. If Mercury already had its complex model, but with $e = e' = 2;30$, then the elongations would be 19;28° $\approx 19\frac{1}{2}°$ at apogee and the same 23;2° at $\bar{\kappa} = 180°$. Note that these elongations are all near to whole or half degrees that suggest fairly rough observational estimates.

and the equation of center c_3 then follows from

$$\tan c_3 = \frac{e \sin \bar{\kappa}}{EC + e \cos \bar{\kappa}}. \tag{62}$$

The distance OC from the earth to the center of the epicycle is

$$OC = (EC^2 + e^2 + 2\,EC\,e \cos \bar{\kappa})^{1/2}, \tag{63}$$

the minimum value of which occurs at $\bar{\kappa} \approx \pm 120°$.[6] Correcting the mean anomaly $\bar{\alpha}$ to the true anomaly α by $\alpha = \bar{\alpha} + c_3$, where c_3 is positive for $\bar{\kappa} \leq 180°$ and negative for $\bar{\kappa} \geq 180°$, the equation of the anomaly c is found from

$$\tan c = \frac{r \sin \alpha}{OC + r \cos \alpha}, \tag{64}$$

and c is positive for $\alpha \leq 180°$ and negative for $\alpha \geq 180°$. The apparent elongation Δ from the mean sun is then

$$\Delta = c + c_3, \tag{65}$$

regarding the sign of c and taking c_3 as negative for $\bar{\kappa} \leq 180°$ and positive for $\bar{\kappa} \geq 180°$. We have given these rules since the fundamental purpose of the model is to compute elongations, longitudes being reached by adding the elongation to the longitude of the mean sun, that is, $\lambda = \bar{\lambda}_{\odot} + \Delta$.

The principle underlying the model of Ibn ash-Shāṭir and Copernicus is a separation of the eccentricities producing the equation of center from the motion of the center of the eccentric in the small circle. However, the model cannot be developed *a priori*, as was the case for the other planetary models, because the structural details of Ptolemy's model, as well as the numerical parameters, are specifically derived to fit the observations. Hence in what follows we shall have to work numerically as well as theoretically in order to reach a model like that of Ibn ash-Shāṭir and Copernicus, and parts of our exposition will in fact correspond to steps in the *Almagest* and *De revolutionibus*, although without taking the resulting model itself as given.[7]

In Fig. 69 Mercury's epicycle is shown at three locations, A and B at each end of the apsidal line, A apogee and B perigee, and C at 90° of mean eccentric anomaly. The observations of the greatest elongations in these locations will be considered later; here we require only the numerical results which are:

$$c_a = 19;3°, \qquad c_b = 23;15°, \qquad c_c = 23;15°, \qquad c_3 = 3;0°.$$

[6] The minimum and its precise location depend upon the values of e and e', and are quite sensitive to small changes. Hartner (1955), 109–18, gives a detailed discussion of the problem for the condition $e = e'$, and Whiteside (1974), 4 n. 7, a somewhat briefer analysis. For $e = e' = 3;0$, Hartner (1974a), 7 n. 8, finds the minimum at $120;28,19° \pm 0;0,2°$, and Whiteside at $120;26\frac{1}{2}°$ (limiting a series expansion to the second power). Our own check, $\sim 120;28,20°$, agrees with Hartner. For the parameters of the *Planetary Hypotheses*, $e = 3;0$, $e' = 2;30$, there is a noticeable change of more than $3\frac{1}{2}°$ to $\sim 124;6,5°$.

[7] A somewhat different analysis, with the model and Ptolemy's parameters taken as already established, is given in Hartner (1974a), 10–15, which we have found helpful and informative.

We must first find the distance $OM = e + e'$, from the earth to the center of the eccentric, which is given by

$$e + e' = R\frac{\sin c_b - \sin c_a}{\sin c_b + \sin c_a}. \tag{66}$$

From c_a and c_b, we find that where $R = 60$,

$$e + e' = 5;41,13, \qquad = .09478,$$

the decimal figure to the right being the same quantity where $R = 1$, which will be useful to compare with Copernicus's derivation. The radius of the epicycle r_{ab} follows from

$$r_{ab} = (R + (e + e')) \sin c_a = (R - (e + e')) \sin c_b, \tag{67}$$

leading to

$$r_{ab} = 21;26,23, \qquad = .3573.$$

Thus far we have followed the initial steps in the derivations of both Ptolemy and Copernicus, and what we have established is the eccentricity and epicyclic radius that follow from, and in turn will produce, the greatest elongations in the apsidal line where the equation of center is zero.

We now consider the situation at C where $\bar{\kappa} = 90°$ and the equation of center is near maximum, and immediately a problem arises: since $c_c = c_b$, and if r is constant, then $OC = OB$, which is clearly impossible. Hence, we make no assumption about r, and will even be prepared to let it vary, but since $e + e'$ is small, and thus e and e' are separately small, we assume provisionally that $EC = R$, an assumption that will turn out to be strictly true in the resulting model.[8] In this way, we find that

$$e = R \tan c_3 = 3;8,40, \qquad = .05241,$$

$$e' = (e + e') - e = 2;32,33, \qquad = .04237,$$

and e is somewhat larger than a bisection of $e + e'$, specifically $(e + e')/e = 1;50,38/1 \approx 2/1;5$. Still assuming that $EC = R$, the distance OC to the center of the epicycle,

$$OC = (R^2 + e^2)^{1/2} = 60;4,56 \approx 60;5, \qquad = 1.00138,$$

and the radius of the epicycle r_c is now

$$r_c = OC \sin c_c = 23;43,3, \qquad = .3953.$$

But previously $r_{ab} = 21;26,23$, and thus

$$\Delta r = r_c - r_{ab} = 2;16,40, \qquad = .03796,$$

[8] In Ibn ash-Shāṭir's and Copernicus's models (see Figs. 70 and 73), at $\bar{\kappa} = \pm 90°$, $EC = R$, as is also true for Venus, and in the models for the superior planets (see above, p. 296, figure 8), $EP = R$.

shows the amount by which the radius of the epicycle must change in order to produce the observed elongations in a model with a fixed center M.[9] Hence the required alternative to Ptolemy's variation of the distance of the epicycle by the motion of the center of the eccentric in the small circle is a corresponding variation in the radius of the epicycle. In a model of the kind just developed, the distance EC for arbitrary values of $\bar{\kappa}$ is given by

$$EC = (R^2 - e'^2 \sin^2 \bar{\kappa})^{1/2} + e' \cos \bar{\kappa}, \qquad (68)$$

and the equation of center again by (62), although these relations cannot easily be compared with (61) and (62) taken together because the numerical values of e and e' in the two models are necessarily different.

We may now construct a model according to Ibn ash-Shāṭir's principles as shown in Fig. 70.[10] The model is a double epicycle of the same kind used for Venus and the superior planets, except that since the equant lies between the earth at O and the center M of the model in Fig. 69, the radius of the smaller epicycle r_2 is extended away from r_1, that is C is at its greatest distance from C_1, when $\bar{\kappa} = (0°, 180°)$ in the apsidal line. The radii r_1 and r_2 derived from e and e' are

$$r_1 = e + e'/2 = 4;24,56,30, \qquad = .07359,$$

$$r_2 = e'/2 = 1;16,16,30, \qquad = .02118,$$

so that $r_1 \approx 3;28,25 r_2$. The mean radius $\bar{r}$ of the principle epicycle is the mean of the extrema established for $\bar{\kappa} = (0°, 180°)$ and $\bar{\kappa} = \pm 90°$, that is

$$\bar{r} = \tfrac{1}{2}(r_{ab} + r_c) = 22;34,43, \qquad = .3763.$$

Since the true radius r of the principle epicycle increases from minimum to maximum as $\bar{\kappa}$ goes from 0° to 90°, and then decreases back to minimum at 180°, its variation is a function of $2\bar{\kappa}$. The variation could be produced by placing the planet on the circumference of a small epicycle of radius $\frac{1}{2}\Delta r$ that rotates through $2\bar{\kappa}$, but this would cause problems and it is not what Ibn ash-Shāṭir did, although Copernicus at one time attempted it.[11] The method employed by Ibn ash-Shāṭir, and likewise by Copernicus in two of his models, uses either of two Marāgha devices for generating a rectilinear motion from two circular motions, and they are shown in Fig. 71(a) and (b). One (a) consists of two equal circles of radius $\rho = \frac{1}{4}\Delta r$, the second on the circumference of the first, that rotate in opposite directions through $2\bar{\kappa}$ and

[9] By iteration from $EC = (R^2 - e'^2)^{1/2}$, where provisionally $e' = 2;32,33$, it is found that more accurately $EC = 59;56,46$, which now leads to $e = EC \tan c_3 = 3;8,30$ (= .05236), $e' = 2;32,43$ (= .04242), $OC = 60;1,42$ (= 1.0005), $r = 23;41,45$ (= .3949), $\Delta r = 2;15,22$ (= .0376). The differences are not significant, and it should be noted that the original decimal values already round to precisely Copernicus's results (below, pp. 419–420).

[10] Ibn ash-Shāṭir's model is described in Kennedy and Roberts (1959), 231–32, and the parameters, which differ from those established here, are in Abbud (1962), 497.

[11] Below, p. 412.

$4\bar{\kappa}$ respectively, causing the planet P to oscillate about a mean position $\bar{P}$ over a distance

$$r' = 2\rho = \tfrac{1}{2}\Delta r = 1;8,20, \qquad = .01898.$$

The alternate device (b) consists of a circle of radius ρ rolling internally on the circumference of a circle of radius 2ρ, and the result is the same. Ibn ash-Shāṭir appears to use (a), while Copernicus definitely uses (b) in the *Commentariolus*, although with complete spheres.[12] In *De revolutionibus* he says only that the motion along the diameter of the small epicycle upon which the planet oscillates "can also be brought about by uniform circular motions, as was exaplained earlier concerning the precession of the equinoxes" (N 164v22–24) without being more specific. In any case, all he ever illustrates is a circle, shown in Fig. 71(c) and inset in Fig. 70, with the radius $PQ = r'$ and the perpendicular QP, which is all that is needed for computation. The true radius of the epicycle r is given by

$$r = \bar{r} - r' \cos 2\bar{\kappa}. \tag{69a}$$

The computation of the equations from the model will be taken up after the description of Copernicus's final model from *De revolutionibus*.[13]

Copernicus's first description of a model for Mercury, illustrated in Fig. 72, is from the *Commentariolus*. In transforming Ibn ash-Shāṭir's model to a heliocentric form the principal epicycle becomes the orbit of Mercury about the mean sun, and since the radius of the epicycle was variable so the radius of the orbit must vary. In the *Commentariolus* this is done by letting the center C_1 of the larger epicycle oscillate on the diameter of the small epicycle of radius r', as shown inset in the figure. The planet is located at the end of the radius r_2 of the two epicycles producing the equation of center. The model thus has the same problems as the model for Venus, only now they are a bit worse. Note first that in order to measure the equation of center it is necessary to construct the parallelogram $O\bar{S}C_1N$, where the distances $\bar{S}C_1 = ON$ are required to vary as a function of $2\bar{\kappa}$, and then construct an instantaneous apsidal line NEA at N with the equant point at E. The equation of center is then measured at N, not where it ought to be at O. As we have already mentioned in our discussion of Venus, this peculiar construction may be a vestige of Copernicus's use of the eccentric model of the second anomaly, but clearly it is not desirable in the present model.[14] Another problem in the *Commentariolus* is that in finding the parameters Copernicus made $r_1 = 3r_2$, the equivalent of a bisected eccentricity, which is not ap-

---

[12] Swerdlow (1973), 504–05.

[13] Below, pp. 410–411.

[14] See above, pp. 372–373. Note that if N is the center of Mercury's eccentric in such a model (see figure 55), the distance ON must be variable, meaning that it is N that must oscillate on the diameter of the small circle, while the distance $NC_1 = R$ is constant.

plicable to Mercury, and also found meaningless numerical values for r_1 and r_2.[15]

The models in the *Commentariolus* were not intended for practical application—at least not with the crude and incomplete parameters supplied in the text—and at the time of its composition Copernicus was evidently not secure in constructing a model for Mercury. He finally did reach a correct model—correct in the sense of doing what was expected of it—in *De revolutionibus*, where he in fact describes two that differ rather significantly. Further, examination of the manuscript version of V,25, where the principal model is described, shows remains of yet another, and incorrect, method for varying the radius of Mercury's orbit. We shall first describe the final, correct model, and then take up the two alternatives that Copernicus considered.

The model actually used in *De revolutionibus*, the final model, is shown in Fig. 73(a) along with its transformation from the double-epicycle in Fig. 73(b). As can be seen by comparison with Fig. 70, it is properly equivalent to Ibn ash-Shāṭir's model, the two radii r_1 and r_2 carrying the center of the epicycle now being transformed into the eccentricities e_1 and e_2 carrying the center of Mercury's orbit. This was precisely the adjustment that was required for Venus.[16] With respect to an instantaneous geocentric apsidal line OEA, the center of the planet's orbit C appears to move uniformly about the equant E located at the distance $OE = e_1 - e_2$, and the equation of center c_3 is now properly measured at O. As was also true for Venus, the equation of center is a function of the earth's distance from Mercury's "apogee" or "highest apsis" A', the point at which the earth is at its greatest distance from the center C of Mercury's orbit or, geocentrically, the point at which the center C is at its greatest distance from the earth. The motion of e_2 is such that e_2 is extended away from e_1, and C is farthest from the mean sun $\bar{S}$, when the earth is in the apsidal line at $\bar{\kappa} = (0°, 180°)$.

In a modern formulation, the equation of center is given by

$$\tan c_3 = \frac{(e_1 - e_2)\sin\bar{\kappa}}{R + (e_1 + e_2)\cos\bar{\kappa}}. \tag{70}$$

When compared to the equant model of Fig. 69 with fixed center M, letting $e_1 - e_2 = e$ and $e_1 + e_2 = e + e'$, equation (70) differs from (68) combined with (62) by about $e'^2 \sin^2 \bar{\kappa}/2R$ in the lower term which, at its maximum with Copernicus's parameters, amounts to about $0.0009R$, a very small quantity that will not affect c_3. On the other hand, the differences from Ptolemy's model, while still quite small, are more noticeable and more

[15] There are different reconstructions of the derivation of the parameters in Swerdlow (1973), 505–09, and Hartner (1974a), 18–22, both concluding that something is wrong. In V,25 (N 164v29) Copernicus specifies the equivalent $e_1 = 3e_2$, but he does not follow it in deriving the correct values in V,27.

[16] Above, p. 373.

complicated, and will be taken up after reviewing the derivation of the parameters.[17] Copernicus's method of solving the equation of center is first to draw OM, dividing c_3 into components a_1 and a_2 as shown in Fig. 74. Analogously with (6–10), he finds

$$OM = (R^2 + e_1^2 + 2Re_1 \cos \bar{\kappa})^{1/2}, \tag{71}$$

leading to

$$a_1 = \sin^{-1} (e_1 \sin \bar{\kappa}/OM). \tag{72}$$

Then, forming $\bar{\kappa} + a_1$, he finds

$$OC = (OM^2 + e_2^2 + 2\,OM\,e_2 \cos (\bar{\kappa} + a_1))^{1/2}, \tag{73}$$

with which

$$a_2 = \sin^{-1} (e_2 \sin (\bar{\kappa} + a_1)/OC). \tag{74}$$

Then the equation of center c_3 and the true eccentric anomaly κ are

$$c_3 = a_1 - a_2, \tag{75}$$

$$\kappa = \bar{\kappa} + c_3, \tag{76}$$

where c_3 is negative for $\bar{\kappa} \leq 180°$ and positive for $\bar{\kappa} \geq 180°$. The radius of Mercury's orbit, as in the previous models, is

$$r = \bar{r} - r' \cos 2\,\bar{\kappa}, \tag{69a}$$

although Copernicus prefers to let $r_m = \bar{r} - r'$, and solve

$$r = r_m + r'(1 - \cos 2\,\bar{\kappa}). \tag{69b}$$

Finding the true anomaly $\alpha = \bar{\alpha} + c_3$, where c_3 is positive for $\bar{\kappa} \leq 180°$ and negative for $\bar{\kappa} \geq 180°$, the equation of the anomaly is then

$$\tan c = \frac{r \sin \alpha}{OC + r \cos \alpha}, \tag{64}$$

c being positive for $\alpha \leq 180°$ and negative for $\alpha \geq 180°$. The elongation Δ from the mean sun

$$\Delta = c + c_3, \tag{65}$$

here taking c_3 negative for $\bar{\kappa} \leq 180°$ and positive for $\bar{\kappa} \geq 180°$, and finally the sidereal and tropical longitudes of the planet are

$$\lambda^* = \bar{\lambda}^*_{\odot} + \Delta, \tag{77a}$$

$$\lambda = \bar{\lambda}'_{\odot} + \Delta, \tag{77b}$$

where $\bar{\lambda}^*_{\odot}$ is the mean sidereal longitude and $\bar{\lambda}'_{\odot}$ the adjusted mean tropical longitude of the sun.

---

[17] Below, pp. 422–423.

This is the model that Copernicus uses in deriving the elements of Mercury's orbit. However, originally in M (176) he had described a model in which the planet moves, apparently in the positive sense, on the circumference of the small circle rather than in a rectilinear oscillation along its diameter, for the movement on the diameter appears only in the form of several marginal corrections and additions.[18] Since in the *Commentariolus* he already had a rectilinear variation of Mercury's orbit, it is not obvious why he would have temporarily departed from it, only to return at a later date. The original arrangement is shown inset in Fig. 73(a), and it is immediately evident that it has two undesirable consequences. The first is that P is displaced from the direction $C\bar{P}$ by an angle k given by

$$\tan k = \frac{r' \sin 2\,\bar{\kappa}}{\bar{r} - r' \cos 2\,\bar{\kappa}}, \tag{78}$$

and k will produce an additional, and unmotivated, correction to the mean anomaly, as though Mercury had a third inequality. With Copernicus's parameters, $\bar{r} = 3763$ and $r' = 190$, k can reach $k_{\max} = \sin^{-1}(r'/\bar{r}) = 2;54°$, which is almost as large as $c_{3\,\max} = 3;2°$. It then follows that the true anomaly would be $\alpha = \bar{\alpha} + c_{3(\bar{\kappa})} + k_{(2\bar{\kappa})}$, the alternate phases of k periodically increasing, decreasing, or even cancelling c_3. Further, the effective radius of the orbit $r = CP$ will now be

$$r = (\bar{r}^2 + r'^2 - 2\,\bar{r}r' \cos 2\bar{\kappa})^{1/2} \tag{79}$$

instead of the former $r = \bar{r} - r' \cos 2\bar{\kappa}$ (69a).

The correction k and the difference in the radius of the orbit will vanish in the apsidal line and at $\pm 90°$ from the apsidal line, and while elsewhere they will alter the position of the planet their effect will not be particularly noticeable.[19] Copernicus was still using this model when he wrote V,27, on Mercury's eccentricities and orbital radius, as is shown by a deleted remark (M 178v3) that the radius of the orbit varies "as in the larger epicycle of the moon" where the variation is produced by motion on the circumference of the smaller epicycle. But in V,28, on the greatest elongations at $\bar{\kappa} = \pm 120°$, the planet is considered to be moving on the diameter of the small circle. Hence, he had made the correction between two chapters, fairly soon one may suppose, even though it happens that the incorrect model has only the slightest effect on the demonstration in V,28.[20]

The cancelled model, however, was not simply an error, for it is identical in effect to what appears to be an entirely different model for Mercury described by Copernicus in V,32 as what appears to be an afterthought.

---

[18] The entire passage is translated, with the marginal additions set off from the original text, in Swerdlow (1975b), 146–48.

[19] See below, p. 414.

[20] See below, p. 421, n. 35.

At the end of his exposition he remarks that the model will be of use in Mercury's latitude theory, and he refers to it again at the end of VI,2 following his description of the component of Mercury's latitude called the *deviatio* (deflection). The model is illustrated in Fig. 75. The eccentricities for producing the equation of center are the same as before, but C is now the center of a circle of radius $2r'$, twice the radius of the small circle controlling the variation of the radius of the orbit in the former models. The diameter KL of this circle rotates about C in the positive sense such that the distance of L from the apsidal direction CA is the mean anomaly $\bar{\alpha}$.[21] Now, extend KL to meet the planet's orbit at I, and imagine that point I and the entire orbit rotate about C through the mean anomaly $\bar{\alpha}$ with respect to direction CA, or through $\bar{\alpha} - \bar{\kappa}$ with respect to the direction ECT of the mean geocentric apogee T. Then a point C', the center of Mercury's orbit, oscillates along the diameter KL as a function of $\bar{\kappa}$ such that $CC' = 2r' \cos \bar{\kappa}$. The radius of the orbit must now be set at $\bar{r} + r'$, where $\bar{r}$ is the mean radius of the orbit and r' the radius of the small circle in the former models, although Copernicus fails to mention this. Finally, the planet P moves about C' through $\bar{\kappa}$ measured from the direction CI, with the result that its total motion with respect to CA is $\bar{\alpha} + \bar{\kappa}$, and with respect to ECT the mean anomaly $\bar{\alpha}$. Note that what Copernicus has done is to break up the motion of P into two components which, measured from CA, are respectively $ACI = \bar{\alpha}$, belonging to the entire orbit, and $IC'P = \bar{\kappa}$, belonging to the planet itself.

We have remarked that the cancelled model in V,25 is identical in effect to this alternate model, and this is easily proved. In order to determine the radius of the orbit, we have (see inset in Fig. 75)

$$CC' = 2r' \cos \bar{\kappa}, \qquad CQ' = 2r' \cos \bar{\kappa} \sin \bar{\kappa} = r' \sin 2\,\bar{\kappa},$$

$$C'Q' = 2r' \cos \bar{\kappa} \cos \bar{\kappa} = r'(\cos 2\,\bar{\kappa} + 1),$$

and since $C'P = \bar{r} + r'$,

$$Q'P = \bar{r} + r' - r'(\cos 2\,\bar{\kappa} + 1) = \bar{r} - r' \cos 2\,\bar{\kappa},$$

so that the effective radius of the orbit

$$CP = r = (\bar{r}^2 + r'^2 - 2\,\bar{r}r' \cos 2\,\bar{\kappa})^{1/2}, \tag{79}$$

[21] By error, Copernicus gives this motion as "about 2;7° per day from the apogee of the eccentric of the star, that is, the amount by which the motion of parallax of the star exceeds the motion of the earth in the zodiac" (N 172v2–4). But 2;7° $= \bar{\alpha} - \bar{\kappa}$ is the motion of L with respect to the mean geocentric apogee T lying on ECT, and this error suggests that Copernicus first worked out the model geocentrically, probably because it was still clearer for him to envision motions geocentrically. Note that in his descriptions of the motions of the inferior planets, the motion of the planet with respect to its own apsidal line is not mentioned, since all that is ever of concern is the motions of the anomaly from the mean or true geocentric apogees of the orbit T or V.

exactly as in the cancelled model in which the planet moved on the circumference of the small circle of radius r'. Further, an additional correction k is introduced that is given by

$$\tan k = \frac{CQ'}{Q'P} = \frac{r' \sin 2\bar{\kappa}}{\bar{r} - r' \cos 2\bar{\kappa}}, \tag{78}$$

which is identical to the unwanted correction in the cancelled model. This means that the cancelled model and the alternate model in V,32 are identical in their effect on the apparent motion of Mercury.

Copernicus is aware of the correction k introduced by the alternate model, and presumably also the cancelled model where it is even more obvious, and explains it away by pointing out that it is zero for $\bar{\kappa} = (0°, \pm 90°, 180°)$ and more important, that where it can have its maximum effect the planet will be too close to the sun to be visible. This is in fact true, for k will have its greatest effect on the equation of the anomaly c when the radius of the orbit CP is directed toward or away from the earth and the planet is thus at inferior or superior conjunction.[22] On the other hand, when the planet is near greatest elongation and clearly visible, even though k can have a large effect on the anomaly α, near greatest elongation c changes very slowly as a function of α, and will thus not be noticeably affected. Copernicus does not address the question of the difference between the radii of the orbit in the two models, but it turns out to be quite small. The maximum difference between relation (69) for the radius in the standard model and (79) for the radius in the alternate and cancelled models is about $r'^2/2R\bar{r}$, which, with Copernicus's parameters, amounts to about $0.0005R$ at $2\bar{\kappa} = 90° - \frac{1}{2}k_{max} = 88;33°$, which in turn can affect c by less than 0;1°.

Copernicus had said that the alternate model would be useful for Mercury's latitude theory, and at the end of VI,2 he says that while the inequality of the motion in longitude was demonstrated by means of a small epicycle, meaning the small circle producing the variation in the radius of the orbit, the latitude is demonstrated by an eccentric of an eccentric, referring to the alternate model. In fact it is only the third component of Mercury's motion in latitude, the *deviatio*, which we translate as "deflection", that is related to the alternate model. The deflection is a combined rotation and variable inclination of the orbital plane of an inferior planet that is intended as an equivalent to the variable inclination of the plane of the eccentric in Ptolemy's model for latitude. The deflection will be considered in detail in the following chapter,[23] but here we shall explain enough to show its relation to the alternate model.

The model for the deflection is shown in Fig. 76 in which the line of intersection of the plane of the deflection and the plane of the earth's orbit passes through the center of the orbit C'. This is not strictly correct since the nodal

---

[22] This is analogous to the small effect of the second lunar inequality near syzygy discussed above, pp. 215–216.

[23] Below, pp. 509–511.

line of Mercury's orbit coincides with the apsidal line, but the deflection is treated as though measured from the plane of the ecliptic. Let us assume that at some time the earth is in Mercury's apsidal line at $\bar{\kappa} = 0°$. Then the point of maximum deflection I_0 coincides with the planet P_0, and the orbital plane is inclined to the south by $-i_{\max}$. Now the earth moves through $\bar{\kappa}$ to O, the point of maximum deflection, which can be thought of as carrying with it the entire orbit, moves through $\bar{\alpha}$ to I, and Mercury moves through $\bar{\alpha} + \bar{\kappa}$, or moves beyond I by $\bar{\kappa}$, to P. Thus we have exactly the division of Mercury's motion into the components, $\bar{\alpha}$ moving the entire orbit and $\bar{\kappa}$ the motion of the planet itself, that occurred in the alternate model, and indeed the point of maximum deflection I is the same point lying on KL extended to meet the orbit at I in Fig. 75. The angle of the deflection $-i$ goes to zero when the earth reaches $\bar{\kappa} = 90°$, at which point P is 90° ahead of I and has no deflection. Then, as $\bar{\kappa}$ exceeds 90° and P approaches I', the inclination of I becomes positive and I' negative, so the planet again is deflected to the south, reaching the maximum deflection when $\bar{\kappa} = 180°$. The return to I takes place in the same manner, the result being that Mercury's deflection always carries the planet to the south. In principle, the inclination $-i$ is given by

$$-i = -i_{\max} \cos \bar{\kappa},$$ [24]

while the motion of the center of the orbit C' in Fig. 75 was

$$CC' = 2r' \cos \bar{\kappa},$$

so both are functions of $\bar{\kappa}$ and have the same period, as Copernicus mentions at the end of VI,2. His idea in uniting the two models is that while the center of the orbit C' librates between K and L on the diameter of the small circle, the orbital plane librates from $-i_{\max}$ to $+i_{\max}$ (although P always is located in the negative semicircle of the orbit). However, as clever as his combination of the models for longitude and latitude may be—and there is no denying that it took a good deal of ingenuity to come up with them—his explanation of their relation in V,32 and VI,2 is far from clear. This lack of clarity is also evident in his description of the alternate model, and the inherent difference between the alternate model, supposedly intended for latitude, and the standard model used in the longitude theory makes one wonder just how he thought they could coexist in the same heaven.

Observations

Copernicus uses ten observations of Mercury of which seven (1–7) are from the *Almagest* and three (8–10) are modern, made by Bernhard Walther

---

[24] There is some ambiguity in Book VI about whether the deviation depends upon the mean or true eccentric anomaly, but in associating the alternate model for longitude with the deviation component of latitude, Copernicus must be considering only the mean eccentric anomaly $\bar{\kappa}$, as we do here.

in 1491 and 1504.[25] Observations (1–6) are of the planet near greatest elongation with the mean sun at specific locations of the same type used for Venus. These were used by Ptolemy to determine the direction of the apsidal line, the eccentricities, and the radius of the epicycle, and Copernicus uses them to adapt Ptolemy's elements to his own model. Observations (8–10) as a group are used to confirm a modern location of Mercury's apsidal line that Copernicus estimated, and to find the value of the mean anomaly for comparison with (7). (7) and (10) are then used to correct the mean anomaly and estimate the motion of the apsidal line.

The observations are summarized in Tables 7 and 8, of which Table 7 gives the following information:

(1) Number of observation.
(2) Observer.
(3) Egyptian date and time in text: M, morning E, evening.
(4) Julian date and time.
(5) Julian day.
(6) Δt since epoch in Egyptian years and days.

The Julian date and time and Δt since epoch are either expressly stated in the text or, in the case of (3) and (4), apparently used by Copernicus for computing the mean position of the sun. In principle these times are for the meridian of Cracow, but Copernicus gives the time of (7), which was made in Alexandria, and of (8–10), which were made in Nuremberg, without an adjustment of meridian, although he did in fact adjust the meridian for computing the longitude of the mean sun.[26]

Table 8 concerns solar longitude, and the longitude and elongation of the planet, as follows:

(1) Number of observation.
(2) Cop $\bar{\lambda}^*_{\odot}$, Copernicus's sidereal longitude of the mean sun.
(3) $\bar{\lambda}^*_{\odot}$, by computation.
(4) π, precession by computation.
(5) $\bar{\lambda}'_{\odot} = \bar{\lambda}^*_{\odot} + \pi$, adjusted mean tropical longitude.
(6) Ptol $\bar{\lambda}_{\odot}$, report from *Almagest* for (1–7).
Cop $\bar{\lambda}'_{\odot}$, Copernicus's report for (8–10).
(7) Obs λ_p, observed tropical longitude of Mercury, from *Almagest* for (1–7) and from Schoener (1544) for (8–10).

[25] Walther's observations were first published in Schoener (1544), which we have used, and Copernicus's use of them is treated perceptively in Kremer (1981). Ptolemy's observations of Mercury have been considered, in various ways, in Czwalina (1959), 18–35, Gingerich (1971), 57–59, Wilson (1972), 226–32, Moesgaard (1974), 98–101, 256, who also considers Walther's observations, Pedersen (1974), 309–10, *HAMA*, 159–67.

[26] Below, pp. 430–431. Cf. the different times used by Moesgaard (1974), 100, 256. As frequently noted, the Julian date of (3) in the text is MCCCV for CXXXV, and the Egyptian date of (6), Mes 21, is correct in *De revolutionibus*, the *Epitome*, and the 1515 *Almagest*, although the Greek manuscripts, except Heiberg's D, give by error Mes *24*.

(8) Cop λ_p^*, Copernicus's sidereal longitude for (1–7).
Cop λ_p, Copernicus's tropical longitude for (8–10).
(9) Δ, elongation of Mercury stated and used by Copernicus: (+) to the east or (−) to the west of the mean sun.

Discrepancies in this table, particularly between Walther's and Copernicus's reports of λ_p for (8–10), will be discussed along with the applications of the observations.[27]

The various problems discussed earlier for Venus—that the planet was not observed exactly at greatest elongation, that greatest elongation does not necessarily occur exactly when the line to the planet is tangent to the orbit, and that only a correctly chosen pair of equal greatest elongations will be symmetrical to the apsidal line—also apply to Mercury and need not be further considered.[28]

Longitude of the Apsidal Line (V,26)

The demonstration of the direction of the apsidal line is considerably abbreviated from Ptolemy's demonstration in *Almagest* IX,7 and *Epitome* IX,11. Copernicus takes observations (1) and (2) in which

$$\bar{\lambda}^*_{\odot 1} = 63;50°, \qquad \bar{\lambda}^*_{\odot 2} = 303;19°, \qquad \Delta_1 = -\Delta_2 = 26;30°.$$

The configuration is shown in Fig. 77 which is simplified in that C is shown on the apsidal line when it is really symmetrically displaced in the two observations. From the equality of the opposite maximum elongations, Copernicus concludes that one end of the apsidal line is directed toward

$$\lambda^*_{A'} = \tfrac{1}{2}(63;50° + 303;19° - 360°) = 3;34,30° \approx 3;34°,$$

and the other end toward 183;34°. Ptolemy's demonstration had been more detailed. He also considered two other observations in which the mean sun occupied about the same positions as in (1) and (2), while the planet had the equal opposite elongations shown in Fig. 77 as

$$\Delta_2' = -\Delta_1' = 21;15°.$$

The importance of these additional observations is that

$$|\Delta_1 - \Delta_1'| = |\Delta_2 - \Delta_2'| = 26;30° + 21;15° = 47;45°$$

is the maximum sum of greatest elongations, taking place when the mean sun (or the earth) is $\sim \pm 120°$ from the apogee. Note that with Copernicus's figures, if $\lambda^*_A = 183;34°$,

$$\bar{\lambda}^*_{\odot 1} - \lambda^*_A = -119;44°, \qquad \bar{\lambda}^*_{\odot 2} - \lambda^*_A = +119;45°,$$

---

[27] Textual corrections: Cop $\bar{\lambda}'_{\odot 9}$ ♑27;7° for ♒27;7° Cop λ_{p10} 26;55° in M (181r9 marg.) is correct against 26;6° in N (169v16), due to a misreading of *deunce* as *decima*, as frequently noted.
[28] See above, pp. 377, 379, n. 28.

which fulfill the same conditions. However by omitting elongations Δ_1' and Δ_2', Copernicus has removed the observational evidence that the sum of Mercury's greatest elongations is a maximum when $\bar{\kappa} \approx \pm 120°$ rather than 180°, which is of course the entire motivation for Mercury's complex model.[29]

Ptolemy next considers six observations made some four hundred years earlier, four in Alexandria dated by the Era of Dionysius, two in Babylon dated by the Seleucid Era, in order to show that Mercury's apsidal line has moved about 4° in the intervening 400 years, leading to the conlusion that Mercury's apsidal line is sidereally fixed since 4° in 400 years is exactly the rate of the precession. Without presenting further evidence, he applies this conclusion to all of the planets. Copernicus omits this section, the reason being that he has already demonstrated independent sidereal motions for the apsidal lines of the superior planets, although not for Venus, and he later assumes and attempts to confirm an independent motion for Mercury's apsidal line in V,30.

In order to distinguish the apsides, Copernicus, following Ptolemy in *Almagest* IX,8 and *Epitome* IX,14, uses observations (3) and (4) in which

$$\bar{\kappa} = \bar{\lambda}^*_{\odot 3} - \lambda^*_{\text{A}} = 182;38° - 183;34° = -0;56°, \quad \Delta_3 = -19;3°,$$

$$\bar{\kappa} - 180° = \bar{\lambda}^*_{\odot 4} - \lambda^*_{\text{A}'} = 4;28° - 3;34° = +0;54, \quad \Delta_4 = +23;15°,$$

so that the earth may be considered to be in the apsidal line as shown in Fig. 78. Then since $|\Delta_3| < |\Delta_4|$, it is evident that $O_3C > O_4C$, and thus according to Copernicus's definition of "apogee" as the direction seen from the earth when the center C of an inferior planet's orbit is most distant, the (geocentric) apogee is directed toward A when C appears in the direction $\lambda^*_{\text{A}} = 183;34°$, and the perigee toward A' when C appears in the direction $\lambda^*_{\text{A}'} = 3;34°$. However, Copernicus now simply subtracts a precession of 6;40° from Ptolemy's apogee at ♎10° to give

$$\lambda^*_{\text{A}} = 190° - 6;40° = 183;20°,$$

which he then uses, not only for the time of Ptolemy, but also for four hundred years earlier.[30] Note that in this demonstration it would be preferable to consider the sum of the greatest elongations at each location, particularly because the maximum sum of Mercury's elongations does not occur in the apsidal line. But Ptolemy does not do so, nor does he remark upon such observations as he did for Venus.[31] The reason is that by Ptolemy's theory of greatest elongation from the true sun in *Almagest* XII,9–10 and his visibility theory in XIII,8–10, Mercury is only barely visible as an evening star in Libra, the opposite elongation to $-\Delta_3$, and not visible at all as a morning star in Aries, the opposite elongation to $+\Delta_4$. Hence, the required observations were simply not possible.[32]

---

[29] See below, pp. 421–422.

[30] Below, p. 426.

[31] Above, p. 381.

[32] As has likewise been noted in Gingerich (1971), 57–58.

Radius of Orbit and Eccentricities (V,27)

The method of deriving the eccentricities and variable radius of the orbit of Mercury has already been discussed in our analysis of the motivation of Ibn ash-Shāṭir's and Copernicus's model.[33] Our previous analysis was done for a geocentric model, but its adaptation to Copernicus's model is trivial. Here we shall go through Copernicus's derivation of these quantities which, we note in advance, are identical but for roundings to those found earlier. In Fig. 79, from observations (3) and (4),

$$|\Delta_3| = 19;3^\circ, \qquad |\Delta_4| = 23;15^\circ,$$

in agreement with Ptolemy. Hence, where $AC = 100{,}000$,

$$CP_3 = AC \sin \Delta_3 = 32639,$$

and where $BC = 100{,}000$,

$$CP_4 = BC \sin \Delta_4 = 39474.$$

But where $AC = 100{,}000$,

$$BC = (\sin \Delta_3/\sin \Delta_4)\, AC = 82685,$$

$$A\bar{S} = \tfrac{1}{2}(AC + BC) = 91342\tfrac{1}{2} \approx 91342,$$

$$\bar{S}C = AC - A\bar{S} = 8658.$$

Now, letting $R = A\bar{S} = 100{,}000$, the radius of the orbit r and the maximum eccentricity $e_1 + e_2$ are

$$r = CP_3\,(R/A\bar{S}) = 35733, \qquad e_1 + e_2 = \bar{S}C\,(R/A\bar{S}) = 9479,$$

or equivalently, where $R = 1;0$,

$$r = 0;21,26, \qquad e_1 + e_2 = 0;5,41.$$

We next consider observations (5) and (6) in which

$$\bar{\lambda}^*_{\odot 5} = 93;30^\circ, \qquad \bar{\lambda}^*_{\odot 6} = 93;39^\circ, \qquad |\Delta_5| = 26;15^\circ, \qquad |\Delta_6| = 20;15^\circ.$$

Using the accurate $\lambda^*_A = 183;34^\circ$,

$$\bar{\kappa}_5 = \bar{\lambda}^*_{\odot 5} - \lambda^*_A = -90;4^\circ \approx -90^\circ,$$

$$\bar{\kappa}_6 = \bar{\lambda}^*_{\odot 6} - \lambda^*_A = -89;55^\circ \approx -90^\circ,$$

and $O\bar{S}$ may be taken as perpendicular to the apsidal line at $\bar{\kappa} = -90^\circ$ as shown in Fig. 80. Since OC bisects $\Delta_5 + \Delta_6$,

$$2c = \Delta_5 + \Delta_6 = 46;30^\circ, \qquad c = 23;15^\circ,$$

and the equation of center

$$c_3 = \Delta_5 - c = c - \Delta_6 = 3;0^\circ.$$

[33] Above, pp. 406–409.

Thus, where $R = 10{,}000$,

$$e_1 - e_2 = \bar{S}C = R \tan c_3 = 524,$$

which Copernicus can find equivalently. Hence it is established that where $R = 10{,}000$,

$$e_1 + e_2 = 948, \quad e_1 - e_2 = 524, \quad e_1 = \bar{S}M = 736, \quad e_2 = MC = 212.$$

In order to find the radius r of the orbit, we have

$$OC = (R^2 + (e_1 - e_2)^2)^{1/2} = 10014,$$

$$r = OC \sin c = 3953.$$

But in the previous demonstration, when the earth was in the apsidal line $r = 3573$. Hence, we have for the minimum and maximum radii

$$r_m = 3573, \qquad r_M = 3953, \qquad \Delta r = 380,$$

and, as shown inset in the figure, the mean radius $\bar{r}$ and the radius of the small circle r',

$$\bar{r} = C\bar{P} = \tfrac{1}{2}(r_m + r_M) = 3763, \qquad r' = \bar{P}P = \tfrac{1}{2}\Delta r = 190.$$

It should be noted that the demonstration of the radius of the orbit in the apsidal line and at $\pm 90°$ from the apsidal line is also a proof that in a model of the kind used by Copernicus, the radius of the orbit must be variable.

Elongations at $\bar{\kappa} = \pm 120°$ (V,28)

The derivation of the fundamental elements of the model is now nearly complete. What remains to be found are the correction of the mean anomaly, the positions at epoch, and whatever can be determined about the motion of the apsidal line, all of which require modern observations. However, it has not yet been shown that the sum of the greatest elongations produced by the model is larger at $\bar{\kappa} = \pm 120°$ than at 180°. Ptolemy takes this up in *Almagest* IX,9 and *Epitome* IX,20, and in *Epitome* IX,21 Regiomontanus adds a demonstration that the center of Mercury's epicycle is closer to the earth at $\bar{\kappa} = \pm 120°$ than at 180°. Copernicus, however, reviews only Ptolemy's demonstration since Regiomontanus's proof does not apply to Copernicus's model, that is, the center of Mercury's orbit is not closer to the earth at $\bar{\kappa} = \pm 120°$ than 180°, for it is in fact farther away, while the larger elongations at $\pm 120°$ are due to the larger size of the orbit.

The elements of the orbit have thus far been established using observations with the earth either in the apsidal line or at 90° from the apsidal line. The sum of the greatest elongations at each position are:

$$\bar{\kappa} = 0°, \qquad 2\Delta_3 = 38;6°,$$

$$\bar{\kappa} = \pm 90°, \qquad \Delta_5 + \Delta_6 = 46;30°,$$

$$\bar{\kappa} = 180°, \qquad 2\Delta_4 = 46;30°.$$

The identity of the sum of the elongations at $\pm 90°$ and $180°$ has been used to show that the radius of Mercury's orbit must be larger at $\pm 90°$ than at $180°$. But the largest sum of elongations occurs somewhere in between. In demonstrating the direction of the apsidal line, Ptolemy had used two pairs of observations, which we have called Δ_1, Δ'_1 and Δ_2, Δ'_2 in which

$$\bar{\kappa} = \pm 120°, \qquad |\Delta_1 - \Delta'_1| = |\Delta_2 - \Delta'_2| = 26;30° + 21;15° = 47;45°,$$

exceeding $2\Delta_4$ at $\bar{\kappa} = 180°$ by 1;15°. Copernicus omits an important part of the observational evidence, since he does not mention Δ'_1 or Δ'_2, but now he sets out to show that the sum of the elongations at $\bar{\kappa} = \pm 120°$ exceeds the sum at 180°.[34]

The demonstration is straightforward. Figure 81 shows the model in the configuration for $\bar{\kappa} = 120°$. It has already been established that where $R = 10{,}000$,

$$e_1 = 736, \qquad e_2 = 212, \qquad \bar{r} = 3763, \qquad r' = 190.$$

For $\bar{\kappa} = 120°$, the equation of center,

$$c_3 = 3;47° - 1;4° = 2;44°,$$

which seems to show that Copernicus must have computed to seconds since accurately, from (71–75),

$$c_3 = 3;47,9° - 1;3,31° = 2;43,38° \approx 2;44°,$$

and the distances

$$OM = 9655, \quad (\text{Corr. } 9653) \qquad OC = 9540. \quad (\text{Corr. } 9537)$$

For the radius of the orbit (see inset), letting $r_m = \bar{r} - r'$, from (69b)

$$r = 3573 + 285 = 3858,$$

and thus the maximum equation of the anomaly,

$$c = \sin^{-1}(r/OC) = \sin^{-1}(3858/9540) = 23;52°.^{35}$

Hence, the sum of the elongations

$$\Delta + \Delta' = 2c = 47;44°,$$

although Copernicus says $2c = 47;45°$, in exact agreement with Ptolemy's observations (even though he failed to mention half of Ptolemy's pertinent

---

[34] Since the elongations $\Delta' = 21;15°$—described in *Almagest* IX,7 and *Epitome* IX,11—are essential to establishing the model for Mercury, Copernicus's omission is curious. See below, p. 422.

[35] Corr. 23;51,13° although in fact $\sin^{-1}(3858/9537) = 23;51,41° \approx 23;52°$. M 178v–179r contains a number of errors and alterations in this demonstration, some of which are noted in Copernicus (1978), 432 n. to 283:30. Were the cancelled model of V,25 in M or the alternate model of V,32 used in this demonstration, the result would be $r = 3861.5$ and $c = 23;53°$, an insignificant difference.

observations), and then points out that this exceeds $2c$ at $\bar{\kappa} = 180°$ by 1;14°. Note, however, that the separate elongations, which Copernicus does not consider, are

$$\Delta = c + c_3 = 26;36°, \qquad \Delta' = c - c_3 = 21;8°,$$

while Ptolemy observed 26;30° and 21;15° respectively. It is perhaps for this reason that Copernicus does not discuss the separate elongations, either from Ptolemy's observations or as produced by his model. Yet the agreement can hardly be called poor. For the same configuration, Ptolemy finds

$$2c = 47;46°, \qquad c = 23;53°, \qquad c_3 = 2;41°,$$

leading to

$$\Delta = 26;34°, \qquad \Delta' = 21;12°,$$

in somewhat better agreement with the observations.

Comparison of Equations and Elongations in the Models of Ptolemy and Copernicus

Copernicus has correctly shown that the sum of the elongations at $\bar{\kappa} = \pm 120°$ exceeds the sum at $\bar{\kappa} = 180°$. And as is true in Ptolemy's model, the values at 120° are exceedingly close to the maximum. Since $\Delta + \Delta' = 2c$, we may find the maximum simply by computing c for various values of $\bar{\kappa}$ using Ptolemy's and Copernicus's models respectively, and we find that the maxima are:

	$\bar{\kappa}$[36]	c_{max}
Ptolemy	120;28,20°	23;53,20°
Copernicus	120;47,28	23;51,45

That Copernicus's model agrees so well with Ptolemy's is quite remarkable since both models and their parameters were based upon elongations at $\bar{\kappa} = (0°, \pm 90°, 180°)$, and nothing was specifically derived from the elongations at 120°.

The maximum values of c as a function of $\bar{\kappa}$, $c_{\max(\bar{\kappa})}$, in Copernicus's model are shown in Fig. 82(a) as the graph marked r. The differences between Copernicus's model and Ptolemy's for both the maximum equation of the anomaly, $\Delta c_{\max(\bar{\kappa})}$, and the equation of center, $\Delta c_{3(\bar{\kappa})}$, as a function of

[36] Since c_{max} changes slowly as a function of $\bar{\kappa}$, it is difficult to give a precise value of $\bar{\kappa}$ without computing to a large number of places.

$\bar{\kappa}$ are shown in Fig. 82(b), and it can be seen that they are quite small. The graph $\Delta c_{\max(\bar{\kappa})}$ gives the differences

$$\Delta c_{\max(\bar{\kappa})} = c_{\max(\bar{\kappa})C} - c_{\max(\bar{\kappa})P}, \qquad \Delta_{\max} = \pm 0;4,31^\circ \qquad (\bar{\kappa} = \pm 50^\circ),$$

where C and P refer respectively to Copernicus's model and Ptolemy's model. In the same way, the differences of the equations of center, $\Delta c_{3(\bar{\kappa})}$, are graphed as

$$\Delta c_{3(\bar{\kappa})} = c_{3(\bar{\kappa})C} - c_{3(\bar{\kappa})P}, \qquad \Delta_{\max} = \pm 0;6,36^\circ \qquad (\bar{\kappa} = \pm 147^\circ).$$

Since $c_{\max}$ and c_3 can be combined in any combination in finding the greatest elongation from the mean sun, the greatest elongations in the two models can differ by up to about $\pm 0;11^\circ$ near $\bar{\kappa} = \pm 45^\circ$ where, of course, no observations were used in deriving either the models or their parameters.

It is of interest that the greater part of the difference between the two models is due to the equation of center rather than to the equation of the anomaly. That the formulas for the equation of center, (61–62) and (70), differ—most significantly by the additional $e' \cos 2\bar{\kappa}$ term in (61)—is fairly obvious. But that the variable radius of the orbit in Copernicus's model so effectively reproduces the effect of the variable distance of the center of the epicycle in Ptolemy's is not so obvious, particularly when one considers the limited locations of the elongations used to derive the elements and, in the case of Ptolemy's model, the structure of the model. Since Copernicus's model was designed to fit the same limited locations, Ibn ash-Shāṭir, or whoever originally invented the principles of the model, must have carried out a careful analysis or been quite lucky.[37] Figure 82(a) also shows the graphs of the maximum equations of the anomaly that would be produced if r were held at its minimum r_m, mean $\bar{r}$, or maximum r_M. Compared to the variable radius r, the fixed radii r_m and r_M produce discrepancies of over 2°, while $\bar{r}$ still leaves errors of well over 1° (since the graph of r may be taken as equivalent to Ptolemy's model for the purpose of these comparisons). Hence, the difference of only about $0;4\frac{1}{2}^\circ$ in Copernicus's model is quite remarkable.

The purpose of the model is of course to compute elongations from the mean sun since it is by adding these elongations to the longitude of the mean sun that the longitude of Mercury is computed. And by then taking the elongations from the true sun, along with Mercury's latitude, the conditions for the planet's visibility are determined. The greatest elongations as a function of $\bar{\kappa}$ produced by Copernicus's model are graphed in Fig. 83 in which the vertical scale is the absolute number of degrees of elongation from the mean sun. If we define the equation of center as

$$c_3 \leq 0^\circ \quad \text{for } \bar{\kappa} \leq 180^\circ, \qquad c_3 \geq 0^\circ \quad \text{for } \bar{\kappa} \geq 180^\circ,$$

[37] For other models designed to duplicate Ptolemy's, see Whiteside (1974), 4–5, on Quṭb ad-Dīn ash-Shīrāzī and Viète, and Swerdlow (1975a), 197–206, on Viète, who had considerably more than luck.

then the morning elongations, when Mercury is west of the mean sun, will be $-\Delta_M = -c_{\max} - c_3$, and the evening elongations, when the planet is east of the mean sun, will be $+\Delta_E = +c_{\max} + c_3$. The behavior of the elongations is interesting. The minimum occurs, not in the apsidal line at $\bar{\kappa} = 0°$, but at $\bar{\kappa} \approx \pm 30°$ for evening and morning respectively. Likewise, the maximum is not at $\bar{\kappa} \approx \pm 120°$, but at $\bar{\kappa} \approx \pm 105°$ for morning and evening respectively. Nevertheless, it is true that the sum of the elongations is a minimum at $\bar{\kappa} = 0°$ and a maximum at $\bar{\kappa} \approx \pm 120°$, for the algebraic sum $+\Delta_E - \Delta_M = 2c_{\max}$, and it can be seen in the graph of $c_{\max}$ that the maximum is at $\bar{\kappa} \approx \pm 120°$.

Finally, we have indicated in the figure the tropical longitudes of $\bar{\kappa} = (0°, \pm 120°, 180°)$ at the time of Ptolemy and Copernicus. Ptolemy located the apogee at ♎10°, and in V,30 Copernicus assumes a tropical longitude for his own time of ♏28;30°, an increase of 48;30° of which the sidereal motion of the apsidal line is 28;10° and the precession the remaining 20;20°.[38] Note that the points at $\bar{\kappa} = \pm 120°$ have shifted from ♒10° and ♊10° to ♓28;30° and ♋28;30°. Hence, according to Copernicus's theory, in his own time the sum of Mercury's greatest elongations reaches a maximum near the beginning of Leo and Aries, the latter almost exactly at the vernal equinox.

Correction of the Mean Anomaly and Positions at Epoch (V,29–31)

Copernicus's method for correcting the mean anomaly of Mercury is in essence the same as that used for Venus, that is, finding the mean anomaly at the time of one ancient and one modern observation (in which the planet should not be near greatest elongation), and then computing the annual motion of the mean anomaly from its change in the intervening time. In the case of Venus, he assumed—he evidently could not prove—that its apsidal line had remained sidereally fixed, and for reasons that are not entirely clear he reduced its eccentricity for his own time. In this way he had all that he needed to find the mean anomaly at the time of his observation (11) of 1529. For Mercury he apparently had reason to believe that its apsidal line had moved since the time of Ptolemy, who had shown it to be sidereally fixed for the preceding 400 years, but how far it had moved was not easy to determine since, as we have seen, Copernicus could not make the kinds of observations required to determine its direction.[39] Consequently, he took an alternative course which was, briefly, to assume a direction for the apsidal line, and then attempt to recompute observed positions of Mercury. In order to do this, he used three observations (8–10) by Bernhard Walther, and his procedure gave him a great deal of trouble which we will examine.[40] Having reached an agreement between the computed and observed positions

---

[38] Below, p. 432.

[39] Above, pp. 403–404.

[40] Below, pp. 428–438.

that he found satisfactory—although he had to alter the observed positions to do this—he took the mean anomaly from the latest observation (10) and used it along with an ancient observation (7) from the *Almagest* in order to correct the mean anomaly, at least in principle, by the normal method. In our exposition, we shall consider first the derivation of the mean anomaly from the ancient observation, then Copernicus's application of Walther's observations, and finally his computation of the corrected mean anomaly and epoch positions.

1. *Ancient Observation* (V,29)

For correcting the mean motion of an inferior planet, it is best to use observations as far as possible from greatest elongation, preferably toward inferior conjunction, so that small changes in the elongation will correspond to large changes in the anomaly. Finding such observations for Venus should not be difficult since the planet is often visible very close to inferior conjunction, but for Mercury it is not easy since the planet is generally visible, if at all, only rather near its greatest elongation. Ptolemy uses two observations of Mercury, from 139 and −264, in the later of which the planet was between greatest elongation and superior conjunction with an elongation of about +25°, and in the earlier between greatest elongation and inferior conjunction with an elongation of −17;30°. The earlier is better suited to finding the mean anomaly, and Copernicus uses it for his ancient observation (7). The observation is dated in the *Almagest* according to the Era of Dionysius (and was perhaps made by an Alexandrian astonomer of the third century B.C. named Dionysius). Ptolemy equates the date with Era Nabonassar 484, which is E.A. 60, Thoth 18/19 (−264 Nov 15). Since Copernicus had used an observation for Venus (9) dated to Ptolemy Philadelphus 13 or E.A. 52, he converts E.A. 60 to Ptolemy Philadelphus 21 and dates observation (7) accordingly.

The description of the observation in *Almagest* IX,10 and *Epitome* IX,22, and also in *De revolutionibus*, is not without problems. According to Ptolemy, Mercury was seen as a morning star, east of the straight line connecting the northern and middle stars in the forehead of Scorpio (β and δ Scorpii) by one lunar diameter, and north of the northern star (β Sco) by two lunar diameters. The moon, a bit before last quarter at about ♌4°, was easily visible for comparison just west of the meridian. The configuration is shown in Fig. 84. Since the observation is about 400 years before the epoch of his star catalogue, Ptolemy reduces the longitudes in the catalogue by a precession of −4°. Ptolemy's tropical coordinates of the stars and the sidereal coordinates in Copernicus's catalogue are:

	Ptol λ	Cop λ*	β
β Sco	♏2;20°	209;40°	+1;20°
δ Sco	♏1;40	209;0	−1;40°

Taking the diameter of the moon $d_☾ \approx \frac{1}{2}°$, if Mercury is $2d_☾$ north of β Sco, it will have a latitude of $+2;20°$, and if it is $d_☾$ east of the straight line connecting the stars, it will fall at A with a longitude of about ♏3;4°. But Ptolemy says that the longitude is ♏3;20°, which would be at B, about $1\frac{1}{2}d_☾$ east of the line, but exactly $2d_☾$ east of β Sco at ♏2;20°. Evidently, Ptolemy's reduction of the report was really to consider Mercury $2d_☾$ north and $2d_☾$ east of the star, i.e. 1° north and 1° east of β Sco.[41]

Copernicus, perhaps, drew a figure much like ours, and found the report puzzling, which it is. His solution was to reverse the description, making the planet $2d_☾$ east of the line and $d_☾$ north of the star, but in fact measuring the difference in longitude from the star rather than the line, just as Ptolemy did. This nicely gives a longitude of ♏3;20°, Ptolemy's figure, but the latitude is now $+1;50°$ instead of $+2;20°$, which places Mercury at C. Copernicus's sidereal coordinates are then

$$\lambda_p^* = \lambda_*^* + 2d_☾ = 209;40° + 1° = 210;40°, \qquad \beta = +1;50°.$$

The sidereal longitude is equivalent to Ptolemy's tropical longitude, indeed, for $\pi = 2;40°$, $\lambda = 210;40° + 2;40° =$ ♏3;20°, but the latitude, which plays no part in the demonstration, is 0;30° less than Ptolemy's. If we now compute the positions of the stars and Mercury from modern theory, we find to sufficient accuracy:[42]

β Sco	$\lambda =$ ♏1;45°	$\beta = +1;16°$
δ Sco	♏1; 9	$-1;41$
Mercury	♏2;50	$+2;45$

The positions of the stars are shown in the figure as β' and δ', and Mercury is at D, in fact about $1\frac{1}{2}d_☾$ east of the line, a little over $2d_☾$ east of the star, but nearly $3d_☾$ north of the star. Here is another example of the difficulty of estimating distances by lunar diameters.[43]

The time elapsed since the Epoch of Alexander, $\Delta t = 59^y\,17;45^d$, for which, according to Copernicus $\bar{\lambda}_☉^* = 228;8°$ and $\lambda_A^* = 183;20°$, the value for the time of Ptolemy, which implies that for four hundred years the apsidal line remained sidereally fixed. Now, this is what Ptolemy believed, and it is also what Copernicus assumed for Venus. Since, however, in V,30 he assumes a sidereal motion of 28;10° for the apsidal line of Mercury between Ptolemy and the beginning of the sixteenth century, there is something contradictory about his assumption here, for it implies that the apsidal line remained at rest for 400 years and then moved nearly 30° in the following 1350 years. However, if Copernicus is to use the observation from -264, there is little

[41] Cf. the analysis in *HAMA*, 166–67.

[42] Using P. V. Neugebauer (1912) and Tuckerman (1962).

[43] Cf. above, pp. 364–365, n. 84.

else he can do since his goal is to find a mean anomaly at the observation identical or at least very close to the value found by Ptolemy, and attributing any further motion to the apsidal line would make this unlikely if not impossible.

In any case, at the time of the observation,

$$\bar{\lambda}_{\odot}^{*} = 228;8°, \qquad \lambda^{*} = 210;40°, \qquad \Delta = -17;28°,^{44}$$
$$\lambda_{A}^{*} = 183;20°, \qquad \bar{\kappa} = 44;48°.$$

The configuration is shown in Fig. 85 in which, since according to the report of the observation the elongation of the planet was still increasing, P is located on the close side of the orbit. For this demonstration, Copernicus uses, where $R = 10{,}000$,

$$e_1 = 736\tfrac{1}{2}, \qquad e_2 = 211\tfrac{1}{2}, \qquad \bar{r} = 3763, \qquad r' = 190.$$

Why Copernicus has altered e_1 and e_2 by $+\frac{1}{2}$ and $-\frac{1}{2}$ is not clear, for it makes absolutely no difference in the demonstration. From (71–75), the equation of center,

$$c_3 = 2;49° - 0;50° = 1;59°,$$

and the distances,

$$OM = 10534, \quad \text{(Corr. 10535)} \qquad OC = 10678. \quad \text{(Corr. 10679)}$$

The radius of the orbit, letting $r_m = \bar{r} - r'$, from (69b), is

$$r = 3573 + 189 = 3762.$$

Hence, in triangle OCP,

$$c = \Delta - c_3 = 15;29°,$$
$$\gamma_1 = \sin^{-1}(OC \sin c/r) = 130;44°,$$
$$\gamma_2 = 180° - (c + \gamma_1) = 33;46°. \quad \text{(Corr. 33;47°)}$$

Consequently,

$$\gamma_2 - c_3 = \bar{\alpha} - 180° = 31;47°,$$
$$\bar{\alpha}_7 = 211;47°,$$

the mean anomaly at the time of the observation. Ptolemy had found 212;34°, and the discrepancy of $-0;47°$ is the largest of any of Copernicus's recomputations of observations used by Ptolemy.[45]

[44] Ptolemy finds $-17;30°$, and with the correct $\bar{\lambda}_{\odot}^{*} = 228;11°$ from Copernicus's solar theory, $\Delta = -17;31°$, in better agreement. Copernicus's Δt = E.A. 59^{y} $17;45^{d}$ corresponds to 6^{h} after midnight, apparently the time of the observation in Alexandria, but perhaps he actually reduced it to 5^{h} to change the meridian to Cracow without noting the reduced time in his text, i.e. Δt = E.A. 59^{y} $17;42,30^{d}$, which would give $\bar{\lambda}_{\odot}^{*} = 228;8,30° \approx 228;8°$.

[45] It is the result of an accumulation of small differences, e.g. Ptolemy had $\Delta = 17;30°$ and $c_3 = 1;54°$ while Copernicus has 17;28° and 1;59°, and the elongation is great enough that $\bar{\alpha}$ is sensitive to such differences. Note that in Fig. 82(b), near $\bar{\kappa} = 45°$, as in this observation, the sum of the differences between Ptolemy's and Copernicus's models are near maximum (above, p. 423).

2. *Modern Observations* (V,30)

In order to find the mean anomaly at his own time, Copernicus must know the other elements of the planet's orbit, and for the superior planets he used his own observations to establish modern elements, the correct procedure. However, it is evident that he did not have suitable observations to do this for the inferior planets, so he had to substitute assumptions for derivations, and confirm them as well as he could by trial and error methods. For Venus he assumed that the apsidal line was sidereally fixed, but that the eccentricity had decreased since Ptolemy, and he then used a single observation (11) from 1529 to find the mean anomaly, although his assumptions remained untested, at least in the text. For Mercury he tries something more elaborate. He takes three modern observations (8–10) by Bernhard Walther, one (8) from 1491 and two (9–10) from 1504, and he assumes that all elements of Mercury's orbit except the direction of the apsidal line have remained unchanged since Ptolemy. He then assumes a value for the longitude of the apsidal line, and using it along with the constant elements, computes the mean anomaly for the first (8) of the three observations. Then, using the mean anomaly at (8) as epoch, and keeping the other elements the same, he computes the elongations of Mercury from the mean sun in observations (9) and (10). If the results agree with the observations, he considers that his original assumption of the longitude of the apsidal line was correct, as was the value of the mean anomaly derived from (8). If not, he must start with a new assumption for the longitude of the apsidal line and repeat the entire procedure, continuing to do so until he finds agreement. And he never does, but ends up altering the original observations, in fact all three of them, to agree with his computed elongations.[46]

Indeed, there was no way, short of a remarkable coincidence, that he could have succeeded, for the method is hopelessly flawed *ab initio* in that it assumes that all elements except the direction of the apsidal line are not only unchanged since Ptolemy, but are correct. Copernicus was attempting to test one element, the longitude of the apsidal line, when in fact his procedure was testing all the elements with no way of distinguishing which one was at fault. Further, the method was extremely laborious and aimlessly repetitious, for each tedious solution was based upon only a guess, although perhaps a narrower guess, about the longitude of the apsidal line. Unlike Ptolemy's iterative procedure for finding the elements of the orbits of the superior planets, there was no step in which corrections were applied that would cause the successive solutions to converge upon a correct result—correct in the sense that it would properly fit the observations under investigation. All Copernicus could do was to keep guessing. Hence, after wasting an unknown amount of time on an unknown number of trials—although he

[46] Kremer (1981) is a perceptive study of the alterations and of Copernicus's procedure.

really may have done only one—he wrote out a demonstration in his manuscript, and changed the observed longitudes and elongations of Mercury to fit it.

Observations (8–10) were, as far as one knows, made by Bernhard Walther in Nuremberg, although for unknown reasons Copernicus attributes (9) and (10) to Johann Schoener, who had access to Walther's papers and prepared the observations for their first printing in Nuremberg in 1544. Obviously, Copernicus must have received at least some part of them earlier, but when and how he received them, how many he possessed, and in what form, are all unknown.[47] The printed reports of the three observations are translated, rather literally, as follows:[48]

(8) 1491 Sep 9
9 September, 5 hours past midnight, I found Mercury in ♍13;23° with a northern latitude of 1;50°. Note how great the difference from the tables. Mercury appeared very faint, and without doubt near the limit of its heliacal setting (*occultationis*). It is also observed that the ray [was] refracted because it was little distant from the horizon, as is clear.

In a following observation of Sept 11, Walther notes that Mercury was still visible, but too faint to be observable, that is, with any precision, in the clearest sky. Since it was approaching morning setting, it was moving toward superior conjunction. The remark about the great difference from the tables can be understood by recomputing the position of Mercury with the *Tabulae resolutae*, which were based upon the *Alfonsine Tables* and were commonly in use in the fifteenth century.[49] The computed longitude is ♍7;57°, a difference of −5;26° compared with the observation, fully justifying Walther's comment.

(9) 1504 Jan 9
9 January, when the midheaven was ♏10°, [found] by Regulus, Mercury at ♑3;15° with a northern latitude of 0;45°, very carefully [observed]. Mercury was very faint.

In the observation for Jan 10 in the morning, Mercury is reported at ♑4°, and thus the observation of Jan 9 is also in the morning. Although no time is given, it can be found from the distance between the midheaven and the solar longitude computed for the date, and it is $6;31^h$ after midnight. By computing the position of the mean sun for Jan 9 and Jan 10, it can be

[47] There is considerable speculation on these matters, some of it acrimonious. For one possibility, see below, pp. 437–438.
[48] From Schoener (1544), 55r (misnumbered 59), 58r, 60r.
[49] We have used an edition published by Schoener (1536) for the meridian of Nuremberg.

determined that the elongation of Mercury is decreasing so that it is moving toward superior conjunction.

(10) 1504 Mar 18
18 March in the morning, the midheaven was ♑26°, when the sun was located at ♈6;53°, Venus at ♒20;15° with a northern latitude of 1;30°. And the same day after noon, through Aldebaran at ♊2;45°, Mercury at ♈26;30° with a northern latitude of 3;0°. The midheaven was ♋25°.

From the longitude of the sun and of the midheaven, the time of the observation can be determined as 7;21^h after noon. In the report for after noon of Mar 17, Walther notes that by aligning the armillary on Aldebaran at ♊2;45°, he found Mercury at ♈25;30°, thus moving at about 1° per day, with a northern latitude of 2;53°. On Mar 24 after noon, Mercury was observed at ♉0;5° with a northern latitude of 3;30°. By computing the position of the mean sun for Mar 17/18/24, it can be found that Mercury's elongation is decreasing, and since it is east of the sun it is moving toward inferior conjunction.

In order to show how the times for (9) and (10) were established, we compute, as Copernicus would, the longitude of the true sun for an approximate time, say, 6^h after midnight for (9) and 6^h after noon for (10). Then, from the spherical astronomy of Book II, the apparent time of each observation can be found from the difference between the right ascension of the true sun and the right ascension of the midheaven given in Walther's reports. In this way, from midnight,

$$\Delta t^h = \Delta t^\circ/15 = (\alpha_M + 180^\circ - \alpha_\odot)/15,$$

from which we find:

	λ_M	$\lambda_\odot$	α_M	$\alpha_\odot$	Δt°	Δt^h
(9)	220°	297;45°	217;35°	299;50°	97;45°	6;31^h ≈ $6\frac{1}{2}^h$
(10)	115	7;25	116;57	6;49	290;8	19;21 ≈ $19\frac{1}{2}$

Copernicus gives the following times:

$$(8)\ 5^h, \qquad (9)\ 6\tfrac{1}{2}^h, \qquad (10)\ 7\tfrac{1}{2}^h \text{ after noon, i.e. } 19\tfrac{1}{2}^h.$$

The time for (8) is that given by Walther, and the times for (9) and (10) agree with our computations. However, it should be noted that Frauenburg is about 8;35° = 0;34,20^h east of Nuremberg, so about $\frac{1}{2}^h$ should be added to the time, which can make a difference of a little over +0;1° in the longitude of the sun.

Although Copernicus gives the times applicable to Nuremberg, he actually computed his solar longitudes for a time fully 1^h later, overestimating the difference in longitude between Nuremberg and Frauenburg, and using the difference without adjusting the times in his text.[50] Note that the computed value of $\bar{\lambda}^*_{\odot 8} = 149;45°$ is 0;3° less than Copernicus's reported 149;48°. If we compute for 1^h later, that is 9 Sep 1491, 6^h, we find $\bar{\lambda}^*_{\odot 8} = 149;48°$, and adding the values of $\Delta\bar{\lambda}^*_{\odot}$ between the observations that he gives later, we have

$$\bar{\lambda}^*_{\odot} \qquad (8)\ 149;48°, \qquad (9)\ 270;2°, \qquad (10)\ 338;34°.$$

If we now add a precession of 26;59° for (8) and 27;5° for (9) and (10), we have the adjusted mean tropical longitudes

$$\bar{\lambda}'_{\odot} \qquad (8)\ 176;47°, \qquad (9)\ 297;7°, \qquad (10)\ 5;39°,$$

which are Copernicus's values. This must be the way he computed the values in the text, and since $\bar{\lambda}^*_{\odot 8}$ was for a time 1^h later than the reported time, all are really for a time 1^h later than the time given in the text.

Copernicus's treatment of the observations, which is not altogether straightforward, is shown by a number of corrections in M.[51] It can be seen in Table 8 that the observed longitudes of Mercury and the longitudes reported by Copernicus differ, but the situation is actually more complicated, for the reported values are the last of a series of changes in both the longitude and elongation made in M and then carried over to N. Taking Copernicus's computed $\bar{\lambda}'_{\odot}$ and Walther's observed λ_p, we should have:

	$\bar{\lambda}'_{\odot}$	λ_p	Δ
(8)	176;47°	163;23°	−13;24°
(9)	297;7	273;15	−23;52
(10)	5;39	26;30	+20;51

Copernicus's original and revised longitudes and elongations are:[52]

	λ_p Orig.	Δ	λ_p Rev.	Δ	
(8)	~163;24°	−13;23°	163;30°	~ −13;15°	(Corr. 13;17°)
(9)	273;15	−23;52	273;20	−23;47	
(10)	26;30	+20;51	26;55	+21;17	(Corr. 21;16°)

[50] The 1492 edition of the *Alfonsine Tables* gives a difference of 14° = $0;56^h \approx 1^h$ between Nuremberg and Cracow.

[51] 180v–182r, 195, the last is part of a separate quire (195–202), containing the end of V,30, and V, 31–33,35–36, that is inserted out of place in the manuscript.

[52] An intermediate value, $\lambda_{p8} = 163;15°$, was written and cancelled, but has no corresponding Δ. By scribal error, Copernicus altered Δ_9 from 23;52° to 23;42°, although he definitely intended 23;47° (he left out a *v*). The revised λ_{p10} in N is 26;5° for the intended 26;55° in M (*deunce* misread as *decima*). The revised Δ_8 correctly implies $\lambda_{p8} = 163;32°$ and Δ_{10} implies $\lambda_{p10} =$ 26;56°, but Copernicus definitely intended the values given. Cf. further discussion in Kremer (1981).

The original values are the same as Walther's except for the change of 0;1° in (8), and the revised values of Δ contain small errors. The reasons for the revisions will be discussed in our exposition of Copernicus's application of the observations.

For the intervals of time between the observations, Copernicus gives

$$\Delta t_{89} = 12^y\ 125;3,45^d, \qquad \Delta t_{910} = 69;31,45^d, \quad (\text{Corr. } 69;32,30^d)$$

and with these intervals, from the tables in III,14,

$$\Delta\bar{\lambda}^*_{\odot 89} = 120;14°, \qquad \Delta\bar{\lambda}^*_{\odot 910} = 68;32°.$$

Copernicus gives as the differences in mean anomaly

$$\Delta\bar{\alpha}_{89} = 316;1°, \qquad \Delta\bar{\alpha}_{910} = 216;0°,$$[53]

which agree exactly with Copernicus's tables, but differ from the tables in the *Almagest* or the *Alfonsine Tables*, $\Delta\bar{\alpha}_{89}$ by about +0;9°. This is odd since in principle Copernicus's tables of mean motions cannot be made up until after the following derivation of the corrected mean anomaly, while here $\Delta\bar{\alpha}$ should be estimated from some earlier source. All we can conclude is that there was a derivation of Mercury's mean anomaly that preceded the writing of V,30 in M, perhaps from an earlier trial using these observations, that Copernicus used the value so derived for the demonstration in V,30, and that possibly the tables of Mercury's mean anomaly are based upon that earlier trial.[54]

Copernicus then sets out the assumptions that will underlie the following demonstration:

> Now, from these three observations we wish to examine the motion of Mercury for the present time, in which we think it must be granted that the proportions (*commensurationes*) of the circles have continued from Ptolemy even to the present, since likewise in the other planets, they are not found to have deceived earlier reliable authors in this particular. And if along with these we also have the position of the apsides of the eccentric, nothing further will be required in [the examination of] the apparent motion of this star also. We have now assumed (*assumpsimus*) the position of the highest apsis in 211½°, that is, in Scorpio 28½°, for it could not be taken as less without prejudice to the observations. (N 171v27–36)

How Copernicus reached the apsidal longitude 211;30° is not clear, but his statement suggests an assumption tested by some kind of trial and error, presumably successive repetitions of the following demonstration with the three observations.[55] The conversion of 211;30° to the tropical longitude

[53] M 181r17–20 contains other cancelled values of $\Delta\bar{\alpha}$ of which we can make no sense. The differences are so large that they must be erroneous.

[54] It will be seen that the tabulated $\Delta\bar{\alpha}$ does not agree with or follow from the values of $\bar{\alpha}$ established for the observations, below, pp. 439–440.

[55] There is a suggested explanation in Moesgaard (1974), 293–98, but it is based on the revised values of Δ that Copernicus altered to fit 211;30° (including the scribal error $\Delta_9 = 26;42°$), and makes rather a large number of assumptions. We have no idea how Copernicus reached 211;30°, and later we shall even question whether he used successive trials by the method set out in V,30 (below, pp. 436–438).

♏28;30° implies a precession of exactly 27° which precisely fits observation (8) of 1491. The apsidal line is then considered fixed for observations (9–10) thirteen years later. The eccentric anomaly at each observation then follows from

$$\bar{\kappa}_8 = \bar{\lambda}^*_{\odot 8} - \lambda^*_{\mathrm{A}} = 298;15^\circ, \qquad \bar{\kappa}_9 = \bar{\kappa}_8 + \Delta\bar{\lambda}^*_{\odot 89} = 58;29^\circ,$$

$$\bar{\kappa}_{10} = \bar{\kappa}_9 + \Delta\bar{\lambda}^*_{\odot 910} = 127;1^\circ.$$

The configuration for (8) is shown in Fig. 86,[56] in which $-\bar{\kappa}_8 = 360^\circ - \bar{\kappa}_8 = -61;45^\circ$, and as in the demonstration for (7), where $R = 10{,}000$,

$$e_1 = 736\tfrac{1}{2}, \qquad e_2 = 211\tfrac{1}{2}, \qquad \bar{r} = 3763, \qquad r' = 190, \qquad r_m = 3573.$$

From (71–75), the equation of center,

$$c_3 = 3;35^\circ - 1;5^\circ = 2;30^\circ, \quad (\text{Corr. } 3;35^\circ - 1;3^\circ = 2;32^\circ)$$

and the distances,

$$OM = 10369, \qquad OC = 10371, \quad (\text{Corr. } 10459)$$

the error for OC being rather serious, and the radius of the orbit, from (69b),

$$r = 3573 + 295 = 3868.$$

Thus, in triangle OCP, taking $\Delta = 13;15^\circ$, the revised value,[57]

$$c = \Delta + c_3 = 15;45^\circ, \qquad \gamma_1 = (OC \sin c/r) = 49;8^\circ,$$

the latter being another serious error since correctly, if $OC = 10371$, $\gamma_1 = 46;42^\circ$, or with the correct $OC = 10459$, $\gamma_1 = 47;13^\circ$. Copernicus continues

$$\gamma_2 = c + \gamma_1 = 64;53^\circ, \qquad \gamma_2 - c_3 = 360^\circ - \bar{\alpha}_8 = 62;23^\circ,$$

$$\bar{\alpha}_8 = 297;37^\circ.$$

Taking correctly $c_3 = 2;32^\circ$ and $OC = 10459$, one finds

$$c = 15;47^\circ, \qquad \gamma_1 = 47;21^\circ, \qquad \gamma_2 = 63;8^\circ,$$

$$\gamma_2 - c_3 = 60;36^\circ, \qquad \bar{\alpha}_8 = 299;24^\circ,$$

---

[56] Since Mercury was past greatest elongation and moving toward superior conjunction (above, p. 429), it is on the farther side of its orbit as seen from the earth. It is, however, of interest that in drawing the figure for this demonstration in M (181v), Copernicus originally put the planet and the radius of the orbit on the close side of the orbit, which would alter the derived $\bar{\alpha}$ by over 80°. Perhaps too much should not be made of a correction in a drawing, but such an error suggests that this was the first time Copernicus had worked through the demonstration, since it would be a strange mistake if in earlier trials he had located the planet correctly, and had merely to copy an earlier drawing.

[57] If Copernicus ever used the original $\Delta = 13;23^\circ$ following from $\lambda_p = 163;24^\circ$ (M 180v27), it was for some kind of trial prior to writing out this demonstration, which appears on the very next page. Perhaps between writing these two pages he carried out a number of trials with the three observations using the original Δ and different values of λ^*_{A}. But that seems a strange, and uncertain, way to work.

so Copernicus's error is $-1;47°$.[58] This error is carried through to the remaining observations, and thus into the final correction of the mean anomaly.

Having found a value of the mean anomaly at observation (8), $\bar{\alpha}_8 = 297;37°$, Copernicus next tries to show that, along with the assumed longitude of the apsidal line, $\lambda_A^* = 211;30°$, observations (9) and (10) are correctly reproduced. Thus the use of observations (9) and (10) is distinctly different from the use of observation (8), for (8) was used to find $\bar{\alpha}$, while (9) and (10) are used to confirm that λ_A^* and $\bar{\alpha}$ have been correctly determined. The problem in (9) and (10) is thus the straightforward one: given the model with all its elements, to compute the elongation of Mercury from the mean sun for a specified time. Copernicus probably uses two observations for confirmation since agreement with a single observation might be fortuitous. But as we shall see, the two he uses do not confirm that the model correctly reproduces the observations.

The configuration for (9) is shown in Fig. 87. We are given that

$$\bar{\kappa}_9 = \bar{\kappa}_8 + \Delta\bar{\lambda}^*_{\odot 89} = 58;29°, \qquad \bar{\alpha}_9 = \bar{\alpha}_8 + \Delta\bar{\alpha}_{89} = 253;38°,$$

and the remaining elements are unchanged.[59] What follows is a trigonometric computation of an elongation from the mean sun. From (71–75), the equation of center

$$c_3 = 3;28° - 1;1° = 2;27°,$$

the distances,

$$OM = 10404, \qquad OC = 10505,$$

and from (69b), the radius of the orbit,

$$r = 3573 + 276 = 3849,$$

are all computed correctly. The true anomaly

$$\alpha = \bar{\alpha} + c_3 = 256;5°,$$

and thus in triangle OCP,

$$OP = (OC^2 + r^2 + 2\,OCr\cos\alpha)^{1/2} = 10282,$$

$$c = \sin^{-1}(r\sin\alpha/OP) = 21;19°, \quad \text{(Corr. 21;18°)}$$

and the elongation

$$\Delta = c + c_3 = 21;19° + 2;27° = 23;46°.$$

---

[58] All the preceding errors have been noted in Kremer (1981), 131 n.15. The cancelled $\Delta = 13;23°$ leads to $\bar{\alpha}_8 = 299;25°$ with Copernicus's c_3 and OC, or to 298;45° with the correct c_3 and OC. We shall see (below, p. 436) that Copernicus's $\bar{\alpha}_8 = 297;37°$ is already too high to reproduce elongations (9) and (10), so these values would make the situation even worse.

[59] The text reads $e_1 = 736$ rather than $736\frac{1}{2}$, but the equation of center follows from $736\frac{1}{2}$.

Now Walther's observation (9) gave $\Delta = 23;52^\circ$, which Copernicus originally wrote, but now he alters it to 23;47° (although he mistakenly wrote 23;42°, i.e. he left out a *v*), and he remarks that his computed 23;46° differs little from the observation. We can only remark that had he honestly reported Walther's observation as 23;52°, the difference of −0;6° would hardly constitute a poor agreement. In fact it would be quite good for the planet is in the worst possible position for the kind of test he was making. An elongation of 23;52° puts the planet so close to greatest elongation that it takes a very large change in the anomaly to produce a small change in the elongation, and in fact the correct elongation of 23;52° would necessitate reducing $\bar{\alpha}$ by at least 2;46° to 250;52°, which is not compatible with the mean anomaly derived from (8). With Copernicus's value of Δ, the tropical longitude of the planet

$$\lambda = \bar{\lambda}'_{\odot} - \Delta = 297;7^\circ - 23;46^\circ = 273;21^\circ,$$

i.e. ♑3;21°, and for this reason he alters Walther's ♑3;15° to ♑3;20°.

Copernicus came within 0;6° of reproducing observation (9). But he was not so lucky for (10), the configuration for which is Fig. 88. It is given that

$$\bar{\kappa}_{10} = \bar{\kappa}_9 + \Delta\bar{\lambda}^*_{\odot 910} = 127;1^\circ, \qquad \bar{\alpha}_{10} = \bar{\alpha}_9 + \Delta\bar{\alpha}_{910} = 109;38^\circ,$$

and the remaining parameters are as before. From (71–75), the equation of center,

$$c_3 = 3;31^\circ - 0;59^\circ = 2;32^\circ,$$

the distances

$$OM = 8575, \qquad OC = 9440, \quad \text{(Corr. 9439)}$$

and from (69b), the radius of the orbit,

$$r = 3573 + 242 = 3815.$$

The true anomaly

$$\alpha = \bar{\alpha} + c_3 = 112;10^\circ,$$

and in triangle *OCP*, in the same way as the previous computation,

$$OP = 8746, \qquad c = 23;50^\circ.$$

The elongation and the tropical longitude are thus

$$\Delta = c - c_3 = 21;18^\circ, \qquad \lambda = \bar{\lambda}'_{\odot} + \Delta = 26;57^\circ,$$

and consequently Copernicus changes the elongation implied by Walther's observation from 20;51° to 21;17°, and changes Walther's observed tropical longitude from ♈26;30° to ♈26;55°, covering all but 0;2° of the error of +0;27° in his result. He then says that his elongation agrees aproximately with the observation, referring to the difference of 0;1° between 21;17° and 21;18° (not to the real difference of 0;27°).

None of this is reassuring, but once again Copernicus's position was hopeless because the planet was so close to greatest elongation that no reasonable change in $\bar{\alpha}$ alone could remedy the defect.[60] In the case of (9), correcting an error of $-0;6°$ in Δ required a change of $-2;46°$ in $\bar{\alpha}$, but here in (10) the error of $+0;27°$ requires a reduction of $\bar{\alpha}$ by at least 9;28° to 100;10° in order to reach agreement with the observed $\Delta = 20;51°$. Thus neither (9) nor, even more so, (10) can be recomputed correctly from the mean anomaly derived in (8), and for that reason, with the original observations left unaltered, Copernicus's assumption for the longitude of the apsidal line has not been confirmed. In fact, at the time of (8), from modern theory $\bar{\alpha}_8 = 292;38°$, so Copernicus's derivation of 297;37° is in error by $+4;59°$.[61] If now we recompute (9) and (10) keeping $\bar{\kappa}$ the same but using a corrected value of $\bar{\alpha}$, that is, Copernicus's value reduced by 4;59°, we find:

	$\bar{\alpha}$	c_3	c	Δ	Obs Δ	Δ − Obs Δ
(9)	248;39°	2;27°	21;28°	23;55°	23;52°	+0;3°
(10)	104;39	2;33	23;41	21;8	20;51	+0;17

Note that the error for (9) is in the opposite direction to Copernicus's error of $-0;6°$ but in better agreement, while the error in (10), although reduced by 0;10°, is still not negligible, for $\bar{\alpha}$ would have to be reduced another $4\frac{1}{2}°$ to reproduce the observed elongation. And this analysis has, of course, considered only the mean anomaly by itself, not taking into account all the remaining untested elements of Copernicus's model that could contribute to the errors.

Comment. We have set out Copernicus's attempt at confirming the longitude of the apsidal line and deriving the mean anomaly under the assumption, apparently supported by the procedure he presents in the text, that it was the last of a number of trials with different estimates of λ_A^*, or λ_A since the precession makes them equivalent, that resulted in different values of $\bar{\alpha}_8$ to be tested by observations (9) and (10). However, we have already noted two pieces of evidence, the correction in the figure for (8) and the revision of the original $\Delta_8 = 13;23°$, that suggest that the exposition in V,30 may be practically the entirety of Copernicus's work on this problem.[62] There is in fact more evidence for this conclusion, and it tells us something of the way Copernicus was working when he wrote this section of his book.

---

[60] As has been noted by Kremer (1981), 126.

[61] $\bar{\alpha}$ is found from taking the difference between the mean heliocentric longitude of Mercury and the mean longitude of the sun in P. V. Neugebauer (1914). The correctly computed $\bar{\alpha}_8 = 299;24°$ increases the error to $+6;46°$.

[62] Above, p. 433, n. 56,57.

If Copernicus carried out a number of trials prior to writing V,30, and wrote down only his last, albeit unsuccessful, attempt in M, does it not appear strange that he mentions Walther's original observed values of Mercury's longitude and the resulting elongations at all? It would mean that he wrote 180v–181r of M, in which the original values are given, in the hope that his demonstrations would reproduce them, but without yet knowing what the demonstrations would yield or even how many attempts by trial and error they would involve. This seems odd. He would then have stopped writing, and carried out any number of trials with different estimates of λ_A^* and, presumably, the original $\Delta_8 = 13;23°$. After finding that Δ_9 and Δ_{10} could not be correctly reproduced, the reason being that $\Delta_8 = 13;23°$ gives $\bar{\alpha}_8$ too high,[63] he would then have reduced Δ_8, which would necessitate more trials and start him on the path of altering Walther's observations.[64] At this point he was, of course, no longer testing λ_A^*, he was now simply altering observation (8) to find an $\bar{\alpha}$ that would reproduce observations (9) and (10). But we know that he was still unsuccessful, and so ended up altering (9) and (10) also.

All of this seems rather a lot of work to carry out between writing folios 180v–181r, in which the original observations were recorded, and writing the remainder of the demonstration, and suggests that Copernicus may have done a good deal less, specifically, that the apsidal longitude $\lambda_A^* = 211;30°$ was not arrived at and tested by a series of trials using these observations, and that all Copernicus ever did with them was to try one or two values of Δ_8 in order to find $\bar{\alpha}_8$, and then alter all three observations to agree with his results. When combined with the fact that Copernicus was already using the tabulated values of $\Delta\bar{\alpha}$ for the intervals between the observations,[65] one is left with the suspicion that both the longitude of the apsidal line and the rate of the mean anomaly were already determined, and determined by unknown methods, before the supposed confirmation of the former and derivation of the latter was carried out. And this suggests considerable haste and a deliberate disregard of proper method in the writing of V,30, the contents of which are quite superfluous to both the origin and confirmation of the two parameters.

A question that therefore arises is when did Copernicus write V,30? The evidence indicates that it was quite late, even after the arrival of Rheticus in 1539, and was thus part of Copernicus's late revision of his book. First of all, Rheticus makes no mention of the use of Walther's observations in the *Narratio prima*, which was written to Schoener, who apparently possessed the manuscripts of Walther's observations and later published them. One

---

[63] Above, p. 434, n. 58.

[64] Perhaps the intermediate, cancelled $\lambda_{p8} = 163;15°$ (above, p. 431, n. 52), implying $\Delta_8 = 13;32°$, is a relic of this stage (if it was ever used at all). Since, however, it increases Δ_8, and thus would increase $\bar{\alpha}_8$, it is a step in the wrong direction.

[65] Above, p. 432.

would think that if Rheticus could have mentioned to Schoener that Copernicus was using Walther's observations, or as Copernicus evidently believed, two of Schoener's observations, he would have done so. Yet this argument *ex silentio* is hardly conclusive. But there is internal evidence in M that indicates a late date. We have noted that V,30 continues on to a separate quire (ff. 195–202) inserted out of place and containing V,31–33, 35–36, the last two chapters on retrograde arcs. The organization, or disorganization, of this part of M is quite confusing, and it is difficult to draw certain conclusions about the exact order in which its parts were written. Yet, the last nine pages of this displaced quire must have been written after Copernicus received the 1538 Basel edition of the Greek text of the *Almagest* from Rheticus, for in V,35 the lemma to Apollonius's Theorem on stationary points is referred to as a *lemmation*, which Copernicus transliterated directly from the Greek text, while it appears in none of his other sources for the theorem.[66]

If V,30 was indeed written after Rheticus's arrival, certain problems become explicable. Copernicus almost certainly did receive Walther's observations from or through Rheticus, since there is no evidence of his having any earlier communication with Schoener. The fact that the mean anomaly used between the observations already follows from Copernicus's tables, while the observations themselves are not the source of the tabulated mean anomaly, shows that this parameter, and perhaps the modern location of the apsidal line, were already established by a previous demonstration not, or no longer, in M. What then was Copernicus attempting to do with Walther's observations? Evidently to confirm by a fairly elaborate test parameters that he had earlier derived in some way that he considered less certain, perhaps pure trial and error or even by untested assumption. Unfortunately, Copernicus's test failed to confirm his earlier work, and rather than use the observations to correct the location of the apsidal line and mean anomaly, he altered the observations to fit his previously established values.

3. *Corrected Mean Anomaly and Positions at Epoch* (V,30–31)

Copernicus concludes his demonstration with this remarkable statement:

> Therefore, these three [computed] positions agreeing so well (*sic*) with the observed positions confirm without doubt that the exact position (*ipsum . . . locum*) of the highest apsis of the eccentric is what we assumed, $211\frac{1}{2}°$ under the sphere of the fixed stars in our time, and consequently that what follows is certain, namely, the mean anomaly of parallax, in the first position 297;37°, in the second 253;38°, in the third 109;38°, which was to be investigated. (N 171v13–27)

It is, perhaps, not out of place to point out that only two of the observations (9–10) were used for confirmation, and that any reader who acquired Schoener's 1544 printing of Walther's observations might have reason to

[66] As discovered by I. N. Veselovsky; cf. Copernicus (1978) 436 n. to 302:32, and as evidence for Rheticus's bringing Copernicus Walther's observations, 433 n. to 285:22. The word occurs in M, 197v33, and in N 179v11–12 misprinted as *lemation*.

doubt whether anything had been confirmed. Copernicus, nevertheless, seems quite content with what he has done, and goes on to show that the mean anomaly, however he may have reached it, is correct.

To do this, he uses the latest of the observations (10) from 1504 along with the ancient observation (7) from −264. The interval of time between the observations,

$$\Delta t = 1768^{y}\,200^{d}\,13;30^{h} = 1768^{y}\,200;33,45^{d},$$

which Copernicus gives as $1768^{y}\,200;33^{d}$.[67] From the previous demonstrations, it is known that at observations (7) and (10),

$$\bar{\alpha}_7 = 211;47^\circ, \qquad \bar{\alpha}_{10} = 109;38^\circ, \qquad \Delta\bar{\alpha} = 5570^{r} + 257;51^\circ,$$

$$\lambda^*_{A7} = 183;20^\circ, \qquad \lambda^*_{A10} = 211;30^\circ, \qquad \Delta\lambda^*_A = 28;10^\circ.$$

Note that λ^*_A is assumed constant from 1491 to 1504. The number of whole revolutions of $\Delta\bar{\alpha}$ is found from (36) or from dividing Δt by Mercury's approximate synodic period, i.e.

$$(1768^{y}\,200^{d})/115;53^{d} = 5570^{r} + \cdots,$$

which Copernicus does, although in a rather roundabout way. He then says that $\Delta\bar{\alpha} = 5570^{r} + 257;51^\circ$ agrees with the numbers set out in his tables. But this is simply untrue for, from the tables of mean anomaly,

$$\Delta t = 1768^{y}\,200;33^{d}, \qquad \Delta\bar{\alpha} = 258;0,51^\circ \approx 258;1^\circ,$$

a difference of +0;10°, which is not inconsiderable and shows that the tables were not derived directly from the mean anomaly established for observations (7) and (10) in the text.[68]

In fact, Copernicus's intervals of time and mean anomaly lead to the following values. Given that

$$\Delta t = 1768^{y}\,200;33^{d}, \qquad \Delta\bar{\alpha} = 5570^{r} + 257;51^\circ,$$

from (37), the annual motion,

$$\bar{\alpha}^{y} = 3^{r} + 53;57,22,47,45,33\ldots^\circ,$$

from (38), the period of the anomaly,

$$P_\alpha = 115;52,38,55,15,34\ldots^{d},$$

and taking the sidereal year of $365;15,24,10^{d}$, from (39),

$$145\,P_\alpha = 16802;14,3,32^{d} = 46^{y} + 0;25,31,52^{d},$$

[67] The difference, $-0;18^{h}$, is probably not due to the equation of time, which would reduce the interval by only $1;53^\circ = 0;7,32^{h}$. Since the time corresponding to the solar longitude of (10) was really 1^{h} later (above, p. 431), the correct interval is $1768^{y}\,200;36,15^{d}$, differing by $+1;18^{h}$ from Copernicus's interval.

[68] The additional $0;0,45^{d}$ would increase the computed anomaly to 258;3°, increasing the difference to +0;12°.

so that $c \approx +0;25,32^{d}$, all of which agree better with V,1 in N than in M (see Tables 1(a) and 1(b)), but in no case is the agreement perfect.[69]

There is, however, no reason to believe that the mean anomaly was in fact derived from observations (7) and (10). First, we have seen that the tabulated $\bar{\alpha}$ was already used for finding $\Delta\bar{\alpha}_{89}$ and $\Delta\bar{\alpha}_{910}$, and thus must have preceded the application of (10) in the text.[70] Second, it is really pointless to make any use of (10) in correcting the mean anomaly since $\bar{\alpha}_{10}$ was found merely by adding a $\Delta\bar{\alpha}$ to $\bar{\alpha}_8$, while it is $\bar{\alpha}_8$, in principle derived from observation, that should have been used with $\bar{\alpha}_7$ to correct the mean anomaly. But it does not appear that Copernicus did this either—at least, we have been no more successful at recomputing his values of $\bar{\alpha}^{y}$, P_{α}, and c with $\bar{\alpha}_8$ than with $\bar{\alpha}_{10}$, Hence, as was true for the other planets, the precise origin of Copernicus's mean anomaly is uncertain.

The motion of the apsidal line is treated simply. Copernicus says that the motion of 28;10° compared to the time in which the apogee has moved leads to a motion of 1° in 63 years, "if only," he says, "the motion were uniform." Here he is probably referring to Ptolemy's evidence that the apogee had been sidereally fixed for 400 years. The rate is a simple rounding of

$$(1768^{y}\ 200^{d})/28;10^{\circ} = 62;47^{y}/1^{\circ} \approx 63^{y}/1^{\circ},$$

and a more precise statement would be

$$\lambda_{A}^{*\circ/y} = 28;10^{\circ}/(1768^{y}\ 200^{d}) = 0;0,57,20^{\circ/y}.$$

The interval of time from the Epoch of Christ to observation (10) of 1504 is correctly $1504^{y}\ 87;48,45^{d}$, although Copernicus gives $1504^{y}\ 87;45^{d}$, for which, from the tables in V,1,

$$\Delta\bar{\alpha} = 63;13,29^{\circ} \approx 63;13^{\circ},$$

and thus, at the Epoch of Christ

$$\bar{\alpha}_{0G} = \bar{\alpha}_{10} - \Delta\bar{\alpha} = 109;38^{\circ} - 63;13^{\circ} = 46;25^{\circ},$$

although Copernicus gives $\bar{\alpha}_{0G} = 46;24^{\circ}$.[71] For the interval to the First Olympiad,

$$\Delta t = 775^{y}\ 12\tfrac{1}{2}^{d}, \qquad \Delta\bar{\alpha} = 95;3,31^{\circ} \approx 95;4^{\circ},$$

but Copernicus gives 95;3°, and then finds

$$\bar{\alpha}_{0A} = 46;24^{\circ} - 95;3^{\circ} = 311;21^{\circ}.$$

The Era of Alexander, $\bar{\alpha}_{0C}$ is computed correctly from $\bar{\alpha}_{0A}$, but the Era of Caesar, D., is omitted.

[69] A larger Δt (above, n. 67, 68) would make everything worse.

[70] Above, p. 432.

[71] $\Delta t = 1504^{y}\ 87;48,45^{d}$ gives $\bar{\alpha}_{0G} = 46;21^{\circ}$, not an improvement. If the observation is taken 1^{h} later, as it should be (above, p. 431), $\bar{\alpha}_{0G} = 46;14^{\circ}$. Perhaps Copernicus rounded $\Delta\bar{\alpha} = 63;13,29^{\circ}$ to 63;14°, but this would not be characteristic.

Determining the longitude of the apsidal line at epoch poses a problem. If we let $\lambda^*_{A10} = 211;30°$, and compute back at the rate $0;0,57,20^{\circ/y}$ to Ptolemy's observation (1), we find $\lambda^*_{A1} = 189;44°$. But in V,26 Copernicus used (1) and (2) to find 183;34°, which he then reduced to ♎10° − 6;40° = 183;20° for the time of Ptolemy, 6;24° less than the value just computed. Copernicus does not say what to do about this problem. If we take $\Delta\lambda^*_A = 28;10°$ to apply between observations (1) and (10), then the annual rate is

$$\lambda^{*\circ/y}_A = 28;10°/(1366^y\,264^d) = 0;1,14,11^{\circ/y} \approx 1°/48\tfrac{1}{2}^y,$$

which is compatible with Copernicus's treatment of the apsidal longitude in his time and at Ptolemy's time. We have thought it best to provide two sets of epochs for the apsidal line: (1) computed from $0;1,14,11^{\circ/y}$ back from $\lambda^*_{A10} = 211;30°$ in 1504, which is compatible with Ptolemy's location and is recommended, and (2) from $0;0,57,20^{\circ/y}$, which is compatible with reaching 183;20° at the time of the early observation of −264, but seems otherwise worthless. Both sets are given here along with the epoch values of $\bar{\alpha}$ and the mean tropical longitude $\bar{\lambda}_{A0} = \lambda^*_{A0} + \bar{\pi}$.

Epoch	$\bar{\alpha}_0$	(1) λ^*_{A0}	(1) $\bar{\lambda}_{A0}$	(2) λ^*_{A0}	(2) $\bar{\lambda}_{A0}$	
A. Olympiad	311;21°	164;32°	159;16°	175;12°	169;56°	
C. Alexander	213;3	173;50	174;52	182;24	183;26	(80)
G. Christ	46;24	180;30	186;2	187;33	193;5	

The mean sidereal and tropical longitudes at each epoch are those of the sun given in III,19.[72] Since Copernicus omitted epoch D., and since it is never really of interest, we have likewise omitted it.

Conclusion to the Inferior Planets

It is evident that Copernicus's treatment of the inferior planets is less satisfactory and less independent than his treatment of the superior planets. The most obvious reason is that he could not obtain the proper observations to carry out his own determination of elements by Ptolemy's methods. For the superior planets, the required observations are principally oppositions, which occur regularly and are easily observable. However, for the inferior planets, the observations are of particular configurations, with the mean sun at selected locations with respect to the planet's apsidal line and the planet rather close to greatest elongation. In the case of Venus, satisfactory observations of all configurations are not necessarily obtainable during one

[72] Above, p. 165.

lifetime. In the case of Mercury, Copernicus had the additional disadvantage of living at so high a latitude that the planet was frequently invisible at greatest elongation, due to the low inclination of the ecliptic to the horizon and the rather high inclination of Mercury's orbit to the ecliptic. Kepler, of course, managed to use Tycho's observations of Mercury, which were made at an even higher latitude, but Kepler was not confined to Ptolemy's procedures. But Copernicus had little choice except to rely on Ptolemy's observations of both planets for establishing the elements of their orbits, and then carry out whatever sorts of tests were possible to him to determine whether the elements were still correct. Unfortunately, his reporting of his procedures is so slight that there is little evidence for what sorts of tests he actually performed.

For Venus, he says that all of Ptolemy's parameters remained unchanged, including the location of the apsidal line, except that the eccentricity has decreased from 416 to 350 "as many observations teach us." The eccentricity referred to is the double eccentricity to the equant, and Copernicus applies the entire reduction to e_1 in his model, leaving e_2 unchanged, just as he had reduced Mars's e_1 while leaving r' the same. The most one can say about this reduction is that it is conformable with the reduction of Venus's eccentricity in medieval sources, where it is made equal to the eccentricity of the sun, and that Copernicus might have found or confirmed it using observations of the mean sun 90° from the apsidal line with the planet near greatest opposite elongations, for such configurations were obtainable during the period he was observing.

Why he does not explain what he did is not certain. One possibility is that his determinations gave only rough values—he did once use 332 or 333—and he ended up merely guessing at 350 as lying in between values that seemed either too large or too small. Why he chose to leave the apsidal line sidereally fixed when he found the apsidal lines of the other planet's to have moved is also not obvious. Here it appears that he simply had no information for either finding the direction on his own or showing that Ptolemy's direction was no longer correct. While he speculates in V,4 that the reduction of Venus's eccentricity, like that of Mars, has something to do with the motion of the center of the earth's orbit, there is really no way of reconciling the fixed apsidal line with the fact that the motion takes place in a small circle, the center of which has itself a circular motion. These could hardly combine to produce a reduction of the eccentricity of Venus along a straight line in a fixed direction, and to give the apsidal line of Venus a compensating motion would strain all credulity.

In the case of Mercury, he adapts all of Ptolemy's parameters to his own model through the use of Ptolemy's observations, but then advances the apsidal line through some 28° of sidereal longitude for his own time, although how he did this is not clear. He then uses three observations by Bernhard Walther to find the mean anomaly and confirm the direction of the apsidal line, and in principle his procedure could be used to find and correct the direction by successive approximation. Since, however, he does little more than alter the observations to agree with what appear to be predetermined values of both the mean anomaly and the direction of the apsidal line, the use of Walther's observations has nothing to do with either the origin or confirmation of these parameters.

What finally is most notable about Copernicus's treatment of the inferior planets is that he adapts his models and parameters to Ptolemy's even if it might appear peculiar to do so. The equation of center in Ptolemy's models depends upon the distance of the center of the epicycle from the apsidal line, and so in Copernicus's heliocentric conversion of Ptolemy's model upon the distance of the earth from the planet's apsidal line. This appears odd, at least to us for we know better, but it evidently did not appear odd to Copernicus, who assumed that Ptolemy's models produced essentially correct apparent motions that he wished to reproduce in his own. Hence, the coincidence went without question. And the same applies, perhaps more so, to the sum of the greatest opposite elongations of Mercury reaching a maximum at $\pm 120°$ from the apogee of the eccentric. This was indeed shown by Ptolemy's observations, and Copernicus follows it so unquestioningly that he even fails to mention an indispensable half of Ptolemy's observational evidence. But we should not expect anything else, first because Copernicus had no suitable observations of his own for examining the phenomenon, and second because it did not occur to him that he had reason to do so. Since he managed with his own model to reproduce very closely all the effects of Ptolemy's, he had reason only to be satisfied. This was the principal goal of Copernicus's planetary theory, and he managed for the most part very well. When, however, he attempted to do the same in his latitude theory, he was not so successful, for it is only in the latitude theory of the inferior planets that Copernicus found himself unable to devise heliocentric models observationally equivalent to Ptolemy's. But even these problems he recognized only partially, while in the longitude theory, whatever we may think, given his intentions and limitations, there was really nothing to alert him to the fact that something might have been wrong.

5. The Tables and Their Use

Arrangement and Computation of the Tables (V,1,33)

The tables of mean motion in V,1 are arranged in the usual form for 1–60 Egyptian years and 1–60 days. For each planet Copernicus tabulates only the mean anomaly $\bar{\alpha}$, which he calls the "motion of parallax" (*motus commutationis*). The mean anomaly can be combined with the mean sidereal or mean tropical longitude of the sun to give the mean heliocentric longitude of the planet, i.e. for a superior planet,

$$\bar{\lambda}_p^{*\prime} = \bar{\lambda}_{\odot}^{*} - \bar{\alpha}, \tag{81a}$$

$$\bar{\lambda}_p^{\prime} = \bar{\lambda}_{\odot} - \bar{\alpha}, \tag{81b}$$

and for an inferior planet,

$$\bar{\lambda}_p^{*\prime} = \bar{\lambda}_{\odot}^{*} + \bar{\alpha}, \tag{82a}$$

$$\bar{\lambda}_p^{\prime} = \bar{\lambda}_{\odot} + \bar{\alpha}. \tag{82b}$$

Copernicus uses relation (81) only indirectly in computing planetary positions, and never uses (82) at all. Since for the inferior planets the equation of center is a function of the motion of the earth through the mean eccentric anomaly $\bar{\kappa}$, only the (essentially geocentric) quantity $\bar{\alpha}$ is required for computing the position of the planet. The values of $\bar{\alpha}^y$ and $\bar{\alpha}^d$ underlying the tables in *N* are as follows:

Planet	$\bar{\alpha}^y$	$\bar{\alpha}^d$	
Saturn	5,47;32,3,9,40°	0;57,7,44,5°	
Jupiter	5,29;25,8,15,6	0;54,9,3,49	
Mars	2,48;23,30,36,4	0;27,41,40,22	(83)
Venus	3,45;1,45,3,40	0;36,59,28,35	
Mercury	$(3^r +)$ 53;57,23,6,30	3;6,24,13,40	

In using the tables, one enters with multiples of 60 years and single years, multiples of 60 days, single days, and sexagesimal fractions of a day, and takes the sum together with the position at epoch for some era, of which only Alexander and Christ are of any real use.

The tables of corrections in V,33 are arranged in six columns:

(1 & 2) Arguments for the entry of the remaining columns.
(3) $c_{3(\bar{\kappa})}$, equation of center.
(4) $c_{4(\bar{\kappa})}$, coefficient of interpolation to correct $c_{6(\alpha)}$ for $\bar{\kappa} > 0°$.
(5) $c_{5(\alpha)}$, least equation of the anomaly for $\bar{\kappa} = 0°$.
(6) $c_{6(\alpha)}$, excess of the greatest equation of the anomaly for $\bar{\kappa} = 180°$ (Mercury $\bar{\kappa} = \pm 120°$).

Copernicus calls these columns:

(1 & 2) *Numeri communes*, common numbers.
(3) *Prosthaphaereses* or *aequatio eccentri*, corrections or equation of the eccentric.
(4) *Scrupula proportionum*, proportional minutes.
(5) *Parallaxes orbis*, parallaxes of the sphere.
(6) *Excessus parallaxeos*, excesses of the parallax.

In M (183r only) he gives for (5) and (6) the following headings:

(5) *Parallaxes orbis magni in summa abside*, parallaxes of the great sphere in the highest apsis.
(6) *Excessus parallaxeos in infima abside*, excesses of the parallax in the lowest apsis.

For the superior planets c_5 is indeed the angle subtended at the planet by the radius of the "great sphere" of the earth, but for the inferior planets c_5 is really the angle subtended at the earth by the radius of the sphere of the planet. Probably for this reason Copernicus dropped this description of c_5.[1]

Columns 1 and 2 are the arguments tabulated at intervals of 3° for 3°–180° and 180°–357°. Ptolemy's arguments in the correction tables in *Almagest* XI,9 were at intervals of 6° for 6°–90°, 270°–354°, and 3° for 90°–180°, 180°–270°. In the *Handy Tables* he extended these to single degrees by linear interpolation in the remaining columns, and all the descendants of the *Handy Tables*, such as the correction tables in the *Alfonsine Tables*, are likewise at intervals of 1°.

Copernicus made some direct use of the correction tables in both the *Almagest* and the *Alfonsine Tables* in making up his own, which is quite understandable. The computation of the correction tables was a formidable task, and Copernicus can hardly be blamed if, wherever possible, he lightened his work by copying or adapting corrections in earlier tables. He could do this because the equations themselves, aside from the established parameters, give identical or nearly identical results in Ptolemy's and Copernicus's planetary theory (except for Mercury where differences are unavoidable).[2]

Column 3, $c_{3(\bar{\kappa})}$, the equation of center. In the correction tables in the *Almagest* Ptolemy gives the equation of center in two components. The first, $c_{3(\bar{\kappa})}$, is computed as if the center of the epicycle were located on a circle around the equant of eccentricity $2e$, and the second, $c_{4(\bar{\kappa})}$, corrects for the proper location on a circle of eccentricity e. The two components are added together with the proper sign of c_4 to give the equation of center $c_3' = c_3 + c_4$. In the *Handy Tables* he simply gives the sum c_3', and this practice, which was standard in the middle ages, is followed by the *Alfonsine Tables* and by Copernicus.[3] In principle, Copernicus computed $c_{3(\bar{\kappa})}$ from formulas (6–10) for all the planets except Mercury, and from (71–75) for Mercury. However, it turns out that for Saturn he copied c_3 directly from the *Alfonsine Tables*,

---

[1] It has been restored and applied indiscriminately to all the planets in some of the critical editions and translations.

[2] For other examples of Copernicus's copying and adapting tables for spherical astronomy and lunar theory, see above, pp. 111, 227, 229.

[3] In M 183r, Copernicus calls c_3 *prosthaphaereses eccentri collectae*, the last word probably referring to the sum $c_3 = a_1 + a_2$ of formula (10a) in his method of computing the equation of center.

and for Jupiter he seems to have adapted c_3 from $c_3 + c_4$ in the *Almagest*.[4] The equation of center appears in three applications:

(1) $\kappa = \bar{\kappa} + c_3$, correction of mean to true eccentric anomaly,

$$c_3 \leq 0° \quad \text{for } \bar{\kappa} \leq 180°, \qquad c_3 \geq 0° \quad \text{for } \bar{\kappa} \geq 180°.$$

(2) $\alpha = \bar{\alpha} + c_3$, correction of mean to true anomaly,

$$c_3 \geq 0° \quad \text{for } \bar{\kappa} \leq 180°, \qquad c_3 \leq 0° \quad \text{for } \bar{\kappa} \geq 180°.$$

(3) $\Delta = c + c_3$, elongation of an inferior planet from the mean sun,

$$c_3 \leq 0° \quad \text{for } \bar{\kappa} \leq 180°, \qquad c_3 \geq 0° \quad \text{for } \bar{\kappa} \geq 180°.$$

Columns 4, 5, and 6 are used together to find the equation of the anomaly $c_{(\alpha,\bar{\kappa})}$ as a function of the true anomaly α and the mean eccentric anomaly $\bar{\kappa}$. Column 5, $c_{5(\alpha)}$, gives the least equation at $\bar{\kappa} = 0°$ (a) for a superior planet at its greatest distance from the mean sun, (b) for an inferior planet when the center of its orbit is at its greatest distance from the earth. Column 6, $c_{6(\alpha)}$, gives the excess of the greatest equation that must be added (a) for a superior planet at its least distance from the mean sun at $\bar{\kappa} = 180°$, (b) for Venus when the center of its orbit is at its least distance from the earth at $\bar{\kappa} = 180°$, (c) for Mercury at the eccentric anomaly $\bar{\kappa} = \pm 120°$. Column 4, $c_{4(\bar{\kappa})}$, is a coefficient of interpolation from 0;0 to 1;0 for $\bar{\kappa} > 0°$. The equation of the anomaly is given by

$$c_{(\alpha,\bar{\kappa})} = c_{5(\alpha)} + c_{4(\bar{\kappa})} \cdot c_{6(\alpha)}. \tag{84}$$

Ptolemy tabulated the equation of the anomaly differently, using four columns: $c_{6(\alpha)}$ the equation for the center of the epicycle at mean distance, $c_{5(\alpha)}$ a negative correction for greatest distance, $c_{7(\alpha)}$ a positive correction for least distance, $c_{8(\bar{\kappa})}$ a coefficient of interpolation from $-1;0$ to $+1;0$ for intermediate distances that also indicates whether $c_{5(\alpha)}$ or $c_{7(\alpha)}$ is to be combined with $c_{6(\alpha)}$. The *Handy Tables* and medieval tables, such as the *Alfonsine Tables*, use the same arrangement except that all functions are tabulated at 1° intervals, and the coefficient of interpolation is $c_{4(\kappa)}$, a function of the true eccentric anomaly $\kappa = \bar{\kappa} + c_{3(\bar{\kappa})}$.

Copernicus's tabulation of the equation in three columns, a simplification at the cost of some precision of interpolation, was used earlier by Ibn ash-Shāṭir, and is also the arrangement of Ptolemy's and Copernicus's lunar correction tables.[5] Given the same parameters, a simple relation should hold between c_5, c_6, and c_7 in the *Almagest* and the *Alfonsine Tables* and Copernicus's corresponding c_5 and c_6. Designating the former A and the latter C, we have

$$c_{5C} = c_{6A} - c_{5A}, \tag{85a}$$

$$c_{6C} = c_{5A} + c_{7A}, \tag{85b}$$

---

[4] Below, pp. 450–451.

[5] Ibn ash-Shāṭir's tables are described in Abbud (1962), and Copernicus's lunar tables are described above, pp. 226–228.

relations that turn out to be useful in Copernicus's tables for Saturn and Jupiter.[6] We now consider the computation of each column in the order c_5, c_6, c_4.

Column 5, $c_{5(\alpha)}$, the least equation of the anomaly, is the equation for any value of α at $\bar{\kappa} = 0°$, that is, for a superior planet at the greatest distance from the mean sun, and for an inferior planet at the greatest distance of the center of its orbit from the earth. The situation is shown in Fig. 89(a) for a superior planet, 89(b) for Venus, and Fig. 89(c) for Mercury. The greatest distance R_0 is:

Superior planets	$R_0 = R + (e_1 - r')$	(86a)
Venus	$R_0 = R + (e_1 - e_2)$	(86b)
Mercury	$R_0 = R + (e_1 + e_2)$	(86c)

The radius r, of the earth's orbit in Fig. 89(a) and of Venus's orbit in Fig. 89(b), is constant, while the radius r_0 of Mercury's orbit in Fig. 89(c) is the minimum value $r_0 = \bar{r} - r'$. $c_{5(\alpha)}$ is given most directly by

$$c_{5(\alpha)} = \tan^{-1}(r \sin \alpha/(R_0 + r \cos \alpha)). \tag{87}$$

and with only a sine function by

$$\rho_0 = (R_0^2 + r^2 + 2R_0 r \cos \alpha)^{1/2}, \tag{88}$$

$$c_{5(\alpha)} = \sin^{-1}(r \sin \alpha/\rho_0). \tag{89}$$

Column 6, $c_{6(\alpha)}$, the excess of the equation of the anomaly, is the difference between the least equation for any value of α at $\bar{\kappa} = 0°$ and the greatest equation for any value of α at $\bar{\kappa} = 180°$ (Mercury $\bar{\kappa} = \pm 120°$). For superior planets the greatest equation occurs when the planet is at its least distance from the mean sun, for Venus when the center of its orbit is at its least distance from the earth, both at $\bar{\kappa} = 180°$, and for Mercury when the eccentric anomaly $\bar{\kappa} = \pm 120°$. Again, the situation is shown in Fig. 89(a) for a superior planet, Fig. 89(b) for Venus, and Fig. 89(c) for Mercury, in which the distances are

Superior planets	$R_{180} = R - (e_1 - r')$	(90a)
Venus	$R_{180} = R - (e_1 - e_2)$	(90b)
Mercury	$R_{120} = (R^2 + e_1^2 + e_2^2 - R(e_1 + e_2) - e_1 e_2)^{1/2}$	(90c)

The radius of Mercury's orbit $r_{120} = \bar{r} + r'/2$.

In order to compute $c_{6(\alpha)}$, we first compute, except for Mercury, the equation $c'_{5+6(\alpha)}$ for $\bar{\kappa} = 180°$, i.e.

$$c'_{5+6(\alpha)} = \tan^{-1}(r \sin \alpha/(R_{180} + r \cos \alpha)), \tag{91}$$

[6] Below, p. 451. Likewise, for $c_{4(\bar{\kappa})}$ in Copernicus's tables and $c_{8(\bar{\kappa})}$ in the *Almagest* (but not $c_{4(\kappa)}$ in the *Handy Tables* or *Alfonsine Tables*), the relation $c_{4(\bar{\kappa})C} \approx \frac{1}{2}(1;0 + c_{8(\bar{\kappa})A})$, taking the correct sign of $c_{8(\bar{\kappa})A}$, will hold if the parameters are the same. But Copernicus does not use it for any planet.

or with a sine function

$$\rho_{180} = (R_{180}^2 + r^2 + 2R_{180}r\cos\alpha)^{1/2}, \tag{92}$$

$$c'_{5+6(\alpha)} = \sin^{-1}(r\sin\alpha/\rho_{180}), \tag{93}$$

and for Mercury, the equation at $\bar{\kappa} = \pm 120°$,

$$c'_{5+6(\alpha)} = \tan^{-1}(r_{120}\sin\alpha/(R_{120} + r_{120}\cos\alpha)). \tag{94}$$

or with only the sine,

$$\rho_{120} = (R_{120}^2 + r_{120}^2 + 2R_{120}r_{120}\cos\alpha)^{1/2}, \tag{95}$$

$$c'_{5+6(\alpha)} = \sin^{-1}(r_{120}\sin\alpha/\rho_{120}). \tag{96}$$

Then $c_{6(\alpha)}$ is given by the difference

$$c_{6(\alpha)} = c'_{5+6(\alpha)} - c_{5(\alpha)}. \tag{97}$$

Column 4, $c_{4(\bar{\kappa})}$, is the coefficient of interpolation to adjust $c_{6(\alpha)}$ for $\bar{\kappa} > 0°$. Since $c_{5(\alpha)}$ is the least equation at R_0, and $c_{6(\alpha)}$ the excess of the greatest equation at R_{180} (Mercury R_{120}, r_{120}), $c_{6(\alpha)}$ can be corrected for all values of R, that is, for all values of $\bar{\kappa}$, by a coefficient representing the variation of the excess of the equation of the anomaly for any value of α due to the variation of the distance $R_{(\bar{\kappa})}$ in the interval $0° < \bar{\kappa} < 180°$. In the case of Mercury, one must also take into account the variation of the radius of the orbit $r_{(\bar{\kappa})}$ as a function of $2\bar{\kappa}$. The coefficient $c_{4(\bar{\kappa})}$ is, however, computed only for the maximum equation $c_{\max(\bar{\kappa})} = \sin^{-1}(r/R_{(\bar{\kappa})})$, and then assumed to hold for all other values of $c_{(\alpha,\bar{\kappa})}$.[7]

In order to compute $c_{4(\bar{\kappa})}$, one must first find the distance $R_{(\bar{\kappa})}$, which is the distance $P\bar{S}$ in Fig. 90(a) for a superior planet, and the distance OC in Fig. 90(b) and (c) for Venus and Mercury respectively. The distance follows from relations (6–8) for every planet but Mercury, and from (71–73) for Mercury, which were already found in computing $c_{3(\bar{\kappa})}$.[8] The radius of Mercury's orbit, $r_{(\bar{\kappa})} = \bar{r} \pm \Delta r$ in Fig. 90(c), is

$$r_{(\bar{\kappa})} = \bar{r} - r'\cos 2\bar{\kappa}. \tag{69a}$$

Then the maximum equation is

$$c_{\max(\bar{\kappa})} = \sin^{-1}(r/R_{(\bar{\kappa})}), \tag{98a}$$

and for Mercury

$$c_{\max(\bar{\kappa})} = \sin^{-1}(r_{(\bar{\kappa})}/R_{(\bar{\kappa})}). \tag{98b}$$

Call the extremal values of $c_{\max(\bar{\kappa})}$ for $\bar{\kappa} = 0°$ and $\bar{\kappa} = 180°$ (Mercury $\pm 120°$) c_{m} and c_{M}, so that

$$c_{\mathrm{m}} = c_{\max(0)} = \sin^{-1}(r/R_0) \approx c_{5(\alpha)\max}, \tag{99a}$$

$$c_{\mathrm{M}} = c_{\max(180)} = \sin^{-1}(r/R_{180}) \approx (c_{5(\alpha)} + c_{6(\alpha)})_{\max}, \tag{99b}$$

---

[7] On the justification of this assumption, see *HAMA*, 94–95.

[8] More direct formulas for $R_{(\bar{\kappa})}$ can easily be derived from relations (2) and (70) for the equation of center, and these can be reduced to the formulas in Hartner (1974a), 10.

and for Mercury,

$$c_{\mathrm{M}} = c_{\max(120)} = \sin^{-1}(r_{120}/R_{120}) \approx (c_{5(\alpha)} + c_{6(\alpha)})_{\max}. \tag{99c}$$

Then the coefficient of interpolation $c_{4(\bar{\kappa})}$ is the ratio

$$c_{4(\bar{\kappa})} = \frac{c_{\max(\bar{\kappa})} - c_{\mathrm{m}}}{c_{\mathrm{M}} - c_{\mathrm{m}}}. \tag{100}$$

Copernicus tabulates $c_{4(\bar{\kappa})}$ to seconds from 0;0,0 to 1;0,0 for every planet except Saturn, for which he tabulates only to minutes.

Numerical Evaluation of the Correction Tables

Column 3, $c_{3(\bar{\kappa})}$, the equation of center. This correction is based upon Copernicus's parameters e_1 and r' for superior planets and e_1 and e_2 for inferior planets. We give them here along with their sum $2e$ (for Mercury their difference e) and the corresponding $2e$ (for Mercury e) from the *Almagest* and the *Alfonsine Tables* in units of $R = 10{,}000$.

Planet	e_1	e_2, r'	$2e$	*Alm* $2e$	*Alf* $2e$	
Saturn	854	285	1139	1139	1139	
Jupiter	687	229	917	917	1039	
Mars	1460	500	1960	2000	2000	(101)
Venus	246	104	350	416	378	
Mercury	736	212	e 524	e 500	e 500	

Only for Saturn and Jupiter are Copernicus's parameters essentially the same as the earlier sources. For Mars and Venus he has used the values applying to his own time, and even though his parameters for Mercury are chosen to come as close as possible to Ptolemy's, the equation of center in the two models is not quite equivalent. Note that Jupiter and Venus in the *Alfonsine Tables* differ from the *Almagest*, Venus because it is given the solar eccentricity, Jupiter for reasons that are unclear (even though $2e$ suggests a new determination of the eccentricity from observation, this seems very unlikely).

Columns 5 and 6, $c_{5(\alpha)}$ and $c_{6(\alpha)}$, the equation of the anomaly for $\bar{\kappa} = 0°$ and the excess for $\bar{\kappa} = 180°$ (Mercury $\pm 120°$). The parameters for these columns are R_0, R_{180}, and r for all planets except Mercury, and R_0 R_{120}, r_0, and r_{120} for Mercury. They are given here with the corresponding values from the *Almagest* and *Alfonsine Tables*, which are in every case the same.[9]

[9] Strictly, the distances R_0 and R_{180} and the equations of the anomaly for Jupiter and Venus in the *Alfonsine Tables* should differ from the *Almagest* and the *Handy Tables* because the eccentricities used in computing the equations of center have been changed. But in good medieval fashion such inconsistencies are disregarded, and Copernicus does much the same in taking over the Alfonsine equations of the anomaly for Saturn.

Planet		*De revolutionibus*			*Almagest & Alfonsine*		
		r	R_0	R_{180}	r	R_0	R_{180}
Saturn		1090	10569	9431	1083	10569	9431
Jupiter		1916	10458	9542	1917	10458	9542
Mars		6580	10960	9040	6583	11000	9000
Venus		7193	10142	9858	7194	10208	9792
Mercury	r_0	3573	10948	—	3750	11500	—
	r_{120}	3858		R_{120} 9540[10]			R_{120} 9261

(102)

Copernicus's corrections for each planet are graphed in Figs. 91–94, $c_{3(\bar{\kappa})}$ in Fig. 91, $c_{5(\alpha)}$ in Fig. 92, $c_{6(\alpha)}$ in Fig. 93, and $c_{4(\bar{\kappa})}$ in Fig. 94. If we call the scale in Fig. 92 of $c_5 = 1$, then in Fig. 91 $c_3 = 3$, and in Fig. 92 for Mars and Mercury $c_6 = 2$ and for Saturn Jupiter, and Venus $c_6 = 4$. A comparison of recomputations of c_3, c_5, and c_6 with Copernicus's tables shows that, of the 60 entries in each column for each planet, the number of errors is as follows (errors $>0;1°$ are in parentheses):[11]

Planet	$c_{3(\bar{\kappa})}$	$c_{5(\alpha)}$	$c_{6(\alpha)}$
Saturn	39 (11)	48 (18)	40 (16)
Jupiter	40 (11)	33 (6)	34 (2)
Mars	23 (6)	45 (14)	35 (16)
Venus	20 (0)	26 (2)	40 (20)
Mercury	21 (1)	51 (25)	28 (9)

(103)

In all, 523 out of 900 entries show errors, of which 366 are 0;1°, and many of these could result from roundings, but 157 are from 0;2° to 0;7°. However, the errors for Saturn and Jupiter are not due to Copernicus's computations, but are the result of his copying or adapting the corrections from the *Alfonsine Tables* and the *Almagest*. c_3 for Saturn agrees exactly with c_3 in the *Alfonsine Tables*, and when c_3 for Jupiter is compared with $c_3 + c_4$ in the *Almagest*—

[10] Corr. 9537, but in V,28 Copernicus computed 9540 (above, p. 421 and n. 35), and he probably used it to compute his tables (although the errors in the tables are so numerous that the results cannot be distinguished).

[11] Textual corrections in c_3, c_5, c_6: Saturn: none. Jupiter: c_5 132°, read 8;46° for 8;56°. Mars: c_3 117°, 10;21° for 10;11°; c_5 141°, 35;20° for 35;25°; 156°, 28;25° for 28;5°; c_6 138°, 10;44° for 10;47°; 156°, 13;3° for 13;28°. Venus: c_6 129°, 1;30° for 1;50°; 156°, 2;24° for 2;34°; 165°, 2;22° for 2;27°; 180°, 0;0° for 0;16°. Mercury: none. The preceding emendations are made for cases in which the distinction between textual and computational errors is obvious, either because of an error of 0;10°, or through very obvious errors of line to line difference, or through use in computing $c_{4(\bar{\kappa})}$. There may well be a good many more scribal errors, and of course it is trivial to correct anything to its correct value, but we have seen no reason to do so since most errors are in Copernicus's computation.

with linear interpolation for 3° intervals below 90°—only two entries disagree by 0;1°.[12] Comparing c_5 and c_6 for Saturn with the *Alfonsine Tables* using relations (85ab) shows one disagreement of 0;1° for c_5 and five of 0;1° for c_6.[13] The same comparison for Jupiter shows perfect agreement for c_5 and five discrepancies of 0;1° for c_6. Since it is not unusual for Copernicus to make small computational errors when adapting other sources—witness his errors in adapting Ptolemy's star catalogue—it appears certain that Copernicus simply adapted c_3, c_5, and c_6 for Saturn and Jupiter from the *Alfonsine Tables* and the *Almagest*. The number of discrepancies for Mars, Venus, and Mercury is not substantially lower than for Saturn and Jupiter, but for these three planets Copernicus had to compute the corrections himself—for Mars and Venus because his eccentricities differ from the earlier sources, for Mercury because his model necessarily produces slightly different equations than Ptolemy's.[14]

Column 4, $c_{4(\bar{\kappa})}$, the coefficient of interpolation. An evaluation of $c_{4(\bar{\kappa})}$ is difficult because its computation is quite lengthy and subject to small errors or roundings in subsidiary steps. In addition to all the parameters required for computing $c_{\max(\bar{\kappa})}$, the coefficient requires c_m at $\bar{\kappa} = 0°$, c_M at $\bar{\kappa} = 180°$ (Mercury ±120°), computed accurately or extracted from $c_{5(\alpha)}$ and $c_{6(\alpha)}$ by relation (99), and their difference $c_M - c_m$. Accurate computation with Copernicus's parameters leads to a set of values that we give unrounded to seconds followed in parentheses by the values extracted from the tables.[15]

Planet	c_m	c_M	$c_M - c_m$	
Saturn	5;55,10° (5;53°)	6;38,12° (6;38°)	0;43,2° (0;45°)	
Jupiter	10;33,24 (10;34)	11;35,0 (11;35)	1;1,36 (1;1)	
Mars	36;53,45 (36;54)	46;42,31 (46;43)	9;48,46 (9;49)	(104)
Venus	45;10,19 (45;10)	46;51,28 (46;52)	1;41,9 (1;42)	
Mercury	19;2,52 (19;3)	23;51,41 (23;53)	4;48,49 (4;50)	

[12] Using the 1515 *Almagest*, the only discrepancies are at 123° and 132° (after making six obvious corrections of 10 to 50 in the minutes column, due originally to misreadings in the Arabic text underlying Gerard's translation). Curiously, George of Trebizond's translation agrees with Gerard's. Copernicus could not use c_3 for Jupiter from the *Alfonsine Tables* since $2e$ is quite different.

[13] However, because $c_{5(\alpha)}$ and $c_{7(\alpha)}$ in the *Alfonsine Tables* contain small errors, their addition produces more serious errors in Copernicus's $c_{6(\alpha)}$ that are obvious in the graph in Fig. 93, in which the correct values lie on the broken curve. That Copernicus would independently make the same errors as the *Alfonsine Tables* is very unlikely. Further, Copernicus's $r = 1090$ gives $c_{5\,\max} = 5;55°$, while his table gives 5;53°, which is correct for $r = 1083$ in the *Alfonsine Tables* and the *Almagest*.

[14] Above, pp. 422–423.

[15] The tabular c_m for Saturn corresponds to $r = 1083$, as in the *Alfonsine Tables* and *Almagest*, and thus 1083 rather than Copernicus's $r = 1090$ must be used in computing $c_{\max(\bar{\kappa})}$. The tabular c_M for Mars follows from reading, at 138°, $c_6 = 10;44°$ for 10;47°. The accurate c_M for Mercury follows from $R_{120} = 9537$ rather than Copernicus's 9540, but for computing $c_{4(\bar{\kappa})}$ he appears to have used the tabular $c_M = 23;53°$.

We have recomputed $c_{4(\bar{\kappa})}$ using both the accurate and the approximate values, and Copernicus's computation is definitely closer to the latter. However, it is very inaccurate, and almost every entry is in error—sometimes by a few seconds, often by 20 or 30 seconds, and not uncommonly by more than a minute. It can be seen of even the few points graphed in Fig. 94 that some cannot be placed on smooth curves, which were all that could be drawn under the circumstances. Detailed comparisons show that in the case of Saturn, which admittedly is tabulated only to minutes, nearly every entry above 90° is 0;1, occasionally 0;2, too high. Similarly for Jupiter, every entry from 39°–108° is over 0;1 too high, which could be a systematic transcription error. Using the accurate values for Saturn changes the errors to 0;1 or 0;2 too low, while Jupiter remains the same. Mars shows errors reaching −0;0,20, making it the least inaccurate of the coefficients, while Venus has many errors exceeding ±0;0,30 and Mercury contains errors of nearly −0;2, particularly for $\bar{\kappa} > 120°$. The situation is not improved by computing with rounded numbers or by varying the parameters—indeed, any plausible changes only make matters worse. One could suggest a variety of textual emendations, e.g. displacing the minutes one line up for Saturn, one line down for Jupiter, but this only removes the long runs of systematic errors. In any case, the errors, whether of computation or copying, are due to Copernicus, and should probably be left alone.[16]

Calculation of Longitudes from the Tables (V,34)

Copernicus's recommended steps for the use of the tables to compute longitude, considering superior and inferior planets separately, are as follows:

Superior planets (Fig. 95): For a given time $t_0 + \Delta t$ since epoch, find the mean sidereal longitude of the sun and the mean anomaly of the planet,

$$\bar{\lambda}^*_{\odot} = \bar{\lambda}^*_{\odot 0} + \Delta\bar{\lambda}^*_{\odot}, \tag{105a}$$

$$\bar{\alpha} = \bar{\alpha}_0 + \Delta\bar{\alpha}, \tag{105b}$$

and although Copernicus does not mention that the apsidal line moves, the sidereal longitude of the highest apsis,

$$\lambda^*_{\mathrm{A}} = \lambda^*_{\mathrm{A}0} + \Delta\lambda^*_{\mathrm{A}}. \tag{105c}$$

The mean eccentric anomaly $\bar{\kappa}$ is then

$$\bar{\kappa} = \bar{\lambda}^*_{\odot} - \lambda^*_{\mathrm{A}} - \bar{\alpha}, \tag{106a}$$

which is true since, calling the mean heliocentric longitude of the planet $\bar{\lambda}^{*\prime}$, a quantity that Copernicus does not use by itself,

$$\bar{\lambda}^{*\prime} = \bar{\lambda}^*_{\odot} - \bar{\alpha}, \qquad \bar{\kappa} = \bar{\lambda}^{*\prime} - \lambda^*_{\mathrm{A}}.$$

[16] By contrast, Reinhold's $c_{4(\bar{\kappa})}$ in the *Prutenic Tables* usually agrees with our computation with the accurate parameters to within one or two seconds where we have checked, and appears to be computed very carefully.

With $\bar{\kappa}$, find the equation of center $c_{3(\bar{\kappa})}$ and the coefficient of interpolation $c_{4(\bar{\kappa})}$, and form the true or corrected anomaly

$$\alpha = \bar{\alpha} + c_{3(\bar{\kappa})}, \qquad c_{3(\bar{\kappa})} \gtrless 0^\circ \quad \text{for } \bar{\kappa} \lessgtr 180^\circ. \tag{107}$$

With α find the least equation of the anomaly $c_{5(\alpha)}$ and the excess of the greatest equation of the anomaly $c_{6(\alpha)}$, and compute the equation of the anomaly $c_{(\alpha, \bar{\kappa})}$ from

$$c_{(\alpha, \bar{\kappa})} = c_{5(\alpha)} + c_{4(\bar{\kappa})} \cdot c_{6(\alpha)}. \tag{84}$$

The next step is rather odd, although it works well enough. Let the positive elongation of the planet from the mean sun as seen from the earth be Δ. In the figure, consider $V'OP'$ drawn parallel to $V\bar{S}P$, with respect to which the negative elongation $\Delta' = 360^\circ - \Delta$ is

$$\Delta' = 360^\circ - \Delta = \alpha + c_{(\alpha, \bar{\kappa})}, \qquad c_{(\alpha, \bar{\kappa})} \lessgtr 0^\circ \quad \text{for } \alpha \lessgtr 180^\circ, \tag{108a}$$

so that the true geocentric sidereal longitude of the planet λ^* is

$$\lambda^* = \bar{\lambda}^*_{\odot} - \Delta' = \bar{\lambda}^*_{\odot} + \Delta. \tag{109}$$

This is equivalent to

$$\begin{aligned} \lambda^* &= \bar{\lambda}^*_{\odot} - (\bar{\alpha} + c_{3(\bar{\kappa})} - c_{(\alpha, \bar{\kappa})}) \\ &= \bar{\lambda}^{*\prime} - c_{3(\bar{\kappa})} + c_{(\alpha, \bar{\kappa})} = \lambda^{*\prime} + c_{(\alpha, \bar{\kappa})}, \end{aligned} \tag{110}$$

where $\bar{\lambda}^{*\prime}$ is the mean heliocentric longitude and $\lambda^{*\prime}$ the true heliocentric longitude, quantities that, as mentioned, Copernicus avoids using directly. Finally, the tropical longitude is

$$\lambda = \lambda^* + \pi, \tag{111}$$

where the true precession π has been found from the theory of Book III.

Inferior planets (Fig. 96): Again, for $t_0 + \Delta t$ since epoch, find

$$\bar{\lambda}^*_{\odot} = \bar{\lambda}^*_{\odot 0} + \Delta\bar{\lambda}^*_{\odot}, \qquad \bar{\alpha} = \bar{\alpha}_0 + \Delta\bar{\alpha}, \qquad \lambda^*_{\text{A}} = \lambda^*_{\text{A}0} + \Delta\lambda^*_{\text{A}}. \tag{105}$$

The eccentric anomaly is now

$$\bar{\kappa} = \bar{\lambda}^*_{\odot} - \lambda^*_{\text{A}}, \tag{106b}$$

and, finding the equation of center $c_{3(\bar{\kappa})}$ and the coefficient of interpolation $c_{4(\bar{\kappa})}$, form the corrected anomaly

$$\alpha = \bar{\alpha} + c_{3(\bar{\kappa})}, \qquad c_{3(\bar{\kappa})} \gtrless 0^\circ \quad \text{for } \bar{\kappa} \lessgtr 180^\circ. \tag{107}$$

With α, find $c_{5(\alpha)}$ and $c_{6(\alpha)}$, and compute the equation of the anomaly

$$c_{(\alpha, \bar{\kappa})} = c_{5(\alpha)} + c_{4(\bar{\kappa})} \cdot c_{6(\alpha)}. \tag{84}$$

The elongation from the mean sun is

$$\Delta = c_{(\alpha, \bar{\kappa})} + c_{3(\bar{\kappa})}, \qquad \begin{aligned} c_{(\alpha, \bar{\kappa})} &\gtrless 0^\circ \quad \text{for } \alpha \lessgtr 180^\circ, \\ c_{3(\bar{\kappa})} &\lessgtr 0^\circ \quad \text{for } \bar{\kappa} \lessgtr 180^\circ, \end{aligned} \tag{108b}$$

and the true geocentric sidereal longitude is

$$\lambda^* = \bar{\lambda}^*_{\odot} + \Delta, \tag{109}$$

observing the proper sign of Δ. The tropical longitude is again

$$\lambda = \lambda^* + \pi. \tag{111}$$

The procedure for both superior and inferior planets may be summarized as follows:

---

1. $\bar{\lambda}^*_{\odot} = \bar{\lambda}^*_{\odot 0} + \Delta\bar{\lambda}^*_{\odot}, \qquad \bar{\alpha} = \bar{\alpha}_0 + \Delta\bar{\alpha}, \qquad \lambda^*_A = \lambda^*_{A0} + \Delta\lambda^*_A.$

	Superior planets	Inferior planets
2.	$\bar{\kappa} = \bar{\lambda}^*_{\odot} - \lambda^*_A - \bar{\alpha}$	$\bar{\kappa} = \bar{\lambda}^*_{\odot} - \lambda^*_A$

3. $$\alpha = \bar{\alpha} + c_{3(\bar{\kappa})},$$
$$c_{3(\bar{\kappa})} \gtrless 0^\circ \quad \text{for } \bar{\kappa} \lessgtr 180^\circ,$$

4. $$c_{(\alpha,\bar{\kappa})} = c_{5(\alpha)} + c_{4(\bar{\kappa})} \cdot c_{6(\alpha)}.$$

	Superior planets	Inferior planets
5.	$\Delta' = \alpha + c_{(\alpha,\bar{\kappa})}$, $c_{(\alpha,\bar{\kappa})} \lessgtr 0^\circ$ for $\alpha \lessgtr 180^\circ$,	$\Delta = c_{(\alpha,\bar{\kappa})} + c_{3(\bar{\kappa})}$, $c_{(\alpha,\bar{\kappa})} \gtrless 0^\circ$ for $\alpha \lessgtr 180^\circ$, $c_{3(\bar{\kappa})} \lessgtr 0^\circ$ for $\bar{\kappa} \lessgtr 180^\circ$,
6.	$\lambda^* = \bar{\lambda}^*_{\odot} - \Delta'$,	$\lambda^* = \bar{\lambda}^*_{\odot} + \Delta.$

7. $$\lambda = \lambda^* + \pi.$$

Example. As an example of the use of the tables, we shall compute the longitudes of Mars and Mercury for 19 Feb 1473 at 4;48^h after noon. This is the time of Copernicus's birth given in his horoscope, and appears to be the principle evidence for the exact date of his birth.[17] The time of day, 4;48^h = 0;20^d, is obviously schematic, based perhaps upon somone's (Copernicus's?) knowledge that he was born in the late afternoon, but is not the correct time of the horoscope.[18] The date, 19 Feb 1473, 16;48^h after midnight (j.d. 2259, 121.7), follows the Epoch of Christ by

$$\Delta t = 1473^y\ 52^d\ 16;48^h = 24,33^y\ 52;42^d.$$[19]

---

[17] The horoscope, reproduced in Biskup (1973), plate 22, is from Cod. lat. Monac. 27003, a collection of horoscopes of famous people, dated, apparently, to the 1540's. There is another horoscope for Copernicus in Cod. lat. Monac. 10667 containing the birth date 10 Feb. 1473, 4;38^h after noon. Cf. Birkenmajer (1900), 406–12. Most of the positions in the two horoscopes are the same, and the correct date of the horoscope, given by the faster bodies, is definitely 19 Feb.

[18] The midheaven ♊23° and the horoscopus ♍25° combined with the sun at ♓11° indicate a time of about 6;40^h, i.e. 1;20^h after sunset at 5;20^h, and the sun is shown under the horizon in the sixth house.

[19] We are computing for the stated time of 4;48^h and ignore a meridian reduction of $-0;5^\circ$ = 0;0,20^h between Frauenburg–Cracow and Torun.

For this time, we find from the precession theory of Book III that $\pi = 26;47°$, and from the tables in III,14 and V,1:

Δt	$\bar{\lambda}^*_{\odot}$	$\bar{\alpha}_{\text{♂}}$	$\bar{\alpha}_{\text{☿}}$
$24{,}0^{y}$	5,55;38,49°	5,24;14,25°	4,57;14,36
33^{y}	5,31;39, 0,53	2,39;40,49,50	5,40;33,42,34
52^{d}	51;15, 5,50	24; 0, 6,59	2,41;32,59,50
$0;42^{d}$	0;41,23,43	0;19,23,10	2;10,28,57
	39;14,19,26	2,28;14,44,59	1,21;31,47,21
A.D. 1	4,32;31	3,58;22	46;24
	5,11;45	26;37	2, 7;56

Strictly, the apsidal line of Mars moves nonuniformly, but we can estimate its position in 1473 by computing back with a uniform motion of $0;0,28,10^{\circ/y}$ from Copernicus's observation (6) of 22 Feb 1523.[20] The longitude of Mercury's apsidal line is computed forward at the rate of $0;1,14,11^{\circ/y}$ from the Epoch of Christ.[21] In this way, we find

$$\lambda^*_{A\text{♂}} = 119;40° - 0;23° = 119;17°,$$

$$\lambda^*_{A\text{☿}} = 180;30° + 30;21° = 210;51°.$$

The mean eccentric anomaly of each planet is thus

$$\bar{\kappa}_{\text{♂}} = \bar{\lambda}^*_{\odot} - \lambda^*_{A} - \bar{\alpha} = 165;51°,$$

$$\bar{\kappa}_{\text{☿}} = \bar{\lambda}^*_{\odot} - \lambda^*_{A} = 100;54°.$$

From the tables of corrections in V,33:

Mars	Mercury
$c_{3(\bar{\kappa})} = 3;2°, \quad c_{4(\bar{\kappa})} = 0;58,2$	$c_{3(\bar{\kappa})} = 2;59°, \quad c_{4(\bar{\kappa})} = 0;56,52$
$\alpha = \bar{\alpha} + c_{3(\bar{\kappa})} = 29;39°$	$\alpha = \bar{\alpha} + c_{3(\bar{\kappa})} = 130;55°$
$c_{5(\alpha)} = 11;3,18°, \quad c_{6(\alpha)} = 1;23,57°$	$c_{5(\alpha)} = 17;26,45°, \quad c_{6(\alpha)} = 5;9°$
$c_{(\alpha,\bar{\kappa})} = c_{5(\alpha)} + c_{4(\bar{\kappa})} \cdot c_{6(\alpha)}$	$c_{(\alpha,\bar{\kappa})} = c_{5(\alpha)} + c_{4(\bar{\kappa})} \cdot c_{6(\alpha)}$
$= 11;3,18° + 1;21,11°$	$= 17;26,45° + 4;52,52°$
$= 12;24°$	$= 22;20°$
$\Delta' = \alpha - c_{(\alpha,\bar{\kappa})} = 17;15°$	$\Delta = c_{(\alpha,\bar{\kappa})} - c_{3(\bar{\kappa})} = +19;21°$
$\lambda^* = \bar{\lambda}^*_{\odot} - \Delta' = 294;30°$	$\lambda^* = \bar{\lambda}^*_{\odot} + \Delta = 331;6°$
$\lambda = \lambda^* + \pi = 321;17°$	$\lambda = \lambda^* + \pi = 357;53°$

[20] Above, p. 363.

[21] Above, p. 441.

We can compute the longitudes of the other planets directly from Copernicus's theory, and we find the following results:

Planet	λ_A^*	$\bar{\alpha}$	$\bar{\kappa}$	$c_{3(\bar{\kappa})}$	$c_{(\alpha,\bar{\kappa})}$	λ^*	
Saturn	239;46°	253;53°	178;6°	0;10°	6;21°	51;21°	
Jupiter	158;50	99;57	52;58	4;4	10;42	218;26	
Mars	119;17	26;37	165;51	3;2	12;24	294;30	(112)
Venus	48;20	67;14	263;25	2;0	26;40	340;25	
Mercury	210;51	127;56	100;54	2;59	22;20	331;6	

The tropical longitudes follow from adding the precession $\pi = 26;47°$. In order to complete the comparison with Copernicus's horoscope, we compute from Copernicus's theory the longitudes of the sun and moon, and we compare the horoscope with computation from Copernicus's theory, the *Tabulae resolutae*, which are based on the *Alfonsine Tables* and give about the same results as using the *Alfonsine Tables* directly, and a modern computation:[22]

Planet	Hor λ		Cop λ	*Alf* λ	Mod λ	
Saturn	♊21° =	81°	78;8°	80;10°	78;35°	
Jupiter	♐ 4	244	245;13	243;23	244;33	
Mars	♒22	322	321;17	322;35	321;43	
Venus	♈ 7	7	7;12	7;59	7;10	(113)
Mercury	♈ 0	0	357;53	0;0	356;30	
Sun	♓11	341	340;14	340;37	340;11	
Moon	♐ 7	247	245;39	245;55	245;13	

On the whole, the horoscope agrees somewhat better with the Alfonsine computation than with Copernicus, and the exact agreement for Mercury is quite striking. Since, however, the longitudes are given only to integer degrees, they may have been computed roughly rather than rounded from more precisely computed values. The modern computation is added to show that the horoscope and the longitudes computed with Copernicus's theory and with the *Tabulae resolutae* are all more-or-less consistent with the date

[22] We have included an equation of time of $+0;8^h$ for computing the position of the moon, without which its longitude is 245;34°. The *Tabulae resolutae*, printed in Schoener (1536) for the meridian of Nuremberg, have been reduced to Torun by $-0;30^h$. The modern computation is from Tuckerman (1964), even for the moon since no great accuracy is required for the comparison. The longitude of Mercury, 356;30°, is from graphical interpolation near first station.

19 Feb 1473, although the moon at ♐ 7° indicates a time about 2^h later than $4;48^h$, as do the midheaven and the horoscopus.[23]

Verification of the Planetary Theory

The simplest and, we believe, the most appropriate test of Copernicus's theory is using it to recompute the observations that form its basis, the same test we performed earlier for the sun and moon. It must be stressed that what may seem the best test, comparison with modern computed positions, is not really of interest since we are uncertain, not only of the accuracy, but even of the existence of observations performed by Copernicus, or by any contemporary, to test his planetary theory.[24] Indeed, prior to the time that accurate systematic observations of planets were made that could be used as a test of planetary theory—and that really means Tycho's observations at the earliest since Walther's hardly count—comparisons of positions computed with older and modern theory are simply anachronistic, and give a positively misleading idea of the contemporary judgment that is of real historical interest.[25] Further, although the comparison of positions computed for, say, the first half of the sixteenth century can show, however unhistorically, the accuracy of Copernicus's theory for his own time, it completely overlooks that what was of the first importance to Copernicus was not just whether his theory was correct in the first half of the sixteenth century, but whether it was also correct for the observations cited in the *Almagest*. It was, after all, for this reason that he devoted so much attention to secular changes in the precession, the tropical year, the obliquity, the eccentricities of the earth, Mars, and Venus, and the motions of apsidal lines.

A. Superior Planets

Copernicus gives seven observations of each superior planet, six of which are oppositions that are correctly recomputed if $\alpha = 180°$ and $c_{(\alpha, \bar{\kappa})} = 0°$. Only the seventh observation, of the planet outside opposition, provides a

[23] Above, p. 454, n. 18. The other bodies will advance less than 0;5°.

[24] There is evidence that his lunar and eclipse theory were tested by the observation of eclipses, which is far simpler, and that rather crude observations of conjunctions of planets were made and compared with his theory. An early example is in a letter of 17 Feb 1545 from Matthias Lauterwalt to Rheticus edited in Burmeister (1967–68) 3, 59–64. Lauterwalt observed the total lunar eclipse of 29 Dec 1544 (the letter gives 28 Dec), and a conjunction of Jupiter and Mercury on 21 Dec 1544 (although from Tuckerman (1964), the conjunction took place on 19 Dec), and compared them with computation from Copernicus's tables and eclipse theory and, for the conjunction, also with the *Alfonsine Tables*.

[25] This is not to say that it is not often useful to compare early *observations* with modern theory, specifically, observations of eclipses, occultations, and conjunctions for which the circumstances of the observations are known, for in this way one can discover what was there to be observed. But to compare bare longitudes with modern theory, whether they were observed or computed, is an uninformative exercise that has, on the whole, done more harm than good since it has led mostly to specious conclusions about the accuracy of early observations.

test of the radius of the planet's eccentric and the equation of the anomaly. In order to test these further, we have added two additional observations of each planet from the *Almagest* that were known to Copernicus but not used in *De revolutionibus*. These two are the observation (8) used by Ptolemy to find the radius of the epicycle and an early observation (9) used to correct the mean motion. The observations are given in Table 9 in which $\bar{\lambda}^*_\odot$ and π are computed from Copernicus's theory, and the observed λ_p for the early observations (9) is corrected by the computed π. The reason is that Ptolemy found λ_p from positions of fixed stars reduced by $\pi = 1°/100^y$ from the positions in his star catalogue. We have substituted the precession from Copernicus's theory, as Copernicus would have done in principle.[26]

In order to compare longitudes computed with Copernicus's theory with the observations, we have computed the sidereal longitude λ^*_p, and then added the precession from Tables 2–4 and 9 to give the tropical λ_p. For the observations in the *Almagest*, λ_p is compared directly with the observed λ_p reported, with the adjustment just mentioned for (9). For the oppositions observed by Copernicus, the comparison is with $\lambda_p = \lambda^*_p + \pi$ based upon the correctly computed $\lambda^*_p = \bar{\lambda}^*_\odot + 180°$ rather than with Copernicus's reported λ^*_p. The reason is that his report is sometimes in error due to a miscomputation of $\bar{\lambda}^*_\odot$, while the recomputation necessarily contains the correct $\bar{\lambda}^*_\odot$. For each computation we have used the apsidal longitude appropriate to the period of the observation, and for Mars we have used $2e = 2000$ for the observations in the *Almagest* and $2e = 1960$ for Copernicus's observations.[27] The results are summarized in Table 10 which gives the following:

(1) Number of observation, (1) to (9).
(2) λ^*_p, by computation.
(3) $\lambda_p = \lambda^*_p + \pi$, computed tropical longitude.
(4) Obs λ_p, observed tropical longitude.
(5) Error, (3)–(4).

The only outstanding error, for Jupiter, is in the additional early observation (9) of -240. The cause is principally a difference in the mean anomaly $\bar{\alpha}$ at the time of the observation; Ptolemy found 77;2° while Copernicus's tables give 76;21°. Substitution of $\bar{\alpha} = 77;2°$ gives $\lambda^*_p = 94;33°$ and $\lambda_p = 97;26°$, over-reducing the error from $+0;29°$ to $-0;7°$. The small errors for all three planets are within reason, and due principally to small deviations of 0;2° or 0;3° from $\alpha = 180°$ at opposition that, for Mars, can produce an equation of the anomaly of 0;4° or 0;5° when it should be 0;0°. And in

[26] The sources of the observations are: Saturn, *Almagest* XI,6–7, *Epitome* XI,16–17; Jupiter, *Almagest* XI,2–3, *Epitome* XI,7–8; Mars, *Almagest* X,8–9, *Epitome* X,23–24. Ptolemy definitely considers observation (9) of Mars to be at 6^h in Alexandria, but he computes $\bar{\lambda}_\odot$ for 7^h, and we have thus treated the observation as 6^h in Cracow. The early time will reduce the longitude of Mars by 0;1°.

[27] For observation (9): $\lambda^*_{A♄} = 222;45°$, $\lambda^*_{A♃} = 153;7°$, $\lambda^*_{A♂} = 105;37°$, (improperly) assuming a uniform motion. If $\lambda^*_{A♂} = 108;50°$, the longitude is increased by 0;1°.

computing λ_p^*, these errors are added together. On the whole, Copernicus's theory of the superior planets is quite good at reproducing these groups of observations.

B. Inferior Planets

The critical value to recompute for the inferior planets is the elongation $\Delta = c_{(\alpha,\bar{\kappa})} + c_{3(\bar{\kappa})}$ from the mean sun. The longitude $\lambda_p = \bar{\lambda}'_{\odot} + \Delta$ will be affected by any errors in $\bar{\lambda}'_{\odot} = \bar{\lambda}^*_{\odot} + \pi$, so is not as informative. We have added three additional observations of Mercury from the *Almagest* that are shown in Table 9. The first two, (11) and (12), are the greatest elongations $\Delta =$ 21;15° at $\bar{\kappa} = \pm 120°$ that Copernicus omitted, and the third (13) was used by Ptolemy together with (7) from −264 for correcting the mean anomaly.[28] Mercury's apogee has been held at $\lambda^*_A = 183;20°$ for all the ancient observations, even for (7), since that is what Copernicus used and it gives better results than if the apogee is moved back to an earlier theoretical position. For (8–10) the apogee is at 211;30°, and the times are taken one hour later than given in the text, as actually used by Copernicus for computing $\bar{\lambda}^*_{\odot}$.[29] Needless to say, the comparisons of (8–10) are with Walther's observations rather than with Copernicus's adjustments. For the ancient observations of Venus $2e = 416$, and for Copernicus's observation (11) $2e = 350$, and the apogee is fixed at $\lambda^*_A = 48;20°$.

The observations and computations are compared in Table 11, which gives the following:

(1) Number of observation.
(2) Δ, computed elongation,
(3) Obs Δ, observed elongation.
(4) Error, (2)–(3).
(5) λ_p, computed tropical longitude.
(6) Obs λ_p, observed tropical longitude.
(7) Error, (5)–(6).

Only the errors in Δ require comment since any further changes of λ_p, for better or worse, are due to $\bar{\lambda}'_{\odot}$. Among the observations (1–8) of Venus near greatest elongation are three large errors of about 1° in (2), (5), and (6). These occur because none of the eight observations was used with a determinate value of $\bar{\alpha}$, which was established only for (9) and (11). It turns out, for example, that (6), which shows the largest error, is about ten days too early, and $\bar{\alpha}$ about 7° too low, to produce the observed elongation of +47;20°.[30]

---

[28] Above, pp. 417, 425.

[29] Above, p. 431.

[30] However, the greatest elongation, of no less than 48;7°, is not reached until fully 25 days after the observation. Wilson (1972), 216, finds from Tuckerman (1964) that the true greatest elongation occurs 21 days after the observation, although it is only 47;26°.

Hence, what needs explaining is not why these errors are large, but why the other five errors, particularly for (7) and (8), are smaller. For the most part this is due to the fact that near greatest elongation Δ changes slowly as a function of α, and thus of $\bar{\alpha}$, and at these five observations $\bar{\alpha}$ happens to be close enough to the required value to produce something close to greatest elongation. And in the case of (7) and (8), if $\bar{\alpha}$ is nearly correct for one, it must be nearly correct for the other because of the relation between $\bar{\alpha}$ and $\bar{\lambda}_{\odot}$ that necessitates opposite and equal greatest elongations six years apart.[31] Observations (9) and (11) were used to correct $\bar{\alpha}$, and thus ought to be accurately reproduced, which they are, while (10) was the observation in the original version of V,23 in M that led to different values of $\bar{\alpha}^{\circ/y}$ and $\bar{\alpha}_0$, and thus it shows an expected error.[32]

Mercury is much the same although the errors are smaller. Observations (1–6, 11–12) are independent of $\bar{\alpha}$ so the errors are largely fortuitous. Observations (8–10) by Walther contain errors on the order of Copernicus's alterations, which is not surprising. That (7), the observation of −264 used to correct $\bar{\alpha}$ shows an error is also not surprising since $\Delta\bar{\alpha}$ between (7) and (10) from the tables differed by +0;10° from the value derived from the observations.[32a] The additional observation (13) of 139 was used by Ptolemy to correct $\bar{\alpha}$. The error of +0;24° is a consequence of a difference of +3;25° between $\bar{\alpha} = 99;27°$ derived by Ptolemy from the observation and $\bar{\alpha} = 102;52°$ computed from Copernicus's tables. One might conclude that it is a wonder that Copernicus's theory of Mercury reproduces the observations as well as it does.

6. Stations and Retrogradations

The most noticeable effects of the second anomaly are the periodic stations and retrograde motion of the planets. It is frequently said that one of the virtues, if not the principal virtue, of Copernicus's heliocentric theory is that it "explains" retrograde motion with a simplicity that earlier geocentric theory was lacking. The explanation is certainly simple enough since in the heliocentric theory apparent retrograde motion is the result of the earth's passing the slower superior planets or being passed by the faster inferior planets. Yet it must be pointed out that in geocentric theory retrograde motion is likewise given a simple and obvious cause, namely, that the apparent velocity of the planet in the lower part of its epicycle exceeds the apparent velocity of the center of the epicycle. And of course, as is true of all effects of the second anomaly, the results are geometrically and observationally indistinguishable.

---

[31] Above, p. 388.

[32] Above, pp. 401–403.

[32a] Above, p. 439.

The qualitative explanation of retrograde motion in both theories may be simple, but as soon as one inquires into the specific lengths and times of retrograde arcs the subject becomes considerably more complicated, although fortunately still equivalent in both theories. And because of the equivalence in both heliocentric and geocentric theories, Copernicus can take over virtually unchanged the methods of treating stations and retrogradations in the *Almagest* and *Epitome*. These fall into two parts:

(1) the proof of Apollonius's Theorem on the location of the stationary points that divide direct and retrograde motion,
(2) the application of Apollonius's Theorem for finding the anomaly at which the stations occur, and the length and time of the retrograde arc.

Following a remark in *Almagest* XII,1 and *Epitome* XII,2, Copernicus says that the mathematicians, and particularly Apollonius of Perga, considered stations and retrogradations only for one inequality, the one with respect to the sun, that is, the second anomaly, which Copernicus here calls the "parallax due to the motion of the great sphere of the earth." Ptolemy gives a proof of Apollonius's Theorem that is followed in the *Epitome* and by Copernicus, and then treats in detail the numerical application to each planet, taking into account the effects of the first anomaly at greatest and least distance, and finally he prepares a table of the true anomaly at first and second station as a function of the mean eccentric anomaly.[33] The exposition in the *Epitome* is abbreviated, although quite sound, and is confined to explaining Ptolemy's procedures without the numerical applications to individual planets. Copernicus's treatment is shorter still, consisting only of the proof of Apollonius's Theorem and a single numerical example of Mars at mean distance. However, in M he had originally given further examples of Mars at greatest and least distance, and he seems to have intended to give similar examples for an inferior planet. These demonstrations were somewhat more complicated in that the effects of the first anomaly on the velocity and distance of the planet had to be taken into account. But then he deleted the demonstrations for greatest and least distance of Mars, did not even write a demonstration for an inferior planet, and left only the simpler case of the mean distance of Mars for inclusion in N.

Apollonius's Theorem (V,35)

Copernicus proves Apollonius's Theorem for an inferior planet, presumably because he can stay closer to the form of the proof for a geocentric configuration, regarding the orbit of an inferior planet as an epicycle. The first anomaly is neglected in the proof. Thus, in Fig. 97 the earth is at *O* and *P*

[33] On Ptolemy's treatment of the entire subject, see *HAMA*, 190–206, 267–70, Pedersen (1974), 329–51.

is an inferior planet moving around center C which, since the first anomaly is not included, may be identified with the mean sun. For a superior planet, the the earth and planet are interchanged, i.e. the planet is at (P) and the earth at (O). Let a line $OP = \rho$ be drawn and extended to meet the orbit at L, and bisecting PL at K, let $PK = \sigma$. Let the velocity of C with respect to O be v_c, for an inferior planet the earth's motion about the mean sun, and the velocity of P with respect to the direction OC be v_p, the planet's mean anomaly. In the case of a superior planet, v_c is the motion of the planet about the mean sun and v_p, as before, the planet's mean anomaly. Apollonius's Theorem states that if

$$\frac{\sigma}{\rho} = \frac{v_c}{v_p}, \tag{114}$$

the planet will appear stationary from O.

A simple modern proof using vectors is shown in Fig. 98.[34] The motion of P in the direction of v_c will be the vector ρv_c perpendicular to OP. The motion of P in the direction of v_p will be the vector rv_p perpendicular to r. The planet will appear stationary from O when the resultant of the vectors lies on OP. Let angle $KCP = \eta$, and when P appears stationary, the parallelogram of vectors at P will be divided by OP into two right triangles similar to KCP (since the angle in the parallelogram at P is $180° - (90° - \eta) - 90° = \eta$). Consequently,

$$\frac{\rho v_c}{r v_p} = \frac{\sigma}{r} = \sin \eta, \qquad \frac{\sigma}{\rho} = \frac{v_c}{v_p},$$

which proves the theorem. Note that as long as the first anomaly is not considered, the vector rv_p is constant in magnitude and the vector ρv_c is variable with the distance ρ. It follows that when the planet is closer to the apogee, ρv_c is increased, $\sigma/\rho < v_c/v_p$, and the planet will appear to move forward. When the planet is closer to the perigee, ρv_c is decreased, $\sigma/\rho > v_c/v_p$, and the planet will appear to move retrograde. An obvious condition for retrogradation is therefore that

$$(\sigma/\rho)_{\max} = r/(R - r) > v_c/v_p,$$

for if $r/(R - r) \leq v_c/v_p$, there will either be no station or a single instantaneous station at perigee.[35]

In *Almagest* XII,1 Ptolemy provides an indirect demonstration of the theorem for both an epicyclic and equivalent eccentric model of the second

[34] Cf. *HAMA*, 191, Pedersen (1974), 331–32.

[35] This condition is fulfilled for all the planets; cf. *HAMA*, 191. It is of interest to note that it is not fulfilled for the moon as seen from the sun, either with the older distances of Ptolemy and Copernicus or with the correct distances, so the moon's heliocentric motion is always direct.

anomaly that may be Apollonius's original proof.[36] Regiomontanus proves the theorem in still greater detail in *Epitome* XII,3–5, although only for an epicyclic model, and Copernicus's proof appears to rely closely on the *Epitome*. The proof is equivalent to showing (see Fig. 98) that when the angle subtended at O by ρv_c is greater than the angle subtended by the projection of rv_p perpendicular to OP at P, the planet moves forward, when less the planet moves retrograde, and therefore when equal the planet appears stationary.

Application of Apollonius's Theorem (V,36)

The application of Apollonius's Theorem to finding the length and time of the retrograde arc is shown in Fig. 99 for a superior planet P and the earth O moving about center C. We are given the parameters R and r and the ratio v_c/v_p, which may be either mean or instantaneous values. From the proof of the theorem, it is known that when P appears stationary at O,

$$\frac{OK}{OP} = \frac{\sigma}{\rho} = \frac{v_c}{v_p}, \tag{114}$$

and consequently, we can find the ratio

$$\frac{PL}{OP} = \frac{\rho + 2\sigma}{\rho} = \frac{v_p + 2v_c}{v_p} = z. \tag{115}$$

From Euclid III,36, $OP \cdot PL = BP \cdot PA$, so that

$$\rho(\rho + 2\sigma) = (R - r)(R + r) = R^2 - r^2, \tag{116}$$

and thus

$$\frac{R^2 - r^2}{z} = \frac{\rho(\rho + 2\sigma)}{\left(\dfrac{\rho + 2\sigma}{\rho}\right)} = \rho^2, \tag{117}$$

or

$$\rho = \left(\frac{R^2 - r^2}{z}\right)^{1/2}. \tag{118}$$

Therefore, the three sides R, r, and ρ of triangle OCP are known, and the angles $\alpha' = 180° - \alpha$ and c, the equation of the anomaly, are given by

$$\cos \alpha' = \frac{R^2 + r^2 - \rho^2}{2Rr}, \qquad \cos c = \frac{R^2 + \rho^2 - r^2}{2R\rho}, \tag{119}$$

[36] Cf. *HAMA*, 267–70, Pedersen (1974), 331–38.

or, having found either one by the law of cosines, the other follows from

$$\sin \alpha' = (\rho \sin c)/r, \qquad \sin c = (r \sin \alpha')/\rho.^{37} \tag{120}$$

During the time that O moves through α' to B, the planet's heliocentric motion $\Delta\lambda'$ is

$$\Delta\lambda' = \alpha'(v_c/v_p), \tag{121}$$

and half the retrograde arc $-\Delta\lambda$ observed from O between the station and opposition is

$$-\Delta\lambda = c - \Delta\lambda'. \tag{122}$$

Finding the mean values $\bar{\alpha}'$ and $\Delta\bar{\lambda}'$ corresponding to α' and $\Delta\lambda'$,[38] the time of half the retrograde arc is

$$\Delta t = \bar{\alpha}'/\bar{v}_p, \qquad \text{or} \qquad \Delta t = \Delta\bar{\lambda}'/\bar{v}_c, \tag{123}$$

where $\bar{v}_p$ and $\bar{v}_c$ are the mean values of v_p and v_c. If the first anomaly is disregarded, the total arc is $2(-\Delta\lambda)$ and the total time $2\Delta t$.

The preceding demonstration is completely general, although when the first anomaly is taken into account, it becomes more complicated, the reason being that v_c/v_p cannot be taken as the ratio of mean motions, $\bar{\lambda}_p/\bar{\alpha}$ for a superior planet, $\bar{\lambda}_\odot/\bar{\alpha}$ for an inferior, but must be corrected for the true motions, which vary as a function of the mean eccentric anomaly $\bar{\kappa}$, and the true distance $R_{(\bar{\kappa})}$ must be substitued for the mean distance R. Ptolemy, as mentioned, carries out the demonstration for mean, greatest, and least distance for all the planets, and Copernicus attempted to do so for Mars, and seems to have intended to do so for an inferior planet. But then he deleted the demonstrations for greatest and least distance written in M. We shall first review his demonstration for mean distance as it appears in N, and then take up the cancelled demonstrations in M.

In *Almagest* XII,1 & 4 Ptolemy sets the ratio of the mean velocities $\bar{v}_c/\bar{v}_p$ of Mars at

$$\bar{v}_c/\bar{v}_p = 42/37 = 1/0;52,51,$$

for which Copernicus takes the reciprocal

$$\bar{v}_c/\bar{v}_p = 1/0;52,51 = 1;8,7/1.$$

[37] (116–120) are essentially the steps of Copernicus's alternate solution to finding the angles of a triangle of given sides in I,13,vii (N, 20v–21r), which is in part similar to Regiomontanus's *De triangulis* I,45. (117–118) are an application of the solution to the problem: given the area of a rectangle and the ratio of its sides, to find the sides. The problem is considered by Regiomontanus in *Epitome* XII,6, and in more detail in *De triangulis* I,18. Ptolemy treats the solution for α' and c differently, avoiding a triangle of three given sides, by first finding η and $\eta + \alpha'$ in right triangles OKC and PKC, so that $\alpha' = (\eta + \alpha') - \eta$, and c then follows from (120).

[38] On finding the mean motions, see below, p. 469.

Thus, in Fig. 100, (114–115),

$$\frac{\sigma}{\rho} = \frac{\bar{v}_c}{\bar{v}_p} = \frac{1;8,7}{1}, \qquad z = \frac{\rho + 2\sigma}{\rho} = \frac{3;16,14}{1}.$$

It has been established that

$$\frac{r}{R} = \frac{6580}{10000} = \frac{39;29}{60}, \qquad \frac{R + r}{R - r} = \frac{99;29}{20;31},$$

and thus, from (116–118),

$$R^2 - r^2 = 2041;4, \qquad \rho^2 = 2041;4/3;16,14 = 624;4,$$

$$\rho/R = 24;58,52/60 = 4163\tfrac{1}{2}/10000. \quad (\text{M } 4163\tfrac{1}{2}, \text{ N } 4163)$$

Thus, in triangle OCP, the sides are

$$R = 10000, \qquad r = 6580, \qquad \rho = 4163\tfrac{1}{2},$$

and by (119–120),

$$\bar{\alpha}' = 16;50^\circ, \qquad c = 27;15^\circ. \quad (\text{Corr. } 27;14^\circ)$$

During half the retrogradation, the motion of the planet, from (121),

$$\Delta\bar{\lambda}' = 16;50^\circ/0;52,51 \approx 19;6,39^\circ, \quad (\text{Accur. } 19;6,38,29^\circ)$$

the length of half the retrograde arc, by (122),

$$-\Delta\lambda = 27;15^\circ - 19;7^\circ = 8;8^\circ,$$

and the elapsed time, by (123),

$$\Delta t = 16;50^\circ/0;27,41,40^{\circ/\text{d}} = 36;28^{\text{d}} \approx 36\tfrac{1}{2}^{\text{d}}.$$

Hence, the entire arc and time are

$$2(-\Delta\lambda) = 16;16^\circ, \qquad 2\Delta t = 73^{\text{d}},$$

while Ptolemy had found $2(-\Delta\lambda) = 16;18,44^{\text{d}}$ and $2\Delta t = 73^{\text{d}}$.

Copernicus concludes by remarking that the demonstration can be done similarly for locations other than mean distance, but it is then necessary to consider the planet's true velocity for the particular location. And the same demonstration can be done for the inferior planets, in which the positions of the earth and planet are reversed, "and therefore," he says, "lest we repeat the same song (*cantilenam*) again and again, let these things suffice." But then he doubts whether this is even a useful procedure for finding stations, for since the variable velocities make their determination difficult, a difficulty that is not removed by Apollonius's Theorem, perhaps it would be better to find stations by interpolation in nearby computed true longitudes, just as is done for finding oppositions to the mean sun or conjunctions of any planets.

Original Version of V,36[39]

A curious vestige of the original version of this chapter remains in the introductory remarks that are unchanged in M and N. Copernicus says that if the spheres of the planets were homocentric with the great sphere, it would be easy to carry out the demonstrations since the velocities would be constant. But since the spheres are eccentric, it is necessary to use corrected motions and velocities except at mean distance where alone the mean motions apply. He continues: "Let us now show these things [first in the three superior planets], taking Mars as an example [because it is carried with a greater inequality than the others], from which example the retrogradations of the remaining planets will also become more clear."[40] Evidently, Copernicus also intended to give an example of an inferior planet following a demonstration for Mars showing the effect of the first inequality, which has a pronounced effect on the retrograde arcs of Mars because of its large eccentricity. He did not write the demonstration for an inferior planet, but for Mars he wrote in M demonstrations for the retrogradation at mean, greatest, and least distance, just as Ptolemy did in *Almagest* XII,4. M also contains a deleted earlier, and less complete, demonstration for mean distance which, however, was already revised before the two further demonstrations were written.

The earlier version for mean distance is as follows (refer to Fig. 100): Copernicus first converts Ptolemy's ratio of $\bar{v}_c/\bar{v}_p$ into a decimal,

$$\bar{v}_c/\bar{v}_p = 1/0;52,51 = 10000/8808 = \sigma/\rho,$$

and thus

$$z = (\rho + 2\sigma)/\rho = 28808/8808.$$

It has been established that

$$r/R = 6580/10000, \qquad (R + r)/(R - r) = 16580/3420.$$

so that from (116–118),

$$R^2 - r^2 = 56703600, \qquad \rho^2 = 17337035, \qquad \rho = 4164.$$

All the computations have so far been correct, but now Copernicus has trouble. In triangle OCP,

$$R = 10000, \qquad r = 6580, \qquad \rho = 4164,$$

and correctly, by (119–120),

$$c = 27;15°, \qquad \bar{\alpha}' = 16;50°, \qquad \bar{\alpha} = 180° - \bar{\alpha}' = 163;10°.$$

But Copernicus gives

$$c = 27;3°, \qquad \bar{\alpha}' = 17;2°, \qquad \bar{\alpha} = 180° - \bar{\alpha}' = 162;58°,$$

---

[39] M 199v–201v.

[40] M 199v8–11; deleted phrases are in brackets.

and he further says that at second station the anomaly will be

$$\bar{\alpha} + 2\bar{\alpha}' = 180° + \bar{\alpha}' = 197;2°.$$

He concludes that $\bar{\alpha}'$ will show the time between first station and opposition, and $2\bar{\alpha}'$ the time of the entire retrograde arc, but computes neither the length nor time of the arc.[41]

He then wrote another demonstration for mean distance, taking Ptolemy's $\bar{v}_c/\bar{v}_p = 1/0;52,51$, but with an erroneous $z = 2;32,15$ instead of the correct 3;16,14, and this demonstration is corrected in the manuscript into the final form that appears in N. However, probably before he made the corrections, he wrote the demonstrations for greatest and least distance that were deleted. In these demonstrations, the mean velocities $\bar{v}_c$ and $\bar{v}_p$ must be corrected to the true velocities v_c and v_p at greatest and least distance. The correction depends upon the rate of change of the equation of center $c_{3(\bar{\kappa})}$ over an interval of 1°, i.e.

$$\Delta c_{3(\bar{\kappa})} = c_{3(\bar{\kappa}+1°)} - c_{3(\bar{\kappa})}, \tag{124}$$

which can be computed directly or found by interpolation in the correction tables. Then the true velocities are

$$v_c = \bar{v}_c - \Delta c_{3(\bar{\kappa})} \cdot \bar{v}_c, \tag{125a}$$

$$v_p = \bar{v}_p + \Delta c_{3(\bar{\kappa})} \cdot \bar{v}_c, \tag{125b}$$

observing the sign of $\Delta c_{3(\bar{\kappa})}$ that is given by $c_{3(\bar{\kappa}+1°)} - c_{3(\bar{\kappa})}$. Taking the ratio v_c/v_p and dividing by $\bar{v}_c$, at greatest distance,

$$\frac{v_c}{v_p} = \frac{\bar{v}_c - \Delta c_{3(\bar{\kappa})} \cdot \bar{v}_c}{\bar{v}_p + \Delta c_{3(\bar{\kappa})} \cdot \bar{v}_c} = \frac{1 - \Delta c_{3(\bar{\kappa})}}{\bar{v}_p/\bar{v}_c + \Delta c_{3(\bar{\kappa})}}, \tag{126a}$$

and at least distance,

$$\frac{v_c}{v_p} = \frac{\bar{v}_c + \Delta c_{3(\bar{\kappa})} \cdot \bar{v}_c}{\bar{v}_p - \Delta c_{3(\bar{\kappa})} \cdot \bar{v}_c} = \frac{1 + \Delta c_{3(\bar{\kappa})}}{\bar{v}_p/\bar{v}_c - \Delta c_{3(\bar{\kappa})}}. \tag{126b}$$

Ptolemy's demonstrations for greatest and least distance are arranged such that the opposition between the two stations falls in the apsidal line. Thus, in order to find $\bar{\kappa}$ at the station, he estimates that $\bar{\kappa} \approx \Delta\bar{\lambda}'$ of the previous demonstration for mean distance, and in the case of Mars he also computes the distance $R_{(\bar{\kappa})}$ from the earth to the center of the epicycle accordingly. Copernicus seems to have missed this point since he takes the station, rather than the opposition between the stations, as occurring when the planet is in the apsidal line. Consequently, while the previous demonstration would lead to the estimates $\bar{\kappa} = \Delta\bar{\lambda}' \approx 19°$ and 180° − 19°, Copernicus simply takes

[41] There follows (M 200r20–32) a fragmentary demonstration for greatest distance that begins with the ratio $v_c/v_p = \sigma/\rho = 8917/10000 = 0;53,30/1$, which is very wrong for greatest distance, and breaks off after finding $\rho = 4441$, which is also very wrong.

$\bar{\kappa} = 0°$ and 180°, and the distances $R_{(\bar{\kappa})}$ are then R_0 and R_{180}. It is possible that when he caught these errors he decided to delete the two demonstrations rather than do them over again correctly, but it is also true that the demonstrations contain a number of internal errors that he may later have noticed.

The demonstration for greatest distance is as follows: Taking Ptolemy's ratio of the velocities for mean distance, $\bar{v}_c/\bar{v}_p = 1/0;52,51$, we must first find the true velocities at greatest distance. From Copernicus's correction tables, for $\bar{\kappa} = 0°$, $\Delta c_{3(0°)} = 0;10,40°$, or solving the equation of center directly, $\Delta c_{3(0°)} = 0;10,44°$, Taking the latter, more accurate, value, from (126a)

$$\frac{v_c}{v_p} = \frac{1 - 0;10,44}{0;52,51 + 0;10,44} \approx \frac{0;46,29}{1},$$

but Copernicus gives $v_c/v_p = 0;46,20,6/1$, which follows exactly from $\bar{v}_c/\bar{v}_p = 1/0;52,50$ and $\Delta c_{3(0°)} = 0;10,50$, i.e.,

$$\frac{v_c}{v_p} = \frac{1 - 0;10,50}{0;52,50 + 0;10,50} = \frac{0;46,20,6}{1}.$$

Curiously, the value $\Delta c_{3(0°)} = 0;10,50°$ comes from the table for Mars in the *Almagest*, where $c'_{3(6°)} = 1;5°$ so that $c'_{3(1°)} = \frac{1}{6}\,1;5° = 0;10,50°$, and interpolation in the *Alfonsine Tables* (rather than taking the rounded value for 1°) will give the same result. One can only wonder why Copernicus did not use his own table since Ptolemy's is for the larger eccentricity $2e = 2000$. In any case, in Fig. 101, from (114–115)

$$\sigma/\rho = 0;46,20,6/1, \qquad z = (\rho + 2\sigma)/\rho = 2;32,40,12/1.$$

At greatest distance, $R_0 = R + (e_1 - r') = 10960$, so that

$$\frac{r}{R_0} = \frac{6580}{10960} \approx \frac{36;1,20}{60}, \qquad \frac{R_0 + r}{R_0 - r} = \frac{96;1,20}{23;58,40},$$

and by (116–118),

$$R_0^2 - r^2 = 2302;23,58, \qquad \rho^2 = 904;51,12, \qquad \rho = 30;4,51.$$

Hence, in triangle OCP, where $R_0 = 100{,}000$,

$$r = 36;1,20R_0/60 = 60037, \qquad \rho = 30;4,51R_0/60 = 50135,$$

and from (119–120)

$$c = 27;18,40°, \quad (\text{Corr. } 27;18,49°)$$

$$\alpha' = 22;9,50°, \quad (\text{Corr. } 22;31;51°)$$

the latter error of about $-0;22°$ affecting both the length and time of the retrograde arc.

From (121), the true motion of the planet from the station to opposition would be

$$\Delta\lambda' = 22;9,50° \cdot 0;46,20,6 = 17;6,58°,$$

but Copernicus gives $\Delta\lambda' = 17;19,3°$.[42] In Ptolemy's demonstration $\Delta\lambda'$ is the motion of the center of the epicycle between first station and opposition in the apsidal line, so $\kappa = 0°$ at opposition and at the station $\kappa = \Delta\lambda'$. In Copernicus's demonstration, the station takes place in the apsidal line where $\kappa = 0°$, and the opposition is at $\kappa = \Delta\lambda'$ from the apsidal line. Hence, in both cases the mean motion in longitude $\Delta\bar{\lambda}'$ between station and opposition is

$$\Delta\bar{\lambda}' = \bar{\kappa} = \kappa - c_{3(\kappa)}, \qquad c_{3(\kappa)} \lesseqgtr 0° \quad \text{for } \kappa \lesseqgtr 180°, \tag{127}$$

where $c_{3(\kappa)}$ is the equation of center as a function of the true eccentric anomaly κ, and is found by inverse use of the correction table, entering with $\kappa = c_{1(\bar{\kappa})} - c_{3(\bar{\kappa})}$ and interpolating for $c_{3(\kappa)}$. In this case, taking the correct $\kappa = \Delta\lambda' \approx 17;7°$, inverse use of Copernicus's correction table for Mars gives $c_{3(\kappa)} = -3;38,30° \approx -3;39°$,[43] which is Copernicus's result since he gives for the mean motion in longitude

$$\Delta\bar{\lambda}' = 17;19,3° + 3;39° = 20;58,3°.$$

Another method is given by Regiomontanus in *Epitome* XII,9, although his explanation does not seem complete. Assume that $\Delta\lambda' = \Delta\bar{\kappa}_0 = \bar{\kappa}_0$, and in the normal way find the equation of center $c_{3(\bar{\kappa}_0)}$. Then a series of better approximations to $\bar{\kappa}$ will be

$$\bar{\kappa}_1 = \Delta\lambda' - c_{3(\bar{\kappa}_0)}, \qquad \bar{\kappa}_2 = \Delta\lambda' - c_{3(\bar{\kappa}_1)}, \ldots, \qquad \bar{\kappa} = \Delta\lambda' - c_{3(\kappa)}, \tag{128}$$

and finally,

$$\Delta\bar{\lambda}' = \Delta\bar{\kappa} = \Delta\lambda' - c_{3(\kappa)}.^{44} \tag{129}$$

For $\bar{\kappa}_0 = \Delta\lambda' \approx 17;7°$, we find,

$$c_3\colon\ (\bar{\kappa}_0) = -3;1°, \qquad (\bar{\kappa}_1) = -3;32°, \qquad (\bar{\kappa}_2) = -3;38°, \qquad (\bar{\kappa}_3) = -3;39°,$$

and since further iterations do not change the minutes, using the correct value of $\Delta\lambda' = 17;6,58°$,

$$\Delta\bar{\lambda}' = 17;6,58° + 3;39° = 20;45,58° \approx 20;46°.$$

We may now inquire into the origin of Copernicus's $\Delta\lambda' = 17;19,3°$, although what follows is only a guess. If the station took place at $\bar{\kappa} = 0°$ and the opposition at $\bar{\kappa} \approx 20;46°$, the velocity of the planet has not remained at the original value $v_{c(0)}$, but has increased in the ratio

$$\frac{v_{c(\bar{\kappa})}}{v_{c(0)}} = \frac{1 - \Delta c_{3(\bar{\kappa})}}{1 - \Delta c_{3(0)}}, \tag{130}$$

[42] On the possible source of this value, see below.

[43] Interpolating in $c_{3(\bar{\kappa})}$ computed to seconds gives $c_{3(\bar{\kappa})} = -3;38,49°$.

[44] Regiomontanus appears to stop with the first solution, although the iterations are necessary to get a reasonable result and seem to follow almost by common sense the first solution with $\bar{\kappa} = \Delta\lambda'$.

and therefore the true motion in longitude has increased to

$$\Delta\lambda'_{(\bar{\kappa})} = \Delta\lambda'_{(0)}\left(\frac{1 - \Delta c_{3(\bar{\kappa})}}{1 - \Delta c_{3(0)}}\right). \qquad (131)$$

It was earlier found for computing v_c/v_p at $\bar{\kappa} = 0°$ that a strict computation gave $\Delta c_{3(0°)} = 0;10,44°$, although Copernicus's table for Mars gives 0;10,40° and he used 0;10,50°. From the correction table for $\bar{\kappa} = 20;46° \approx 21°$, $\Delta c_{3(\bar{\kappa})} = 0;10,10°$, and for the following computation let $\Delta c_{3(0°)} = 0;10,45°$, Then, by (131),

$$\Delta\lambda' = 17;6,58°\left(\frac{1 - 0;10,10}{1 - 0;10,45}\right) \approx \frac{17;6,58°}{0;59,18} = 17;19,5°,$$

which is close to Copernicus's 17;19,3°, although it is far from certain that this is what he did. Finally, the corrected mean motion is, from (129),

$$\Delta\bar{\lambda}' = 17;19,3° + 3;39° = 20;58,3°.$$

The remainder of the demonstration is straightforward. Half of the retrograde arc between station and opposition is, by (122),

$$-\Delta\lambda = 27;18,40° - 17;19,3° = 9;59,37°,$$

and the time, from (123),

$$\Delta t = \Delta\bar{\lambda}'/\bar{\lambda}^{\circ/\mathrm{d}} = 20;58,3°/0;31,26,31^{\circ/\mathrm{d}} = 40;0,42^{\mathrm{d}} \approx 40^{\mathrm{d}}.$$

If we now assume, as Ptolemy did, that the opposition rather than the station takes place in the apsidal line, then the total arc and time between stations are

$$2(-\Delta\lambda) = 19;59,14°, \qquad 2\Delta t = 80^{\mathrm{d}},$$

compared with Ptolemy's results of 19;53,32° also in 80^{d}. Note that this last step of doubling the arc and time is not valid if, as Copernicus arranged his computation, the station takes place in the apsidal line, for then the arcs and the times between the stations and opposition will not be equal.

For least distance, we again begin with the ratio for mean distance, $\bar{v}_p/\bar{v}_c = 1/0;52,51$, and since for $\bar{\kappa} = 180°$, by computation or from Copernicus's correction table $\Delta c_{3(180°)} = 0;13°$, the ratio of the true velocities, from (126b), is

$$\frac{v_c}{v_p} = \frac{1 + 0;13}{0;52,51 - 0;13} = \frac{1;49,54}{1},$$

although Copernicus gives $v_c/v_p = 1;50,40/1$, which again follows from $\bar{v}_c/\bar{v}_p = 1/0;52,50$ and $\Delta c_{3(180°)} = 0;13,10°$, i.e.,

$$\frac{v_c}{v_p} = \frac{1 + 0;13,10}{0;52,50 - 0;13,10} = \frac{1;50,40,20}{1}.$$ [45]

Thus, in figure 102, from (114–115),

$$\sigma/\rho = 1;50,40/1, \qquad z = (\rho + 2\sigma)/\rho = 4;41,20/1,$$

although Copernicus gives 4;41,21, which follows if $v_c/v_p = 1;50,40,20/1$, leading to $z = 4;41,20,40/1$. At least distance $R_{180} = R - (e_1 - r') = 9040$, so that

$$\frac{r}{R_{180}} = \frac{6580}{9040} = \frac{43;40,21}{60}, \qquad \frac{R_{180} + r}{R_{180} - r} = \frac{103;40,21}{16;19,39},$$

and from (116–118).

$$R_{180}^2 - r^2 = 1692;42,45.$$ [46]

$$\rho^2 = 1692;42,45/4;41,21 = 360;59,1, \qquad \rho = 18;59,58.$$

Thus, in triangle OCP, where $R_{180} = 100,000$,

$$r = 43;40,21 R_{180}/60 = 72787, \quad \text{(Accur. 72787.5)}$$

$$\rho = 18;59,58 R_{180}/60 = 31665, \quad \text{(Accur. 31665.7)}$$

and by (119–120),

$$c = 25;45,16°, \quad \text{(Corr. 25;44,15°)}$$

$$\alpha' = 10;53,13°, \quad \text{(Corr. 10;53,22°)}$$

The true motion of the planet between the station and opposition is, from (121),

$$\Delta\lambda' = 10;53,13° \cdot 1;50,40 = 20;4,49°,$$

but Copernicus gives $\Delta\lambda' = 19;44,58°$. Letting $\kappa = 180° + 19;45°$, from inverse use of the equation table $c_{3(\kappa)} = +3;27,51°$, and if it is assumed instead that $\bar{\kappa}_0 = 199;45°$, from iterations with (128), we find on the fourth iteration that $c_{3(\bar{\kappa})} = +3;27,25°$, while Copernicus uses $+3;27,27°$. However, the true motion $\Delta\lambda' = 19;44,58°$ requires further investigation. From the correct $\Delta\lambda' \approx 20;5°$, if $\kappa = 200;5°$, then from inverse use of the correction table $\bar{\kappa} \approx 200° - 4° = 196°$, for which $\Delta c_{3(\bar{\kappa})} = 0;12,20°$. For finding v_c/v_p

---

[45] In this case, from the *Almagest* $\Delta c_{3(180°)} = 0;13,20°$, so 0;13,10° is halfway between the values from Copernicus's and Ptolemy's tables (which may be of no significance). Copernicus must have used 0;13,10° since the computation is extremely sensitive to changes of even 0;0,1° in $\Delta c_{3(\bar{\kappa})}$.

[46] The text (M 201v4) reads 1672;42,52, and several cancelled numbers in later steps follow from this error. But Copernicus must then have computed 1692;42,45 correctly, although he forgot to enter it, for it is confirmed by his corrections of ρ^2 and ρ.

at $\bar{\kappa} = 180°$, Copernicus had used $\Delta c_{3(180)} = 0;13,10°$, and thus, from (131), the corrected motion in longitude is

$$\Delta\lambda' = 20;4,49°\left(\frac{1 - 0;13,10}{1 - 0;12,20}\right) \approx \frac{20;4,49°}{1;1} = 19;45,3°,$$

which differs by $+0;0,5°$ from Copernicus's 19;44,58° (and, it must be noted, is not necessarily found by his method, whatever that may have been). Then, from $c_{3(\bar{\kappa})} = +3;27,27°$, the mean motion in longitude, from (129), is

$$\Delta\bar{\lambda}' = 19;44,58° - 3;27,27° = 16;17,31°.$$

The final steps are again trivial. From (122), half the retrograde arc between station and opposition is

$$-\Delta\lambda = 25;45,16° - 19;44,58° = 6;0,18° \approx 6°,$$ [47]

and the time of half the retrogradation, from (123),

$$\Delta t = \Delta\bar{\lambda}'/\bar{\lambda}^{°/d} = 16;17,31°/0;31,26,31^{°/d} = 31;5,22^d \approx 31\tfrac{1}{12}.^d$$

Therefore, if the opposition were in the apsidal line, the total arc and time between stations would be

$$2(-\Delta\lambda) \approx 12;1°, \qquad 2\Delta t = 62\tfrac{1}{6}^d,$$

which differ somewhat from Ptolemy's 11;12,14° in $64\frac{1}{2}^d$. However, as mentioned previously, the simple doubling of the arc and time is not valid in Copernicus's demonstration since he took the station rather than the opposition to fall in the apsidal line. It seems reasonable to guess that this error, which significantly affects the entire demonstration for greatest and least distance, was the reason that he deleted the entire section in M and published only the simpler, and essentially correct, demonstration for mean distance in N.

Appendix: The Distances of the Planets and Cosmology

There appears to be little doubt that to Copernicus himself, the principal argument for the heliocentric theory was that it gave a certain and straightforward demonstration of the order and distances of the planets, without the necessity of making additional assumptions, however plausible, about the arrangement of the planetary spheres. He emphasizes this point about order and distance in his dedicatory preface to Paul III, in the discussion of the order of the planets in I,10, and he returns to it briefly in the introduction to Book V, all of which are well-known and require no further comment. In the course of

---

[47] From $\Delta\lambda' = 20;4,49°$, it follows that $-\Delta\lambda = 5;40,27°$, which is much closer to Ptolemy's $-\Delta\lambda = 5;36,7°$.

Book V, he derives for each planet the ratio of the radius of its sphere to the radius of the great sphere of the earth, since this ratio is required for computing the equation of the anomaly, and for the superior planets he also gives the greatest, least, and mean distances from the mean sun, taking the radius of the earth's orbit as the unit. However, what he does not do anywhere is to bring these distances together for a general description of the dimensions of the planetary system, or convert them into the standard unit of terrestrial radii used to measure the distances of the moon and sun in Book IV.

It is not surprising that he did not do so. Ptolemy did nothing of the kind in the *Almagest*, since the distances are unknown and only the ratios of the radii of the eccentrics and epicycles, not distances in a common measure, are required for computing the positions of the planets. In *De revolutionibus*, even though the distances are known, the conversion of planetary distances to a common unit is equally superfluous since likewise all that is required for computation are the ratios of the distances. Copernicus computes his correction tables for each planet simply from the relative parameters r, e_1, etc., measured in units of R, and never explicitly makes use of the heliocentric distances of the planets. By contrast, the distance of the moon and, if it is not too large, the distance of the sun are required in units of terrestrial radii because the diurnal parallax must be taken into account in computing apparent positions of the moon and eclipses of the sun. However, the distances of the planets, although some may not be so far away as the sun, are all considered to be sufficiently great, or planetary theory sufficiently imprecise, that diurnal parallax has an insignificant effect upon computed apparent positions, and may therefore be ignored along with the necessity of converting distances into terrestrial radii. Hence, the conversions are justifiably absent in the *Almagest* and *De revolutionibus*, both of which are concerned with the theory and application of mathematical astronomy rather than the less certain subject of cosmology.

Since Copernicus determined the distance between the earth and sun in terrestrial radii, and since, from the maximum equation of the anomaly, the relative distances of the planets from the mean sun are easily found in units of the radius of the earth's orbit, it is of course trivial to compute the heliocentric distances of the planets in the common measure of terrestrial radii. This, however, is not true of the distances of the planets from the earth in a geocentric theory. The eccentricities and epicyclic radii of the planets are known only in terms of the radius of each planet's eccentric, and there is no common unit for the eccentrics of different planets as the distance between the earth and sun provides in the heliocentric theory whether Copernican or Tychonic in form. Hence, in a geocentric theory, one can determine the change of relative distance for any one planet, but cannot state the distance of one planet in terms of another, or of all the planets in terms of a common measure, without some further assumption about the arrangement of the planets.

Such an assumption, and it is fundamentally a physical assumption, was made by Ptolemy, not in the *Almagest*, but in his later work the *Planetary*

Hypotheses, where he presents a detailed description of the physical constitution of the heavens in which the spheres of the planets are contiguous to each other with no empty space between them. In Book II of the *Hypotheses*, Ptolemy describes complete spherical models for the planets of the kind illustrated earlier in Fig. 5.[48] It will be noted that in addition to the eccentric sphere 3 and epicyclic sphere 4 in the figure, there are two further spheres, an inner sphere 1 with its inner surface concentric to the earth at O, and an outer sphere 2 with its outer surface likewise concentric to O. If it is now assumed, as Ptolemy does, that the spheres of successive planets are contiguous, then the next higher planet has its least distance equal to the greatest distance of the planet shown in the figure, and the inner surface of sphere 1 of the higher planet will be contiguous to the outer surface of sphere 2 of the lower. Since the variation of relative distance of each planet is known, it is then possible to compute the greatest distance, that is, the distance to the outer surface of sphere 2, of the higher planet, and the procedure may be continued until one runs out of planets.

The result is the series of contiguous planetary spheres extending outward from the moon to the sphere of the fixed stars generally known as the Ptolemaic System, the foundation of the more sophisticated cosmology of the middle ages and the received physical representation of the heavens until Tycho showed convincingly that the heavens could not be filled with contiguous spheres or indeed with any spheres at all. There is even reason to believe that, although the principle was no longer necessary to him as a means of finding the distances of the planets, Copernicus himself still believed, not merely that the planets were carried by spheres—the fundamental assumption underlying his insistence upon uniform circular motion—but that the successive spheres of the planets were contiguous, extending through some indefinite distance so as to fill all space in the planetary system. The evidence for this conclusion is the well-known diagram of the planetary system in *De revolutionibus* I,10, which appears to depict a series of contiguous spheres contained between the concentric circles surrounding the sun.[49]

In order to find the distances of the planets through the assumption of contiguous spheres, the distance of any one planet must be known independently in some unit. But Ptolemy already knew the distances of two, the moon and the sun, that he had found in terrestrial radii in *Almagest* V,13,15. Hence, in the *Planetary Hypotheses*, beginning with the greatest distance of the moon, he finds the greatest distances of Mercury and Venus—and significantly finds that there is sufficient space between the greatest distance of the moon and the least distance of the sun to contain Mercury and Venus, but not Mars, thereby confirming his order of the planets. And beginning with the greatest distance of the sun, he finds the greatest distances of Mars, Jupiter, and Saturn, assuming at the end, as seems only reasonable, that the

---

[48] Above, p. 40.

[49] See above, p. 58.

sphere of the fixed stars is contiguous to the outer surface of the sphere of Saturn. The system extends from the least distance of the moon, at 33^{tr}, to the sphere of the fixed stars 19865^{tr}, which he rounds to $20{,}000^{tr}$. But if, he remarks, there are in fact spaces between the spheres, which he considers improbable on the principle that nature does nothing in vain, then the distances found in this way are at least the minimum distances of the planets.

Ptolemy's exposition of the distances, and also the sizes, of the planets is at the end of Book I of the *Planetary Hypotheses*, which was known in Arabic and Hebrew translations in the middle ages, and became a standard subject for Arabic astronomers and philosophers, who frequently recomputed the distances, sometimes with minor variations such as leaving space for the bodies of the planets. But that Ptolemy was himself responsible for the theory of contiguous spheres and distances was unknown in modern times until Willy Hartner made the astute conjecture that the distances given in Arabic sources originated, at least in principle, in the *Planetary Hypotheses*, a conjecture that was confirmed when Bernard Goldstein discovered that the section on distances survives in the Arabic and Hebrew translations, although the Greek text breaks off just before it.[50]

The distances from the *Hypotheses* were unknown in Europe in the sixteenth century, but two more-or-less typical sets of distances computed from its principles and contained in the medieval Latin translations of al-Farghānī and al-Battānī had been known since the twelfth century and were available in printed editions.[51] And Copernicus himself discusses and criticizes in I,10 a partial exposition of the theory, extending only from the moon, through Mercury and Venus, to the sun, from Proclus's *Hypotyposis*, which was published in a Latin translation by Giorgio Valla in 1501, although Copernicus's source was more likely Valla's encyclopedic *De expetendis et fugiendis rebus* (1501) that we have had occasion to mention earlier as a source for Copernicus's star catalogue.[52] Of the complete sets of distances, Farghānī's was the more widely known in the thirteenth through fifteenth centuries, but Battānī's, since his work contains not only the distances but the method of

[50] Hartner (1964), Goldstein (1967), containing a translation of the last part of Book I. The surviving Greek text of Book I, first published in a remarkably learned edition by John Bainbridge in 1620, is edited by Heiberg along with a German translation from the Arabic of Books I and II, but curiously omitting the last part of Book I, in Heiberg (1907), 69–145. An exposition of Ptolemy's theory of distances and sizes is in *HAMA*, 917–22 and a general history of the dimensions of the planetary system from antiquity to the early eighteenth century is given in a forthcoming study, *Measuring the Universe*, by A. van Helden.

[51] Both are printed in Battānī (1537), Farghānī's at f. 20 of the first foliation, Battānī's in cap. 50 at ff. 76v–78v of the second. Copernicus probably had not seen this edition, but he certainly knew Farghānī's distances from earlier printings, and he also knew the distances given by Proclus. Battānī's distances are in Nallino 1, 120–23, with notes on distances and sizes, 286–92.

[52] Above, pp. 93, 123. Copernicus's use of *De expetendis* for the distances (N 8r) rather than Valla's separate translation of Proclus is shown by Rosen in Copernicus (1978), 355–56, and Copernicus's use of Valla's work is well known. He would of course prefer a classical to a medieval Arabic source for the theory, and does not refer to Farghānī or Battānī.

computing them, seems to have been taken more seriously in the sixteenth century—at least, his set is specifically cited by Tycho and Kepler.[53] Battānī's distances are in any case more suitable for our purposes to compare with distances computed from Copernicus's theory because their solar distances are fairly close, that is, Battānī's greatest solar distance is 1146^{tr} and Copernicus's mean distance 1142^{tr}.[54]

In order to compare the distances, we shall set out Battānī's least, greatest, and mean distances in terrestrial radii, and compute the mean distance where the radius of the sun's orbit is the unit. Battānī's distances are computed by Ptolemy's method, which is as follows: From the parameters of each model, we can find the ratio M'/m' of the greatest to least relative distance, i.e.,

$$\frac{M'}{m'} = \frac{R + e + r}{R - e - r}.^{55} \tag{132}$$

Then, given the least distance of a planet m_1, the greatest distance M_1 is

$$M_1 = m_1(M'/m'), \tag{133}$$

and from the assumption of contiguity, the least distance of the next higher planet $m_2 = M_1$, and so on. The mean distance $\mu = \frac{1}{2}(m + M)$, and the mean distance in units of the radius of the sun's orbit is $\mu/\mu_{\odot}$. Copernicus's distances are of course simpler to compute. The mean distance in units of the radius of the earth's orbit, that is, the distance between the earth and the mean sun, is $\mu/\mu_{\odot}$, which is equal to r/R for an inferior planet and R/r for a superior planet, and the mean distance in terrestrial radii is $\mu = (\mu/\mu_{\odot})\, 1142^{tr}$. The results are shown in Table 12.[56]

Some care must be exercised in comparing the distances since they differ, most obviously in that Battānī's are measured from the earth and Copernicus's from the mean sun, but more significantly because the methods of computation are so different. Battānī's result from a series of multiplications through which the distances of lower planets affect the distances of higher planets, while Copernicus's depend only on the ratio $\mu/\mu_{\odot}$ that is determined independently for each planet. Nevertheless, the mean distances μ and the ratios $\mu/\mu_{\odot}$ are in a sense comparable, at least for the superior planets, since these are essentially the radii R of the eccentrics without the variations produced by r and e. For the inferior planets, Battānī's mean distances correspond to R,

[53] *Progymnasmata* II,7 in Brahe (1913–29) 2, 417, where Farghānī's diameters of the planets are given along with Battānī's; Kepler, in his attempted reconstruction of III,14 of Ptolemy's *Harmonics*, ed. by Frisch in Kepler (1858–71) 5, 392.

[54] Above, pp. 245–248.

[55] Battānī's ratios are rounded, and for Mercury, which is more complicated, he simply rounds Ptolemy's incorrect ratio of 88/34 ≈ 2;35. The ratio for the sun is $(R + e)/(R - e)$, but the distances of the sun were determined independently in cap. 40, Nallino 1, 61.

[56] Battānī's distances are from Nallino 1, 120–22. Those in Plato of Tivoli's Latin translation, printed in Battānī (1537), 76v–78v, contain a number of errors, although their emendation is trivial.

but Copernicus's to the radius of the planet's heliocentric orbit r. The corresponding epicyclic radii in Battānī's system of distances would be $(r/R)\mu$, which is strictly true for Venus and approximately true for Mercury, and gives for Mercury $0;22,30 \cdot 115^{\text{tr}} = 43^{\text{tr}}$ and for Venus $0;43,10 \cdot 618^{\text{tr}} = 445^{\text{tr}}$. In I,10 Copernicus notes that the epicycle of Venus in this system is much larger than the entire region occupied by the earth, moon, and Mercury, which he considers implausible.[57]

The most remarkable fact brought out by the comparison is that the size of the planetary system contracts, and contracts by about one-third, in the distances resulting from Copernicus's theory. If the outer boundaries of the systems are compared, the contraction is even more severe. With Copernicus's parameters, the greatest distance of Saturn is

$$M = \frac{R + (e_1 - r')}{r}\mu_{\odot} = 9;41,46 \cdot 1142^{\text{tr}} = 11073^{\text{tr}},$$[58]

while Battānī gives $M = 18094^{\text{tr}} = 16;19,49\mu_{\odot}$, so the system has contracted by about 40% from over $16\mu_{\odot}$ to less than $10\mu_{\odot}$. That Copernicus's theory results in a smaller planetary system than Ptolemy's is not usually recognized, but it is a fact, and is true no matter what value of $\mu_{\odot}$ is used since the multiplication of the ratios M'/m' for Mars, Jupiter, and Saturn will always give Saturn a greatest distance of about $16\mu_{\odot}$, while with Copernicus's theory the greatest distance of Saturn is necessarily less than $10\mu_{\odot}$. Consequently, in changing from the Ptolemaic to the Copernican theory, the size of the planetary system became considerably smaller.

Of course, when one considers the distances to the stars things are quite different. Battānī assumed, as Ptolemy did, that the sphere of the fixed stars is located just outside the sphere of Saturn, placing the stars at a distance of about $19{,}000^{\text{tr}}$. Copernicus is forced by his theory of the annual motion of the earth to assume that the stellar sphere is so far away that the effects of annual parallax are essentially unobservable. If it is assumed, to take the example used by Tycho, that there is an annual displacement of 0;1°, which evidently would not have been detectable, corresponding to an annual parallax of 0;0,30°,[59] then the distance to the stars would be

$$\mu_{*} = \mu_{\odot}/\sin 0;0,30^{\circ} = 1142^{\text{tr}}/\sin 0;0,30^{\circ} = 7{,}851{,}814^{\text{tr}},$$

---

[57] N 8r31–8v2, where the reference is to the numerical values of the distances from Valla and Proclus, above, p. 475, n. 52.

[58] The outer limit of the sphere of Saturn, the outermost physical boundary of Copernicus's system of spheres, would be

$$\mu_{\odot}(R + e_1 + r')/r = 1142^{\text{tr}} \cdot 10;13,9 = 11670^{\text{tr}},$$

or let us say, less than $12{,}000^{\text{tr}}$.

[59] *Progymnasmata* II,7 in Brahe (1913–29) 2, 430:19–26, That Tycho is estimating a parallax of 0;0,30° is shown by his stating the result that the distance of the fixed stars is more than 700 times the greatest distance of Saturn. Tycho himself takes the distance of the fixed stars as about $14{,}000^{\text{tr}}$ (*ibid.*, 430:28 ff.), which is 1100^{tr} beyond the outer limit of Saturn's epicycles that he sets at $12{,}900^{\text{tr}}$, although Saturn reaches only $12{,}300^{\text{tr}}$.

which is somewhat larger than Battānī's $19{,}000^{tr}$. This remarkable distance leaves a vast space between the greatest distance of Saturn and the sphere of the fixed stars of no less than $7{,}840{,}741^{tr}$, more than seven hundred times the distance between the mean sun and the greatest distance of Saturn, "and that," as Tycho says, "altogether devoid of stars and intended for no conceivable purpose, which is absurd to believe."

Quite as remarkable as the effect on the distance of the stars is the effect on their size. Battānī, following Ptolemy's method in the *Planetary Hypotheses*, computes the true diameter of a star of the first magnitude as follows:[60] Let the ratio of the apparent diameter of the star to the apparent diameter of the sun be $s'_*/s'_\odot = \frac{1}{20}$, the canonical value from the *Hypotheses*, the ratio of the distance of the star to the mean distance of the sun be $\mu_*/\mu_\odot = 19000/1108$, and the diameter of the sun in terrestrial diameters be $s_\odot = 5\frac{1}{2}^{td}$. Then, the true diameter of the star is

$$s_* = \frac{s'_*}{s'_\odot} \cdot \frac{\mu_*}{\mu_\odot} \cdot s_\odot = \frac{1 \cdot 19000}{20 \cdot 1108} \cdot 5;30^{td} = 4;43^{td},$$

and for whatever the point may have been worth at the time for philosophical speculation, the star is roughly the size of the sun. It should be noted that prior to the invention of the telescope, there was nothing unreasonable about taking the apparent diameter of a first-magnitude star to be $\frac{1}{20}$ the apparent diameter of the sun, that is, about $\frac{1}{20} \cdot 0;30^\circ = 0;1,30^\circ$, for Tycho himself estimates the diameter of a first-magnitude star to be about $0;2^\circ \pm 0;0,15^\circ$.[61] Now let us examine the size of the star under the assumption of the distance implied by Copernicus's theory. In IV,20 Copernicus found the diameter of the sun to be $5;27^{td}$.[62] Hence, using the standard ratio $s'_*/s'_\odot = \frac{1}{20}$ and, from the distance μ_* just found, the ratio of the distances $\mu_*/\mu_\odot = 7{,}851{,}814/1142$, we shall have

$$s_* = \frac{1 \cdot 7{,}851{,}814}{20 \cdot 1142} \cdot 5;27^{td} = 1873;34^{td},$$

which, were it placed at the mean sun, would entirely comprehend the mean distance of Mars at 1736^{tr}, and reach nearly to its greatest distance of $M = (10960/6580)\mu_\odot = 1902^{tr}$.

An astronomer or philosopher of the sixteenth century would have good reason to wonder what this enormous object, no longer in any way comparable to the sun or the planets, could possibly be, and he could hardly be blamed

---

[60] Nallino 1, 122–23.

[61] *Supra*, n. 59, 430–31. Tycho's apparent diameters of the stars extend from $0;2^\circ$ for first magnitude to $0;0,20^\circ$ for sixth, and the apparent diameters of the planets (*ibid.*, 424–25) from $0;3\frac{1}{4}^\circ$ for Venus to $0;1,50^\circ$ for Saturn. Cf. Dreyer (1890), 191. All are of course much too large, but there is no doubt that to the unaided eye, planets and stars give the illusion of having a detectable apparent diameter related to their brightness (and perhaps color).

[62] Above, p. 249.

for doubting whether the motion of the earth about the sun was worth having to accept the consequences of such preposterously large stellar distances and sizes, such vast spaces and things, as Tycho said, "intended for no conceivable purpose." Of course, the problems were removed early in the seventeenth century when Galileo found that the number of stars revealed by the telescope was beyond counting and that the apparent sizes of stars were not increased under telescopic observation, leading to the conclusion that the appearance of stars is of points of light of indeterminate size at indeterminate but immense distances. Galileo's discovery, far more than the heliocentric theory, fundamentally transformed the universe into the fathomless spaces and countless stars of modern astronomy and cosmology. But until the consequences of the early telescopic observations were understood and accepted, the distances and sizes of the stars implicit in the assumption of the annual motion of the earth were not a small obstacle to the Copernican theory. Yet ironically, it is now known that the distances, thought by Tycho to be absurdly large, are really much too small, and there really are some stars even larger than the orbit of Mars. Copernicus, while entirely vindicated, would doubtless be amazed.

6

Planetary Theory of Latitude

Introduction

"Copernicus, ignorant of his own riches, took it upon himself for the most part to represent Ptolemy, not nature, to which he had nevertheless come the closest of all." In this famous, and just, assessment of Copernicus, Kepler was referring to the latitude theory of Book VI, specifically to the "librations" of the inclinations of the planes of the eccentrics, not in accordance with the motion of the planet, but (*quod monstri simile est*) by the unrelated motion of the earth. This improbable connection between the inclinations of the orbital planes and the motion of the earth was the result of Copernicus's attempt to duplicate the apparent latitudes of Ptolemy's models in which the inclinations of the epicyclic planes were variable. In a way this is nothing new since Copernicus was also forced to make the equation of center of the inferior planets depend upon the motion of the earth rather than the planet. But the image of the orbital planes of all the planets dutifully inclining one way and another in synchronization with the motion of the earth was to Kepler an incongruous combining of unrelated motions that he found incredible even before he was able to show, from Tycho's observations, that it was false.

But Copernicus's representation, or better, imitation of Ptolemy in Book VI goes farther than the structure of his models. The "observational" basis of Copernicus's parameters was neither his own observations nor even Ptolemy's but the extremal latitudes in Ptolemy's tables. And when he found, in the case of Mars, that his model could not reproduce the extrema of the tables, he changed the "observations" to the closest values that his model would fit, or at least he attempted with only partial success to do so. Still more remarkable, even though his model for the latitude of the inferior planets is equivalent to Ptolemy's only when the earth is in the apsidal line or 90° from the apsidal line, Copernicus nevertheless copied out Ptolemy's tables and applied them indifferently to all locations of the earth, although neither the tables nor the method of computing latitudes from them are applicable to his model.[1]

Part of the problem with the latitude theory is not really Copernicus's fault. If we consider his previous treatment of longitude, we see that he attempted, and to some extent succeeded, in deriving elements from observations and computing tables appropriate to his parameters and his models even though, for the inferior planets, he had to rely upon Ptolemy's observations rather than his own. That he relies upon Ptolemy for the observational basis of the latitude theory is likewise understandable. The observations of extremal latitudes, upon which Ptolemy's theory is based, can be made only at particular configurations of the earth and planet that occur rarely, and there is little doubt that when he wrote the *Almagest* Ptolemy was

[1] The incompatibility of Copernicus's model and tables has been noted in Riddell (1980), 129 on the basis of an analysis in Riddell (1978), 96–98.

himself relying at least as much upon conventional estimates of extremal latitudes as upon his own observations. And it was precisely the inaccuracies of these observations that forced upon Ptolemy the variable inclinations of the epicyclic planes and in turn forced upon Copernicus the variable inclinations of the orbital planes that Kepler so distrusted. So in this most mischievous example of representing Ptolemy rather than nature, Copernicus can be defended on the grounds that he had no other description of nature to represent.

On the other hand, the internal problems of Copernicus's latitude theory, the consequential damages of representing Ptolemy, cannot be passed over so lightly. First there is the point that so disturbed Kepler, which is not so much that the inclinations of the orbital planes vary, but that they vary in accordance with the motion of the earth. Copernicus believed that his theories were physically correct descriptions of the planetary system according to certain fundamental rules, e.g. uniform rotations of all component spheres, and not, as the much maligned Osiander would have it, congeries of geometrical models that need not be true, or even probable, so long as they suffice for computation. But it is far from clear what physical connection the motion of the earth, or the rotation of the Great Sphere of the earth, could possibly have with the librations of the orbital planes of the several planets, each with its own period and magnitude. Copernicus does not venture a guess, which is reasonable enough considering the incongruity of the whole situation, but neither does he so much as remark upon the plausibility of his theory, an omission that seems more characteristic of theology than astronomy.

If it be granted for the sake of argument that the models do in some way represent the physical structure of the heavens, then another problem arises in that they are not compatible with Copernicus's fundamental principle of uniform circular motion. The librations of the orbital planes are presumably produced by the same device of Naṣir ad-Dīn aṭ-Ṭūsī used for the inequality of the precession and the variation of the obliquity of the ecliptic, that is, two rotating spheres that together produce an oscillation along the arc of a great circle. But in the case of the superior planets, the libration corresponds to the true anomaly of the planet, and in the case of the inferior planets to the true eccentric anomaly, that is, the true motion of the earth with respect to the planet's apsidal line, and both of these motions are nonuniform. Consequently, the spheres producing the librations must rotate nonuniformly. Further, in Copernicus's model for the third component of latitude of the inferior planets, called the *deviatio*, the planet must move so that it is always ahead of the point of maximum inclination of its orbital plane by the true eccentric anomaly. But this means that the planet itself is moving nonuniformly, or more physically, is being carried nonuniformly by its sphere. And since the motion of the planet in the corresponding motion for longitude is uniform, the actual positions of the planet in the models for longitude and latitude differ by the equation of center, which reaches 2° for Venus and 3° for Mercury. Now this is something that never happens in Ptolemy's models, i.e. the planet occupying two different positions at the same time. About all of these problems Copernicus likewise says nothing, which seems hardly fair after his berating Ptolemy for not adhering strictly to uniform circular motion.

So let us further concede, again for the sake of argument and in benefit to Copernicus, that his models need not make physical sense of the structure of the heavens nor conform to principles of uniform circular motion, but only provide assumptions by which the motions can be correctly calculated *ex geometriae principiis*, as Osiander says. Then we ask, not whether Copernicus's theory allows a correct calculation of the latitudes of the planets—for that would be unfair and asking too much—but only if his tables and the latitudes computed from his tables are compatible with his models. For the superior planets, aside from some inaccuracies in the tables, which are not here at issue, they are compatible so there is no problem. However, for the inferior planets the tables, which were copied from Ptolemy, and thus computed originally for Ptolemy's models, are not compatible, for direct computation with Copernicus's models will produce differences amounting to as much as 1° for Venus. Hence Copernicus's models, which were intended to represent Ptolemy, and his tables, which were copied from Ptolemy, will not produce the same latitudes. This is a serious failure since it undermines what is after all, to both Ptolemy and Copernicus, the fundamental task of mathematical astronomy, that is, to represent the motions of the planets by appropriate models and then to compute planetary positions from the models by means of appropriately computed tables. Copernicus's latitude theory for the inferior planets does neither.

Copernicus had essentially developed his latitude theory by the time he wrote the *Commentariolus*, and in the period of perhaps thirty years between his first description of the theory and its final publication it is not unreasonable to expect that he would have given some thought to these difficulties. Although in *De revolutionibus* he says not a word that would indicate that he was even aware of them, it seems remarkable that he could have passed the entire period in complete ignorance. One may venture any number of guesses for Copernicus's silence, but it is hard enough to explain what Copernicus did, let alone what he did not do.[2]

Of course the fundamental problem was the librating planes. If the inclinations could remain fixed, the other problems would not have occurred. There would be no inexplicable relation to the motion of the earth, no violation of uniform circular motion, and no contradiction between straightforward models and equally straightforward tables. We have mentioned

[2] It is possible that Book VI in its present form was written in haste and only after Copernicus had decided to prepare *De revolutionibus* for publication. See below, pp. 535–537. There is little literature on Copernicus's latitude theory. Brief accounts may be found in Delambre (1821) 1, 136–38, Herz (1887–94) 2, 94–99, Riddell (1980), 127–30. The latitude theory of the *Commentariolus* is considered in Swerdlow (1973), 482–89, 494–99, 509–10.

that the reason Copernicus had to make the orbital planes librate was to reproduce, in so far as possible, the effects of the variable inclinations of the epicyclic planes in Ptolemys models. This was Ptolemy's latitude theory in the *Almagest*. The variable inclinations were the result of faulty observations or, as seems more likely, of faulty conventional estimates of extremal latitudes that Ptolemy must have used because at the time he had nothing better. However, the *Almagest* does not contain Ptolemy's final latitude theory, for he continued to refine it, at least in part on the basis of better observations, in the *Handy Tables* and the *Planetary Hypotheses*.[3] And in the last work he eliminated the oscillations of the epicyclic planes altogether and gave them fixed inclinations, for the superior planets parallel to the ecliptic, for the inferior planets inclined to the ecliptic at the angle used previously for the latitudes at greatest elongation.[4] These modifications must have been based upon newer observations because the extremal latitudes they produce differ noticeably from the extremal latitudes of the theory in the *Almagest* and are in fact much improved.

Unfortunately, Ptolemy's later latitude theory seems to have been virtually unknown. It does not appear as though the latitude tables from the *Handy Tables* were translated into Arabic, although most of the rest of the *Handy Tables* was. And while the *Planetary Hypotheses* was translated into Arabic, as far as is known medieval Arabic astronomers simply noted that the *Almagest* and the *Hypotheses* contained different latitude theories without inquiring into the reason for the differences or whether the theory of the *Hypotheses* might not be preferable.[5] For all practical purposes, Ptolemy's latitude theory remained the theory of the *Almagest*, as though his later revisions never existed. And of course this was true for Copernicus since the *Handy Tables* and the *Planetary Hypotheses* were all but unknown in Europe at his time. It is thus ironic that had he been able to represent heliocentrically the latitude theory of the *Planetary Hypotheses*, with its epicyclic planes of fixed inclinations, he would not only have been spared all the problems that plague his latitude theory, but would have had a simpler and more accurate latitude theory all around.

General Considerations

The apparent motion of a planet in latitude is, like the apparent motion in longitude, affected by two anomalies. The planet moves in an orbit inclined to the plane of the ecliptic by a small angle, not exceeding 7° for Mercury and

[3] The latitude theory of the *Almagest* is treated in Pedersen (1974), 355–86, *HAMA*, 206–26, Riddell (1978) for the inferior planets. On the latitude theory of the *Planetary Hypotheses* and the *Handy Tables*, see *HAMA*, 908–13, 1006–16.

[4] See below, p. 515, n. 37.

[5] There may of course be exceptions. But Ibn al-Haytham, to take an example, in his *Doubts concerning Ptolemy* (unpublished translation by D. Voss), finds the differences to be contradictions worthy of criticism. Ibn ash-Shāṭir, in Roberts (1966), describes separate models for the theory of the *Almagest* and the *Hypotheses*, but without considering why they differ.

less than half that amount for the remaining planets visible to the unaided eye. The plane of the orbit intersects the plane of the ecliptic along a line, called the nodal line, that passes through the true sun. Since the orbit is eccentric to the true sun, and since the nodal line does not pass through the center of the planet's orbit, the periods that the planet is on either side of the ecliptic are unequal, and thus the rates of change of latitude at equal distances from the nodal line on either side of the ecliptic are unequal. This is the first anomaly of latitude, and it corresponds to the nonuniform description of arcs in the first anomaly of longitude. The latitude affected by the first anomaly alone is the planet's heliocentric latitude. The geocentric latitude is subject to a further anomaly due to the motion of the earth in the plane of the ecliptic, which changes the distance between the earth and the planet, and thus varies the apparent angle under which the heliocentric latitude is observed at the earth. This is the second anomaly of latitude, and it corresponds to the variation of geocentric longitude due to the motion of the earth in the second anomaly of longitude. And like the second anomaly of longitude, the second anomaly of latitude has the equivalent of stations and retrogradations in that it can cause the planet periodically to appear to remain at a constant latitude and even reverse the direction of its heliocentric motion of latitude.

The orbits of the earth and a planet are shown in Figs. 1(a) and 1(b) for a superior and inferior planet respectively. The common focus of the orbits is the true sun S, the center of the earth's orbit is the mean sun $\bar{S}$ and of the planet's orbit M, the perihelia of the planet and the earth are π and π_0, and the ascending node is ☊. Since the nodal line ☊S does not pass through M, the planet's orbit is divided unequally by the nodal line, and this produces the first anomaly in heliocentric latitude, which depends upon the eccentricity e and the angle ☊$S\pi$ between the nodal and apsidal lines. The first anomaly is most pronounced for Mars because its eccentricity is large and its nodal line nearly perpendicular to its apsidal line, and it is less noticeable in the other planets because either e or ☊$S\pi$ is small. It is also obvious from the figure that the motion of the earth will produce considerable changes in the distance OP between the earth and planet, and this accounts for the second anomaly in geocentric latitude. The variation of latitude due to the second anomaly, like the variation of longitude, depends upon the ratio of the radius of the earth's orbit to the radius of the planet's orbit—r/R for a superior planet, R/r for an inferior—and the closer the ratio approaches to unity, the greater will be the variation of distance and of geocentric latitude. Thus it is greatest for Mars and Venus, and less for the other planets, just as the second anomaly of longitude.

The method of computing latitude is shown in Fig. 2 which is simplified by the omission of eccentricities. The earth O is shown in two locations, inside the orbit of a superior planet and outside the orbit of an inferior planet. The inclination of the orbital plane of the planet to the plane of the ecliptic is i, the heliocentric latitude b, the geocentric latitude β, and the true angular distance of the planet from the ascending node is ω', called the argument of latitude. Then the heliocentric latitude b is given by

$$b = \sin^{-1} (\sin i \sin \omega'), \tag{1}$$

and the geocentric latitude β by

$$\beta = \sin^{-1} ((SP/OP) \sin b) = \sin^{-1} ((SP/OP) \sin i \sin \omega'), \tag{2a}$$

where the relative distances SP and OP have been determined previously in computing the planet's longitude, which is also required to find ω'.

This is a characteristically "heliocentric" method of computing latitude in that one first finds the heliocentric latitude b and then reduces it to the geocentric latitude β. However, it is not Copernicus's procedure since he follows Ptolemy in the arrangement of his latitude tables. For a superior planet he first computes directly the geocentric latitude of the planet at the northern or southern limit of latitude, $\pm 90°$ from the ascending node, given by $\sin^{-1} ((SP/OP) \sin i)$, and then corrects the result for other locations of the planet by multiplying by $\cos \omega$, where $\omega = \omega' - 90°$ is the argument of latitude measured from the northern limit.[6] That Copernicus's procedure is still geocentric is not unlike his fundamentally geocentric method of computing longitude, likewise following Ptolemy, in which the heliocentric longitude does not appear as a step in the computation. Nevertheless, the resulting geocentric latitude is nearly identical to the previous computation. If $\sin \omega'$ is substituted for $\cos \omega$, Copernicus's computation is equivalent to

$$\beta = \sin^{-1} ((SP/OP) \sin i) \sin \omega', \tag{2b}$$

which differs little from (2a) since i is small.

In order to show the transformation between a heliocentric and geocentric model for latitude, we consider a simplified model in which the nodal line passes through the mean sun. This is the model used by Copernicus, just as in his model for longitude the apsidal line passes through the mean sun. In Fig. 3 the plane of the orbit $\bar{S}P$ intersects the plane of the ecliptic at the mean sun $\bar{S}$ with an inclination i. The planet at P is seen from the earth O in the direction OP with the latitude β. If we now complete parallelogram $\bar{S}PCO$, as was done for longitude in Fig. 1 of Chapter 5, the planet will be located on an epicycle CP that remains parallel to the plane of the ecliptic, while the orbital plane of the center of the epicycle will be OC, intersecting the plane of the ecliptic at the earth with the same inclination i. The planet is seen along the diagonal of the parallelogram OP, and the models are in every way equivalent. This was essentially the transformation that Copernicus applied, although the other way around, from the orbital plane of the center of the epicycle OC passing through the earth to the orbital plane of the planet $\bar{S}P$ passing through the mean sun. But since in

[6] The procedure for the inferior planets is more complicated, and even more geocentric throughout.

Ptolemy's model the inclination of the plane of the epicycle is variable and does not remain parallel to the ecliptic, that is, angle OCP is not constant, so in Copernicus's model angle $O\bar{S}P$ is not constant, and since the plane of the ecliptic is taken to be fixed, the inclination of the orbital plane of the planet is variable.

Ever since Kepler criticized the oscillating planes of Ptolemy's and Copernicus's latitude models in *Astronomia nova* 14, and showed that the orbital plane of Mars passing through the true sun has a fixed inclination, it has been presumed that the reason the orbital plane must oscillate in Copernicus's model is that it passes through the mean sun rather than the true sun. But this turns out to be incorrect. It is true that the substitution of the mean sun for the true sun does necessitate an oscillation of the orbital plane, but the resulting oscillation differs significantly from that in Copernicus's model in that it is minuscule and takes place in different directions on each side of the ecliptic.

The actual effect of moving the orbital plane from the true sun to the mean sun is shown in Fig. 4. The plane of the orbit intersects the ecliptic at the true sun S' with an inclination i, and the planet P^+ is at any distance $S'P^+$ north of the ecliptic and P^- at an equal distance $S'P^-$ south of the ecliptic (although unequal distances would also do for the demonstration). Now, imagine that S', P^+, and P^- have been projected into the plane perpendicular to the ecliptic passing through the mean sun $\bar{S}$ and the northern and southern limits of latitude, and let e'_0 be the projection of the earth's eccentricity into the same plane. Consider first the planet at P^+. At opposition it is seen from the earth O_1 at the large latitude β_o and at the conjunction it is seen (in principle) from O_2 at the smaller latitude β_c. If the planet is now to lie in a plane passing through the mean sun, we shall take distances equal to SP^+ from $\bar{S}$, and the planet will be at P_o^+ at opposition and P_c^+ at conjunction, the inclinations of the planes $\bar{S}P_o^+$ and $\bar{S}P_c^+$ differing by Δi. Note that the inclination of $\bar{S}P_o^+$ is greater than the inclination of $\bar{S}P_c^+$, that is, a greater inclination is required at opposition than at conjunction. We do the same for the planet at P_o^-, and on planes passing through $\bar{S}$ the planet will be at P_o^- at opposition and P_c^- at conjunction, the inclinations of the planes differing by $\Delta i'$, which is not necessarily equal to Δi. And note that this time the inclination of $\bar{S}P_o^-$ is less than $\bar{S}P_c^-$, that is, a smaller inclination is required at opposition than at conjunction.

Now, this model with oscillating planes is a correct transformation from the intersection of the orbital planes at the true sun to an intersection at the mean sun. It is completely general, and can be applied as well to the inferior planets. However, it is not Copernicus's model for either the superior or inferior planets. In his model for the superior planets, the maximum inclination always occurs at opposition and the minimum at conjunction, while we have found that the inclinations are reversed for opposition and conjunction on different sides of the ecliptic, which correspond to different distances from the true sun. And in his model for the inferior planets, the inclination depends only upon the position of the earth with respect to the planet's nodal line, which is taken to coincide with the apsidal line, while the position of the earth with respect to the planet is of no effect. Hence, the transformation of the orbital plane from the true sun to the mean sun does not account for the oscillation of the planes in Copernicus's models.

Further, as we have remarked, the change of inclination caused by this transformation is minute. We have checked it for Mars, using the correct values of the required parameters for Ptolemy's time, and the result is that if the inclination through the true sun $i = 1;52°$, then, calling the inclination at opposition i_o and at conjunction i_c, to the nearest 0;0,5°:

northern limit	$i_o = +1;53,45°$	$i_c = +1;52,25°$	$\Delta i = 0;1,20°$
southern limit	$i_o = -1;49,10°$	$i_c = -1;51,30°$	$\Delta i' = 0;2,20°$

Note the reverse of the larger and smaller inclination at each limit. It is obvious that these minute variations have nothing to do with the oscillations of 1;42° of the orbital plane of Mars in Copernicus's model and of $\pm 1;15°$ in the plane of the epicycle in Ptolemy's model.

And if the oscillation due to substituting the mean sun for the true sun is small for Mars, it is vanishing for Jupiter and Saturn. For the superior planets, the total oscillation Δi is of the order

$$\Delta i \approx \frac{2ie'_0 r}{R^2 - r^2}, \qquad (3a)$$

where i is the inclination through the true sun, e'_0 the projection of the earth's eccentricity on to the plane perpendicular to the ecliptic passing through the limits of latitude, and R the mean radius of the planet's orbit in units of r, the mean radius of the earth's orbit. For the inferior planets, the corresponding relation is

$$\Delta i \approx \frac{2ie'_0 r}{r^2 - R^2}. \qquad (3b)$$

We have computed approximate values of Δi for all five planets. In each case e'_0 is found from $e'_0 = e_0|\cos \omega_{A\odot}|$, where $e_0 = 0.0175$ and $\omega_{A\odot} = \lambda_{Np} - \lambda_{A\odot}$ is the longitudinal difference between the northern limit of the planet and the apogee of the sun according to modern theory for the time of Ptolemy.

Planet	i	$\omega_{A\odot}$	e'_0	R	Δi
Saturn	2;33°	117°	0.0079	9.54	0;0,1,36°
Jupiter	1;25	101	0.0033	5.2	0;0,1,18
Mars	1;52	55	0.0100	1.524	0;1,42
Venus	3;23	80	0.0030	0.723	0;2,33
Mercury	6;58	46	0.0122	0.387	0;12,0

(4)

These results are only approximations, but they do show the order of magnitude of the effect, and it is obviously quite small. Only for Mercury is there any noticeable oscillation, but this plays no part in Ptolemy's or Copernicus's latitude theories in which, when an inferior planet is at either inferior or superior conjunction at either limit of latitude, the orbital plane has the same inclination to the ecliptic. Clearly, something other than shifting the orbital planes from the true sun to the mean sun must account for the variable inclinations in Ptolemy's and Copernicus's models, and, as we have mentioned, the problem is in the observations—which are probably conventional estimates—of apparent latitudes used by Ptolemy for deriving the parameters of his models. These forced upon him the variable inclinations of the epicyclic planes, and since Copernicus's models are based upon the extrema of Ptolemy's tables, and thus upon the same faulty observations, a corresponding oscillation of the planet's orbital plane is a direct consequence.

Another relic of Ptolemy's latitude theory that has a significant effect on Copernicus's theory is a sharp division between the latitudes of superior and inferior planets. These appear somewhat different in that the superior planets, with orbits surrounding the orbit of the earth, reach their extremal latitudes in opposite parts of the zodiac and cross the ecliptic at nodes midway between the extrema. The motion of the earth affects their apparent latitudes by changing the distance between the earth and planet, but does not change the locations of the extremal latitudes or the nodes. In the case of the inferior planets, the orbits of which are surrounded by the orbit of the earth, the location of the earth in itself produces notable differences in the latitudinal motion of the planet. When the earth is in the planet's nodal line, the orbit is seen perpendicular to the line of sight and extremal latitudes on opposite sides of the ecliptic occur at opposite greatest elongations, while at inferior and superior conjunctions the planet is in the nodal line and thus in the plane of the ecliptic. On the other hand, when the earth is 90° from the nodal line, the orbit is inclined in the line of sight so that extremal latitudes on either side of the ecliptic are at inferior and superior conjunctions, the planet is in the plane of the ecliptic at 90° of anomaly from the conjunctions, and the latitudes at greatest elongations are of no special significance. Hence, the apparent locations of the extremal latitudes and the absence of latitude depend upon the position of the earth. Further, in the latitude theory of the *Almagest*, Ptolemy gives the epicycle different inclinations at locations corresponding to the nodal line and 90° from the nodal line, and treats these two positions as limiting cases of two distinct components of latitude, altogether different from the simpler behavior of the superior planets, and he finds yet a third component when the epicycle is at the nodal line. Copernicus does the same in relation to the position of the earth with respect to the planet's nodal line, and thus develops different latitude theories for the superior and inferior planets. In our exposition we shall follow Copernicus in considering first the superior planets and then the considerably more complicated, and more troublesome, theory of the inferior planets.

1. Superior Planets

Development of the Model (VI,1–2)

Ptolemy's model for the latitude of the superior planets is designed to account for a limited number of phenomena that are described specifically in *Almagest* XIII,1, expanded somewhat in *Epitome* XIII,1, and in turn adapted to a heliocentric form by Copernicus in VI,1. According to the *Almagest* and the *Epitome*, the phenomena and the conclusions to be drawn from them are as follows:

1. The planet reaches its maximum northern latitude in the vicinity of the apogee of the eccentric and its maximum southern latitude in the vicinity of the perigee, indicating that the eccentric is inclined to the north in the direction of the apogee and to the south in the direction of the perigee.
2. When the center of the epicycle is about 90° of true longitude from the northern and southern limits, and the planet about 90° of true anomaly from the apogee of the epicycle, the planet has no latitude, which shows that at the nodal line, the epicycle lies in the plane of the ecliptic, and not in the plane of the eccentric.
3. When the center of the epicycle is near the apogee and perigee of the eccentric, and thus near the northern and southern limits of latitude, the planet appears farther from the ecliptic at opposition in the perigee of the epicycle than near conjunction near the apogee. This shows that likewise at the limits of latitude the epicycle does not lie in the plane of the eccentric, but is inclined with its perigee farther from the ecliptic and its apogee closer than if it were in the plane of the eccentric.

The precise locations of the limits of latitude, and the inclinations of the eccentric to the ecliptic and of the epicycle to the eccentric, remain to be determined from more precise observations. Ptolemy's description of the phenomena is compatible, for example, with an epicycle that is always parallel to the plane of the ecliptic. However, from observations of apparent latitudes at opposition and near conjunction, it turns out that when the epicycle is at the northern or southern limit, it is not parallel to the plane of the ecliptic, but is inclined such that its perigee is farther from the ecliptic than its apogee.

Ptolemy's model is shown in Fig. 5. The plane of the eccentric intersects the plane of the ecliptic along a nodal line ☊O☋ passing through the earth at O, and the eccentric is inclined to the ecliptic at an angle i_1. The distance $OM = e'$ is the projection of the eccentricity on to the line joining the northern limit of latitude N and the southern limit S. In the case of Mars,

NS coincides with the apsidal line and $e' = e$, while for Jupiter and Saturn NS is displaced from the apsidal line and thus $e' < e$. When the center of the epicycle is at the nodal line, the epicycle lies in the plane of the ecliptic, and the planet has no latitude wherever it may be located on the epicycle. When the center of the epicycle is at the northern or southern limit, the perigee of the epicycle is inclined away from the plane of the ecliptic and the apogee toward the plane of the ecliptic by an inclination i_2 from a plane parallel to the ecliptic, and thus by $(i_1 + i_2)$ from the plane of the eccentric. Consequently, the latitudes $\pm\beta_o$ of the planet at opposition P_o will be greater than the latitudes $\pm\beta_c$ of the planet at conjunction P_c, and the differences between β_o and β_c will be greater than if the epicycle were parallel to the plane of the ecliptic. As the epicycle moves away from the northern limit, its inclination changes by $i_2 \cos \omega$, where ω is the true distance in longitude from the northern limit. It will thus lie in the plane of the ecliptic at the node where $\omega = 90°$ and reach the opposite inclination i_2 at the southern limit where $\omega = 180°$, and in each zodiacal period of the center of the epicycle i_2 will complete one cycle of oscillation.

A direct transformation of Ptolemy's model to a heliocentric arrangement by the parallelogram of Fig. 3 would require that the plane $O\bar{S}$ remain parallel to the plane of the epicycle CP, But if CP has a variable inclination, $O\bar{S}$ cannot be made to move out of the ecliptic since it is the plane of the earth's orbit and thus defines the ecliptic. Hence, the variable inclination must be transferred to the orbital plane $\bar{S}P$, and this is the principle underlying the transformation from Ptolemy's model with a variable inclination of the plane of the epicycle to Copernicus's model with a variable inclination of the orbital plane of the planet. The transformation is shown in Fig. 6(a) for the planet at opposition and Fig. 6(b) at conjunction. Let the plane of the eccentric intersect the plane of the ecliptic at the earth O with the inclination i_1, and let the center of the epicycle C be at the northern limit. Then if the plane of the epicycle were parallel to the ecliptic, completing parallelogram $OCP\bar{S}$ gives a plane $\bar{S}P$ passing through the mean sun and the planet with the same inclination i_1, just as in Fig. 3. But the epicycle is inclined to the plane parallel to the ecliptic at an angle i_2 so that the planet P_o at opposition is seen with the latitude β_o in the direction OP_o, and the planet P_c at conjunction is seen at the latitude β_c in the direction OP_c. Let a distance equal to OC be taken from $\bar{S}$ that will meet OP_o at P'_o and OP_c at P'_c, and the plane $\bar{S}P'_o$ will be inclined to $\bar{S}P$ by δ_i and $\bar{S}P'_c$ will be inclined to $\bar{S}P$ by δ'_i. Hence, the orbital plane through the mean sun oscillates such that its inclination i_o is greater than i_1 at opposition and its inclination i_c is less than i_1 at conjunction.

Note that as shown in the figure δ_i and δ'_i are unequal, as they must be if i_2 is an appreciable angle. However as i_2 is quite small, always less than 2°, δ_i and δ'_i are very nearly equal. It then follows that $\frac{1}{2}(i_o + i_c) = i_1$, a relation that holds precisely for Ptolemy's and Copernicus's parameters. Further, since the inclination δ_i of the eccentric of radius R is about equivalent to the inclination i_2 of the epicycle of radius r, it follows that $\delta_i/i_2 \approx r/R$, which holds very nearly provided that i_2 is small. But it is also clear from these approximations that the transformation is only possible with small angles, for all symmetries of i_o and i_c with i_1 break down if i_2 is large.[7]

Copernicus had developed his latitude model for the superior planets completely when he wrote the *Commentariolus*, the model in *De revolutionibus* differing only in its more carefully derived parameters. The model is shown in Fig. 7 for the planet P_o at opposition with the earth at O_o, and P_c at conjunction with the earth at O_c. The distance $\bar{S}M = e'$ is the component of the eccentricity in the direction of the northern limit, which is N_o at opposition and N_c at conjunction. The plane of the orbit oscillates over an angle $\Delta i = \delta_i + \delta_i'$ between the greatest inclination i_o at opposition and the least inclination i_c at conjunction. Hence the orbital plane completes one oscillation, or "libration" as Copernicus calls it, in the planet's true synodic period between two successive oppositions or conjunctions.

The oscillation of the orbital plane raises two of the problems mentioned earlier.[8] In Ptolemy's model the period of the variation of the inclination of the epicycle is the zodiacal period of the center of the epicycle, so whatever may be thought of oscillating planes in general—and we note that Ptolemy eventually eliminated them in the *Planetary Hypotheses*—the motion at least belongs to the planet's own model, and there is exactly one oscillation in each true zodiacal period. Since the epicycle inclines so that its perigee is farther from the ecliptic than its apogee, the motion of the planet on the inclined epicycle then accounts for the larger latitudes at opposition and smaller latitudes at conjunction than would be produced by an epicycle that remained parallel to the ecliptic. However, in order to produce the same effect, Copernicus requires the planet's orbital plane to oscillate from maximum to minimum inclination between each true opposition and conjunction of the planet with the mean sun, which of course depends upon the motion of the earth with respect to the planet. Kepler found this strange relation objectionable—he called it a "monstrosity," meaning that it was unnatural—and the problem is now so well known as not to require further discussion.

The oscillation of the epicyclic plane in Ptolemy's model is produced by a mechanism, described by Ptolemy in *Almagest* XIII,2, consisting of small vertical circles with equant points that carry the diameter of the epicycle passing through apogee and perigee up and down from its mean position of parallelism to the plane of the eccliptic. Since the oscillation corresponds to the true zodiacal period with respect to the nodes or limits of latitude of the eccentric, it is nonuniform, and for this reason the small circles are

---

[7] A numerical analysis of the relations is given below, p. 503.

[8] Above, pp. 483–484.

given equant motion.[9] The libration of the orbital plane in Copernicus's model is completed in the planet's true synodic period, and so it too depends upon a nonuniform motion. Copernicus does not describe the mechanism of the libration in *De revolutionibus*—he says only that it is like the libration of the precession of the equinoxes, "but simple and commensurable with the motion of parallax," that is, with the true anomaly. In the *Commentariolus* he describes the two concentric spheres of Naṣir ad-Dīn aṭ-Ṭusī that in this case must rotate through α and 2α respectively, and he doubtless has the same device in mind in his later work as indicated by his remark about the libration used for the precession.[10] Since α does not increase uniformly, the rotation of the spheres is nonuniform, and the principle of uniform circular motion is violated in the models for latitude. Copernicus does not comment on this difficulty, but in terms of his own physical principles, in which uniform circular motion is essential to all spheres in the heavens, it must be counted as a serious failing.

Derivation of Parameters (VI,3)

Ptolemy's parameters for the latitude are all simple, rounded numbers reflecting the difficulty of extracting anything more precise from the observations, which we note again are most likely conventional estimates, of apparent latitudes that he had at his disposal when he wrote the *Almagest*. He does not give specific, dated observations of latitudes because the configurations he requires—the center of the epicycle at the northern and southern limits and the planet as close as possible to apogee and perigee of the epicycle—occur seldom, and it appears that the best he could do at the time was to adopt conventional estimates of extremal latitudes in different parts of the zodiac and assume that they occurred under the required conditions. And he evidently knew that it would be unreasonable to expect more than acceptable approximations out of such rough observations. For example, it is very difficult to determine precisely where the northern or southern limit lies since the center of the epicycle, like the sun near the solstice, is moving nearly linearly for a considerable distance on either side of the limit, and the effect of the motion of the planet on the epicycle far exceeds the minute change of latitude as the center of the epicycle moves through the limit. It may have been easier to find the location of the nodes since, when the planet crosses the ecliptic, the center of the epicycle must be at the node and its position may be computed for the time of crossing. The limits are then taken to be 90° from the nodes. Ptolemy's longitude of the northern limit

[9] See the descriptions of these mechanisms for the inferior planets in Riddell (1978). It should be noted that the purpose of Ibn ash-Shāṭir's latitude theory, like his longitude theory, was the development of models that preserve uniform circular motion. While his longitude theory can easily be adapted to a heliocentric form, his latitude theory cannot, and Copernicus's theory has no relation to it.

[10] The model in the *Commentariolus* is described in Swerdlow (1973), 483, 488–89.

λ_N is given with the longitude of the apogee λ_A, the difference of longitude, $\omega_A = \lambda_A - \lambda_N$, which is also the true argument of latitude of the apogee, and the correct value of λ_N for Ptolemy's time:

Planet	λ_A	λ_N	ω_A	Corr λ_N	
Saturn	♏23°	~♎ 0°	$-53° \approx -50°$	♎7°	
Jupiter	♍11	~♎ 0	$+19° \approx +20°$	♍22	(5)
Mars	♋$25\frac{1}{2}$	~♋30	$+4\frac{1}{2} \approx 0$	♌5	

Note that λ_N is given by Ptolemy simply as near the beginning of Libra for Saturn and Jupiter and near the end of Cancer for Mars, and that ω_A is rounded severely. The values of ω_A seem to have become canonical; they are still used in the *Alfonsine Tables* and are retained by Copernicus.

Ptolemy derives the inclination of the eccentric i_1 and the maximum inclination of the epicycle $(i_1 + i_2)$ from the apparent latitudes of Saturn and Jupiter at opposition and near conjunction at both limits, that is $\pm\beta_o$ and $\pm\beta_c$, and of Mars at opposition alone at both limits since Mars is invisible for such long periods before and after conjunction that there were evidently no reliable estimates of its latitude at conjunction. The periods of invisibility are of some interest here because only if they are short can β_c be estimated. At the latitude of Alexandria in Ptolemy's time, the superior planets have the following intervals from last visibility Ω to conjunction, and from conjunction to first visibility Γ at Ptolemy's locations of the northern limit λ_N and southern limit λ_S.[11]

Planet	λ_N	Ω	Γ	Ω → Γ	λ_S	Ω	Γ	Ω → Γ	
Saturn	♎0°	17^d	13^d	30^d	♈0°	13^d	31^d	44^d	
Jupiter	♎0	14	12	26	♈0	10	25	35	(6)
Mars	♌0	46	45	91	♒0	99	116	215	

The periods of invisibility of Saturn and Jupiter are relatively short, and because the planets move slowly, observations near last and first visibility provide the basis for an estimate of the latitude at conjunction. Yet, it is important to note that near last and first visibility the planet is seen only close to the horizon where measurements of position are notoriously inaccurate. Hence, the best that can be done is to interpolate between two uncertain observations, and the round values of the latitudes given by Ptolemy, $\pm 2°$ for Saturn and $\pm 1°$ for Jupiter, are both underestimated. It also happens that the latitudes he gives for opposition at the limits, $\pm 3°$ for Saturn and $\pm 2°$ for Jupiter, are both overestimated. It appears obvious

[11] From P. V. Neugebauer (1938).

that he was using conventional estimates of both extremal latitudes. The result is that the variation of latitude between opposition and conjunction is too large, and this leads to the inclination of the plane of the epicycle to the plane of the ecliptic at the limits of latitude when it ought to be parallel.

In the case of Mars conditions are far worse since the periods of invisibility are so long, nearly 100^d at the northern limit and over 200^d at the southern, that any attempt to estimate the latitude at conjunction from observations near first and last visibility is futile. Ptolemy therefore gives apparent latitudes only for opposition at each limit, $+4;20°$ at the northern limit and $-7°$ at the southern, the difference being due to Mars's large eccentricity so that the planet is considerably farther at the apogee than at the perigee of the eccentric. Here again the difference between the latitudes is too large, the northern latitude is underestimated and the southern overestimated, and this likewise leads to the inclination of the plane of the epicycle to the plane of the ecliptic.[12]

The apparent latitudes β_o and β_c at opposition and conjunction at the limits, and the inclinations i_1, $(i_1 + i_2)$, and i_2 derived from them are as follows:

Planet	Apparent Latitudes		i_1	$(i_1 + i_2)$	i_2
Saturn	$\beta_o \pm 3°$	$\beta_c \pm 2°$	$2\frac{1}{2}°$	$4\frac{1}{2}°$	$2°$
Jupiter	$\beta_o \pm 2$	$\beta_c \pm 1$	$1\frac{1}{2}$	$2\frac{1}{2}$	1
Mars	$+\beta_o + 4\frac{1}{3}$	$-\beta_o - 7$	1	$2\frac{1}{4}$	$1\frac{1}{4}$

(7)

Correctly i_2 ought to be zero, and in the *Planetary Hypotheses* Ptolemy does just that for he gives the same inclination to the plane of the eccentric and the plane of the epicycle that is equal to $2\frac{1}{2}°$ for Saturn, $1\frac{1}{2}°$ for Jupiter, and $1;50°$ for Mars, nearly the correct values in each case.[13]

Copernicus's parameters are adapted as closely as he could manage from Ptolemy's within the limits of the slightly different behavior of his model. In VI,1 he says that the longitudes of the northern limits have changed between Ptolemy and his own time, "just as the apogees have also shifted up to our time, for the inclinations and cardinal points of the latitude follow the very same motion (*ipsum motum*) of their spheres."[14] This means that Copernicus takes the intervals $\omega_A = \lambda_A - \lambda_N$ to have remained constant since Ptolemy. Consequently, λ_N, which is given as a tropical longitude, follows from

$$\lambda_N = \lambda_A^* + \pi + \omega_A, \qquad (8)$$

[12] There is a further problem for Mars in that the maximum apparent latitudes do not take place exactly at the limits on account of the large eccentricity of its orbit.

[13] *HAMA*, 908.

[14] N 182v11–13.

i.e. taking λ_A^* and π from Tables 2–4 of Chapter 5, that is, from observations (6) of the derivations in Book V, we find the following values of λ_N which we compare with those given by Copernicus:[15]

Planet	Date	λ_A^*	π	λ_A	ω_A	λ_N	Cop λ_N
Saturn	1527	240;21°	27;21°	♐ 27;42°	−50°	♏ 7;42°	♏ 7°
Jupiter	1529	159;1	27;22	♎ 6;23	+20°	♎ 26;23	♎ 27
Mars	1523	119;40	27;18	♌ 26;58	0	♌ 26;58	♌ 27

(9)

While the roundings may be a bit off, this appears to be the way Copernicus found the values of λ_N given in VI,1. In any case they hardly matter, for in the instructions in VI,9 for computing latitudes from the tables, he simply gives Ptolemy's rule from *Almagest* XIII,6, which is equivalent to $\lambda_N = \lambda_A + \omega_A$, where for Saturn $\omega_A = -50°$, for Jupiter $+20°$, and for Mars $0°$.

Copernicus derives the inclinations of the planes of the orbits, not from his own observations, nor exactly from Ptolemy's, but in so far as possible, from the extrema of Ptolemy's tables in *Almagest* XIII,5. The extrema from the tables are as follows:

Planet	$+\beta_o$	$+\beta_c$	$-\beta_o$	$-\beta_c$
Saturn	+3;2°	+2;4°	−3;5°	−2;2°
Jupiter	+2;4	+1;7	−2;8	−1;5
Mars	+4;21	+0;8	−7;7	−0;4

(10a)

Copernicus gives, and as we shall see, uses only the southern latitudes for his derivations. Originally in M (192v11–15) he wrote exactly Ptolemy's values, but then he changed three of them for reasons that will become clear from the derivations. His values are as follows, with the original numbers from M in parentheses:

Planet	$-\beta_o$	$-\beta_c$
Saturn	−3;5°	−2;2°
Jupiter	−2;7 (−2;8)	−1;5
Mars	−7;0 (−7;7)	−0;5 (−0;4)

(10b)

[15] In a page of notes bound in with his copy of the *Alfonsine Tables*, Copernicus gives precisely these dates and the values of λ_A^* and λ_N from *De revolutionibus*. See the facsimile of the page in Biskup (1973), plate 11, and the discussion in Swerdlow (1973), 427.

In describing the latitudes at conjunction, Copernicus shows that he is aware that they can only be estimated from observations near last and first visibility, and he is probably also aware that the long periods of invisibility of Mars make even an estimate next to impossible. Ptolemy is quite clear about this since he does not report observational estimates of Mars's latitude at conjunction, and the latitudes in his tables for conjunction, which are actually more than 1° too low, are simply the result of computation using the inclinations of the eccentric and epicycle established from the apparent latitudes at opposition. Ptolemy could ignore the latitudes of Mars at conjunction since in his model the inclination of the eccentric is fixed and the inclination of the epicycle, depending only upon the nodal distance, is the same at opposition and conjunction at the same nodal distance. Copernicus, however, must use the latitudes at conjunction, even for Mars, in order to derive the minimum inclination of the orbital plane. And thus he uses Ptolemy's tabulated latitudes of Mars at conjunction as though having some kind of observational basis.

Copernicus shows the derivation of the maximum and minimum inclinations only for Mars, and then gives the results for Saturn and Jupiter without demonstration. The derivation is shown in Fig. 8. We are given the larger latitudes β_o at opposition at each limit, from which we derive the maximum inclination i_o, and the smaller latitudes β_c at conjunction at each limit, from which are derived the minimum inclination i_c. Actually, we have more information than needed since only one β_o and β_c is required for each inclination, and it is possible that two will result in different maximum and minimum inclinations at each limit. This turns out to be true for all three planets, and is particularly troublesome for Mars. Copernicus's solution for Mars is to alter Ptolemy's β_o and β_c to values compatible with the same i_o and i_c at both limits, although he still does not succeed for i_o. For Jupiter and Saturn he derives the inclinations from the southern latitudes alone, and then computes the resulting northern latitudes.

From the parameters of the longitude theory, we know the ratio R/r and the eccentricity $e = e_1 - r'$, from which we may find e', the component of the eccentricity in the direction of the northern limit, by $e' = e \cos \omega_A$. In the figure are four triangles $\bar{S}PO$ in which $\bar{S}P^+ = R + e'$, $\bar{S}P^- = R - e'$, and because the inclinations will be small, it is adequate to assume that

$$OP_o^+ = R + e' - r, \qquad OP_o^- = R - e' - r,$$

$$OP_c^+ = R + e' + r, \qquad OP_c^- = R - e' + r.$$

Consequently, we may find the ratios

$$\begin{aligned} \frac{OP_o}{\bar{S}P} &\approx \frac{(R \pm e') - r}{R \pm e'} = 1 - \frac{r}{R \pm e'}, \\ \frac{OP_c}{\bar{S}P} &\approx \frac{(R \pm e') + r}{R \pm e'} = 1 + \frac{r}{R \pm e'}, \end{aligned} \tag{11}$$

and then find the inclinations

$$\sin i_o = \sin \pm\beta_o\left(1 - \frac{r}{R \pm e'}\right), \qquad \sin i_c = \sin \pm\beta_c\left(1 + \frac{r}{R \pm e'}\right).^{16} \qquad (12)$$

Just before his demonstration, Copernicus reports Ptolemy's observed latitudes of Mars at opposition, $-\beta_o = 7°$ at perigee and $+\beta_o = 4;20°$ at apogee.[17] Since the northern limit of Mars coincides with the apogee, $e' = e$, and where $R = 10{,}000$, $r = 6580$ and $e' = 960$. Therefore, from (11),

$$1 - r/(R - e') = 0;16,20, \qquad 1 - r/(R + e') = 0;24,$$

and the inclinations i_o derived from β_o at each limit are, from (12),

$$-i_o = \sin^{-1}(\sin 7° \cdot 0;16,20) = 1;54°,$$
$$+i_o = \sin^{-1}(\sin 4;20° \cdot 0;24) = 1;44°,$$

which differ by 0;10° and are not compatible. Evidently $-\beta_o$ is too large and $+\beta_o$ too small to give the same inclination. Copernicus does not show these computations, but he must have done them, or something similar, and then decided to decrease $-\beta_o$ and increase $+\beta_o$ until they fit the same i_o, although he was only partially successful. In the text he takes $-\beta_o = 6;50°$ and $+\beta_o = 4;30°$, changing Ptolemy's values by 0;10°, perhaps because of the discrepancy of 0;10° in the previous computations. He then finds

$$i_o = \sin^{-1}(\sin 6;50° \cdot 0;16,20) = 1;51,12° \approx 1;51°,^{18}$$

and then he claims that $+\beta_o = 4;30°$ will result at the northern limit. This, however, is untrue for

$$+\beta_o = \sin^{-1}(\sin 1;51°/0;24) = 4;38°,$$

while $+\beta_o = 4;30°$ actually implies an inclination of

$$i_o = \sin^{-1}(\sin 4;30° \cdot 0;24) = 1;48°.$$

And in fact it appears that Copernicus used $i_o = 1;48°$ in computing the northern latitudes in his table from about 120° – 180° of the argument of latitude.[19]

Nevertheless, Copernicus seems content with what he has done, and moves on to the minimum inclination i_c given by β_c. This time, as Copernicus

[16] Copernicus gives the ratios $(R \pm e')/r$ for Mars, which suggests a less direct procedure.

[17] Originally in M 193r28 he gave $+\beta_o$ as 4;15°, "or as others [report] 4;20°," but then deleted 4;15° and *siue ut alii*. The value 4;15° is given in *Epitome* XIII,5 as Ptolemy's, which is odd since 4;20° is given correctly in the Gerard of Cremona translation.

[18] In M 193r1–4, Copernicus originally found 1;50°, which is used in VI,4.

[19] Below, p. 528.

must have originally, we consider the latitudes in Ptolemy's table, $-\beta_c = 0;4°$ and $+\beta_c = 0;8°$, which have no observational basis, but follow from computation. In this case, from (11),

$$1 + r/(R - e') = 1;44,40, \qquad 1 + r/(R + e') = 1;36,$$

and the minimum inclinations i_c, from (12),

$$-i_c = \sin^{-1}(\sin 0;4° \cdot 1;44,40) = 0;7°,$$

$$+i_c = \sin^{-1}(\sin 0;8° \cdot 1;36) = 0;13°,$$

which are again incompatible. So Copernicus changes $-\beta_c$ to 0;5°, and finds

$$i_c = \sin^{-1}(\sin 0;5° \cdot 1;44,40) = 0;8,43° \approx 0;9°,$$

which he takes as the minimum inclination at conjunction. He next computes that at the northern limit at conjunction

$$+\beta_c = \sin^{-1}(\sin 0;9°/1;36) = 0;5,37° \approx 0;6°,$$

differing a bit from 0;8° in Ptolemy's table, and finally he points out that the libration of the orbital plane is

$$\Delta i = 1;51° - 0;9° = 1;42°.$$ [20]

For Jupiter and Saturn, Copernicus gives the inclinations without showing how they were derived, but the derivations can be reconstructed using the same procedure as for Mars. In the case of Saturn, where $R = 10{,}000$, $r = 1090$, $e = e_1 - r' = 569$, and since $\omega_A = -50°$, $e' = 569 \cos 50° = 366$. Hence, at opposition, from (11),

$$1 - r/(R - e') = 0;53,13, \qquad 1 - r/(R + e') = 0;53,41,$$

and taking the latitudes at opposition from Ptolemy's tables, the maximum inclinations are, from (12),

$$-i_o = \sin^{-1}(\sin 3;5° \cdot 0;53,13) = 2;44°,$$

$$+i_o = \sin^{-1}(\sin 3;2° \cdot 0;53,41) = 2;43°.$$

Copernicus finds 2;44°, evidently from $-\beta_o$ that he reported previously. For the least inclination, from (11),

$$1 + r/(R - e') = 1;6,47, \qquad 1 + r/(R + e') = 1;6,19,$$

and from the latitudes in Ptolemy's tables, the minimum inclinations are, from (12),

$$-i_c = \sin^{-1}(\sin 2;2° \cdot 1;6,47) = 2;16°,$$

$$+i_c = \sin^{-1}(\sin 2;4° \cdot 1;6,19) = 2;17°.$$

[20] N 186r33–34 gives 1;41° and for $\frac{1}{2}\Delta i$ about 0;50,30° which follow from the original $i_o = 1;50°$ mentioned in n. 18. 1;50° and 0;50,30° are used in VI,4 (below, pp. 504–505).

Copernicus gives 2;16° so again he has used $-\beta_c$. The variation of the inclination is

$$\Delta i = 2;44° - 2;16° = 0;28°. \quad (\text{M } 0;18°, \text{ N } 0;19°)$$

It then follows that at the northern limit,

$$+\beta_c = \sin^{-1} (\sin 2;16°/1;6,19) = 2;3°,$$

while Ptolemy's tables gives 2;4°.

For Jupiter, where $R = 10{,}000$, $r = 1916$, $e = e_1 - r' = 458$, and since $\omega_A = +20°$, $e' = 458 \cos 20° = 430$. Therefore, at opposition, from (11),

$$1 - r/(R - e') = 0;48, \qquad 1 - r/(R + e') = 0;49,$$

and from the latitudes at opposition in Ptolemy's tables, the maximum inclinations are, from (12),

$$-i_o = \sin^{-1} (\sin 2;8° \cdot 0;48) = 1;42°,$$

$$+i_o = \sin^{-1} (\sin 2;4° \cdot 0;49) = 1;41°.$$

Copernicus gives 1;42° from $-\beta_o$, although he also changed the text from 2;8° to 2;7°, perhaps by

$$-\beta_o = \sin^{-1} (1;42°/0;48) = 2;7,30° \approx 2;7°,$$

since he usually rounds . . . 30 down. For the minimum inclination at conjunction, from (11),

$$1 + r/(R - e') = 1;12, \qquad 1 + r/(R + e') = 1;11,$$

and from the latitudes in Ptolemy's tables, the inclinations, from (12), are

$$-i_c = \sin^{-1} (\sin 1;5° \cdot 1;12) = 1;18°,$$

$$+i_c = \sin^{-1} (\sin 1;7° \cdot 1;11) = 1;19°.$$

Copernicus uses the result from $-\beta_c$, so that $i_c = 1;18°$ and the variation

$$\Delta i = 1;42° - 1;18° = 0;24°.$$

Finally, the apparent latitude at the northern limit,

$$+\beta_c = \sin^{-1} (\sin 1;18°/1;11) = 1;6°,$$

compared with 1;7° in Ptolemy's tables.

This completes the derivation of the parameters for the superior planets, which are summarized in the following table.

Planet	ω_A	i_o	i_c	Δi
Saturn	−50°	2;44°	2;16°	0;28°
Jupiter	+20	1;42	1;18	0;24
Mars	0	1;51	0;9	1;42

(13)

It was noted earlier that because of the relation between Copernicus's and Ptolemy's models, and because their parameters were derived essentially from the same apparent latitudes, some simple relations should hold between the inclinations in both models. Specifically, the mean inclination of the orbital plane in Copernicus's model should be equal to the fixed inclination i_1 of the eccentric in Ptolemy's, i.e.

$$\tfrac{1}{2}(i_o + i_c) = i_1. \tag{14}$$

Further, since the libration of the orbital plane Δi in Copernicus's model is about equivalent to the variable inclination i_2 of the epicyclic plane in Ptolemy's, and since the angles are small,

$$\tfrac{1}{2}\Delta i(R/r) \approx i_2. \tag{15}$$

As the following table shows, relation (14) is exactly fulfilled and relation (15) very nearly fulfilled by the inclinations in the two models.

Planet	$\frac{1}{2}(i_o + i_c)$	i_1	R/r	$\frac{1}{2}\Delta i(R/r)$	i_2	
Saturn	2;30°	2;30°	9;10	2;8°	2;0°	
Jupiter	1;30	1;30	5;13	1;3	1;0	(16)
Mars	1;0	1;0	1;31	1;17	1;15	

Fulfilling relation (15) exactly would require setting Δi equal to 0;26° for Saturn, 0;23° for Jupiter, and 1;39° for Mars. The resulting inclinations i_o and i_c would change β_o and β_c by at most 0;1° for Jupiter and Saturn, considerably less than the difference between the observations used by Ptolemy and the tabulated latitudes of both Ptolemy and Copernicus. However, for Mars one would find $-\beta_o = 6;43°$ and $+\beta_o = 4;34°$, while Ptolemy had used 7° and 4;20° respectively and Copernicus's inclinations lead strictly to 6;50° and 4;38°. The comparisons are of interest in showing that, by its very structure, Copernicus model cannot produce both extremal latitudes $\pm\beta_o$ that differ as much as they do in Ptolemy's observations and tables if the extrema are as great as they are for Mars.

Computation from the Model (VI,4)

Copernicus next explains how to compute latitudes from the model for the purpose of making up tables of the same form used by Ptolemy, which are therefore essentially geocentric in that the heliocentric latitude plays no role in their composition. The tables contain three columns in addition to the columns of entry c_1 and c_2. Column 3, $c_{3(\alpha)}$, is the apparent latitude of the planet when located at the northern limit as seen from the earth in any location, and is thus a function of the true anomaly α. Column 4, $c_{4(\alpha)}$, is similarly the apparent latitude at the southern limit as a function of α. Finally, $c_{5(\omega)} = |\cos \omega|$ is a coefficient of interpolation for all other locations of the planet in its orbit.[21] In VI,4 Copernicus explains the procedure for computing $c_{3(\alpha)}$ and $c_{4(\alpha)}$, and gives an example for Mars at the southern limit and the earth at 225° of anomaly, i.e. 45° past true opposition. Since the inclination of the planet's orbit varies as a function of α, it is in the computation of $c_{3(\alpha)}$ and $c_{4(\alpha)}$ that the proper inclination must be determined.

The configuration is shown in Fig. 9 in which the planet P is shown at the northern and southern limits, and the earth O at $\alpha' = 180° \pm \alpha$ measured from opposition rather than conjunction. Since the maximum inclination i_o takes place at opposition, the inclination i at any location is

$$i = i_o - \tfrac{1}{2}\Delta i(1 - \cos \alpha'). \qquad (17)$$

Let $e' = e \cos \omega_A$ be the component of the eccentricity in the direction of the northern limit, and in triangles $P\bar{S}P'$, at the northern limit P^+ and at the southern limit P^-,

$$\bar{S}P = R + e', \qquad \bar{S}P' = (R + e')\cos i, \qquad PP' = (R + e')\sin i,$$

where e' is positive at P^+ and negative at P^-. In triangles $P'\bar{S}O$,

$$OQ = r \sin \alpha', \qquad \bar{S}Q = r \cos \alpha',$$

$$P'Q = \bar{S}P' \pm \bar{S}Q = (R + e')\cos i - r \cos \alpha',$$

$$OP' = (P'Q^2 + OQ^2)^{1/2}$$

$$= ((R + e')^2 \cos^2 i + r^2 - 2(R + e')r \cos i \cos \alpha')^{1/2}.$$

Finally, in triangles $OP'P$, the geocentric latitude,

$$\beta = \tan^{-1}(PP'/OP') = \sin^{-1}(PP'/(OP'^2 + PP'^2)^{1/2}).$$

For Jupiter and Saturn it would be adequate to approximate the last step by $\beta \approx \sin^{-1}(PP'/OP')$, but this will not do for the large latitudes of Mars near opposition.

In Copernicus's example, Mars is at the southern limit P^- and the earth is past opposition at $+\alpha' = 45°$. The inclination to the south is computed from values cancelled in VI,3 in M, that is,

$$-i = 1;50° - 0;50,30°(1 - \cos 45°) = 1;50° - 0;15° = 1;35°.$$ [22]

Since Mars's northern limit coincides with its apogee, $e' = e = 960$ where $R = 10{,}000$. Thus, in triangle $P\bar{S}P'$,

$$\bar{S}P = 9040, \qquad \bar{S}P' = 9036, \quad \text{(Corr. 9036.54)}$$

$$PP' = 249\tfrac{1}{2}. \quad \text{(Corr. 249.78)}$$

[21] The computation of $c_{5(\omega)}$ and the use of the tables will be taken up below, pp. 528–529.

[22] Above, notes 18, 20.

Then, in triangle $P'\bar{S}O$, where $r = 6580$,

$$OQ = 4653, \qquad \bar{S}Q = 4653, \qquad P'Q = 9036 - 4653 = 4383,$$
$$OP' = (4383^2 + 4653^2)^{1/2} = 6392.$$

And in triangle OPP', the apparent latitude

$$-\beta = \tan^{-1}(249\tfrac{1}{2}/6392) = 2;14,13°,$$

although Copernicus gives 2;15°, and this is the value placed in the table for $\alpha = 180° \pm \alpha'$, i.e. 225° and 135°.[23]

We next consider the planet at the northern limit P^+ with the earth in the same location. Consequently, $-\alpha' = 135°$, and the inclination to the north, again using the cancelled values of i_0 and Δi, is

$$+i = 1;50° - 0;50,30°(1 - \cos 135°) = 1;50° - 1;26° = 0;24°.$$

Therefore, in triangle $P\bar{S}P'$,

$$\bar{S}P = 10960, \qquad \bar{S}P' = 10960, \qquad PP' = 75\tfrac{1}{2},$$

and in triangle $P'\bar{S}O$,

$$OQ = 4653, \qquad \bar{S}Q = -4653, \qquad P'Q = 9040 + 4653 = 13693,$$
$$OP' = (13693^2 + 4653^2)^{1/2} = 14462.$$

Hence, the apparent latitude

$$+\beta = \tan^{-1}(75\tfrac{1}{2}/14462) = 0;18°.$$

At $\alpha = (45°, 315°)$ Copernicus's table gives 0;19°.

2. Inferior Planets

Development of the Model (VI,1–2)

The specific phenomena of the latitudes of the inferior planets that determine Ptolemy's model are described in *Almagest* XIII,1 and *Epitome* XIII,2, and Copernicus adapts them to a heliocentric form in VI,1. It was noted earlier that Ptolemy found for the superior planets that the northern limits of latitude were in the vicinity of the apogees. In the case of the inferior planets, it happens that at Ptolemy's time the nodal lines are rather close to

[23] If, as revised in VI,3, $i_0 = 1;51°$, then correctly $i = 1;36°$ and $-\beta = 2;15°$, but Copernicus's computation in the text is for $i_0 = 1;50°$. In M 194v23, he originally wrote 2;12° following a series of errors, but corrected it to 2;15°.

the apsidal lines found by Ptolemy, specifically, the ascending node of Venus and the descending node of Mercury are near the apogees:

Venus	Ptol $\lambda_A = 55°$	$\lambda_{☊} = 60°$	$\Delta = 5°$
Mercury	Ptol $\lambda_A = 190°$	$\lambda_{☋} = 206°$	$\Delta = 16°$

It is thus not surprising that Ptolemy finds that the apsidal line of the eccentric and the equivalent of the nodal line of the epicycle have the same direction, which corresponds in a heliocentric arrangement to the apsidal line and nodal line of the orbit coinciding, as is true in Copernicus's model.

Ptolemy describes the following phenomena for the latitudes of the inferior planets:

1. When the center of the epicycle is $\pm 90°$ of true eccentric anomaly from the apsidal line, the maximum latitudes on either side of the ecliptic occur near inferior and superior conjunction, the larger latitude near inferior conjunction. At $+90°$ the latitude of Venus at inferior conjunction is south and at superior conjunction north, and at $-90°$ the reverse. For Mercury at $+90°$ the latitude at inferior conjunction is north and at superior conjunction south, and at $-90°$ the reverse.
2. When the center of the epicycle is in the apsidal line, the planet reaches approximately equal maximum latitudes on either side of the ecliptic at opposite greatest elongations. At apogee, the evening elongation of Venus is north and the morning elongation south, and at perigee the reverse. For Mercury, at apogee the evening elongation is south and the morning north, and this is likewise reversed at perigee.
3. When the center of the epicycle is in the apsidal line and the planet near inferior or superior conjunction, Venus always has a small northern latitude and Mercury a small southern latitude.

Omitting for the moment the third phenomenon, Ptolemy reaches the model shown in Fig. 10 for Venus. The earth is at O, the center of the eccentric is M, and the epicycle is shown at apogee at $\kappa = 0°$ and 90° from the apsidal line at $\kappa = 270°$. At $\kappa = \pm 90°$, the epicycle is inclined at an angle i_1 along a nodal line that lies in the plane of the eccentric perpendicular to the direction from the earth to the center of the epicycle. The planet reaches its maximum latitude β_1 on one side of the ecliptic at apogee P_a, its maximum latitude β_1' at perigee P_b, and obviously $|\beta_1'| > |\beta_1|$. As the epicycle moves to $\kappa = 0°$, i_1 decreases to zero and the epicycle takes on a second inclination i_2 along a nodal line coinciding with the direction from the earth to the center of the epicycle, i_2 reaching its maximum when $\kappa = (0°, 180°)$. The planet will then have its maximum latitude β_2 on one side of the ecliptic at greatest elongation P_e and its maximum latitude β_2' on the other side at greatest morning elongation P_m, and under the conditions shown $|\beta_2| = |\beta_2'|$. As the epicycle moves from $\kappa = 0°$ to $\kappa = 90°$, i_2 decreases to zero and i_1 increases to maximum, and thus at any value of κ, the inclinations will be:

Venus	$i_{1(\kappa)} = i_1 \cos(\kappa + 90°)$	(18a)
Mercury	$i_{1(\kappa)} = i_1 \cos(\kappa - 90°)$	(18b)
Venus	$i_{2(\kappa)} = i_2 \cos \kappa$	(19a)
Mercury	$i_{2(\kappa)} = i_2 \cos(\kappa + 180°)$	(19b)

Ptolemy's mechanism to produce the changes of inclination again consists of small vertical circles. This time there are two sets of small circles located at the ends of diameters of the epicycle perpendicular to the nodal lines, with their centers in the plane of the eccentric and their motion uniform with respect to equant points located within the circles.[24] Their object is to make the inclinations change, and change nonuniformly as a function of κ. The inclinations i_1 and i_2 are determined by the apparent latitudes β_1 and β_2, and they turn out to be unequal. Ptolemy treats them as separate components of the latitude of the planet, and tabulates them separately.

Copernicus's transformation of Ptolemy's model follows from noting that the two nodal lines for the inclinations i_1 and i_2 in Fig. 10, when considered in the limiting positions shown, in the apsidal line and $\pm 90°$ from the apsidal line, may be regarded as a single nodal line that maintains a fixed direction parallel to the apsidal line while the epicycle is carried around the earth, as shown in Fig. 11. This is not really a correct representation of the two variable and independent inclinations of the epicycle on two perpendicular nodal lines, which are not parallel to the apsidal line in other locations of the center of the epicycle, but it is a simplification made by Copernicus in order to reach his model.[25] If the epicycle is then transformed to the heliocentric orbit of the inferior planet, there follows the model shown in Fig. 12, in which the nodal line coincides with the apsidal line and the different latitudes β_1 and β_2 are the result of the position of the earth with respect to the interior orbit that inclines on a single fixed nodal line from i_1 at $\kappa = \pm 90°$ to i_2 at $\kappa = (0°, 180°)$.[26] When the earth is $\pm 90°$ from the apsidal line, and thus from the nodal line, the maximum latitudes on either side of the ecliptic are again β_1 and β'_1, the planet being at the geocentric apogee P_a and geocentric perigee P_b respectively. When the earth is in the apsidal line, and thus also in the nodal line, the maximum opposite latitudes are β_2 and β'_2, with the planet at greatest evening elongation P_e and greatest morning elongation P_m.

[24] These are described in detail in Riddell (1978).

[25] That the transformation is not equivalent was pointed out in Riddell (1978), 96–98 as an objection to a description of the transformation in Swerdlow (1973), 495–96.

[26] The figure is simplified in that the center of the orbit C is shown in the same position for $\kappa = 0°$ and $\kappa = \pm 90°$.

If in Ptolemy's model i_1 and i_2 were equal, the new model would have a single fixed inclination accounting for both latitudes β_1 and β_2. When he wrote the *Commentariolus*, Copernicus believed that two inclinations were unnecessary, for in describing the model for the first two components of latitude he says that they result from looking at the same inclination in two ways and he gives only one inclination for the orbit of each planet, nowhere indicating that the inclination is variable.[27] Having reduced the two perpendicular nodal lines of the epicycle to a single fixed nodal line of the orbit, perhaps he also believed that the different distances of the orbit from the earth, in the apsidal line and $\pm 90°$ from the apsidal line, would account for the changes in latitude without changing the inclination. However, Ptolemy found i_2 to be greater than i_1, and Copernicus must later have realized that in his model also the apparent latitudes required a variable inclination. Hence, in the description in *De revolutionibus*, when the earth is at $\kappa = \pm 90°$ the orbit must take the same inclination i_1 as in Ptolemy's model, and at $\kappa = (0°, 180°)$ the inclination i_2.

In principle, the inclination i of the orbit at any value of κ is

$$i = i_2 - \tfrac{1}{2}(i_2 - i_1)(1 - \cos 2\kappa), \tag{20}$$

although the instantaneous inclination i is not used by Copernicus, who both derives i_1 and i_2 from β_1 and β_2, and tabulates β_1 and β_2 separately just as Ptolemy does. In fact, he simply copies Ptolemy's tables for β_1 and β_2, which is not correct since, as we have noted, the single variable inclination about a single nodal line of fixed direction in his model is not equivalent to the two variable inclinations about two perpendicular nodal lines in Ptolemy's. This is the third of the serious problems of Copernicus's latitude theory mentioned earlier, and the latitude theory of the inferior planets also suffers from the first two.[28] First, the variation of the inclination between i_1 and i_2 depends upon the motion of the earth through the true eccentric anomaly κ, and although the equation of center of the inferior planets also depends upon the motion of the earth through the mean eccentric anomaly $\bar{\kappa}$, it is hardly obvious why there should be such a relation, except of course as a direct conversion of Ptolemy's models. Then, the "libration" of the inclination is supposed to be produced by the same device used for the inequality of precession and the variation of the obliquity of the ecliptic, that is, the two spheres of Naṣir ad-Dīn aṭ-Ṭūsī used also for the libration of the orbital planes of the superior planets. And again, as was true for the superior planets, the principle of uniform circular motion is violated since

[27] For Venus he gives Ptolemy's i_1 and for Mercury i_2, which may show some confusion, and in enumerating the circles required for the longitude and latitude models of each planet, he does not give Venus and Mercury enough to account for a variation of the inclination of the orbital plane from i_1 to i_2. See Swerdlow (1973), 494–96, 509–10. Curiously, in the *Narratio prima* Rheticus likewise omits the variation of this inclination, concerning which see below, pp. 536–537.

[28] Above, pp. 483–485.

the libration is a function of the true eccentric anomaly κ which increases nonuniformly. Since the inclination is i_2 when $\kappa = (0°, 180°)$ and i_1 when $\kappa = \pm 90°$, the spheres must rotate nonuniformly through 2κ and 4κ. Copernicus does not mention this difficulty, and in VI,2 he even describes the libration as though it depended upon the mean motion of the earth, although in explaining the computation of latitudes in VI,9 he makes it clear that it depends upon the true motion, which of course is nonuniform.[29]

The third component of latitude described by Ptolemy is the small northern latitude of Venus and southern latitude of Mercury when the center of the epicycle is in the apsidal line and the planet near inferior or superior conjunction. Since the latitude is the same whether the planet is near apogee or perigee of the epicycle, Ptolemy accounts for it by inclining the plane of the eccentric, as shown in Fig. 13 for Venus, in which the inclination takes place to the north. The nodal line of the second component of latitude β_2 lies in the plane of the eccentric, and thus whether the planet is at apogee P_a or perigee P_b of the epicycle, it will be removed from the ecliptic by the latitude β_3 which, at $\kappa = (0°, 180°)$, is equal to the inclination i_3 of the plane of the eccentric to the plane of the ecliptic. As the center of the epicycle moves to $\kappa = 90°$, i_3 decreases by $i_{3(\kappa)} = i_3 \cos \kappa$, and since if i_3 were constant the latitude of the center of the epicycle would also be $i_3 \cos \kappa$, the latitude with decreasing i_3 is given by

$$\beta_3 = i_{3(\kappa)} \cos \kappa = i_3 \cos^2 \kappa, \tag{21}$$

a point that Copernicus seems not to have understood.[30] As the center of the epicycle continues beyond $\kappa = 90°$, the opposite half of the eccentric inclines to the north, and at $\kappa = 180°$ i_3 and β_3 again reach their maximum. The change of inclination and latitude in the remaining semicircle, from $\kappa = 180°$ to $\kappa = 0°$, is identical. Figure 13 is shown for Venus, in which i_3 and β_3 are to the north of the ecliptic; for Mercury they must be taken to the south, and everything else is the same.

It is not immediately obvious how to convert this model into a heliocentric form. A direct transformation would require—if one can look at Fig. 12 and imagine such a thing—inclining the nodal line of the orbit, and thus the orbit itself, entirely out of the plane of the ecliptic, and rotating it so that the nodal line always points to the earth. But even to Copernicus this must have seemed implausible. Instead, he attempted to reproduce only the two extremal conditions of Ptolemy's model, namely, that when the center of the epicycle is in the apsidal line, i_3 is maximum and thus β_3 is maximum wherever the planet is located on the epicycle, and when the epicycle is 90° from the apsidal line, i_3 and thus β_3 are both zero. Unfortunately, only the second condition is met by his model, for although i_3 takes on its maximum value when the earth is in the apsidal line, β_3 is

[29] This has been noted in Riddell (1980), 129–30.

[30] See below, pp. 523–524.

affected by the distance between the earth and planet, a function of the true anomaly, and can have a fairly wide range. Copernicus originally developed the model in the *Commentariolus*, and there seemed unaware of its problems. In *De revolutionibus* he takes the consequences of his model literally, turns necessity into a virtue, and tabulates the variation of β_3, but makes a strange computational error for Venus that diminishes the effect of the change of distance to a fraction of its true range.

The model for the third component of latitude has already been discussed and illustrated earlier in connection with the alternate model for Mercury's longitude in V,32.[31] It is shown now in somewhat more detail for Venus in Fig. 14. In our earlier description, we considered the model as though it depended upon mean motions, and Copernicus's description in VI,2 is unclear about whether the motions are mean or true. However, β_3 in Ptolemy's model uses the true eccentric anomaly, and Copernicus's instructions in VI,9 for computing β_3 from the tables shows that in his model likewise one must use the true eccentric anomaly. Therefore, when the earth is in the planet's apsidal line at $\kappa = 0°$, the orbital plane is inclined at an angle i_3 such that the point of maximum inclination I_0 coincides with the planet P_0 wherever the planet may be, and P_0 is seen from the earth with the latitude β_{30}. Let us assume that initially the ascending node of i_3 is in the apsidal line at ☊$_0$. As the earth moves through κ to O, the nodal line and the point of maximum inclination move at the rate of the mean anomaly $\Delta\bar{\alpha}$ to ☊ and I respectively, and the planet moves ahead of I through the true eccentric anomaly κ to P and is seen from O with the latitude β_3. Then, when the earth has reached $\kappa = 90°$, P will have reached the node where β_3 is zero. The inclination i_3 is also variable as a function of κ such that $i_{3(\kappa)} = i_3 \cos \kappa$, as in Ptolemy's model, and at $\kappa = 90°$, i_3 is also zero. As the earth continues from 90° to 180°, the opposite side of the orbit will cross the ecliptic, just as the eccentric did in Ptolemy's model, and at $\kappa = 180°$ P will again be at the point of maximum inclination I' and on the same side of the ecliptic as at I. The remaining semicircle back to I is traversed in the same way. The model for Mercury is the same except that i_3 and β_3 always incline to the south.

In order to show the behavior of β_3 and its relation to Ptolemy's β_3, we may substitute arcs for sines since all angles are small. When the inclination i_3 is maximum and the planet at mean distance R, the apparent latitude is $(r/R)i_3$, and thus at any distance $OP_0 = \rho$, the apparent latitude β_{30} will be $(r/\rho)i_3$ which can vary considerably. In Ptolemy's model, whenever i_3 is maximum β_3 is also at its maximum value. As the planet moves away from the point of maximum inclination, if the inclination were constant the latitude would be $(r/\rho)i_3 \cos \kappa$, but since the inclination varies as $i_{3(\kappa)} = i_3 \cos \kappa$, the latitude with variable i_3 is

$$\beta_3 = (r/\rho)i_{3(\kappa)} \cos \kappa = (r/\rho)i_3 \cos^2 \kappa. \tag{22}$$

[31] Above, pp. 414–415. In Copernicus's description of the model in VI,2, at N 185r22 for *terra in A uel B* read *terra in A uel C*.

There are two differences between Copernicus's and Ptolemy's models, both depending upon distance. First, i_3 cannot be the same quantity in each since in Ptolemy's model it was the inclination of the eccentric of radius R and in Copernicus's the inclination of the orbit of radius r. This in itself is not a problem since i_3 can be chosen to produce the same β_3 at mean distance R as in Ptolemy's model. But what cannot be avoided is that Copernicus's model introduces the ratio of the distances r/ρ while Ptolemy's is independent of distance. Since ρ varies greatly, the latitudes produced by the two models really have little to do with each other except for the single condition that when $\kappa = \pm 90°$ in both cases $\beta_3 = 0°$. At all other values of κ, the differences can be very large, and Copernicus therefore tabulates β_3 separately as a function of the true anomaly although, as noted before, he makes so serious an error for Venus that the table hardly represents the correct behavior of the model.

The model also contains two violations of uniform circular motion. The first is that since the variation of i_3 is a function of the true eccentric anomaly κ, the spheres producing the libration of the orbital plane must rotate nonuniformly through κ and 2κ. This, of course, is already a familiar problem, but there is a stranger violation in this model that is more serious. The planet must move in its orbit, or more physically, be carried by its sphere, so that it is always ahead of the point of maximum inclination I by the true eccentric anomaly κ, and thus its heliocentric motion with respect to a fixed direction, e.g. of the apsidal line, is $(\bar{\alpha} + \kappa)$ and its motion about C is itself nonuniform. But in the model for longitude, the planet's motion is $(\bar{\alpha} + \bar{\kappa})$, which is uniform and therefore differs by the equation of center $c_{3(\bar{\kappa})}$ that reaches 2° for Venus and 3° for Mercury. The result is that the planet occupies two different positions in the longitude and latitude models, a serious inconsistency. The problem is even more obvious if one compares the latitude model with the alternate model of Mercury in V,32 that Copernicus says is intended to be applicable to the latitude theory.[32] It can be seen that in the alternate model the planet is ahead of I by $\bar{\kappa}$, while here it must be ahead by κ. And Copernicus has no choice but to use κ in the model for latitude, for since the variable inclination i_3 is a function of κ, if the planet exceeded the motion of I by $\bar{\kappa}$, it would not reach the node at the same time that i_3 becomes zero, and would not then return to the same side of the ecliptic precisely as it passes the node. An alternative would be to make both the inclination and the motion of the planet depend upon $\bar{\kappa}$, but this would no longer be equivalent to Ptolemy's model, although considering the other differences between the models and the serious problems introduced here, such a solution would probably offer the lesser evil.

Finally, there is the question of how the models for the components of latitude of Figs. 12 and 14 are to be combined. According to Copernicus's description, the model for β_3 must be placed in the inclined orbital plane

---

[32] Above, pp. 412–413.

of β_1 and β_2 for in this way the moveable nodal line and point of maximum inclination of β_3 can rotate through $\Delta\bar{\alpha}$ without disturbing the direction of the fixed nodal line and limits of latitude of β_1 and β_2, and indeed this is the only sensible arrangement. But this will introduce into the combined latitude model the problem of the location of *P*. In the model of Fig. 12, *P* falls just where it does in the longitude theory since it is moving uniformly about *C*. But in the model of Fig. 14 the motion about *C* is nonuniform, the difference being of course the equation of center. There thus appears to be no way of combining the two latitude models consistently as long as the inclination i_3 and the motion of *P* with respect to *I* are both functions of κ.

Note on Technical Terms

The technical names for the three components of latitude of the inferior planets have a varied history extending from the Greek text of the *Almagest*, through the medieval Arabic translations to the Gerard of Cremona Latin version and translations of other Arabic astronomical works. The Latin terms are contained in standard works used by Copernicus, such as the *Epitome*, Peurbach's *Theoricae novae planetarum*, and the *Alfonsine Tables*. Copernicus's own terminology is taken mostly from this tradition, and in one case probably from George of Trebizond's translation of the *Almagest* directly from the Greek. The purpose of this note is to sketch the background to the technical terms.

1. The first component of latitude β_1 and the inclination i_1. Ptolemy calls the latitude β_1 ἔγκλισις, "inclination," and the angle i_1 ἔγκλισις τοῦ ἐπικύκλου, the "inclination of the epicycle." The corresponding Arabic terms are *mayl* and *mayl falak at-tadwīr* that in Gerard's translation are *declinatio* and *declinatio orbis revolutionis*. The *Alfonsine Tables* call β_1 *declinatio*, while Peurbach uses both *inclinatio* and *declinatio*, and in the *Epitome* Regiomontanus uses only *inclinatio*. George of Trebizond uses *declinatio* and *inclinatio*, and Copernicus likewise gives both words, but generally uses *declinatio*. To represent its meaning in the context of the latitude theory, *declinatio* must be translated as "inclination." The inclination reaches its maximum when $\kappa = \pm 90°$, that is, when the earth is at 90° of true eccentric anomaly from the apsidal line, or geocentrically when the center of the epicycle is at 90° of true eccentric anomaly from the apsidal line. Copernicus, following the Gerard of Cremona *Almagest*, calls these locations "at mean distances" (*in mediis longitudinibus*), and he calls the corresponding points of the planet's orbit 90° from the apsidal line "at mean apsides" (*ad medias absidas*).[33]

[33] The term *longitudo media* is discussed by Regiomontanus in *Disputationes contra cremonensia deliramenta* where it is taken properly to mean the points on the eccentric at the distance *R* from the observer. The use of the term to mean the points at which $\kappa = \pm 90°$ is specifically criticized as incorrect, but this is nevertheless the way Copernicus uses the term in the latitude theory.

2. The second component of latitude β_2 and the inclination i_2. Ptolemy calls β_2 λόξωσις, "slant" or "obliquity," and angle i_2 λόξωσις τοῦ ἐπικύκλου, the "slanting of the epicycle." The distinction between ἔγκλισις and λόξωσις as Ptolemy uses the terms in the latitude theory is that the first refers to an inclination in the line of sight and the second to an inclination across the line of sight. The inclination of the epicycle in the theory of the superior planets, which takes place in the line of sight, is referred to only as ἔγκλισις. In Arabic the slant β_2 is *warāb*, but preferably *inḥirāf*, and the slanting of the epicycle *inḥirāf falak at-tadwīr*, from which comes Gerard's translation *reflexio*, a "slanting," for β_2 and *reflexio orbis revolutionis* for i_2. The *Alfonsine Tables*, the *Theoricae novae*, and the *Epitome* call β_2 *reflexio*, while George of Trebizond translates λόξωσις as *obliquatio*. Copernicus says that β_2 is called *obliquatio*, and by others *reflexio*, *obliquatio* being by far his preferred term, and he calls i_2 *obliquitas orbis* and *inflexio*. *Obliquatio* is in any case a good equivalent of λόξωσις, and is best translated "slant" or "obliquity" since "obliquation" is an obsolete word.

3. The third component of latitude β_3 and the inclination i_3. Although not specifically using the term, Ptolemy considers β_3 an ἔγκλισις, "inclination," one may presume because it takes place in the line of sight, and he calls i_3 ἔγκλισις τοῦ ἐκκέντρου, these being the same terms used for the single latitude and the inclination of the eccentric in the latitude theory of the superior planets, both of which take place in the line of sight. The corresponding Arabic for β_3 is *mayl* and for i_3 *mayl falak al-ᶜārij al-markaz*, "inclination of the eccentric sphere," which Gerard translates as *declinatio* and *declinatio orbis egredientis centri* respectively. Peurbach calls β_3 *deviatio* and i_3 *deviatio deferentis*, and Regiomontanus likewise calls β_3 *deviatio* and i_3 *deviatio eccentri*. In XIII,3 he explains *deviatio* as *deviatio sive inclinatio eccentri ad superficiem ecliptici*. We have not found β_3 referred to as *deviatio* prior to Peurbach, but do not know whether the term was his invention. Copernicus calls β_3 *deviatio* and i_3, which is seldom referred to, the *obliquitas circuli deviationis*. As long as it is understood as a specific technical term, *deviatio* can simply be transliterated as "deviation," although "deflection" is a better descriptive term for the small effect of β_3, and we shall us it.[34]

In order to keep some kind of consistency, we shall use the following translations in referring to both Ptolemy's and Copernicus's latitude theories:

β_1, i_1: ἔγκλισις, *declinatio*, "inclination",

β_2, i_2: λόξωσις, *obliquatio*, "slant",

β_3, i_3: ἔγκλισις, *deviatio*, "deflection".

---

[34] As a suitable English equivalent, *deflection* is used in the *Oxford English Dictionary* art. *deviation* 1. b.

Derivation of Parameters and Computation from the Model

The parameters still to be derived for the latitude theory are the maximum values of the three inclinations of the orbital planes: the inclination i_1, the slant i_2, and in principle the deflection i_3, although Copernicus substitutes the maximum apparent latitude of the deflection at mean distance, $(r/R)i_3$, which is taken as equal to i_3 in Ptolemy's model. As was true for the superior planets, Copernicus's derivations are not based upon his own observations of apparent latitudes, but on a combination of the apparent latitudes reported by Ptolemy and the extremal apparent latitudes in Ptolemy's tables. In a sense Copernicus's derivations are unnecessary since he simply takes over Ptolemy's values, and his demonstrations of how to compute the tabulations of the inclination and slant are likewise superfluous since he simply copies Ptolemy's tables. Only for the deflection does he independently compute a tabulation, and this is because his model is so different from Ptolemy's that he could not avoid taking account of the effect of the variation of distance upon the apparent latitude.

Ptolemy derived the inclinations from observations of apparent latitudes near inferior and superior conjunction and near greatest elongation, and must, at least in part, have been relying upon conventional estimates as he did for the superior planets. The isolated effects of the inclination and deflection (Ptolemy of course does not use that term) are seen near inferior and superior conjunction, and the isolated effect of the slant at greatest elongation. These are further localized in that the maximum inclination is observed when the center of the epicycle is at $\kappa = \pm 90°$ from the apsidal line, and the maximum slant and deflection when the center of the epicycle is in the apsidal line at $\kappa = (0°, 180°)$. Hence, in order to find the inclination and the deflection, the planet must be observed as close as possible to inferior and superior conjunction at all four quadrants of eccentric anomaly, and this is far from easy.

Ptolemy places the apogee of Venus at 55° and of Mercury at 190°. It was explained earlier that during a period of forty or so years Venus reaches its greatest elongations when the mean sun is in only five regions of the zodiac about 72° apart.[35] This is a consequence of its highly regular cycle of five synodic periods in $8^y - 2^d$, and since in four years Venus's anomaly changes by about

$$4(\tfrac{5}{8} \cdot 360°) = 2^r + 180°,$$

an inferior and superior conjunction will take place at about the same longitude four years apart, and there will likewise be only five locations of inferior and superior conjunctions about 72° apart during an observer's lifetime. And of course the latitudes at greatest elongations present the same problem in that they too can be observed in only five locations. This

[35] Above, pp. 387–388.

greatly limited Ptolemy's opportunity to make the required observations of Venus, and thus he had no choice but to use earlier conventional estimates of its latitude in different parts of the zodiac.

The problem is not so serious for Mercury since its elongations and conjunctions are distributed throughout the zodiac during its 46-year cycle of 145 synodic periods, but the planet's invisibility between heliacal setting and rising poses difficulties. Taking Ptolemy's longitude of Mercury's apogee, 190°, at inferior conjunction in the apsidal line Mercury is invisible not less than 25^d and at superior conjunction in the apsidal line not less than 27^d at the latitude of Alexandria.[36] At 90° from the apsidal line, the period of invisibility at inferior conjunction is not less than 12^d and at superior conjunction not less than 27^d. And it is from uncertain observations near last and first visibility that the latitudes at conjunctions must be estimated in order to find the inclination and deflection. For Venus the corresponding periods of invisibility are as short as 1^d at inferior conjunction $-90°$ from the apsidal line and 7^d in the apsidal line at perigee, to long periods of 59^d to 67^d at superior conjunction in the four locations. Thus it is not surprising that Ptolemy's observations near greatest elongation used to find the slant are considerably better than the latitudes taken for conjunctions. The later improvements of his latitude theory in the *Handy Tables* and *Planetary Hypotheses* appear to follow from his finding that the larger inclinations of the epicycle established for the slant at greatest elongation apply also at conjunctions.[37] In his later latitude theory, the epicycles of Venus and Mercury are given a single fixed inclination and are made to rotate such that the nodal line maintains a fixed direction. Heliocentrically, this would produce an orbit of fixed inclination and fixed nodal line.

Using the notation of Figs. 10 and 13, i.e. the maximum apparent latitudes at superior and inferior conjunction are β_1 and β_1' leading to the inclination i_1, the maximum apparent latitudes at greatest elongation are β_2 and β_2' leading to the slant i_2, and the maximum apparent latitude β_3 at conjunctions is equal to the deflection i_3, Ptolemy's observations and resulting inclinations may be summarized as follows:

Planet	$\pm\beta_1$	$\mp\beta_1'$	i_1	$\pm\beta_2 \approx \mp\beta_2'$	i_2	$\beta_3 = i_3$
Venus	~1;0°	~6;20°	2;30°	~2;30°	3;30°	+0;10°
Mercury	~1;45	~4;0	6;15	~2;30	7;0	−0;45

(23)

[36] Using P. V. Neugebauer (1938).

[37] For Venus the inclination is 3;30°, exactly that of i_2 in the *Almagest*, but for Mercury it is 6;30°, lying between $i_1 = 6;15°$ and $i_2 = 7°$ in the *Almagest*, the latter being in fact the correct inclination of Mercury's orbit.

Note that all apparent latitudes except β_3 are given by Ptolemy as approximate. With regard to β_2, he says that near apogee for Venus $\beta_2 - \beta_2'$ is a bit less than 5°, and at perigee a bit more, and for Mercury the difference is about $\frac{1}{2}$° less than 5° at apogee and $\frac{1}{2}$° more at perigee. These differences are shown to result from the distances of the center of the epicycle at apogee and perigee respectively. Copernicus derives the same inclinations from the same apparent latitudes, and by the same methods, as Ptolemy. He then copies Ptolemy's tables for the inclination and slant, and computes an additional column for the effect of the variation of the planet's distance from the earth on the deflection. We shall now consider separately Copernicus's exposition of each component in VI,5–8.

A. Inclination, β_1 and i_1 (VI,5)

The inclination is isolated when the earth is "at mean distances," that is, at $\kappa = \pm 90°$ from the apsidal line and thus at $\pm 90°$ from the nodes of the planet's orbit.[38] Copernicus takes as the extremal latitudes the values of β_1 and β_1' from Ptolemy's table, and finds exactly Ptolemy's values of i_1.

Planet	$\pm\beta_1$	$\pm\beta_1'$	i_1
Venus	1;2°	6;22°	2;30°
Mercury	1;45	4;5	6;15

(24)

He does not show the derivations, but remarks that the inclinations are made clear by the tables of equations, meaning the correction tables for longitude in Book V. The reference is to the method used by Ptolemy in *Almagest* XIII,3, which is shown in Fig. 15.[39] Imagine that the plane of the planet's orbit has been rotated 90° so that it is perpendicular to the plane of the ecliptic. Then the longitude correction tables may be used to find the inclination i_1 from the observed latitude β_1. Let $\beta_1 = c_{(\alpha)}$, the equation of the anomaly at a distance α from apogee, and by inverse use of the tables find $i_1 = \alpha$ near apogee, and in the same way from β_1' find $i_1' = \alpha'$ near perigee. Since the earth is taken to be $\kappa = 90°$ from apogee, one must interpolate in Copernicus's tables with $c_{(\alpha)} = c_{5(\alpha)} + 0;30c_{6(\alpha)}$ for Venus and

[38] Copernicus's statement of these conditions at N 187v3–6 is not a little confusing. He says that the inclination is isolated "near mean distances and near the nodes" (*circa medias longitudines, circaque nodos*), but the earth is 90° from the nodes when the inclination is isolated. Perhaps he is thinking of the nodes of the eccentric in Ptolemy's model, which are at $\kappa = \pm 90°$, that is, at what Copernicus calls means distances.

[39] *Epitome* XIII,4 gives a strict trigonometrical procedure for finding i_1, but Copernicus refers only to Ptolemy's simpler method using the correction tables.

with $c_{(\alpha)} = c_{5(\alpha)} + 0;53c_{6(\alpha)}$ for Mercury. Thus, letting β_1 and β_1' each be equal to $c_{(\alpha)}$ at apogee and perigee respectively,

$$i_1 = 3°(\beta_1/c_{3°}), \qquad i_1' = 3°(\beta_1'/c_{183°}). \tag{25}$$

Using Copernicus's tables for Venus and Mercury in Book V, we find:

Planet	β_1	$c_{3°}$	i_1	β_1'	$c_{183°}$	i_1'
Venus	1;2°	1;15°	2;29°	6;22°	7;39°	2;30°
Mercury	1;45	0;51	6;11	4;5	1;58	6;14

(26)

Copernicus gives 2;30° for Venus and 6;15° for Mercury in agreement with Ptolemy, and it is possible that he did not even bother carrying out the procedure we have just described.

Next, Copernicus shows how to compute β_1 for different distances between the planet and the earth due to the motion of the planet in its orbit, and thus as a function of the true anomaly α. The inclination of the orbital plane is taken as the maximum value of i_1 for the earth at mean distances, $\kappa = \pm 90°$, and the resulting latitude $\beta_{1(\alpha)}$ is entered into the table of latitudes as column 3, the inclination (*declinatio*) $c_{3(\alpha)}$. The method of computation is based upon Ptolemy's examples for Venus and Mercury in *Almagest* XIII,4 and the procedure described in *Epitome* XIII,8 even down to the lettering of the diagram. Essentially, it is the same as the computation of the single latitude of the superior planets. In Fig. 16 the planet at P is taken as $\alpha' = 180° \pm \alpha$ from the geocentric perigee of the orbit, and its projection into the plane of the ecliptic is P'. When $\kappa = 90°$, $\bar{\kappa} \approx 90° + c_{3(\bar{\kappa})\max}$ and the distance $R_{(\bar{\kappa})} = OC$ from the earth to the center of the orbit is, for Venus

$$R_{(\bar{\kappa})} = ((R + (e_1 - e_2)\cos\bar{\kappa})^2 + (e_1 + e_2)^2 \sin^2\bar{\kappa})^{1/2},$$

and for Mercury

$$R_{(\bar{\kappa})} = ((R + (e_1 + e_2)\cos\bar{\kappa})^2 + (e_1 - e_2)^2 \sin^2\bar{\kappa})^{1/2}.$$

Letting $CP = r$, the radius of the orbit,

$$PQ = P'Q' = r\sin\alpha', \qquad CQ = r\cos\alpha',$$

$$PP' = OQ' = r\sin i_1 \cos\alpha', \qquad Q'C = r\cos i_1 \cos\alpha',$$

$$OQ' = OC - Q'C = R_{(\bar{\kappa})} - r\cos i_1 \cos\alpha',$$

$$OP' = (OQ'^2 + P'Q'^2)^{1/2} = ((R_{(\bar{\kappa})} - r\cos i_1\cos\alpha')^2 + r^2\sin^2\alpha')^{1/2},$$

$$\beta_{1(\alpha)} = \tan^{-1}(PP'/OP') = \sin^{-1}(PP'/(OP'^2 + PP'^2)^{1/2}).$$

Following *Almagest* XIII,4 and *Epitome* XIII,9, Copernicus also checks the reduction to the ecliptic of the equation of the anomaly in the inclined orbit for the purpose of showing that it is sufficiently small that longitudes may be computed as though the orbit were in the plane of the ecliptic. Ptolemy had also done this for the superior planets, but Copernicus did not do so because the inclinations are so small that the reduction, already not greater than 0;1° in Ptolemy's model, would be even less in his. The equation of the anomaly reduced to the plane of the ecliptic is

$$\text{angle } Q'OP' = \sin^{-1}(P'Q'/OP').$$

In order to find the equation $Q'OP$ in the inclined orbit, Copernicus uses a different method than Ptolemy used. He takes the diagonal PQ', which is given by

$$PQ' = (PP'^2 + P'Q'^2)^{1/2},$$

and since the distance from the earth to the planet OP is

$$OP = (OQ'^2 + PQ'^2)^{1/2},$$

the equation of the anomaly in the inclined orbit is

$$\text{angle } Q'OP = \sin^{-1}(PQ'/OP).$$

The difference $Q'OP - Q'OP'$ is the reduction to the ecliptic.

Copernicus's examples, following Ptolemy, are for $\alpha' = 45°$. First for Venus, $i_1 = 2;30°$ and using the parameters applicable to Ptolemy's time, where $R = 10{,}000$,

$$e_1 = 312, \qquad e_2 = 104, \qquad r = 7193.$$

At $\kappa = 90°$, $\bar{\kappa} \approx 92;23°$ and the distance of the center of the orbit, $R_{(\bar{\kappa})} \approx$ 10,000. Then, following the procedure just given,

$$PQ = 5086, \qquad CQ = 5086, \qquad PP' = 221, \quad \text{(Corr. 222)}$$

$$Q'C = 5081, \qquad OQ' = 4919, \qquad OP' = 7075, \quad \text{(Corr. 7075.6)}$$

$$\beta_{1(\alpha)} = 1;47°,$$

while Ptolemy found 1;48°. In principle $c_{3(\alpha)} = 1;47°$ should be put in the table for $\alpha = (135°, 225°)$, but Copernicus's table is copied from Ptolemy's, and so gives 1;48°. To check the reduction to the ecliptic, since $P'Q' = PQ$,

$$\text{angle } Q'OP' = \sin^{-1}(5086/7075) = 45;57,40° \approx 45;58°,$$

is the equation of the anomaly in the plane of the ecliptic. Copernicus next finds that

$$PQ' = 5091, \qquad OP = 7079, \qquad \text{angle } Q'OP = 45;59°,$$

which differs by only 0;1°.[40]

[40] By error here M 204r1–2 gives $Q'OP'$ as 45;57° and the difference as 0;2°. N 188r34–36 is even more confusing since it gives $Q'OP' = 45;57°$, $Q'OP = 45;58°$, and a difference of 0;2°. Ptolemy found $Q'OP' = 46;0°$, $Q'OP = 46;2°$, and a difference of 0;2°.

For Mercury $i_1 = 6;15°$, and where $R = 10{,}000$,

$$e_1 = 748, \qquad e_2 = 212, \qquad r = \bar{r} + r' = 3953.$$

At $\kappa = 90°$, $\bar{\kappa} \approx 93°$ so that $R_{(\bar{\kappa})} = 9964$, and thus

$$PQ = 2795, \qquad CQ = 2795, \qquad PP' = 304,$$

$$Q'C = 2778, \qquad OQ' = 7186, \qquad OP' = 7710,$$

$$\beta_{1(\alpha)} = 2;16°,$$

in agreement with Ptolemy. The equation of the anomaly in the plane of the ecliptic,

$$\text{angle } Q'OP' = \sin^{-1}(2795/7710) = 21;16°, \quad (\text{Corr. } 21;15°)$$

and since

$$PQ' = 2811, \qquad OP = 7716,$$

the equation in the inclined plane,

$$\text{angle } Q'OP = \sin^{-1}(2811/7716) = 21;22°,$$

although Copernicus gives 21;23°, which follows from taking OP' for OP, that is, $\sin^{-1}(2811/7710) = 21;23°$. He then says that the difference from 21;16° is about 0;7°.

B. Slant, β_2 and i_2 (VI,6–7)

The slant has its maximum effect when the earth is in the apsidal line, which is also the nodal line, and the orbital plane of the planet is inclined across the line of sight. But the deflection also has its maximum effect when the earth is in the apsidal line, and thus it is necessary to distinguish the slant and the deflection. According to Ptolemy's description of the deflection, when the planet is at apogee or perigee of the epicycle it has a small latitude β_3, for Venus $+0;10°$ and for Mercury $-0;45°$. At other points on the epicycle β_3 will be added algebraically to β_2, and consequently the northern latitudes of Venus can exceed the southern latitudes by $2\beta_3 = +0;20°$, and the southern latitudes of Mercury can exceed the northern by $2\beta_3 = -1;30°$. It is, however, clear from Ptolemy's description in *Almagest* XIII,3 that these differences were not specifically observed, but are a consequence of combining the models for β_2 and β_3. What he reports is that when the center of the epicycle is in the apsidal line and the planet at greatest elongations, where β_2 reaches its maximum, the difference between the northern and southern latitudes of both planets is about 5°, which is thus $2\beta_2$. More specifically, for Venus at apogee the difference is slightly less than 5° and at perigee slightly more, and for Mercury there is a difference of about $\frac{1}{2}°$ at apogee and perigee, i.e. $5° - \frac{1}{2}°$ at apogee and $5° + \frac{1}{2}°$ at perigee, these differences being due to the distances of the center of the epicycle. Hence, when the effect of the variation of distance is removed $2\beta_2 \approx 5°$, and when the effect of the deflection is removed $\beta_2 = \beta_2' \approx 2\frac{1}{2}°$.

Copernicus explains all this at the beginning of VI,6, and then turns to the proof that the maximum value of the slant, $\beta_{2\,\max}$, takes place at greatest elongation, and that at any other location of the planet the ratio of β_2 to $\beta_{2\,\max}$ is equal to the ratio of the equation of the anomaly $c_{(\alpha)}$ in the same location to the maximum equation of the anomaly $c_{(\alpha)\max}$ at greatest elongation. This proof is of some importance since the equality of the ratios is the basis for the tabulation of β_2. The demonstration appears in *Almagest* XIII,4 and *Epitome* XIII,11,19, and Copernicus follows his sources closely. In Fig. 17 the planet is at any position P including greatest elongation P_e. For any position of P,

$$\sin i_2 = b/a, \qquad \sin c = a/d, \qquad \sin \beta_2 = b/d,$$

and thus

$$\sin i_2 \sin c = b/d = \sin \beta_2, \qquad \sin \beta_2 = \sin i_2 \sin c.$$

Therefore, at greatest elongation $c_{\max} > c$,

$$\sin \beta_{2\,\max} = \sin i_2 \sin c_{\max} > \sin i_2 \sin c = \sin \beta_2,$$

which proves the proposition. It also follows that

$$\sin \beta_2/\sin \beta_{2\,\max} = \sin c/\sin c_{\max}, \tag{27a}$$

which gives an easy way of computing β_2 since $c_{\max}$ and c may be taken from the table of corrections in Book V and $\beta_{2\,\max}$ is known from observation. However, Ptolemy and Copernicus claim that

$$\beta_2/\beta_{2\,\max} = c/c_{\max}, \tag{27b}$$

and although the substitution of arc for sine is acceptable for β_2, which is small, it is not acceptable for c, which reaches 22° for Mercury and 46° for Venus at mean distance for which β_2 is computed.[41]

The derivation of i_2 from $\beta_{2\,\max}$ is straightforward, and is shown in Fig. 18 in which the planet P_e is at greatest elongation. Ptolemy and Copernicus find i_2 in four steps:

$$d = (R^2_{(\bar{K})} - r^2)^{1/2}, \tag{28a}$$

$$a = d(r/R_{(\bar{K})}), \tag{28b}$$

$$b = d \sin \beta_{2\,\max}, \tag{28c}$$

$$i_2 = \sin^{-1}(b/a), \tag{28d}$$

which can be reduced to the direct solution

$$\sin i_2 = \sin \beta_{2\,\max}(R_{(\bar{K})}/r). \tag{29a}$$

[41] This has been pointed out in Pedersen (1974), 380–81.

Since $R_{(\bar{\kappa})}$ is variable, Ptolemy and Copernicus first find i_2 for mean distance R, and then check the variation of $\beta_{2\,\max}$ for the distance R_0 at apogee and R_{180} at perigee by finding d and a as before and then

$$b = a \sin i_2, \tag{28e}$$

$$\beta_{2\,\max} = \sin^{-1}(b/d). \tag{28f}$$

Again, (29a) will serve for a more direct solution in the form

$$\sin \beta_{2\,\max} = \sin i_2(r/R_{(\bar{\kappa})}). \tag{29b}$$

First, consider Venus for which $\beta_{2\,\max} = 2;30°$ and, using the parameters for the time of Ptolemy, where $R = 10{,}000$, $r = 7193$ and $e_1 - e_2 = 208$. Hence, at mean distance, from (28a–d),

$$d = 6947, \qquad a = 4997, \qquad b = 303, \qquad i_2 = 3;29°,$$

while Ptolemy gives 3;30°. Since the inclination $i_1 = 2;30°$ at $\kappa = \pm 90°$, the variation of the inclination of the orbital plane is 1°. At greatest distance $R_0 = 10208$, and computing correctly from (28ab–ef),

$$d = 7243, \qquad a = 5104, \qquad b = 310, \qquad \beta_{2\,\max} = 2;27°.$$

Copernicus makes a series of small errors, and finds

$$d = 7238, \qquad a = 5102, \qquad b = 309, \qquad \beta_{2\,\max} = 2;27°,$$

so the result is not affected. At least distance $R_{180} = 9792$, and thus

$$d = 6644, \qquad a = 4883, \qquad (\text{Corr. } 4881), \qquad b = 297, \qquad \beta_{2\,\max} = 2;34°.$$

Hence, exactly as Ptolemy found, $\beta_{2\,\max}$ can differ from 2;30° by $-0;3°$ at apogee and $+0;4°$ at perigee, which Copernicus says are too small to be measured by means of "astrolabic instruments," i.e. the armillary, and therefore 2;30° may be retained as the maximum value of β_2.

In the case of Mercury, the variation of the radius of the orbit must be taken into account, i.e. $r = \bar{r} - r' \cos 2\bar{\kappa}$, and since the slant is isolated in the apsidal line at $\bar{\kappa} = (0°, 180°)$, $r = \bar{r} - r'$, the minimum value. Hence, where $R = 10{,}000$, $r = 3573$ and $e_1 + e_2 = 948$. Since $\beta_{2\,\max} = 2;30°$, just as for Venus, at mean distance, from (28a–d),

$$d = 9340, \qquad a = 3337, \qquad b = 407, \qquad i_2 = 7;0°,$$

in agreement with Ptolemy, and since $i_1 = 6;15°$ at $\kappa = \pm 90°$, the change of inclination is 0;45°. At greatest distance $R_0 = 10948$, and correctly from (28ab–ef),

$$d = 10349, \qquad a = 3378, \qquad b = 412, \qquad \beta_{2\,\max} = 2;17°.$$

Here Copernicus makes an error in d, and finds

$$d = 9452, \qquad a = 3085, \qquad b = 376, \qquad \beta_{2\,\max} = 2;17°,$$

the final result being the same since the four steps are equivalent to (29b) and d is cancelled out. Then at least distance $R_{180} = 9052$, and

$$d = 8317, \qquad a = 3283, \qquad b = 400, \qquad \beta_{2\max} = 2;45°.$$

Hence, $\beta_{2\max}$ can differ from 2;30° by −0;13° at apogee and +0;15° at perigee, while Ptolemy found −0;13° and +0;16°, and following Ptolemy, Copernicus rounds the difference to $\pm\frac{1}{4}°$.

Ptolemy and Copernicus also compute for mean distance the reduction to the ecliptic of the maximum equation of the anomaly $c_{\max}$. In Fig. 18 the equation in the inclined orbit is $c_{\max}$ and reduced to the ecliptic $c'_{\max}$, which are given by

$$c_{\max} = \sin^{-1}(a/d), \qquad (30a)$$
$$c'_{\max} = \sin^{-1}(a'/d'), \qquad (30b)$$

where

$$a' = (a^2 - b^2)^{1/2}, \qquad (31a)$$
$$d' = (d^2 - b^2)^{1/2}. \qquad (31b)$$

The results are as follows:

Planet	a	d	$c_{\max}$	b	a'	d'	$c'_{\max}$
Venus	4997	6947	46;0°	303	4988	6940	45;57°
Mercury	3337	9340	20;56	407	3314[42]	9331	20;48

(32)

Hence the reduction $c_{\max} - c'_{\max}$ is 0;3° for Venus and 0;8° for Mercury. Ptolemy had found reductions of 0;1° and 0;6° respectively.

In order to compute β_2 for any value of the true anomaly α, Ptolemy and Copernicus use the relation

$$\beta_{2(\alpha)} = (\beta_{2\max}/c_{\max})c_{(\alpha)}, \qquad (33)$$

where $c_{(\alpha)}$ is the equation of the anomaly for mean distance, and the result is entered in the latitude tables of the inferior planets as $c_{4(\alpha)}$. As noted earlier, the substitution of arc $c_{\max}$ and c for sine $c_{\max}$ and c leads to errors that can amount to 0;6° for Venus and only 0;1° for Mercury. Since $c_{(\alpha)}$ is for mean distance, using the longitude correction tables in *Almagest* XI,11, $c_{(\alpha)} = c_{6(\alpha)}$. With Copernicus's tables in Book V, for Venus $c_{(\alpha)} = c_{5(\alpha)} + 0;30c_{6(\alpha)}$, and for Mercury, in order to approximate Ptolemy's $c_{6(\alpha)}$ one would use $c_{(\alpha)} = c_{5(\alpha)} + 0;37c_{6(\alpha)}$. Copernicus says that at mean distance $c_{\max}$ is 46° for Venus and about 22° for Mercury. Since in both cases $\beta_{2\max}$ is 2;30°, the coefficient $(\beta_{2\max}/c_{\max})$ is 0;3,15 for Venus and 0;6,49 for Mercury, and in principle the slant $c_{4(\alpha)}$ in Copernicus's latitude tables is computed from:

Venus	$c_{4(\alpha)} = 0;3,15(c_{5(\alpha)} + 0;30c_{6(\alpha)})$
Mercury	$c_{4(\alpha)} = 0;6,49(c_{5(\alpha)} + 0;37c_{6(\alpha)})$

[42] Corr. 3312.

However, since Copernicus simply copied $c_{4(\alpha)}$ from the latitude tables in *Almagest* XIII,5, these formulas were never used, and the columns in both Ptolemy's and Copernicus's tables were computed from

Venus	$c_{4(\alpha)} = 0;3,15c_{6(\alpha)}$
Mercury	$c_{4(\alpha)} = 0;6,48c_{6(\alpha)}$

where $c_{6(\alpha)}$ is from *Almagest* XI,11 and the coefficients follow from Ptolemy's $(\beta_{2\,\max}/c_{\max})$.

C. Deflection, β_3 and i_3 (VI,8)

The component of latitude that Copernicus calls the deflection (*deviatio*) is in the *Almagest* the latitude resulting from the variable inclination i_3 of the eccentric. We have noted earlier that $i_{3(\kappa)} = i_3 \cos\kappa$, and as the point of maximum inclination is always in the apsidal line, the latitude β_3 of the planet at any location of the center of the epicycle is

$$\beta_3 = i_{3(\bar{\kappa})} \cos\kappa = i_3 \cos^2\kappa, \qquad (21)$$

where for Venus $i_3 = +0;10^\circ$ and for Mercury $i_3 = -0;45^\circ$. Ptolemy tabulates the function $|\cos\kappa|$ sexagesimally to two places, and in XIII,6 he explains that having already found the "sixtieths" for computing β_2, take from them the same quantity of the sixtieths, i.e. $\cos\kappa \cos\kappa$, and then take the sixth part for Venus, always to the north, and the three-fourths part for Mercury, always to the south. This is exactly the same as the relation given above.

Copernicus, however, seems to have misunderstood the instructions, and is thereby led to a misunderstanding of Ptolemy's inclination of the eccentric, for he writes in VI,8:[43]

> Nevertheless, it is not sufficiently clear whether they [earlier astronomers] intended such an inclination of the spheres to be always equal and the same, for this is what their computations indicate when they order that the sixth part of the proportional minutes always be taken for the deflection of Venus, but the three-fourths part for the deflection of Mercury. This cannot be so unless the angle of the inclination always remains the same, just as the proportion of those minutes, upon which [the computations] are based, requires. And further, if the angle always remains the same, it will not be possible to understand how this latitude of those stars suddenly leaps back from the intersection [i.e. the nodal line] into the same latitude that it previously left unless you were to say that it takes place like the bending [reflection] of light rays (*per modum refractionis luminum*) as in optics. But here we are concerned with a motion that is not instantaneous, but by its own nature is proportional to time.

Evidently, Copernicus believes that the latitude is found from $\beta_3 = i_3 \cos\kappa$, instead of $\cos^2\kappa$, and this would indeed indicate an eccentric plane of fixed inclination and make it hard to understand how the planet can return

[43] N 191v30–192r7 with corr. *tempori* for *ipsi* at 192r6 as in M 207v34.

to the same side of the ecliptic after passing the node. The source of his problem must be that he is relying upon the instructions for computing β_3 by Santritter in his edition of the *Alfonsine Tables*, for in Propositions 25–26 on the latitudes of Venus and Mercury, the instructions say simply to take for Venus a sixth part, and for Mercury a half plus a quarter part, of the proportional minutes used for finding β_2, i.e. $\frac{1}{6}\cos\kappa$ and $\frac{3}{4}\cos\kappa$ respectively, which is incorrect since the coefficient should be $\cos^2\kappa$.[44] Nevertheless, Copernicus does figure out that his own model for the deflection requires $\cos^2\kappa$, and he tabulates the coefficient as $c_{7(\kappa)}$ of his latitude tables for the inferior planets.

The most important difference between Ptolemy's inclination of the eccentric and Copernicus's inclination of the orbital plane of the planet for producing the deflection is that Ptolemy's model, by applying i_3 to the plane of the eccentric surrounding the earth, gives the planet the same latitude β_3 wherever it is located on the epicycle, while Copernicus's model, by inclining the interior orbital plane of the planet, makes β_3 also depend upon the distance between the earth and the planet, and thus upon the location of the planet in its orbit. For example, an inclination i_3 that will produce at mean distance a latitude $\beta_{30} \approx i_3(r/R)$, will produce at geocentric apogee a latitude $\beta_3 \approx \beta_{30}(R/(R+r))$ and at geocentric perigee a latitude $\beta_3 \approx \beta_{30}(R/(R-r))$. In the case of Venus, where $r \approx .72R$, the result is an apparent latitude six times greater at perigee than at apogee. Hence, Copernicus's model is in no way equivalent to Ptolemy's, but produces instead an altogether different component of latitude that is not supported by any cited observations. In particular, it directly contradicts Ptolemy's observations that the planet has equal latitudes at inferior and superior conjunctions, that is, at geocentric perigee and apogee, for these are precisely the locations where the distances, and thus the latitudes produced by Copernicus's model, differ the most.

At the time he first described the model for the deflection in the *Commentariolus*, Copernicus appeared unaware of these problems since he explains the behavior of the inclination of the orbital plane in considerable detail without any reference to the effect of the changing distance of the planet on the apparent latitude. By the time of *De revolutionibus*, he realized that he had to take the variation of distance into account, and that is the principle subject of VI,8, in which he shows how to compute β_3 as a function of the distance of the planet, which then appears as column 5, $c_{5(\alpha)}$, of his latitude tables, giving the apparent latitude of the point of maximum inclination of the orbital plane as a function of the true anomaly α. However, he makes a

[44] Apparently this error was not uncommon. Among the latitude tables copied by Copernicus on pages bound in with his copy of the *Alfonsine Tables* are tables for the deflection of Venus and Mercury that give $i_3 \cos\kappa$, but the accompanying instructions omit the order to multiply this by $\cos\kappa$, which is tabulated on the next page (below, p. 529, n. 51). It would have been better to tabulate directly $\beta_3 = i_3 \cos^2\kappa$, but this was not done. The tables and instructions are printed in Curtze (1875), 235–38, and Prowe (1883–84), 2, 236–40.

remarkable error for Venus that reduces the variation of the latitude from fully 0;30° to only 0;9°, which he then reduces further to 0;7° in the tables. Whether this error proceeded from negligence or deliberation is unclear.

Copernicus first investigates the variation of extremal latitudes at greatest and least distance. In Fig. 19 the planet is shown at P_0 at a mean distance R, and at P_a at greatest distance $R + r$ and P_b at least distance $R - r$. The entire orbit is shown raised out of the ecliptic plane such that at P_0 the apparent latitude β_{30} is equal to the maximum value of $\beta_3 = i_3$ in Ptolemy's inclination of the eccentric. Hence, the elevation of the orbital plane $q = CC'$ is

$$q = R \sin \beta_{30}, \tag{34}$$

and at greatest distance P_a and least distance P_b the apparent latitudes will be

$$\beta_a = \sin^{-1} (q/(R + r)), \tag{35a}$$

$$\beta_b = \sin^{-1} (q/(R - r)), \tag{35b}$$

since the eccentricities of the orbits are neglected. Copernicus does not find the inclination i_3 of the orbital plane, but it is given by

$$i_3 = \tan^{-1} (q/r) \approx \beta_{30}(R/r). \tag{36}$$

Taking first Venus, we have

$$\beta_{30} = 0;10°, \qquad R = 10{,}000, \qquad r = 7193,$$

from which correctly,

$$q = 29, \qquad \beta_a = 0;5{,}48°, \qquad \beta_b = 0;35{,}31°, \qquad i_3 = 0;13{,}52°.$$

Copernicus gives $\beta_a = 0;6°$, but $\beta_b \approx 0;15°$, which is a serious error, and he remarks that the differences of $-0;4°$ and $+0;5°$ from 0;10° are small enough to be generally neglected. He does not give i_3. For Mercury, one must use the radius of the orbit in the apsidal line, the minimum radius $r = \bar{r} - r'$, and thus

$$\beta_{30} = 0;45°, \qquad R = 10{,}000, \qquad r = 3573.$$

Consequently,

$$q = 131, \qquad \beta_a = 0;33°, \qquad \beta_b = 1;10°, \qquad i_3 = 2;6°,$$

which Copernicus gives correctly, although omitting i_3. The differences from 0;45° are $-0;12°$ and $+0;25°$, but Copernicus says that they nearly disappear under the rays of the sun before the planet becomes visible, for which reason the ancients only considered the apparent part of the deflection as if it were simple, i.e. as if the deflection in the visible part of the orbit were not affected by the change of distance. This is not a good excuse even for Mercury, and it is simply untrue for Venus, which has an enormous variation of distance in the visible part of its orbit and can be observed extremely close to inferior conjunction. In any case, it contradicts Ptolemy's description of the phenomena underlying the inclination of the eccentric.

Copernicus originally concluded the chapter here in a cancelled passage in M (208v5–14) in which he says that if anyone wishes to investigate the part of Mercury's path hidden under the rays of the sun, he will expend more labor on it than on any of the other latitudes previously explained. And therefore, he says, he will use the computation of the ancients, which does not disagree much with the truth, lest, as they say, we appear to have a fight "over the shadow of an ass."[45] But then he changed his mind, added an explanation of how to find β_3 at any point in the orbit, and needless to say, deleted his disparaging comment on the exercise.

In fact, the computation is quite simple, and is shown in Fig. 20. The planet P is at $\alpha' = 180° \pm \alpha$ from the perigee of its orbit, which is elevated from the plane of the ecliptic by the distance q that produces the deflection β_{30} at mean distance. Thus, as before $q = R \sin \beta_{30}$, and since eccentricities are neglected, in the plane of the ecliptic,

$$P'Q' = r \sin \alpha', \qquad C'Q' = r \cos \alpha',$$

$$OQ' = R - r \cos \alpha', \qquad OP' = (P'Q'^2 + OQ'^2)^{1/2},$$

and thus the latitude,

$$\beta_3 = \tan^{-1}(q/OP') = \sin^{-1}(q/(OP'^2 + q^2)^{1/2}).$$

Since the angles are all very small, it would be adequate to find $\beta_3 \approx \beta_{30}(R/OP')$. Copernicus gives as an example Mercury at $\alpha' = 45°$. We are given that

$$\beta_{30} = 0;45°, \qquad R = 10{,}000, \qquad r = 3573,$$

and therefore

$$q = 131, \qquad P'Q' = 2526, \qquad C'Q' = 2526,$$

$$OQ' = 7474, \qquad OP' = 7889, \qquad \beta_3 = 0;57°.$$

Copernicus fails to mention the resulting β_3, but in the tables for $\alpha = (135°, 225°)$, $c_{5(\alpha)} = 0;57°$.

The chapter concludes with a lengthy explanation of the motivation and computation of the coefficient of interpolation for the deflection, $c_{7(\kappa)} = \cos^2 \kappa$. A possible reason that Copernicus devotes so much attention to something so simple is that, owing to his misunderstanding of the computation of β_3 from Ptolemy's model, he thought he was doing something new and different by introducing $\cos^2 \kappa$ as a coefficient of interpolation.

---

[45] M 208v5–14. . . . *de umbra, quod aiunt, asini*, from a common Greek proverb, cf. Aristoph. *Vespae* 191 *et schol.* which gives a fanciful account of its origin.

3. The Tables and Their Use (VI,9)

Except for the deflection component of the latitude of the inferior planets, Copernicus's tables are of the same form as those in the *Almagest*, and to a greater extent than is warranted are copied from the *Almagest*.[46] The first two colums, c_1 and c_2, are the arguments of entry at intervals of 3° for 3°–180° and 180°–357°. The remaining columns differ for the superior and inferior planets, which we consider separately.

A. Superior Planets

Each planet is given three columns in addition to the columns of entry:

Column 3, $c_{3(\alpha)}$, latitudes at the northern limit, $\omega = 0°$.
Column 4, $c_{4(\alpha)}$, latitudes at the southern limit, $\omega = 180°$.
Column 5, $c_{5(\omega)} = |\cos \omega|$, coefficient of interpolation for $0° \leq \omega \leq 180°$.

Columns 3 and 4, $c_{3(\alpha)}$ and $c_{4(\alpha)}$, give the apparent latitude of the planet when located at the limit as a function of the true anomaly α since the distance between the earth and the planet varies as a function of α. Fig. 21 shows graphs of the northern latitudes $c_{3(\alpha)}$ marked *N* and the southern latitudes $c_{4(\alpha)}$ marked *S*. The method of computing the columns was explained earlier in VI,4, and the parameters required are as follows:[47]

Planet	i_o	$\frac{1}{2}\Delta i$	ω_A	e'	$R + e'$	$R - e'$	r	
Saturn	2;44°	0;14°	−50°	366	10366	9634	1090	
Jupiter	1;42	0;12	+20	430	10430	9570	1916	(37)
Mars	1;51	0;51	0	960	10960	9040	6580	

A comparison of recomputations of $c_{3(\alpha)}$ and $c_{4(\alpha)}$ with Copernicus's tables shows that in so far as possible he avoided computation by copying from the tables in the *Almagest*. The following remarks may be made on each planet:

Saturn: Compared with recomputation, both columns show errors of 0;1°, but the agreement with $c_{3(\alpha)}$ and $c_{4(\alpha)}$ in the *Almagest*, using linear interpolation for 3° intervals below 90°, is perfect.

---

[46] The 1515 *Almagest* gives the latitudes at 6° intervals throughout rather than 6° up to 90° and 3° over 90° as in other texts of the *Almagest*. But Copernicus must have used a complete version, and interpolated only for 3° intervals below 90°. Possibilities are the 1528 printing of George of Trebizond's translation or even the Greek text printed in 1538 if Copernicus wrote the tables after Rheticus gave it to him in 1539. Concerning the possibility that the latitude theory was written so late, see below, pp. 535–537, and on the peculiar origin of $c_{5(\omega)}$, p. 529.

[47] Above, pp. 503–505.

Jupiter:[48] $c_{3(\alpha)}$ shows errors of 0;1°, but agrees perfectly with the *Almagest*: $c_{4(\alpha)}$ contains errors of 0;1°, occasionally 0;2°, and agrees with the *Almagest* below 120°. From 120°–180° Copernicus seems to have computed himself, and most entries are in error by −0;1°.

Mars:[49] Since Copernicus's inclination of Mars's orbit is intended to reproduce the extremal latitudes −6;50° and +4;30°, while Ptolemy derived inclinations from −7;0° and +4;20°, the columns for Mars cannot be copied from the *Almagest*. However, there is no way of reconciling the columns with a single value of i_o since 1;51° gives an extreme northern latitude of +4;38° rather than the required +4;30° in the table. On the whole, the best agreement for $c_{3(\alpha)}$ is found by using 1;51° up to 117° and 1;48° from 120°–180°, and in fact it was found earlier that the extremal latitude +4;30° leads to an inclination of 1;48°.[50] However, the same variation, $\frac{1}{2}\Delta i = 0;51°$, appears to have been used throughout. Even so, errors of 0;3° and 0;4° occur in the earlier part of the column, and while this seems odd, isolated entries may have been copied from the *Almagest*. $c_{4(\alpha)}$ for southern latitudes was computed from 1;51° consistently, and has only small errors of 0;1° or 0;2° up to 141°, after which the errors increase to as much as 0;15°.

Column 5, $c_{5(\omega)}$, is a coefficient of interpolation for the planet at locations other than the northern or southern limits. If we define a heliocentric latitude b as the latitude of the planet seen from the mean sun as it moves along its inclined orbit, then $b = \sin^{-1}(\sin i \cos \omega)$, where ω is the angular distance from the northern limit. Since the geocentric latitude, equivalent to $\sin^{-1}((\bar{S}P/OP)\sin i)$, has already been tabulated for the planet at the northern limit in $c_{3(\alpha)}$ and the southern limit in $c_{4(\alpha)}$, it is only necessary to multiply each of these by $\cos \omega$ to find the geocentric latitude for other locations of the planet. Both Ptolemy and Copernicus provide a column, $c_{5(\omega)} = |\cos \omega|$, for this purpose. The column is given sexagesimally to two places from +1;0 to −1;0, and is applicable to all the planets. The computation of the coefficient is explained in *Almagest* XIII,4, *Epitome* XIII,21, and by Copernicus is VI,7. The latitude of the moon was tabulated in *Almagest* V,8 as

$$\beta_{☾} = c_{7(\omega)} = \sin^{-1}(\sin 5° \cos \omega) \approx 5° \cos \omega,$$

and thus, multiplying by 0;12,

$$c_{5(\omega)} = 0;12 c_{7(\omega)} = 0;12 \cdot 5 \cos \omega = \cos \omega.$$

---

[48] Textual corrections to N: c_3 120°, 1;45° for 1;44°; 135°, 1;51° for 1;52°. Both are correct in M.

[49] Textual correction: c_3 150°, 1;51° in both M and N appears to be a scribal error for 1;59°. There are many other errors, but they are probably computational.

[50] Above, p. 500. The tables do not appear to use $i_o = 1;50°$ and $\frac{1}{2}\Delta i = 0;50,30°$ that were used in VI,4. Cf. above, notes 18, 20.

Copernicus did not compute $c_{5(\omega)}$ from either his own or Ptolemy's lunar latitude column, nor did he copy it directly from the *Almagest*. Rather, he extracted it at 3° intervals from a table called *Tabula minutorum proportionabilium quinque planetarum* at 1° that he had copied, along with an extensive set of latitude tables also at 1° intervals, on pages bound in with his copy of the *Alfonsine Tables*.[51] The original table was derived from linear interpolation in the corresponding column of Ptolemy's tables, but in a version at 6° intervals throughout rather than at 6° below 90° and 3° for 90°–180°. Such versions at 6° intervals are printed in the *Alfonsine Tables* and oddly, in the 1515 *Almagest*, and interpolating at 3° intervals gives results identical to Copernicus's column and its source, but differing from, and somewhat less accurate than, Ptolemy's complete column.

The computation of the latitude is quite simple. From the earlier computation of longitude, one has the true eccentric anomaly $\kappa = \bar{\kappa} + c_{3(\bar{\kappa})}$ and the true anomaly $\alpha = \bar{\alpha} + c_{3(\bar{\kappa})}$, observing the opposite signs of $c_{3(\bar{\kappa})}$ in each case. The true argument of latitude from the northern limit is $\omega = \kappa - \omega_A$, so that

$$\text{Saturn: } \omega = \kappa + 50°, \qquad \text{Jupiter: } \omega = \kappa - 20°, \qquad \text{Mars: } \omega = \kappa. \tag{38}$$

With ω as argument, find the coefficient $c_{5(\omega)}$, and then with the true anomaly α find $c_{3(\alpha)}$ or $c_{4(\alpha)}$ and form the geocentric latitude

$$\begin{aligned} +\beta_{(\alpha,\omega)} &= c_{5(\omega)} \cdot c_{3(\alpha)} \qquad \text{for } 270° \leq \omega \leq \ \ 90°, \\ -\beta_{(\alpha,\omega)} &= c_{5(\omega)} \cdot c_{4(\alpha)} \qquad \text{for } \ \ 90° \leq \omega \leq 270°. \end{aligned} \tag{39}$$

The latitude will be northern or southern depending upon whether $c_{3(\alpha)}$ or $c_{4(\alpha)}$ is used according to the preceding rule.

Example. Compute the latitude of Mars for its position in Copernicus's horoscope of 19 Feb 1473, 16;48^h.[52] From the computation of the longitude,

$$\kappa = 165;51° - 3;2° = 162;49°, \qquad \alpha = 26;37° + 3;2° = 29;39°.$$

Since for Mars $\omega = \kappa$, and $90° \leq \omega \leq 270°$, from the tables,

$$c_{5(\omega)} = 0;57,13, \qquad c_{4(\alpha)} = -0;11°, \qquad \beta_{(\alpha,\omega)} = -0;10°.$$

---

[51] Printed in Curtze (1875), 236, and Prowe (1883–84) 2, 237. This is the table, used as cos κ, that is referred to in n. 44. In M 210 the entries in $c_{5(\omega)}$ for 69°–81° are wrong by a series of dittography errors, although the corresponding entries for 99°–111° are correct. Curiously, in N 193v–194r, 69°–81° are correct and 99°–111° contain the same errors. Characteristically, all the critical editions and translations have adopted the wrong numbers, and some have even "emended" the correct numbers to agree with them. The correct sequence for 69°–81° and 111°–99° is 21,24/18,24/15,24/12,24/9,24 instead of the erroneous 21,21/18,18/15,15/12,12/9,9. Copernicus's $c_{7(\kappa)}$ for the inferior planets is computed from the correct values by $c_{7(\kappa)} = c^2_{5(\omega)}$. See below, p. 531.

[52] Above, pp. 454–457.

In the same way, we compute the latitudes of Saturn and Jupiter for the positions in the horoscope, and the results are summarized in the following table along with a modern computation.[53]

Planet	κ	α	ω	$c_{5(\omega)}$	$c_{(\alpha)}$	β	β mod
Saturn	177;56°	254;3°	227;56°	0;40,3	c_4 −2;38°	−1;45°	−1;4°
Jupiter	48;54	104;1	28;54	0;52,29	c_3 +1;37	+1;25	+0;55
Mars	162;49	29;39	162;49	0;57,13	c_4 −0;11	−0;10	−1;7

(40)

The large error for Mars, which is near enough to conjunction to be invisible, is a notable flaw in Ptolemy's latitude theory in the *Almagest*, and thus in Copernicus's theory. The inclination i_2 of the epicycle in Ptolemy's model brings Mars practically into the ecliptic near conjunction, when in fact it can have a latitude of over 1°. The elimination of i_2 in the *Planetary Hypotheses* and the corresponding increase of i_1 to 1;50° gives Mars nearly correct latitudes even at conjunction. Likewise, the elimination of i_2 for Saturn and Jupiter in the same work also improves their latitudes to essentially the correct values. Unfortunately, Copernicus knew nothing of Ptolemy's later latitude theory.

B. Inferior Planets

The inferior planets require the use of five columns in addition to the columns of entry:[54]

Column 3, $c_{3(\alpha)}$, inclination (*declinatio*) β_1, $\kappa = \pm 90°$.
Column 4, $c_{4(\alpha)}$, slant (*obliquatio*) β_2, $\kappa = (0°, 180°)$.
Column 5, $c_{5(\alpha)}$, deflection (*deviatio*) β_3, $\kappa = (0°, 180°)$.
Column 6, $c_{6(\kappa)} = |\cos \kappa|$, found as $c_{5(\omega)}$ for superior planets, coefficient of interpolation for $c_{3(\alpha)}$ and $c_{4(\alpha)}$.
Column 7, $c_{7(\kappa)} = \cos^2 \kappa$, coefficient of interpolation for $c_{5(\alpha)}$.

Columns 3, 4, and 5 are graphed in Fig. 22. In principle, the columns are computed by the methods explained in VI,5–8, but $c_{3(\alpha)}$ and $c_{4(\alpha)}$ are simply copied from the same columns in the *Almagest* with linear interpolation for 3° intervals below 90°.[55] The columns for the deflection, $c_{5(\alpha)}$, are

[53] Tuckerman (1964).

[54] The order of the columns was changed between M 211 and N 194v–95r, but the headings were incorrectly left in the original order.

[55] Two entries that differ must be scribal errors: Venus c_3 144° is 2;25° for 2;23°, Mercury c_3 48° is 1;15° for 1;16°. See above, p. 527, n. 46.

inconsistent, particularly for Venus where a correct computation would have the latitude increase to 0;36° at $\alpha = 180°$, while Copernicus holds it to a false limit of 0;14°.[56] Mercury is not computed accurately either, but does stay within the limits 0;33° and 1;10° found in VI,8. What we are calling $c_{6(\kappa)}$ is the column $c_{5(\omega)}$ for the superior planets, and thus is here used as $|\cos \kappa|$. Finally, $c_{7(\kappa)} = \cos^2 \kappa$ is computed from the column $c_{5(\omega)}$ for the superior planets, that is, $c_{7(\kappa)} = c^2_{5(\omega)}$.[57]

The computation of latitude for an inferior planet is considerably more complicated than for a superior planet. There are three separate components that are computed from the tables using the true eccentric anomaly κ and the true anomaly α with a bewildering combination of rules for when the latitudes are northern or southern, when the components are added or subtracted. Further, Copernicus's tables and method of computation are applicable to Ptolemy's model, where the inclination β_1 and slant β_2 are separate components produced by the separate inclinations i_1 and i_2 on perpendicular diameters of the epicycle, but not to Copernicus's model, where there is a single nodal line of fixed direction and a single inclination that varies between i_1 and i_2. In fact, Copernicus here missed an opportunity to make an improvement in his latitude theory simply by developing tables for computing from his own model rather than Ptolemy's. In order to demonstrate the differences between the two models, we shall first set out Copernicus's rules for computing from his tables, and then in the following section show the proper method of finding latitudes from his model.

From the preceding computation of the planet's longitude, one has the true eccentric anomaly κ and the true anomaly α. Each component of latitude is computed as follows (see Figs. 10 and 12):

1. β_1, inclination. β_1 is computed from $c_{3(\alpha)}$ which applies to $\kappa = \pm 90°$ where the inclination of the epicycle or the orbital plane in the line of sight is i_1. For other positions of the center of the epicycle or the earth, β_1 is computed from

$$\beta_{1(\alpha,\kappa')} = c_{6(\kappa')} \cdot c_{3(\alpha)}, \qquad \kappa' = \kappa \begin{cases} +90° & \text{for Venus,} \\ -90° & \text{for Mercury,} \end{cases} \tag{41}$$

since the geocentric apogee of Venus's epicycle or orbit is inclined to the north at $\kappa = +90°$ and the geocentric apogee of Mercury's epicycle or orbit is north at $\kappa = -90°$. We have now to determine the signs of $c_{3(\alpha)}$ and $c_{6(\kappa')}$. Since at $\kappa' = 0°$ the northern limit of latitude is at apogee where $\alpha = 0°$, the sign of $c_{3(\alpha)}$ is

$$c_{3(\alpha)} \geq 0° \quad \text{for } 270° \leq \alpha \leq 90°, \qquad c_{3(\alpha)} \leq 0° \quad \text{for } 90° \leq \alpha \leq 270°.$$

---

[56] Reinhold does the same in the *Prutenic Tables*.

[57] There are six errors at 9°, 21°, 36°, 57°, 87°, but all entries for 69°–81° follow from the correct $c_{5(\omega)}$ (above, p. 529, n. 51).

The coefficient $c_{6(\kappa')}$ represents the variable inclination $i_{1(\kappa)} = i_1 \cos \kappa'$, and thus has the sign of a cosine function, i.e.

$$c_{6(\kappa')} \geq 0^\circ \quad \text{for } 270^\circ \leq \kappa' \leq 90^\circ, \qquad c_{6(\kappa')} \leq 0^\circ \quad \text{for } 90^\circ \leq \kappa' \leq 270^\circ.$$

2. β_2, slant. β_2 is computed from $c_{4(\alpha)}$ which applies to $\kappa = (0^\circ, 180^\circ)$ where the inclination of the epicycle or the orbital plane across the line of sight is i_2. For Mercury $c_{4(\alpha)}$ is a mean value since $\beta_{2\,\max} = 2;30^\circ$ was found to differ at $\kappa = 0^\circ$ and $\kappa = 180^\circ$ by $\mp 0;15^\circ$. Hence, we shall consider Venus and Mercury separately. First, in the case of Venus, for any position of the center of the epicycle or the earth

$$\beta_{2(\alpha,\kappa'')} = c_{6(\kappa'')} \cdot c_{4(\alpha)}, \qquad \kappa'' = \kappa, \tag{42a}$$

since the positive (western, evening) semicircle of Venus's epicycle or orbit is inclined to the north at $\kappa = 0^\circ$. For Mercury, we note that the difference of latitude at apogee and perigee, 0;15°, is $\frac{1}{10}\beta_{2\,\max}$, and further, that the positive semicircle of Mercury's epicycle or orbit is inclined to the north at $\kappa = 180^\circ$. Hence, for any position of the center of the epicycle or the earth

$$\beta_{2(\alpha,\kappa'')} = c_{6(\kappa'')} \cdot c'_{4(\alpha)}, \qquad \kappa'' = \kappa + 180^\circ, \tag{42b}$$

$$c'_{4(\alpha)} = \tfrac{9}{10} c_{4(\alpha)} \quad \text{for } 270^\circ \leq \kappa'' \leq 90^\circ,$$

$$c'_{4(\alpha)} = \tfrac{11}{10} c_{4(\alpha)} \quad \text{for } 90^\circ \leq \kappa'' \leq 270^\circ.$$

The signs of $c_{4(\alpha)}$ and $c_{6(\kappa'')}$ follow from noting that at $\kappa'' = 0^\circ$, the positive semicircle is north of the ecliptic, so that

$$c_{4(\alpha)} \geq 0^\circ \quad \text{for } 0^\circ \leq \alpha \leq 180^\circ, \qquad c_{4(\alpha)} \leq 0^\circ \quad \text{for } 180^\circ \leq \alpha \leq 360^\circ,$$

and since $c_{6(\kappa'')} = \cos \kappa''$ corresponds to the inclination $i_{2(\kappa)} = i_2 \cos \kappa''$, $c_{6(\kappa'')}$ has the sign of a cosine function, i.e.

$$c_{6(\kappa'')} \geq 0^\circ \quad \text{for } 270^\circ \leq \kappa'' \leq 90^\circ, \qquad c_{6(\kappa'')} \leq 0^\circ \quad \text{for } 90^\circ \leq \kappa'' \leq 270^\circ.$$

3. β_3, deflection (see Fig. 14). β_3 is computed from $c_{5(\alpha)}$ which applies to $\kappa = (0^\circ, 180^\circ)$ where the inclination i_3 is maximum, and is always northern for Venus and southern for Mercury. For other values of κ, β_3 is computed from

$$\beta_{3(\alpha,\kappa)} = c_{7(\kappa)}(\pm c_{5(\alpha)}), \qquad \begin{matrix} + \\ - \end{matrix} \begin{cases} \text{for Venus,} \\ \text{for Mercury,} \end{cases} \tag{43}$$

where $c_{7(\kappa)} = \cos^2 \kappa$ represents both the variation of the inclination $i_{3(\kappa)} = i_3 \cos \kappa$ and the change of latitude as the planet moves from the point of maximum inclination to the node, which is also a function of $\cos \kappa$.

Finally, the three components are added algebraically to give the latitude

$$\beta = \beta_{1(\alpha,\kappa')} + \beta_{2(\alpha,\kappa'')} + \beta_{3(\alpha,\kappa)}. \tag{44}$$

Example. Compute the latitude of Mercury for its position in Copernicus's horoscope of 19 Feb 1473, 16;48^h.[58] From the computation of the longitude,

$$\kappa = 100;54^\circ - 2;59^\circ = 97;55^\circ, \qquad \alpha = 127;56^\circ + 2;59^\circ = 130;55^\circ.$$

1. β_1, inclination: $\kappa' = \kappa - 90^\circ = 7;55^\circ$,

$$c_{3(\alpha)} = +2;2^\circ, \qquad c_{6(\kappa')} = +0;59,17, \qquad \beta_{1(\alpha,\kappa')} = +2;1^\circ.$$

2. β_2, slant: $\kappa'' = \kappa + 180^\circ = 277;55^\circ$,

$$c'_{4(\alpha)} = 0;54c_{4(\alpha)} = 0;54(+2;21^\circ) = 2;7^\circ,$$

$$c_{6(\kappa'')} = +0;8,19, \qquad \beta_{2(\alpha,\kappa'')} = +0;18^\circ.$$

3. β_3, deflection:

$$c_{5(\alpha)} = -0;56^\circ, \qquad c_{7(\kappa)} = 0;1,11, \qquad \beta_{3(\alpha,\kappa)} = -0;1^\circ.$$

Hence, the latitude

$$\beta = \beta_{1(\alpha,\kappa')} + \beta_{2(\alpha,\kappa'')} + \beta_{3(\alpha,\kappa)} = +2;18^\circ,$$

and were the computation carried to seconds, $\beta = +2;17^\circ$. According to Tuckerman (1964), the latitude is $+2;44^\circ$.

The position of Venus in the horoscope may be computed more concisely.

$$\alpha = 65;14^\circ, \qquad \kappa = 265;25^\circ, \qquad \kappa' = 355;25^\circ, \qquad \kappa'' = 265;25^\circ.$$

1. $c_{3(\alpha)} = +0;30^\circ$, $\quad c_{6(\kappa')} = +0;59,42$, $\quad \beta_1 = +0;30^\circ$.
2. $c_{4(\alpha)} = +1;27^\circ$, $\quad c_{6(\kappa'')} = -0;4,53$, $\quad \beta_2 = -0;7^\circ$,
3. $c_{5(\alpha)} = +0;9^\circ$, $\quad c_{7(\kappa)} = 0;0,26$, $\quad \beta_3 = 0;0^\circ$.

$$\beta = \beta_1 + \beta_2 + \beta_3 = +0;23^\circ.$$

According to Tuckerman (1964), the latitude is $+0;44^\circ$.

Correct Computation of β_1 and β_2 from Copernicus's Model

We have noted that the computation of the inclination β_1 and the slant β_2 from Ptolemy's tables is not a correct representation of the latitudes produced by Copernicus's model.[59] The two models are equivalent only for the extremal cases used to derive the inclinations i_1 and i_2, that is, at $\kappa = \pm 90^\circ$ where β_1 is isolated and β_2 zero, and conversely at $\kappa = (0^\circ, 180^\circ)$ where β_2 is isolated and β_1 is zero. For all other values of κ the models behave differently. The reason is that the variation of a single inclination between the limits i_1 and i_2 on a single nodal line of fixed direction in Copernicus's model is not equivalent to the two separate variable inclinations i_1

[58] Above, pp. 454–457.

[59] This was first noted in Riddell (1980), 129.

and i_2, each of which can become zero, on two perpendicular nodal lines of changing direction in Ptolemy's model. In order to show the differences, we shall first develop a correct method of computation for Copernicus's model, and then show its relation to a corresponding formulation for Ptolemy's model. The latitude we wish to find is $\beta = \beta_1 + \beta_2$, the sum of the inclination and the slant, since the deflection β_3 is a totally separate component not affected by the inclinations i_1 and i_2.

In Fig. 23, which is drawn for Venus, the earth O and the planet P are shown at arbitrary positions, and the ascending node is on the side of the orbit opposite $\kappa = 0°$. If the figure were drawn for Mercury, the descending node would be opposite $\kappa = 0°$. The inclination i varies between i_1 at $\kappa = \pm 90°$ and i_2 at $\kappa = (0°, 180°)$, and for any value of κ is given by

$$i = i_2 - \tfrac{1}{2}(i_2 - i_1)(1 - \cos 2\kappa) = i_1 \sin^2 \kappa + i_2 \cos^2 \kappa, \qquad (45)$$

the second formulation being of significance for comparison with Ptolemy's model. Since i is small, angles in the orbital plane are shown projected into the plane of the ecliptic. Letting ω' be the argument of latitude of P from the ascending node,

$$b = r \sin i \sin \omega'.$$

Let $\kappa' = \kappa - 180°$, and regarding ω' as measured in the plane of the ecliptic,

$$\text{angle } OCP' = \omega' - \kappa' = \alpha' = \alpha - 180°,$$

so that

$$\omega' - \kappa = \omega' - (\kappa' + 180°) = \alpha,$$

and the geocentric distance of the planet reduced to the ecliptic, $OP' = \rho$, is

$$\rho = (R_{(\kappa)}^2 + r^2 + 2R_{(\kappa)} r \cos(\omega' - \kappa))^{1/2}, \qquad (46)$$

where $R_{(\kappa)}$ is the distance OC from the earth to the center of the orbit.[60] The apparent latitude is then

$$\beta = \tan^{-1}(b/\rho) = \tan^{-1}((r/\rho) \sin i \sin \omega'), \qquad (47)$$

or substituting i for $\sin i$,

$$\beta \approx (r/\rho) i \sin \omega' = (r/\rho) \sin \omega' (i_1 \sin^2 \kappa + i_2 \cos^2 \kappa). \qquad (48)$$

Now consider the latitudes β_1 and β_2 in Fig. 24 computed with Ptolemy's and Copernicus's tables for the inclinations i_1 at $\kappa = \pm 90°$ and i_2 at $\kappa = (0°, 180°)$ with interpolation for arbitrary values of κ. Note that in transforming the model implied by the tables, which is of course Ptolemy's model, to a heliocentric form with a moving earth, the orbit of the planet

[60] In finding ρ the reduction of r to the ecliptic is omitted.

must rotate along with the motion of the earth through the true eccentric anomaly κ, which is nonuniform, so that the inclination i_1 is always directed toward the earth at O and the slant i_2 is always perpendicular to the line of sight from the earth. This is a curious model, but it is what is actually required by Copernicus's adoption of Ptolemy's tables. Since i_1 is zero at $\kappa = 0°$ and maximum at $\kappa = 90°$, and conversely i_2 is maximum at $\kappa = 0°$ and zero at $\kappa = 90°$, the inclinations at any value of κ are $i_1 \sin \kappa$ and $i_2 \cos \kappa$ respectively. Hence, if latitudes β_1 and β_2 are computed independently as a function of the mean anomaly $\alpha = \omega' - \kappa$, by the same steps as before we reach

$$\beta_1 = \tan^{-1}((r/\rho) \sin i_1 \cos(\omega' - \kappa) \sin \kappa), \qquad (49a)$$

$$\beta_2 = \tan^{-1}((r/\rho) \sin i_2 \sin(\omega' - \kappa) \cos \kappa). \qquad (49b)$$

Again substituting i for $\sin i$, the apparent latitude β, the sum of both components is, after some reduction,

$$\beta = \beta_1 + \beta_2 \approx (r/\rho)[\sin \omega'(i_1 \sin^2 \kappa + i_2 \cos^2 \kappa) + \tfrac{1}{2}(i_1 - i_2) \cos \omega' \sin 2\kappa]. \qquad (50)$$

But by a correct computation from Copernicus's model,

$$\beta \approx (r/\rho) \sin \omega'(i_1 \sin^2 \kappa + i_2 \cos^2 \kappa), \qquad (48)$$

and thus the latitudes computed from the two models differ by

$$\Delta\beta \approx \tfrac{1}{2}(r/\rho)(i_1 - i_2) \cos \omega' \sin 2\kappa. \qquad (51)$$

Since $\sin 2\kappa$ is zero at multiples of 90°, $\Delta\beta$ is likewise zero and the two computations are identical for the extremal cases of $\kappa = \pm 90°$ and $\kappa = (0°, 180°)$ that are tabulated as $c_{3(\alpha)}$ and $c_{4(\alpha)}$. However, for intermediate values of κ they differ, and in the case of Venus can differ rather greatly. We have checked latitudes for various configurations, computing accurately with $\sin i$, and the greatest differences occur near, but not at, inferior conjunction where r/ρ is largest. For Venus r/ρ can amount to over 5/2, and the difference of latitude can reach about 1°, but for Mercury r/ρ is never more than 2/3, and the differences are less than 0;12°.

Concluding Remarks

The latitude theory of the planets is certainly the most flawed part of *De revolutionibus*. Rather than constructing his theory upon independent observations—which were in any case probably unobtainable in the required locations of the earth and each planet—Copernicus set the goal of reproducing as directly as possible in a heliocentric form the latitudes of Ptolemy's tables in the *Almagest*. But even this more modest task was beyond Copernicus's reach, and led to a series of problems, the most notable being the inexplicable relation of the oscillating orbital planes of the planets to the motion of the earth, the violations of uniform circular motion, and finally the incompatibility of the model for the inferior planets with the very tables they were intended to duplicate. Considering that the latitude theory was already described in nearly its complete form in the *Commentariolus*, it seems odd that Copernicus could not have done more to improve it in the intervening period of nearly thirty years. There is, however, reason to believe that during most of this period Copernicus hardly modified his original theory at all, and that even the changes found in the present form of Book VI were a very late part of his work and may have been done in haste with little thought for the many difficulties remaining in his final theory.

The evidence is provided by Rheticus's *Narratio prima*, which can be used, although with some caution, as a guide to the state of Copernicus's work as of 1539. It was noted earlier that in the *Commentariolus* Copernicus believed that a single fixed inclination of the orbital planes of Venus and Mercury was adequate to account for the inclination and slant components of their latitude.[61] Rheticus likewise describes a model with a single inclination, and describes it in considerable detail, the different effects of the inclination and the slant being attributed solely to the different positions of the earth with respect to the orbit without any libration of the orbital plane.[62] It could be argued that Rheticus did not yet understand Copernicus's latitude theory at the time since he himself says at the beginning of the *Narratio* that "I have just begun to understand the models of the remaining Books," referring to Books V and VI.[63] But Rheticus's description appears thorough, and he correctly and completely describes the libration of the orbital planes of the superior planets and the libration of the deflection component of the inferior planets, although without noting the effect of the variation of distance. Since he is so meticulous in describing these librations, which he seems to find especially interesting parts of Copernicus's models, it is hard to believe that he could have missed the libration between the inclination and the slant if it were present in the text of *De revolutionibus* upon which the *Narratio prima* was based.

Hence, if Rheticus's omissions are to be taken seriously, and it appears that they must, the version of Book VI that he had studied in 1539 was significantly different from the present text contained in M. In particular, the description of the model for the inclination and slant in VI,2 must have omitted the libration, and the derivation of i_1 and i_2 in VI,5–7, which necessitates the libration, must have been very different or not even written. Nor does it seem as though Copernicus had yet treated the effect of distance

---

[61] Above, p. 508.

[62] The latitude theory is described in the edition of the *Narratio prima* printed in Kepler (1937–) 1, 122–26, and the inclination and slant are at 123–24. These are translated in Rosen (1971), 178–85, 180–82. Rheticus's description of only a single inclination has been noted in Hugonnard-Roche and Verdet (1982), 189–90.

[63] In Kepler (1937–) 1, 88:39–40, *reliquorum vero hypotheses primum animo complexus sunt.*

upon the deflection in VI,8.[64] In fact, the latitude theory described by Rheticus is in all essentials that of the *Commentariolus*, which it thus appears Copernicus had retained more-or-less unchanged as late as 1539. Consequently, Book VI must have been revised into its present form during the period that Copernicus was preparing *De revolutionibus* for publication, and in his haste he may have been less than careful about the undesirable consequences of his attempt to represent Ptolemy rather than nature.

[64] Since VI,2 in M 190r–192r is written on paper of watermark D, this would indicate that Copernicus was still using papers D and E together after September of 1539, which is surprisingly late. Further, the addition to VI,8 on 208–09, showing the computation of $c_{5(\alpha)}$ and $c_{7(\kappa)}$, can be dated to as late as 1541 from the use of paper F for 209. See above, p. 88.